MW00996795

THE HISTORY OF PHYSICS

THE HISTORY OF PHYSICS

ISAAC ASIMOV

Walker and Company • New York

This edition first published in the United States of America in 1984
by the Walker Publishing Company, Inc.

First published in the United States of America
in 1966 by the Walker Publishing Company, Inc., as three separate volumes
under the title *Understanding Physics*, volumes I, II and III.

Published simultaneously in Canada by John Wiley & Sons
Canada, Limited, Rexdale, Ontario.

Library of Congress Cataloging in Publication Data

Asimov, Isaac, 1920-
The history of physics.

Bibliography: p.
Includes index.
1. Physics. I. Title.
QC23.A8 1983 530 83-6478
ISBN 0-8027-0751-3

Library of Congress Catalog Card Number: 83-6478

PRINTED IN THE UNITED STATES OF AMERICA

10 9 8 7 6 5 4 3 2 1

Contents

Introduction *xiii*

Part One: Motion, Sound, and Heat

1–The Search for Knowledge **1**

From Philosophy to Physics 1
The Greek View of Motion 3
Flaws in Theory 7

2–Falling Bodies **10**

Inclined Planes 10
Acceleration 13
Free Fall 18

3–The Laws of Motion **23**

Inertia 23
Forces and Vectors 26
Mass 30
Action and Reaction 33

4–Gravitation **37**

Combination of Forces 37
The Motion of the Moon 41
The Gravitational Constant 47

5–Weight **53**

The Shape of the Earth 53
Beyond the Earth 57
Escape Velocity 61

6–Momentum **65**

Impulse 65
Conservation of Momentum 67
Rotational Motion 72
Torque 75
Conservation of Angular Momentum 79

7–Work and Energy **84**

The Lever 84
Multiplying Force 89
Mechanical Energy 93
The Conservation of Energy 97

8–Vibration **101**

Simple Harmonic Motion 101
Period of Vibration 103
The Pendulum 108

9–Liquids **115**

Pressure 115
Buoyancy 119
Cohesion and Adhesion 125
Viscosity 130

10–Gases **135**

Density 135
Gas Pressure 138
Boyle's Law 143

11–Sound **148**

Water Waves 148
Sound Waves 154
Loudness 158

12–Pitch **162**

The Velocity of Sound 162
The Musical Scale 167
Modification of Pitch 173
Reflection of Sound 177

13–Temperature **181**

Hot and Cold 181

Temperature Scales 185
Expansion 188
Absolute Temperature 192

14–Heat **197**

The Kinetic Theory of Gases 197
Diffusion 204
Real Gases 208
Specific Heat 212
Latent Heat 215

15–Thermodynamics **220**

The Flow of Heat 220
The Second Law of Thermodynamics 226
Entropy 229
Disorder 234

Part Two: Light, Magnetism, and Electricity

16–Mechanism **241**

The Newtonian View 241
Action at a Distance 243

17–Light **247**

Transmission 247
Reflection 252
Curved Mirrors 255
Refraction 263

18–Lenses **269**

Focus by Transmission 269
Spectacles 273
Cameras 277
Magnification 281
Microscopes and Telescopes 285

19–Color **290**

Spectra 290
Reflecting Telescopes 294
Spectral Lines 297
Diffraction 300

20–Light Waves **306**

Interference 306
The Velocity of Light 310
The Doppler-Fizeau Effect 316
Polarized Light 320

21–The Ether **326**

Absolute Motion 326
The Michelson-Morley Experiment 331
The FitzGerald Contraction 336

22–Relativity **341**

The Special Theory 341
Mass-Energy Equivalence 347
Relative Time 352
The General Theory 357
Gravitation 359

23–Quanta **365**

Black-Body Radiation 365
Planck's Constant 368
Photoelectric Effect 372
Photons 375

24–Magnetism **380**

Magnetic Poles 380
Magnetic Domains 384
The Earth as a Magnet 388
Magnetic Fields 391

25–Electrostatics 397

Electric Charge 397
Electrons 402
Electromotive Force 406
Condensers 409

26–Electric Currents 417

Continuous Electron Flow 417
Chemical Cells 420
Resistance 423
Electric Power 428
Circuits 431
Batteries 436

27–Electromagnetism 441

Oersted's Experiment 441
Applications of Electromagnetism 446
Measurement of Current 452
Generators 455

28–Alternating Current 460

The Armature 461
Impedance 464
Transformers 469
Motors 472

29–Electromagnetic Radiation 476

Maxwell's Equations 476

Part Three: The Electron, Proton, and Neutron

30–The Atom 481

Origin of Atomism 482
The Chemical Elements 484
The Modern Atomic Theory 487

The Periodic Table 493
The Reality of Atoms 498

31–Ions and Radiation **503**

Electrolysis 503
Particles of Electricity 507
The Radiation Spectrum 511

32–The Electron **516**

The Discovery of the Electron 516
The Charge of the Electron 520
Electronics 521
Radio 524
Television and Radar 528

33–Electrons Within Atoms **533**

The Photoelectric Effect 533
The Nuclear Atom 536
Characteristic X Rays 540
Atomic Numbers 543
Electron Shells 546

34–Electrons and Quanta **551**

Spectral Series 551
The Bohr Atom 555
Sub-shells 559
Transition Elements 563

35–Electron Energy Levels **570**

Semiconductors 570
Solid-State Devices 573
Masers and Lasers 577
Matter-Waves 581

36–Radioactivity **588**

Uranium 588
Alpha Particles 591

Particle Detection 595
The Neutron 598
New Radioactive Elements 600

37–Isotopes **604**

Atomic Transformation 604
Radioactive Series 608
Half-Lives 613
Stable Isotopes 617

38–Nuclear Chemistry **625**

Mass Number 625
Radioactive Dating 630
Nuclear Reactions 632
The Electron-Volt 635
Particle Accelerators 637

39–Artificial Radioactivity **643**

Radioisotopes 643
The Biochemical Uses of Isotopes 645
Units of Radioactivity 649
Neutron Bombardment 651
Synthetic Elements 654

40–Nuclear Structure **660**

Nucleons, Even and Odd 660
Packing Fractions 665
Nuclear Energy 669
Nuclear Fission 672

41–Nuclear Reactors **675**

Uranium-235 675
The "Atomic Pile" 678
The "Atomic Age" 682
Nuclear Fusion 685
Radiation Sickness 689
Fusion Power 693

42–Anti-Particles **696**

Cosmic Rays 696
The Positron 701
Matter Annihilation 704
Anti-Baryons 708
Antimatter 711

43–Other Particles **714**

The Neutrino 714
Neutrino Interactions 718
The Muon 722
The Pion 727
The Frontier 732

Appendix ***737***

Suggested Further Reading ***743***

Index ***745***

Introduction

The History of Physics was published under the title *Understanding Physics* in 1966 in three volumes by Walker and Company.

In the eighteen years that have since elapsed, it has sold pleasantly and a great many kind words have been said about it. This has warmed my heart, to say nothing of the large, conglomerate heart of Sam and Beth Walker and all the nice people who work for them.

Therefore, it occurred to the Walkers to put out a new edition of *The History of Physics*, one in which the three volumes would be put within a single book cover. This will lend new life to what they are pleased to call a "classic" (a term that my own well-known modesty would never allow me to use myself; my courtly sense of proper behavior, however, does not allow me to contradict others lightly, so we'll let it stand). It will also make the book easier to use and handle.

It did raise a question, though. Eighteen years have passed. Is it possible that the book needs to be rewritten?

Surprisingly enough, the answer is, "Not really."

This is not because science has not advanced. Of course it has—and enormously, too. It is because of the way in which I had chosen to write the book.

It is my custom in dealing with any subject in science to take up the matter historically. I do this because it seems to me that a series of flat statements of this "fact" followed by that "law," with an additional announcement of a "rule" is unbearably dull—and I have written enough fiction to know that dullness is the most important no-no in writing.

To avoid the dullness I considered the matter in this fashion:

If we have an intelligent layman who feels a need to read about physics and find out what it says about the universe, and if physics has reached its present state because a number of intelligent people felt a need to collect and organize the observations and thoughts that led to the present state of our knowledge about the universe, then surely there seems an interesting match between the two needs. Why not let the intelligent layman find out about the universe in the same manner that the scientists did? What satisfied the need of the latter might well satisfy the need of the former, too.

Professional scientists of any kind enjoy "doing science" not because of what science tells them *now*, but because of the problems that still exist and the manner in which science teaches them to play the game of question and answer with the universe in order to squeeze *new* information out of it.

If the scientists enjoy the getting there more than the being there, then why shouldn't anyone else? You, for instance. Besides, teaching physics by the historical method is, to my way of thinking, not only more interesting, but more philosophically correct. There is danger in presenting the conclusions of physics flatly as though they were all delivered to us in letters of fire carved into stone on Mount Sinai. That sort of presentation not only reduces knowledge to memorization, but also encourages us to forget that science is forever correcting and advancing itself. Someone who is under the impression that "Science says—" is final is studying theology and not science.

By learning how scientific knowledge has progressed through chains of incomplete (and sometimes wrong) observations, theories, and conclusions, we learn that we are still only part way along the endless chain of discovery and that any conclusion, however certain it may seem, has within it the potentiality of extension, intensification, revision, and perhaps even total abolition. When we learn *that* we learn something more important than any individual "conclusion" or combination of "conclusions."

The interesting side effect of all this is that when physics is written in the historical fashion, little of it gets out of date. No matter what we find out about motion now, the manner in which Galileo upset Aristotelian notions of falling bodies does not change, and what Galileo taught the world remains valuable forever.

For the most part, then, here is *The History of Physics* exactly as it was. For the *most* part. In subatomic physics, recent discoveries are too important to go unmentioned. A new world of hadrons, quarks and gluons, of leptons, tauons and neutrinos, of "grand unified theories" and visions of universal origin, face us now—and these are addressed in an appendix.

So here is *The History of Physics* and I hope it helps you understand.

PART ONE

Motion, Sound, and Heat

CHAPTER 1

The Search for Knowledge

From Philosophy to Physics

The scholars of ancient Greece were the first we know of to attempt a thoroughgoing investigation of the universe—a systematic gathering of knowledge through the activity of human reason alone. Those who attempted this rationalistic search for understanding, without calling in the aid of intuition, inspiration, revelation, or other nonrational sources of information, were the *philosophers* (from Greek words meaning "lovers of wisdom").*

Philosophy could turn within, seeking an understanding of human behavior, of ethics and morality, of motivations and responses. Or it might turn outside to an investigation of the universe beyond the intangible wall of the mind—an investigation, in short, of "nature."

Those philosophers who turned toward the second alternative were the *natural philosophers,* and for many centuries after the palmy days of Greece the study of the phenomena of nature continued to be called natural philosophy. The modern word that is used in its place—*science,* from a Latin word meaning "to

* Undoubtedly there were wise men, and even rationalists, before the Greeks, but they are not known to us by name. Furthermore, the pre-Greek rationalists labored in vain, for it was only the Greek culture that left behind it a rationalistic philosophy to serve as ancestor to modern science.

know"—did not come into popular use until well into the nineteenth century. Even today, the highest university degree given for achievement in the sciences is generally that of "Doctor of Philosophy."

The word "natural" is of Latin derivation, so the term "natural philosophy" stems half from Latin and half from Greek, a combination usually frowned upon by purists. The Greek word for "natural" is *physikos,* so one might more precisely speak of *physical philosophy* to describe what we now call science.

The term *physics,* therefore, is a brief form of physical philosophy or natural philosophy and, in its original meaning, included all of science.

However, as the field of science broadened and deepened, and as the information gathered grew more voluminous, natural philosophers had to specialize, taking one segment or another of scientific endeavor as their chosen field of work. The specialties received names of their own and were often subtracted from the once universal domain of physics.

Thus, the study of the abstract relationships of form and number became mathematics; the study of the position and movements of the heavenly bodies became astronomy; the study of the physical nature of the earth we live upon became geology; the study of the composition and interaction of substances became chemistry; the study of the structure, function, and interrelationships of living organisms became biology, and so on.

The term physics then came to be used to describe the study of those portions of nature that remained after the above-mentioned specialties were subtracted. For that reason the word has come to cover a rather heterogeneous field and is not as easy to define as it might be.

What has been left over includes such phenomena as motion, heat, light, sound, electricity, and magnetism. All these are forms of "energy" (a term about which I shall have considerably more to say later on), so that a study of physics may be said to include, primarily, a consideration of the interrelationships of energy and matter.

This definition can be interpreted either narrowly or broadly. If it is interpreted broadly enough, physics can be expanded to include a great deal of each of its companion sections of science. Indeed, the twentieth century has seen such a situation come about.

The differentiation of science into its specialties is, after all, an artificial and man-made state of affairs. While the level of knowledge was still low, the division was useful and seemed natural. It was possible for a man to study astronomy or biology without reference to chemistry or physics, or for that matter to study either chemistry or physics in isolation. With time and accumulated information, however, the borders of the specialties approached, met, and finally overlapped. The techniques of one science became meaningful and illuminating in another.

In the latter half of the nineteenth century, physical techniques made it possible to determine the chemical constitution and physical structure of stars, and the science of "astrophysics" was born. The study of the vibrations set up in the body of the earth by quakes gave rise to the study of "geophysics." The study of chemical reactions through physical techniques initiated and constantly broadened the field of "physical chemistry," and the latter in turn penetrated the study of biology to produce what we now call "molecular biology."

As for mathematics, that was peculiarly the tool of physicists (at first, much more so than that of chemists and biologists), and as the search into first principles became more subtle and basic, it became nearly impossible to differentiate between the "pure mathematician" and the "theoretical physicist."

In this book, however, I will discuss the field of physics in its narrow sense, avoiding consideration (as much as possible) of those areas that encroach on neighboring specialties.

The Greek View of Motion

Among the first phenomena considered by the curious Greeks was motion. One might initially suspect that motion is an attribute of life; after all, men and cats move freely but corpses and stones do not. A stone can be made to move, to be sure, but usually through the impulse given it by a living thing.

However, this initial notion does not stand up, for there are many examples of motion that do not involve life. Thus, the heavenly objects move across the sky and the wind blows as it wills. Of course, it might be suggested that heavenly bodies are pushed by angels and that wind is the breath of a storm-god, and indeed such explanations were common among most societies and through most centuries. The Greek philosophers, however, were committed to explanations that involved only that portion of the universe that could be deduced by human reason from phenomena apparent to human senses. That excluded angels and storm-gods.

Furthermore, there were pettier examples of motion. The

smoke of a fire drifted irregularly upward. A stone released in midair promptly moved downward, although no impulse in that direction was given it. Surely not even the most mystically-minded individual was ready to suppose that every wisp of smoke, every falling scrap of material, contained a little god or demon pushing it here and there.

The Greek notions on the matter were put into sophisticated form by the philosopher Aristotle (384–322 B.C.). He maintained that each of the various fundamental kinds of matter ("elements") had its own natural place in the universe. The element "earth," in which was included all the common solid materials about us, had as its natural place the center of the universe. All the earthy matter of the universe collected there and formed the world upon which we live. If every portion of the earthy material got as close to the center as it possibly could, the earth would have to take on the shape of a sphere (and this, indeed, was one of several lines of reasoning used by Aristotle to demonstrate that the earth was spherical and not flat).

The element "water" had its natural place about the rim of the sphere of "earth." The element "air" had its natural place about the rim of the sphere of "water," and the element "fire" had its natural place outside the sphere of "air."

While one can deduce almost any sort of scheme of the universe by reason alone, it is usually felt that such a scheme is not worth spending time on unless it corresponds to "reality"—to what our senses tell us about the universe. In this case, observation seems to back up the Aristotelian view. As far as the senses can tell, the earth is indeed at the center of the universe; oceans of water cover large portions of the earth; the air extends about land and sea; and in the airy heights there are even occasional evidences of a sphere of fire that makes itself visible during storms in the form of lightning.

The notion that every form of substance has its natural place in the universe is an example of an *assumption*. It is something accepted without proof, and it is incorrect to speak of an assumption as either true or false, since there is no way of proving it to be either. (If there were, it would no longer be an assumption.) It is better to consider assumptions as either useful or useless, depending on whether or not deductions made from them corresponded to reality.

If two different assumptions, or sets of assumptions, both lead to deductions that correspond to reality, then the one that explains more is the more useful.

On the other hand, it seems obvious that assumptions are the weak points in any argument, as they have to be accepted on faith in a philosophy of science that prides itself on its rationalism. Since we must start somewhere, we must have assumptions, but at least let us have as few assumptions as possible. Therefore, of two theories that explain equal areas of the universe, the one that begins with fewer assumptions is the more useful. Because William of Ockham (1300?–1349?), a medieval English philosopher, emphasized this point of view, the effort made to whittle away at unnecessary assumptions is referred to as making use of "Ockham's razor."

The assumption of "natural place" certainly seemed a useful one to the Greeks. Granted that such a natural place existed, it seemed only reasonable to suppose that whenever an object found itself out of its natural place, it would return to that natural place as soon as given the chance. A stone, held in the hand in midair, for instance, gives evidence of its "eagerness" to return to its natural place by the manner in which it presses downward. This, one might deduce, is why it has weight. If the supporting hand is removed, the stone promptly moves toward its natural place and falls downward. By the same reasoning, we can explain why tongues of fire shoot upward, why pebbles fall down through water, and why bubbles of air rise up through water.

One might even use the same line of argument to explain rainfall. When the heat of the sun vaporizes water ("turns it into air" a Greek might suppose), the vapors promptly rise in search of their natural place. Once those vapors are converted into liquid water again, the latter falls in droplets in search of their natural place.

From the assumption of "natural place," further deductions can be made. One object is known to be heavier than another. The heavier object pushes downward against the hand with a greater "eagerness" than the lighter object does. Surely, if each is released the heavier object will express its greater eagerness to return to its place by falling more rapidly than the lighter object. So Aristotle maintained, and indeed this too seemed to match observation, for light objects such as feathers, leaves, and snowflakes drifted down slowly, while rocks and bricks fell rapidly.

But can the theory withstand the test of difficulties deliberately raised? For instance, an object can be forced to move away from its natural place, as when a stone is thrown into the air. This is initially brought about by muscular impulse, but once the stone leaves the hand, the hand is no longer exerting an impulse upon it. Why then doesn't the stone at once resume its natural motion and fall to earth? Why does it continue to rise in the air?

Aristotle's explanation was that the impulse given the stone was transmitted to the air and that the air carried the stone along. As the impulse was transmitted from point to point in the air, however, it weakened, and the natural motion of the stone asserted itself more and more strongly. Upward movement slowed and eventually turned into a downward movement until finally the stone rested on the ground once more. Not all the force of an arm or a catapult could, in the long run, overcome the stone's natural motion. ("Whatever goes up must come down," we still say.)

It therefore follows that forced motion (away from the natural place) must inevitably give way to natural motion (toward the natural place) and that natural motion will eventually bring the object to its natural place. Once there, since it has no place else to go, it will stop moving. The state of *rest*, or lack of motion, is therefore the natural state.

This, too, seems to square with observation, for thrown objects come to the ground eventually and stop; rolling or sliding objects eventually come to a halt; and even living objects cannot move forever. If we climb a mountain we do so with an effort, and as the impulse within our muscles fades, we are forced to rest at intervals. Even the quietest motions are at some cost, and the impulse within every living thing eventually spends itself. The living organism dies and returns to the natural state of rest. ("All men are mortal.")

But what about the heavenly bodies? The situation with respect to them seems quite different from that with respect to objects on earth. For one thing, whereas the natural motion of objects here below is either upward or downward, the heavenly bodies neither approach nor recede but seem to move in circles about the earth.

Aristotle could only conclude that the heavens and the heavenly bodies were made of a substance that was neither earth, water, air, nor fire. It was a fifth "element," which he named "ether" (a Greek word meaning "blazing," the heavenly bodies being notable for the light they emitted).

The natural place of the fifth element was outside the sphere of fire. Why then, since they were in their natural place, did the heavenly bodies not remain at rest? Some scholars eventually answered that question by supposing the various heavenly bodies to be in the charge of angels who perpetually rolled them around the heavens, but Aristotle could not indulge in such easy explanations. Instead, he was forced into a new assumption to the effect that the laws governing the motion of heavenly bodies were different from those governing the motion of earthly bodies. Here the natural state was rest, but in the heavens the natural state was perpetual circular motion.

Flaws in Theory

I have gone into the Greek view of motion in considerable detail because it was a physical theory worked out by one of history's greatest minds. This theory seemed to explain so much that it was accepted by great scholars for two thousand years afterward; nevertheless it had to be replaced by other theories that differed from it at almost every point.

The Aristotelian view seemed logical and useful. Why then was it replaced? If it was "wrong," then why did so many people of intelligence believe it to be "right" for so long? And if they believed it to be "right" for so long, what eventually happened to convince them that it was "wrong"?

One method of casting doubt upon any theory (however respected and long established) is to show that two contradictory conclusions can be drawn from it.

For instance, a rock dropping through water falls more slowly than the same rock dropping through air. One might deduce that the thinner the substance through which the rock is falling the more rapidly it moves in its attempt to return to its natural place. If there were no substance at all in its path (a *vacuum,* from a Latin word meaning "empty"), then it would move with infinite speed. Actually, some scholars did make this point, and since they felt infinite speed to be an impossibility, they maintained that this line of argument proved that there could be no such thing as a vacuum. (A catch-phrase arose that is still current: "Nature abhors a vacuum.")

On the other hand, the Aristotelian view is that when a stone is thrown it is the impulse conducted by the air that makes it possible for the stone to move in the direction thrown. If the air were gone and a vacuum were present, there would be nothing to move the stone. Well then, would a stone in a vacuum move at infinite speed or not at all? It would seem we could argue the point either way.

Here is another possible contradiction. Suppose you have a one-pound weight and a two-pound weight and let them fall. The two-pound weight, being heavier, is more eager to reach its natural place and therefore falls more rapidly than the one-pound weight. Now place the two weights together in a tightly fitted sack and drop them. The two-pound weight, one might argue, would race downward but would be held back by the more leisurely fall of the one-pound weight. The overall rate of fall would therefore be an intermediate one, less than that of the two-pound weight falling alone and more than that of the one-pound weight falling alone.

On the other hand, you might argue, the two-pound weight and the one-pound weight together formed a single system weighing three pounds, which should fall more rapidly than the two-pound weight alone. Well then, does the combination fall more rapidly or less rapidly than the two-pound weight alone? It looks as though you could argue either way.

Such careful reasoning may point out weaknesses in a theory, but it rarely carries conviction, for the proponents of the theory can usually advance counter-arguments. For instance, one might say that in a vacuum natural motion becomes infinite in speed, while forced motion becomes impossible. And one might argue that the speed of fall of two connected weights depends on how tightly they are held together.

A second method of testing a theory, and one that has proved to be far more useful, is to draw a necessary conclusion from the theory and then check it against actual phenomena as rigorously as possible.

For instance, a two-pound object presses down upon the hand just twice as strongly as a one-pound object. Is it sufficient to say that the two-pound object falls more rapidly than the one-pound object? If the two-pound object displays just twice the eagerness to return to its natural place, should it not fall at just twice the rate? Should this not be tested? Why not measure the exact rate at which both objects fall and see if the two-pound object falls at just twice the rate of the one-pound object? If it doesn't, then surely the Greek theories of motion will have to be modified. If, on the other hand, the two-pound weight does fall just twice as rapidly, the Greek theories can be accepted with that much more assurance.

Yet such a deliberate test (or *experiment*) was not made by Aristotle or for two thousand years after him. There were two types of reasons for this. One was theoretical. The Greeks had had their greatest success in geometry, which deals with abstract concepts such as dimensionless points and straight lines without width. They achieved results of great simplicity and generality that they could not have obtained by measuring actual objects. There arose, therefore, the feeling that the real world was rather crude and ill-suited to helping work out abstract theories of the universe. To be sure, there were Greeks who experimented and drew important conclusions therefrom; for example, Archimedes (287?–212 B.C.) and Hero (first century A.D.). Nevertheless, the ancient and medieval view was definitely in favor of deduction from assumptions, rather than of testing by experimentation.

The second reason was a practical one. It is not as easy to experiment as one might suppose. It is not difficult to test the speed of a falling body in an age of stopwatches and electronic methods of measuring short intervals of time. Up to three centuries ago, however, there were no timepieces capable of measuring small intervals of time, and precious few good measuring instruments of any kind.

In relying on pure reason, the ancient philosophers were really making the best of what they had available to them, and in seeming to scorn experimentation they were making a virtue of necessity.*

The situation slowly began to change in the late Middle Ages. More and more scholars began to appreciate the value of experimentation as a method of testing theories, and here and there individuals began trying to work out experimental techniques.

The experimentalists remained pretty largely without influence, however, until the Italian scientist Galileo Galilei (1564–1642), came on the scene. He did not invent experimentation, but he made it spectacular and popular. His experiments with motion were so ingenious and conclusive that they not only began the destruction of Aristotelian physics but demonstrated the necessity, once and for all, of experimentalism in science. It is from Galileo (he is invariably known by his first name only) that the birth of "experimental science" or "modern science" is usually dated.

* And yet we can regret that the Greek philosophers did not conduct certain simple experiments that required no instruments. For instance, a sheet of thin papyrus falls slowly. The same sheet, crumpled into a small, tight ball, drops at a clearly greater speed. Since its weight hasn't changed as a result of the crumpling, why the change in the rate of fall? A question as simple as this might have been crucial in modifying Greek theories of motion in what we would now consider the proper direction.

CHAPTER 2

Falling Bodies

Inclined Planes

Galileo's chief difficulty was the matter of timekeeping. He had no clock worthy of the name, so he had to improvise methods. For instance, he used a container with a small hole at the bottom out of which water dripped into a pan at, presumably, a constant rate. The weight of water caught in this fashion between two events was a measure of the time that had elapsed.

This would certainly not do for bodies in "free fall"—that is, falling downward without interference. A free fall from any reasonable height is over too soon, and the amount of water caught during the time of fall is too small to make time measurements even approximately accurate.

Galileo, therefore, decided to use an inclined plane. A smooth ball will roll down a smooth groove on such an inclined plane at a manifestly lower speed than it would move if it were dropping freely. Furthermore, if the inclined plane is slanted less and less sharply to the horizontal, the ball rolls less and less rapidly; with the plane made precisely horizontal, the ball will not roll at all (at least, not from a standing start). By this method, one can slow the rate of fall to the point where even crude time-measuring devices can yield useful results.

One might raise the point as to whether motion down an

inclined plane can give results that can fairly be applied to free fall. It seems reasonable to suppose that it can. If something is true for every angle at which the inclined plane is pitched, it should be true for free fall as well, for free fall can be looked upon as a matter of rolling down an inclined plane that has been maximally tipped—that is, one that makes an angle of 90° with the horizontal.

For instance, it can be easily shown that relatively heavy balls of different weights would roll down a particular inclined plane at the same rate. This would hold true for any angle at which the inclined plane was tipped. If the plane were tipped more sharply, the balls would roll more rapidly, but all the balls would increase their rate of movement similarly; in the end all would cover the same distance in the same time. It is fair to conclude from that alone that freely falling bodies will fall through equal distances in equal times, regardless of their weight. In other words, a heavy body will *not* fall more rapidly than a light body, despite the Aristotelian view.

(There is a well-known story that Galileo proved this when he dropped two objects of different weight off the Leaning Tower of Pisa and they hit the ground in a simultaneous thump. Unfortunately, this is just a story. Historians are quite certain that Galileo never conducted such an experiment but that a Dutch scientist, Simon Stevinus (1548–1620), did something of the sort a few years before Galileo's experiments. In the cool world of science, however, careful and exhaustive experiments, such as those of Galileo with inclined planes, sometimes count for more than single, sensational demonstrations.)

Yet can we really dispose of the Aristotelian view so easily? The observed equal rate of travel on the part of balls rolling down an inclined plane cannot be disputed, but on the other hand neither is it possible to dispute the fact that a soap bubble falls far more slowly than a ping-pong ball of the same size, and that the ping-pong ball falls rather more slowly than a solid, wooden ball of the same size.

We have an explanation for this, however. Objects do not fall through nothing; they fall through air, and they must push the air aside, so to speak, in order to fall. We might take the viewpoint that to push the air aside consumes time. A heavy body pressing down hard pushes the light air to one side with no trouble and loses virtually no time. It doesn't matter whether the body is one pound or a hundred pounds. The one-pound weight experiences so little trouble in pushing the air to one side that the hundred-pound weight can scarcely improve on it. Both weights therefore fall through equal distances in equal times.* A distinctly light body such as a ping-pong ball would press down so softly that it would experience considerable trouble in pushing the air out of the way, and it would fall slowly. A soap bubble, for the same reason, would scarcely fall at all.

Can this use of air resistance as an explanation be considered valid? Or is it just something concocted to explain the failure of Galileo's generalization to hold in the real world? Fortunately, the matter can be checked. First, suppose that of two objects of equal weight one is spherical and compact while the other is wide and flat. The wide, flat object will make contact with air over a broader front and have to push more air out of the way in order to fall. It will therefore experience more air resistance than the spherical, compact one, and will fall more slowly, even though the two bodies are of equal weight. This turns out to be so, when tested. In fact, if a piece of paper is crumpled into a pellet, it falls more quickly because it suffers less air resistance. I referred to this experiment on page 9 as being one the Greeks might easily have performed, and from which they might have discovered that there must be something wrong with the Aristotelian view of motion.

An even more unmistakable test would be to get rid of air and allow bodies to fall through a vacuum. With no resistance to speak of, all bodies, however light or heavy they might be, ought to fall through equal distances in equal times. Galileo was convinced this would be so, but in his time there was no way of creating a vacuum to test the matter. In later years, when vacuums could be produced, the experiment of causing a feather and a lump of lead to fall together in a vacuum, and noting the fact that both covered an equal distance in an equal time, became commonplace. Air resistance is therefore real and not just a face-saving device.

Of course, this raises the question of whether one is justified, for the sake of enunciating a simple rule, in describing the universe in nonreal terms. Galileo's rule that all objects of whatever weight fell through equal distances in equal times could be expressed in very simple mathematical form. The rule is true, however, only in a perfect vacuum, which, as a matter of fact, does not exist. (Even the best vacuums we can create, even the vacuum of interstellar space, are not perfect.) On the other hand, Aristotle's view

* Actually, there is a small difference. This does not show up in falls of reasonable length, but would become visible if both weights were dropped from an airplane. In such case, the lighter weight would be held up a bit and lag behind a trifle.

that heavier objects fall more rapidly than light ones is true, at least to a certain extent, in the real world. However, it cannot be reduced to as simple a mathematical statement, for the rate of fall of particular bodies depends not only upon their weight but also upon their shapes.

One might suppose that reality must be held to at all costs. However, though that may be the most moral thing to do, it is not necessarily the most useful thing to do. The Greeks themselves chose the ideal over the real in their geometry and demonstrated very well that far more could be achieved by consideration of abstract line and form than by a study of the real lines and forms of the world; the greater understanding achieved through abstraction could be applied most usefully to the very reality that was ignored in the process of gaining knowledge.

Nearly four centuries of experience since Galileo's time has shown that it is frequently useful to depart from the real and to construct a "model" of the system being studied; some of the complications are stripped away, so a simple and generalized mathematical structure can be built up out of what is left. Once that is done, the complicating factors can be restored one by one, and the relationship suitably modified. To try to achieve the complexities of reality at one bound, without working through a simplified model first, is so difficult that it is virtually never attempted and, we can feel certain, would not succeed if it were attempted.

It is useless then to try to judge whether Galileo's views are "true" and Aristotle's "false" or vice versa. As far as rates of fall are concerned there are observations that back one view and other observations that back the other. What we can say, however, as strongly as possible, is that Galileo's views of motion turned out to explain many more observations in a far simpler manner than did Aristotle's views. The Galilean view was, therefore, far more useful. This was recognized not too long after Galileo's experiments were described, and Aristotelian physics collapsed.

Acceleration

If we were to measure the distance traversed by a body rolling down an inclined plane, we would find that the the body would cover greater and greater distances in successive equal time intervals.

Thus, a body might roll a distance of 2 feet in the first second. In the next second it would roll 6 feet, for a total distance of

8 feet. In the third second it would roll 10 feet, for a total distance of 18 feet. In the fourth second it would roll 14 feet, for a total distance of 32 feet.

Clearly the ball is rolling more and more rapidly with time.

This in itself represents no break with Aristotelian physics, for Aristotle's theories said nothing about the manner in which the velocity of a falling body changed with time. In fact, this increase in velocity might be squared with the Aristotelian view, for one might say that as a body approached its natural place its eagerness to get there heightened, so its velocity would naturally increase.

However, the importance of Galileo's technique was just this: he took up the matter of change of speed, not in a qualitative way but in a quantitative way. It is not enough to simply say, "Velocity increases with time." One must say, if possible, by just how much it increases and work out the precise interrelationship of velocity and time.

For instance, if a ball rolls 2 feet in one second, 8 feet in two seconds, 18 feet in three seconds, and 32 feet in four seconds, it would appear that there was a relationship between the total distance covered and the square of the time elapsed. Thus, 2 is equal to 2×1^2, 8 is equal to 2×2^2, 18 is equal to 2×3^2, and 32 is equal to 2×4^2. We can express this relationship by saying that the total distance traversed by a ball rolling down an inclined plane (or by an object in free fall) after starting from rest is directly proportional* to the square of the time elapsed.

Physics has adopted this emphasis on exact measurement that Galileo introduced, and other fields of science have done likewise wherever this has been possible. (The fact that chemists and biologists have not adopted the mathematical attitude as thoroughly as have physicists is no sign that chemists and biologists are less intelligent or less precise than physicists. Actually, this has come about because the systems studied by physicists are simpler than those studied by chemists and biologists and are more easily idealized to the point where they can be expressed in simple mathematical form.)

Now consider the ball rolling 2 feet in one second. Its average *velocity* (distance covered in unit time) during that one-

* When we say that *a* is "directly proportional" to *b*, we mean that as *b* increases, *a* increases as well. Sometimes, a relationship is such that as *b* increases, *a* decreases. (For instance, as the price of an object increases, the number of sales may decrease.) We then say that *a* is "inversely proportional" to *b*.

second interval is 2 feet divided by one second. It is easy to divide 2 by 1, but it is important to remember that we must divide the units as well, the "feet" by the "second." We can express this division of units in the usual fashion by means of a fraction. In other words, 2 feet divided by one second can be expressed as $\frac{2 \text{ feet}}{1 \text{ second}}$, or 2 feet/second. This can be abbreviated as 2 ft/sec and is usually read as "two feet per second." It is important not to let the use of "per" blind us to the fact that we are in effect dealing with a fraction. Its numerator and denominator are units rather than numbers, but the fractional quality remains nevertheless.

But to return to the rolling ball . . . In one second it covers 2 feet, for an average velocity of 2 ft/sec. In two seconds, it covers 8 feet, for an average velocity over the entire interval of 4 ft/sec. In three seconds it covers 18 feet, for an average velocity over the entire interval of 6 ft/sec. And you can see for yourself, the average velocity for the first four seconds is 8 ft/sec. The average velocity, all told, is in direct proportion to the time elapsed.

Here, however, we are dealing with average velocities. What is the velocity of a rolling ball at a particular moment? Consider the first second of time. During that second the ball has been rolling at an average velocity of 2 ft/sec. It began that first second of time at a slower velocity. In fact, since it started at rest, the velocity at the beginning (after 0 seconds, in other words) was 0 ft/sec. To get the average up to 2 ft/sec, the ball must reach correspondingly higher velocities in the second half of the time interval. If we assume that the velocity is rising smoothly with time, it follows that if the velocity at the beginning of the time interval was 2 ft/sec less than average, then at the end of the time interval (after one second), it should be 2 ft/sec more than average, or 4 ft/sec.

If we follow the same line of reasoning for the average velocities in the first two seconds, in the first three seconds, and so on, we come to the following conclusions: at 0 seconds, the velocity is 0 ft/sec; at one second, the velocity (at that moment) is 4 ft/sec; at two seconds, the velocity is 8 ft/sec; at three seconds, the velocity is 12 ft/sec; at four seconds, the velocity is 16 ft/sec, and so on.

Notice that after each second, the velocity has increased by exactly 4 ft/sec. Such a change in velocity with time is called an *acceleration* (from Latin words meaning "to add speed"). To determine the value of the acceleration, we must divide the gain in velocity during a particular time interval by that time interval. For instance at one second, the velocity was 4 ft/sec while at four seconds it was 16 ft/sec. Over a three-second interval the velocity increased by 12 ft/sec. The acceleration then is 12 ft/sec divided by three seconds. (Notice particularly that it is *not* 12 ft/sec divided by 3. Where units are involved, they *must* be included in any mathematical manipulation. Here you are dividing by three seconds and not by 3.)

In dividing 12 ft/sec by three seconds, we get an answer in which the units as well as the numbers are subjected to the division—in other words $4 \frac{\text{ft/sec}}{\text{sec}}$. This can be written 4 ft/sec/sec (and read four feet per second per second). Then again, in algebraic manipulations a/b divided by b is equal to a/b multiplied by $1/b$, and the final result is a/b^2. Treating unit-fractions in the same manner, 4 ft/sec/sec can be written 4 ft/sec^2 (and read four feet per second squared).

You can see that in the case just given, for whatever time interval you work out the acceleration, the answer is always the same: 4 ft/sec^2. For inclined planes tipped to a greater or lesser extent, the acceleration would be different, but it would remain constant for any one given inclined plane through all time intervals.

This makes it possible for us to express Galileo's discovery about falling bodies in simpler and neater fashion. To say that all bodies cover equal distances in equal times is true; however, it is not saying enough, for it doesn't tell us whether bodies fall at uniform velocities, at steadily increasing velocities, or at velocities that change erratically. Again, if we say that all bodies fall at equal velocities, we are not saying anything about how those velocities may change with time.

What we can say now is that all bodies, regardless of weight (neglecting air resistance), roll down inclined planes, or fall freely, at equal and constant accelerations. When this is true, it follows quite inevitably that two falling bodies cover the same distance in the same time, and that at any given moment they are falling with the same velocity (assuming both started falling at the same time). It also tells us that the velocity increases with time and at a constant rate.

Such relationships become more useful if we introduce mathematical symbols to express our meaning. In doing so, we introduce nothing essentially new. We would be saying in mathematical symbols exactly what we have been trying to say in words, but more briefly and more generally. Mathematics is a shorthand language in which each symbol has a precise and agreed-upon meaning. Once the language is learned, we find that it is only a form of English after all.

For instance, we have just been considering the case of an acceleration (from rest) of 4 ft/sec^2. This means that at the end of one second the velocity is 4 ft/sec, at the end of two seconds it is 8 ft/sec, at the end of the three seconds it is 12 ft/sec, and so on. In short, the velocity is equal to the acceleration multiplied by the time. If we let v stand as a symbol for "velocity" and t for "time," we can say that in this case v is equal to $4t$.

But the actual acceleration depends on the angle at which the inclined plane is tipped. If the plane is made steeper, the acceleration increases; if it is made less steep, the acceleration decreases. For any given plane, the acceleration is constant, but the particular value of the constant can vary greatly from plane to plane. Let us not, therefore, commit ourselves to a particular numerical value for acceleration, but let this acceleration be represented by a. We can then say:

$$v = at \qquad \text{(Equation 2–1)}$$

It is important to remember that included in such equations in physics are units as well as numerals. Thus a, in Equation 2–1, does not represent a number merely, say 4, but a number and its units—4 ft/sec^2—the unit being appropriate for acceleration. Again, t, for time, represents a number and its units—three seconds let us say. In evaluating at, then, we multiply 4 ft/sec^2 by three seconds, multiplying the units as well as the numerals. Treating the units as though they were fractions (in other words, as though we were to multiply a/b^2 by b) the product is 12 ft/sec. Thus, multiplying acceleration (a) by time (t) does indeed give us velocity (v), since the units we obtain, ft/sec, are appropriate to velocity.

In any equation in physics, the units on either side of the equals sign must balance after all necessary algebraic manipulation is concluded. If this balance is not obtained, the equation does not correspond to reality and cannot be correct. If the units of one symbol are not known, they can be determined by deciding just what kind of unit is needed to balance the equation. (This is sometimes called *dimensional analysis.*)

With that out of the way, let us consider a ball starting from rest and rolling down an inclined plane for t seconds. Since the ball starts at rest, its velocity at the beginning of the time interval is 0 ft/sec. According to Equation 2–1, at the end of the interval, at time t, its velocity v is at ft/sec. To get the average velocity, during this interval of smoothly increasing velocity, we take the sum of the original and final velocity $(0 + at)$ and divide by 2. The average velocity during the time interval is therefore $at/2$. The distance (d) traversed in that time must be the average velocity multiplied by the time, $at/2 \times t$. We therefore conclude that:

$$d = \frac{at^2}{2} \qquad \text{(Equation 2–2)}$$

I will not attempt to check the dimensions for every equation presented, but let's do it for this one. The units of acceleration (a) are ft/sec^2 and the units of time (t) are sec (seconds). Therefore, the units of at^2 are ft/sec^2 $\times$ sec $\times$ sec, which works out to $\frac{\text{ft-sec}^2}{\text{sec}^2}$ or simply ft. Dividing at^2 by 2 does not alter the situation for in this case 2 is a "pure number"—that is, it lacks units. (Thus if you divide a foot-rule in two, each half has a length of 12 inches divided by 2, or 6 inches. The unit is not affected.) Thus the units of $at^2/2$ are ft (feet), an appropriate unit for distance (d).

Free Fall

As I said earlier, the value of the acceleration (a) of a ball rolling down an inclined plane varies according to the steepness of the plane. The steeper the plane, the greater the value of a.

Experimentation will show that for a given inclined plane the value of a is in direct proportion to the ratio of the height of the raised end of the plane to the length of the plane. If you represent the height of the raised end of the plane by H, and the length of the plane by L, you can express the previous sentence in mathematical symbols as $a \propto H/L$, where the symbol $\propto$ means "is directly proportional to."

In such a direct proportion the value of the expression on

one side changes in perfect correspondence with the value of the expression on the other. If H/L is doubled, a is doubled; if H/L is halved, a is halved; if H/L is multiplied by 2.529, a is multiplied by 2.529. This is what is meant by direct proportionality. But suppose that for a particular value of a, the value of H/L happens to be just a third as large. If the value of a is changed in any particular way, the value of H/L is changed in a precisely corresponding way, so it is still one third the value of a. In this particular case then, a is three times as large as H/L not for any one set of values but for all values.

This is a general rule. Whenever one factor, x, is directly proportional to another factor, y, we can always change the relationship into an equality by finding some appropriate constant value (usually called the *proportionality constant*) by which to multiply y. Ordinarily, we don't know the precise value of the proportionality constant to begin with, so it is signified by some symbol. This symbol is very often k (for "Konstant"—using the German spelling). Therefore, we can say that if $x \propto y$, then $x = ky$.

It is not absolutely necessary to use k as the symbol for the proportionality constant. Thus, the velocity of a ball rolling from rest is directly proportional to the time during which it has been rolling, and the distance it traverses is directly proportional to the square of that time; therefore, $v \propto t$ and $d \propto t^2$. In the first case, however, we have the special name "acceleration" for the proportionality constant, so we symbolize it by a; while in the second case, the relationship to acceleration is such that we symbolize the proportionality constant as $a/2$. Therefore $v = at$ and $d = at^2/2$.

In the case now under discussion, where the value of the acceleration (a) is directly proportional to H/L, it will prove convenient to symbolize the proportionality constant by the letter g. We can therefore say:

$$a = gH/L \qquad \text{(Equation 2–3)}$$

The quantities H and L are both measured in feet. In dividing H by L, feet are divided by feet and the unit cancels. The result is that the ratio H/L is a pure number and possesses no units. But the units of acceleration (a) are ft/sec^2. In order to keep the units in balance in Equation 2–3, it is therefore necessary that the units of g also be ft/sec^2, since H/L can contribute nothing in the way of units. We can conclude then that the proportionality constant in Equation 2–3 has the units of acceleration and therefore must represent an acceleration.

We can see what this means if we consider that the steeper we make a particular inclined plane, the greater the height of its raised end from the ground—that is, the greater the value of H. The length of the inclined plane (L) does not change, of course. Finally, when the plane is made perfectly vertical, the height of the raised end is equal to the full length of the plane, so that H equals L, and H/L equals 1.

A ball rolling down a perfectly vertical inclined plane is actually in free fall. Therefore, in free fall, H/L becomes 1, and Equation 2–3 becomes:

$$a = g \qquad \text{(Equation 2–4)}$$

This shows us that g is not only an acceleration, but is the particular acceleration undergone by a body in free fall. The tendency of a body to have weight and fall toward the earth is the result of a property called *gravity* (from the Latin word for "weighty"), and the symbol g is used because it is the abbreviation of "gravity."

If the actual acceleration of a body rolling down any particular inclined plane is measured, then the value of g can be obtained. Equation 2–3 can be rearranged to yield $g = aL/H$. For a particular inclined plane, the length (L) and height (H) of the raised end are easily measured, and with a known, g can be determined at once. Its value turns out to be equal to 32 ft/sec^2 (at least at sea level).

Now so far, for the sake of familiarity, I have made use of feet as a measure of distance. This is one of the common units of distance used in the United States and Great Britain, and we are accustomed to it. However, scientists all over the world use the *metric system* of measure, and we have gotten far enough into the subject, I think, to be able to join them in this.

The value of the metric system is that its various units possess simple and logical relationships among themselves. For instance, in the common system, 1 mile is equal to 1760 yards, 1 yard is equal to 3 feet, and 1 foot is equal to 12 inches. Converting one unit into another is always a chore.

In the metric system, the unit of distance is the "meter." Other units of distance are always obtained by multiplying the meter by 10 or a multiple of 10. Thanks to our system of writing numbers, this means that conversions of one unit to another within the metric system can be carried out by mere shifts of a decimal point.

Furthermore, standardized prefixes are used with set meaning. The prefix "deci-" always implies 1/10 of a standard unit, so a decimeter is 1/10 of a meter. The prefix "hecto-" always implies 100 times a standard unit, so a hectometer is 100 meters. And so it is for other prefixes as well.

The *meter* itself is 39.37 inches long. This makes it the equivalent, roughly, of 1.09 yards, or 3.28 feet. Two other metric units commonly used in physics are the *centimeter* and the *kilometer*. The prefix "centi-" implies 1/100 of a standard unit, so a centimeter is 1/100 of a meter. It is equivalent to 0.3937 inches, or approximately 2/5 of an inch. The prefix "kilo-" implies 1000 times the standard unit, so a kilometer is equal to 1000 meters or 100,000 centimeters. The kilometer is 39,370 inches long, which makes it just about 5/8 of a mile. The abbreviations ordinarily used for meter, centimeter, and kilometer are m, cm, and km, respectively.

Seconds, as a basic unit of time, are used in the metric system as well as in the common system. Therefore, if we want to express acceleration in metric units, we can use "meters per second per second" or m/sec^2 for the purpose. Since 3.28 feet equal 1 meter, we divide 32 ft/sec^2 by 3.28 and find that in metric units the value of g is 9.8 m/sec^2.

Once again, consider the importance of units. It is improper and incorrect to say that "the value of g is 32" or "the value of g is 9.8." The number by itself has no meaning in this connection. One must say either 32 ft/sec^2 or 9.8 m/sec^2. These last two values are absolutely equivalent. The numerical portions of the expression may be different, taken by themselves, but with the units added they are identical values. One is by no means "more true" or "more accurate" than the other; the expression in metric units is merely more useful.

We must know at all times which units are being used. In free fall, a is equal to g, so Equation 2–1 can be written $v = 32t$, if we are using common units; and $v = 9.8t$, if we are using metric units. In the shorthand of equations, the units are not included, so there is always the chance of confusion. If you try to use common units with the equation $v = 9.8t$, or metric units with the equation $v = 32t$, you will end up with results that do not correspond to reality. For that reason, the rules of procedure must be made perfectly plain. In this book, for instance, it will be taken for granted henceforward that the metric system will be used at all times, except where I specifically say otherwise.

Therefore, we can say that for bodies in free fall, from a starting position at rest:

$$v = 9.8t \qquad \text{(Equation 2–5)}$$

In the same way, for such a body, Equation 2–2 becomes $d = gt^2/2$ or:

$$d = 4.9t^2 \qquad \text{(Equation 2–6)}$$

At the end of one second, then, a falling body has dropped 4.9 m and is falling at a velocity of 9.8 m/sec. At the end of two seconds, it has fallen through a distance of 19.6 m and is falling at a velocity of 19.6 m/sec. At the end of the three seconds, it has fallen through a distance of 44.1 m and is falling at a velocity of 29.4 m/sec, and so on.*

* Since this book is not intended as a formal text, I am not presenting you with problems to be solved. I hope, nevertheless, that you have had enough experience with algebra to see that equations in physics not only present relationships in brief and convenient form, but also make it particularly convenient to solve problems—that is, to find the value of a particular symbol when the values of the other symbols in the equation are known or can be determined.

CHAPTER 3

The Laws of Motion

Inertia

Galileo's work on falling bodies was systematized a century later by the English scientist Isaac Newton (1642–1727), who was born, people are fond of pointing out, in the year of Galileo's death.

Newton's systematization appeared in his book *Philosophiae naturalis principia mathematica* (Mathematical Principles of Natural Philosophy) published in 1687. The book is usually referred to simply as the *Principia*.

Aristotle's picture of the physical universe had been lying shattered for nearly a hundred years, and it was Newton who now replaced it with a new one, more subtle and more useful. The foundations of the new picture of the universe consisted of three generalizations concerning motion that are usually referred to as Newton's Three Laws of Motion.*

* The important generalizations of science are brief descriptions of the behavior of the universe that are known to cover all observed cases. It is strongly felt that they will also cover all unobserved cases, here or anywhere, now or at any time. Such generalizations are sometimes called "laws of nature." This is actually a poor phrase because it seems to draw an analogy with man-made law, as something that is imposed and can be repealed, that can be violated at the cost of a penalty, and so on. All such analogies are misleading. It would be better therefore to speak of "Newton's generalizations concerning

His first law of motion may be given thus:

A body remains at rest or, if already in motion, remains in uniform motion with constant speed in a straight line, unless it is acted on by an unbalanced external force.

As you can see, this first law runs counter to the Aristotelian assumption of "natural place" with its corollary that the natural state of an object is to be at rest in its natural place.

The Newtonian view is that there is no natural place for any object. Wherever an object happens to be at rest† without any force acting upon it, it will remain at rest. Furthermore, if it happens to be in motion without any force acting upon it, it will remain in motion forever and show no tendency at all to come to rest. (I am not defining "force" just yet, but you undoubtedly already have a rough idea of what it means, and a proper definition will come eventually; see page 26.)

This tendency for motion (or for rest) to maintain itself steadily unless made to do otherwise by some interfering force can be viewed as a kind of "laziness," a kind of unwillingness to make a change. And indeed the first law of motion is referred to as the principle of *inertia,* from a Latin word meaning "idleness" or "laziness." (The habit of attributing human motivations or emotions to inanimate objects is called "personification." This is a bad habit in science, though quite a common one, and I indulged in it here only to explain the word "inertia.")

At first glance, the principle of inertia does not seem nearly as self-evident as the Aristotelian assumption of "natural place." We can see with our own eyes that moving objects do indeed tend to come to a halt even when, as nearly as we can see, there is nothing to stop them. Again, if a stone is released in midair it starts moving and continues moving at a faster and faster rate,

---

motion." However, everybody calls them the "laws" of motion, and if I did otherwise, I would merely seem eccentric. Nevertheless, by this footnote you are warned.

† In Aristotle's time the earth was considered a motionless body fixed at the center of the universe; the notion of "rest" therefore had a literal meaning. What we ordinarily consider "rest" nowadays is a state of being motionless with respect to the surface of the earth. But we know (and Newton did, too) that the earth itself is in motion about the sun and about its own axis. A body resting on the surface of the earth is therefore not really in a state of rest at all. In fact, the whole problem of what is really meant by "rest" and "motion" forced a new view of the universe in the form of what is called the "theory of relativity," advanced by Albert Einstein in 1905. In this book, however, we will run into no complications if we accept the fact that by "rest" and "motion" we really mean "rest with respect to the earth's surface" and "motion with respect to the earth's surface."

even though, as nearly as we can see, there is nothing to set it into motion.

If the principle of inertia is to hold good, we must be willing to admit the presence of subtle forces that do not make their existence very obvious.

For instance, a hockey puck given a sharp push along a level cement sidewalk will travel in a straight line, to be sure, but will do so at a quickly decreasing velocity and soon come to a halt. If the same puck is given the same sharp push along a smooth layer of ice, it will travel much farther, again in a straight line, but this time at only a slowly decreasing velocity. If we experiment sufficiently, it will quickly become clear that the rougher the surface along which the puck travels, the more quickly it will come to a halt.

It would seem that the tiny unevennesses of the rough surface catch at the tiny unevennesses of the hockey puck and slow it up. This catching of unevennesses against unevennesses is called *friction* (from a Latin word meaning "rub"), and the friction acts as a force that slows the puck's motion. The less the friction, the smaller the frictional force and the more slowly the puck's velocity is decreased. On a very smooth surface, such as that of ice, friction is so low that a puck would travel for great distances. If one could imagine a horizontal surface with no friction at all, then the hockey puck would travel in a straight line at constant velocity forever.

The Newtonian principle of inertia therefore holds exactly only in an imaginary ideal world in which no interfering forces exist: no friction, no air resistance.

Next consider a rock held in midair. It is at rest, but the instant we let go it begins to move. Clearly, then, there must be some force that makes it move, since the principle of inertia requires that in the absence of a force it remain at rest. Since the motion of the rock, if merely released, is always in the direction of the earth, the force must be exerted in that direction. Since the property that makes a rock fall had long been spoken of as "gravity," it was natural to call the force that brought about the motion *gravitational force* or the *force of gravity*.

It would therefore seem that the principle of inertia depends upon a circular argument. We begin by stating that a body will behave in a certain way unless a force is acting on it. Then, whenever it turns out that a body does not behave in that way, we invent a force to account for it.

Such circular argumentation would be bad indeed if we set

about trying to prove Newton's first law, but we do not do this. Newton's laws of motion represent assumptions and definitions and are not subject to proof. In particular, the notion of "inertia" is as much an assumption as Aristotle's notion of "natural place." There is this difference between them, however: The principle of inertia has proved extremely useful in the study of physics for nearly three centuries now and has involved physicists in no contradictions. For this reason (and not out of any considerations of "truth") physicists hold on to the laws of motion and will continue to do so.

To be sure, the new relativistic view of the universe advanced by Einstein makes it plain that in some respects Newton's laws of motion are only approximations. At very great velocities and over very great distances, the approximations depart from reality by a considerable amount. At ordinary velocities and distance, however, the approximations are extremely good.*

Forces and Vectors

The term *force* comes from the Latin word for "strength," and we know its common meaning when we speak of the "force of circumstance" or the "force of an argument" or "military force." In physics, however, force is defined by Newton's laws of motion. A force is that which can impose a change of velocity on a material body.

We are conscious of such forces, usually (but not always), as muscular effort. We are conscious, furthermore, that they can be exerted in definite directions. For instance, we can exert a force on an object at rest in such a way as to cause it to move away from us. Or we can exert a similar force in such a way as to cause it to move toward us. The forces are clearly exerted in different directions, and in common speech we give such forces two separate names. A force directed away from ourselves is a *push;* one directed toward ourselves is a *pull.* For this reason, a force is sometimes defined as "a push or a pull," but this is actually no definition at all, for it only tells us that a force is either one kind of force or another kind of force.

* It is sometimes said that Einstein's view of the universe "disproved" Newton's view. This is too simple a view. Actually, Einstein's view is more useful over a wider range of circumstances. Under ordinary circumstances, however, the Einsteinian view works out to be just about identical with the Newtonian view. In this book, ordinary circumstances only will be involved, and it will not be necessary to introduce relativity.

A quantity that has both size and direction, as force does, is a *vector quantity,* or simply a *vector.* One that has size only is a *scalar quantity.* For instance, distance is usually treated as a scalar quantity. An automobile can be said to have traveled a distance of 15 miles regardless of the direction in which it was traveling.

On the other hand, under certain conditions direction does make a difference when it is combined with the size of the distance. If town B is 15 miles north of town A, then it is not enough to direct a motorist to travel 15 miles to reach town B. The direction must be specified. If he travels 15 miles north, he will get there; if he travels 15 miles east (or any direction other than north), he will not. If we call a combination of size and direction of distance traveled *displacement,* then we can say that displacement is a vector.

The importance of differentiating between vectors and scalars is that the two are manipulated differently. For instance, in adding scalars it is sufficient to use the ordinary addition taught in grade school. If you travel 15 miles in one direction, then travel 15 miles in another direction, the total distance you travel is 15 plus 15, or 30 miles. Whatever the directions, the total mileage is 30.

If you travel 15 miles north, then another 15 miles north, the total displacement is, to be sure 30 miles north. Suppose, however, that you travel 15 miles north, then 15 miles east. What is your total displacement? How far, in other words, are you from your starting point? The total distance traveled is still 30 miles, but your final displacement is 21.2 miles northeast. If you travel 15 miles

Scalar and vector

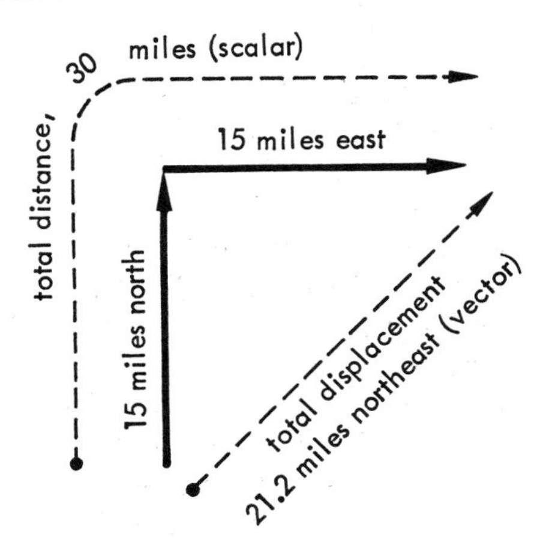

north and then 15 miles south, you have still traveled 30 miles altogether, but your total displacement is 0 miles, for you are back at your starting point.

So there is both ordinary addition, involving scalars, and *vector addition*, involving vectors. In ordinary addition $15 + 15$ is always 30; in vector addition, $15 + 15$ can be anything from 0 to 30, depending on circumstances.

Since force is a vector, two forces are added together according to the principles of vector addition. If one force is applied to a body in one direction and an exactly equal force is applied in the opposite direction, the sum of the two forces is zero; in such a case, even though forces are involved, a body subjected to them does not change its velocity. If it is at rest, it remains at rest. In fact, in every case where a body is at rest in the real world, we can feel certain that this does not mean there are no forces present to set it into motion. There are *always* forces present (the force of gravitation if nothing else). If there is rest, or unchanging velocity, that is because there is more than one force present and because the vector sum of all the forces involved is zero.

If the vector sum of all the forces involved is *not* zero, then there is an unbalanced force (mentioned in my definition of Newton's first law), or a net force. Whenever I speak of a force exerted on a body, it is to be understood that I mean a net force.

A particular force may have one of several effects on a moving body. The force of gravity, for instance, is directed downward toward the ground, and a falling body, moving in the direction of the gravitational pull, travels at a greater and greater velocity, undergoing an acceleration of 9.8 m/sec^2.

A body propelled upward, however, is moving in a direction opposite to that of the force of gravity. Consequently, it seems to be dragged backward by the force, moving more and more slowly. It finally comes to a halt, reverses its direction, and begins to fall. Such a slowing of velocity may be called "deceleration" or "negative acceleration." However, it would be convenient if a particular force was always considered to produce a particular acceleration. To avoid speaking of negative acceleration, we can instead speak of negative velocity.

In other words, let us consider velocity to be a vector. This means that a movement of 40 m/sec downward cannot be considered the same as a movement of 40 m/sec upward. The easiest way to distinguish between opposed quantities is to consider one positive and the other negative. Therefore, let us say that the downward motion is +40 m/sec and the upward one is −40 m/sec.

Since a downward force produces a downward acceleration* (acceleration being a vector, too), we can express the size of the acceleration due to gravity not as merely 9.8 m/sec, but as +9.8 m/sec.

If a body is moving at +40 m/sec (downward, in other words), the effect of acceleration is to increase the size of the figure. Adding two positive numbers by vector addition gives results similar to those of ordinary addition; therefore, after one second, the body is moving +49.8 m/sec, after another second, +59.6 m/sec, and so on. If, on the other hand, a body is moving at −40 m/sec (upward), the vector addition of a positive quantity resembles ordinary subtraction, as far as the figure itself is concerned. After one second, the body will be traveling −30.2 m/sec; after two seconds, −20.4 m/sec; and after four seconds −0.8 m/sec. Shortly after the four-second mark, the body will reach a velocity of 0 m/sec, and at that point it will come to a momentary halt. It will then begin to fall, and after five seconds its velocity will be +9.0 m/sec.

As can be seen, the acceleration produced by the force of gravity is the same whether the body is moving upward or downward, and yet there is something that is different in the two cases. The body covers more and more distance each second of its downward movement, but less and less distance each second of its upward movement. The amount of distance covered per unit time can be called the velocity or speed of the body.

In ordinary speech speed and velocity are synonymous, but not so in physics. Speed is a scalar quantity and does not involve direction. An object moving 16 m/sec north is traveling at the same speed as one moving at 16 m/sec east, but the two are traveling at different velocities. In fact, it is possible under certain circumstances to arrange a force so as to cause it to make a body move in circles. The speed, in that case, might not change at all, but the velocity (which includes direction) would be constantly changing.

Of the two terms, velocity is much more frequently used by physicists, for it is the broader and more convenient term. For

* We know from experience that if we push an object away from us, it moves away from us; if it is already moving, it moves away more rapidly. In the same way, to stop a moving body we always exert a force in the direction opposite to its motion. Experience tells us that the acceleration produced by a force is in the same direction as the force.

instance, we might define a force as "that which imposes a change in the speed of a body, or its direction of motion, or both." Or we might define it as "that which imposes a change in the velocity of a body," a briefer but as fully meaningful a phrase.

Since a change in velocity is an acceleration, we might also define a force as "that which imposes an acceleration on a body, the acceleration and force being in the same direction."

Mass

Newton's first law explains the concept of a force, but something is needed to allow us to measure the strength of a force. If we define a force as something that produces an acceleration, it would seem logical to measure the size of a force by the size of the acceleration it brings about. When we restrict ourselves to one particular body, say a basketball, this makes sense. If we push the basketball along the ground with a constant force, it moves more and more quickly, and after ten seconds it moves with a velocity, let us say, of 2 m/sec. Its acceleration is 2 m/sec divided by 10 seconds, or 0.2 m/sec^2. If you start from scratch and do not push quite as hard, at the end of ten seconds the basketball may be moving only 1 m/sec; it will therefore have undergone an acceleration of 0.1 m/sec^2. Since the acceleration is twice as great in the first case, it seems fair to suppose that the force was twice as great in the first case as in the second.

But if you were to apply the same forces to a solid cannonball instead of a basketball, the cannonball will not undergo anything like the previously noted accelerations. It might well take every scrap of force you can exert to get the cannonball to move at all.

Again, when a basketball is rolling along at 2 m/sec, you can stop it easily enough. The velocity change from 2 m/sec to 0 m/sec requires a force to bring it about, and you can feel yourself capable of exerting sufficient force to stop the basketball. Or you can kick the basketball in mid-motion and cause it to veer in direction. A cannonball moving at 2 m/sec, however, can only be stopped by great exertion, and if it is kicked in mid-motion it will change its direction by only a tiny amount.

A cannonball, in other words, behaves as though it possesses more inertia than a basketball and therefore requires correspondingly more force for the production of a given acceleration. Newton used the word *mass* to indicate the quantity of inertia possessed by a body, and his second law of motion states:

The acceleration produced by a particular force acting on a body is directly proportional to the magnitude of the force and inversely proportional to the mass of the body.

Now I have already explained that when x is said to be directly proportional to y, then $x = ky$ (see page 19). However, in saying that x is inversely proportional to another quantity, say z, we mean that as z increases x decreases by a corresponding amount and vice versa. Thus, if z is increased threefold, x is reduced to 1/3; if z is increased elevenfold x is reduced to 1/11, and so on. Mathematically, this notion of an inverse proportion is most simply expressed as $x \propto 1/z$, for then when z is 3, x is 1/3; when z is doubled to 6, x is halved to 1/6, and so on. We can change the proportionality to an equality by multiplying by a constant, so that if x is inversely proportional to z, $x = k/z$. If x is both directly proportional to y and inversely proportional to z, then $x = ky/z$.

With this in mind, let's have a represent the acceleration, f the magnitude of the force and m the mass of the body. We can then represent Newton's second law of motion as:

$$a = \frac{kf}{m} \qquad \text{(Equation 3–1)}$$

Let us next consider the units in which we will measure each quantity, turning to mass first, since we have not yet taken it into account in this book. You may think that if I say a cannonball is more massive than a basketball, I mean that it is heavier. Actually, I do not. "Massive" is not the same as "heavy," and "mass" is not the same as "weight," as I shall explain later in the book (see page 53). Nevertheless, there is a certain similarity between the two concepts and they are easily confused. In common experience, as bodies grow heavier they also grow more massive, and physicists have compounded the chance of confusion by using units of mass of a sort which nonphysicists usually think of as units of weight.

In the metric system, two common units for mass are the *gram* (gm) and the *kilogram* (kg). A gram is a small unit of mass. A quart of milk has a mass of about 975 grams, for example. The kilogram, as you might expect from the prefix, is equal to 1000 gm and represents, therefore, a trifle more than the mass of a quart of milk.

(In common units, mass is frequently presented in terms of "ounces" and "pounds," these units also being used for weight. In this book, however, I shall confine myself to the metric system as far as possible, and shall use common units, quarts, for example, only when they are needed for clarity.)

In measuring the magnitude of a force, two quantities must be considered: acceleration and mass. Using metric units, acceleration is most commonly measured as m/sec^2 or cm/sec^2, while mass may be measured in gm or kg. Conventionally, whenever distance is given in meters, the mass is given in kilograms, both being comparatively large units. On the other hand, whenever distance is given in the comparatively small centimeters, mass is given in the comparatively small grams. In either case, the unit of time is the second.

Consequently, the units of many physical quantities may be compounded of centimeters, grams, and seconds in various combinations; or of meters, kilograms, and seconds in various combinations. The former is referred to as the *cgs system,* the latter is the *mks system.* A generation or so ago, the cgs system was the more frequently used of the two, but now the mks system has gained in popularity. In this book, I will use both systems.

In the cgs system, a unit force is described as one that will produce an acceleration of 1 cm/sec^2 on a mass of 1 gm. A unit force is therefore 1 cm/sec^2 multiplied by 1 gm. (In multiplying the two algebraic quantities *a* and *b,* we can express the product simply as *ab.* We manipulate units as we would algebraic quantities, but to join words together directly would be confusing, so I will make use of a hyphen, which, after all, is commonly used to join words.) The product of 1 cm/sec^2 and 1 gm is therefore 1 gm-cm/sec^2—the magnitude of the unit force. The unit of force, gm-cm/sec^2, is frequently used by physicists, but since it is an unwieldy mouthful, it is more briefly expressed as the *dyne* (from a Greek word for "force").

Now let's solve Equation 3–1 for *k*. This works out to:

$$k = \frac{ma}{f} \qquad \text{(Equation 3–2)}$$

The value of *k* is the same for any consistent set of values of *a, m* and *f,* so we may as well take simple ones. Suppose we set *m* equal to 1 gm and *a* equal to 1 cm/sec^2. The amount of force that corresponds to such a mass and acceleration is, by our definition, 1 gm-cm/sec^2 (or 1 dyne).

Inserting these values into Equation 3–2, we find that:

$$k = \frac{1\ \text{cm/sec}^2 \times 1\ \text{gm}}{1\ \text{gm-cm/sec}^2} = \frac{1\ \text{gm-cm/sec}^2}{1\ \text{gm-cm/sec}^2} = 1$$

In this case, at least, *k* is a pure number.

Since k is equal to 1, we find that Equation 3–2 can be written as $ma/f = 1$, and, therefore:

$$f = ma \qquad \text{(Equation 3–3)}$$

provided we use the proper sets of units—that is, if we measure mass in gm, acceleration in cm/sec^2, and force in dynes.

In the mks system of measurement, acceleration is measured in m/sec^2 and mass in kg. The unit of force is then defined as that amount of force which will produce an acceleration of 1 meter per second per second when applied to 1 kilogram of mass. The unit force in this system is therefore 1 m/sec^2 multiplied by 1 kg, or 1 kg-m/sec^2. This unit of force is stated more briefly as 1 *newton,* in honor of Isaac Newton, of course. Equation 3–3 is still true, then, for a second combination of consistent units—where mass is measured in kg, acceleration in m/sec^2, and force in newtons.

From the fact that a kilogram is equal to 1000 grams and that a meter is equal to 100 centimeters, it follows that 1 kg-m/sec^2 is equal to (1000 gm) (100 cm)/sec^2, or 100,000 gm-cm/sec^2. To put it more compactly, 1 newton = 100,000 dynes.

Before leaving the second law of motion, let's consider the case of a body subject to no net force at all. In this case we can say that $f = 0$, so that Equation 3–3 becomes $ma = 0$. But any material body must have a mass greater than 0, so the only way in which ma can equal 0, is to have a itself equal 0.

In other words, if no net force acts on a body, it undergoes no acceleration and must therefore either be at rest or traveling at a constant velocity.

This last remark, however, is an expression of Newton's first law of motion. It follows, then, that the second law of motion includes the first law as a *special case.* If the second law is stated and accepted, there is no need for the first law. The value of the first law is largely psychological. The special case of $f = 0$, once accepted, frees the mind of the "common-sense" Aristotelian notion that it is the natural tendency of objects to come to rest. With the mind thus freed, the general case can then be considered.

Action and Reaction

A force, to exist, must be exerted by something and upon something. It is obvious that something cannot be pushed unless something else is pushing. It should also be obvious that something cannot push unless there is something else to be pushed. You cannot imagine pushing or pulling a vacuum.

A force, then, connects two bodies, and the question arises as to which body is pushing and which is being pushed. When a living organism is involved, we are used to thinking of the organism as originating the force. We think of ourselves as pushing a cannonball and of a horse as pulling a wagon, not of the cannonball as pushing us or the wagon as pulling the horse.

Where two inanimate objects are concerned, we cannot be so certain. A steel ball falling upon a marble floor is going to push against the floor when it strikes and therefore exert a force upon it. On the other hand, since the steel ball bounces, the floor must have exerted a force upon the ball. Whereas the force of the ball was exerted downward onto the floor, the force of the floor was exerted upward onto the ball.

In this and in many other similar cases there would seem to be two forces, equal in magnitude and opposite in direction. Newton made the generalization that this was always and necessarily true in all cases and expressed it in his third law of motion. This is often stated very briefly: "For every action, there is an equal and opposite reaction." It is for that reason that the third law is sometimes referred to as the "law of action and reaction."

Perhaps, however, this is not the best way of putting it. By speaking of action and reaction, we are still thinking of a living object exerting a force on some inanimate object that then responds automatically. One force (the "action") seems to be more important and to precede in time the other force (the "reaction").

But this is not so. The two forces are of exactly equal importance (from the standpoint of physics) and exist simultaneously. Either can be viewed as the "action" or the "reaction." It would be better, therefore, to state the law something like this:

Whenever one body exerts a force on a second body, the second body exerts a force on the first body. These forces are equal in magnitude and opposite in direction.

So phrased, the law can be called the "law of interaction."

The third law of motion can cause confusion. People tend to ask: "If every force involves an equal and opposite counterforce, why don't the two forces always cancel out by vector addition, leaving no net force at all?" (If that were so, then acceleration would be impossible and the second law would be meaningless.)

The answer is that two equal and opposite forces cancel out by vector addition when they are exerted on the same body. If a force were exerted on a particular rock and an equal and opposite force were also exerted on that same rock, there would be no net force; the rock, if at rest, would remain at rest no matter how large each force was. (The forces might be large enough to crush the rock to powder, but they wouldn't move the rock.)

The law of interaction, however, involves equal and opposite forces exerted on *two separate bodies.* Thus, if you exert a force on a rock, the equal and opposite force is exerted by the rock on you; the rock and you each receive a single unbalanced force. If you exert a force on a rock and let go of it at the same time, the rock, in response to this single force, is accelerated in the direction of that force—that is, away from you. The second force is exerted on you, and you in turn accelerate in the direction of that second force—that is, in the direction opposite to that in which the rock went flying. Ordinarily, you are standing on rough ground and the friction between your shoes and the ground (accentuated, perhaps, by muscular bracing) introduces new forces that keep you from moving. Your acceleration is therefore masked, so the true effect of the law of interaction may go unnoticed. However, if you were standing on very smooth, slippery ice and hurled a heavy rock eastward, you would go sliding westward.

In the same way, the gases formed by the burning fuel in a rocket engine expand and exert a force against the interior walls of the engine, while the walls of the engine exert an equal and opposite force against the gases. The gases are forced into an acceleration downward, so that the walls (and the attached rocket) are forced into an acceleration upward. Every rocket that rises into the air is evidence of the validity of Newton's third law of motion.

In these cases, the two objects involved are physically separate, or can be physically separated. One body can accelerate in one direction and the other in the opposite direction. But what of the case where the two bodies involved are bound together? What of a horse pulling a wagon? The wagon also pulls the horse in the opposite direction with an equal force. Yet horse and wagon do not accelerate in opposite directions. They are hitched together and both move in the same direction.

If the forces connecting wagon and horse were the only ones involved, there would indeed be no overall movement. A wagon and horse on very slippery ice would get nowhere, no matter how the horse might flounder. On ordinary ground, there are frictional effects. The horse exerts a force on the earth and the earth exerts a counterforce on the horse (and its attached wagon). Consequently, the horse moves forward and the earth moves backward.

The earth is so much more massive than the horse that its acceleration backward (remember that the acceleration produced by a force is inversely proportionate to the mass of the body being accelerated) is completely unmeasurable. We are aware only of the horse's motion, and so it seems to us that the horse is pulling the wagon. We find it hard to imagine that the wagon is also pulling at the horse.

CHAPTER 4

Gravitation

Combination of Forces

Newton had already turned his attention to an important and very profound question while still in his twenties. Did the laws of motion apply only to the earth and its environs, or did they apply to the heavenly bodies as well? The question first occurred to him on his mother's farm when he saw an apple fall from a tree* and began to wonder whether the moon was in the grip of the same force as the apple was.

It might seem at first thought that the moon could *not* be in the grip of the same force as the apple, since the apple fell to earth and the moon did not. Surely, if the same force applied to both, the same acceleration would affect both, and therefore both would fall. However, this is an oversimplification. What if the moon is indeed in the grip of the same force as the apple and therefore moving downward toward the earth; in addition, what if the moon also undergoes a second motion? What if it is the combination of two motions that keeps the moon circling the earth and never quite falling all the way?

This notion of an overall motion being made up of two or more component motions in different directions was by no means

* It did *not* hit him on the head, despite the hundreds of cartoons drawn by hundreds of cartoonists.

an easy concept for scientists to accept. When Nicholas Copernicus (1473–1543) first suggested that the earth moved about the sun (rather than vice versa), some of the most vehement objections were to the effect that if the earth rotated on its axis and (still worse) moved through space in a revolution about the sun, it would be impossible for anything movable to remain fixed to the earth's surface. Anyone who leaped up in the air would come down many yards away, since the earth beneath him would have moved while he was in the air. Those arguing in this manner felt that this point was so obvious as to be unanswerable.

Those who accepted the Copernican notion of the motion of the earth had to argue that it was indeed possible for an object to possess two motions at once: that a leaping man, while moving up and down, could also move with the turning earth and therefore come down on the same spot from which he leaped upward.

Galileo pointed out that an object dropped from the top of the mast of a moving ship fell to a point at the base of the mast. The ship did not move out from under the falling object and cause it to fall into the sea. The falling object, while moving downward, must also have participated in the ship's horizontal motion. Galileo did not actually try this, but he proposed it as what is today called a "thought experiment." Even though it was proposed only in thought, it was utterly convincing; ships had sailed the sea for thousands of years, and objects must have been dropped from mast-tops during all those years, yet no seaman had ever reported that the ship had moved out from under the falling object. (And of course, we can flip coins on board speeding jets these days and catch them as they come down without moving our hand. The coin participates in the motion of the jet even while also moving up and down.)

Why then did some scholars of the sixteenth and seventeenth centuries feel so sure that objects could not possess two different motions simultaneously? Apparently it was because they still possessed the Greek habit of reasoning from what seemed valid basic assumptions and did not always feel it necessary to check their conclusions against the real universe.

For instance, the scholars of the sixteenth century reasoned that a projectile fired from a cannon or a catapult was potentially subject to motions resulting from two causes—first the impulse given it by the cannon or catapult, and secondly, its "natural motion" toward the ground. Assuming, to begin with, that an object could not possess two motions simultaneously, it would seem necessary that one motion be completed before the second began. In other words, it was felt that the cannonball would travel in a straight line in whatever direction the cannon pointed, until the impulse of the gunpowder explosion was used up; it would then at once fall downward in a straight line.

Galileo maintained something quite different. To be sure, the projectile traveled onward in the direction in which it left the cannon. What's more, it did so at constant velocity, for the force of the gunpowder explosion was exerted once and no more. (Without a continuous force there would be no continuous acceleration, Newton later explained.) *In addition,* however, the cannonball began dropping as soon as it left the cannon's mouth, in accordance with the laws of falling bodies whereby its velocity downward increased with a constant acceleration (thanks to the continuous presence of a constant force of gravity). It was easy to show by geometric methods that an object that moved in one direction at a constant speed, and in another at a speed that increased in direct proportion with time, would follow the path of a curve called a "parabola." Galileo also showed that a cannonball would have the greatest range if the cannon were pointed upward at an angle of 45° to the ground.

A cannon pointed at a certain angle would deliver a cannonball to one place if the early views of the cannonball's motions were correct, and to quite another place if Galileo's views were correct. It was not difficult to show that it was Galileo who was correct.

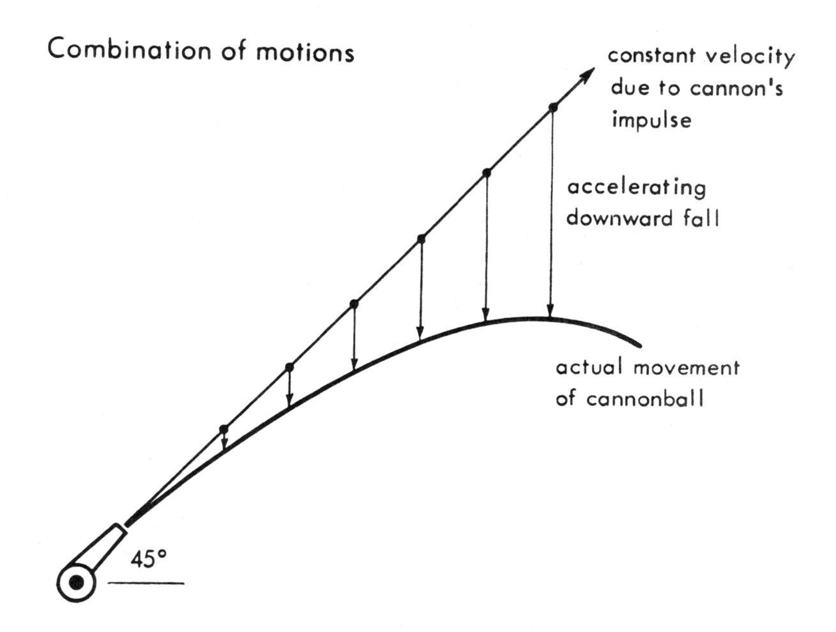

Indeed, the gunners of the time may not have dabbled much in theory, but they had long aimed their weapons in such a way as to take advantage of a parabolic motion of the cannonball.* In short, the possibility of a body's possessing two or more motions at once was never questioned after the time of Galileo.

How can separate motions be added together and a resultant motion obtained? This can be done by vector addition, according to a method most easily presented in geometric form. Consider two motions in separate directions, the two directions at an angle α to each other. (The symbol α is the Greek letter "alpha." Greek letters are often used in physics as symbols, in order to ease the overload on ordinary letters of the alphabet.) The two motions can then be represented by two arrows set at angle α, the two arrows having lengths in proportion to the two velocities. (If the velocity of one is twice that of the other, then its corresponding arrow is twice as long.) If the two arrows are made the sides of a parallelogram, the resultant motion built up out of the two component motions is

* The correct aim, especially nowadays when shells are hurled for miles, requires more than the idealized parabola of Galileo. Many factors of the real world—as, for instance, the curvature of the earth's surface, the manner in which its speed of rotation varies with latitude, the amount of air resistance (which varies with height and temperature), the strength and direction of the wind, the motion of the object aimed at and the object carrying the cannon (if both are ships, for instance), affect the situation. All these effects merely serve to modify the parabola, however, and do not affect the basic worth of Galileo's argument, which serves only to present a greatly simplified but nevertheless vastly useful model of the real situation.

Parallelogram of forces

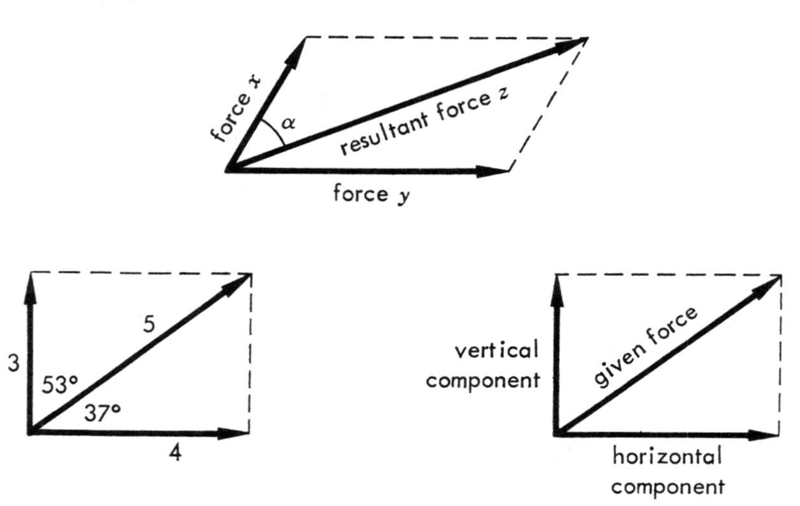

represented by the diagonal of the parallelogram, the one that lies in a direction intermediate between those of the two components.

Given the values of the two velocities and the angle between them, it is possible to calculate the size and direction of the resultant velocity even without the geometric construction, although the latter is always useful to lend visual aid. For instance, if one velocity is 3 m/sec in one direction, and the other is 4 m/sec in a direction at right angles to the first, then the resultant velocity is 5 m/sec in a direction that makes an angle of just under 37° with the larger component and just over 53° with the smaller.

In the same way, a particular velocity can be separated into two component velocities. The particular velocity is made the diagonal of a parallelogram, and the adjacent sides of the parallelogram represent the component velocities. This can be done in an infinite number of ways, since the line representing a velocity or force can be made the diagonal of an infinite number of parallelograms. As a matter of convenience, however, a velocity is divided into components that are at right angles to each other. The parallelogram is then a rectangle.

This device of using a parallelogram can be employed for the combination or resolution of any vector quantity. It is very frequently used for forces, as a matter of fact, so one usually speaks of this device as involving a *parallelogram of force.*

The Motion of the Moon

Now let us return to the moon. It travels about the earth in an elliptical orbit. The ellipse it describes in its revolution about the earth is not very far removed from a circle, however. The moon travels in this orbit with a speed that is almost constant.

Although the moon's speed is approximately constant, its velocity certainly is not. Since it travels in a curved path, its direction of motion changes at every instant and, therefore, so does its velocity. To say that the moon is continually changing its velocity is, of course, to say that it is subject to a continuing acceleration.

If the moon is viewed as traveling at a constant speed along a uniformly circular path (which is at least approximately true), it can be considered to be changing its direction of motion by precisely the same amount in each successive unit of time. It is therefore undergoing a constant acceleration and must be subject to a constant force, according to Newton's second law of motion. Since the shift in the direction of motion is always toward the earth, the acceleration, and therefore the force, must be directed toward the earth.

Certainly, if there is a force attracting the moon to the earth, it might well be the same as the force attracting the apple to the earth. However, if that were so, and the moon were undergoing a constant acceleration toward the earth under the pull of a constant force, why does it not fall to the earth as an apple would?

We can see why if we break the moon's motion into two component motions at mutual right angles. One of the components can be drawn as an arrow pointing toward the earth, along a radius

Motion of the moon

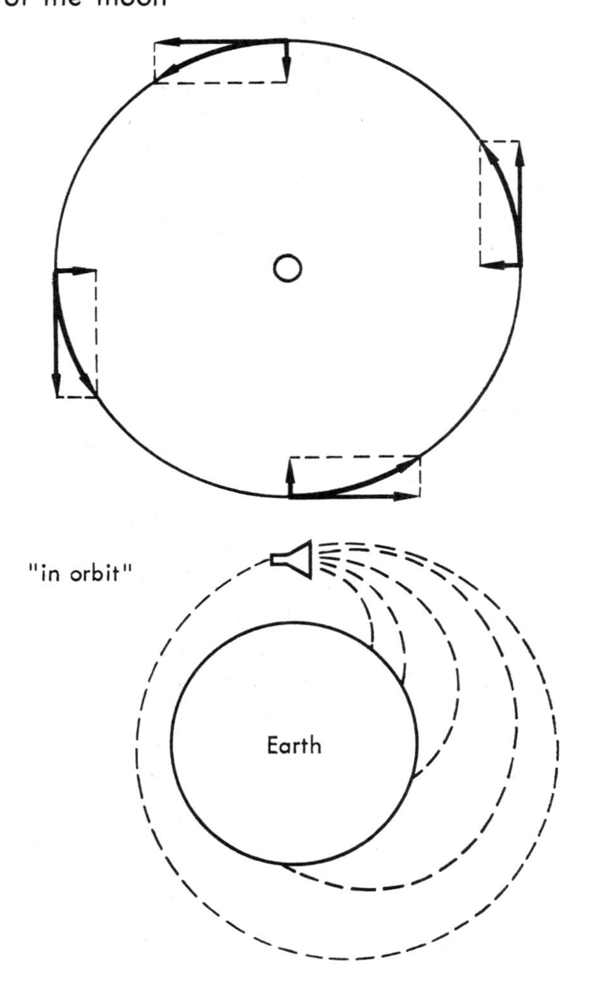

of the moon's circular orbit. This represents the motion in response to the force attracting the moon to the earth. The other component is drawn at right angles to the first and is therefore tangent to the circle of the moon's orbit. This tangential motion represents that which the moon would experience if there were no force attracting it to the earth. The actual motion lies between the two, and the tangential component carries the moon to one side just far enough in a unit of time to make up for the motion toward the earth in that same unit of time. The moon, in other words, is always falling toward the earth, but it also "sidesteps."

In a sense, this "sidestep" means that the earth's surface curves away from the moon just as fast as the moon approaches by falling, and the distance between earth and moon remains the same. This can be made plain if one supposes a projectile fired horizontally from a mountaintop on earth with greater and greater velocity. The greater the velocity, the farther the projectile travels before striking the ground. The farther it travels, the more the surface of the spherical earth curves away from it, thus adding to the distance the projectile covers. Finally, if the projectile is shot forward with sufficient velocity, its rate of fall just matches the rate at which the earth's surface curves away, and the projectile "remains in orbit." It is in this fashion that satellites are placed in orbit, and it is in this fashion that the moon remains in orbit.

In considering the moon's motion, therefore, we need only consider that component which is directed toward the earth, and we can ask ourselves whether that component is the result of the same force that attracts the apple. Let's first concentrate on the apple and see how to interpret the force between it and the earth in the light of the laws of motion.

In the first place, all apples fall with the same acceleration regardless of how massive they are. But if one apple has twice the mass of a second apple, yet falls at the same acceleration, the first apple must be subjected to twice the force, according to the second law of motion. The force attracting the apple to the earth (often spoken of as the *weight* of the apple) must be proportional to the mass of the apple.

But according to the third law of motion, whenever one body exerts a force on a second, the second is exerting an equal and opposite force on the first. This means that if the earth attracts the apple with a certain downward force, the apple attracts the earth with an equal upward force.

That seems odd. How can a tiny apple exert a force equal

to that exerted by the tremendous earth? If it did, one would expect the apple to attract other objects as the earth does, and the apple most certainly does not. The logical way to explain this is to suppose that the attractive force between apple and earth depends not only on the mass of the apple but on the mass of the earth as well. It cannot depend on the sum of the masses, for when the mass of the apple is doubled, the sum of the mass of the apple and the earth remains just about the same as before, and yet the force of attraction doubles. Instead it must depend upon the product of the masses.

If we multiply the masses, the small mass has just as much effect on the final product as the large one. Thus, the minute quantity a multiplied by the tremendous quantity b yields the product ab. If a is now doubled, it becomes equal to $2a$. If that is multiplied by b, the product is $2ab$. Thus doubling one of two factors in a multiplication, however small that factor may be, doubles the product. And doubling the mass of the apple doubles the size of the force between the apple and the earth.

Furthermore, the apple does not measurably attract any other object of ordinary size because the product of the masses of two ordinary objects is an infinitesimal fraction of the product of the mass of either object and that of the vast earth. The attractive force between two objects of ordinary size is correspondingly smaller, and while the force does exist, it is far too small to be noticed in the ordinary course of events.

Since the earth attracts all material objects to itself (even the gaseous atmosphere is held firmly to the planet through gravitational force) it would seem that the force is produced by mass in whatever form the latter occurs. In that case, the earth need not be involved. Any two masses ought to interact gravitationally, and if we notice the force only when the earth is involved, that is only because the earth itself is the only body in our neighborhood massive enough to produce a gravitational force sufficient to obtrude itself on our notice.

Such is the essence of Newton's contribution. He did not discover the law of gravity merely in the sense that all earthly objects are attracted to the earth. (This limited concept is at least as old as Aristotle and the word "gravity" was used in that sense for many centuries before Newton.) What Newton pointed out was that *all* masses attracted *all other* masses, so that the earth's attraction was not unique. Because Newton maintained that there was a gravitational attraction between any two material bodies in the universe, his generalization is called the *law of universal gravitation*. The adjective "universal" is the most important word of the phrase.*

If this were all, we would now be able to decide the size of that component of the moon's motion that is directed toward the earth. All bodies on earth fall with the same acceleration, and therefore it might be decided that the moon, if it were in the grip of the same force, ought to do the same. In one second it ought to fall some 4.9 meters toward the earth. Actually, the earthward component of the moon's motion is much smaller than that.

To account for this, one might suppose that the earth's gravitational force weakens with distance, and certainly this seems a reasonable supposition. It is common experience that many things weaken with distance. Such is the case with light and sound, to name two common phenomena with which man has always been familiar.

And yet is such weakening supported by experimental evidence? At first blush it might seem that it wasn't. A stone dropped from a height of 100 meters falls with an acceleration of 9.8 m/sec^2, and one dropped from a height of 200 meters falls with the same acceleration. If the gravitational force decreased with distance from the earth, ought not the fall from a greater height involve a smaller acceleration? In fact, ought not the acceleration increase steadily as the stone approached the earth, instead of remaining constant, as it does?

But Newton's view was that *all* bodies attracted *all other* bodies. A falling rock is attracted not only by the portion of the earth making up the surface immediately under it, but also by the portions deep underneath, all the way to the center and beyond—to the antipodes, 12,740 kilometers (8000 miles) distant. It should also be attracted by portions at all distances to the north, east, south, west and points in between.

It would seem reasonable that for a body like the earth, which has nearly the symmetrical shape of a sphere, we could simplify matters. The pull from the north would balance the pull from the south; the pull from the west would balance the pull from the east; the distant pull of the antipodes would balance the nearby pull of the surface directly beneath. In consequence, we might suppose that the net effect is that the overall pull of the earth would be concentrated exactly at its center.†

* It is possible to contrive an exception. If the earth were hollow, there would be no net gravitational force within the hollow. A body within the hollow would not be attracted by the earth. However, this is a highly artificial exception that has no practical significance, for any body that is large enough to have an important gravitational field is too large to support a hollow structure.

† It is so easy to say "it would seem reasonable" and end with the happy

The radius of the earth is about 6370 kilometers (3960 miles). An object falling from a height of 100 meters (0.1 kilometers) begins its fall, therefore, from a point 6370.1 kilometers from the earth's center, while one falling from a height of 200 meters begins its fall from a point 6370.2 kilometers from the center. The difference is so insignificant that the gravitational attraction can be considered constant over that small distance. (Actually, modern instruments can measure the difference in the strength of the gravitational field over even such small distances with considerable accuracy.)

However, the moon's distance from the earth (center to center) is, on the average, 384,500 kilometers (239,000 miles). This is 60.3 times as far from the earth's center as is an object on the earth's surface. With a sixtyfold increase in distance, gravitational force might indeed decrease considerably.

But how much is "considerably"?

The earth attracts bodies on every portion of its surface; therefore, the gravitational force may be considered as radiating outward from the earth in all directions. If the force does this, it can be viewed as occupying the surface of a sphere that is inflating to a larger and larger size as it recedes from the earth. If a fixed amount of gravitational force is stretched out over the surface of such a growing sphere, then the intensity of the force at a given spot on the surface ought to decrease as the total surface area grows larger.

From solid geometry it is known that the surface area of a sphere is directly proportional to the square of its radius. If one sphere has three times the radius of another, it has nine times the surface area. As distance between two bodies increases, then the gravitational force between them ought to be inversely proportional to the square of that distance. (This relationship is familiarly known as the *inverse-square law*. Not only gravitation but also such phenomena as the intensity of light, the intensity of magnetic attraction, and the intensity of electrostatic attraction weaken as the square of the distance.)

In comparing the motion of the moon to the motion of an

---

conclusion that the gravitational force of the earth seems to originate at its center. Yet it took the transcendent genius of Newton eighteen years to convince himself of this fact, and he had to invent that branch of mathematics now known as the calculus, before he could prove it to his own satisfaction and that of others. Throughout this book, I say "it would seem reasonable" and "it is clear" and "it is easy to see" when I am reaching conclusions that in actual fact were attained only through great ingenuity and hard labor. In doing so, my conscience hurts—but in an introductory book I have no alternative.

apple on the earth's surface, we must remember that the moon is 60.3 times as far from the earth's center as the apple is and that the gravitational force on the moon is weaker by a factor of 60.3 times 60.3, or 3636. Whereas an apple falls 4.9 meters in the first second of fall, the moon should fall 1/3636 that distance, or 0.0013 meters, in a second of fall. (A thousandth of a meter is a *millimeter,* so that 0.0013 meters is equal to 1.3 millimeters.)

Indeed, astronomical measurements show that the moon in its course about the earth does indeed deviate from a straight line course by about 1.3 millimeters in one second. This alone would have been sufficient to make it strongly probable that the same force that held the apple held the moon. However, Newton went on to show how gravitational force on a universal scale would account for the fact that the orbit of the moon about the earth is an ellipse with the earth at one focus; that the planets revolved about the sun in a similar elliptical manner; that the tides took place as they did; that the precession of equinoxes took place, and so on. The one simple and straightforward generalization explained so much that it had to be triumphantly accepted by the scientific community.

A century after Newton's death, the German-English astronomer William Herschel (1738–1822) discovered instances of far distant stars that revolved about each other in strict accordance with Newton's law of universal gravitation, which thus seemed universal indeed. Unseen planets were eventually discovered through the tiny gravitational effects produced by their otherwise unsuspected presence. It is no wonder that Newton's working out of the law of universal gravitation is often considered as the greatest single discovery in the history of science.*

The Gravitational Constant

Newton succeeded in establishing the generalization that any two bodies in the universe attract each other with a force (F) that is directly proportional to the product of the masses (m

* Nevertheless, Newton's generalization concerning gravity is only an approximation and is not absolutely correct. Already in the mid-nineteenth century, it was discovered that the planet Mercury had one small component of its motion that could not be explained by Newton's law. It remained unexplained until Albert Einstein advanced his "General Theory of Relativity" in 1915. This theory, more advanced, powerful and controversial than the Special Theory of 1905, involved a broader view of the universe than was implicit in Newton's laws. In all ordinary cases, the two views were just about equivalent. At certain extremes, however, the two views diverge, and when such extremes are tested it appears that Einstein's view, rather than Newton's, carries the day.

and m') of the bodies, and inversely proportional to the square of the distance (d) between them. To convert the proportionality to an equality, it is necessary, of course, to introduce a constant. The one introduced in this case is usually referred to as the *gravitational constant* and is symbolized as G. Newton's law of universal gravitation can then be expressed as:

$$F = \frac{Gmm'}{d^2} \qquad \text{(Equation 4–1)}$$

One problem left unsolved by Newton was the value of G.

To see why it was left unsolved, let's consider the famous case of Newton's falling apple and try to substitute values in Equation 4–1, in the mks system of units. We know the value of the distance from the apple to the earth's center and can set d equal to 6,370,-000 meters. There are ways of measuring the mass of the apple and m can be set at, let us say, 0.1 kilograms. As for the strength of the gravitational force (F) between the apple and the earth, it is equal (see Equation 3–3) to the product of the mass of the apple and the acceleration to which it is subjected by the action of gravity, according to Newton's second law of motion. The value of F, then is 0.1 kg times 9.8 m/sec^2, or 0.98 kg-m/sec^2.

This leaves us with two items still undetermined: G, the gravitational constant, and m', the mass of the earth. If we knew either one, we could calculate the other at once, but Newton knew neither, nor did anyone else in his time.

(You might wonder whether we could not eliminate the constant in Equation 4–1, as we did the constant in Equation 3–3. That, however, was done by a proper choice of units. We could do so here by inventing a unit called "earth-unit," for example, and saying that the earth had a mass of 1 earth-unit. We could further invent similar arbitrary units for the mass of the apple and the distance of the apple from the center of the earth. Such tricks would be of limited value, however. It is unsatisfying to be told that the earth has a mass of 1 earth-unit, and that is all we would find out in this way. What we want is the mass of the earth in terms of familiar objects—that is, in the units of the mks system. And for that we must know the value of G in the mks system.)

The law of universal gravitation implies that the value of G is the same under all conditions. Therefore, if we could measure the gravitational force between two bodies of known mass, separated by a known distance, we could at once determine G and from that the mass of the earth.

Unfortunately, the force of gravity is just about the weakest known force in nature. It takes a body with the enormous size of the earth to produce enough gravitational force to bring about an acceleration of 9.8 m/sec^2. The puny forces that can be produced by a few pounds of muscle can and do counter all that gravitational force whenever we chin ourselves, do pushups, jump upward, or climb a mountain.

For bodies that are large, though less massive than the earth, the decline in gravitational force has drastic effects. The earth maintains a firm grip on its atmosphere through the force of its gravity, but the planet Mars, which has only 1/10 the earth's mass, can hold only a thin atmosphere. The moon has an enormous mass by ordinary standards; nevertheless it is only 1/81 as massive as the earth and has a gravitational force too weak to hold any atmosphere at all.

Where bodies of ordinary size are concerned, the gravitational forces produced are completely insignificant. The mass in a mountain exerts a gravitational attraction on you, but you are aware of no difficulty in stepping away from such a mountain.

The problem, therefore, is how to measure so weak a force as that of gravity. We might speculate on possible ways of measuring the gravitational forces between two neighboring mountains, but the masses of individual mountains are not much easier to determine than the mass of the earth. Furthermore, the mountains are of irregular shape and the gravitational force is concentrated at some "center," a position that would be difficult to determine.

We would therefore have to measure the gravitational forces originating in symmetrical bodies small enough to be handled easily in the laboratory, and the measurement of the tiny gravitational forces to which such bodies would give rise might well be considered too difficult to lie within the realm of the possible.

The beginning of a solution to the problem came about, however, in the time of Newton himself, thanks to the work of the English scientist Robert Hooke (1635–1703).

As a preliminary to explaining Hooke's work, let us keep in mind that when forces are applied to a body, that body will change its shape as a result. If a plank of wood is suspended across two supports, and someone sits down on the center, that plank will bend under the load. If a rubber band is pulled at both ends in opposite directions, it stretches. If a sponge is clenched in a fist, it compresses, and if rotated at each end in opposite directions, it will twist. If pushed to the right at one end and to the left at the other, without being allowed to rotate, it will shear.

All these types of deforming forces can be referred to as *stresses*. The deformation undergone by the body under stress is a *strain*.

When an object undergoes deformation as the result of a stress, the original shape may be restored when the stress is removed. The wooden plank, after you stand up, unbends; the rubber band, after the pull is released, contracts to its normal size; the sponge once released from the compressive, twisting, or shearing force, springs back. Again, a steel ball flattens upon striking the ground, a baseball on striking the bat, and a golf ball on striking the club. When the deforming force is gone all are spheres once more. This tendency to return to the original shape after deformation under stress is called *elasticity*.

There is a limit to the elasticity of any substance, a point beyond which stress will produce a permanent deformation. For a substance such as wax, this point is easily reached and even light stresses will cause a lump of wax to change its shape permanently. (It is "plastic" rather than elastic.) A wooden plank will break if too great a force is exerted on its unsupported center. A rubber band will snap under too great a pull. A steel ball will be permanently flattened under too great a compression.

However, if one works with forces not strong enough to surpass this limit, one can arrive, as Hooke did, at a useful generalization that can be briefly expressed as follows:

The strain is proportional to the stress.

This is called *Hooke's law*. One would expect, from Hooke's law, that if a force x stretches a spring through a distance y, then a force of $2x$ will stretch it through a distance of $2y$, and a force of $x/2$ will stretch it through a distance of $y/2$. Suppose then that the amount of stretch produced by a known force is measured. Any force of unknown size (within the elastic limit) can then be measured by simply measuring the strain it produced.

This principle can be applied to any other form of stress that produces an easily measured strain; for instance, it can be applied to the twisting, or *torsion,* of an elastic rod or fiber. When torsion is used to measure the size of an unknown stress by the amount of strain produced, the set-up is called a *torsion balance*. If an extremely thin fiber is used, one that can be twisted by very small forces, it becomes conceivable that even tiny gravitational forces may be measured.

In 1798, the English scientist Henry Cavendish (1731–1810) made use of a delicate torsion balance for just this purpose. His torsion balance consisted of a light rod suspended at the middle by a delicate wire approximately a yard long. At each end of the light rod was a lead ball about two inches in diameter. Imagine a force applied to each lead ball in opposite directions and at right angles to both the rod and the delicate wire. The wire would twist, and extremely small forces would be sufficient to produce such a twist.

As a preparatory step Cavendish applied tiny forces to determine the amount of twist that would result. Next, carefully shielding his apparatus from air currents he brought two larger lead balls, each about eight inches in diameter, almost in contact with the small lead balls, but on opposite sides. The gravitational force between the lead balls now produced a twist in the fiber and from the total angle of twist Cavendish could measure the strength of the force between the small and large lead ball. (It turned out to be about 1/2,000,000 of a newton.)

Suppose, we rearrange Equation 4–1 as follows:

$$G = \frac{Fd^2}{mm'} \qquad \text{(Equation 4–2)}$$

With the value of F determined as I have just described, it is a simple matter to measure the mass of the lead balls (m and m') and the distance between them (d), center to center. Once all the values of the symbols on the right hand side of the equation are known, it is simple arithmetic to calculate the value of G. (Since the units of F in the mks system are kg-m/sec^2, those of d^2 are m^2, and those of mm' are kilograms times kilograms, or kg^2; the units

Cavendish experiment

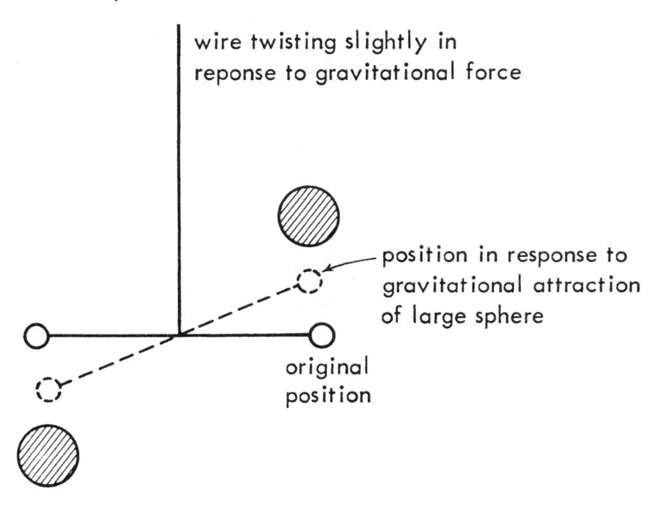

of G work out, by Equation 4–2, to [(kg-m/sec^2) m^2]/kg^2, or m^3/kg-sec^2.)

The best modern determination of G gives it a value of 0.0000000000667 m^3/kg-sec^2, certainly a tiny enough value. It is a tribute to Cavendish's great talents as an experimenter that in his first determination he obtained a value very close to this.

Suppose, now, we arrange Equation 4–1 as follows:

$$m' = \frac{Fd^2}{Gm} \qquad \text{(Equation 4–3)}$$

and try once more to determine the mass of the earth (m'). We already have, in the mks system, a value of 0.98 for F, one of 6,370,000 for d, and one of 0.1 for m. If we now add the value of 0.0000000000667 for G, it is simple arithmetic to solve for m', the mass of the earth. You can see that m' is equal to (0.98) (6,370,000) (6,370,000) divided by (0.0000000000667) (0.1), or just about 6,000,000,000,000,000,000,000,000 kilograms.

Physicists customarily express large numbers as powers of 10. Thus, 1,000,000 is usually written 10^6, which signifies the product of six 10's. The exponent (for numbers larger than 1) signifies the number of 0's in the number. It follows that 6,500,000 is 6.5×10^6. Negative exponents signify numbers less than 1, so that 10^{-6} is equal to $1/10^6$, or 1/1,000,000 or 0.000001. Again, 0.00000235 is 2.35×10^{-6}.

Using such *exponential notation,* the value of G is 6.67×10^{-11} m^3/kg-sec^2, and the mass of the earth is 6×10^{24} kg. (In the cgs system, the value of G is 6.67×10^{-8} cm^3/gm-sec^2 and the mass of the earth is 6×10^{27} gm.)

CHAPTER 5

Weight

The Shape of the Earth

By determining the value of G, Cavendish had, in effect, determined the mass of the earth. For this reason, Cavendish is often said to have "weighed the earth," but this is *not* what he had done.

In common language, "weight" and "mass" are often spoken of as though they were the same things, and a body may be spoken of as "heavy" or "massive" interchangeably; even physicists sometimes fall into the trap. However, consider what weight is. The weight of a body is the force with which it is attracted to the earth. To repeat, weight is a force and has the units of a force!

A simple way of measuring the weight of an object is to suspend it from a coiled spring. In accordance with Hooke's law the force by which the body is attracted to the earth will extend the spring, the amount of extension (or strain) being proportional to the force (or stress). A weight-measuring device of this sort is a *spring balance.*

The mass of a body, on the other hand, is the quantity of inertia it possesses. By Newton's second law $m = f/a$; it is a force divided by an acceleration. Weight, which is a force, must by the same law be a mass multiplied by an acceleration. In the case of weight, which is the force of earth's gravitational field upon a body, the particular acceleration involved is, naturally, that produced by the earth's gravitational field.

The weight (w) of a body, in other words, is equal to the mass (m) of that body times the acceleration (g) due to the pull of earth's gravity:

$$w = mg \quad \text{(Equation 5–1)}$$

Since the value of g is, under ordinary circumstances, just about constant, weight may be said to be directly proportional to mass. To say that A is 3.65 times as massive as B is equivalent to saying that under ordinary circumstances A is 3.65 times as heavy, or as weighty, as B. Since the two statements are usually equivalent, there is a strong temptation to consider them synonymous, and there lies the source of confusion between mass and weight.

The confusion is made worse because of the common units used for weight. A body with a mass of one kilogram is commonly said to have a weight of one kilogram, too.* In the mks system, however, the units of m are kilograms (kg), and the units of g are m/sec^2. Since weight is equal to mass times gravitational acceleration (mg), the units of weight are kg-m/sec^2 or newtons. A kilogram of mass therefore exerts (under ordinary circumstances) 9.8 newtons of force.

A kilogram of weight (which may be abbreviated as kg(wt.) to distinguish it from a kilogram of mass) is, therefore, *not* equal to 1 kg but to 9.8 newtons. In the cgs system, g is equal to 980 cm/sec^2. The weight of a body with a mass of 1 gm is therefore 1 gm multipled by 980 cm/sec^2, or 980 gm-cm/sec^2. Consequently 1 gm (wt.) equals 980 dynes.

All this may strike you as unnecessarily puristic and refined —as making a great deal out of a distinction without a difference. After all, if weight and mass always vary in the same way, why bother so much about which is which?

The point is that mass and weight do not always vary in the same way. They are related by g, and the value of g is not a constant under all conditions.

The gravitational force (F) exerted by the earth upon a particular body is equal to mg, as indicated by Equation 5–1. It

* The units of weight (pound, ounce, etc.) were in use long before Newton established the concept of mass. The units of weight were borrowed and applied to mass, which was a mistake—but one which is now beyond retrieval.

is also equal to Gmm'/d^2 as shown in Equation 4–1. Therefore $mg = Gmm'/d^2$; or dividing through by m:

$$g = \frac{Gm'}{d^2} \qquad \text{(Equation 5–2)}$$

Of the quantities upon which the value of g depends in Equation 5–2, the gravitational constant (G) and the mass of the earth (m') may be considered as constant. The value of d, however, which is the distance of the body from the center of the earth is certainly not constant, and g varies inversely as the square of that distance.

An object at sea level, for instance, may be 6370 km from the center of the earth, but at the top of a nearby mountain it may be 6373 km from the center, and a stratoliner may take it to a height of 6385 km from the center.

Even if we confine ourselves to sea level, the distance to the center of the earth is not always the same. Under the action of gravity alone, the earth would be a perfect sphere (barring minor surface irregularities)—a fact pointed out by Aristotle—and then the distance from sea level to the earth's center would be the same everywhere. A second factor is introduced, however, by the fact that the earth rotates about its axis. This rotation means, as Newton was the first to recognize, that the earth cannot be a perfect sphere.

As the earth rotates about its axis, the surface of the earth is continually undergoing an acceleration inward toward the center of the earth (just as the moon does in revolving about the earth). If this is so, then Newton's third law (see page 34) comes into play. The earth's center exerts a constant force on the earth's outer layers to maintain that constant inward acceleration as the planet rotates; the outer layers must, therefore, by action and reaction, exert a force outward on the earth's center. The force directed inward is usually called a *centripetal force,* and the one directed outward is called a *centrifugal force* (the words coming from Latin phrases meaning "move toward the center" and "flee from the center," respectively).

The two forces are oppositely directed and the result is a stretching of the earth's substance. If you were to imagine a rope stretching from the earth's surface to the earth's center, with the earth's surfaces pulling outward at one end of the rope and the earth's center pulling inward at the other end, you would expect the rope to stretch by a certain amount; the earth's substance does exactly that.

If every point on the earth's surface were rotating at the same speed, the stretch would be the same everywhere and the earth would be perfectly spherical still. However, the earth rotates about an axis, and the nearer a particular portion of the earth's surface is to the axis, the more slowly it rotates. At the poles, the earth's surface touches the axis and the speed of rotation is zero. At the equator, the earth's surface is at a maximum distance from the axis and the speed of rotation is highest (just over 1600 kilometers an hour).

The interacting forces are zero at the poles, therefore, and increase smoothly as the equator is approached. The "stretch" increases, too, and a bulge appears in the earth, which reaches maximum size at the equator. Because of this equatorial bulge, the distance from the center of the earth to sea level at the equator is 21 km (13 miles) greater than the distance from the center of the earth to sea level at either pole.

The earth, therefore, is not a sphere, but an *oblate spheroid.*

To be sure, 21 km in a total distance of 6370 km is not much, but it is enough to introduce measurable differences in the value of *g*. What with the equatorial bulge and local differences in

Centrifugal force and equatorial bulge

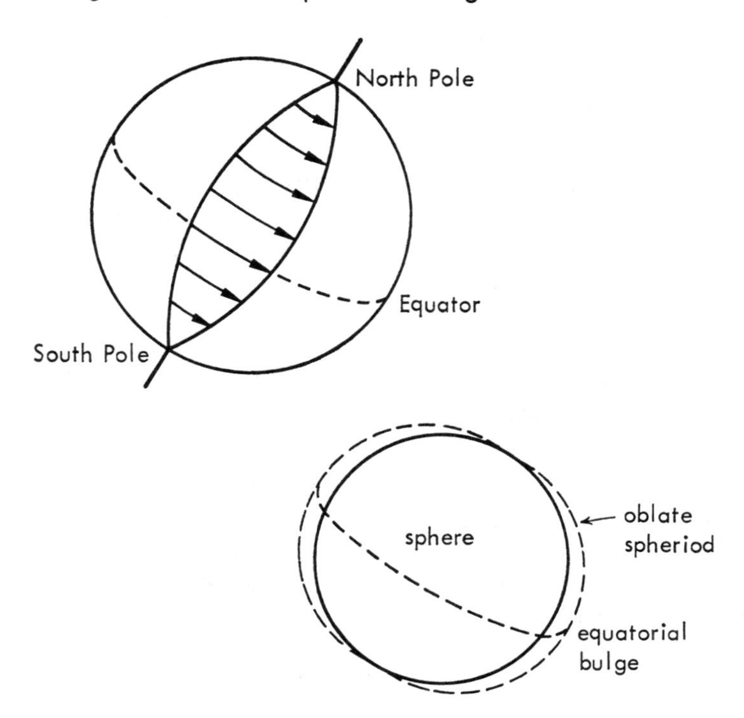

altitude, there are points in Alaska where the value of g is over 9.82 m/sec^2, whereas at the equator it is barely higher than 9.78 m/sec^2. That represents a difference of nearly one-half of one percent and is reflected in weight.

In other words, the weight of an object changes measurably from place to place on the earth's surface, as a spring balance would show. A man who weighs 200 pounds at the poles would weigh 199 pounds at the equator. To a chemist or physicist interested in the mass of an object (many properties depend on the mass), the measurement of weight as a substitute for mass would introduce serious inaccuracies.

What the scientist usually does when he "weighs" an object is to make use of a balance consisting of two pans suspended from opposite ends of a beam that is pivoted at the center. Objects of known weight are placed on one pan, the object to be weighed is placed in the other. The known weights are adjusted to the point of balance. The force of gravity is then the same on both pans; (if it were greater on one of them, that pan would sink downward while the other would rise upward).

If the weights are the same on both sides, then mg is the same on both sides. However much g might vary from point to point on the earth's surface, it is the same for two neighboring balance-pans at one particular point on the earth's surface. Therefore the mass (m) is the same for the objects in both pans. The mass of the unknown object is therefore equal to the mass of the weights (which is known).*

Beyond the Earth

Naturally, the small changes in the value of g become large ones if we alter greatly the distance of the body from the earth's center. A further complication is introduced if in removing the body to a great distance from the earth we bring it close to some other sizable conglomeration of mass. This situation is most likely to become important in connection with the moon, for man-made objects have already landed there, and living men may be standing on the surface of the moon before many years have passed.

An object on the moon's surface is still within the earth's gravitational field, which extends not only to the moon but, in principle, through all the universe. However, the moon also has

* Mass is not completely constant, by the way. However, the variation in mass becomes important only at extreme velocities not likely to be met with in ordinary life.

a gravitational field. That field is much smaller than the earth's, for the moon is much the less massive of the two. Nevertheless, an object on the moon's surface is much closer to the moon's center than to the earth's center; the moon's gravitational attraction would therefore be far greater than that of the distant earth, and a man standing on the surface of the moon would be conscious of the moon's pull only.

But the moon's pull on an object on its surface is not the same as the earth's pull on an object on its surface. To see how the two forces compare, let us refer back to Equation 4–1, which states that $F = Gmm'/d^2$. This F refers to the intensity of the earth's pull upon an object upon its surface. The moon's pull upon an object upon its surface, we can call F_m.

Now an object has the same mass whether it is on the surface of the earth or the surface of the moon, so m remains unchanged. The value of G is also unchanged, for it is constant throughout the universe. The mass of the moon, however, is known to be 1/81 the mass (m') of the earth. The mass of the moon, consequently, is $m'/81$. The distance from the surface of the moon to its center is 1737 km or just about 3/11 that of the 6370 km distance (d) from the surface of the earth to the center. Consequently, we can set the distance of the moon's surface to its center as $3d/11$.

We can now modify Equation 4–1, using the mass and radius of the moon to get an expression of the gravitational force of the moon for an object on its surface. This is:

$$F_m = \frac{Gm(m'/81)}{(3d/11)^2} \qquad \text{(Equation 5–3)}$$

If we now divide Equation 5–3 by Equation 4–1, we find that F_m/F (the ratio of the moon's gravitational force to that of the earth) is equal to 1/81 divided by $(3/11)^2$, or almost exactly 1/6. Thus, the gravitational force we would experience on the moon's surface would be 1/6 that to which we are accustomed on the surface of the earth. A 180-pound man who weighed himself on a spring balance would find he weighed 30 pounds.

But though the weight was decreased so drastically, the mass of an object would remain unchanged. This means that the force required to accelerate a particular object at a given rate would be the same on the moon as on the earth. We could lift a 180-pound friend without much trouble, for the sensation of the effort involved in the lift would be like that of lifting 30 pounds on earth. We could not, however, lift the man any more rapidly on the moon than on the earth. Now we could manipulate something that felt 30 pounds with a certain amount of ease on the earth. On the moon, something that felt 30 pounds would have six times the "normal" quantity of mass, and it could only be moved slowly. For this reason, maneuvering objects on the moon would give one the feel of "slow motion" or of pushing through molasses.

Again, if we jump on the moon, the force of our muscles will be countered by only 1/6 the gravitational force to which we are accustomed on the earth. The center of our body will therefore rise to six times the height it would on the earth. Having reached this unusual height, we would then fall toward the surface, but at 1/6 the usual acceleration (1.63 m/sec^2). This means we would seem to fall downward slowly and "like a feather." By the time we reached the surface again, however, having dropped at 1/6 the usual acceleration for six times the distance, we would be landing at just the same velocity that we would be landing at from a similar jump (equal exertion, but reaching a much lower height) on the earth.

To bring ourselves to a halt from that velocity would require as much force on the moon as it would on the earth, for it is mass that counts in this respect, not weight, and the mass remains unchanged on the moon. If we are seduced by our slow-paced fall into thinking we are indeed a feather, and try to land gracefully on one big toe, we are likely to break that toe.

The situation can be made even more radically unusual without having to go to the moon.

The subjective sensation we call "weight" arises from the fact that we are physically prevented from responding to the force of gravitation with an acceleration. Standing on the surface of the earth, we are prevented by the substance of the earth itself from an accelerated fall toward earth's center. It is the force exerted upon us against the stubborn opposition of the ground we stand on that we interpret as "weight."

If we were falling at precisely the acceleration imposed upon us by the gravitational acceleration (free fall), we would feel no weight. If we were in an elevator that had broken loose and was dropping without restraint, or if we were in an airplane that was in an unpowered fall, our sensation of weight would be gone. We could not press against the floor of the elevator or the airplane since that floor would be falling as rapidly as we were. If we were in midair within the elevator, we could not drop to its floor, for the floor would be moving as fast as we were. We would therefore seem to remain floating in midair and to be weightless.

Such examples of free fall are imperfect. Neither an elevator nor an airplane could fall for long without coming to disaster and ruining the experiment. Furthermore, the falling elevator or airplane would be slowed somewhat by the resistance of the air it was rushing through, and slowed to a greater extent than the man within the the elevator or airplane would be slowed by the quiet air about him. There would therefore be the sensation of some slight weight.

For the true feeling of free fall, we would need to be beyond the major portion of the atmosphere, say at a height of 160 km or more above the surface of the earth. To keep at that height it would be best if there were also a sideways motion that would keep one in orbit about the earth, in the same way that a combination of inward and sideways forces keeps the moon in orbit about the earth (see page 42).

This is exactly the situation in a manned orbiting satellite. Such a satellite is in free fall and can continue in free fall for long periods. The astronaut within has no sensation of weight. This is *not* because he is "beyond the pull of earth's gravity" as some news announcers maintain. It is only because he is in free fall, so the satellite and everything in it are falling at precisely the same acceleration.

The earth itself is in free fall in an orbit that takes it around the sun. Although its mass is huge (see page 52), its weight is zero. Cavendish, therefore, did not "weigh the earth," for he did not need to; its weight was understood to be zero from Newton's time. What Cavendish did was to determine the earth's mass.

Even in free fall, where weight is zero, the mass of a particular body remains unchanged. Astonauts building a space station will be moving huge girders that will have no weight. They will be able to balance such girders on one finger, if girder and finger are motionless with respect to each other. If a girder must be set in motion, however, or if it is already moving and must either be stopped or have its direction of motion altered, the effort will be precisely as great as it would be on the earth. A man trapped between two girders moving toward each other may well be crushed to death by two weightless but not massless objects.

The distinction between mass and weight, which seems so trivial on the surface of the earth, is therefore anything but trivial in space, and can easily become a matter of life and death.

Escape Velocity

As an object is dropped from a greater and greater height above the ground, it takes longer and longer to fall and strikes the ground with a higher and higher velocity. If we use Equations 2–1 and 2–2 (see page 17-18), letting the symbol a in those equations be set equal to 9.8 m/sec^2 (the acceleration in free fall), we can make some easy calculations. A body dropped from a height of 4.9 meters will strike the ground in one second and be moving, at the moment of impact, at 9.8 m/sec. If it were dropped from a height of 19.6 meters, it would strike after two seconds, moving then at 19.6 m/sec. If it were dropped from 44.1 meters, it would strike after three seconds, moving then at 29.4 m/sec.

It would seem that if you could only drop an object from a great enough height, you could make the velocity of impact as high as you pleased. Certainly this would seem so if the value of g were the same for all heights.

But the value of g is not constant; it decreases with height. The value of g varies inversely as the square of the distance from the earth's center. A point 6370 kilometers above the earth's surface would be 12,740 kilometers from the earth's center—twice as far from the center as a point on the surface would be. The value of g at that height would therefore be just 1/4 what it is at the surface.

An object falling from an initial state of rest 6370 kilometers above the earth's surface would in the first second attain a velocity of only 2.45 m/sec, instead of the 9.8 m/sec it would attain after a one-second drop in the immediate vicinity of the earth's surface.

As the body continued to drop and approach the earth, the value of g would, of course, increase steadily and approach 9.8 m/sec^2 at the end. However, the falling body would not strike the earth's surface with as high a velocity of impact as it would have done if the value of g had been 9.8 m/sec^2 all the way down.

Imagine a body dropped first from a height of 1000 kilometers, then from 2000 kilometers, then from 3000 kilometers, and so on. The drop from 1000 kilometers would result in a velocity of impact, v_1. If the value of g were constant all the way up, then a drop from 2000 kilometers would involve a gain in velocity in the first 1000 kilometers equal to the gain in the second 1000 kilometers, so the final velocity of impact would be $v_1 + v_1$ or $2v_1$. However, the upper 1000 kilometers represents a region where g is smaller than in the lower 1000 kilometers. Less velocity is added in the upper than in the lower half of the drop, and the final velocity of impact is $v_1 + v_2$, where v_2 is smaller than v_1. The same argument can be repeated all the way up, so a fall from a height of 10,000 kilometers would result in a velocity of impact of $v_1 + v_2 + v_3 + v_4$ and so on, up to v_{10}. Here each symbol represents the portion of the final velocity contributed by a higher and higher 1000 kilometer region of drop, and the value of each symbol is less than that of the preceding one.

Whenever one is faced with a series of numbers each smaller than the one before, there is the possibility of a *converging series.* In such a series, the sum of the numbers never surpasses a certain fixed value, the *limiting sum,* no matter how many numbers are added. The best-known case of such a converging series is $1 + 1/2 + 1/4 + 1/8 + 1/16$, where each number is half the one before. The sum of the first two numbers is 1.5; the sum of the first three numbers is 1.75; the sum of the first four numbers is 1.875; the sum of the first five numbers is 1.9325, and so on. As more and more numbers in the series are added, the sum grows larger and larger, and approaches closer and closer to 2 without ever quite reaching it. The limiting sum of this particular series is 2.

It turns out that the numbers representing increments of velocity resulting from falls from regularly increased heights do indeed form a converging series. As a body is dropped from a greater and greater height, the final velocity of impact does not increase without limit; instead it tends toward a limiting velocity it cannot surpass.

This limiting velocity of impact (v_l) depends on the value of g and on the radius (r) of the body that is the source of the gravitational field. The importance of the radius rests on the fact that the larger its value, the more slowly does the value of g fade off with distance. Suppose a body has a radius of 1000 kilometers. At 10,000 kilometers from its center, a falling body is ten times as far from the center as an object on the surface is, and the value of g is therefore 1/100 the value at the surface. Suppose that a body has a radius of 2000 kilometers, however; at a distance of 10,000 kilometers from its center, a falling body is then only five times as far from the center as an object on the surface is, and the value of g is 1/25 the value at the surface. Through all heights, therefore, the value of g would decline more rapidly for the small body than for the large body, and the final velocity of impact would be less for the smaller body even though its surface value of g might be the same as for the larger body.

It turns out that:

$$v_i = \sqrt{2gr} \qquad \text{(Equation 5–4)*}$$

In the mks system, the value of g is 9.8 m/sec^2 and that of r is 6,370,000 m, so that $2gr$ is equal to about 124,800,000 m^2/sec^2. In taking the square root of this number, we must also take the square root of the unit. Since the square root of a^2/b^2 is ab, it should be clear that the square root of m^2/sec^2 is m/sec. The square root of 124,800,000 m^2/sec^2 is about 11,200 m/sec. This limiting velocity of impact is equal to 11.2 km/sec (or just about seven miles per second). No object, falling to earth *from rest,* could ever strike with an impact of more than 11.2 km/sec. (Of course, if an object such as a meteor happens to be speeding in the direction of the earth, so that its own speed is added to the velocity produced by earth's gravitational field, it will strike with an impact of more than 11.2 km/sec.)

For the moon, with its smaller values for both g and r, the maximum velocity of impact is only 2.4 km/sec (or 1.5 miles per second).

Suppose we now turn the matter around. Instead of a falling body, consider one that is propelled upward. For a body moving upward, g represents the amount by which its speed is diminished each second (see page 29). The situation is symmetrically reversed to that of a body moving downward; that is, if a body initially at rest falls from a height h and attains a velocity v at the moment of impact, then a body hurled upward with a velocity v will attain a height h before coming to rest (and beginning its fall back to the earth).

But a body dropped from any height, however great, can never attain a velocity of impact greater than 11.2 km/sec. This means that if a body is hurled upward with a velocity of 11.2 km/sec or more it will never reach a point of rest and, therefore, never fall back to the earth (barring the interference of gravitational fields of other bodies).

The limiting velocity of impact is consequently also the velocity at which a body hurled upward will escape from the earth permanently; it is therefore called the *escape velocity.* The escape velocity at the surface of the earth is 11.2 km/sec and the escape velocity at the surface of the moon is 2.4 km/sec.

* As this is an introduction to physics, I shall not always give the derivation of the equations used—since at times these would involve concepts not yet explained or mathematical techniques I would prefer not to use.

A body that orbits the earth has not escaped from the earth. It is falling toward it, and only its sideward motion prevents it from finally making impact. A smaller velocity is therefore required to place an object in orbit than to cause it to escape from the earth altogether. For a circular orbit, the velocity must be equal to $\sqrt{gr}$, where r is the distance of the orbiting body from the earth's center and g is the value of the gravitational acceleration at that distance. In the immediate vicinity of the earth's surface, this comes out to 7.9 km/sec (or 4.9 miles per second). Orbiting satellites travel at this velocity and complete the 40,000 kilometer circumnavigation of the earth in a minimum time of 85 minutes.

As the distance from the earth's center increases, the value of r increases, of course, while the value of g decreases, varying as $1/r^2$. The variation of $\sqrt{gr}$ (which is the orbital velocity) as distance increases is as $\sqrt{(1/r^2)(r)}$ or $\sqrt{1/r}$. In other words, the orbital velocity of a body varies inversely as the square root of its distance from the object around which it is in orbit.

Thus, the distance of the moon from the earth's center is 382,400 kilometers. This is 60.3 times the distance from the center of a satellite orbiting just above the atmosphere. The orbital velocity of the moon is therefore less than that of the satellite by a factor equal to the $\sqrt{60.3}$. The moon's orbital velocity, in other words, is $7.9/\sqrt{60.3}$, or just about 1 km/sec.

Consider, also, a satellite in orbit 42,000 kilometers from the earth's center (about 35,600 kilometers above its surface). Its distance from the earth's center would be 6.6 times that of an object on earth's surface. Its orbital velocity would therefore be $7.9/\sqrt{6.6}$, or not quite 3.1 km/sec. The length of its orbit would be about 264,000 kilometers, and at its orbital velocity it would take the satellite just 24 hours to complete one revolution. It would, therefore, just keep pace with the surface of the rotating earth and would seem to hang motionless in the sky. Such apparently motionless satellites serve admirably as communication relays.

CHAPTER 6

Momentum

Impulse

Let's consider a falling body again.

An object held at some point above the ground is at rest. If it is released, it begins to fall at once. Motion is apparently created where it did not previously exist. But the word "created" is a difficult one for physicists (or for that matter philosophers) to swallow. Can anything really be created out of nothing? Or is one thing merely changed into a second, so the second comes into existence only at the expense of the passing into nonexistence of the first? Or perhaps one object undergoes a change (from rest to motion, for instance) because, and only because, another object undergoes an opposing change (from rest to motion in the opposite direction, for instance). In this last case, what is created is not motion but motion plus "anti-motion," and if the two together cancel out to zero, there is perhaps no true creation at all.

To straighten this matter out, let's start by trying to decide exactly what we mean by motion.

We can begin by saying that a force certainly seems to create motion. Applied to any body initially at rest, say to a hockey puck on ice, a force initiates an acceleration and sets the puck moving faster and faster. The longer the force acts, the faster the hockey puck moves. If the force is constant, then the velocity at any given time is proportional to the amount of the force multiplied by the time during which it is applied. The term *impulse* (I) is applied to this product of force (f) and time (t):

$$I = ft \qquad \text{(Equation 6–1)}$$

Since a force produces motion, we might expect that a given impulse (that is, a given force acting over a given time) would always produce the same amount of motion. If this is so, however, then the amount of motion cannot be considered a matter of velocity alone. If the same force acts upon a second hockey puck ten times as massive as the first, it will produce a smaller acceleration and in a given time will bring about a smaller velocity than in the first case. The quantity of motion produced by an impulse must therefore involve mass as well as velocity.

That this is indeed so is actually implied by Equation 6–1. By Newton's second law we know that a force is equal to mass times acceleration ($f = ma$). We can therefore substitute ma for f in Equation 6–1 and write:

$$I = mat \qquad \text{(Equation 6–2)}$$

But by Equation 2–1 (see page 17), we know that for any body starting at rest the velocity (v) produced by a force is equal to the acceleration (a) multiplied by time (t), so that $at = v$. If we substitute v for at in Equation 6–2, we have:

$$I = mv \qquad \text{(Equation 6–3)}$$

It is this quantity, mv, mass times velocity, that is really the measure of the motion of a body. A body moving rapidly requires a greater effort to stop it than does the same body moving slowly. The increase in velocity adds to its total motion, therefore. On the other hand, a massive body moving at a certain velocity requires a greater effort to stop it than does a light body moving at the same velocity. The increase in mass also adds to total motion. Consequently, the product mv has come to be called *momentum* (from a Latin word for "motion").

Equation 6–3 means that an impulse (ft) applied to a body at rest causes that body to gain a momentum (mv) equal to the impulse. More generally, if the body is already in movement, the application of an impulse brings about a change of momentum equal to the impulse. In brief, impulse equals change of momentum.

The units of impulse must be those of force multiplied by those of time, according to Equation 6–1, or those of mass multiplied by those of velocity, according to Equation 6–3. In the mks system, the units of force are newtons, so impulse may be measured in newton-sec. The units of mass are kilograms, however, and the units of velocity are meters per second, so the units of impulse (mass times velocity) are kg-m/sec. However, a newton has been defined as a kg-m/sec^2. A newton-sec, therefore, is a kg-m-sec/sec^2, or a kg-m/sec. Thus the units of I considered as ft are the same as the units of I considered as mv. In the cgs system, it is easy to show, the units of impulse are dyne-sec, or gm-cm/sec, and these are identical also.

Conservation of Momentum

Imagine a hockey puck of mass m speeding across the ice at a velocity, v. Its momentum is mv. Imagine another hockey puck of the same mass moving at the same speed but in the opposite direction. Its velocity is therefore $-v$ and its momentum is $-mv$. Momentum, you see, is a vector, since it involves velocity, and not only has quantity but direction. Naturally, if we have two bodies with momenta in opposite directions, we can set one momentum equal to some positive value and the other equal to some negative value.

Suppose now that the two hockey pucks are rimmed with a layer of glue powerful enough to make them instantly stick together on contact. And suppose they do make contact head-on. When that happens, they would come to an instant halt.

Has the momentum been destroyed? Not at all. The total momentum of the system* was $mv + (-mv)$, or 0, before the collision and $0+0$, or (still) 0, after the collision. The momentum was distributed among the parts of the system differently before and after the collision, but the total momentum remained unchanged.

Suppose that instead of sticking when they collided (an inelastic collision) the two pucks bounced with perfect springiness (an elastic collision). It would then happen that each puck would reverse directions. The one with the momentum mv would now have the momentum $-mv$ and vice versa. Instead of the sum $mv+(-mv)$, we would have the sum $(-mv)+mv$. Again there would be a change in the distribution of momentum, but again the total momentum of the system would be unchanged.

* By a "system" is meant the entire collection of bodies being discussed, in this case, the two hockey pucks, considered in isolation from the rest of the universe.

If the collision were neither perfectly elastic nor completely inelastic, if the pucks bounced apart but only feebly, one puck might change from mv to $-0.2mv$, while the other changed from $-mv$ to $0.2mv$. The final sum would still be zero.

This would still hold true if the pucks met at an angle, rather than head-on, and bounced glancingly. If they met at an angle, so their velocities were not in exactly opposite directions, the two momenta would not add up to zero, even though the velocities of the two pucks were equal. Instead the total momentum of the system would be arrived at by vector addition of the two individual momenta. The two pucks would then bounce in such a way that the vector addition of the two momenta after the collision would yield the same total momentum as before. This would also be true if a moving puck struck a puck at rest a glancing blow. The puck at rest would be placed in motion, and the originally moving puck would change its direction; however, the two final momenta would add up to the original.

Matters would remain essentially unchanged even if the two pucks were of different masses. Suppose one puck was moving to the right at a given speed and had a momentum of mv, while an-

Conservation of momentum

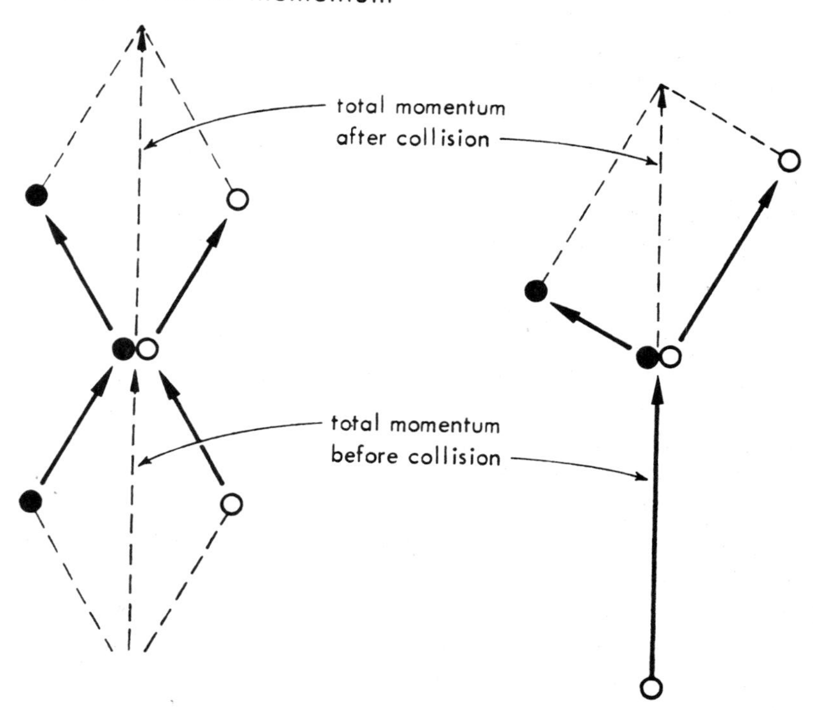

other, three times as massive, was moving at the same speed to the left and had, therefore, a speed of $-3mv$. If the two stuck together after a head-on collision, the combined pucks (with a total mass of $4m$) would continue moving to the left—the direction in which the more massive puck had been moving—but at half the original velocity ($-v/2$). The original momentum of the system was $mv + (-3mv)$, or $-2mv$. The final momentum of the system was $(4m)(-v/2)$, or $-2mv$. Again, the total momentum of the system would be unchanged.

And what if momentum is seemingly created? Let us consider a bullet initially at rest—and with a momentum, therefore, of 0—which is suddenly fired out of a gun and moves to the right at high velocity. It now has considerable momentum (mv). However, the bullet is only part of the system. The remainder of the system, the gun, must gain $-mv$ by moving in the opposite direction. If the gun has n times the mass of the bullet, it must move in the opposite direction with $1/n$ times the velocity of the speeding bullet. The momentum of the gun (minus the bullet) would then be $(nm)(-v/n)$, or $-mv$. (If the gun were suspended freely when it was fired, its backward jerk would be clearly visible. When fired in the usual manner its backward motion is felt as "recoil.") The total momentum of gun plus bullet was therefore 0 before the gun was fired and 0 after it was fired, though here the distribution of momentum among the parts of the system varied quite a bit before and after firing.

In short, all the experiments we can make will bring us to the conclusion that:

The total momentum of an isolated system of bodies remains constant.

This is called the *law of conservation of momentum.* (Something that is "conserved" is protected, guarded, or kept safe from loss.)

Of course, it is impossible to prove a generalization by merely enumerating isolated instances. No matter how often you experiment and find that momentum is conserved, you cannot state with certainty that it will *always* be conserved. At best, one can only say, as experiment after experiment follows the law and as no experiment is found to contradict it, that the law is increasingly probable. It would be far better if one could show the generalization to be a consequence of another generalization that is already accepted.

For instance, suppose two bodies of any masses and moving at any velocities collide at any angle with any degree of elasticity.

At the moment of collision, one body exerts a force (f) on the second. By Newton's third law, the second body exerts an equal and opposite force ($-f$) on the first. The force is exerted only while the two bodies remain in contact. The time (t) of contact is obviously the same for both bodies, for when the first is no longer in contact with the second, the second is no longer in contact with the first. This means that the impulse of the first body on the second is ft, and that of the second on the first is $-ft$.

The impulse of the first body on the second imparts a change in momentum mv to the second body. But the impulse of the second body on the first, being exactly equal in quantity but opposite in sign, must impart a change in momentum $-mv$ to the first. The changes in momentum may be large or small depending on the size of the impulse, the angle of collision, and the elasticity of the material; however, whatever the change in momentum of one, the change in the other is equal in size and opposition in direction. The total momentum of the system must remain the same.

Thus, the law of conservation of momentum can be derived from Newton's third law of motion. In actual fact, however, it was not, for the law of conservation of momentum was first enunciated by an English mathematician, John Wallis (1616–1703), in 1671, a dozen years before Newton published his laws of motion. One could, indeed, work it the other way, and derive the third law of motion from the law of conservation of momentum.

At this point you might feel that if the physicist proves the conservation of momentum from the third law of motion, and then proves the third law of motion from the conservation of momentum, he is actually arguing in a circle and not proving anything at all. He would be if that were what he is doing, but he is not.

It is not so much a matter of "proving" as of making an assumption and demonstrating a consequence. One can begin by assuming the third law of motion and then showing that the law of conservation of momentum is a consequence of it. Or one can begin by assuming the law of conservation of momentum and showing that the third law is the consequence of that.

The direction in which you move is merely a matter of convenience. In either case, no "proof" is involved and no necessary "truth." The whole structure rests on the fact that no one in nearly three centuries has been able to produce a clearcut demonstration that a system exists, or can be prepared, in which either the third law of motion or the law of conservation of momentum is not obeyed. Such a demonstration may be made tomorrow, and the foundations of physics may have to be modified as a consequence; but by now it seems very unlikely* that this will happen.

And yet it may be that with a little thought we might think of cases where the law is not obeyed. For instance, suppose a billiard ball hits the rim of the billiard table squarely and rebounds along its own line of approach. Its velocity v becomes $-v$ after the rebound, and since its mass remains unchanged, its original momentum mv has become $-mv$. Isn't that a clear change in momentum?

Yes, it is, but the billiard ball does not represent the entire system. The entire system includes the billiard table that exerted the impulse that altered the billiard ball's momentum. Indeed, since the billiard table is fixed to the ground by frictional forces too large for the impact of the billiard ball to overcome, it includes the entire planet. The momentum of the earth changes just enough to compensate for the change in the momentum of the billiard ball. However, the mass of the earth is vastly larger than that of the billiard ball, and its change in velocity is therefore correspondingly smaller—far too small to detect by any means known to man.

Yet one might assume that if enough billiard balls going in the same direction were bumped into enough billiard tables, at long, long last, the motion of the earth would be perceptibly changed. Not at all! Each rebounding billiard ball must strike the opposite rim of the table, or your hand, or some obstacle. Even if it comes to a slow halt through friction, that will be like striking the cloth of the table little by little. No matter how the billiard ball moves it will have distributed its changes in momentum equally in both directions before it comes to a halt, if only itself and the earth are involved.

A more general way of putting it is that the distribution of momentum among the earth and all the movable objects on or near its surface may vary from time to time, but the total momentum, and therefore the net velocity of the earth *plus* all those movable objects (assuming the total mass to remain constant), must remain the same. No amount or kind of interaction among the components of a system can alter the total momentum of that system.

And now the solution to the problem of the falling body with which I opened the chapter is at hand. As the body falls it gains momentum (mv), this momentum increasing as the velocity increases. The system, however, does not consist of the falling body

* Please remember that "unlikely" does not mean "impossible."

alone. The gravitational force that brings about the motion involves both the body and the earth. Consequently, the earth must gain momentum ($-mv$) by rising to meet the body. Because of the earth's huge mass, its upward acceleration is vanishingly small and can be ignored in any practical calculation. Nevertheless, the principle remains. Motion is not created out of nothing when a body falls. Rather, both the motion of the body and the anti-motion of the earth are produced, and the two cancel each other out. The total momentum of earth and falling body, with respect to each other, is zero before the body starts falling, is zero after it completes its fall, and is zero at every instant during its fall.

Rotational Motion

So far, I have discussed motion as though it involved the displacement of an object as a whole through space with the different parts of the object maintaining their mutual orientation unchanged. This is *translational motion* (from Latin words meaning "to carry across").

It is possible, however, for a body not to be displaced through space as a whole, and yet still be moving. Thus, the center of a wheel may be fixed in place so that the wheel as a whole does not change its position; nevertheless, the wheel may be spinning about that center. In similar fashion, a sphere fixed within a certain volume of space may yet spin about a fixed line, the axis. This kind of motion is *rotational motion* (from the Latin word for "wheel"). (It is, of course, possible for a body to move in a combination of these two types of motion, as when a baseball spins as it moves forward, or when the earth rotates about its axis as it moves forward in its orbit about the sun.)

Rotational motion is quite analogous to translational motion, but it requires a change of viewpoint. For instance, we are quite used to thinking of the speed of translational motion in terms of miles per hour or centimeters per second. Furthermore, we take it for granted that if one part of a body has a certain translational velocity so has every other part of the body. The tail of an airplane, in other words, moves forward just as rapidly as its nose.

In the case of rotational motion, matters are different. A point on the rim of a turning wheel is moving at a certain speed, a point closer to the center of the wheel is moving at a smaller speed, and a point still closer to the center is moving at a still smaller speed. The precise center of a turning wheel is motionless. To say that a turning wheel moves at so many centimeters per second is therefore meaningless, unless we specify the exact portion of the wheel to which we refer, and this can be most inconvenient.

It would be neater if we could find some method of measuring rotational speed that would apply to the entire rotating body at once. One method might be to speak of the number of revolutions in a unit time. Though various points on the wheel might move at various speeds, every point on the wheel completes a revolution in precisely the same period, since the wheel rotates "all in one piece." We might therefore speak of a wheel or any rotating object as having a speed of so many *revolutions per minute* (usually abbreviated as *rpm*).

Or we might divide one revolution into 360 equal parts called *degrees* and abbreviated as a zero superscript ($^\circ$). In that case 1 rpm would be equal to 360° per minute, or 6° per second. As the wheel sweeps out those degrees, a line connecting the center of the wheel with a point on its rim marks out an angle. A speed given in revolutions per minute or degrees per second is therefore spoken of as *angular speed.*

It is possible for rotational motion to take place in one of two mirror-image fashions. As viewed from a fixed position, a wheel may be observed to be rotating *clockwise*—that is, in the same sense that the hands of a clock move. It could, on the other hand, move *counterclockwise*—that is, in the opposite sense to the moving clockhand.* Therefore, we can speak of *angular velocity* as indicating not only speed but direction as well. (Velocities involved in translational motion can be spoken of as *linear velocity,* since movement is then along a line rather than through an angle.)

Physicists use another unit in measuring rotational velocity: the *radian.* This is an angle that marks out on the rim of a circle an arc that is just equal in length to the radius of the circle. The circumference of the circle is π times the length of the

* It is important to specify "from a fixed position," for clockwise and counterclockwise are not absolute terms. A wheel may seem to be turning clockwise to you, but if you move to the opposite side and view it, it will then seem to be turning counterclockwise. The same is true if you speak of translational motion as being "left" or "right," or "toward" or "away." Those are terms that have meaning only with reference to your own position. However, if you speak of "north," "south," "east," or "west," those are terms that are fixed with respect to the earth and do not depend on your own position.

diameter* and is therefore 2π times the length of the radius. The circumference must therefore also be 2π times the length of the arc marked out by one radian. The entire circumference is marked out in one revolution, so one revolution equals 2π radians or 360°. It follows that one radian equals $360°/2\pi$ or, since π equals 3.14159, one radian is about equal to 57.3°.

Angular velocity is often symbolized by the Greek letter ω ("omega"), since this is the equivalent in Greek of the Latin letter v usually used for linear velocity.

For any given point on a rotating body, angular velocity can be converted to linear velocity. The linear velocity depends not only on the angular velocity but also on the distance (r) of the point in question from the center of rotation. If the distance is doubled for the same angular velocity, the linear velocity of the point is doubled. We can say then that:

$$v = r\omega \qquad \text{(Equation 6–4)}$$

This equation is precisely correct when ω is measured in radians per unit of time. For instance, if the angular velocity is one radian per second, then in one second a given point anywhere on the wheel sweeps out an arc equal to its distance from the center, and $v=r$. If ω equals 2 radians per second then $v=2r$, and so on.

If we were measuring ω in revolutions per unit of time, then Equation 6–4 would have to read $v = 2\pi r\omega$, and if we were

* The Greek letter π ("pi") is used to represent the ratio of the circumference (c) of a circle to its diameter (d); in other words, $c/d = \pi$. Although every circle may have a different value for c and for d, the ratio of the two, c/d, is always the same for all circles. Therefore, π is a constant, and it is approximately equal to 3.14159.

Size of the radian

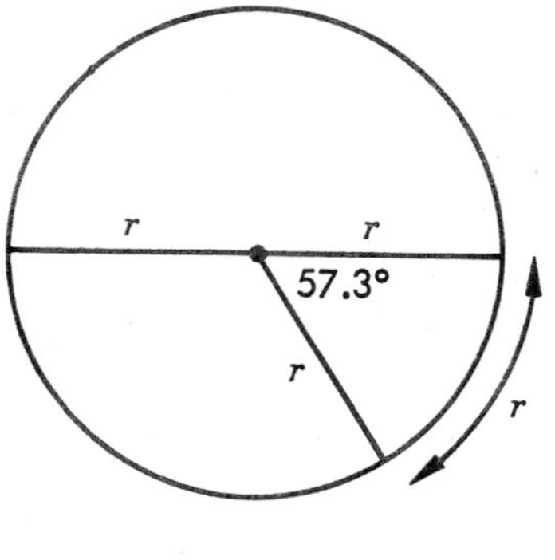

Angular velocity

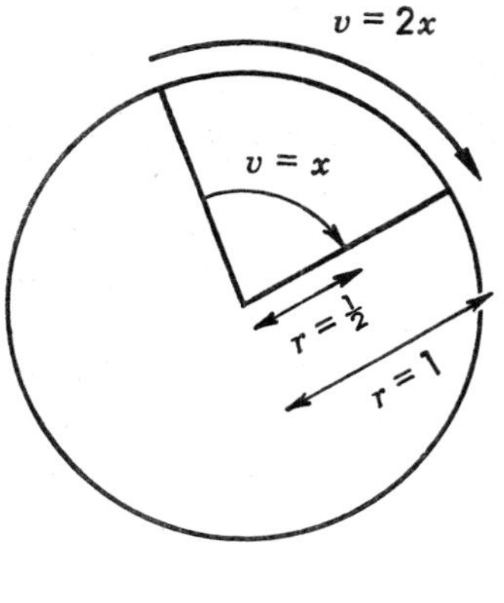

measuring it in degrees per unit time, it would have to read $v = r\omega/57.3$. This is an example of how a unit that may, at first blush, seem to have an odd and inconvenient size can yet turn out to be useful because it allows relationships to be expressed with maximum simplicity.

Torque

It takes a force to set a body at rest into translational motion. Under certain conditions a force can set a body at rest into rotational motion instead. Suppose, for instance, you nailed a long, flat rod loosely to a wooden base at one end. If you pushed the rod, it would not move as a whole, in a translational manner, because one end is fixed. The rod would instead begin to make a rotational movement about the fixed end.

A force that gives rise to such a rotational movement is called a *torque* (from a Latin word meaning "to twist"). To continue the use of Greek letters for rotational motion, a torque may be symbolized by the Greek letter τ ("tau"), which is the equivalent of the Latin t (for "torque," obviously).

A given force does not always give rise to the same torque by any means. In the case of the rod just mentioned, the amount of torque depends on the distance from the point at which the force is applied to the fixed point. A force applied to the fixed point will not itself produce a torque. As one recedes from that point a given force will produce a more and more rapid rotation and will therefore represent a greater and greater torque. In fact, the torque is equal to the force (f) multiplied by the distance (r):

$$\tau = fr \qquad \text{(Equation 6–5)}$$

In the past, torque has been referred to as the *moment of force,* but this phrase is now quite out of fashion.

Nor need a torque be produced only where a portion of a body is fixed in space; one can be produced even if the entire body is free to move.

Consider a body possessing mass but consisting of but a single point. Such a body can only undergo translational motion. A rotating body, after all, spins about some point (or line); if that point is all that exists, then there is nothing to spin and only linear motion is possible. It is to such point-masses that the laws of motion can be made to apply most simply.

In the real universe, however, there are no point-masses. All real massive bodies have extension. Nevertheless, it can be shown that in some ways such real bodies behave as if all their mass were concentrated at some one point. The point at which this seeming concentration is found is the *center of mass.* Where a body is symmetrical in shape, and where it is either uniform in density or has a density that changes in symmetrical fashion, the center of mass is at the geometrical center of the body. For instance, the earth is an essentially spherical body; while it is not uniformly dense, it is most dense at the center, and this density falls off equally in all directions as one approaches the surface. The earth's center of mass therefore coincides with its geometric center, and it is toward that center that the force of gravity is directed.

The concept of the center of mass can explain several things that might otherwise be puzzling. According to Newton's first law of motion, a moving object will continue moving at constant velocity unless acted upon by some outside force. Suppose that a shell containing an explosive is moving at constant velocity through space and that at a certain point it explodes. Fragments of the shell are hurled in all directions, and the various chemical products of the explosion also expand outward. This explosion is an internal force, however, one produced within the system in question, and it should have no effect on the motion of the system, according to the first law. Yet the various fragments of the shell are no longer traveling at the original velocity. Do Newton's laws of motion break down?

Not at all. The laws apply to a system as a whole, and not necessarily to one part or another taken in isolation. As a result of the explosion the system has changed its shape. But has it changed its center of mass? The center of mass might be viewed as the "average point" of the body. If one portion of the shell hurls outward in one direction, it is balanced by another portion hurled in the opposite direction. To be more precise, the vector sum of all the momenta in one direction must be equal to the vector sum of all the momenta in the opposite direction, according to the law of conservation of momentum. This can be shown to imply that no matter how the body changes shape through internal forces, the center of mass remains where it would have been if no change of shape had occurred. In other words, the center of mass of the system moves on at constant velocity regardless of the explosion that hurled bits of the system this way and that.

If a body were under the influence of a gravitational force

and following a parabolic path, its sudden explosion would not prevent the center of mass from continuing smoothly in that parabolic path even though the individual fragments moved all over the lot. (This implies no interference by forces outside the system. If fragments strike other bodies and are halted, the motion of the center of mass changes. Again, the effect of air resistance on the multitude of particles after explosion may not be the same as the effect upon the single shell before explosion; this may change the motion of the center of mass.)

Suppose next that a body is falling toward the earth. Every particle of the body is being pulled by the force of gravity, but the body behaves as if all that force were concentrated at one point within the body; that point is the *center of gravity*. If the body were in a uniform gravitational field, the center of gravity would be identical with the center of mass. However, the lower portion of a body is somewhat closer to the center of the earth than is the upper, and the lower portion is therefore more strongly under gravitational influence. The center of gravity is consequently very slightly below the center of mass; therefore, while the difference under ordinary conditions is so small as to be easily neglected, it is better form not to interchange the two phrases.

The concept of the center of gravity is useful in considering the stability of bodies. Imagine a brick resting on its narrowest base. If it is tipped slightly and then released, it drops back to

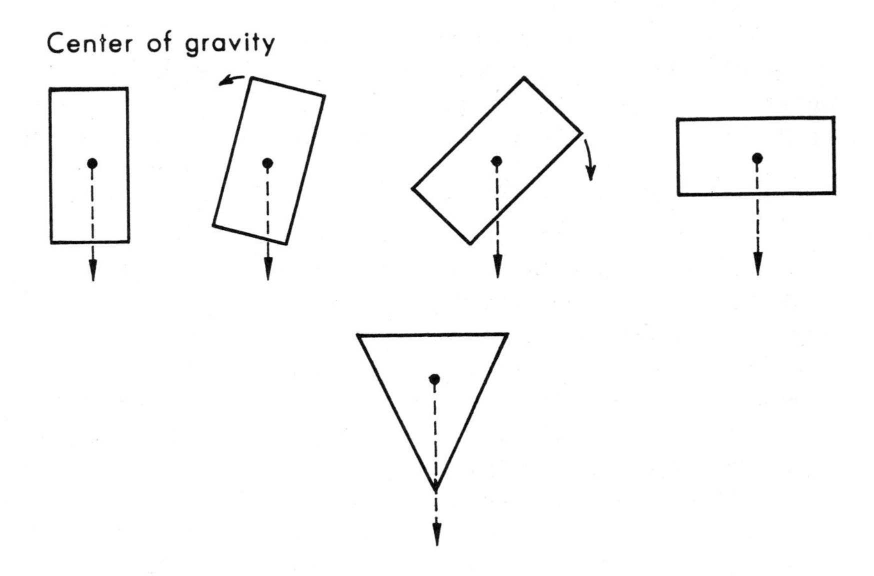

its original position. If it is tipped somewhat more and then released, it drops back again. As it is tipped more and more, however, there comes a point when it flops over onto one of its other bases. At what point does this flop-over come?

We can view the gravitational force as pulling upon the center of gravity of the brick, and upon that point only. As long as the center of gravity is located directly over some portion of the original base, the effect of the gravitational pull is to move the brick back upon that base once the tipping force is removed. If the brick is tipped so much that the center of gravity is located directly over some point outside the original base, the brick drops onto the base over which the point is now located.

Naturally, the wider the base in comparison with the height of the center of gravity, the greater the degree of tipping required before the center of gravity moves beyond that base—the more stable the body, in other words. A brick resting on its broadest base is more stable than one resting upon its narrowest base.

A cone resting on its pointed end may be so adjusted that its center of gravity will be directly above that point. It will then remain in balance. The slightest movement, however, the smallest breath of air, will move the center of gravity beyond the point in one direction or another, and down it will flop onto its side. A juggler keeps objects balanced upon points or, more accurately, upon very small bases, by moving his own body in such a way as to bring the base under the center of gravity again every time the center of gravity moves out of position.

Where a body is not uniform in density, its center of gravity is not located at its geometrical center but is displaced toward the denser portions. An object that is particularly dense in its lowermost portion ("bottom-heavy") has an unusually low center of gravity. Even a large degree of tipping will not bring that low center of gravity beyond the line of the base, and on being released the object will return to its original position. On the other hand, an object that is particularly dense in its uppermost portion ("top-heavy") has an unusually high center of gravity and will flop over after but a slight degree of tipping. Since our common experience is with objects of reasonably uniform density, we are generally surprised at the refusal of a bottom-heavy object to fall over (like those round-bottomed children's toys that spring up again even when forced down to their side), or at the ease with which a top-heavy object topples.

Let us now return to our point-mass which undergoes only translational motion. If we imagine a force directed against a real body in such a way as to intersect its center of mass, then that real body behaves as a point-mass does and undergoes a purely translational motion. Thus, a falling body is subjected to the pull of gravity directly upon the center of gravity (usually equivalent to the center of mass). Therefore (provided a torque is not applied to the body at the moment of its release and that the effect of wind and air resistance is neglected), a body will fall in a purely translational manner.

If, however, a force is applied to a body in such a way as to be directed to one side or another of the center of mass, a torque is produced as well. Such bodies, even when forced into translational motion by the force, undergo rotational motion also. The manner in which footballs, baseballs and similar objects spin as they move is well known to all of us. It is so difficult to center the force upon the center of mass that it is virtually impossible to keep them from spinning.

Naturally, the further the force from the center of mass, the greater is the rotational motion compared with the translational motion. A coin standing on edge can be flicked into a spin by snapping a finger against its rim, and it will turn very rapidly while moving forward only slowly.

It is thought that the stars and planets originated by accretion and that small fragments struck the growing nuclei of the bodies. Astronomers have labored to devise schemes whereby these colliding bodies would be shown as tending to strike more often to one side of the center of mass than the other, setting up torques that do not average out to zero. Thus heavenly bodies, however they may move translationally, also rotate about some axis.

Conservation of Angular Momentum

There is rotational inertia as well as the more familiar translational inertia. If a wheel is rotating about a frictionless axis, it will continue to rotate at a constant angular velocity unless an outside torque is exerted upon it.

The application of a torque will induce an acceleration in the angular motion. This angular acceleration can be represented by α, the Greek letter "alpha," which is the equivalent of the Latin *a*. The units of angular acceleration are radians per second per second or radians/sec^2. Just as linear velocity is equal to angular velocity times distance from the center of rotation (see Equation 6–4) so, by the same line of reasoning, linear acceleration (a) is equal to angular acceleration (α) times distance from the center (r), or:

$$a = r\alpha \qquad \text{(Equation 6–6)}$$

By the second law of motion, we know that force is equal to mass times linear acceleration $(f = ma)$. Combining this with Equation 6–6, we can substitute $r\alpha$ for a, and have:

$$f = mr\alpha \qquad \text{(Equation 6–7)}$$

We have already decided that torque (τ) is equal to force times distance from the center (fr). This was expressed in Equation 6–5 on page 75. Substituting the value for f given in Equation 6–7, we have:

$$\tau = (mr\alpha)\ (r) = mr^2\alpha \qquad \text{(Equation 6–8)}$$

Now according to the laws of motion as applied to translation, the ratio of the force to the acceleration (f/a) is the mass (m) (see Equation 3–3, on page 33). What if we take the analogous ratio in angular motion—that is, the ratio of the torque to the angular acceleration (τ/α)? By rearranging Equation 6–8, we can obtain a value for such a ratio:

$$\frac{\tau}{\alpha} = mr^2 \qquad \text{(Equation 6–9)}$$

In rotational motion, therefore, the quantity mr^2 (mass times the square of the distance from the center of rotation) is analogous to mass alone (m) in translational motion. This introduces interesting differences in the two types of motion.

Consider a body moving in a straight line and made up of a thousand subunits of equal mass. The force required to stop the motion of this body in a given period of time depends only on the total mass. It does not depend on how the subunits are distributed—whether they are packed closely together, arranged in a hollow sphere, in a cubical array, in a straight line or anything else. Only the total mass counts, and the manner in which the subunits are distributed does not change the total mass.

In rotational motion, however, it is not mass alone that counts but mass times the square of the distance from the point (or line) about which the rotation is taking place. Consider a rotating sphere, for instance, made up of a thousand subunits of equal mass. Some of the subunits are close to the axis and some are far away from the axis. Those close to the axis have a small *r,* and therefore a small mr^2, while those far from the axis have a large *r,* and therefore a large mr^2. The body as a whole has some average mr^2, which is called the *moment of inertia,* and this is often symbolized as *I.* The torque required to stop the rotating sphere in a given period of time depends not upon the mass of the sphere but its moment of inertia.

The value of the moment of inertia depends on the distribution of the mass and can be changed without altering the total mass. If instead of a solid sphere we made up a hollow sphere of the same subunits, some of the subunits previously close to the axis would now be located far from the axis. On the other hand, no subunits would have been moved closer to the axis. The average *r* would increase and the moment of inertia (the average mr^2) would increase considerably even though the total mass had not changed. It would require a much larger torque to stop a spinning hollow sphere in a given period of time than it would to stop a solid sphere of the same mass spinning at the same angular velocity.

Thus, gyroscopes and fly-wheels in which it is desired to maintain as even an angular velocity as possible, despite torques of one sort or another, are constructed to have rims as massive as possible and interiors as light as possible. The accelerations produced by given torques are then reduced to a minimum because the moment of inertia has been raised to a maximum.

It is not surprising, considering the analogies between rotational and translational motions, that experiment shows such a thing as a *law of conservation of angular momentum.* By analogy with the law of conservation of momentum in translational motion, this additional law might be stated:

The total angular momentum of an isolated system of bodies remains constant.

But how would we define angular momentum? Ordinary translational momentum is *mv,* mass times velocity. For angular momentum, we must substitute moment of inertia (*I*) for mass, and angular velocity (ω) for translational velocity. Angular momentum, then, is equal to $I\omega$.

Again, however, the moment of inertia (the average value of mr^2) can be altered without altering the total mass, and this produces curious effects.

Suppose, for instance, that you are standing on a frictionless turntable that has been set to spinning; you are holding your arms extended, a heavy weight in each hand.

The axis of rotation is running down the center of your body from head to toe, and the mass of your extended arms is further from that axis than is the rest of you. The weights in either hand are further still. Consequently, your arms and the weights they carry, being associated with large values of *r,* contribute greatly to the mr^2 average and give you a much higher moment of inertia than you might ordinarily possess.

Suppose next that while spinning you lower your arms to your side. The mass content of your arms and the weights they carry is now considerably closer to the axis of rotation, and without any change in total mass, the moment of inertia is greatly decreased. If the moment of inertia (I) is decreased, the angular velocity (ω) must be correspondingly increased to keep the angular momentum ($I\omega$) constant. (In other words, if you are interested in having the product of two numbers always equal 24, then if you start with 8 times 3 and reduce the 8 to 4, you must increase the 3 to 6, to have the new numbers, 4 times 6, still equal 24.)

This, indeed, is what happens. The turntable suddenly increases its rate of spin as you bring your arms to your side. The rate decreases again promptly if you extend your arms once more.

A figure skater makes use of the same device on ice. At first, as rapid a spin as possible is produced with arms extended. The arms are then brought down, and the body spins on the point of one skate with remarkable velocity.

A body that possesses only angular momentum cannot transmit an unbalanced translational momentum to another body, for it has none to transmit. To be sure, the turning wheels of an

Conservation of angular momentum

automobile send it forward and give it translational momentum. There, however, an equal momentum is given the earth in the opposite direction. The two translational momenta add up to zero. No motorist who has ever tried to drive on ice will dismiss that fact. Once friction has decreased to the point where little or no momentum can be transmitted to the earth, the car will itself gain little or no momentum, and the wheels will spin vainly.

CHAPTER 7

Work and Energy

The Lever

Laws of conservation are popular with scientists. In the first place, a conservation law sets limits to possibilities. In considering a new phenomenon, it is convenient to be able to rule out all explanations that would involve a violation of one of the conservation laws (at least until it is found that nothing short of a violation will do). It is then easier to work with the possibilities that remain.

In addition, there is an intuitive feeling that one will not be able to get something for nothing. It therefore seems proper and orderly to suppose that the universe possesses a fixed amount of something or other (such as momentum) and that while this may be distributed among the different bodies of the universe in various ways, the total amount may neither be increased nor decreased.

Consequently, if we observe a situation in which it appears that in some respect something is obtained for nothing, a search is quickly begun for some other factor in the situation which decreases in compensation. It may prove that it is the two factors combined in some fashion that are conserved. In the case of angular momentum, for instance, the moment of inertia can be changed at will and can seemingly be made to appear out of nowhere or disappear into nowhere. The angular velocity, however, always changes in the opposite sense at once, and it is the product of the moment of inertia and the angular velocity that is conserved.

Another case of this sort arises from a consideration of the *lever*. This is any rigid object capable of turning about some fixed point called the *fulcrum*. As a practical example we might consider a wooden plank resting upon a sawhorse—the former being the lever, the latter the fulcrum.

If the fulcrum is directly under the lever's center of gravity, the lever will remain balanced, tipping neither this way nor that. Since the lever, like any other object, behaves as though all its weight were concentrated at the center of gravity, it can then be supported, as a whole, on the narrow edge of the fulcrum. If the lever is of uniform dimensions and density, the center of gravity is at the geometrical center, and it is there that the fulcrum must be placed, as in the well-known children's amusement device, the seesaw.

If a downward force is applied to any point on the lever, the force times the distance of its point of application from the fulcrum represents a torque (see page 75), and the lever takes on rotational motion in the direction of the torque.

Suppose though that a downward force is at the same time applied to the lever on the other side of the fulcrum. If the second force is equal to the first and is applied at the same distance from the fulcrum, the two torques are equal in size but not in direction. The torque on one side of the fulcrum tends to set up a clockwise rotation; the one on the other side tends to set up a counterclockwise rotation. If one torque is symbolized as τ, the other must be $-\tau$. The two torques add up to zero and the lever does not move. It remains in balance.

(On the other hand, if the force is exerted downward on one side of the fulcrum and upward on the other, then both produce a motion in the same direction: both clockwise or both counterclockwise. The torques are then both of the same sign and add up to either 2τ or -2τ. Such a doubled torque is a *couple*, and it naturally is easier to move a lever about a fulcrum by means of a couple than by means of a single torque. It is a couple we use when we wind an alarm clock or manipulate a corkscrew.)

The torques used in connection with levers are often weights that are resting on the ends of the balance, or they are on pans suspended from those ends. We can say that two equal weights will leave a lever in balance if they are placed on opposite sides of the fulcrum and at equal distances from it.

This, in fact, is the principle of the "balance." A balance

has two pans of equal weight suspended from the ends of a horizontal rod that pivots about a central fulcrum. If an object of unknown weight is placed in one pan, combinations of known weights can be put in the other till the two pans balance. We then know that the unknown weight is equal to the sum of the known weights in the other pan. (As explained on page 57, this actually serves to measure mass as well as weight.)

Because of its use in the balance, a lever subjected to equal and opposite torques is said to be in *equilibrium* (from Latin words meaning "equal weights"), and this expression has come to be applied to any system under the stress of forces that produce effects that cancel out and leave the overall condition unchanged.

For a lever to be in equilibrium it must be subjected to equal and opposite torques, and this may be true even if the forces applied are unequal. Consider a downward force (f) applied on one side of a lever at a given distance (r) from the fulcrum. The torque would be fr. Next consider a downward force twice as large ($2f$) applied to the other side of the fulcrum but at a distance only half that of the first ($-r/2$). (The distance is here given a negative sign because it is in the opposite direction from the fulcrum, as compared with the first). This second torque is $(2f)(-r/2)$, or $-fr$. The two torques are equal and opposite, and the lever remains in equilibrium.

If the forces are produced by unequal weights resting on the ends of the lever, it is easy to see that the center of gravity of the system must shift toward the end with the greater weight. To maintain equilibrium, the fulcrum must be directly under the new position of the center of gravity. When this is done, it will be found that its position is such that the product of one weight and its distance from the fulcrum will be equal to the product of the other and its distance from the fulcrum.

Thus if two children of roughly equal weight are on a seesaw, they are right to sit at the ends. If one child is markedly heavier than the other, he should sit closer to the fulcrum. The two should so distribute themselves, in fact, that their own center of gravity plus that of the seesaw remains directly above the fulcrum. (It is also possible, in the case of some seesaws, to shift the board and adjust the position of the fulcrum.)

Because of the fact that torques rather than forces must be equal in order to produce equilibrium, a lever can be put to good use. Suppose a 250-kilogram weight (equivalent to a force of about 2450 newtons) is placed 1 meter from the fulcrum. Next suppose that 10 meters from the fulcrum on the other side of the lever a man applies a downward force of 245 newtons (the equivalent of a 25-kilogram weight). The torque associated with the force (25 × 10) is equal and opposite to that of the torque produced by the weight on the other side of the lever (250 × 1). The lever is then placed in equilibrium and the heavy weight is supported by the light force. If the man applies a somewhat greater force (one that is still considerably less than that produced by the weight on the other side), the lever overbalances on his own side.

A man is not so much conscious of torque as of force (more exactly, of muscular effort). He knows that he cannot apply sufficient force directly to the 250-kilogram weight to lift it. By making use of the lever, however, he can do the job with a force one-tenth that required for direct lifting. By adjusting the differences properly, he could make do with a force one-hundredth, one-thousandth, or indeed any fraction of that required for direct lifting. The usefulness of the lever as a method of multiplying man's lift-

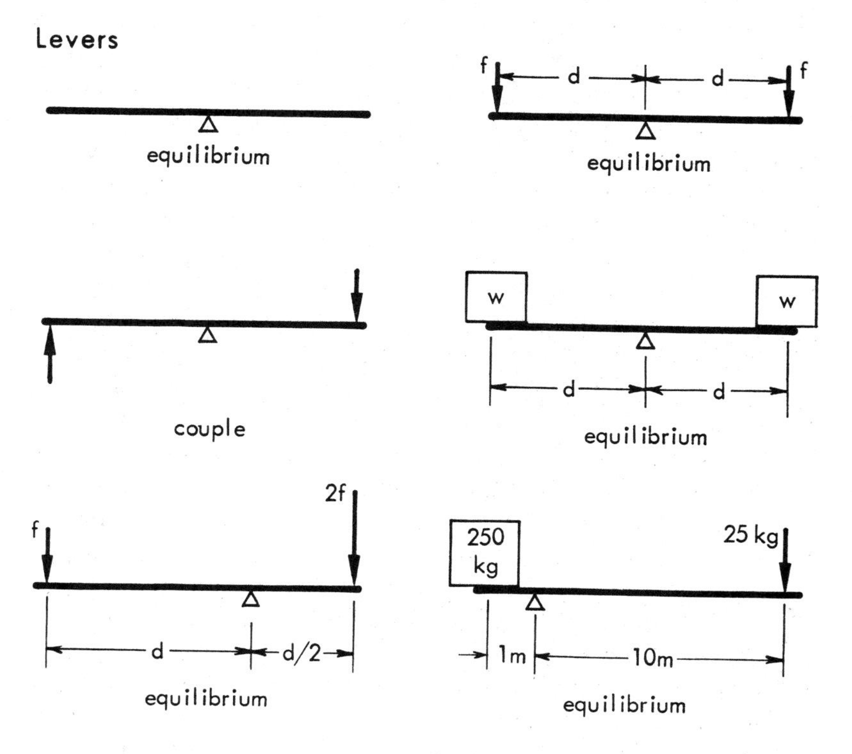

ing ability is evidenced by the very word "lever," which comes from a Latin word meaning "to lift."

No doubt even primitive man had stumbled upon this "principle of the lever," but it was not until the time of the Greek mathematician Archimedes (287–212 B.C.) that the situation was analyzed scientifically. So well did Archimedes appreciate the principle of the lever and its use in unlimited multiplication of force that he said, with pardonable bombast, "Give me a place to stand on and I will move the world."

Any device that transfers a force from the point where it is applied to another point where it is used, is a *machine* (from a Latin word meaning "invention" or "device"). The lever does this, since a force applied on one side of the fulcrum can lift a weight on the other side; it does this in so uncomplicated a fashion that it cannot be further simplified. It is therefore an example of a *simple machine.* Other examples of simple machines are the inclined plane, and the wheel and axle. Some add three other simple machines to the list: the pulley, the wedge, and the screw. However, the pulley can be viewed as a sort of lever, the wedge consists of two inclined planes set back to back, and the screw is an inclined plane wound about an axis.

Virtually all the more complicated machines devised and used by mankind until recent times have been merely ingenious combinations of two or more of these simple machines. These machines depend upon the motions and forces produced by moving bodies through direct contact. As a result, that branch of physics that deals with such motions and forces is called *mechanics.*

That branch of mechanics that specifically deals with motion is called *dynamics,* while that branch that deals with motions in equilibrium is called *statics* (from a Greek word meaning "to cause to stand"). Archimedes was the first great name in the history of statics because of his work with the lever. Galileo, of course, was the first great name in the history of dynamics.

One force that does not seem to be the result of direct contact of one body upon another is gravitation. Gravitation seemingly exerts a force from a distance and produces a motion without involving direct contact between bodies. Such "action at a distance" troubled Newton and many physicists after him. Expedients were worked out to explain this away, and gravitation was included among the mechanical forces. Thus, the study of the motions of the heavenly bodies that result from and are controlled by gravitational forces is called *celestial mechanics.*

Multiplying Force

A machine not only transfers a force, it can often be used to multiply that force, as in the example of the lever described above. Yet this multiplication of force should be approached with suspicion. How can one newton of force do the work of ten newtons just by transmitting it through a rigid bar? Such generosity on the part of the universe is too much to expect, as I pointed out at the beginning of the chapter. Something else must be lost to make up for it.

If we consider the lever lifting the 250-kilogram weight by use of a force equivalent to only 25 kilograms of weight, we can see in the accompanying diagram that we have two similar triangles. The sides and altitude of one are to the corresponding sides and altitude of the other as the distance of the weight from the fulcrum is to the distance of the applied force from the fulcrum.

In other words, if we apply a force at a point ten times as far from the fulcrum as the weight is, then to lift the weight a given distance, we must push down through a distance ten times as great. There is the answer! In lifting a weight by means of a lever, we may adjust distances from the fulcrum in such a way as to make use of a fraction of the force that would be required without the lever, but we must then apply that fractional force through a correspondingly greater distance. The product of the force multiplied by the distance remains the same at either end of the lever.

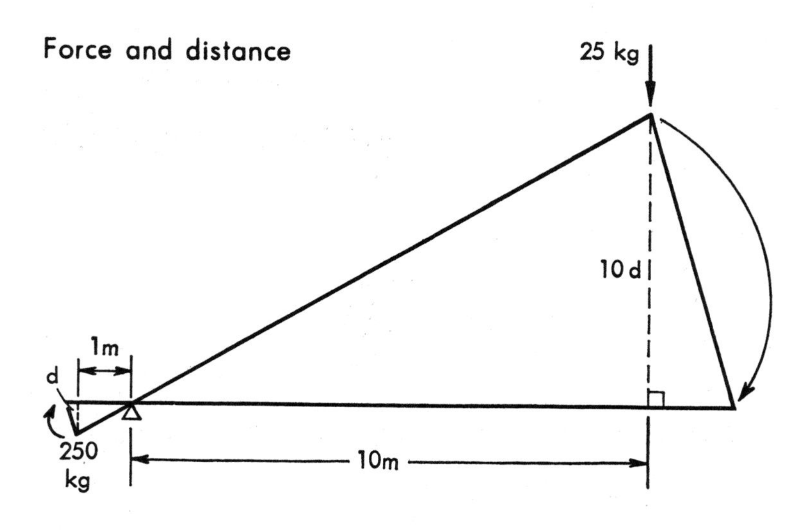

This turns out to be true of any machine that seems to multiply a force. The smaller force performs the task that would require a larger force without the machine, but always at the cost of having to be exerted through a correspondingly longer distance. The product of force and distance in the direction of the force is called *work* and is usually symbolized as w, so that:

$$w = fd \qquad \text{(Equation 7–1)}$$

In a sense, work is an unfortunate term to use in this connection. Anyone will agree that lifting a weight through a distance is work, but in the common use of the term matters are not confined to this alone. In the common language, work is a term applied to the product of any form of exertion. To sit quietly in my chair for half an hour and think of what I am going to say next in this book may strike me as being hard work, but it involves no action of a force through a distance and is not work to a physicist. Again, to stand in one place and hold a heavy suitcase seems hard work, but since the suitcase doesn't move, no work is being done on it. If one walks along with the suitcase, there is still no work being done on it, for although it is moving (horizontally), it is not moving in the direction of the (vertical) force that keeps it from falling.

Nevertheless, the term *work,* signifying a force multiplied by the distance through which a body moves in the direction of the force, is ineradicably established and must be accepted.

The units of work are those of force multiplied by those of distance. In the mks system, the unit of work is the newton-meter, and this is named the *joule* (pronounced "jool") after an English physicist whom I will have occasion to mention later. In the cgs system, the unit of work is the dyne-centimeter, which is called the *erg* (from a Greek word meaning "work"). Since a newton is equal to 100,000 dynes and a meter to 100 centimeters, a newton-meter is equal to 100,000 times 100 dyne-centimeters. In other words, one joule is equal to 10,000,000 ergs.

Since force is a vector quantity, it might seem that work, which is after all the product of a force and a distance, might also be a vector; and that one might speak of a given amount of work to the right and the same amount of work to the left as being equal and opposite. This is not so, however, as we will find if we consider the units of work once more.

A newton is defined as a kilogram-meter per second per second, or $kg\text{-}m/sec^2$. If a joule is a newton-meter, then it is also a kilogram-meter-meter per second per second, or a kg-m²/sec². This last can be written kg-(m/sec)². But m/sec (meters per second) is a unit of velocity, and this means that the unit of work is equal to the unit of mass times the square of the unit of velocity, or $w = mv^2$.

It is true that velocity is a vector quantity, therefore one might speak of $-v$ and $+v$, but the unit of work involves the square of the velocity. The square of a positive number $(+v)(+v)$ and the square of a negative number $(-v)(-v)$ are both positive $(+v^2)$, as we know from elementary algebra. Consequently, the square of the velocity involves no differences in signs, and a unit that includes the square of the velocity is not a vector unit (unless it contains vector units other than velocity, of course).

We conclude then that work is a scalar quantity.

Returning to the lever, we see that the work involved in raising a boulder with a lever is the same as that involved in raising a boulder without a lever, but that the distribution of work between force and distance differs. The same is true where an inclined plane is the device used.

Let us say it is necessary to raise a 50-kilogram barrel through a height of two meters onto the back of a truck. Since a kilogram of weight exerts a downward force of 9.8 newtons, a total upward force of 490 newtons is required to lift the barrel. To exert 490 newtons of force through a distance of two meters in the direction of the force is to do 980 joules of work.

Suppose instead that we lay a sloping plank from the ground to the truck so that the plank makes an angle of 30° with the ground. Under those conditions, the length of the plank from ground to truck is just twice the vertical height from ground to truck, or four meters. The force required to roll the barrel up the plank is 295 newtons, just half the force required for direct lift-

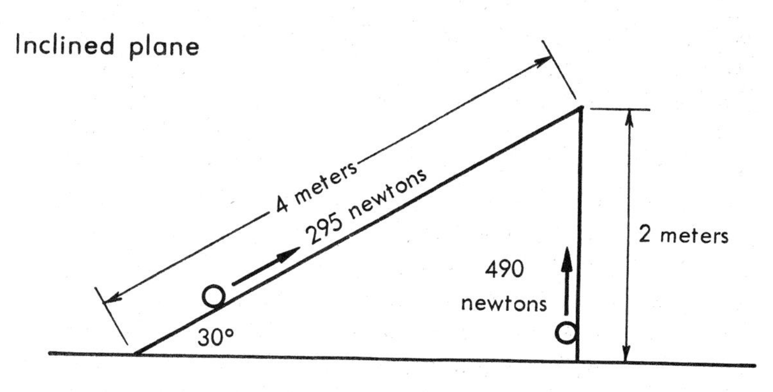

ing. That half-force is exerted through double-distance, however, and 980 joules of work is still done.

The gentler the slope of the inclined plane, the smaller the force required to move the barrel and the longer the distance through which it must be moved. The inclined plane dilutes force —as it diluted velocity for Galileo—by diluting gravitational force (see page 10). Neither the inclined plane, nor the lever, nor any machine, dilutes work. If we stick to work, rather than force, we never get something for nothing.

But if we gain nothing on work, why bother? The answer is that even if we gain nothing directly, we may gain by altering the distribution between force and distance. If it is a question of lifting, by our unaided effort, 250 kilograms two meters directly upward, we must give up. We cannot lift it a meter, a centimeter, or anything at all; we cannot budge it. To move the equivalent of 50 kilograms through ten meters is possible, however, especially if we work slowly; in this way we can do the same work (50×10) that would have been impossible under the previous conditions (250×2). To lift the equivalent of five kilograms through 100 meters may be tedious, but it is quite easy.

Again, if we were asked to shinny up a rope suspended from the roof of a five-story building, we might well decide it to be beyond our capacity unless we were in excellent physical shape. However, a quite ordinary man can lift his weight to a fifth-story roof, if he goes up by way of a ramp, which is an inclined plane that enables him to use less force to lift his body—at the expense of moving it through a longer distance.

It is sometimes convenient to do the opposite: expend extra force in order to gain distance. Thus, a great deal of force is exerted upon the pedals of a bicycle. This is transmitted to a point on the rear wheel near the hub. The spokes of the wheel then act as levers (with the hub the fulcrum), so a much smaller force is applied to the rim of the wheel which, however, moves through a correspondingly larger distance.

The bicycle is therefore a machine that enables the body to convert force into distance (without changing the total work done) more efficiently than it could without the bicycle. It is for this reason that a man on a bicycle can easily outrace a running man, although both are using their leg muscles with equal effort.

The definition of work as the product of a force and the distance through which it acts, says nothing about the time it takes to act. Men usually find it preferable to accomplish a particular amount of work in a short time rather than in a long time and are therefore interested in the rate at which work is done. This rate is spoken of as *power.* The units of power are joules/sec in the mks system and ergs/sec in the cgs system.

A very common unit of power which fits into neither system was originated by the Scottish engineer James Watt (1736–1819). He had improved the steam engine and made it practical toward the end of the eighteenth century, and he was anxious to know how its rate of work in pumping water out of coal mines compared with the rate of work of the horses previously used to operate the pumps. In order to define a *horsepower* he tested horses to see how much weight they could lift through what distance and in what time. He concluded that a strong horse could lift 150 pounds through a height of 220 feet in one minute, so one horsepower was equal to 150 × 200/1, or 33,000 foot-pounds/minute.

This inconvenient unit is equal to 745.2 joules/sec, or 7,452,-000,000 ergs/sec. A joule/sec is defined as a *watt* in James Watt's honor, and so we can also say that one horsepower is equal to 745.2 watts. The watt, however, is most commonly used in electrical measurements. In mechanical engineering (at least in Great Britain and the United States) it is still horsepower all the way. The power of our automobile engines, for instance, is routinely given in horsepower.

Mechanical Energy

It is neat and pleasant to see that the work put into one end of a lever is equal to the work coming out of the other end, and we might fairly suspect that there was such a thing as "conservation of work."

Unfortunately, such a possible conservation law runs into a snag almost at once. After all, where did the work come from that was put into the lever? If one end of the lever was manipulated by a human being who was using the lever to lift a weight, the work came from that done by the moving human arm.

And where does the work of the moving arm come from? A man sitting quietly can suddenly move his arm and do work where no work had previously seemed to exist. This runs counter to the notion of conservation in which the phenomenon being conserved can be neither created nor destroyed.

If one is anxious to set up a conservation law involving work, therefore, one might suppose that work, or something equivalent to work, could be stored in the human body (and perhaps in other objects) and that this work-store could be called upon at need and converted into visible, palpable work.

At first blush such a work-store might have seemed to be particularly associated with life, since living things seemed filled with this capacity to do work, whereas dead things, for the most part, lay quiescent and did not work. The German philosopher and scientist Gottfried Wilhelm Leibnitz (1646–1716), who was the first to get a clear notion of work in the physicist's sense, chose to call this work-store *vis viva* (Latin for "living force").

However, it is clearly wrong to suppose that work is stored only in living things; as a matter of fact, the wind can drive ships and running water can turn millstones, and in both cases force is being exerted through a distance. Work, then, was obviously stored in inanimate objects as well as in animate ones. In 1807, the English physician Thomas Young (1773–1829) proposed the term *energy* for this work-store. This is from Greek words meaning "work-within" and is a purely neutral term that can apply to any object, living or dead.

This term gradually became popular and is now applied to any phenomenon capable of conversion into work. There are many varieties of such phenomena and therefore many forms of energy.

The first form of energy to be clearly recognized as such, perhaps, was that of motion itself. Work involved motion (since an object had to be moved through a distance), so it was not surprising that motion could do work. It was moving air, or wind, that drove a ship, and not still air; moving water that could turn a millstone, and not still water. It was not air or water that contained energy then, but the motion of the air or water. In fact, anything that moved contained energy, for if the moving object, whatever it was, collided with another, it could transfer its momentum to that second object and set its mass into motion; it would thus be doing work upon it, for a mass would have moved through a distance under the urging of a force.

The energy associated with motion is called *kinetic energy,* a term introduced by the English physicist Lord Kelvin (1824–1907) in 1856. The word "kinetic" is from a Greek word meaning "motion."

Exactly how much kinetic energy is contained in a body moving at a certain velocity, v? To determine this, let us assume that in the end we are going to discover that there exists a conservation law for work in all its forms—stored and otherwise. In that case, we can be reasonably confident that if we find out how much work it takes to get a body moving at a certain velocity, v, then that automatically will be the amount of work it can do on some other object through its motion at that velocity. In short, that would be its kinetic energy.

To get a body moving in the first place takes a force, and that force, by Newton's second law, is equal to the mass of the moving body multiplied by its acceleration: $f = ma$. The body will travel for a certain distance, d, before the acceleration brings it up to the velocity, v, which we are inquiring into. The work done on the body to get it to that velocity is the force multiplied by the distance. Expressing the force as ma we have:

$$w = mad \qquad \text{(Equation 7–2)}$$

Now much earlier in the book, in discussing Galileo's experiments with falling bodies, we showed that $v = at$—that velocity, in other words, is the product of acceleration and time. This is easily rearranged to: $t = v/a$. We also pointed out in discussing Galileo's experiments that where there is uniform acceleration, $d = \frac{1}{2}at^2$, where d is the distance covered by the moving body. If, in place of t in the relationship just given, the quantity v/a is substituted, we have:

$$d = \frac{1}{2}\, a\left(\frac{v}{a}\right)^2 = \frac{1}{2}\,\frac{v^2}{a} \qquad \text{(Equation 7–3)}$$

Let us now substitute this value for d in Equation 7–2, which becomes:

$$w = \frac{1}{2}\,\frac{mav^2}{a} = \frac{1}{2}\,mv^2 \qquad \text{(Equation 7–4)}$$

This is the work that must be done upon a body of mass m to get it to move at a velocity v, and it is therefore the kinetic energy contained by the body of that mass and with that velocity. If we symbolize kinetic energy as e_k, we can write:

$$e_k = \frac{1}{2}\,mv^2 \qquad \text{(Equation 7–5)}$$

I have already pointed out that work has the units of mass multiplied by those of velocity squared and, as is clear from Equation 7–5, so has kinetic energy. Therefore, kinetic energy can be measured in joules or ergs, as can work. Indeed, all forms of energy can be measured in these units.

We might now imagine that we can set up a conservation

law in which kinetic energy can be converted into work and vice versa, but in which the sum of kinetic energy and work in any isolated system must remain constant. Such a conservation law will not, however, hold water, as can easily be demonstrated.

An object thrown up into the air has a certain velocity and therefore a certain kinetic energy as it leaves the hand (or the catapult or the cannon). As it climbs upward, its velocity decreases because of the acceleration imposed upon it by the earth's gravitational field. Kinetic energy is therefore constantly disappearing and, eventually, when the ball reaches maximum height and comes to a halt, its kinetic energy is zero and has therefore entirely disappeared.

One might suppose that the kinetic energy has disappeared because work has been done on the atmosphere, and that therefore kinetic energy has been converted into work. However, this is not an adequate explanation of events, for the same thing would happen in a vacuum.

One might next suppose that the kinetic energy had disappeared completely and beyond redemption, without the appearance of work, and that no conservation law involving work and energy was therefore possible. However, after an object has reached maximum height and its kinetic velocity has been reduced to zero, it begins to fall again, still under the acceleration of gravitational force. It falls faster and faster, gaining more and more kinetic energy, and when it hits the ground (neglecting air resistance) it possesses all the kinetic energy with which it started.

Rather than lose our chance at a conservation law, it seems reasonable to assume that energy is not truly lost as an object rises upward, but that it is merely stored in some form other than kinetic energy. Work must be done on an object to lift it to a particular height against the pull of gravity, even if once it has reached that height it is not moving. This work must be stored in the form of an energy that it contains by virtue of its position with respect to the gravitational field.

Kinetic energy is thus little by little converted into "energy of position" as the object rises. At maximum height, all the kinetic energy has become energy of position. As the object falls once more, the energy of position is converted back into kinetic energy. Since the energy of position has the potentiality of kinetic energy, the Scottish engineer William J. M. Rankine (1820–1872) suggested, in 1853, that it be termed *potential energy*, and this suggestion was eventually adopted.

To lift a body a certain distance (d) upward, a force equal to its weight must be exerted through that distance. The force exerted by a weight is equal to mg, where m is mass and g the acceleration due to gravity (see Equation 5–1 on page 54). If we let potential energy be symbolized as e_p then, we have:

$$e_p = mgd \qquad \text{(Equation 7–6)}$$

If all the kinetic energy of a body is converted into potential energy, then the original e_k is converted into an equivalent e_p, or combining Equations 7–5 and 7–6:

$$\frac{1}{2} mv^2 = mgd$$

or simplifying, and assuming g to be constant.

$$v^2 = 2gd = 19.6d \qquad \text{(Equation 7–7)}$$

From this relationship one can calculate (neglecting air resistance) the height to which an object will rise if the initial velocity with which it is propelled upward is known. The same relationship can be obtained from the equations arising out of Galileo's experiments with falling objects.

Kinetic energy and potential energy are the types of energy made use of by machines built up out of levers, inclined planes and wheels, and the two forms may therefore be lumped together as *mechanical energy*. As long ago as the time of Leibnitz it was recognized that there was a sort of "conservation of mechanical energy," and that (if such extraneous factors as friction and air resistance were neglected) mechanical energy could be visualized as bouncing back and forth between the kinetic form and the potential form, or between either and work, but not (taken in all three forms) as appearing from nowhere or disappearing into nowhere.

The Conservation of Energy

Unfortunately, the "law of conservaton of mechanical energy," however neat it might seem under certain limited circumstances, has its imperfections, and these at once throw it out of court as a true conservation law.

An object hurled into the air with a certain kinetic energy, returns to the ground without quite the original kinetic energy. A small quantity has been lost through air resistance. Again, if an elastic object is dropped from a given height, it should (if mechanical energy is to be truly conserved) bounce and return to exactly its original height. This it does not do. It always returns to somewhat less than the original height, and if allowed to drop again and bounce and drop again and bounce, it will reach lower and lower heights until it no longer bounces at all. Here it is not only the air resistance that interferes but also the imperfect elasticity of the body itself. Indeed, if a lump of soft clay is dropped, its potential energy is converted to kinetic energy, but at the moment it strikes the ground with a nonbouncing splat that kinetic energy is gone—and without any re-formation of potential energy. To all appearances, mechanical energy disappears in these cases.

One might argue that these losses of mechanical energy are due to imperfections in the environment. If only a frictionless system were imagined in a perfect vacuum, if all objects were completely elastic, then mechanical energy would be conserved.

However, such an argument is quite useless, for in a true conservation law the imperfections of the real world do not affect the law's validity. Momentum is conserved, for instance, regardless of friction, air resistance, imperfect elasticity or any other departure from the ideal.

If we still want to seek a conservation law that will involve work, we must make up our minds that for every loss of mechanical energy there must be a gain of something else. That something else is not difficult to find. Friction, one of the most prominent imperfections of the environment, will give rise to heat, and if the friction is considerable, the heat developed is likewise considerable. (The temperature of a match-head can be brought to the ignition point in a second by rubbing it against a rough surface.)

Conversely, heat is quite capable of being turned into mechanical energy. The heat of the sun raises countless tons of water vapor kilometers high into the air, so that all the mechanical energy of falling water (where as rain, cataracts or quietly flowing rivers) must stem from the sun's heat. Futhermore, the eighteenth century saw man deliberately convert heat into mechanical energy by means of a device destined to reshape the world. Heat was used to change water into steam in a confined chamber, and this steam was then used to turn wheels and drive pistons. (Such a device is, of course, a steam engine.)

It seemed clear, therefore, that one must add the phenomenon of heat to that of work, kinetic energy and potential energy, in working out a true conservation law. Heat, in short, would have to be considered another form of energy.

But if that is so, then any other phenomenon that could give rise to heat would also have to be considered a form of energy. An electric current can heat a wire and a magnet can give rise to an electric current, so both electricity and magnetism are forms of energy. Light and sound are also forms of energy, and so on.

If the conservation law is to encompass work and all forms of energy (not mechanical energy alone), then it had to be shown that one form of energy could be converted into another quantitatively. In other words, in such energy-conversions all energy must be accounted for; no energy must be completely lost in the process, no energy created.

This point was tested thoroughly over a period of years in the 1840's by an English brewer named James Prescott Joule (1818–1889), whose hobby was physics. He measured the heat produced by an electric current, that produced by the friction of water against glass, by the kinetic energy of turning paddle wheels in water, by the work involved in compressing gas, and so on. In doing so, he found that a fixed amount of one kind of energy was converted into a fixed amount of another kind of energy, and that if energy in all its varieties was considered, no energy was either lost or created. It is in his honor that the unit of work and energy in the mks system is named the "joule."

In a more restricted sense, one can consider that Joule proved that a certain amount of work always produced a certain amount of heat. Now the common British unit of work is the "foot-pound" —that is, the work required to raise one pound of mass through a height of one foot against the pull of gravity. The common British unit of heat is the "British thermal unit" (commonly abbreviated "Btu") which is the amount of heat required to raise the temperature of one pound of water by 1° Fahrenheit. Joule and his successors determined that 778 foot-pounds are equivalent to 1 Btu, and this is called the *mechanical equivalent of heat.*

It is preferable to express this mechanical equivalent of heat in the metric system of units. A foot-pound is equal to 1.356 joules, so 778 foot-pounds equal 1055 joules. Furthermore, the most common unit of heat in physics is the *calorie,* which is the amount of heat required to raise the temperature of one gram of water by 1° Centigrade.* One Btu is equal to 252 calories. Therefore, Joule's mechanical equivalent of heat can be expressed as 1055 joules equal 252 calories, or 4.18 joules = 1 calorie.

Once this much was clear, it was a natural move to suppose that the law of conservation of mechanical energy should be con-

* There will be more to say about Fahrenheit degrees, Centigrade degrees, calories, and other items of the sort later in the book; see chapters 13 and 14.

verted into a *law of conservation of energy,* in the broadest sense of the word—including under "energy," work, mechanical energy, heat, and everything else that could be converted into heat. Joule saw this, and even before his experiments were far advanced, a German physicist named Julius Robert von Mayer (1814–1878) maintained it to be true. However, the law was first explicitly stated in form clear enough and emphatic enough to win acceptance by the scientific community in 1847 by the German physicist and biologist Hermann von Helmholtz (1821–1894), and it is he who is generally considered the discoverer of the law.

The law of conservation of energy is probably the most fundamental of all the generalizations made by scientists and the one they would be most reluctant to discard. As far as we can tell it holds through all the departures of the real universe from the ideal models set up by scientists; it holds for living systems as well as nonliving ones; and for the tiny world of the subatomic realm as well as for the cosmic world of the galaxies. At least twice in the last century phenomena were discovered which seemed to violate the law, but both times physicists were able to save matter by broadening the interpretation of energy. In 1905, Albert Einstein showed that mass itself was a form of energy; and in 1931, the Austrian physicist Wolfgang Pauli (1900–1958) advanced the concept of a new kind of subatomic particle, the neutrino, to account for apparent departures from the law of conservation of energy.

Nor was this merely a matter of saving appearances or of patching up a law that was springing leaks. Each broadening of the concept of conservation of energy fit neatly into the expanding structure of twentieth-century science and helped explain a host of phenomena; it also helped predict (accurately) another host of phenomena that could not have been explained or predicted otherwise. The nuclear bomb, for instance, is a phenomenon that can only be explained by the Einsteinian concept that mass is a form of energy.

CHAPTER 8

Vibration

Simple Harmonic Motion

The law of conservation of energy serves to throw light on a type of motion that we have not yet considered.

So far, the motions that have been discussed, whether translational or rotational, have progressed (unless disturbed) in one direction continuously. It is, however, also possible for motion to progress alternately, first in one direction, then in another, changing direction sometimes after long intervals and sometimes after short intervals—even very short intervals. Such an alternate movement in opposite directions is called a *vibration* or *vibratory motion* (from a Latin word meaning "to shake").

This type of motion is very common, and we are constantly aware of the swaying or trembling of branches and leaves in the wind, for instance; or of the rapid trembling of machinery in operation, such as that of an automobile with its motor idling; even of the chattering of our teeth or the shaking of our hand under conditions of cold or of nervous tension.

The form of vibration that first came under scientific scrutiny was that of a taut string when plucked. Such strings were used in musical instruments known even to the ancients; the plucked strings give rise to musical sounds for reasons involving the vibratory motions lent by the vibrating strings to the air itself (see the chapters on sound beginning on page 148). The first to study such vibrations was the ancient Greek mathematician and philosopher Pythagoras of Samos (sixth century B.C.). His interest lay entirely in the relationship of these vibrations to music, and as a result, vibratory motion is frequently called *harmonic motion.*

Most vibratory motion is of a complicated nature and does not readily lend itself to easy mathematical analysis. The particular type exemplified by the taut, vibrating string is, however, an exception. It can be analyzed with comparative ease and is therefore called *simple harmonic motion* (sometimes abbreviated SHM).

In simple harmonic motion, it has been found that Hooke's law (see page 50) holds at every stage of the movement. If we pull a taut string out of its original equilibrium position, the amount of the displacement from that equilibrium position is proportional to the force tending to restore it to the equilibrium position.

If the string is released after being pulled to the right, let us say, the restoring force accelerates it in the direction of the equilibrium position. In other words, the string snaps back to equilibrium, moving faster and faster as it does so.

As it approaches the equilibrium position, its displacement from that position becomes continually less, and the restoring force becomes continually less in proportion. As the restoring force decreases so, naturally, does the acceleration it imparts; therefore, although the string moves more and more rapidly as it approaches the equilibrium position, the rate of gain of velocity becomes less and less. Finally, when it has reached the equilibrium position the restoring force has become zero and so has acceleration. The string can gain no more velocity and its rate of motion is at a maximum.

But although it is no longer gaining velocity it *is* moving rapidly, and it cannot remain at equilibrium position, but must move past it. Only a force can stop it once it is moving (Newton's first law), and at equilibrium position there is no force to do so. As it goes past the point of equilibrium to the left, however, it is displaced once more, and a restoring force comes into being again; this force produces an acceleration that serves to diminish its velocity of movement (which is now in the direction opposed to the force). As the string continues to move leftward, the displacement and the restoring force continue to increase, and the velocity diminishes at a faster and faster rate until it reaches zero. The string is now motionless at a point of maximum leftward displacement that is equal to the original extent of the rightward displacement.

Under the influence of the restoring force, the string moves to the right again, passes through the equilibrium position, and out to the original maximum rightward displacements. Then it goes back to the left, then back to the right, and so on.

If there were no air resistance and no friction at the points where the string is held taut, the maximum displacements to left and right would always be the same and the vibration would continue indefinitely. As it is, the vibrations do not, after all, quite reach the maximum, but with each rightward (or leftward) motion attain a point of displacement not quite equal to the point reached at the previous motion in that direction. The vibrations are "damped" and slowly die out.

In all cases of simple harmonic motion, the crucial fact is that velocity changes smoothly at all times, never abruptly. Suppose one imagines a falling body passing through the surface of the earth and the solid substance of the planet. The gravitational force upon it would grow continually less as more and more of the substance of the planet lay above the falling body and less and less below it. The body would accelerate as it fell, but by a smaller and smaller amount. By the time it reached the center of the earth, there would be no force upon it at all (at that point), and its velocity would be at a maximum. It would then pass beyond the earth's center and begin to emerge through the opposite portion of the planet, its velocity decreasing as the gravitational force grew larger and larger, until it emerged from the surface at the antipodes and rose as high above it as it had been (on the other side) in the beginning. It would then repeat this movement, returning to its original position, then to the opposite position, and so on. This, too, would be an example of simple harmonic motion.

In actual fact, however, the falling body is interrupted by the surface of the earth, and its velocity is abruptly changed at the moment of contact with that surface. The resultant series of bounces, while an example of a vibratory or harmonic motion, is not a simple harmonic motion.

The Period of Vibration

A particular point of interest in any vibratory motion is the time it takes to move from the extreme point on one side to the extreme point on the other and back. The time taken to complete this motion (or any particular motion, for that matter) is the *period* of that motion.*

Whenever a motion goes through a series of repetitive submotions, each with a period of its own, the motion is said to be a *periodic motion,* particularly when the individual periods are equal. Motion about a circle or any closed curve can be viewed as made up of successive returns to an original point with each single movement about the curve; it is hence a series of repetitive submotions and may be a periodic motion. A vibration also represents a series of returns to an original point, though by way of a forward-and-back motion rather than by motion in a closed curve, and a vibration can also be a periodic motion.

To determine the period of a vibrating object, even when it is vibrating in accordance with the laws governing simple harmonic motion, is rather complicated if the vibration is dealt with directly. In such a vibration, neither velocity nor acceleration is constant, but both are changing with position at every instant. One therefore searches for a way of representing a vibration by means of some sort of motion involving a constant acceleration.

This can be done by switching from vibration to another form of periodic motion—that of motion in a circle. An object can be pictured as moving in a circle under constant inward acceleration, and hence as moving along the circumference of the circle at a constant speed.

If the circle in question has a radius of length a, then its circumference is $2\pi a$. If the point is moving at a speed v, then the time, t, it takes to make a complete revolution (the period of the circular motion) is:

$$t = \frac{2\pi a}{v} \qquad \text{(Equation 8–1)}$$

Now if we imagine the circle casting a shadow edge-on upon the wall, its shadow will be that of a straight line. The point moving about the circle will seem in the shadow to be moving back and forth on the straight line. As the point moves once about the circle, the point on the shadow will seem to move once back and forth upon the straight line. The period of the motion about the circle (Equation 8–1) will also be the period of the shadow-vibration.

* The word "period" comes from Greek words meaning "round path" or "circle," because the first motion to interest mankind from the standpoint of the time it took was, of course, the apparent circular motion of the sun across the sky from one sunrise to the next.

At either extremity of the shadow-line, the point will seem to be moving very slowly, for its motion on the circle will be more or less at right angles to the shadow-line, and there will be very little sideways motion. (And only sideways motion will show up on the shadow.) As the point travels into intermediate parts of the circle, more and more of its motion is sideway and less and less toward or away from the line, so the point on the shadow-line seems to move faster and faster the further it is from the extremity. At the very center, the point on the circle is moving quite parallel to the line and all its motion is sideway. At the center of the shadow line, therefore, the point seems to be moving fastest. The motion of the point on the shadow-line seems to resemble that of a body in simple harmonic motion and, indeed, the motion can be shown to *be* that of a body in simple harmonic motion. Consequently, Formula 8–1 represents the period (t) of a simple harmonic motion.

Equation 8–1 still represents a difficulty for it involves v, a velocity, and while the point travels about the circle at uniform speed, it moves on the shadow-line with a constantly changing one. We must, therefore, find something to substitute for v, if we can.

In any simple harmonic motion, the maximum velocity comes at the midpoint, between the two extremes. A body undergoing such motion is then at equilibrium position, where it would

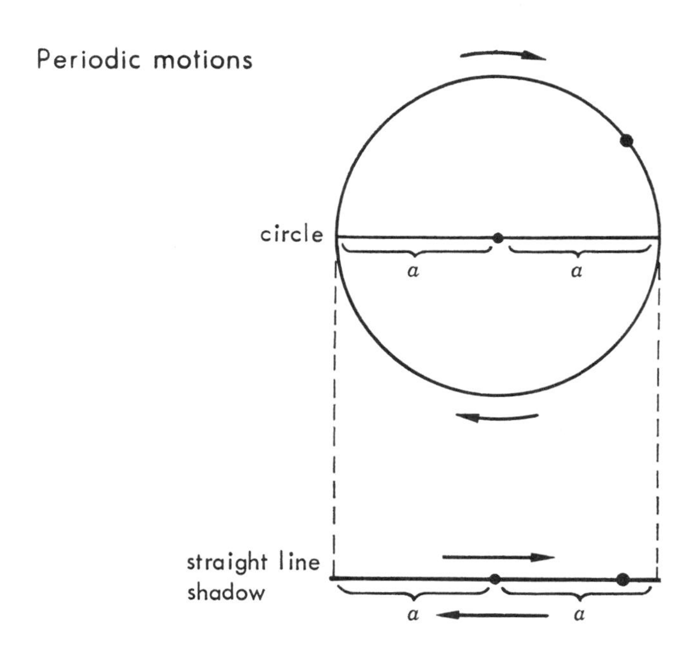

remain if it were at rest. It has no energy of position at that point, and whatever energy it has is all energy of movement, or (see page 94) kinetic energy. As the object moves away from its equilibrium position, it loses velocity and therefore loses kinetic energy. However, it moves into a position it would not take up but for the kinetic energy, and so it gains energy of position, or (see page 96) potential energy. At the extreme position, the body comes to a momentary halt, and all its energy is in the form of potential energy. A body in simple harmonic motion demonstrates a periodic shift from kinetic energy to potential energy and back again, and (barring the damping effect of friction and air resistance) is an excellent example of a conservation of mechanical energy.

Now I have already said that, by Hooke's law, the restoring force on a body undergoing simple harmonic motion is proportional to its displacement from equilibrium position. That is $f = kd$, where f is the restoring force and d is the displacement. The restoring force is least at the position of equilibrium (which is at the center of our straight-line shadow). There is no displacement at that point and the force is equal to 0. The restoring force is at a maximum at the point of maximum displacement, which is, of course, at the extremity of the straight-line shadow. That extremity is a distance of a (the radius of the circle that casts the straight-line shadow) from the center, or equilibrium position, so the force at its maximum is equal to ka.

As the body moves from equilibrium position to the extremity, it moves against a force that begins at 0 and increases smoothly to ka, and the average force against which it moves is therefore ka plus 0 divided by two, or $ka/2$.

The work done on the body to bring it from its equilibrium position to the extremity is equal to force times the distance through which the force is exerted. This comes to $ka/2$ times a, or $ka^2/2$. At the extreme point, all this work is stored as potential energy, and therefore the maximum potential energy of the body moving under the conditions of simple harmonic motion is $ka^2/2$.

At the same time, the maximum kinetic energy of the body comes at the equilibrium point where all the potential energy has been converted into motion and where velocity reaches a maximum. The kinetic energy is then equal to $mv^2/2$, where m is the mass of the body and v its maximum velocity.

Since the potential energy and the kinetic energy are interconverted constantly during simple harmonic motion, without significant loss, the maximum potential energy and the maximum kinetic energy must be equal, so:

$$mv^2/2 = ka^2/2 \qquad \text{(Equation 8–2)}$$

We can easily rearrange this equation to:

$$\frac{a}{v} = \sqrt{\frac{m}{k}} \qquad \text{(Equation 8–3)}$$

We can substitute $\sqrt{m/k}$ for a/v in Equation 8–1 and we have:

$$t = 2\pi\sqrt{\frac{m}{k}} \qquad \text{(Equation 8–4)}$$

This is an astonishing result, for the period of simple harmonic motion turns out to depend only on the mass of the moving body and on the proportionality constant between stress and strain. Both can easily be determined for a particular body, and the period can then be calculated at once.

The period, it should be noted, does *not* depend on the velocity of the body moving in simple harmonic motion, nor on the amount by which it is displaced from equilibrium position, since both v and a have disappeared from Equation 8–4. This means that if a string is pulled out from its equilibrium position by a certain amount, it will attain a certain maximum velocity at midpoint of its swings and will have a certain period of vibration. If it is pulled out a greater distance, or a lesser distance, it will gain a greater maximum velocity, or a lesser one, respectively; in either case the change in velocity will be just enough to make up for the change in distance of displacement, so the period will remain the same.

This constant period of vibration offers mankind a great boon; it is a means of measuring time quite accurately by counting vibrations, even damped vibrations.

In theory, any periodic motion makes this possible. The first periodic motion to serve mankind as a timepiece was the earth itself, for each turn of the planet on its axis marks off one day and night and each turn of the planet about the sun marks off one cycle of seasons. Unfortunately, the earth's movements do not offer a good means of measuring times of less than a day.

During ancient times, mankind made use of nonperiodic motions broken up into (as was hoped) equal parts. These included the motion of a shadow along a background, the movement of sand through a narrow orifice, the dripping of water through an orifice, the shrinking of a burning candle, and so on. All that was obtained in this fashion were rather poor approximations of equal times; not until the mid-seventeenth century was it possible to tell time to closer than an hour or so, or to measure units of time less than an hour with any reasonable accuracy.

It was not until periodic motions with short periods of vibration were put to use that modern time-telling devices became possible—and with them, to a very large extent, modern science.

The Pendulum

Galileo himself suffered greatly from the inability to measure short intervals of time accurately. (He made use of his pulse on occasions, and though this was a periodic phenomenon, it was not, unfortunately, a very steady one.) Nevertheless, although he was himself not to benefit directly from it, he was the first to discover a periodic motion that was eventually to be put to use for the purpose of time-telling.

In 1583, Galileo was a teen-age medical student at the University of Pisa and one day went to the cathedral to pray. Even his devotion to prayers (and Galileo was always a pious man) could not keep his agile mind from working. He could not help but notice the chandelier swaying in the draft. At times, thanks to the vagary of the wind, it swayed in large arcs, at times in smaller ones, but it seemed to Galileo that the period of swing was always the same, regardless of the length of the arc. He interrupted his prayers and checked this conjecture by timing the swing against his pulse.

Back at his quarters, Galileo went on to set up small experimental "chandeliers" by suspending heavy weights ("bobs") from strings attached to the ceiling and letting them swing to and fro. (Such suspended weights are called *pendulums* from a Latin word meaning "hanging" or "swinging"). Galileo was able to show that the period of swing did not depend on the weight of the bobs, but only upon the square root of the length of the string. In other words, a pendulum with a string four feet long would have a period twice as long as one with a string one foot long.

Consider the pendulum, now. If the bob is suspended vertically from its support, it will remain motionless. That is its equilibrium position. If, however, the bob is pulled to one side, the pull of the string forces it to move in the arc of a circle so that it is raised to a higher level. If it is now released, the pull of gravity will cause it to move downward with an accelerating velocity, back along the arc of the circle to the bottommost position.

The gravitational force that brings about this fall is only that part not balanced by the upward pull of the string. As the bob drops, the string becomes more and more nearly vertical and balances more and more of the gravitational force. The unbalanced gravitational force constantly decreases as the bob drops, and the acceleration to which the bob is subjected also decreases. When the bob is at the bottom of the arc, the pendulum is perfectly vertical and the string balances all the gravitational pull. There is no unbalanced gravitational pull at that point and no acceleration. The bob is moving at maximum velocity.

Because of inertia, the bob passes through the point of equilibrium and begins to mount the arc in the other direction. Now there is an unbalanced gravitational force that slows its motion. The higher it climbs the greater the unbalanced gravitational force and the more quickly is the motion of the bob slowed down. Eventually its motion is slowed to zero and it reaches a maximum displacement. Down it comes again, through the equilibrium point, to a maximum displacement on the other side, and so on.

This is very much like the discription of simple harmonic motion (see page 102), except that where the plucking of a string involves motion back and forth in a straight line, that of the pendulum involves motion back and forth along a circular arc. This in itself would not seem to be an essential difference, since there seems no reason why there should not be vibratory rotational motion as well as vibratory translational motion; indeed there are cases of both varieties of simple harmonic motion.

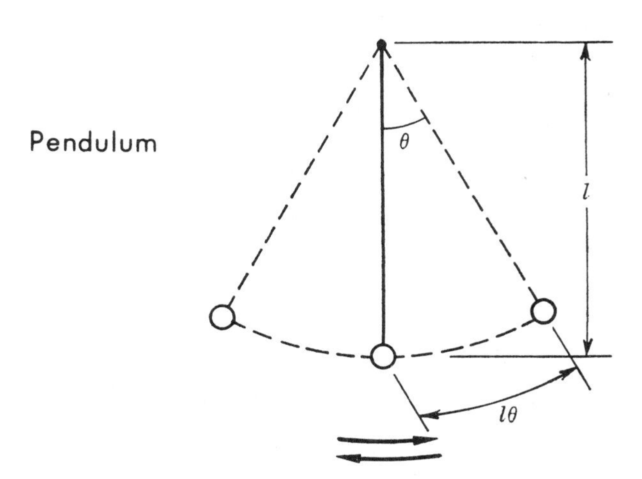

But is the pendulum truly one of them? In all cases of simple harmonic motion such as the vibration of a string, the twisting and untwisting of a wire, the up and down movement of a stretched string, and the opening and closing of a spirally coiled spring, the restoring force lies within the material itself; it is the product of its elasticity. In the case of the pendulum, the restoring force lies outside the system in the form of an unbalanced gravitational pull. This may well introduce a fundamental difference. To check on whether the pendulum swings according to simple harmonic motion, we must see whether the restoring force of gravity is indeed proportional to the amount of displacement, which is what would be required if Hooke's law (characteristic of simple harmonic motion) is to hold.

Let us begin with the displacement. This is the length of the circular arc through which the pendulum has moved in reaching a particular position. The length of this arc depends both on the length (l) of the string and on the size of the angle (θ)* through which it moves. The displacement (D) is, in fact, equal to the length of the string times the angle through which the weight moves:

$$D = l\theta \qquad \text{(Equation 8–5)}$$

Now what about the restoring force? That depends upon the force of gravity, of course. The full pull of gravity, directed downward, would be equal to mg (see page 54), where m is the mass of the bob and g is the gravitational acceleration.† However, the bob is not being pulled directly downward, but to one side. It moves as though it were sliding down an inclined plane that changes its slope at every point.

The situation is similar to that on page 18, where we were also involved with inclined planes. Imagine the bob of a pendulum at a certain point of its movement, where the suspending string makes an angle θ with the vertical. At that point, the bob is acting as though it were sliding down an inclined plane that made a

* The Greek letter, "theta" (θ) is often used to represent angles.

† Actually, the string also has a mass, however light that may be, so there is mass distributed all along the line of the pendulum from the bob up to the very support. At each point in the string there is a little bit of mass suspended by a different length of string. This is also true of the bob itself, different portions of which are different distances from the point of support. Ideally, a pendulum should consist of a massive bob with zero volume attached by a weightless string to the point of support. Such a device is an ideal or *simple pendulum,* which naturally doesn't exist in the actual world. However, by using a dense bob and a light string, a real pendulum can be made to approach the properties of a simple pendulum.

tangent to the arc of swing at that point. We could draw such an inclined plane as part of a right triangle. The inclined plane would have a length L and would be at a height H above the horizontal line. The angle made by the inclined plane to the horizontal could be shown by ordinary geometry to be equal to the angle of displacement, and it, too, can be marked as θ.

As on page 19, the maximum gravitational force would have to be multiplied by the ratio of H to L, so the restoring force (F) would be equal to $mg(H/L)$. The ratio of H to L is usually thought of as the sine* of angle θ, and is symbolized as "sin θ." We can therefore represent the restoring force as:

$$F = mg(\sin\theta) \qquad \text{(Equation 8–6)}$$

The ratio of the restoring force to the displacement in the case of the swinging pendulum is therefore (combining Equations 8–5 and 8–6):

$$\frac{F}{D} = \frac{mg(\sin\theta)}{l\theta} \qquad \text{(Equation 8–7)}$$

Now the question is whether this ratio is a constant, as it must be if the swinging pendulum is to be considered an example of simple harmonic motion. The mass (m) of the bob and the

* The ratio of one side of a right triangle to another varies according to the size of the angles of the right triangle. For a given angle, these ratios are fixed, and each is given a name of its own. Since such ratios are studied in that branch of mathematics called "trigonometry" ("the measurement of triangles" is the meaning of the Greek term from which that expression is derived), such ratios are called *trigonometric functions.* Sines are an example of a trigonometric function. We don't have to go into these in detail. Suffice it to say that it is easy to obtain tables that will give the sine, or any of various other trigonometric functions, for angles of any size.

Restoring force of pendulum

$$\sin\theta = \frac{H}{L}$$

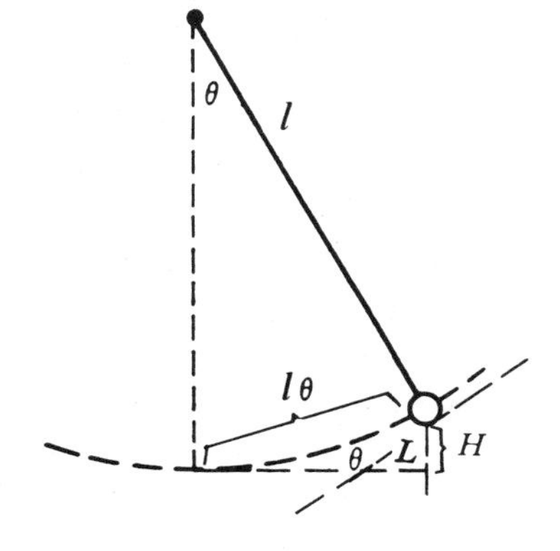

length of the string (l) do not change as the pendulum swings, and the value of g is constant for any given point of the earth's surface, so the quantity mg/l may be considered constant. It remains only to determine whether the quantity $(\sin \theta)/\theta$ is likewise constant. If it is, we are set.

Unfortunately, the ratio is not constant. As we can easily determine, the sine of 30° is 1/2, while the sine of 90° is 1. The angle has tripled, in other words, while the sine of the angle has only doubled. This means that $(\sin \theta)/\theta$ is not a constant, that the restoring force of a pendulum is not proportional to the displacement, and that the swinging of a pendulum is *not* an example of simple harmonic motion.

Nevertheless, if the ratio $(\sin \theta)/\theta$ is not constant, it is nevertheless almost constant for small angles of 10° or less. Therefore, if the pendulum swings back and forth in moderate arcs, it is *almost* an example of simple harmonic motion.

In fact, for small angles $(\sin \theta)/\theta$ is not only constant, it is about equal to unity. For that reason (provided we remember that we are dealing with pendulums swinging through small arcs only), we can eliminate $(\sin \theta)/\theta$ in Equation 8–7 and write:

$$\frac{F}{D} \approx \frac{mg}{l} \qquad \text{(Equation 8–8)}$$

where the symbol $\approx$ signifies "is approximately equal to."

(You may wonder why we are willing to bother with an approximate equality when science should concern itself with exact relationships. The answer is that by being satisfied with an approximation, we can treat the pendulum as an example of simple harmonic motion and make certain other calculations quite simple, if not quite exact.)

For instance, we have already determined that the period (t) of an object undergoing simple harmonic motion is equal to $2\pi\sqrt{m/k}$ (see Equation 8–4).

The symbol k represents the ratio of the restoring force to the displacement for which, in the case of the pendulum, we have found a value in Equation 8–8, where it is set approximately equal to mg/l. Combining Equations 8–4 and 8–8 (and retaining the symbol for approximate equality) we can state that the period of a moderately swinging pendulum is:

$$t \approx 2\pi \sqrt{\frac{m}{mg/l}} \approx 2\pi \sqrt{\frac{l}{g}} \qquad \text{(Equation 8–9)}$$

As you see, the period of a moderately swinging pendulum is

independent of the mass of the bob and depends (at least to a very close approximation) upon the square root of the length of the string—as Galileo had determined by experiment back in the sixteenth century.

The presence of g, the acceleration due to gravity, is of great importance. If we solve Equation 8–9 for g, we obtain:

$$g \approx \frac{4\pi^2 l}{t^2} \qquad \text{(Equation 8–10)}$$

This gives us a far easier method for measuring g than by trying to measure the velocity of free fall directly. The length of a pendulum is easily determined, and so is its period. The use of pendulums in Newton's time showed the manner in which g varied slightly with latitude and added experimental proof to Newton's suggestion that the earth was an oblate spheroid (see page 56).

Since the period of a moderately swinging pendulum is virtually constant, it can also be used to measure time. If a pendulum is connected to geared wheels in such a way that with each oscillation of the pendulum the wheel is pushed forward just one notch, the motion can then easily be scaled down in such a way as to push one pointer around a dial in exactly one hour (the minute hand) and another pointer around the same dial in exactly twelve hours (the hour hand). The addition of weights can keep the pendulum going so that the effects of friction and air resistance do not damp its motion to zero.

In his old age, Galileo had a vision of this application of his youthful discovery, but it was first brought to fruition by the Dutch scientist Christian Huygens (1629–1695) in 1673. Huygens even allowed for the imperfections of the pendulum. He showed how to take into account the fact that an actual pendulum is not a simple pendulum but has a bob of finite volume suspended from a string or rod of finite mass. He also showed that if a pendulum swung in a curve that was not the arc of a circle but the arc of a rather more complicated curve called a cycloid, the period would then be truly constant. He showed, furthermore, how the pendulum could be made to swing in such a cycloidal arc.

Since his time, ingenious methods have also been used to take into account the fact that the length of a pendulum (and therefore its period as well) changes slightly with changes in temperature.

Other examples of simple harmonic motion can also be used to measure time. Hooke (of Hooke's law) devised the "hair-spring," a fine spiral spring which can be made to expand and contract in simple harmonic motion. The fine spring is driven by the uncoiling of a larger "main-spring" that is periodically tightened by mechanical winding. Such hair-springs are used in wristwatches, where there is obviously no room for a pendulum and where (even if room existed) the movements of the arm would throw a pendulum into confusion at once.

In recent years, the vibrations of the atoms moving within molecules in accordance with the rules of simple harmonic motion have been used to measure time. Such "atomic clocks" are far more regular and accurate than any clock based on supra-atomic phenomena can be.

CHAPTER 9

Liquids

Pressure

So far I have assumed that the "bodies" which have been under discussion are *solid*—that is, that they are more or less rigid and have a definite shape. They resist any force tending to alter or deform that shape (though if the force increases without limit, a point is eventually reached where even the most rigid solid shape will deform or break). Solids behave all-in-a-piece, so if part of a solid moves, all of it moves, and in such a way as to maintain the shape.

There are bodies, however, which do not have a definite shape and do not resist deformation. If a stretch or shear, even a small one, is exerted upon them, they alter shape in response. In particular, they will respond to the force of gravity and alter their shape in such a way as to reduce their potential energy to a minimum. In response to gravity, such bodies will move downward and flatten out as much as possible; in so doing, they will take on the shape of any container in which they might be. If the container is open at the top and is tipped, or if an opening is made at the bottom, the material will pour out, under the influence of gravity, to take up a new position of still lower potential energy on the table-top, the floor or in a hole. It is this ability to pour or flow that gives such bodies the name of *fluids* (from a Latin word meaning "to flow").

Fluids fall into two classes. In one class, the downward force of gravity is paramount, so the fluid, while taking on the shape of the container, collects in the lowermost portions and does not necessarily fill it. Such fluids have a definite volume, if not a definite shape, and are called *liquids* (also from a Latin word meaning "to flow"). Water is, of course, the most familiar liquid.

In the other class of fluids, the downward force of gravity is countered by other effects to be discussed in later chapters. In this class there is a certain concentration toward the bottom of a container but not enough to notice under ordinary conditions. On the whole, such a fluid spreads itself more or less evenly through a confined space and has no definite volume of its own. Such fluids without either a definite shape or a definite volume are *gases.** Air is the most familiar gas.

I will take up each variety of fluid separately and will begin with liquids.

The weight of an object, as I explained earlier, is a downward force exerted by that object in response to the gravitational pull. In the case of solids, this force makes itself evident through whatever portion of its nether surface makes contact with another body. Since the nether surface is usually rough (even if only on a microscopic scale) the force is uneven, being exerted at those points where contact is actually made and not at others where contact is not made. For this reason, it is usually convenient to speak only of the total downward force exerted by a solid body, and this is indeed done when we speak of its weight.

In the case of a liquid, however, the contact between its nether surface and the object it rests upon is quite smooth and evenly distributed, so that all portions receive their equal share.† For fluids, therefore, it becomes convenient to speak of weight (or, more properly, force) per unit area. This force per unit area is termed *pressure*.

It is common to use as a unit of pressure "pounds per square

* The word "gas" was coined about 1600 by a Flemish chemist, Jan Baptista van Helmont (1577–1644), who supposedly derived it from the Greek word "chaos."

† If we get sufficiently submicroscopic, unevenness does show up, to be sure. This is because matter is not really continuous but is composed of discrete particles called atoms. We don't have to worry about this right now, but will consider it later (see page 143).

inch" (sometimes abbreviated "psi") where pounds are a unit of weight in this connection and *not* units of mass.

In the metric system the proper units of pressure are newtons per square meter in the mks system, and dynes per square centimeter in the cgs system. Since a newton equals 100,000 dynes and a square meter equals 10,000 square centimeters, 1 newton/m^2 is the equivalent of 100,000 dynes per 10,000 square centimeters, or 10 dynes/cm^2. Translating into metric units, one pound per square inch is equal to 6900 newtons/m^2, and one gram per square centimeter is equal to 98 newtons/m^2.

Suppose we consider a square centimeter of the bottom of a container filled with liquid to a height *n*. The pressure (dynes /cm^2) depends on the weight of liquid resting on that square centimeter. The weight depends, in part at least, upon the volume of the column, one square centimeter in cross-sectional area and *n* centimeters high. The volume of that column is *n* cubic centimeters.

It does not follow, however, that in knowing the volume of a substance we also know its weight. It is common knowledge that the weight of a body of given volume varies according to the nature of the substance making up the body. We are all ready to admit, for instance, that iron is "heavier" than aluminum. By that, of course, is meant that a given volume of iron is heavier than the same volume of aluminum. (If we remove this restriction to equal volumes, we will be faced by the fact that a large ingot of aluminum is much heavier than an iron nail.)

For any object the weight per unit volume is its *density*, and in the metric system the units of density are usually expressed as grams (of weight) per cubic centimeter, or kilograms (of weight) per cubic meter. We should, therefore, say that iron is "denser," rather than "heavier," than aluminum.

If the height of a column of liquid resting upon a unit area determines its volume, and the density of that liquid gives the weight of a unit volume, then the total weight on the unit area, or pressure (p), is equal to the height of the liquid column (h) multiplied by its density (d):*

$$p = hd \qquad \text{(Equation 9–1)}$$

* This assumes that the density does not vary along the height of the column, and as far as liquids are concerned, any variation of density with depth is small enough to be ignored for small pressures. This will not be so for gases (see page 145).

The pressure of a liquid on the bottom of a container therefore depends only upon the height and density of the liquid, and not upon the shape of the container or the total quantity of liquid in the container. This means that the various containers shown in the accompanying figure, with bottoms of equal areas but with different shapes and containing different quantities of liquid, will have their bottoms placed under equal pressure.

It is easy to see that the container with the expanded upper portion ought to experience the same pressure at the bottom, for the weight of the additional liquid is clearly supported by the upper horizontal portion of the container. It may seem not at all logical, however, that the container with the contracted upper portion should also experience the same pressure at the bottom. The missing liquid (not present because of the contraction) has no weight to contribute to the pressure. How then does the pressure remain as great as if the missing liquid were there?

To explain that, we must realize that pressure is exerted differently in liquids as compared with solids. A solid resists the deforming influence of its own weight. A large pillar of marble may rest solidly on a stone floor and transmit a great deal of pressure to that floor, but it will itself remain unmoved under its own weight. The pillar will not, for instance, belly out in the middle,

Pressure and shape

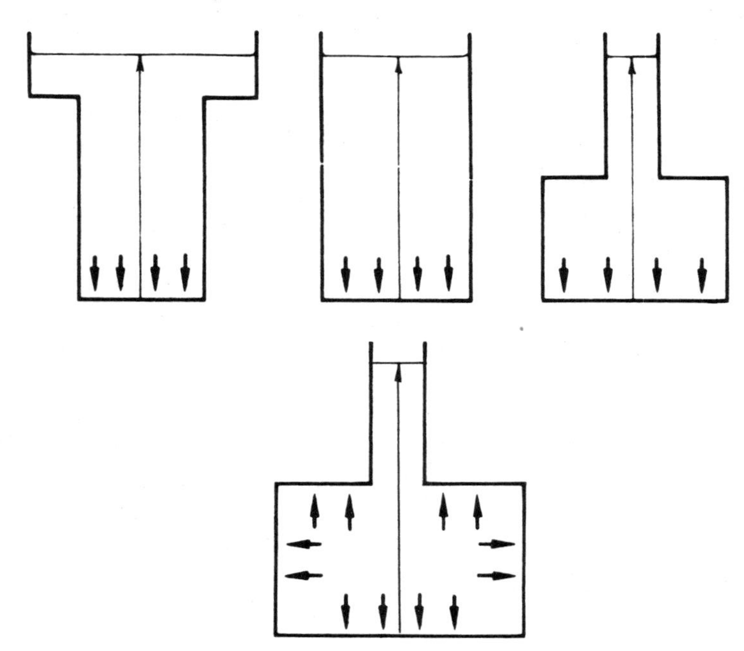

and if we place our hands on the side of the pillar we will be aware of no pressure thrusting out sideways.

Imagine, however, a similar pillar made of water. It could not remain in existence for more than a fraction of a second. Under the force of its own weight it would belly outward at every point and collapse. If a pillar of water is encased in a restraining cylinder of aluminum, the outward-bellying tendency of the water will evidence itself as a sidewise force. If a hole is punched in the aluminum cylinder, water will spurt out sidewise under the influence of that force. This same line of reasoning would show that a liquid would exert a pressure against a diagonally slanted wall with which it made contact.

A fluid, indeed, exerts pressure in all directions and particularly in a direction perpendicular to any wall with which it may make contact. The amount of pressure exerted at any given point depends upon the height of liquid above that particular point. Thus if a hole is punched in a cylindrical container of water, the liquid will spurt out with more force if the hole is near the bottom (with a great height of liquid above) than if it is near the top (with but a small height of liquid above).

In the container with a contracted upper portion, then, there is a pressure of the fluid up against the horizontal section, as indicated in the diagram. The amount of this pressure depends upon the height of the liquid above that horizontal section. By Newton's third law, the upper horizontal section exerts an equal pressure down upon the liquid. The downward pressure of the horizontal section is equal to that which would be produced by the missing liquid if it were there, and so the pressure at the bottom of the container remains the same.

Buoyancy

The generalization concerning pressure, made use of in the previous section, was first clearly stated by the French mathematician Blaise Pascal (1623–1662) and is therefore often referred to as *Pascal's principle.*

This can be expressed as follows: Pressure exerted anywhere on a confined liquid is transmitted unchanged to every portion of the interior and to all the walls of the containing vessel; and is always exerted at right angles to the walls.

This principle can be used to explain the observed fact that if a container of liquid contains two or more openings, to which are connected tubes of various shapes into which the liquid can rise, and if enough liquid is present in the container so that the level will rise into those tubes, the liquid will rise to the same height in each.*

To explain this, let us consider the case of a container with two openings and let us imagine a porous vertical partition dividing the container between the two openings. The pressure against the partition from the left would depend on the height of the liquid on the left, while the pressure from the right would depend on the height of the liquid on the right. If the liquid column is higher on the left, the pressure from the left is greater than that from the right, and there is a net pressure from left to right. Liquid is forced through the partition in that direction, so that the height of the liquid on the left decreases and that on the right increases. When both heights are equal there is no net pressure either left or right, and therefore no further motion.

This effect is part of folk knowledge, as is witnessed by the common saying that "water seeks its own level."

Notice that I am taking for granted here that liquids will move, or flow, in response to a force, and this is actually so. The laws of motion apply to fluids as well as to solids, and the study of mechanics includes, in its broad sense, forces and motions involving fluids as well as solids. However, it is quite common to restrict the use of the term "mechanics" to solid bodies. The mechanics of liquids is then given the special name *hydrodynamics* (from Greek words meaning "the motions of water"), and the mechanics of gases is called *pneumatics* (from the Greek word for "air"). These may be grouped together as *fluid mechanics.*

It is not only the weight of the liquid itself that can be trans-

* This is not strictly true because of capillary action, but this will be taken up later (see page 129).

Water finds its level

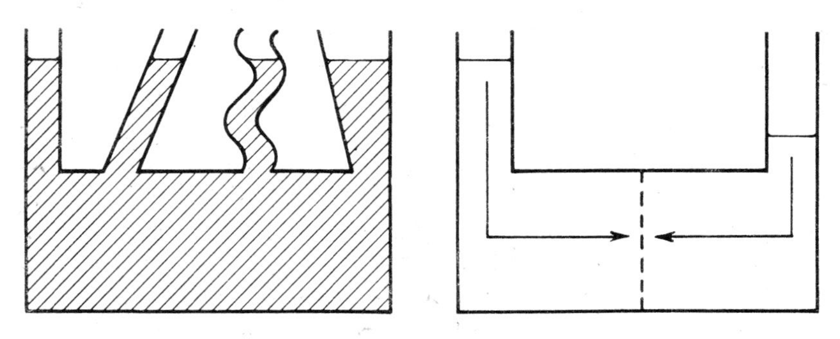

mitted to every part of the liquid as a pressure. Any applied force can be so transmitted.

For instance, suppose a liquid completely fills a container with two necks, each neck being stoppered by a movable piston which we can assume to be weightless. Suppose, furthermore, that the necks are of different width, so the piston in the larger neck has a cross-sectional area of 10 cm^2 while that in the smaller one has a cross-sectional area of only 1 cm^2.

Now imagine that a force of one dyne is exerted downward on the smaller piston. Since the area of the smaller piston is 1 cm^2, the pressure upon it as a result of the applied force is 1 dyne/cm^2. In accordance with Pascal's principle, this pressure is transmitted unchanged through the entire body of liquid and, perpendicularly, to all the walls. It is transmitted, in particular, perpendicularly to that portion of the wall represented by the larger piston. As the small piston moves downward, then, the large piston moves upward.

The upward pressure against the larger piston must be the same as the downward pressure against the smaller piston, 1 dyne /cm^2. The area of the larger piston is, however, 10 cm^2. The total force against the larger piston is therefore 1 dyne/cm^2 multiplied by 10 cm^2, or 10 dynes. The total force has been multiplied tenfold and the weight which the original force would have been capable of lifting has also been multiplied tenfold. It is by "hydraulic presses" based on this effect that heavy weights can be lifted with an expenditure of but a reasonable amount of force.

Are we in this way getting something for nothing? Not at all!

Suppose we press down on the small piston (1 cm^2 in area)

Hydraulic press

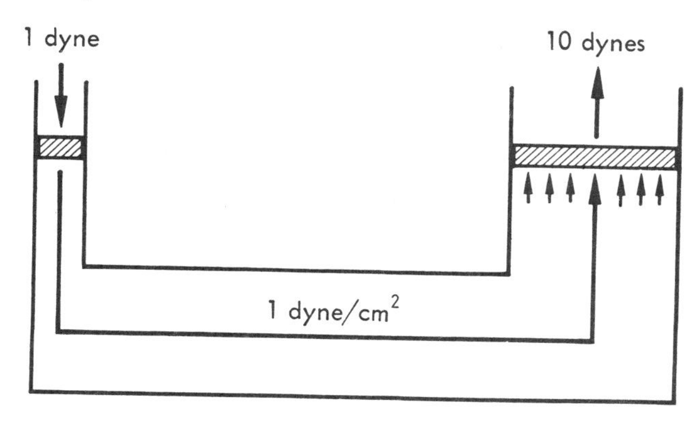

and make it move a distance 1 cm. The volume of liquid it has displaced is 1 cm^2 multiplied by 1 cm, and this comes to one cubic centimeter (1 cm^3). The larger piston (10 cm^2 in area) can only move upward a sufficient distance to make room for the displaced 1 cm^3 of liquid. The distance required is 1 cm^3 divided by 10 cm^2, or 0.1 cm. Thus the situation is the same as it was for the lever (see page 89). The force has been multiplied tenfold, yes, but the distance through which the force has been exerted has been reduced to one-tenth. The total work (force times distance) obtained from the hydraulic press is the same, if we neglect such things as friction, as the total work put into it.

The pressure of a liquid will be transmitted not only to the walls of a container but also (perpendicularly) to the surfaces of any solid object within the liquid. Imagine a cube of iron suspended in liquid so that the top and bottom surface of the cube are perfectly horizontal and the other four surfaces are perfectly vertical. The pressure against each of the four vertical surfaces depends on the height of liquid above them, which is the same for all. For the vertical surfaces, then, we have equal pressures arranged in opposing pairs. There is, consequently, no net sideways pressure in any direction.

But what if we consider the two horizontal surfaces, the one on top and the one on bottom? It is clear that there is a greater height of liquid above the lower surface than above the upper one. There is therefore a comparatively great upward pressure against the lower surface and a comparatively small downward pressure against the upper surface. As a result, a net upward force is exerted by the liquid upon the submerged object. (This is most easily reasoned out in the case of the solid cube, but it can be shown to hold for a solid of any shape or, for that matter, for a submerged drop of liquid or bubble of gas.) This upward force of liquids against submerged objects is called *buoyancy*.

How large is this buoyant force? Consider a solid body dropping into the liquid contents of a vessel. The solid must make room for its own volume by pushing aside, or displacing, an equivalent volume of liquid, and the liquid level in the vessel rises sufficiently to accommodate that displaced volume.

It therefore follows that the submerging solid is exerting a downward force on the liquid, a force large enough to balance the weight of the solid's own volume of liquid. By Newton's third law, it is to be expected that the liquid will in turn exert an upward force on the solid equivalent to the weight of that same quantity of liquid.

The original weight of the submerged body is equal to its volume (V) times its density (D). The weight of the displaced liquid is equal to its volume (which is the same as the volume of the submerged solid, and hence also V) times its density (d). The weight of the body after submersion (W) is equal to its original weight minus the weight of the displaced water:

$$W = VD - Vd \qquad \text{(Equation 9–2)}$$

Solving for D, the density of the submerged solid, we have:

$$D = \frac{W + Vd}{V} \qquad \text{(Equation 9–3)}$$

The weight of the immersed body (W) can be directly measured, the volume of the displaced fluid (V) is obtained at once from the rise in water level and the cross-sectional area of the container, and the density of the fluid (d) is also easily measured. With this data in hand, the density of the immersed body can be calculated easily from Equation 9–3.

This method of measuring density was first made use of by the Greek mathematician Archimedes in the third century B.C. The story is that King Hiero of Syracuse, having received a gold crown from the goldsmith, suspected graft. The goldsmith had, the king felt, alloyed the gold with cheaper silver and had pocketed the difference. Archimedes was asked to tell whether this had been done, without, of course, damaging the crown.

Archimedes knew that a gold-silver alloy would have a smaller density than would gold alone, but he was at a loss for a method of determining the density of the crown. He needed both its

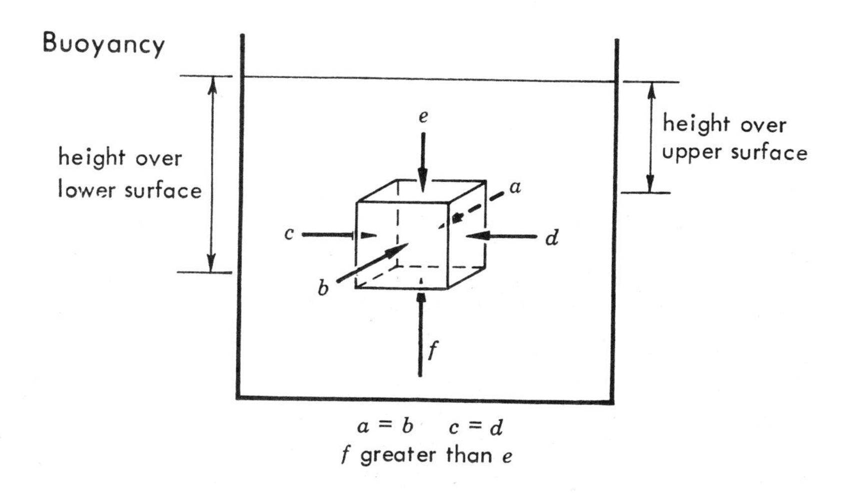

weight and its volume for that, and while he could weigh it easily enough, he could not estimate the volume without pounding it into a cube or sphere or some other shape for which the volume could then be worked out by the geometry of the time. And pounding the crown would have been frowned on by Hiero.

The principle of buoyancy is supposed to have occurred to Archimedes when he lowered himself into a full bathtub and noted the displaced water running over the sides. He ran naked through the streets of Syracuse (so the story goes) shouting, "Eureka! Eureka!" ("I've got it! I've got it!"). By immersing the crown in water and measuring the new weight together with the rise in water level, and then doing the same for an equal weight of pure gold, he could tell at once that the density of the crown was considerably less than that of the gold; the goldsmith was suitably punished. The principle of buoyancy is sometimes called *Archimedes' principle* as a result.

If an immersed body has a greater density than that of the fluid in which it is immersed, then D is greater than d, and VD is naturally greater than Vd. From Equation 9–2, we see that in that case the weight (W) of the immersed body must be a positive number. The weight of the body is decreased, but it is still larger than zero and falls through the fluid. (Thus a solid iron or aluminum object will fall through water.)

However, if the immersed body has a smaller density than that of the fluid, then D is smaller than d, VD is smaller than Vd, and the immersed body has a weight that is a negative number. With a negative weight (so to speak), it moves upward rather than downward in response to a gravitational field. (Thus a piece of wood or a bubble of air submerged in water will "fall upward" if left free to move.)

A solid body less dense than the fluid that surrounds it will

Floating

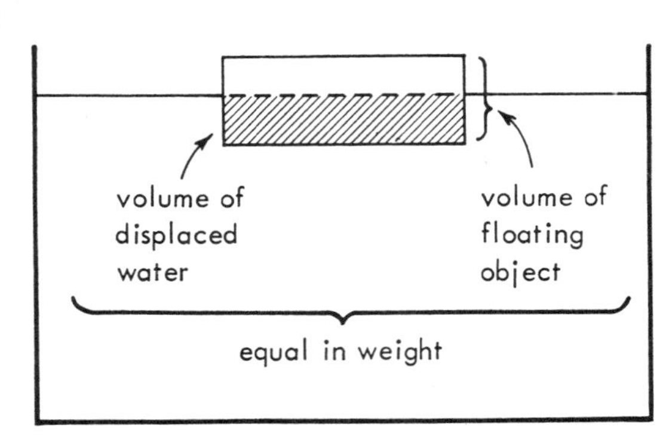

float, partly submerged, on the surface of the fluid under conditions where the weight of water it displaces is equal to its own original weight; in such a case, its weight in water is zero, and it neither rises nor falls. The solid body floats when it has displaced just enough water (less than its own volume) to equal its own original weight.

It is not to be supposed though that because a steel ship floats the density of steel is less than that of water. It is not the steel of the ship alone that displaces water. The ship is hollow, and as it sinks into the water the enclosed air displaces water just as the steel does. The density of the steel-plus-enclosed-air is less than the density of water, though the density of steel alone most certainly is not, and so the steel ship floats.

The force of buoyancy, by the way, is not a matter of calculations and theory alone; it can easily be felt. Lift a sizable rock out of the water and its sudden gain in weight as it emerges into the air can be staggering. Float a sizable block of wood on a water surface and try to push it down so that it will be completely submerged, and you will feel the counterforce of buoyancy most definitely.

Cohesion and Adhesion

Solids, as I said at the beginning of the chapter, act all-in-a-piece. Each fragment of a solid object clings firmly to every other fragment, so if you seize one corner of a rock and lift, the entire substance of the rock rises. This sticking-together is called *cohesion* (from Latin words meaning "to stick to").

Fluids have nothing like the kind of cohesion that exists in solids. If you dip your hand into water and try to lift a piece of it in the hopes that the entire quantity will rise out of its con-

Surface tension

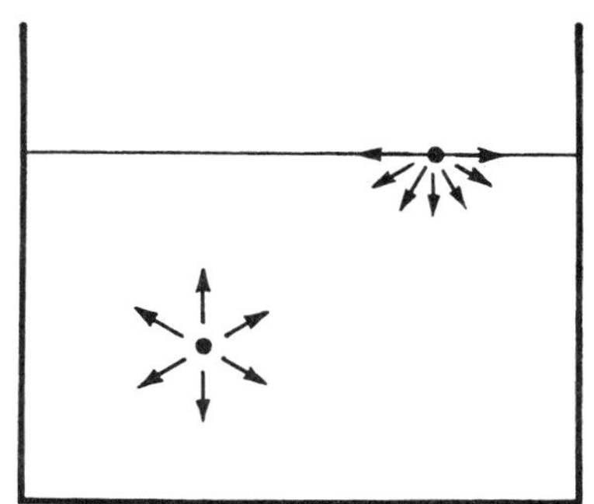

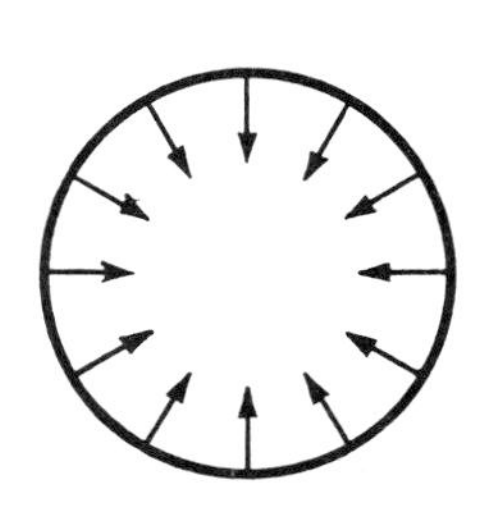

tainer, you will only get your fingers wet. Nevertheless, one should not conclude from this that the force of cohesion in liquids is completely absent. The force is much smaller in most liquids than in solids, but it is not entirely zero. This is most clearly seen at the surface of a liquid.

In the body of a liquid, even a short way below the actual surface, a given portion of the liquid is attracted by cohesive forces in all directions equally by the other portions of the liquid that surround it. There is no net unbalanced force in any particular direction.* At the surface of the liquid, however, the cohesive forces are directed only inward toward the body of the liquid and not outward where there is no liquid to supply cohesive forces. (Most often, there is only air on the other side of the liquid surface, and the attractive forces between air and the liquid are so small that they can be ignored.) The resultant of this semisphere of cohesive forces about a particle of liquid in the surface is a net inward force exerted perpendicularly from that surface.

To keep liquid in the surface against this inward force requires work, so the surface represents a form of energy of position or potential energy. This particular form is usually called *surface energy*.

Such surface energy is distributed over an area of surface, so its units are those of work per area. In the mks system this would be joules per square meter (joules/m^2), and in the cgs system it would be ergs per square centimeter (ergs/cm^2). In the case of surface energy, the cgs system is the more convenient and is generally used. One erg is equal to one dyne-cm or 1 g-cm^2/sec^2 so 1 erg/cm^2 is equal to 1 (g-cm^2/sec^2)/cm^2. If we cancel one of the centimeter units, this becomes 1 (g-cm/sec^2)/cm, or 1 dyne/cm, and as a matter of fact the units of surface energy are most often presented with the last-named unit, dynes per centimeter.

Left to itself, surface energy reduces to a minimum in a way analogous to that in which gravitational potential energy is reduced whenever a ball high in the air falls to the ground, or in which a column of liquid flattens and spreads out if the container is broken. A small quantity of liquid suspended in air will take up the shape of a sphere, for a sphere has for its volume the smallest area of surface, therefore, surface energy is then reduced to a

* In solids, the various particles constituting the substance are lined up in fixed and orderly positions (rather than moving about freely as in liquids). For that reason, the cohesive forces between neighboring particles in solids are oriented in definite directions and are very noticeable.

minimum. Such a sphere of liquid is, however, distorted into a "tear-shaped" object by the unbalanced downward pull of gravity. If it is falling through air, as a raindrop does, for instance, it will be flattened at the bottom through the upward force of air resistance. The smaller the droplet of liquid, the smaller the relative effects of gravity and air resistance, and the more nearly spherical it is. Soap bubbles are hollow, liquid structures that are so light for their volume (because of enclosed air) that the forces of gravity (unusually low in this case) and of air resistance (unusually high) cancel each other. Soap bubbles therefore drift about slowly and show themselves to be virtually perfect spheres.

A sizable quantity of liquid flattens out. The necessity of minimizing the gravitational potential energy rises superior to that of minimizing the surface potential energy, and the surface of an undisturbed pail of water (or pond of water) seems to be a plane. Actually, it is a segment of a sphere, but a large one; one that has a radius equal to that of the earth. Look at the Pacific Ocean on a globe of the earth and you will see that its surface almost forms a semisphere.

If energy in any form is added to a liquid, some may well go into increasing the surface energy by extending the surface area beyond its minimum. Thus, wind will cause the surface of an ocean or lake to become irregular and therefore increase in surface area. The surface in a glass of water will froth if the glass is shaken.

Because the surface is stretched into a larger area by such an input of energy, and because it pulls back to the minimum when the energy input ceases, the analogy between the liquid surface and an elastic skin under tension (a very thin film of stretched rubber, for instance) is unmistakable. The surface effects are therefore frequently spoken of as being caused by *surface tension,* rather than by surface energy.

The same sort of cohesive forces that act to hold different portions of a liquid together, via surface tension, act also to hold a portion of a liquid in contact with a portion of a neighboring solid. In the latter case, where the attractive force is between solid and liquid (unlike particles) rather than between liquid and itself (like particles), the phenomenon is called *adhesion* (also, like cohesion, from Latin words meaning "to stick to"). Adhesive forces may be as great as, or even greater than, cohesive forces. In particular, the adhesion of water to clean glass is greater than the cohesion of water to itself.

This has an effect on the shape of the liquid surface of water in a glass container. Where the water meets the glass, the attraction of the glass for water is large enough to overcome water's cohesive forces. As a result, the water surface rises upward so as to increase the water-glass contact (or "interface") as much as possible at the expense of the weaker water-water forces. If there were no countering forces, water would rise to the top of the container and over. However, there is the countering force of gravity. There comes a point where the weight of the raised water, added to the cohesive forces of water, just balances the upward pull of the adhesive forces, and a point of equilibrium is reached after the water level has been raised by a moderate degree.

If the container is reasonably wide, this upward-bending of the surface is restricted only to the neighborhood of the water-glass contact. The water surface in the interior remains flat. Where the container is a relatively narrow one, however, the surface of the liquid is all in the region of water-glass contact, and the liquid surface is then nowhere plane; instead it forms a semisphere bending down to a low point in the center of the tube. Viewed from the side, the surface resembles a crescent moon and, indeed, it is spoken of as a *meniscus* (Greek for "little moon").

Cohesive forces may well be larger than adhesive forces in particular cases. For instance, the cohesive forces in liquid mercury are much larger than those in water; they are also larger than the adhesive forces between mercury and glass. If we look at mercury in a glass tube, we see that at the interface where mercury meets glass, the mercury pulls away from the glass, reducing the mercury-glass interface. The mercury meniscus in such a tube bends downward at the edges and rises to a maximum height at the center of the tube. The same is true even for water if the glass container has a coating of wax, since the adhesive forces between water and wax are less than the cohesive forces within water.

If water is spilled onto a flat surface of glass, it will spread out into a thin film so as to make the greatest possible contact, adding to the total adhesive force at the expense of the weaker cohesive force. The water, in other words, wets the glass. Mercury, however, when spilled on glass (or water on a waxed surface), makes as little contact with the glass as possible, drawing itself into a series of small gravity-distorted spheres, and adding to the total cohesive force at the expense of the weaker adhesive force. Mercury does not wet glass, and water does not wet wax. In all these events, the effect is to reduce the total surface energy (that of the liquid/air interface plus that of the liquid/solid interface) to a minimum.

Where a water-containing tube attached to a water reservoir is narrow, the rise in water level brought about by the upward force of adhesion is considerable, and the water rises markedly above its "natural level" (see page 120).

It is possible to calculate what the raised height (h) of the water level must be in a particular tube. Adhesion is a form of surface tension (which we can represent as the Greek letter "sigma," σ) acting around the rim of the circle where water meets the glass of the tube. This circle has a length of $2\pi r$, where r is the radius of the tube. The total upward force brought about by adhesion is therefore the surface tension of the water-glass interface, σ dynes/cm, multiplied by the length of the circle where water and glass meet, $2\pi r$ cm, so that the total force is $2\pi r\sigma$ dynes.

Countering this upward force is the downward force of gravitation, which is equal to the weight (mg dynes, see page 54) of the raised water. The mass of the column of water raised by adhesion is equal to its volume (v) times its density (d). Substituting vd for m, we see that the weight of the water is vdg dynes. Since the raised column of water in the tube is in the form of a cylinder, we can make use of the geometrical formula for the volume of a cylinder and say that the volume of the raised water is equal to the height of the column (h) multiplied by the cross-sectional area (πr^2), where r is the radius of the column. Substituting $\pi r^2 h$ for v, we see that the weight of the water is $\pi r^2 hdg$ dynes.

When the water in the narrow tube has been raised as high as it will go, the upward adhesive force is balanced by the downward gravitational force, so we have:

$$2\pi r\sigma = \pi r^2 hdg \qquad \text{(Equation 9–4)}$$

Solving for h:

$$h = \frac{2\sigma}{rdg} \qquad \text{(Equation 9–5)}$$

The acceleration due to gravity (g) is fixed for any given point on the earth; and for any particular liquid, the surface tension (σ) and the density (d) are fixed for the particular conditions of the experiment. The important variable is the radius of the tube (r). As you see, the height to which a column of water is drawn upward in a narrow tube is inversely proportional to the radius of the tube. The narrower the tube, the greater the height to which the liquid is lifted. Consequently, the effect is most noticeable in tubes (natural or artificial) of microscopic width. These are capillary tubes (from a Latin expression meaning "hair-like"), and the rise of columns of water in such tubes is called *capillary action.* It is through capillary action that water rises through the narrow interstices of a lump of sugar or a piece of blotting paper, and it is at least partly through capillary action that water rises upward through the narrow tubes within the stems of plants.

Again, if we know the value of the density of a liquid and the extent of its rise in a tube of known radius (both rise and radius being easily measured), it follows that since the value of g is also known, the value of the surface tension (σ) can be calculated from Equation 9–5.

In the case of mercury, where the adhesive forces with glass are exerted downwards, the level is pulled below the "natural level." The degree to which the level is lowered is increased as the radius of the tube is decreased.

Viscosity

We are accustomed to the notion of friction as a force that is exerted opposite to that which brings about motion when one solid moves in contact with another. Such friction tends to slow,

Capillary action

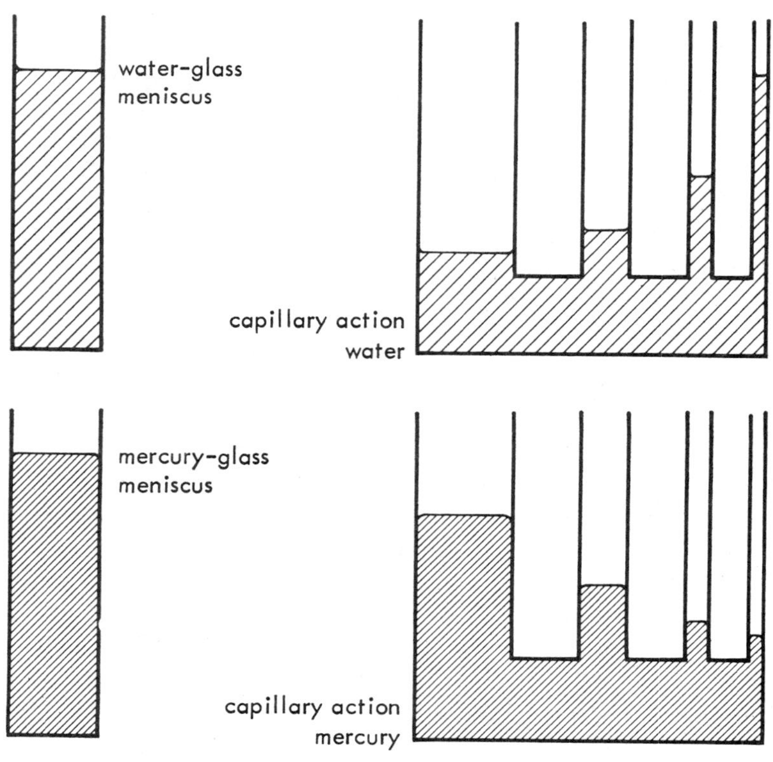

water-glass meniscus

capillary action water

mercury-glass meniscus

capillary action mercury

and eventually stop, motion unless the propulsive force is vigorously maintained.

There is also friction where a solid moves through a fluid, as when a ship plows through water. For all that water seems so smooth and lacking in projections to catch at the ship, the ship once set in motion will speedily come to a halt, its energy absorbed in overcoming the friction with the water, unless the propulsive force is vigorously maintained there, too.

This friction arises from the fact that it is necessary to expend energy to pull the water apart against its own cohesive forces in order to make room for the ship or other object to pass through. The energy expended varies with the shape of the object moving through the fluid. If the fluid is pulled apart in such a way as to force it into eddies and other unevennesses of motion (*turbulence*), the energy expended is multiplied and the motion stops the sooner; to prevent a stop the propulsive force must be increased. If, instead, the fluid is pulled apart gradually by the forward edge of the moving object and allowed to come together even more gradually behind, so turbulence is held at a minimum, the energy expended is reduced considerably, and the force required to maintain motion is likewise reduced. A "streamlined shape" consisting of a bluntly curving fore and a narrowly tapering rear is the "teardrop" shape of waterdrops falling through air, and of fish, penguins, seals, and whales moving through water. It is used in human devices, too, where motion through a liquid medium with maximum economy is desired—such use was enforced by hit-and-miss practice long before it was explained by theory.

The friction between a moving solid and a surrounding liquid increases with velocity. Thus, an object falling through water is accelerated by the gravitational pull against the resistance of friction with the water. However, as the velocity of the falling body increases, the resisting friction increases, too; the force of gravity, of course, remains constant. Eventually, the resisting force of friction increases to the point where it balances the force of gravity, so acceleration is then reduced to zero. Once that happens, the body falls through the liquid at a constant *terminal velocity*.

We ourselves are easily made aware of the friction of solids moving through liquids. Anyone trying to walk while waist-deep in water cannot help but be conscious of the unusual consumption of energy required and of the "slow-motion" effect.

The friction makes itself evident even when the liquid itself is the only substance involved. When a liquid moves, it does not move all-in-one-piece as a solid does. Instead, a given portion will move relative to a neighboring portion, and an "internal friction" between these two portions will counter the motion. Where the cohesive forces that impose this internal friction are low, as in water, we are not ordinarily very conscious of this. Where they are high, as in glycerol or in concentrated sugar solutions, the fluid pours slowly; so slowly indeed that, accustomed as we are to the comparatively rapid water flow, we tend to grow impatient with it. The internal friction is higher at low temperatures than at high temperatures. The folk-saying "as slow as molasses in January" points up our impatience.

A slowly-pouring liquid is said to be "viscous," from the Latin word for a sticky species of birdlime that had this property. The internal friction that determines the manner in which a liquid will pour is called the *viscosity*. There are liquids that are so viscous that the pull of gravity is not sufficient to make them flow against the strong internal friction. Glass is such a liquid and its viscosity is such that it seems a solid to the ordinary way of thinking.*

To consider the measurement of viscosity, imagine two parallel layers of liquid, each in the form of a square of a given area a and separated by a distance d. To make one of these squares move with respect to the other at velocity v against the resisting internal friction requires a force f. It turns out that the relationship among these properties can be expressed by the following equation:

$$\frac{fd}{va} = \eta \qquad \text{(Equation 9–6)}$$

where η (the Greek letter "eta") is a constant at a given temperature and represents the measure of viscosity.

The unit of viscosity can be determined from Equation 9–6. The expression fd in the numerator of the fraction in Equation 9–6 represents force multiplied by distance, or work. The unit of work in the cgs system is dyne-cm or gm-cm^2/sec^2. The expression va in the denominator of the fraction represents volume (centimeters per second) multiplied by area (square centimeters). The unit of va, therefore, is (cm/sec)(cm^2) or cm^3/sec.

* That glass is not a solid, despite its seeming so, is evidenced by its lack of certain properties characteristic of solids. Glass does not have a crystalline structure, for instance, or a sharp melting point. Even so, the case of glass is evidence enough that the distinction between a solid and a liquid is not as clear-cut as might be expected from the most common examples of either. Indeed, most differences and distinctions in science are artificial human conventions imposed on a very complicated universe, and such distinctions cannot help but become fuzzy if viewed with sufficient attention to detail.

To get the unit of viscosity in the cgs system, we must therefore divide the unit of *fd* by that of *va*. It turns out that (gm-cm^2/sec^2)/(cm^3/sec) works out by ordinary algebraic manipulation to gm/cm-sec, or grams per centimeter-second. One gm/cm-sec is defined as one *poise* in honor of the French physician Jean Louis Marie Poiseuille (1799–1869), who in 1843 was the first to study viscosity in a quantitative manner. (As a physician, he was primarily interested in the manner in which that viscous fluid, blood, moved through the narrow blood-vessels.)

The poise is too large for convenience in dealing with most liquids, so the *centipoise* (one-hundredth of a poise) and even the *millipoise* (one-thousandth of a poise) are commonly used. Thus, the viscosity of water at room temperature is just about one centipoise. At the same temperature, the viscosity of diethyl ether (the common anesthetic) is 0.23 centipoises, or 2.3 millipoises, while the viscosity of glycerol is about 1500 centipoises, or 15 poises.

The motion of a fluid has an effect upon its pressure. Imagine a column of water flowing through a horizontal tube of fixed diameter. The water is under pressure or it would not be moving, and the pressure (force per unit area) is the same at all points, for the water is flowing at the same velocity at all points. This could be demonstrated if the pipe were pierced at intervals and a

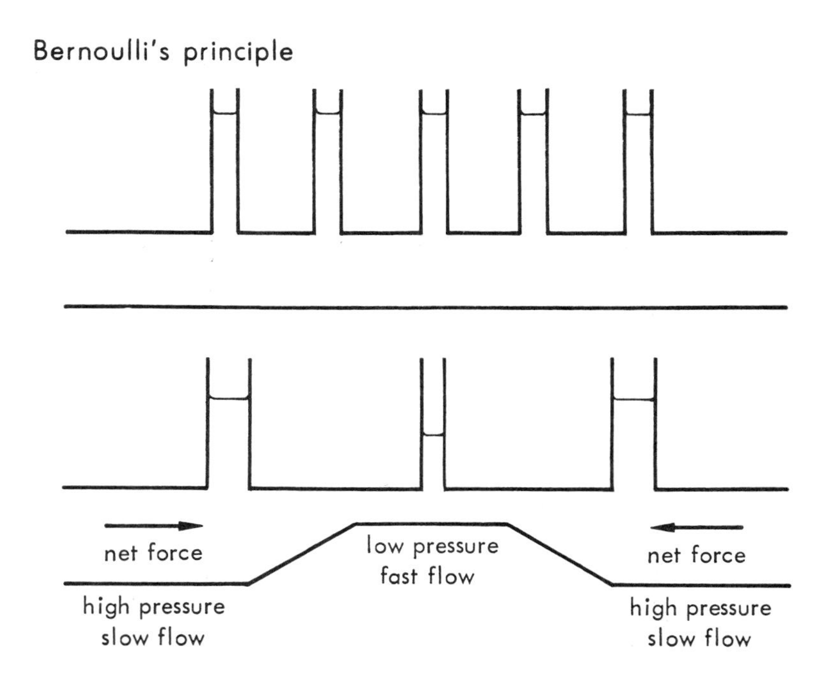

tube inserted into each orifice. The water would rise to the same level in each tube.

But suppose the pipe had a constricted area in the middle. The same volume of water would have to pass through the constricted area in a given time as would have to pass through an equal length of unconstricted area. If that were not the case, water would pile up at the entrance to the constriction, which, of course, it does not. (If the constriction were narrow enough to prevent flow altogether, flow would stop, and the volume of water passing through a given section would be 0 cm^3/sec in the constricted and unconstricted areas alike.)

But in order for the same volume of water to pass through the constricted and unconstricted areas in a given time the flow of water must be more rapid through the constricted area (just as the wide slowly-flowing river becomes a tumbling torrent when passing through a narrow gorge). Since the velocity of water increases as it enters the constricted area, it is subject to an acceleration, and this must be brought about by a force. We can most easily find such a force by supposing a difference in pressure. If the pressure in the unconstricted portion is greater than that in the constricted portion, then there is a net force from the unconstricted portion (high pressure) toward the constricted portion (low pressure), and the liquid is indeed accelerated as it enters the constriction.

Furthermore, when the liquid leaves the constriction and enters a new unconstricted area, its velocity must decrease again. This involves an acceleration again, and there must be a force in the direction opposite to the flow in order to bring about such a slowing of velocity. However, if the new unconstricted area is a region of high pressure again, such a force can be accounted for.

In short, it can be concluded as an important generalization that the pressure of a liquid (or a fluid, generally) falls as its velocity increases. This is called *Bernoulli's principle,* after the Swiss mathematician Daniel Bernoulli (1700–1782), who was the first to study the phenomenon in 1738 and who, on that occasion, invented the term "hydrodynamics."

CHAPTER 10

Gases

Density

The properties of liquids, described in the previous chapter, are important in connection with a fundamental question concerning the ultimate composition of matter, a question that was of great interest to scientists as long ago as the time of the ancient Greeks.

Matter can be subdivided indefinitely, as far as the eye can see. A piece of paper can be torn in half, in quarters, and in eighths —and still remain paper. A drop of water can be divided into two smaller drops or into four still smaller drops—and still remain water. Can such a subdividing process be continued forever? Is matter continuous even to ultimate smallness? There was no way in which the ancient thinkers could test this in actual practice, and they resorted to logical arguments based on what they considered first principles.

Some, notably Democritus of Abdera (fifth century B.C.), maintained that matter could not be subdivided forever, but that eventually a small portion was reached that could not possibly be broken down further. This he called "atomos" (meaning "uncuttable"), and we now speak of his views as representing *atomism,* or an *atomic theory*.

Other Greek philosophers, notably Aristotle, argued against

this notion, however, adducing reasons that made the idea of atoms seem illogical. By and large, the non-atomistic view won out and remained the prevalent belief of scientists for two thousand years.

If one confined oneself to the study of the properties of solids, one could scarcely help but be sympathetic to the Aristotelian view, for there is nothing about a solid that would make it seem logical to consider it to be composed of a conglomeration of small particles. If it were, we would have to suppose the particles to be stuck firmly stogether, since solids acted all-in-one-piece. And if we are going to suppose that particles are stuck firmly together, why not discard the particles altogether and suppose the solid to be all one piece of continuous matter in the first place?

Where liquids are concerned, the situation is quite different. By the very fact that liquids do not move all-in-one-piece, it might reasonably be suggested that they are composed of separate particles. A mass of tiny metal spheres or a heap of powder would take on the overall shape of any container in which they were placed, and they would pour as a fluid would. If the particles were rather sticky, they would pour like a viscous fluid.

In fact, many of the properties of liquids could be explained nicely by supposing them to consist of sub-submicroscopic particles which attract each other somewhat. Surface tension could be explained in this fashion, for instance.

However, all the properties that make liquids suggest atomism more effectively than solids do, are further intensified in gases. And in actual fact, it was the study of gases through the seventeenth and eighteenth centuries that finally forced scientists to reverse the early decision in Aristotle's favor and to take up again, at the start of the nineteenth century, the long-discarded view of Democritus.

Gases differ from liquids most clearly and obviously, perhaps, with respect to density. In comparison with liquids, gases are thin and rarefied.

The density of water is 62.43 pounds per cubic foot in the English system. Using metric units, it is one gram per cubic centimeter (1 gm/cm^3)* in the cgs system and ten kilograms per cubic meter (10 kg/m^3) in the mks system. The least dense liquid at room temperature or below is liquid hydrogen, while the densest is mercury. The former has a density of 0.07 gm/cm^3, the latter a

* This is no coincidence. In setting up the metric system in the 1790's, the French originators defined the gram as the weight of a cubic centimeter of water under set conditions of temperature.

density of 13.546 gm/cm^3. (At elevated temperatures some metals such as platinum would melt to liquids with densities as high as 20 gm/cm^3.)

The density of solids falls for the most part within this range, too. The lightest solid, solid hydrogen, has a density of 0.08 gm/cm^3, while the heaviest, the metal osmium, has one of 22.48 gm/cm^3.

Such densities can be expressed in allied fashion as *specific gravity,* a term dating back to the Middle Ages. Specific gravity may be defined as the ratio of the density of a substance to the density of water. In other words, if the density of mercury is 13.546 gm/cm^3 and that of water is 1 gm/cm^3, then the specific gravity of mercury is $(13.546\ gm/cm^3)/(1\ gm/cm^3)$, or 13.546.

Because the density of water is 1 gm/cm^3, the specific gravity comes out numerically equal to the density in the cgs system, but you must not be misled by this apparent equality, for there is an important difference in the matter of units. In dividing a density by a density, the units (gm/cm^3, in the case cited in the previous paragraph) cancel, so that the figure for specific gravity is a dimensionless number.

The units cancel when specific gravity is calculated, no matter what system of units is used for the densities. In the mks system, the densities of mercury and water are, respectively, 135.46 kg/m^3 and 10 kg/m^3. By taking the ratio, the specific gravity of mercury is 13.546, as before. In the English system of units, the densities of mercury and water are 845.67 pounds per cubic foot and 62.43 pounds per cubic foot, and the ratio is still 13.546.

The convenience of a dimensionless number is just this then: it is valid for any system of units.

The specific gravity of gases is much less than that of either liquids or solids. The most common gas, air, has a specific gravity of 0.0013 under ordinary conditions. The lightest gas, hydrogen, has under ordinary conditions a specific gravity of 0.00009. An example of a very dense gas is the substance uranium hexafluoride, which is a liquid at ordinary temperatures, but if heated gently is converted into a gas with a specific gravity of 0.031.

Thus, under ordinary conditions, even the densest gases are less than half as dense as even the least dense liquids or solids, while a common gas such as air has only about 1/700 the density of a common liquid such as water, and only about 1/2000 the density of common solids such as the typical rocks that make up the earth's crust.

Gas Pressure

Gases share the fluid properties of liquids but in an attenuated form, as is to be expected considering the difference in density. For instance, gases exhibit pressure as well as liquids do, but gas pressure is considerably smaller for a given height of fluid. A column of air a meter high will produce a pressure at the bottom only 1/700 that of a column of water a meter high.

Nevertheless we live at the bottom of an ocean of air many miles high. Its pressure should be considerable and it is; it is equivalent to the pressure produced by a column of water ten meters high. This pressure was first measured by the Italian physicist Evangelista Torricelli (1608–1647) in 1644.

Torricelli took a long tube, closed at one end, and filled it with mercury. He then upended it in a dish of mercury. The mercury in the tube poured out of the tube, of course, in response to the downward pull of gravitational force. There was a counterforce, however, in the form of the pressure of the atmosphere against the mercury surface in the dish. This pressure was transmitted in all directions within the body of the mercury (Pascal's principle, see page 119), including a pressure upward into the tube of mercury.

As the mercury poured out of the tube, the mass of the column, and therefore the gravitational pull upon it, decreased until it merely equalled the force of the upward pressure due to the atmosphere. At that point of balancing forces, the mercury no longer moved. The mercury column that remained exerted a pressure (due to its weight) that was equal to the pressure of the atmosphere (due to its weight). The total weight of the atmosphere is, of course, many millions of times as large as the total weight of the mercury, but we are here concerned with pressure which, be it remembered, is weight (or force) per unit area.

It turns out that the pressure of the atmosphere at sea level is equal to that of a column of mercury just about 30 inches (or 76 centimeters) high; Torricelli had, in effect, invented the barometer. Air pressure is frequently measured, particularly by meteorologists, as so many *inches of mercury* or *centimeters of mercury,* usually abbreviated in Hg or cm Hg, respectively.* It is natural to set 30 in. Hg or 76 cm Hg equal to 1 *atmosphere.* A millimeter of mercury (mm Hg) has been defined as 1 *torricelli,* in honor of the physicist, so one atmosphere is equal to 760 torricellis.

* "Hg" is the chemical symbol for mercury.

Air pressure may also be measured as weight per area. In that case, normal air pressure at sea level is 14.7 pounds of weight per square inch, or 1033 grams of weight per square centimeter ($gm[w]/cm^2$). Expressed in the more formal units of force per area, one atmosphere is equal to 1,013,300 $dynes/cm^2$. One million dynes per square centimeter has been set equal to one *bar* (from a Greek word for "heavy"), so one atmosphere is equal to 1.0133 bars.

Naturally, if it is the pressure of the atmosphere that balances the pressure of the mercury column, then when anyone carrying a barometer ascends a mountain, the height of the column of mercury should decrease. As one ascends, at least part of the atmosphere is below, and what remains above is less and less. The weight of what remains above, and therefore its pressure, is lower and so is the pressure of the mercury it will balance.

This was checked in actual practice by Pascal in 1658. He sent his brother-in-law up a neighboring elevation, barometer in hand. At a height of a kilometer, the height of the mercury column had dropped by ten percent, from 76 centimeters to 68 centimeters.

Furthermore, the atmosphere is not evenly distributed about the earth. There is an unevenness in temperature that sets up air movements that result in the piling up of atmosphere in one place at the expense of another. The barometer reading at sea level can easily be as high as 31 in. Hg or as low as 29 in. Hg. (In the center of hurricanes, it may be as low as 27 in. Hg.) These "highs" and "lows" generally travel from west to east, and their movements can be used to foretell weather. The coming of a high (a rising barometer) usually bespeaks fair weather, while the coming of a low (a falling barometer) promises storms.

For all that air pressure is sizable in quantity (the value, 14.7 pounds of weight per square inch, is most easily visualized by a person used to the common measurements in the United States) it goes unnoticed by us. For thousands of years, men considered air to be weightless. (We still say "as light as air" or "an airy nothing.")

The reason for this is that air exerts its pressure in all directions, as all fluids do. An empty balloon, although supporting the full pressure of miles of air, will rest with its mouth open and its walls not touching, for the air within it has an outward pressure equal to the inward pressure of the air outside. Place the balloon in your mouth, however, and suck out the air within so that the inward push of the air outside is no longer balanced. Now the walls of the balloon will be pushed hard together.

The same factors apply to human beings. The air in our lungs, the blood in our veins, the fluid in our bodies (and living tissue is essentially a thick, viscous fluid) is generally at air pressure and delivers a pressure outward equal to that of the atmosphere inward. The net pressure exerted on us is zero, and we are therefore unaware of the weight of the air.

If we submerge ourselves in water, the pressure from without rapidly increases, and it cannot be matched by pressure from within without damage to our tissues. It is for this reason that an unprotected man, such as a skin diver, is severely limited in the depth to which he can penetrate, regardless of how well equipped with oxygen he may be. On the other hand, forms of life adapted to the deeps exist at the extremest abyss of the ocean, where the water pressure is over a thousand atmospheres. Those life forms are as unaware of the pressure (balanced as it is from within), and as unhampered by it, as we are by air pressure.

Once it was recognized that air had weight and produced a pressure, it was also quickly recognized that this could easily be demonstrated provided it were *not* balanced by an equal pressure from within. In other words, it seemed desirable to be able to remove the air from within a container, producing a *vacuum* (a Latin word meaning "empty") so that the air pressure from without would remain unbalanced by any appreciable pressure from within. Torricelli had formed the first man-made vacuum (inadvertently) when he had upended his tube of mercury. The column of mercury, as it poured out, left behind a volume of nothingness (except for thin wisps of mercury vapor), and this is still called a "Torricellian vacuum."

Just a few years later, in 1650, the German physicist Otto von Guericke (1602–1686) invented a mechanical device that little by little sucked air out of a container. This enabled him to form a vacuum at will and to demonstrate the effects of an unbalanced air pressure. Such air pressure would hold two metal hemispheres together against the determined efforts of two eight-horse teams of horses (whipped into straining in opposite directions) to pull them apart. When the air was allowed to enter the hemispheres once more, they fell apart of their own weight.

Again, air pressure gradually forced a piston into a cylinder being evacuated, even though fifty men pulled at a rope in a vain attempt to keep the piston from entering.

In other respects, too, a gas like air has fluid properties in an attenuated form. It exhibits buoyancy, for instance. We ourselves displace a volume of air equal to our own volume, and the effect is to cause a 150-pound man to weigh some three ounces less than he would in a vacuum. This is not enough to notice ordinarily, of course, but for objects of very low densities the effect is very noticeable.

This is particularly true for substances (such as certain gases) that are lighter than air. Hydrogen gas, for instance, has only 1/14 the density of air. In consequence, hydrogen penned within a container is subjected to an upward force like that of wood submerged in water (see page 124). If the container is light enough, it will be carried upward by this upward force. If enough hydrogen is involved, the force will be sufficient to also carry upward a suspended gondola containing instruments or even men. The first such "balloons" were launched in France in 1783.

When there is relative motion between a solid and a gas there is friction, as there is between a solid and a liquid—though again the effect is much smaller where a gas rather than a liquid is involved. The friction with gas ("air resistance") is enough, however, to slow the velocity of projectiles to the point where it must be allowed for if an artilleryman is to aim correctly. Air resistance also prevents a complete and perfect interchange of kinetic and potential energy by dissipating some of the energy as heat (see page 97).

The downward force of the gravitational field is proportional to the total mass of the body, while the upward force of air resistance is proportional to the area of contact of the moving body with air in the direction of its motion. For compact and relatively heavy bodies, such as stones, bricks and lumps of metal, the gravitational force is high, while the contact with air is over a relatively limited area so that air resistance is low. In such cases motion is close enough to what it would be in a vacuum for Galileo to have been able to draw correct conclusions from his experiments.

For light bodies, the gravitational force is relatively low. If such bodies are also thin and flat (as leaves or feathers are, for instance), they present a relatively large area to the air, and air resistance is relatively high. In such cases, air resistance almost balances the gravitational force, and these light bodies therefore fall slowly (they would fall quickly in a vacuum); this slow rate of fall fooled the ancient Greek observers into believing there was an intrinsic connection between weight and the rate of free fall.

There is a terminal velocity reached in motion through air under the influence of a constant force such as gravity, since air resistance does not remain constant but increases with the velocity of an object through air. As velocity increases, air resistance eventually balances the gravitational force. For heavy, compact objects this terminal velocity is very high, but for light, flat objects it is quite low. Snowflakes quickly reach their low terminal velocity and accelerate no further though they fall for miles. If a compact object is suspended from a light flat one, the two objects together reach a far lower terminal velocity than the compact object would by itself, and this is why a parachute makes it possible to fall safely from great heights.

Again, there is a Bernoulli effect for gases as well as for liquids, and air pressure drops as the velocity of moving air increases. A jet of air moving across an orifice covers that orifice with a low-pressure area (or a "partial vacuum"). If a tube connected with the orifice dips into a liquid under normal atmospheric pressure, that liquid is pushed up the tube and is blown out in a fine spray.

When a baseball or a golf ball spins in the air, one side spins with the motion of the air flowing past the ball as it moves; the other side spins against the motion. The side that spins against

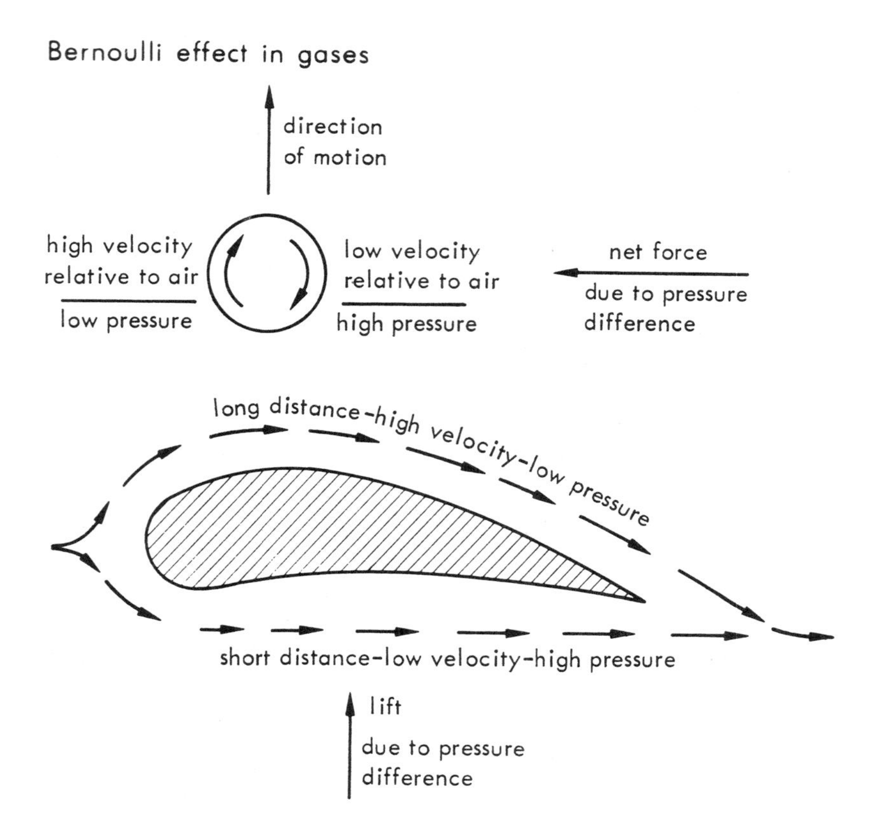

the motion has a greater velocity relative to the air, and the air pressure is less in that direction. The ball is pushed in the direction of lower pressure, so that the baseball curves in its flight (usually desirable, if it is the pitcher throwing the ball), while the golf ball "hooks" or "slices" (usually undesirable).

Where high velocity through air must be maintained with a minimum of force, streamlining is important. This importance increases with velocity, since air resistance also does. Thus, a horse and wagon need display no streamlining and automobiles need display very little. (A trend toward extreme automobile streamlining initiated in the late 1930's was a matter of appearance rather than necessity, and was abandoned.)

Airplanes, however, must be streamlined, and in reaching supersonic speeds it was not so much higher power that had to be developed, but the proper design for minimizing air resistance. Furthermore, airplane wings (themselves streamlined) are so designed that the air must move over a greater distance above than below, and so must move more rapidly above than below. This, by Bernoulli's principle, means there is less pressure above than below, and therefore a net upward force ("lift") that helps support the plane.

Boyle's Law

The properties of gases are of crucial importance with respect to the possibly atomic nature of matter. If matter is nonatomic, then variations in density must be caused by the intrinsic differences in the density of matter itself. Every bit of it, however small, must be as dense as every other bit. There would be no holes or empty space in matter, as there would be if the matter consisted of atoms.

If matter consisted of atoms, there might be space between the atoms, a space containing only vacuum. Matter might be made less dense, then, by pulling the atoms apart in some fashion so as to increase the proportion of empty space within a given volume. Conversely, matter might be made more dense by pushing the atoms together to reduce the proportion of empty space.

Indeed, it might seem that the density of a particular substance could be changed in just this fashion by heating or cooling. Density usually decreases with heating and increases with cooling. Thus, although the density of cold water is 1 gm/cm^3, that of hot water is only about 0.96 gm/cm^3.

Then again, solids melt to liquids if heated sufficiently, and the liquids become solids again if cooled. This change in the *state of matter* is accompanied by a sudden change in density. Thus, ice has a density of 0.92 gm/cm^3, but as soon as it is melted to water the density increases sharply to 1.00 gm/cm^3. Again, solid iron has a density of 7.8 gm/cm^3, but this decreases sharply when iron is melted to the liquid form, which has a density of only 6.9 gm/cm^3. An atomist might point out that a ready explanation for this is that in one state the constituent atoms are more compact than in the other. (Usually it is the solid state which is denser, with water a rather unusual exception.)

However, in all such changes, the density varies by only a few percent, and this is not overwhelmingly convincing. Working against the atomist is the fact that liquids and solids are relatively incompressible. Large increases of pressure (attainable only with specialized equipment) are required to bring about even small decreases in volume. For this reason, there can't be much empty space in ordinary matter, and even atomists must admit that in liquids and solids, atoms, if they exist, are in virtual contact. Since liquids and solids remain incompressible at any temperature, the feeling that atoms are further apart in hot water than in cold, or in liquid iron than in solid iron, seems to be wrong. If it were not wrong, then hot water and liquid iron would be at least moderately compressible, and they are not.

It is another matter entirely, however, when the point of view shifts from solids and liquids to gases. When liquid water is boiled and gaseous steam is formed, the change in density is drastic and dramatic. Where water has a density of 0.96 gm/cm^3 at the boiling point, steam at the same temperature has a density of no more than 0.0006 gm/cm^3. Steam is only about 1/1700 as dense as water.

This can be reasonably explained by adopting an atomistic view. One can suppose that the constituent atoms (or groups of atoms) making up water move far apart in the conversion of the liquid water to the gaseous steam, and that steam is as low in density as it is because it consists mostly of the empty space between atoms. We might generalize and say that whereas in liquids and solids atoms are virtually in contact, in gases they are far apart. This spreading out of atomic particles would account not only for the extremely low density of gases, but also for their low pressure, their small frictional forces, and so on.

If this atomistic view is so, and if the particles of gas are widely spread out, then gases ought to be easily compressible. If pressure is exerted upon a given volume of gas, that volume ought to decrease considerably. This is actually so, and the fact was first clearly presented to the scientific community by the English physicist Robert Boyle (1627–1691) in 1660.

He poured mercury into the open long end of a J-shaped tube and trapped some of the air in the closed short end. By adding additional mercury, he raised the pressure on the trapped air by an amount he could measure through the difference between the inches of mercury in the open and closed sides. He found that doubling the pressure on the trapped gas generally halved its volume; tripling the pressure reduced the volume to a third, and so on.

The trapped gas was always able to support the column of mercury on the other side once its volume had been reduced by the appropriate amount, so that the pressure it exerted was equal to the pressure exerted upon it. (This is to be expected from Newton's third law—which, however, was not yet enunciated in Boyle's time.)

Consequently, we can say that for a given quantity of gas the pressure (P) is inversely related to the volume (V), so as one goes up, the other goes down ($P=k/V$). Therefore, the product of the two remains constant:

$$PV = k \qquad \text{(Equation 10–1)}$$

This relationship is called *Boyle's law.**

Another way of stating Boyle's law is as follows. Suppose that you have a sample of gas with a pressure P_1 and a volume V_1. If you change the pressure, either increasing it or decreasing it, to P_2, you will find that the volume automatically changes to V_2. However, the product of pressure and volume must remain constant by Equation 10–1, so we can say that for a given quantity of gas:

$$P_1V_1 = P_2V_2 \qquad \text{(Equation 10–2)}$$

and that, too, is an expression of Boyle's law.

Indeed, gas is so easily compressed that the pressure of the upper layers of a column of gas will compress the lower layers. Whereas a column of virtually incompressible liquid has a constant density throughout, columns of gases vary considerably in density with height. This is particularly noticeable in the case of the atmosphere itself.

If gas were as incompressible as liquid, and if it were as dense at all heights as it is at sea level, then one could easily calculate

* Boyle's law, as it turned out, is only an approximation (see page 208), but it is a very useful approximation and, in the case of some gases, an approximation very close to the truth.

what the height of the atmosphere ought to be. Air pressure is 1033.2 grams of weight per square centimeter. This means that a column of gas one square centimeter in cross section and extending straight upward to the top of the atmosphere weighs 1033.2 grams. A column with such a cross section, but only one centimeter high, has a weight of 1.3 milligrams. Each additional height of one centimeter added to the column would add an additional 1.3 milligrams, and it would take a total height of about 800,000 centimeters to account for the 1033.2 grams of air pressure. This is a height of just about five miles.

However, this cannot be right, for balloons have found air to exist at heights of over 20 miles, and less direct methods of measurement have shown perceptible quantities of air to exist at heights of over 100 miles.

The point is that the atmosphere is not at constant density. As one moves upward, one finds that a given quantity of gas is under less pressure because the quantity of air above it has become less. By Boyle's law, that given quantity of gas must therefore take up a larger volume. Consequently, as one rises, the amount of atmosphere remaining above, while decreasing rapidly in weight, decreases only very slowly in volume. For that reason, indeed, the atmosphere has no definite upper edge, but fades slowly off for hundreds of miles above earth's surface, decreasing in density until it peters out into the incredibly thin wisps of gas that make up interplanetary space.

By pointing out the atomistic argument first, I was trying to make absolutely clear the importance and significance of Boyle's experiment. It is not to be supposed, however, that that one experiment at once turned scientific opinion toward atomism. It was not until the first decades of the nineteenth century, a century and a half after Boyle's experiment, that the weight of evidence had finally accumulated to the point where scientists could no longer avoid accepting atomism.

The scientist usually given credit for the final establishment of atomism is the English chemist John Dalton (1776–1844). He worked out the "modern atomic theory" in detail, between 1803 and 1808, basing it chiefly on the observations of the properties of gases that had begun with Boyle's experiments. (In fact, one might maintain that Boyle's law made atomism inevitable, and that all that followed merely served to place a finer edge on the concept.)

It is now generally accepted that all matter consists of *atoms*; that these atoms may exist singly, but much more commonly exist in groups of from two to many hundreds of thousands; and that these groups of atoms, called *molecules,* maintain their identity under ordinary circumstances and form the particles of matter.*

It was by considering gases to consist of a collection of widely-spaced molecules (or, occasionally, of widely-spaced individual atoms) that it became possible to view such phenomena as sound and heat in a new and more fundamental manner.

* Under certain circumstances, molecules do alter in nature, and old combinations of atoms shift and change into new combinations. These shifts and changes of molecular combinations are the prime concern of the science of chemistry.

CHAPTER 11

Sound

Water Waves

Fluids can move in the various fashions that solids can move. They can undergo translational motion, as when rivers flow or winds blow. They can undergo rotational motion, as in whirlpools and tornadoes. Finally, they can undergo vibrational motion. It is the last that concerns us now, for a vibration can produce a distortion in shape that will travel outward. Such a moving shape-distortion is called a *wave*. While waves are produced in solids, they are most clearly visible and noticeable on the surface of a liquid. It is in connection with water surfaces, indeed, that early man first grew aware of waves.

If a stone is dropped into the middle of a quiet stretch of water, the weight of the stone pushes down on the water with which it comes into contact and a depression is created. Water is virtually incompressible, so room must be made for the water that is pushed downward. This can only be done by raising the water in the immediate neighborhood of the fallen stone, so the central depression is surrounded by a ring of raised water.

The ring of raised water falls back under the pull of gravity, and its weight acts like the original weight of the stone. It pushes the water underneath downward and throws up a wider ring of water a bit farther away from the original center of disturbance.

This continues, and the ring of upraised water moves farther and farther out from the center. As it moves outward, the total mass of upraised water must be spread out through a larger and larger circumference, and the height of the upraised ring is therefore lower and lower.

Nor is there a single wave emanating from the center of the disturbance. As the initial wall of upraised water immediately about the center of disturbance comes down, it not only pushes up a wall of water beyond itself, but also pushes up the water at the center. This rises and then drops again, acting, so to speak, like a second stone, and setting up a second circular wall of water that spreads outward inside the first wall. This is followed by a third wall, and so on. Each successive wall is lower than the one before, since with each rise and fall of water some of the energy is consumed in overcoming the internal friction of the water and is converted into heat. As a particular wall of water spreads outward, some of its energy is also being continually converted into heat. Eventually, all the waves die out and the pool is quiet again; however, it is very slightly warmer for having absorbed the kinetic energy of the falling stone.

To produce a wave, then, we need an initial disturbance. If this initial disturbance, in correcting itself, disturbs a neighboring region in a fashion similar to the original disturbance, the wave is propagated.

In a propagated wave, if we concentrate our attention on a given point in space, we see that some property waxes and wanes, often periodically. In the case of water waves, for instance, if we view one portion of the water surface and no other, then the varying property is potential energy as that portion of the surface first rises, then falls, then rises again.

It is important to realize that the water is moving up and down only. The disturbance is propagated outward across the surface of the water, and it appears to the casual observer that water is moving outward; however, it is not! Only the disturbance is. A chip of wood floating on water that has been disturbed into ripples will rise and fall with the rise and fall of the water it rests on, but the moving ripples will not carry the wood with it. (To be sure, waves approaching the shore will carry material with them, sometimes even forcefully, as the water dashes on the rocks or beach. These waves, however, are being driven by the horizontal force of the wind and are different from the ripples set up by the vertical force of a falling stone.)

Suppose we imagine a cross section of the surface of the

water that is undergoing the disturbance of a falling stone. Ideally, ignoring loss of height with increasing circumference or loss of energy as heat, we have a steady rise and fall. This rise and fall is what we commonly think of when the word "wave" is spoken, or when we speak of a "wavy line."

In its simplest form, such a wavy line is identical with the type of curve produced if one plots the value of the sine of an angle (see page 111) on graph paper as the size of the angle increases steadily. For an angle of 0°, the sine is 0. As the angle increases, the sine also increases, first quickly, then more and more slowly, till it skims a maximum of 1 at 90°. For still larger angles it begins to decrease, first slowly, then more and more rapidly, reaching 0 again at 180° and passing into negative values thereafter. It skims a minimum of —1 at 270°, then increases again to reach 0 once more at 360°. An angle of 360° can be considered equivalent to one of 0°, so the whole process can be viewed as beginning again and continuing onward indefinitely. In plotting the graph, then, one gets a wave-like figure that can extend outward forever as it oscillates regularly between +1 and —1. It is this wave-like figure (the *sine curve*) that represents the shape of an idealized water wave.

A wave like the water wave, in which the motion of each part is in one direction (up-and-down in this case), and the direction of propagation of the disturbance is at right angles to that direction (outward across the water surface in this case), is a *transverse wave.* (Transverse is from Latin words meaning "lying across"; the motion of the water itself "lies across" the line of propagation.)

The point at which the disturbance is greatest in the upward direction (+1, in the sine curve) is the *crest,* and the point at which it is greatest in the downward direction (—1, in the sine curve) is the *trough.* Between crest and trough are points where the water is momentarily at the level it would be at if the surface were undisturbed (0, in the sine curve); these are *nodes.* There are two kinds of nodes in these water waves, for water may pass through a node on its way down to a trough or on its way up to a crest. We might distinguish these as "descending nodes" and "ascending nodes" (borrowing terms that are used in astronomy for an analogous purpose). The vertical distance from a node to either a crest or a trough is the *amplitude* of the wave.

Two or more points that occupy the same relative positions in the sine curve are said to be *in phase.* For instance, the points on the various crests are all in phase; so are the points on the various troughs. The ascending nodes are all in phase; the descending nodes are all in phase. All points lying a fixed portion of the way between an ascending node and a crest are in phase, and so on. If two waves exist, and if they match up in such a way that the crest of one is even in space or form with the crest of another at the same instant in time, those portions of the two waves are said to be in phase. It is possible that the entire stretch of both waves may be in phase in this fashion, crest for crest and trough for trough.

Naturally, points on a single wave that are not in phase are *out of phase.* And a pair of waves in which the crest in one case does not appear at the same time as the crest of another are out of phase.

A sine curve can be looked upon as consisting of a particular

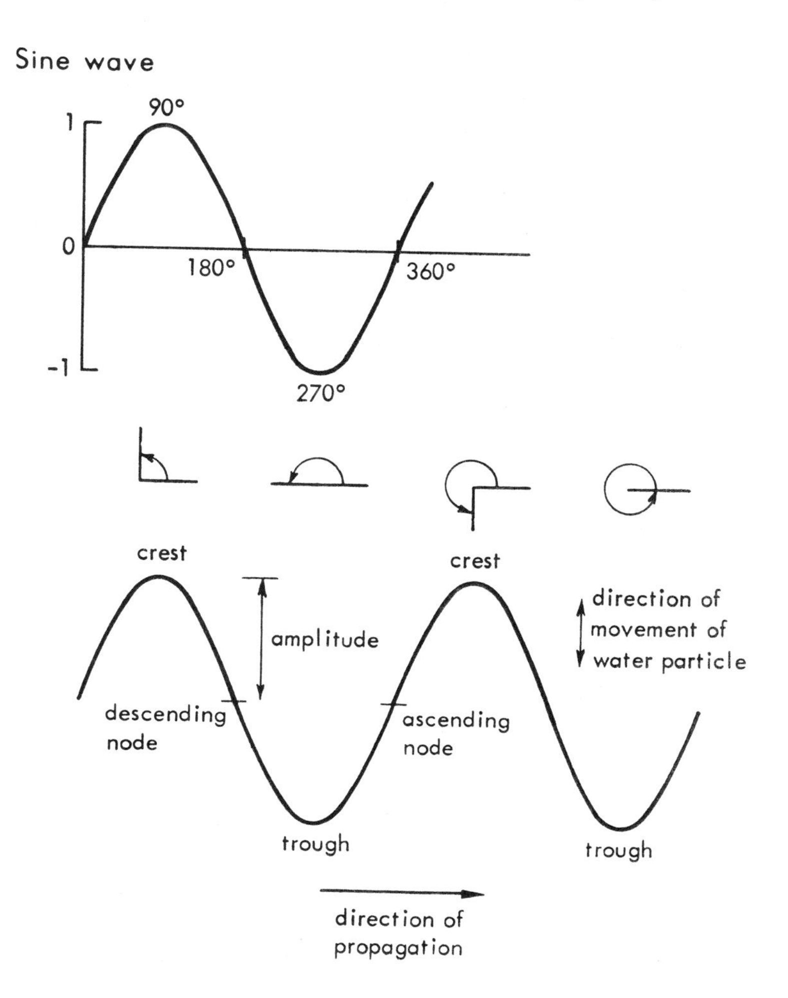

small portion that repeats itself indefinitely. For instance, a portion of the sine curve from one crest to the next can be shaped into a stamp, if you like, and the entire sine curve can be reproduced by stamping that one crest-to-crest portion, then another like it to its right, another like it to the right again, and so on. The same could be done if we took a portion of the sine curve from trough to trough or from ascending node to ascending node or from descending node to descending node, and so on. An appropriate stamp can be made covering the section from any point on the

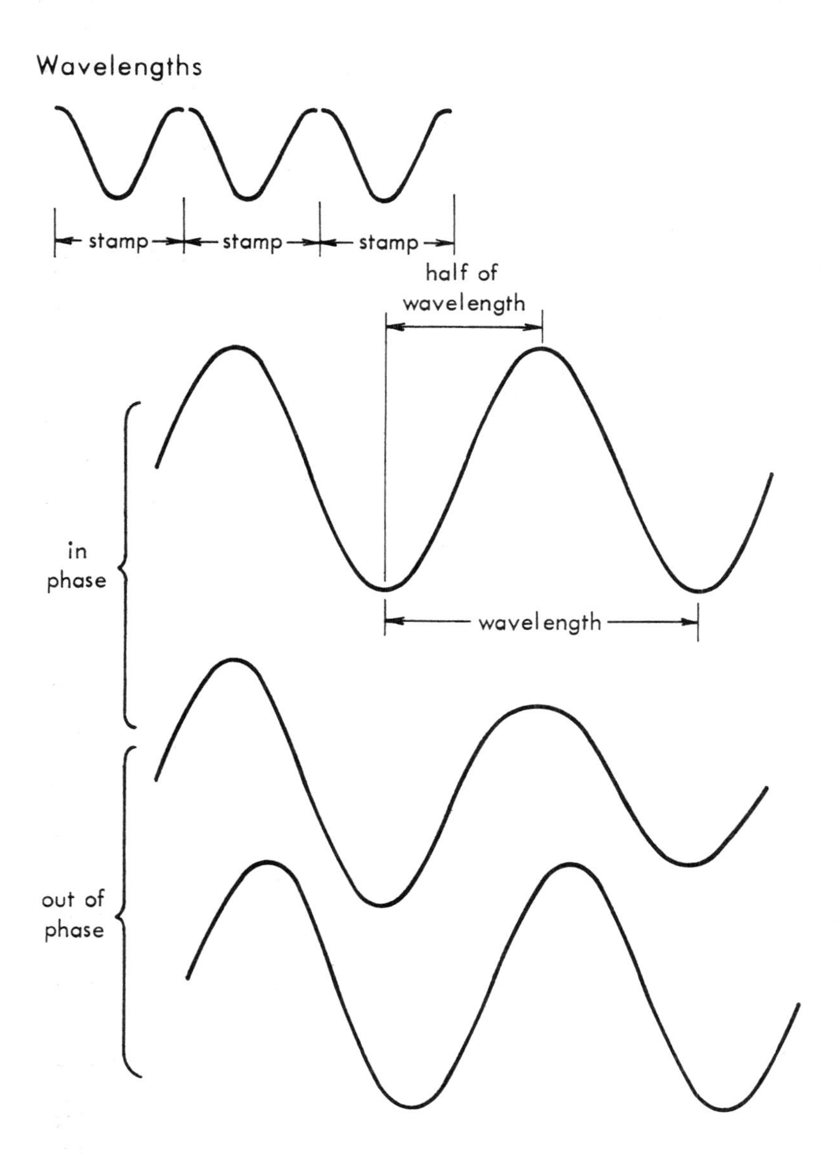

sine curve to the next point in phase. The length from any such point to the next is constant for any particular sine curve. This length (let us say from crest to crest for simplicity's sake) defines a *wavelength.* The wavelength is usually symbolized by the Greek letter "lambda" (λ).

If nodes are not distinguished from each other (and they are usually not in physics), then successive nodes are half a wavelength apart. If crest and trough are lumped together as *antinodes* (as sometimes they are), then successive antinodes are half a wavelength apart.

A particular crest moves outward along the surface of the water (though the water itself, I repeat, does not move outward with it), and the distance it travels in one second is the velocity of the wave.

Let us suppose that the velocity of a particular wave is ten meters per second and that the wavelength (that is, the distance from crest to crest) is two meters. If we fix our attention on a certain point of the water's surface, we will note that a particular crest is at that point. It travels outward, and two meters after it a second crest occupies that point; two meters after that there is is a third crest, and so on. After one second the original crest is ten meters away, and a fifth crest (10 divided by 2) is occupying the original spot.

The number of crests (or the number of troughs, ascending nodes, descending nodes, or any successive points in phase) passing a given point in one second is the *frequency* of the wave. Frequency is usually symbolized by the Greek letter "nu" (ν).

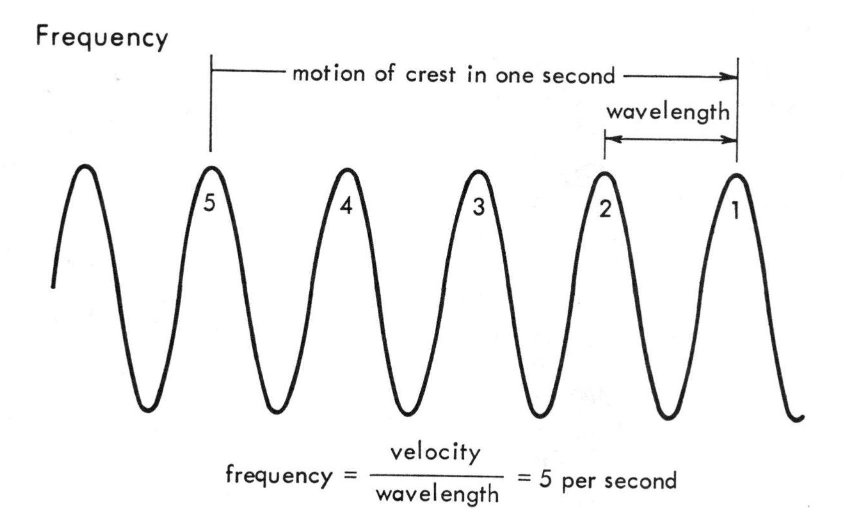

From what I have said, it should be clear that the velocity of a wave divided by its wavelength is equal to its frequency, so:

$$\nu = \frac{v}{\lambda} \qquad \text{(Equation 11–1)}$$

The units of velocity are meters per second in the mks system, and those of wavelengths are meters. The units of frequency are therefore (m/sec)/m, or 1/sec. Since in algebra $1/a$ is said to be the "reciprocal" of a, 1/sec is sometimes spoken of as *reciprocal seconds.* More often it is referred to by the phrase "per second." The frequency of the wave mentioned above as an example can be written 5/sec, and this can be read as "5 per second" or as "5 reciprocal seconds."

Sound Waves

At a comparatively early stage in the quest for knowledge, sound came to be thought of as resulting from a kind of wave motion. The first experiments on sound were conducted by the ancient Greeks, and these were rather remarkable in one way, for the study of sound was one branch of physics in which the Greeks seemed, by modern criteria, to start off in the right direction from the very beginning.

As early as the sixth century B.C., Pythagoras of Samos was studying the sound produced by plucked strings. It could be seen that a string vibrated when plucked. The plucked string's motion was only a blur, but even so, certain facts about that blur could be associated with sound. The width of the blurred motion seemed to correspond to the loudness of the sound. As the vibration died down and the blur narrowed, the sound grew softer. And when the vibration stopped, either by natural slowing or by an abrupt touch of the hand, so did the sound. Furthermore, it could be made out that shorter strings vibrated more rapidly than longer ones, and the more rapid vibration seemed to produce the shriller sound.

By 400 B.C., Archytas of Tarentum (420?–360? B.C.), a member of the Pythagorean school, was suggesting that sound was produced by the striking together of bodies—swift motion producing high pitch and slow motion producing low pitch. By about 350 B.C., Aristotle was pointing out that the vibrating string was striking the air; and that the portion of the air which was struck must in turn be moved to strike a neighboring portion, which in turn struck the next portion, and so on. To Aristotle, then, it seemed that air was necessary as a medium through which sound was conducted, and he reasoned that sound would not be conducted through a vacuum. (In this, Aristotle was correct.)

Since in rapid rhythm a vibrating string strikes the air not once but many times, not one blow, but a long series of blows, must be conducted by the air. The Roman engineer Marcus Vitruvius Pollio, writing in the first century B.C., suggested that the air did not merely move, but vibrated, and that it did so in response to the vibrations of the string. It was these air vibrations, he held, that we heard as sound.

Finally, about 500 A.D., the Roman philosopher Anicius Manilius Severinus Boethius (480?–524?) made the specific comparison of the conduction of sound through the air with the waves produced in calm water by a dropping pebble. While this analogy has its value, and while water waves can be used to this day (and are so used in this book, for instance) to serve as a preliminary to a consideration of sound waves, there are nevertheless important differences between water waves and sound waves.

Transverse waves, such as water waves, can appear only under certain conditions. Such waves represent conditions in which one section of a body moves sideways with respect to another and then reverses that motion. (You can produce a transverse wave in a tall stack of cards by moving each card sideways by the proper amount.) Such a sideways motion is produced by a type of force called a *shear*. For such a force to result in a transverse wave, however, the force producing the shear must be countered by another force that brings the portions of the body back into line.

Within a solid, for instance, a blow may cause a portion of the substance to move sideways with respect to a neighboring portion. The strong cohesive forces between the molecules of a solid, which tend to keep each molecule in place, act to bring the displaced section back. It shoots back, overshoots the mark, shoots back again, overshoots the mark again, and so on. The resulting vibration is propagated just as the waves on a water surface are, and as a result it is possible to have transverse waves through the body of a solid.

The cohesive forces in liquids and gases are, however, very weak in comparison to those in solids and do not serve to restore a shear. If a portion of water or air is shifted sideways with respect to a neighboring portion, additional water or air will simply flow into the region left "empty" by the shifting portion, and the new arrangement of portions will remain. There are therefore no transverse waves through the body of a fluid.

To be sure, transverse waves will travel over the horizontal upper surface of liquids, for there we have the special case of an outside force, gravity, resisting the up-and-down shear. Within the body of the liquid, gravity cannot be counted upon to do this work, for each fragment of water is buoyed up by the surrounding water. Since the density of each bit of water is equal to the density of the surrounding water, each bit of water has a weight of zero (see page 124) and does not respond to gravity. If a portion of water within the body of the liquid is raised by a shear, it remains in the new position, in spite of gravity. Since transverse waves are confined to the surface of a fluid, and since gases have no definite volume and therefore no definite surface, it follows that transverse waves cannot be transmitted by gases under any condition.

Consequently, if sound is transmitted through the air as a wave form (as all the evidence indicates it must be), that wave form cannot be transverse. A logical alternative is that it consists of periodic compressions and rarefactions.

Consider the vibrations of a tuning fork, for instance. The prong of a tuning fork moves right, left, right, left, in a rapid periodic motion. As it moves right, the molecules of air lying immediately to the right are pushed together, forming a small volume of compression. The pressure within the compressed volume is greater than in the neighboring volume of normal air. The molecules in the compressed volume spring apart and push against the neighboring volume, compressing it. The neighboring volume compresses its neighbor as it springs apart, and so on. Thus, a volume of compression is propagated outward in all directions, forming an expanding sphere about the source of the disturbance just as the crest of a water wave forms an expanding circle about *its* source of disturbance. (The atmosphere is a three-dimensional medium, the surface of the water a two-dimensional one, which is why we have an expanding sphere in one case, an expanding circle in the other.)

Meanwhile, the prong of the tuning fork, having moved to the right and set off an expanding volume of compression, next moves to the left. More room is made to the right of the prong and the air in that immediate volume expands and becomes relatively rarefied. Pressure is higher in the neighboring un-rarefied air, which therefore pushes into the rarefied volume and is itself rarefied in the process. In this way, a volume of rarefaction expands outward on the heels of the volume of compression.

Again the prong of the tuning fork moves right, then left, then right, so that volumes of compression and volumes of rarefaction follow each other outward in rapid alternation for as long as the prong continues to vibrate. Each period of the prong (one movement back and forth) sets up one compression/rarefaction combination.

In these waves of alternate compression and rarefaction, the individual molecules of air move in one direction when compressed, then in the reverse direction when rarefied; the volumes of compression and rarefaction move outward and are propagated in a direction parallel to the back and forth motion of the molecules. Such a wave, in which the particles move parallel to the propagation rather than perpendicular to it, is a *longitudinal wave,* or a *compression wave.*

Longitudinal waves are harder to picture and grasp than are transverse waves, for there are no examples out of common experience that we can draw on to illustrate the former in the way we used water waves to illustrate the latter. Nevertheless, having gone into some detail about transverse waves, we can deal with longitudinal waves by analogy.

The points of maximum compression are analogous to the crests of transverse waves, and the points of maximum rarefaction to the troughs. In between there are areas where pressure is momentarily normal, and these correspond to the nodes.

The distance between points of maximum compression (or between points of maximum rarefaction) is the wavelength of the longitudinal wave. The number of points of maximum compression (or of maximum rarefaction) passing a given position in one second is the frequency of the longitudinal wave.

Since the molecules of liquids and solids, as well as those of gases, evolve a restoring counterforce when compressed,

Sound waves

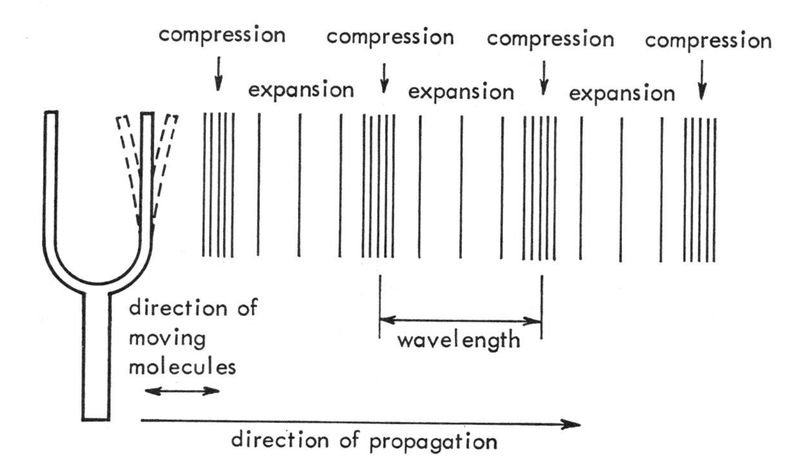

longitudinal waves can be carried through gases, liquids and solids. Sound waves in particular are carried by water and by steel, as well as by air. (Waves produced in the body of the earth by the vibrations induced by earthquakes are of both varieties, transverse and longitudinal. Both can be transmitted by the solid matter of the earth, but it was found that when the waves penetrated a certain depth below the earth's surface, only the longitudinal ones continued onward, while the transverse waves were stopped suddenly and entirely. It was from this that geologists were able to deduce that the earth contains a liquid core, and to measure its diameter with considerable accuracy.)

Sound waves, however, cannot be conducted in the complete absence of molecules. If an electric bell is suspended in a bell jar and set to ringing, it will be heard through the glass (which can carry sound waves). If the bell jar is gradually evacuated, the sound of the bell will become fainter and will eventually fade out altogether. The clapper may continue to strike the bell furiously and the bell may even be seen to vibrate, but no longitudinal waves can be set up among the molecules of an air that does not exist. As a result, sound will not be heard.

(It is frequently stated that the moon, which lacks an atmosphere, is a soundless world. However, sound can be transmitted through the moon's crust, and an astronaut may hear a distant explosion if he makes the proper contact with the moon's surface.)

Loudness

Suppose we consider a sound wave in which the succession of compressions and rarefactions are regular. This would be analogous to a transverse wave that had the form of a regular sine curve. Such a sound wave is heard by us as a steady musical note and is produced by a tuning fork. Indeed, if a pen is attached to the prong of a tuning fork in such a way that it makes contact with a roll of paper being moved at constant velocity in a direction at right angles to the vibration of the prong, a sine wave will be produced.

A tuning fork may produce sounds that differ in loudness. If it is struck lightly, it will emit a soft sound; if struck more heavily it will emit a sound which our ear will detect as identical with the first except for being louder. The lightly struck tuning fork will move back and forth over a comparatively small arc; the more heavily struck one will move back and forth over a larger arc. As is to be expected of simple harmonic motion, the two movements will involve an identical period despite the difference in amplitude, so either way the same number of volumes of compression and of rarefactions are set up per second. The frequency of the sound produced is, therefore, the same in either case.

However, the more heavily struck tuning fork, moving in the larger arc, compresses the air more violently. Therefore, a louder note differs from a softer note in that the compressed volumes of the former are more compressed, and the rarefied volumes more rarefied. The greater difference in extent of compression in a longitudinal wave is analogous to a greater amplitude in a transverse wave. This can easily be visualized if we think again of the tuning fork with the pen attached. A gently vibrating prong would mark off a sine curve of small amplitude; one vibrating through a greater arc, as a result of a heavier blow, would mark off a sine curve of greater amplitude.

To compress air against the resistance of its pressure requires energy, and the compressed air contains a store of energy that it can expend by expanding and pushing whatever is in the neighborhood. For this reason, sound waves can be considered a form of energy.

The more the air is compressed, the more energy it contains and can expend. Another way of looking at it is to consider that the vibrating tuning fork has kinetic energy that is expended in compressing air.* If the prong swings through a greater arc but completes its period in the same time, it moves at a greater average velocity and has more kinetic energy that it can expend in compressing air. Whichever way we look at it, we can come to the conclusion that loudness is a matter of quantity of energy, and that a loud sound contains more energy than a soft one.

The loudness, or *intensity,* of sound is measured in terms of the quantity of energy passing each second through one square centimeter of area, the area being perpendicular to the direction of propagation of the sound. Energy expended per unit time is power, and the amount of power involved in sound is very small. To indicate just how small, let's reconsider some units of power.

A watt is the mks unit of power and is equal to one joule per second. We are familiar with watts in connection with light

* The tuning fork, or any sound-producing device, also rarefies air against its own pressure, which also requires energy. The argument is exactly analogous if rarefaction is considered rather than compression.

bulbs; we all know that a light with a power of 75 watts is none too bright for reading purposes, and that one with as little power as 40 watts is rather dim. Even a night light, just bright enough to dispel the worst of the shadows and enable us to get to the bathroom at night without tripping over the furniture, has a power of 1/4 watt. A microwatt is 1/1,000,000 of a watt, so that such a night light has a power of 250,000 microwatts.

In comparison with that, ordinary conversational sounds carry a power of but 1000 microwatts, and low sounds sink down to bare fractions of a microwatt.

The ear detects differences in loudness by ratios of power rather than by actual differences. Thus, a 2000-microwatt sound will seem a certain amount louder than a 1000-microwatt sound, but a 3000-microwatt sound will not appear louder by as much again. It takes a 4000-microwatt sound to seem louder by as much as a 2000-microwatt sound is louder than a 1000-microwatt sound. To get a sound that is as much louder still than a 4000-microwatt sound, we must rise to an 8000-microwatt sound. The ratios 2000/1000, 4000/2000, and 8000/4000 are all equal even though the differences are not, and it is by ratios that the ear judges.

This means that the ear acts not by the power of a sound, but by the logarithm* of that power. When one sound carries ten times the power of a second sound, the ratio of the power of the first to that of the second is 10, and the logarithm of that ratio is 1. The difference in sound intensity is then said to be one *bel,* so named in honor of Alexander Graham Bell (1847–1922), who studied the physics of sound and invented the telephone. Similarly, if one sound is 100 times as powerful as another, it is two bels

* The common logarithm of a number is its exponent when it is expressed as a power of 10. For instance, 10^2 is (10) (10), or 100; 10^3 is (10)(10)(10), or 1000. Therefore, the logarithm of 100 is 2 and that of 1000 is 3. The use of logarithms converts a geometric series (one in which each number is obtained by multiplying the preceding number by a fixed quantity) into an arithmetic one (where each number is obtained from the preceding by addition). In the series 10—100—1000—10,000—100,000, etc., each number is obtained by multiplying the previous number by 10. If the logarithms of the numbers in the series are written instead, we have 1—2—3—4—5, etc., where each number is obtained by adding 1 to the previous number. Our senses generally work by converting a geometric series to an arithmetic one in this fashion. If one stimulation is 100,000 times as intense as another of the same sort, the sense organ, working by logarithms, detects it as, say, five times as intense. In this way sense organs can be useful over an enormous range of intensity. This is the *Weber-Fechner Law,* so named in honor of two Germans: Ernst Heinrich Weber (1795–1878), who first expressed the law; and Gustav Theodor Fechner (1801–1887), who popularized it.

louder; if it is 1000 times as powerful it is three bels louder, and so on. This kind of unit imitates the logarithmic working of the ear.

The bel is rather too large a unit for convenience. A tenth of a bel is a *decibel*. One sound is a decibel louder than another sound when the first is 1.26 times as powerful as the second, for the logarithm of 1.26 is just about 0.1.

Because of the small amount of energy represented by even loud sounds, sound energy is not something we are usually aware of. The energy of a roll of thunder may be sufficient to cause objects to vibrate noticeably. The telephone is an example of the manner in which human ingenuity has managed to usefully convert sound energy into electrical energy and back to sound energy.

For the most part, however, the sounds that continually surround us, whether created by human beings, by other forms of life, or by inanimate surroundings, simply fade out and are converted into heat.

If sound remained unconverted into other forms of energy, we could easily see how the loudness of sound would fall off with distance from the source. The sound wave moves outward as an expanding sphere from the source, and the total power represented by each sound wave spreads out over that surface. The surface of a sphere is equal to $4\pi r^2$, where r is the radius of the sphere—that is, the distance from the source. If the distance from the center is tripled, the surface area is increased ninefold and only one-ninth as much power passes through any square centimeter on the surface. The intensity of sound would then be expected to vary inversely as the square of the distance from the source. This is how the intensity of gravitational attraction falls off, for instance. However, gravity is not absorbed by matter, whereas sound is easily absorbed by most of the objects with which it makes contact—even by the air itself. As a result, sound falls off more rapidly than one would expect.

CHAPTER 12

Pitch

The Velocity of Sound

A particular object has some natural period of vibration, and in the case of simple harmonic motion at least, this period is proportional to the square root of the mass of the object divided by the restoring force (see Equation 8–4 on page 107). In the case of the pendulum, where the restoring force is gravity (which increases with mass), the period varies as the square root of the length of the pendulum divided by the acceleration due to gravity (see Equation 8–9, page 112).

This means that we will generally expect that of two similar objects the larger and more massive will have the longer period of vibration. It will consequently produce fewer sound waves per unit time, and the individual waves will have a longer wavelength and a lower frequency.

The period of vibration can also be varied by changing the size of the restoring force, the period shortening as the restoring force increases in size. A taut string is more difficult to pull out of its equilibrium position than a slack one is, and from that it is clear that the force tending to restore the string to position is increased as the string grows tauter. Of two strings otherwise alike, the tauter snaps back faster and, if it is a bowstring, shoots the arrow farther. (That is why bowstrings are kept as taut as possible when the bow is in action.) A taut string, snapping back quickly, naturally has a shorter period of vibration than a slack one has and produces sound waves with higher frequency and shorter wavelength.

From experience, however, we know that all the factors that serve to produce a sound wave of low frequency also produce a deep tone, while those that bring about a sound wave of high frequency also produce a shrill tone. Large objects with long periods of vibration produce deep tones, while similar small objects produce shrill ones. Compare the tolling of a church bell with the tinkling of a sleigh bell, the strum of the string on the bass viol with the shrillness of the string on the tenor violin. In the realm of life, compare the trumpeting of the elephant with the squeak of the mouse; the honk of the goose with the tweet of the canary. The voice of a man with his longer vocal cords is deeper than those of women and children with their shorter ones. An individual can vary the shrillness of the sound he produces by adjusting the tautness of his vocal cords (though he is not aware he is doing so), and the sound of a freely vibrating string can be made more shrill as it is made more taut.

This property of shrillness, or depth in a tone, is referred to as the *pitch* of the sound, and it is quite obvious that the ear differentiates the frequencies of sound waves as pitch. As frequency increases, a sound is heard as increasingly shrill. As frequency decreases, a sound is heard as increasingly deep.

It is easy to determine the frequency of a sound wave. The vibrations of a tuning fork can actually be counted in several ways, including (to mention a simple method) having it mark itself by penpoint on a moving scroll of paper and counting the waves produced in a unit time. In this way, frequency and pitch can be matched. For instance, a tuning fork or pitch pipe that produces a "standard A" (the pitch against which musicians standardize their instruments) can be shown to have a frequency of 440 per second.

To calculate the actual wavelength of a sound of a certain pitch, one can make use of Equation 11–1. This tells us that the frequency (ν) is equal to the velocity of the wave (v) divided by the wavelength (λ). Solving Equation 11–1 for λ, we find that:

$$\lambda = \frac{v}{\nu} \qquad \text{(Equation 12–1)}$$

The piece of information we need to make Equation 12–1 useful is the velocity of sound. This velocity may be determined with considerable accuracy by a straightforward experiment which was first carried through successfully in the early seventeenth century.

Suppose a cannon is set up on one hill and observers are stationed on another hill a known distance away. When the cannon is fired, the flash is seen at once (assuming that light travels so quickly that its journey from one hill to the other takes up virtually zero time—which is correct). The sound of the cannon, however, is heard only after a measurable interval of time. The distance between cannon and observers divided by the number of seconds of lag in hearing the cannon (the possession of a good timepiece is assumed) will give the velocity of sound.

To be sure, if there is a wind the compression waves will be hastened onward by the overall movement of the air, or slowed down, depending on the direction of the wind. What can be done, therefore, is to place cannon on both hills and fire each, first one then the other. Whatever effect the wind has in one direction, it has a precisely opposite effect in the other, and averaging the two velocities obtained will give the velocity in quiet air.

The currently accepted velocity of air at ordinary temperatures (say, 20° C or, what is equivalent, 68°F)* is 344 meters per second (or 1130 feet per second, or 758 miles per hour). This velocity varies a bit with temperature. On a cold winter day, it may be as low as 330 meters per second; on a hot summer day, as high as 355 meters per second.

The temperature difference has important effects. During the day, upper levels of the atmosphere are generally cooler than air at ground level. As the upper part of a beam of sound waves penetrates the cooler strata, it slows up; the effect of that is to veer the entire beam upward. (If you are walking, and someone seizes your left arm, slowing that part of your body, you automatically veer leftward.) At night, the situation is reversed, for the upper levels are warmer than the lower levels. The upper part of a beam of sound waves will quicken, and the whole beam will veer downward. It is for this reason that sound can usually be heard more clearly and over greater distances by night than by day.

However, if we confine ourselves to room temperature, we may write Equation 12–1 as:

$$\lambda = \frac{344}{\nu} \qquad \text{(Equation 12–2)}$$

* The question of temperature and temperature scales will be taken up later in some detail (see page 181).

Until recent times, sound traveled at a velocity much greater than that of any man-made vehicle, so for practical purposes the velocity of sound did not concern the traveler. With the invention of the airplane, however, and with the steady increase in the velocities of which it was capable, the velocity of sound became of importance for reasons other than those involving the speed of communication.

It is the speed of the natural rebound of molecules after compression that dictates the rate at which a compressed area restores itself to normal and compresses the next area; so it is this speed of rebound that determines the velocity of sound. It is also the speed of the natural rebound of molecules after striking a speeding plane that makes it possible for air to "get out of the way" of the plane. As the plane approaches the velocity of sound, then, it approaches the velocity with which the air molecules can rebound. The plane begins to "chase after" the rebounding air molecules and, with increasing speed, more and more nearly catches them. Such a plane compresses the air ahead permanently (or at least for as long as it maintains its speed), since the air cannot get out of its way. This volume of compressed air ahead of the plane puts great strains upon the plane's structure; for a time in the 1940's, it was felt that a plane would disintegrate if it approached the speed of sound too closely. Thus, talk began to be heard of a "sound barrier," as though the velocity of sound represented a wall the plane could not break through.

The ratio of the velocity of an object to the velocity of sound in the medium in which the object is traveling is called the *Mach number,* in honor of an Austrian physicist, Ernst Mach (1838–1916), who toward the end of the nineteenth century first investigated the theoretical consequence of motion at such velocities. To equal the velocity of sound is to be moving at "Mach 1," to double it is to be at "Mach 2," and so on. A Mach number does not represent a definite velocity, but depends upon the nature, temperature, and density of the fluid through which the object is traveling. For normal air at room temperature, Mach 1 is 344 meters per second, or 758 miles per hour.

Improved design of planes enabled them to withstand the stresses at high velocities, and on October 14, 1947, a manned plane "broke the sound barrier" by traveling at a velocity of more than Mach 1. Since then velocities of Mach 3 and more have been attained. (An astronaut circling the earth at a speed of five miles per second might be said to be traveling at Mach 25, if the velocity of sound in ordinary air is used as a comparison. However, the astronaut is traveling through a near vacuum across which no significant amount of sound is conducted, and Mach numbers do not really apply to him.)

A plane traveling at *supersonic velocities* (velocities greater than Mach 1) carries its sound waves ahead of it, so to speak, since it travels more quickly than they could alone. The volumes of compression are brought together, and instead of a smooth progression from compression to rarefaction and back, as in ordinary sound waves, there is a sharp dividing line between a volume of strong compressions and the normal surrounding atmosphere. The strong compression streams backward in a cone-shaped band, with an angle depending on the Mach number, and is called a *shock wave*. A similar shock wave streams back from speeding bullets, too; it is also formed by the effect of lightning bolts, for instance, which will energetically expand air at velocities greater than Mach 1. (The shock wave is an example of a wave form that is not periodic.)

If a plane traveling at supersonic velocities slows down or veers off, the shock wave will revert to ordinary sound waves, carrying volumes of unusually strong compression and rarefaction, however. In this train, sound waves expand and weaken as they travel, but if they are fairly close to the ground to begin with, and happen to be directed downward, they will strike the ground with considerable strength, producing the now well-known "sonic boom."

Thunder is the sonic boom produced by lightning, and the crack of a bullwhip is a miniature sonic boom, since it has been established that the tip of such a whip can be made to travel at supersonic velocities.

The velocity of sound, when spoken of simply as such, always implies its velocity through air. However, sound travels through any material body, and its velocity varies with the nature of the body. Intermolecular forces in liquids and solids, stronger than in gases, bring about a much quicker rebound after compression. Consequently, sound travels with greater velocities through liquids and solids than through any gas, and the more rigid the substance (and hence the stronger the intermolecular forces), the greater the velocity of sound through it. In water, sound travels at a velocity of 1450 m/sec (3240 miles per hour), and in steel it travels at a velocity of about 5000 m/sec (or 11,200 miles per hour).

The Musical Scale

Sounds of different pitch can be produced in musical instruments by striking or plucking strings of different length and thickness, as in the piano or harp; by using few strings but altering their effective length by pinning one end with the finger at varying points, as in the case of the violin; by allowing sound waves to fill tubes which may be lengthened or shortened by physical movement, as in the trombone; or by blocking or unblocking certain sections of the tube by stopping a hole with a finger, as in a flute, or depressing a key, as in a trumpet.

When two notes are sounded, either together or one after the other, the combination is sometimes pleasant and sometimes unpleasant. This is partly a subjective and cultural matter, for we like what we are used to and many types of music, such as rock and roll or traditional Japanese, sound unpleasant to the uninitiated but very pleasant to the devotee. Nevertheless, if we confine ourselves to the serious music of the West, we can come to certain conclusions about this.

When two notes are sounded together, the result is not two separate trains of sound waves, each traveling independently through the air. Instead, the two waves add to each other to form a resultant wave.

To make things very simple, suppose that two sound waves are each of the same frequency but are sounded in such a way that one is half a wavelength behind the other. Whenever one sound wave is forming an area of compression at one point, the other is forming an area of rarefaction there, and vice versa. The two effects cancel each other and the air does not move. As a result, the two sounds taken together produce silence, and this phenomenon is called *interference*. It is difficult to picture this if we think of longitudinal waves. However, if the longitudinal waves are pictured as analogous transverse waves (as they invariably are for this purpose) interference is easily pictured. Wherever the sine curve of one sound wave goes up, the sine curve of the other goes down, and if the two are added together a horizontal line (no wave at all) is the result.

On the other hand, if two waves of the same frequency are sounded exactly in phase, they would add to each other, so compressed areas are more compressed and rarefied areas are more rarefied than if either sound had been produced alone. In transverse wave analogy, the crests and troughs of the separate waves would match, and the resulting crests would be higher and troughs deeper of the two waves together than of either alone. The ear would hear one sound of the proper pitch, but louder. This is *reinforcement.*

Actually, perfect interference or reinforcement is unlikely. Instead, two or more waves will combine, reinforcing here, interfering there, and will form resulting patterns of very complicated form that will not at all resemble the regular sine waves of individual notes. However complicated these patterns may be, they will remain periodic. That is, a small unit section of the pattern can be taken, and the entire pattern can be shown to be made up of a succession of these units.

In 1807, the French physicist Jean Baptiste Joseph Fourier (1768–1830), studying wave forms generally, showed that any periodic wave pattern, however complicated it might seem, could be separated by appropriate mathematical techniques into the individual sine waves making it up. The mathematics involved is referred to as *harmonic analysis* because it can be applied to musical sounds. (The wave patterns of musical sounds are composed of separate sine waves that display an orderly set of interrelationships. Where this is not true, but where the component sine waves are chosen and combined at random, so to speak, the result is not

Interference

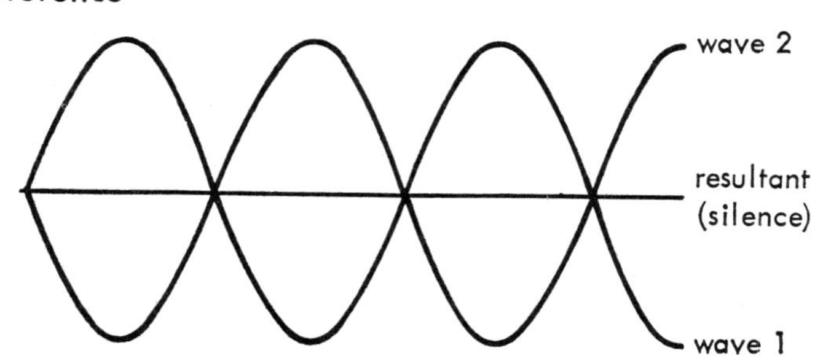

Reinforcement

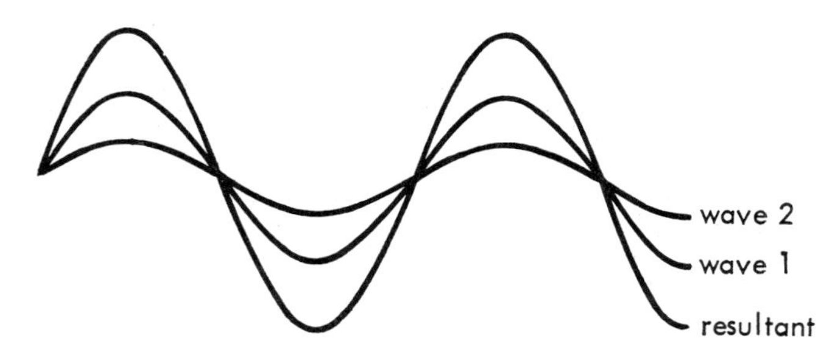

music, but "noise." The difference is analogous to that between an orderly but complicated geometrical figure and the same lines combined in random fashion to produce a scribble. Fourier's methods could be used to analyze the wave patterns of noise, too, so perhaps it should be referred to by the more neutral term *wave analysis.*)

Restricting ourselves to very simple cases, and without inserting appreciable mathematics, let's consider two notes of different pitch, and therefore of different frequency, sounded together. The compressed regions of the sound wave (or the crests, if we wish to speak in the more easily visualized transverse-wave analogy) would be coming at shorter intervals in the case of the note of higher frequency, and they would overtake those of the sound wave with the lower frequency.

Suppose one note has a frequency of 250 per second and another note of 251 per second, and suppose they start in phase. The first crest for both appears simultaneously. The second crest of the 251/sec note appears just a little sooner than the second crest of the 250/sec note. The third crest appears still sooner, and the fourth crest appears still sooner. At the end of one second, however, one note has completed exactly 250 vibrations and the other exactly 251 vibrations. They are back in phase, but the 251/sec note has gained one complete crest.* Each succeeding second, the 251/sec note gains another complete crest.

At the point where the two notes are in phase, crest for crest, there is a short period of complete reinforcement, and the note sounds loudly. As the second progresses and the crests fall more and more out of phase, there is more and more interference and the sound becomes softer. At the half-minute mark, midway between two in-phase periods, the notes are completely out of phase and the crests of one match the troughs of another, and there is a short period of complete interference. The result is a regular swelling and dying of sound, with the maximum loudness coming at second-intervals when the crests match. Such periodically changing loudness when two notes are sounded together is called a beat.

Suppose the two notes had frequencies of 250/sec and 252/sec, respectively. Then, after half a second, one note would have completed 125 vibrations and the other 126 vibrations, and

* The two notes are racing only in connection with the number of crests being produced in a given time, not in terms of velocity. Both notes are traveling through space at the same velocity. Indeed, the velocity of sound does not depend on frequency.

they would be back in phase with crest matching crest. This would be repeated every half-second, and there would be two beats per second. The number of beats per second, where two notes are sounded simultaneously, is generally equal to the difference in the frequencies of those notes.

If beats are infrequent enough to be heard separately, they render the sound combinations unpleasant to the ear. Apparently, 30 beats per second is maximally unpleasant. Where beats are more than 60 per second, however, they melt into each other as far as the ear is concerned, and the combination of sounds seems pleasant or harmonious.

Consider two notes of which one has a frequency exactly double the other. One has a frequency of 220/sec, let us say, and the other 440/sec; the ratio of frequencies is 1:2. The number of beats, when the notes are sounded together, is 440-220, or 220 a second. The beats duplicate the lower note, so the two notes seem to melt into each other and be almost the same note. They go well together.

It was Pythagoras who first noticed that notes that go well together are related by these small whole-number ratios. He had no method of measuring frequency itself, but he considered strings of different lengths. He found that two strings with lengths in a 1:2 ratio produced a pleasant combination; so did strings with a 2:3 ratio and a 3:4 ratio.

(Pythagoras wandered off in mystical fashion from these sound observations—sound in both senses. He assumed that the interplay of small whole numbers in the production of pleasing sounds fit in with his views that all the universe was ruled by number. He and his pupils speculated that the planets themselves produced sounds—the so-called music of the spheres—with notes based on their relative distances from the earth. Science did not free itself of these notions for 2000 years.)

Suppose then that we start with a note of a 440/sec frequency (the standard frequency for musicians) and call it A. A note of twice the frequency sounds so much like it that we can call that A, too, and we can use the letter for a sound of half the frequency, for that matter. In fact, we can have a whole series of such A's, with frequencies of 110/sec, 220/sec, 440/sec, 880/sec, 1760/sec, and so on, extending the range, if we choose, both upward and downward indefinitely.

Between any two successive A's, we can introduce other notes with frequencies that bear some orderly arithmetical relationship to the A-notes and to each other. It is customary to introduce six other notes in the interval; these are lettered B, C, D, E, F, and G. Thus we have, from A to A, the notes: A, B, C, D, E, F, G, A. In passing from A to A, there are eight notes (counting both A's) and seven intervals between notes. The span from A to A is therefore called an *octave* (from a Latin word for "eighth." Other spans are spoken of in plain English. The span from C to G (C, D, E, F, G), involving as it does five notes, is a *fifth,* while the span from C to F is a *fourth.*

The frequencies associated with the notes from the 220/sec A to the 880/sec A are:

A = 220	A = 440	A = 880
B = 247.5	B = 495	
C = 264	C = 528	
D = 297	D = 594	
E = 330	E = 660	
F = 352	F = 704	
G = 396	G = 792	

The range from 220/sec to 440/sec is one octave and that from 440/sec to 880/sec is another octave. Each note in the upper octave is double the corresponding note in the lower octave, so the interval from B to B is an octave; so is the interval from C to C, from D to D, and so on. Remembering to double the frequency for each higher octave and halve it for each lower octave, you can write the frequencies for any note in any octave.

If the successive notes within any octave are sounded, they sound just like the corresponding notes within any corresponding octave lower or higher. The standard piano keyboard covers a range of a little over seven octaves; if the white notes are sounded one after the other, one can easily detect the same "tune" to be repeated seven times, at successively higher pitches.

The frequencies are interrelated by ratios that can be expressed in small whole numbers. The ratio of G to C, for instance, is 396:264, or 3:2; the ratio of F to C is 352:264, or 4:3. It is these simple ratios that Pythagoras studied, and it is the simplicity of the ratios that sets up beats that reinforce the notes themselves

Pattern of octave intervals

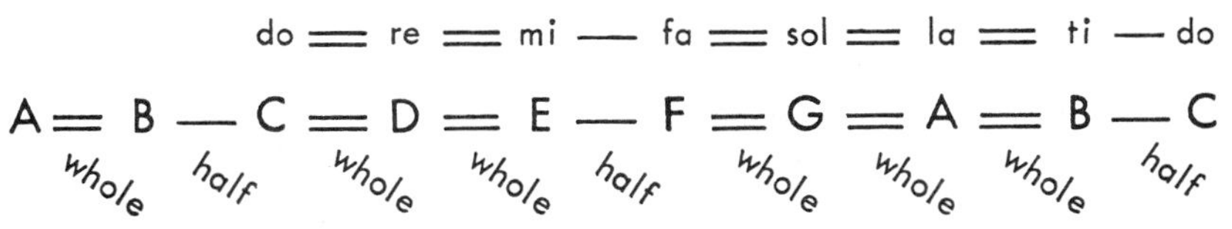

do = re = mi — fa = sol = la = ti — do

A = B — C = D = E — F = G = A = B — C

whole half whole whole half whole whole whole half

and make them blend well together. That is why fifths and fourths are much used as intervals between successive notes.

Then, too, the ratios of C, E and G are 264:330:396, or 4:5:6, and the three notes sounded together as a *major triad* make a pleasing sound-combination, or *chord.* F, A and C also make a major triad, and so do G, B and D. In fact, the note intervals are so designed that every note can be part of one or another of these three major triads.

If the ratio of the frequency of adjacent notes is considered, it turns out that B:A as 9:8. The ratio for D:C and for G:F is also 9:8. The ratio for E:D and A:G is not quite that, but it is close, 10:9. In other words, of the seven intervals between the notes of the octave, five are of roughly equal size, and we can call them "whole intervals."

The frequency ratio of F:E, however, is only half as large, for it is 352:330, or 16:15; this is also true of the ratio of C:B. (This may be easier to see if we express it another way. A ratio of 9:8 represents an increase in frequency of 12.5 percent, and one of 10:9 represents an increase of 11.1 percent. The ratio 16:15, however, represents an increase of only 6.7 percent.) In passing from B to C or from E to F, then, we are traversing only a "half interval."

If we start from A and go up the notes through B, C, and so on, we will be passing intervals in the following pattern: whole, half, whole, whole, half, whole, whole, whole, half, whole, whole, half, and so on. Successive half intervals are separated by two whole intervals, three whole intervals, two whole intervals, three whole intervals, and so on.

When we sing the scale, using the traditional names for the notes (*do, re, mi, fa, sol, la, ti, do*), through long habit, we insist on placing the half-note intervals between *mi* and *fa* and between *ti* and *do.* Any other arrangement sounds wrong to us. We therefore want the seven intervals of the octave to fall into the following pattern: whole, whole, half, whole, whole, whole, half. If you check back, you will see that this particular arrangement can only be brought about if we start with *do* on the note C (it doesn't matter which C). Then *re* becomes D, *mi* becomes E, *fa* becomes F, *sol* becomes G, *la* becomes A, *ti* becomes B, and *do* is C again. The *mi-fa* half interval corresponds to the EF half interval, and the *ti-do* half interval corresponds to the BC half-note interval. The arrangement of notes you sound in singing the scale now matches the successive notes you tap out, beginning at C on the white keys of the piano. If you start on any white key of the piano other than

C and play the successive white keys, the piano and you will sound half intervals at different points in the scale, and the piano (not you, of course) will sound dreadful.

It is desirable to be able to play the scale from any point on the piano keyboard, so that the range of the scale can be adjusted to a particular human voice, for instance. For this reason, in every octave, five black notes are inserted to break up the five whole intervals. This accounts for the familiar black note pattern of two (CD, DE) and three (FG, GA, AB) all along the keyboard. Now the scale can be sounded by beginning at any note on the piano (either white or black) provided you remember to choose your notes carefully and play sometimes black and sometimes white. Only if you start on C, however, can you play the scale by sounding successive white notes only.

It is for this reason that C seems a natural *do* and that the "key of C" is the simplest key to play for beginners (white keys only, for the most part!). "Middle C" is the particular C that is about at the midpoint of the piano keyboard, and it is the C with the frequency 264/sec.*

Modification of Pitch

Pitch will change if the source of sound is moving relative to the hearer. Suppose a distant train, standing motionless, sounds a whistle that has a frequency of 344/sec. In that case, when the sound wave reaches us, 344 compression/rarefaction combinations will strike our eardrum each second. Since sound (at room temperature) travels at 344 m/sec, successive areas of compression are a meter apart.

Suppose next that the train is moving rapidly toward us at

* Physicists often use a frequency of 256/sec for middle C, because as a power of 2, 256 is a particularly easy number to halve and double. It is (2)(2)(2)(2)(2)(2)(2)(2), or 2^8.

Piano keyboard

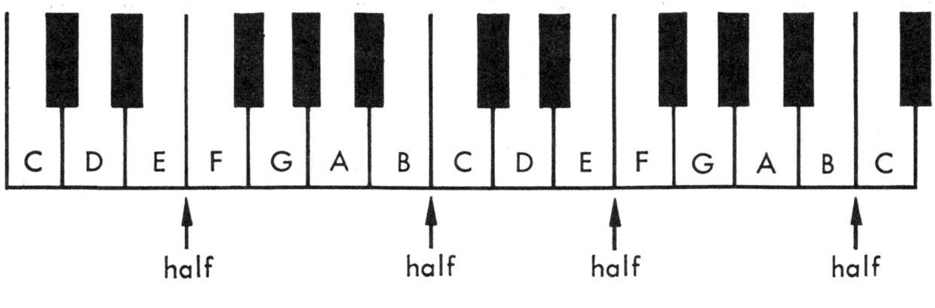

a rate of 34.4 m/sec (75.8 miles an hour), or just one-tenth the velocity of sound. It is still sounding its whistle. One region of compression is moving ahead of it, and by the time it has moved a meter, another region of compression is emitted. By that time, however, the train has moved forward a tenth of a meter and the second region of compression is only 0.9 meters behind the first. This happens for all successive regions of compression if the train maintains a steady pace. For this reason, sound waves from the whistle of the approaching train enter our eardrums 0.9 meters apart and in one second 344/0.9, or 382 of them, strike the eardrums. A person on the train, and therefore moving right along with the whistle, receives 344 regions of compression in one second. The ratio 382:344 is close to 9:8, so the sound is a whole interval shriller (see page 172) for the person watching the train approach than for the person on the train.

On the other hand, if the train were receding, then by the time a region of compression had moved a meter toward the hearer and a new region of compression was due, the train would have moved a tenth of a meter away, and the two areas of compression would be 1.1 meters apart. The frequency would be 344/1.1, or 312 per second. Now it is deeper by nearly a whole interval than it would sound to the person on the train.

If the train passed us at this velocity, the sound we heard would shift suddenly from a frequency of 382/sec as it was approaching and passed, to 312/sec as it passed and receded.

This phenomenon is called the Doppler effect in honor of the Austrian physicist Christian Johann Doppler (1803–1853), who first studied the effect and explained it correctly in 1842.

Pitch can be made to vary in a much more subtle way, too. The same note sounded with the same loudness on the piano, violin, and clarinet sounds different to us. If we have any experi-

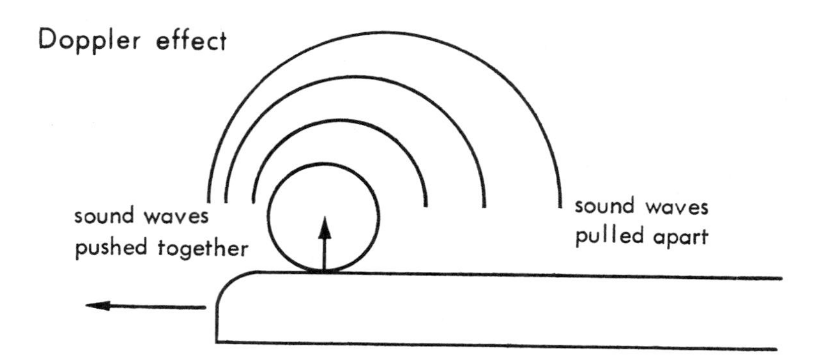

ence at all, we can tell which instrument is sounding the note. This difference in notes that are identical in pitch and loudness is a difference in *quality* or *timbre.*

To explain this, we must consider that the vibrations of a string, or of any sound-producing device, are actually more complicated than I have described them to be. A string, for instance, may indeed vibrate all in one piece to produce a vibration and, therefore, a sound wave of a given frequency. In the transverse-wave analogy, this would be a simple, regular sine curve, and is the *fundamental note.* It is the fundamental note we usually think of when we speak of the frequency of a particular note.

However, the string may also vibrate as two halves: one half moving to the right as the other half moves to the left and vice versa; the midpoint of the string, bounding the two halves, serving as a motionless *node.* Each half of the string vibrates at twice the frequency of the whole string, so a note is sounded with just twice the frequency of the fundamental note. The string may also vibrate in thirds, in fourths, in fifths, and so on, to produce notes with frequencies three times, four times, five times, and so on, that of the fundamental notes. All these notes of higher frequencies are called *overtones.* The fundamental note and the various overtones are sounding simultaneously; the actual motion of the string is a combination of all. The fundamental note remains dominant, but the overtones add their wave forms, and therefore the resulting wave form is far more complicated than a simple sine curve. Furthermore, for strings under different conditions (to say nothing of other sound-producing devices) the overtones may be receiving different proportionate stress; certain overtones may be stronger in some cases than in others, so the final wave form will be different for different instruments. Such a difference imposed on the eardrum is great enough for us to detect.

This difference can be magnified by methods of selecting some overtones from the rest for special magnifications. Let's see how.

A vibrating object may force another to vibrate in unison, so the second object sets up the same sound wave pattern and produces the same sound. If a vibrating tuning fork has its stem placed in contact with a table, its sound is suddenly louder because now the entire table is vibrating in unison.

Such a *forced vibration* need not even be the result of direct physical contact between solids. Indirect contact through air may be sufficient. A given vibration will set the air pulsing in longitudinal waves; these waves will, in turn, set the eardrum vibrating in unison. The eardrum will move inward when a region of compression strikes, and outward when a region of rarefaction does; it moves a greater distance from the equilibrium position as the regions become more compressed and more rarefied. It is through such forced vibrations that the eardrum exactly duplicates the original vibration, and we are able to judge (via a complicated hearing mechanism we will not describe here) the pitch, loudness, and even the timbre of a sound.

There are occasions, however, when one particular frequency is more easily "forced" than another on a second body. Imagine yourself pushing a child on a swing, for instance. The child on a swing represents a form of pendulum and has a natural period of vibration. If you apply successive pushes to the swing at random intervals, you will often push the swing as it is moving back toward you and will cancel what motion it possesses, slowing it. By persisting, you will keep the swing moving in accordance with your pushes, but you will expend a lot of energy doing so. If, however, you timed your pushes to match the natural period of vibration of the swing, you would push each time as the swing begins to move away from you, thus adding to its velocity and increasing it further with each swing and rhythmic push. At the expense of far less energy, you would get a far more rapid and extended swing.

(Marching soldiers crossing a bridge are supposed to break step. Otherwise, if the thud of the footsteps in unison happened to match the natural period of vibration of the bridge, the bridge would swing in wider and wider arcs until it finally broke apart.)

The situation is analogous for sound waves. The sound wave of a particular note would push another object with each region of compression and pull it with each region of rarefaction. If the rhythmic push-and-pull did not match the natural period of the receiving object, the forced vibration could only be obtained at the expense of considerable energy being used to overcome that natural period. If, however, the frequency of the note just matched the natural period of vibration of the receiving object, the latter would begin to vibrate more and more. This is called *resonance* (from Latin words meaning "to sound again").

Any given sound wave would produce far more vibration in a resonating object than in any other kind; in fact only the resonating object might produce sound waves strong enough to be audible. Suppose, for instance, you raise the top of a piano to expose the wires and step on the "loud" pedal to allow all those wires to vibrate freely. Now sing a short, loud note. Only those wires that vibrate at the frequency of that note will resonate, and when you stop singing, you will hear the piano answer back softly in that same note.

Musical instruments depend upon the resonance of the materials making up their structure to strengthen and add richness to the notes produced. Pianos have a "sounding board" just under the wires, and this device can resonate with the various notes. Without that board, the notes sounded by the wires would be quite weak.

Naturally, the resonating portions of each instrument, although resonating to almost all the notes (the resonating portions are complicated in shape and different parts have different natural periods of vibration), do not do so with equal efficiency. The wood of a violin resonates to the notes produced by it, but it may resonate more efficiently to some overtones than to others. No two violins are of exactly the same shape, or of exactly the same wood with the same grain arrangement, or possess exactly the same varnish. As a result, there are subtle differences in resonance from instrument to instrument. The Italian violinmaker Antonius Stradivarius (1644–1737) manufactured violins that are the despair of imitators, for it is almost impossible to duplicate their richness of tone.

The sounds we ourselves make produce resonances in the air filling the hollows in throats, mouths and nasal cavities. The natural vibrations of the air depend on the shape and size of the cavities, and since in no two individuals are these cavities of precisely the same shape and size, voices differ in quality; we usually have no trouble recognizing the voice of a friend from among a large number of others.

Reflection of Sound

A ripple in a water tank is turned back on itself when it strikes the rim of the tank; having progressed, let us say, leftward prior to contact, it proceeds rightward thereafter, much as a billiard ball does that has struck the edge of a pool table head-on. The water wave has been *reflected* (from Latin words meaning "to bend back").

Sound waves can be reflected, too. A mountain wall will reflect them, for instance. A word shouted across a valley is heard almost at once as it leaves the lips, being conducted through the air from lips to ear. It is then heard again seconds later, after the sound wave has reached the mountain wall, been reflected, and crossed the valley a second time. This is the *echo*. If mountain sides are properly arranged, more than one echo may be heard.

Similar echoes may be heard in tunnels, in large empty rooms, and indeed anywhere where hard surfaces reflect sound rather than absorb it. In ellipsoidal rooms a sound uttered at one focus of the ellipse will spread out in all directions; however, in reflecting from various portions of the walls and ceiling, it will concentrate upon the other focus. Two people standing at the foci can converse in whispers, even though separated by a large distance. Such "whispering galleries" always amaze those who have never encountered one before.

In rooms of moderate size, the length of time taken for a sound wave to travel to a wall, be reflected to an opposite wall, be reflected once again to the first wall, and so on, is so short that distinct echoes will not be heard. Instead, a series of very rapid echoes will blend into a dull, hollow rumble that may persist audibly for a considerable time after the original sound is no longer being formed. This persistence of sound is called *reverberation.* The study of the behavior of sound in enclosed places, particularly with regard to such reverberation, is called *acoustics* (from a Greek word meaning "to hear"), a term that is sometimes applied to the study of sound generally.

Reverberation can represent a great inconvenience. A lecturer may find that his words cannot be heard because of the dying sound of his previous words. An orchestra may find its best efforts reduced to discord as previous notes live on past the time when they are wanted or needed. Reverberation can be reduced by draping the walls, using a soft, pulpy material for the ceiling, or even by the presence of an audience in winter clothing. When sound waves enter the small interstices of fabric or other porous material, contact of the moving air molecules with solid material is made over a hugely increased area. Friction is increased and sound energy is converted into heat. The sound waves, in other words, are absorbed rather than reflected.

This can be overdone. If reverberation is reduced to too low a level, there seems a "deadness" to sounds. A reverberation period of one second, or even two if the room is very large, is aimed for.

Sound waves are not always either reflected or absorbed (nor are water waves). There is a third alternative: sound waves (and water waves, too) can bend around obstacles and continue onward. It is because of this that we have no difficulty hearing someone call from behind a tree or from around a corner. This ability to bend around obstacles is not the same for all kinds of waves.

In 1818, the French physicist Augustin Jean Fresnel (1788–1827) was able to show, in connection with his studies of wave motions generally, that whether a wave was reflected or not depended upon the comparative size of the wavelength and the obstacles. When an obstacle was the size of the wavelength or less, it did not reflect the wave, which, instead, bent round the obstacle. If the obstacle was considerably larger than the wavelength, the wave was reflected.

Consider the common sounds we hear about us every day, with frequencies, let us say, through the middle range of the piano from C below low C to C above high C—a range of four octaves. The range of frequencies extends from 66/sec to 1056/sec. If we make use of Equation 12–2, we see that the range of wavelengths over these four octaves is from 5.2 meters down to 0.32 meters (roughly from 1 to 18 feet, in common units). The common obstacles we meet with fall within this range of size and do not reflect such sound to any great degree, so sound bends.

This bending is, of course, more likely for the deeper sounds than for the shriller ones. We judge the direction of a sound by the inequality of loudness in the two ears, automatically turning our head until both ears hear the sound with equal loudness. Our head is large enough to reflect, somewhat, a shrill sound coming from one side. There is then a considerable reduction in the intensity of that sound making its way around our head to the other ear. We have no trouble, therefore, locating a child by its shrill cry. On the other hand, the deep tones of the lower register of an organ move around our head with ease and sound equally intense in both ears. The sound seems to come from all around us; this in itself lends majesty to the swell of the organ.

The full range of the piano covers 7.5 octaves. The lowest note possesses a frequency of 27.5/sec and a wavelength of 12.5 meters. We can hear still deeper sounds; however, the usual extreme in that direction is 15/sec, a sound with a wavelength of 22 meters. The highest note of the piano possesses a frequency of 4224/sec and a wavelength of 0.081 meters, or 8.1 centimeters. The adult human ear can hear sounds with a frequency as high as 15,000/sec (wavelength, 2.2 centimeters), and the child can sometimes hear a frequency as high as 20,000/sec (wavelength, 1.7 centimeters). Such extremely shrill sounds will be reflected quite well by objects too small to reflect sounds in the more common range. The high-pitched creak of a cricket may be reflected so well by various objects that it is next to impossible to tell exactly where the original sound is coming from.

It is, of course, possible for objects to vibrate with frequencies of less than 15/sec and more than 20,000/sec; when this happens sound waves are produced which are not audible. Those that are too deep to be audible are *infrasonic waves* (from Latin words meaning "below sound"), while those that are too shrill to be audible are *ultrasonic waves* (from Latin words meaning "beyond sound").

Infrasonic waves are comparatively unimportant except where they become energetic enough to do physical damage, as in earthquakes. Ultrasonic waves impinge upon us more often and in many ways. For one thing, they are not inaudible to all forms of life; many animals smaller than ourselves can both produce and hear them. The "silent" whistles to which dogs respond produce ultrasonic waves that they can hear though we cannot. The singing canary produces ultrasonic waves that would undoubtedly greatly add to the beauty of the song if we could but hear them. The squeaking mouse also produces them, and the waiting cat can hear them where we cannot—which increases the efficiency of the feline stalk.

Ultrasonic sound waves, with wavelengths even shorter than those of the shrillest sounds we can hear, can be reflected efficiently by quite small objects. Bats take advantage of this fact. They emit a continuous series of ultrasonic squeaks while flying. These have frequencies of from 40,000 to 80,000/sec and, therefore, have wavelengths of from 8 to 4 millimeters. A twig or an insect will tend to reflect such short wavelengths; and the bat, whose squeaks are of extremely short duration, will catch the faint echo between squeaks. It can thus guide its flight by hearing alone and continue flying with perfect efficiency even if blinded. This process is called *echolocation.*

Men duplicate this effect by making use of beams of ultrasonic waves underwater. These are reflected from objects such as the sea bottom, jutting rocks above the sea bottom itself, schools of fish, or submarines. The technique is referred to as *sonar,* an abbreviated form of "*so*und *n*avigation *a*nd *r*anging" (where "ranging" means getting the range of an object—that is, determining its distance).

CHAPTER 13

Temperature

Hot and Cold

Heat has been mentioned several times in the book, notably toward the end of Chapter 7 in connection with the conservation of energy. I have not stopped, however, to consider it in detail, since to do so with proper understanding required first a consideration of the properties of fluids and, in particular, of gases. Enough of these properties have now been described to make it advisable to turn again to the subject of heat.

Heat is most familiar to us as a subjective sensation. We feel something to be "hot" or "cold," and we know what we mean when we say that one object is "hotter" than another. The degree of hotness or coldness of an object is called its *temperature*.

Temperature is of importance to physicists because a great many of the properties of matter with which he deals vary with temperature. In the previous chapter, for instance, I mentioned that the velocity of sound varied with temperature (see page 164). Again, the volume of a given mass of water increases as it is heated to near the boiling point, and so the density decreases. Hot water possesses weaker cohesive forces than does cold water, so the viscosity and surface tension decrease as temperature goes up. Even such seeming unchangeables as the length of an iron rod change with temperature

It follows then that if a physicist is to make proper generalizations concerning the universe, he must know just how the properties of matter change with temperature, and to do that he must be able to measure the temperature accurately. Our subjective feelings are insufficiently fine under the best conditions and are grossly inaccurate at times; therefore they will not do for the purpose. Thus, a polished metal surface exposed to the temperature of freezing water will feel much colder to the touch than a polished wooden surface exposed to the same conditions (for reasons to be discussed on page 225), even though both are at the same temperature. A well-known experiment produces an even greater paradox. If you place one hand in ice water and one in hot water and leave them there for a few moments, and then place both hands in the same container of lukewarm water, you will simultaneously feel the lukewarm water to be warm (with your cold hand) and cold (with your hot hand).

Some objective means of measuring temperature is therefore needed. The logical method is to find some property that changes in an apparently uniform manner with temperature and then associate fixed changes in temperature with fixed changes in that property. Physicists make use of a number of different temperature-dependent properties for the purpose, but the most commonly used property for the temperature range met with in ordinary life is that of volume-change. The volume of a given mass of matter generally increases with rising temperature and decreases with falling temperature. (I say "generally" because there are occasional exceptions to this.)

The change in volume with temperature is, in the case of liquids and solids, quite small and, in fact, unnoticeable to the eye. Thus, a steel rod a meter long will, if brought from the temperature of melting ice to that of boiling water, expand in length by one millimeter—that is, one part in a thousand. Since it is the volume that is expanding, the other dimensions will also increase by one part in a thousand, and if the steel rod has a circular cross section one centimeter in radius, that radius will increase by one-hundredth of a millimeter.

Such changes, while small, are by no means unimportant. Long metal girders such as those used in bridges, or long rails such as those used in railroad tracks, will expand and buckle in the hot summer sun if fixed at both ends. To avoid that, spaces are left between adjoining units so that there will be room to expand. Again, even a trifling change in the length of a clock's pendulum, will alter its period slightly, for that period depends upon its length (see page 112). The error in time-measurement, which depends upon that period, is cumulative and would make it necessary to adjust the clock periodically in summer, although it might run perfectly at cooler temperatures.

Not all substances expand by the same relative amounts when exposed to a given temperature change. An alloy of iron and nickel (in 5 to 3 ratio) will, for instance, only expand to one-tenth the extent that iron or steel will. For this reason, it is a useful alloy out of which to construct measuring tapes, rods of standard length, and so on. Because its length is more nearly invariable than that of most metals, the trade name for the alloy is Invar.

Glass expands with temperature almost to the same extent that steel will. If a glass vessel is exposed to a drastic temperature change, one portion of it may expand (or contract) while another portion, to which the temperature change has not yet penetrated, does not. The expansion or contraction may not be large in an absolute sense, but it is enough to set up internal strains which the cohesive forces of glass are not sufficient to withstand, so the glass cracks.

One way out of this dilemma is to use relatively thin glass so that if one portion is heated (or cooled) the temperature change will penetrate to other portions quickly.* Another and better way is to make use of a boron-containing variety of glass, usually known by the trade name of Pyrex, which will change in volume with a given temperature only by one-third the amount that ordinary glass will. It is therefore far more resistant to cracking under temperature change because smaller strains are set up. Its thickness (and mechanical strength) need not, therefore, be sacrificed to temperature stability. Quartz, with a still smaller tendency to change volume with temperature (less even than that of Invar), is still better for the purpose. A quartz vessel can be heated to red heat and plunged into ice water, and be undamaged by the ordeal.

However startling the effects of trifling changes in volume, those changes remain trifling in actual size. Unless they can somehow be magnified, they would be difficult to use as a measure of temperature. Fortunately, there are simple methods of magnifying small volume changes.

One method is to weld strips of two different metals together, say a strip of brass with one of iron. For a given change of temperature, a brass strip will change in volume (and, therefore, in

* It takes time for temperature change to make its way through glass because glass is a poor conductor (see page 224) of heat.

length) to nearly twice the extent that an iron strip of the same size will. If the two strips of metal were not welded together, the brass would expand under the influence of increasing temperature and slide against the iron, becoming a trifle the longer of the two although they had been equal in length to start with. If the temperature is reduced below the starting point, matters are reversed. Now the brass contracts more than the iron does, sliding against the iron and ending a trifle the shorter of the two.

However, the two strips of metal *are* welded together and the brass cannot slide against the iron. What happens then is that the welded strips (a *bimetallic strip* or a *compound bar*) bend in the direction of the iron if it is heated. The brass would then lie along the outer rim of the curve and the iron along the inner. Since the outer rim is longer than the inner, this allows the brass to be longer than the iron while remaining welded throughout. As the temperature falls again, the curve straightens and becomes entirely straight when the temperature returns to its original value. If the temperature falls lower still, the bimetallic strip bends in the direction of the brass, which now lies on the inner rim while the iron lies on the outer rim.

If such a bimetallic strip is fixed at one end, the other end sways back and forth as temperature changes. The outer rim of the curved strip is very little longer than the inner, and the difference in length increases but slowly with the degree of bending. For that reason, even small changes in temperature, producing very small differences in length between the iron and brass, nevertheless produce a considerable amount of bending.

A device of this sort can be used as a *thermostat*. As the temperature in the house falls, the bimetallic strip begins to curve to the left, let us say, and at a certain temperature the bending is sufficient for it to close an electrical contact that turns on the furnace. As the house heats up, the bimetallic strip bends back and quickly breaks the contact, thereby turning the furnace off. By altering the position of the electrical contact (easily done by hand), we can arrange to have the bimetallic strip turn the furnace on at any temperature we please.

Then, too, the exact position of the free end of such a bimetallic strip can be used as a measure of the temperature. If a pen is attached and a circle of paper is allowed to revolve under it at a fixed speed, the device will automatically make a continuous recording of its position, and temperatures can be deduced from the position of the wavering line.

Again, pendulums can be designed in which the rod is not

a single piece of metal but several strips of two different metals—say, steel and zinc. These can be joined by horizontal bars in such a way that the temperature change in the zinc tends to lengthen the pendulum while the temperature change in the iron tends to shorten it. The combined action tends to leave the pendulum unaltered in length as temperature changes. This is a *compensation pendulum.*

Temperature Scales

A much more common method of magnifying volume change for the purpose of measuring temperature is to make use of liquids rather than solids. Imagine an evacuated spherical container with a long, narrow tube of constant width extending upward. The container holds enough liquid to fill the sphere completely, but the neck remains empty and includes only vacuum. If the liquid is warmed, its volume will increase and there will be no place for the liquid to expand into but the neck. The volume of the water rising into the cylindrical neck can be expressed by the usual formula for the volume of a cylinder, $V = \pi r^2 h$, where r is the radius of the cylindrical neck and h is the height to which the water rises. For a given volume, the smaller the value of the radius of the neck, the greater the height to which the liquid must rise. It follows that even though the additional volume of liquid (due to expansion with temperature) is very small, the change in height can be made quite sizable if only the radius of the tube is made small enough. There is no difficulty in using changes in the height to measure changes in temperature.

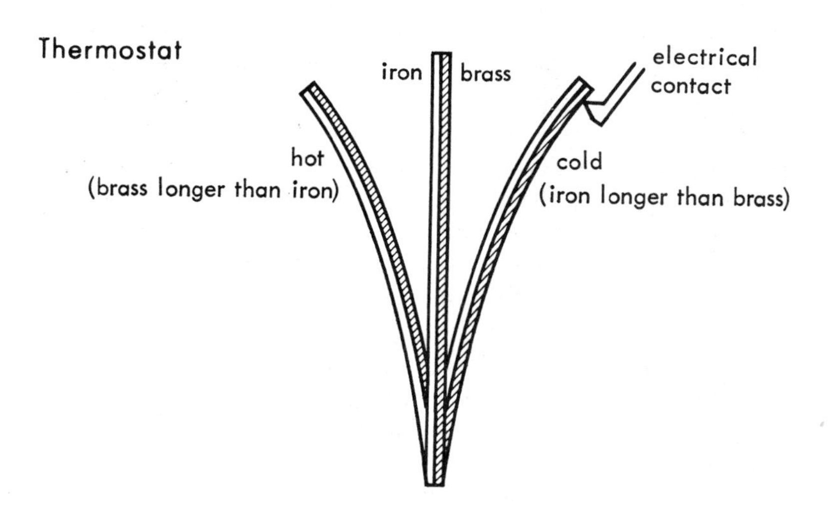

A device making use of a sizable reservoir of fluid and a narrow tube into which that fluid can expand is the most common form of *thermometer* (from Greek words meaning "heat-measure"). Such thermometers were first devised in the seventeenth century and a variety of fluids were used. Water, naturally, was one of the first. Unfortunately, water does not expand uniformly with temperature. In fact, it reaches a point of maximum density and minimum volume at a temperature somewhat above its freezing point. If the temperature is dropped further, water actually expands as the temperature is lowered until it freezes (and in that process expands still further, for ice is less dense than water by nearly ten percent). Furthermore, water remains liquid over a comparatively small temperature range and is useless for temperatures below its freezing point or above its boiling point. Alcohol, also used in thermometers, stays liquid at temperatures far below that at which water freezes; however, it boils at a temperature even lower than that at which water boils.

Furthermore, both water and alcohol wet glass. As the height of liquid sinks with the falling temperature, some remains behind, clings to the glass, and then slowly trickles down. The level of liquid may then be observed to rise slowly, and it would be difficult to decide whether this was because temperature was going up or iiquid was trickling down.

The first to make use of mercury as a thermometric fluid was a German physicist named Gabriel Daniel Fahrenheit (1686–1736), and he did this in 1714. Mercury's freezing point is considerably below that of water and its boiling point is considerably higher; in addition, it expands quite smoothly with temperature change. (One way of judging this is by noting that temperature changes as measured by the change in volume of mercury agree closely with temperature changes as measured by changes in a variety of other properties of matter. It is more reasonable to assume that all these changes are regular than that they all happen to be irregular in just the same way.) Finally, mercury does not wet glass, and the height of the mercury column would not be affected by delayed trickling.

Once the course of temperature change is made easily visible by the rise and fall of the mercury thread in the narrow tube of the thermometer, it is next necessary to associate some definite numerical values with fixed positions.

At atmospheric pressure, for instance, ice melts at a particular temperature and there seems every reason to believe that this temperature is the same in all places and at all times. (At least, there is no reason to believe the contrary.) Similarly, water always boils at a particular temperature at atmospheric pressure in all places and at all times. If a thermometer is placed in melting ice, the level to which the mercury thread rises can be marked; if it is placed in boiling water, a new level is marked. All men would have marks on their own thermometers that were comparable; all thermometers would match and "speak the same language." Once two such marks were set, the distance between could be divided into equal steps, or *degrees.*

Fahrenheit's method of setting his fixed point unfortunately did not involve the freezing point and boiling point of water directly. For his zero point, he used a mixture of ice and salt that produced the lowest freezing point he could get, and for another point, he tried to use the temperature of the human body. He ended by associating the freezing point of pure water with the number 32 and the boiling point of water with 212. (These figures are separated by 180 degrees, you see.) This is the *Fahrenheit scale,* and measurements upon it are given in "degrees Fahrenheit," abbreviated as "°F." Thus, the freezing point of water is 32°F, and the boiling point of water is 212°F. Body temperature is set at 98.6°F, and something like 70°F is considered a comfortable room temperature.

Temperatures below the 0°F mark can be said to be so many "degrees below zero," or a minus sign can be used. Thus alcohol freezes at 179 degrees below zero, Fahrenheit, or at —179°F.

In 1742, the Swedish astronomer Anders Celsius (1701–1744) made use of a different scale, one in which the freezing point of water was associated with 0 and the boiling point with 100, these points therefore being separated by a hundred degrees. This scale is called the *Centigrade scale* (from Latin words meaning "hundred degrees"), but in the 1950's it was decided to honor the inventor by calling it the *Celsius scale.* Whether one speaks of "degrees Centigrade" or "degrees Celsius," however, the abbreviation is "°C." On the Celsius scale, the melting point of ice, or the freezing point of water, is 0°C, and the boiling point of water is 100°C.

The Celsius scale commends itself to scientists because the 0 to 100 stretch fits in with the decimal nature of the metric system and because that stretch of temperature over which water remains liquid is a particularly interesting one, especially to chemists. It is the Celsius scale that is used universally by scientists.

The Fahrenheit scale, however, is used in ordinary affairs in the United States and Great Britain, and it has at least this advantage: the 0 to 100 stretch in the Fahrenheit scale covers the usual range of temperatures in the world. Meterologists using the Celsius scale must frequently descend to negative numbers; those using the Fahrenheit scale need do so but rarely.

Since both scales are used in the United States and Great Britain, it is useful to be able to convert one to the other. Let us begin by noting that in the range between the freezing point and the boiling point of water there are 180 Fahrenheit degrees and 100 Celsius degrees. The Celsius degree is obviously the larger of the two and is equal to 180/100, or 9/5 Fahrenheit degrees. Conversely, a Fahrenheit degree is equal to 100/180, or 5/9 Celsius degrees.

That would be enough if the two scales had their zero point in common, but they don't. If the Celsius reading is multiplied by 9/5, the result is the number of Fahrenheit degrees, not above 0°F, but above 32°F (for 32°F is the equivalent of 0°C). For that reason 32° must be added to the result. In other words:

$$F = \frac{9}{5}C + 32 \qquad \text{(Equation 13–1)}$$

To obtain the reading on the Celsius scale when the Fahrenheit temperature is given, it is only necessary to solve Equation 13–1 for C, and the answer is:

$$C = \frac{5}{9}(F - 32) \qquad \text{(Equation 13–2)}$$

Expansion

Once temperature can be measured with precision, it becomes possible to express temperature-dependent changes accurately. We can decide how much change there is "per degree Celsius" (a phrase that can be abbreviated as "per °C" or as "/°C."

For instance, we can measure the changing length of a rod and determine the increase of length brought about by a definite temperature change. We can then calculate what the relative increase in length is for a rod that has undergone a temperature rise of 1°C. This increase is the *coefficient of linear expansion.*

The size of the coefficient of linear expansion varies from substance to substance, but for solids it is always quite small. For steel, for instance, it is 0.00001/°C, or, expressed in exponential form: 1×10^{-5}/°C. This means that a one-meter rod will expand by 0.00001 meters when temperature goes up 1°C, a one-kilometer rod will expand by 0.00001 kilometers, a one-centimeter rod will expand by 0.00001 centimeters, and so on. (Some other coefficients of linear expansion are 1.9×10^{-5}/°C for brass, 2.6×10^{-5}/°C for aluminum, and only 0.04×10^{-5}/°C for quartz.)

Suppose we represent the coefficient of linear expansion by the Greek letter "alpha" (α). If we start with a rod exactly one meter long at a particular temperature and raise that temperature by 1°C, then the length increases by α meters, and the total length is $1 + \alpha$ meters. If we raise the temperature by 2°C, the expansion is twice as great, so the total length now becomes $1 + 2\alpha$, while for a temperature rise of 3°C it is $1 + 3\alpha$. In short, the value of α is multiplied by the number of degrees by which the temperature is changed.

It is customary in physics and in mathematics to signify a change in a value by the capital form of the Greek letter "delta" (Δ). If we let temperature be symbolized as t, then a temperature change is written Δt, and is usually read "delta t." In other words, we can consider the length of a one-meter rod after a certain rise in temperature to be $1 + \alpha(\Delta t)$.

Naturally, if temperature is allowed to fall instead of rise, Δt is negative and so is $\alpha(\Delta t)$. The expression $1 + \alpha(\Delta t)$ is then smaller than 1, which is reasonable, since with falling temperature the rod contracts.

Suppose, now, that we started with a two-meter rod. We can consider it as consisting of 2 one-meter rods fused together. Each one-meter half has a total length of $1 + \alpha(\Delta t)$ after the temperature has changed, and the total length is therefore $2[1 + \alpha(\Delta t)]$. This can be reasoned similarly for any length. In fact, if we call the length of a rod L, then the new length after a change in temperature is $L[1 + \alpha(\Delta t)]$ or, multiplying this out, $L + L\alpha(\Delta t)$.

We can next ask ourselves what the change in length is as a result of the change in temperature. The change in length, which we can naturally symbolize as ΔL, would be the length after the temperature change minus the original length. This would be $L + L\alpha(\Delta t) - L$, so we can conclude:

$$\Delta L = L\alpha(\Delta t) \qquad \text{(Equation 13–3)}$$

A substance expanding with rise in temperature expands in all directions and not in length only, and the change in volume is often more important than the change in length. In liquids and gases, particularly, it is the expansion in volume that is measured. In solids, however (especially when in the shape of long rods), it is far simpler to measure the linear expansion and calculate the volume expansion from that.

We can begin by assuming that the coefficient of linear expansion for a given substance has the same value for width and height as for length.* Suppose we start with a cubic meter of a substance. Its length, after a 1°C rise in temperature becomes $1 + \alpha$ meters. Its width, however, also expands to $1 + \alpha$ meters, and its height, too. It's volume, which began as 1^3 cubic meters ($1^3 = 1$, of course) is now $(1 + \alpha)^3$ cubic meters. The change in volume with a 1°C rise in temperature is $(1 + \alpha)^3 - 1^3$, or $(1 + \alpha)^3 - 1$, and that is the *coefficient of cubical expansion.*

The quantity $(1 + \alpha)^3$ can be expanded by ordinary algebra to $1 + 3\alpha + 3\alpha^2 + \alpha^3$. We subtract 1 from this and find that the coefficient of cubical expansion is $3\alpha + 3\alpha^2 + \alpha^3$. Where α is very small, as it is in the case of solids and liquids, α^2 and α^3 are much smaller still† and can be ignored as not contributing a significant quantity to the expression. If we throw out the square and cube, then we can say with quite sufficient accuracy that the coefficient of cubical expansion is 3α—three times the coefficient of linear expansion. Thus, if the coefficient of linear expansion is 1×10^{-5}/°C for steel, then we can say that its coefficient of cubical expansion is 3×10^{-5}/°C.

The coefficient of cubical expansion is roughly ten times as high for liquids as for solids, and considerably higher still for gases. It is, indeed, for gases that the coefficient of cubical expansion has proved to have the greatest theoretical significance.

* This is not necessarily strictly true. A single crystal may expand by different amounts in different directions, depending on the orderly arrangement of the atoms and molecules making it up. A crystal may, in this respect and many others, have properties that vary with direction. In these respects, it is *anisotropic*. Common substances about us, however, are often not crystalline or, if they are, are composed of myriads of tiny crystals facing every which way. On the average, then, properties would be the same in every direction, and the substance would be *isotropic*. We tend to think of substances generally as isotropic because this is the less complicated view, but anisotropy is not really a rare phenomenon. We all know that it is much easier to split a wooden plank with the grain than against the grain.

† This may not be at once obvious. If a number is larger than 1, then the square and cube are larger still. The greater the number, the more magnified are the square and cube. Thus, the square of 10 is 100 and the cube is 1000, while the square of 100 is 10,000 and the cube is 1,000,000. The situation is reversed for numbers less than one. Here the square and cube are smaller still, and the smaller the original number the greater is the shrinkage in square and cube. Thus the square of 1/10 is 1/100 and the cube is 1/1000. For a figure like 1/100,000, which is the coefficient of linear expansion for steel, the square is 1/10,000,000,000 and the cube is 1/1,000,000,000,000,000.

Galileo himself realized that gases expand with rising temperature and contract with falling temperature; he even tried to construct a thermometer based on this fact. Taking a warmed glass bulb with an upward stalk, open at the top, he upended it in a trough of water. As the glass bulb cooled, the gas within it contracted and water was drawn part way up the stalk. Later, if the temperature went up, the gas within the bulb expanded, pushing the water level in the stalk downward. If the temperature went down, the water level rose. Unfortunately for Galileo, the water level in the stalk was also affected by changes in air pressure, so the thermometer was not an accurate one. However, the principle of change in gas volume with change in temperature was established.

Since this is so, then a volume of gas trapped under a column of mercury (as in Boyle's experiments) would expand if heated or contract if cooled. This means that if one were studying the manner in which the volume of gas changed with changes in pressure, one would have to be sure to keep the gas at constant temperature. Otherwise changes in volume would take place for which pressure was not responsible. Boyle himself in formulating what we call Boyle's law did not, apparently, take note of this fact. In 1676, however, a decade and a half after Boyle's experiments, a French physicist, Edme Mariotte (1620?–1684), discovered Boyle's law independently, and he did draw attention to the importance of constant temperature. For this reason, on the European continent the relationship of pressure and volume is often called *Mariotte's law* rather than Boyle's law—and with some justice.

The first attempt to study the expansion of gases with temperature change, quantitatively, was in 1699. The French physicist Guillaume Amontons (1663–1705) showed that if the gas were penned in and prevented from expanding as the temperature rose, the pressure increased instead, and that the pressure increased by a fixed amount for a given temperature rise regardless of the mass of gas involved.

Amontons, however, could work only with air, for in his time air was the only gas readily available. All through the eighteenth century, however, a number of gases were produced, distinguished among, and studied. In 1802, the French chemist Joseph Louis Gay-Lussac (1778–1850) not only determined the coefficient of cubical expansion for air but showed that the various common gases such as oxygen, nitrogen and hydrogen all had just about the same coefficient of cubical expansion. (This is quite astonishing, since the coefficient of cubical expansion varies quite a bit from one solid to another and from one liquid to another. Thus, the coefficient of cubical expansion is 77 times as great for aluminum as for quartz and 6 times as great for methyl alcohol as for mercury.)

The coefficient of cubical expansion for gases at 0°C turns out to be 0.00366, or about 300 times the coefficient of cubical expansion for the average solid. We can adapt Equation 13–3 for the expansion of gases. We will substitute volume (V) for length and the coefficient of cubical expansion (0.00366, or 1/273) for the coefficient of linear expansion. If we do this, then for the change in volume of gases (ΔV) with change in temperature from 0°C (Δt), we can write:

$$\Delta V = 0.00366V(\Delta t) = \frac{V(\Delta t)}{273} \qquad \text{(Equation 13–4)}$$

This is one way of expressing *Gay-Lussac's law*. As it happens, the French physicist Jacques Alexandre César Charles (1746–1823) claimed to have reached Gay-Lussac's conclusions as early as 1787. He did not publish them either then or later, and ordinarily a discovery does not count unless it is published. Nevertheless, the relationship is frequently called *Charles's law* because of this.

Absolute Temperature

The fact that objects expand and contract with temperature change raises an interesting point. It is easy to see that an object can expand indefinitely as temperature goes up, but can an object contract indefinitely as temperature goes down? If it continues to contract at a steady rate, will it not eventually contract to zero volume? What then?

The paradox is most acute in the case of gases, which contract more rapidly with falling temperature than do liquids or solids. The volume of a gas after a certain change in temperature from 0°C is the original volume at 0°C plus the change in volume ($V + \Delta V$).

Suppose then that the temperature were to drop 273 degrees below 0°C. In that case, Δt would be -273. From Equation 13–4, we would see that ΔV, in that case, would be equal to $V(-273)/273$, or to $-V$. The new volume ($V + \Delta V$) would become $V - V$, or 0. A strict application of Gay-Lussac's law would indicate that gases would reach zero volume and vanish at −273°C.

Physicists did not panic at this possibility. It seemed quite likely that before −273°C was reached, all gases would be converted to liquid form, and for liquids the coefficient of cubical expansion would then be much smaller. (This turned out to be true.) Even if this were not so, it seemed quite likely that Gay-Lussac's law might not apply strictly at very low temperatures* and that the coefficient of cubical expansion might gradually decrease as temperature dropped, so although volume continued to shrink, it would do so at a slower and slower rate and never reach zero.

Nevertheless, the temperature −273°C was not forgotten. In 1848, William Thomson (later raised to the rank of baron and the title of Lord Kelvin) pointed out the convenience of supposing that −273°C might represent the lowest possible temperature, an *absolute zero.*†

If we let −273°C be zero and count upward from that by Celsius degrees, we would have an *absolute scale* of temperature. Readings on this scale would constitute an *absolute temperature,* and the degrees given in such a reading could be indicated as °A (for "absolute") or, more often, as °K (for Kelvin).

To change a Celsius temperature to one on the absolute scale, it is therefore only necessary to add 273. Since water freezes at 0°C, it does so at 273°K; since it boils at 100°C, it does so at 373°K. To prevent confusion, it is customary to represent temperature readings on the Celsius scale by the symbol t, and temperature readings on the Kelvin scale by the symbol T.** We can write the relationship of the Kelvin scale to the Celsius scale as follows, therefore:

$$T = t + 273 \qquad \text{(Equation 13–5)}$$

* It is important to remember that many scientific generalizations hold true only over limited ranges of pressure, temperature and other such environmental factors. This does not affect the usefulness of the generalization within the proper range, but one must not expect them to be useful outside that range.

† The actual value, according to the best modern determinations, is −273.16°C.

** Confusion cannot be done away with altogether. Thus, t stands not only for Celsius temperature but also, very commonly, for time. Every letter of the Latin and Greek alphabet—and some from Hebrew, Sanskrit and others—in small form, capital form, italics, boldface, and gothic script has been used, and even so there are numerous duplications of symbols. For that reason, in presenting any equation it is always advisable to state the significance of each symbol and never to take it for granted that the meaning of any symbol is self-evident.

The convenience of the absolute scale rests on the fact that certain physical relationships can be expressed more simply by using T rather than t. Thus, suppose we try to express the manner in which the volume of a quantity of gas varies with temperature. We can start at a temperature t_1, with a gas at volume V_1, and when the temperature has changed to t_2 we will find that the gas volume has changed to V_2. The final volume will be the original volume plus the volume change, so $V_2 = V_1 + \Delta V$.

Using Equation 13–4, we see that $\Delta V = V_1(\Delta t)/273$. However, the change in temperature (Δt) is the difference between the final temperature and the original temperature, $t_2 - t_1$. The unit of cubical expansion for gases is determined for a starting temperature of 0°C so $t_2 - t_1$ becomes $t_2 - 0$, or simply t_2. We will therefore substitute t_2 for the Δt in Equation 13–4. Then, in writing $V_2 = V_1 + \Delta V$, we will have:

$$V_2 = V_1 + \frac{V_1 t_2}{273} = V_1\left(1 + \frac{t_2}{273}\right) \qquad \text{(Equation 13–6)}$$

This is easily converted to:

$$\frac{V_2}{V_1} = \frac{273 + t_2}{273} \qquad \text{(Equation 13–7)}$$

Let's consider now what the significance of the number 273 might be. It enters this equation because 1/273 is the coefficient of cubical expansion for a gas at 0°C. Remember, however, that the unit of the coefficient of cubical expansion is "per °C" or "/°C." The number 273 is the reciprocal of that coefficient, and its units should be the reciprocal of the units of the coefficient. The reciprocal of "/°C" is "°C."*

For 273, then, in Equation 13–7, read 273 Celsius degrees. But (see Equation 13–5) adding 273 Celsius degrees to a temperature reading on the Celsius scale gives the reading on the Kelvin scale. Consequently, the final temperature of the gas (t_2) plus 273 is the final temperature on the Kelvin scale; or, in short, $t_2 + 273 = T_2$. Similarly, 273 Celsius degrees represents the freezing point of water on the Kelvin scale, since $0 + 273 = 273$. The initial temperature of the gas was 0°C, so we can let 273 represent T_1, the initial temperature on the Kelvin scale. Consequently, Equation 13–7 becomes:

* Just as a reminder . . . The reciprocal of a is $1/a$, and the reciprocal of $1/a$ is a.

$$\frac{V_2}{V_1} = \frac{T_2}{T_1} \qquad \text{(Equation 13–8)}$$

This is another way of expressing Gay-Lussac's law (or Charles's law), and just about the simplest way. Using any other scale of temperature, the expression would become more complicated. In words, Equation 13–8 can be expressed: *The volume of a given mass of gas is directly proportional to its absolute temperature, provided the pressure on the gas is held constant.*

That last clause is important, because if the pressure on the gas varies, then the volume of the gas will change even though the temperature does not.

So we have Boyle's law which relates volume to pressure, provided temperature is held constant, and now we have Gay-Lussac's law which relates volume to temperature, provided pressure is held constant. Is there any way of relating volume to temperature *and* pressure? In other words, suppose we begin with a quantity of gas with volume V_1, pressure P_1 and temperature T_1, and change both pressure *and* temperature to P_2 and T_2. What will the new volume V_2 be?

Let's begin by changing the pressure P_1 to P_2 while holding the temperature at T_1. With the temperature constant, Boyle's law (see page 145) requires that the new volume (V_x) must fit into the following relationship: $P_2V_x = P_1V_1$. If we solve for V_x, we get:

$$V_x = \frac{P_1V_1}{P_2} \qquad \text{(Equation 13–9)}$$

But V_x is not the final volume we are looking for. It is merely the volume we attain if we alter the pressure. Now let's keep the pressure at the level we have reached, P_2, and raise the temperature from T_1 to T_2. The volume now changes a second time, from V_x to V_2. (The latter is the volume we expect to have when pressure has reached P_2 and temperature has reached T_2.) In going from V_x to V_2, by raising the temperature from T_1 to T_2 and keeping the pressure constant, Gay-Lussac's law must hold, so $V_2/V_x = T_2/T_1$ (see Equation 13–8). By substituting for V_x, the value given in Equation 13–9, we have the relationship:

$$\frac{V_2}{(P_1V_1)/P_2} = \frac{T_2}{T_1} \qquad \text{(Equation 13–10)}$$

This can be rearranged by the ordinary techniques of algebra to:

$$\frac{P_2V_2}{T_2} = \frac{P_1V_1}{T_1} \qquad \text{(Equation 13–11)}$$

We can summarize then by saying that for any given quantity of gas the volume times the pressure divided by the absolute temperature remains constant. The constant here is usually symbolized as R, so we can say: $(PV)/T = R$, or:

$$PV = RT \qquad \text{(Equation 13–12)}$$

Actual measurement, however, shows that Equation 13–12 does not hold exactly for gases (for reasons I shall explain later; see page 208). It would hold under certain ideal conditions that actual gases do not fulfill (though some come pretty close to doing so), and one can imagine an *ideal gas* or *perfect gas* that, if it existed, would follow the relationship shown in Equation 13–12 exactly. For that reason, Equation 13–12 (or its equivalent, Equation 13–11) is called the *ideal gas equation*.

CHAPTER 14

Heat

The Kinetic Theory of Gases

If the atomic theory of gas structure (made inevitable by Boyle's experiments) is to be accepted, it ought to explain the gas laws described in the previous chapter and earlier. The first man to attempt this seriously was Bernoulli (of Bernoulli's principle) in 1738.

If gases are composed of separate particles (atoms or molecules) spaced widely apart, it may reasonably be assumed that these are in constant free motion. If this were not so and the gas molecules were motionless, they would, under the force of gravity, fall to the bottom of a container and remain there. This is indeed the case for liquids and solids, where the atoms do not move freely but are in virtual contact and are constrained to remain so. The assumption that gases are made up of particles in motion, each particle virtually uninfluenced by the presence of the others, is the *kinetic theory of gases* ("kinetic," of course, from a Greek word meaning "to move").

For the moment we will not ask why the particles should be moving but will merely accept the fact that they are. The kinetic energy of the gas particles must far surpass the feeble gravitational force that can be exerted by the earth on so small a particle. (Remember that the force of gravity upon the particle depends in part upon the mass of the earth multiplied by the mass of the particle (see page 44), and the latter is so small a quantity that the total force is minute.)

To be sure, the pull of gravity is not zero and on a large scale it is effective. The earth's atmosphere remains bound to the planet by gravitational force, and most of the particles of the gas surrounding our planet remain within a few miles of the surface. Only thin wisps of gas manage to make their way higher. Nevertheless, for small quantities of gas, for quantities small enough to be contained within man-made structures, the effects of gravity are minute enough to ignore. Consequently, the particles within such containers can be viewed as moving with equal ease in any direction, upward and sideways as easily as downward.

In any given container, the random motion of the particles in any direction keeps the gas evenly spread out. (The even spreading of the gas within the container is enough to show that the motion must be random. If it were not, gas would accumulate in one part or another of the container.) If the same quantity of gas is transferred to a larger container, the random motion of the particles will spread them out evenly within the more spacious confines. Thus a gas expands to fill its container, however large, and (unless the container is so huge that the effect of gravity can no longer be ignored) fills it evenly. On the other hand, if the gaseous contents of a large container are forced into a smaller one, the particles move more closely together and all the gas can be made to fit into the smaller confine. There is none left over.

If we consider the moving gas particles, however, it is clear that no individual particle can move for long without interference. One particle is bound to collide with another sooner or later, and all are bound to collide every now and then with the walls of the container. One must assume these particles have perfect elasticity and bounce without overall loss of energy. If this were not so, the particles would gradually slow and lose energy as they bounced, until finally they were brought to a state of rest or near-rest and fell to the bottom of the container under the pull of gravity. But this does not happen. If we isolate a container of gas as best we can and keep it, that container remains full of gas indefinitely.

Bernoulli pointed out that the bouncing of gas particles off the wall of a container produced an effect that could be interpreted as pressure. As it bounced, each particle subjected the wall to a tiny force, and the total force over a unit area was the pressure. Strictly speaking, what we call pressure then is actually a great many separate pushes. There are so many of these spread so thickly through time, and each separate push is so tiny, that the whole is sensed as a smooth, even pressure. Since the particles move freely and randomly in all directions, pressure is equal in all directions.

Suppose that a gas is in a container topped by a frictionless piston with just enough weights resting upon it to balance the gas pressure (the force of the particles bouncing against the undersurface of the piston). If one of those weights is removed, the external force pressing down upon the upper surface of the piston is decreased. The upward force of the bouncing particles is greater than the downward force that remains, and the piston moves upward.

However, as the piston moves upward, the volume of the container increases. As the volume increases, each particle of the gas has, on the average, a greater distance to travel in order to reach the underside of the piston. Naturally, then, the number of collisions against the wall in any given instant must drop off as each particle spends more time traveling and less time colliding. The pressure decreases in consequence. Eventually, the pressure drops to the point where it is balanced by the fewer weights on the piston, and the piston rises no more. Gas volume has increased as pressure decreased in the manner described by Boyle's law.

Suppose, instead, that additional weights had been added to those originally present on the piston. Now the downward force of gravity moves the piston downward against the force of the particle collisions. As the piston moves downward, the volume decreases. Each particle has, on the average, a smaller distance to travel in order to reach the underside of the piston. The number of collisions in any given instant rises and pressure increases. Eventually, the pressure increases to the point where the additional weight on the piston is balanced. Gas volume has decreased and pressure has increased, again in the manner described by Boyle's law.

A century after Bernoulli's time, when the effect of temperature on the volume and pressure of gases came to be better understood, it was necessary to expand the kinetic theory of gases in order to explain the involvement of temperature.

Imagine a gas in a closed container with immovable walls. If the temperature of the gas is raised, its pressure against the walls increases. This was first observed by Amontons (see page 191) and is to be expected from the ideal gas equation (Equation

13–12), for if volume times pressure is proportional to absolute temperature, and if the volume is held constant, then pressure by itself must be proportional to the absolute temperature.

By kinetic theory, pressure increases only if the number of collisions of gas particles with the walls in any given instant increases. However, since the volume of the container (with its immovable walls) has not changed, the individual particles have the same distance to travel before reaching the walls, after the temperature rise and before. To account for the fact that more of them do reach the walls, and hence raise the pressure, one must conclude that as the temperature rises the particles move more quickly. In that case, they not only strike the wall oftener, but also more energetically. Conversely, with a fall in temperature they move more slowly.

Accepting this, let us consider a sample of gas held under a frictionless, weighted piston. The downward force of the weights is balanced by the upward force of the gas pressure. If the temperature of the gas is raised, the particles making it up move more quickly and their collisions with the underside of the piston are more numerous and energetic. The downward force of the weights is overbalanced, and the piston is raised until the expansion of volume increases the distance that must be traveled by the particles to the point where the number of collisions is so far reduced as to be only sufficient to balance the piston once more. Thus, volume increases with rising temperature. By similar reasoning, we could argue that it would decrease with falling temperature, and thus Gay-Lussac's law is explained.

I have shown how kinetic theory explains the gas laws only in a qualitative manner. In the 1860's, however, the Scottish physicist James Clerk Maxwell (1831–1879) and the Austrian physicist Ludwig Boltzmann (1844–1906) treated the kinetic theory with full mathematical rigor and established it firmly. We can consider some of this.

Let us begin with a container in the form of a parallelepiped

Kinetic theory

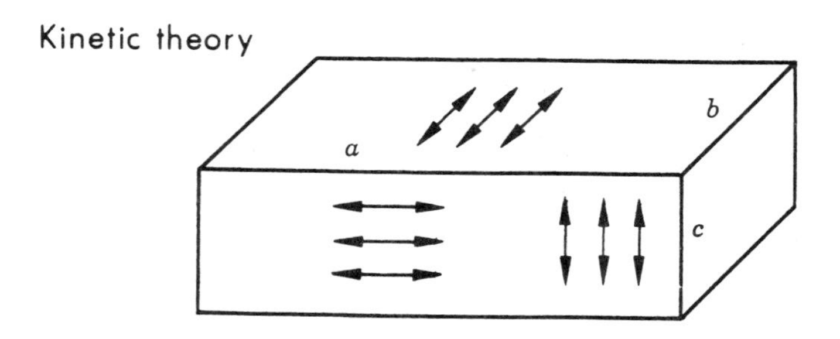

(brick-shaped, in other words), with a length equal to a meters, a width of b meters, and a height of c meters. The volume (V) of the container is equal to abc cubic meters. Suppose next that within this container are a number (N) of particles each with a mass (m), and that all the particles are moving with a velocity of v meters per second.

These particles can be moving in any direction, but such motion can always be viewed as being made up of three components at right angles to each other. (This can be done by setting up a "parallelepiped of force," which is a three-dimensional analog of the parallelogram of force mentioned on page 40). We can arrange the mutually perpendicular components to suit ourselves, and we can select one component parallel with the length of the container, another parallel with the width, and the third parallel with the height.

Since the motions are random and there is no net motion in any one direction (or the whole container would go flying off into space), it is fair to assume that each component contains an equal share of the motion. We suppose then that 1/3 of the total particle motion is parallel to the edge a, 1/3 parallel to the edge b, and 1/3 parallel to the edge c. This means that we are viewing the container of gas as containing three equal streams of particles, one moving left and right in equal amounts, one moving up and down in equal amounts, one moving back and forth in equal amounts.

In reality, of course, all the particles are continually colliding with each other, and bouncing and changing direction. Since the particles are perfectly elastic, this doesn't change the total motion, even though the distribution of motion among the individual particles is constantly changing. To put it as simply as possible, if one particle changes direction in one fashion, another particle changes simultaneously in such a way as to balance the first change. For this reason we can ignore inter-particle collisions.

Let us focus our attention on one particle moving parallel to edge a. It strikes the face bounded by b and c head-on and bounces back at the same speed but in the opposite direction (still parallel to edge a), so its velocity is now $-v$. Its momentum before the collision was mv, and its momentum after the collision is $-mv$. The total change in momentum is $mv-(-mv)$, or $2mv$.

This change in momentum must be balanced by an opposite change in momentum on the part of the wall of the container if the law of conservation of momentum is to be conserved. The wall is therefore pushed in the direction opposite to the rebounded particle, and $2mv$ represents the contribution of that one bounce of that one particle to the force on the face bounded by b and c. For the total force on the face, we need to know how many bounces there are on the entire face in a given unit of time.

The single particle we have been considering, having bounced off the face, travels to the other end of the container, bounces there, comes back, and bounces off the original face a second time; it repeats the process and bounces off a third time, then a fourth time, and so on. In traveling to the other end of the container and back, it travels a distance of $2a$ meters. Since its velocity is v meters per second, the number of its collisions with the face under discussion is $v/2a$ times each second.

The total force delivered to the wall by a single particle in one second is the momentum change in one bounce times the number of bounces per second. This is $2mv$ multiplied by $v/2a$, or mv^2/a. But one third of all the particles in the container $(N/3)$ are moving parallel to edge a and each contributes the same force. The total force delivered in one second by all those particles is therefore $N/3$ multiplied by mv^2/a, or $Nmv^2/3a$.

Pressure is the force exerted against a unit area. The wall we are considering is bounded by lines of dimensions b meters and c meters, so the area of the wall is bc square meters. To get the pressure—that is, force per square meter—one must divide the total force on the wall by the number of square meters. This means we must divide $Nmv^2/3a$ by bc, and we get a pressure equal to $Nmv^2/3abc$. But abc is equal to the volume (V) of the container. We can therefore express the pressure (P) as follows:

$$P = \frac{Nmv^2}{3V} = \frac{N}{3V}(mv^2) = \frac{2N}{3V}\left(\frac{1}{2}mv^2\right) \qquad \text{(Equation 14–1)}$$

But the quantity $\frac{1}{2}mv^2$ represents kinetic energy (e_k) (see page 95). We can therefore rearrange Equation 14–1 as follows:

$$PV = \frac{2N}{3}\left(\frac{1}{2}mv^2\right) = \frac{2N}{3}e_k \qquad \text{(Equation 14–2)}$$

In any given quantity of gas the number of particles is constant, therefore the quantity $2N/3$ is constant. Equation 14–2 tells us, therefore, that for a given sample of gas, the product of its pressure and volume is directly proportional to the kinetic energy of its constituent particles. Equation 13–12 (see page 196) tells us, furthermore, that the product of the pressure and volume of a gas is directly proportional to its absolute temperature.

It is a truism that if x is directly proportional to y and is also directly proportional to z, then y is directly proportional to z. We conclude that if PV is directly proportional both to absolute temperature and to the kinetic energy of the gas particles, then absolute temperature is itself directly proportional to the kinetic energy of the particles of a gas (and, by extension, to the particles of any substance).

To be sure, we have assumed that all the particles in the gas have identical velocities, and that is not so. As the particles collide with each other, momentum will be transferred in a random manner (though the total momentum will always be the same). Briefly, even if the particles had originally been moving at equal velocities, they would soon be moving over a whole range of velocities.

Maxwell derived an equation that would express the distribution of particle velocities at various temperatures. If there is a distribution of velocities, there is also a distribution of kinetic energies. If we know the average velocity, however, and this can be obtained from Maxwell's equation, we know the average kinetic energy.* At any temperature, there will be individual particles with very low energies and others with very high energies. The average kinetic energy per particle, however, keeps precisely in step with the rise and fall of absolute temperature.

By the kinetic theory of gases, then, we can define heat as the internal energy associated with such phenomena as the random motions of the particles (atoms and molecules) that make up matter. Absolute temperature is the measure of the average kinetic energy of the individual particles of a system.

This gives an important theoretical meaning to absolute zero. It is not merely a convenience for simplifying equations, or a point at which the volumes of gases would shrink to zero if they followed Gay-Lussac's law exactly (which they do not). Rather it is the temperature at which the kinetic energy of the particles

* The average here is not the ordinary "arithmetical mean" obtained by adding values and dividing by the number of values. It is rather the "root mean square" (rms), which is the square root of the arithmetical means of the squares of the value. Thus, if we have two values, 4 and 6, the ordinary average is $(4 + 6)/2$, or 5. The rms, however is $\sqrt{\frac{4^2 + 6^2}{2}}$ or $\sqrt{26}$ or 5.1.

of a substance are lowered to an irreducible minimum. Usually this minimum is said to be zero, but that is not completely correct. Modern theories indicate that even at absolute zero, a very small amount of kinetic energy remains present. This amount, however, cannot be further reduced, and temperatures below absolute zero cannot exist.

Diffusion

The motion of gas particles can also be used to explain *diffusion*—that is, the spontaneous ability of two gases to mix intimately, even though originally separate, and even against the pull of gravity. Suppose a container is horizontally partitioned at the center. In the upper portion of the container is hydrogen; in the lower, under equal pressure, nitrogen. If the partition is removed, it might be expected that hydrogen (by far the lighter gas of the two) would remain floating on top, as wood floats on water. Nevertheless, in a short time the two gases are intimately mixed, the nitrogen diffusing upward, the hydrogen downward, in apparent defiance of gravity.

This comes about because the motion of the gas particles is virtually independent of gravity (see page 198). One third of the nitrogen particles (on the average, if we assume random motion) are moving upward at any one instant; and one third of the hydrogen particles are moving downward. Naturally, the two gases mix.

Diffusion also takes place between mutually soluble liquids, although more slowly. For instance, alcohol can be floated on water, which is denser; if one waits, the two liquids will mix evenly. This indicates that while the particles making up liquids must remain in contact, they nevertheless have a certain freedom of movement, slipping and sliding about, so they are able to insinuate themselves among the particles of another liquid.

On the other hand, diffusion between different solids in contact proceeds with excessive slowness, if at all, and this is an indication that the constituent particles in solids are not only in contact but are more or less fixed in place. (This does not mean, however, that the particles of a solid are motionless; all evidence points to the fact that although they have a fixed place, they vibrate about that fixed place with an average kinetic energy corresponding to the absolute temperature of the solid.)

To get a notion of the quantitative relationships that involve diffusion, let us return to Equation 14–1, where P is set equal to

$Nmv^2/3V$, and let us solve for v, the average velocity of the particles. This gives us the relationship:

$$v = \sqrt{\frac{3PV}{Nm}} \qquad \text{(Equation 14–3)}$$

This is easily handled. If a given quantity of, let us say, oxygen is being dealt with, its pressure (P) and volume (V) are easily measured. The quantity Nm represents the number of particles multiplied by the mass of the individual particle and that is, after all, the total mass of the gas, which is easily measured. Without going into the details of the calculation, it turns out that at 0°C (273°K) and 1 atmosphere pressure,* the average velocity of the oxygen molecule is 460 meters per second (0.28 miles per second).

Equation 14–3 can be written: $v = \sqrt{3/Nm}\sqrt{PV}$. For a given quantity of a specific gas (Nm), the total mass, is constant, so the quantity $\sqrt{3/Nm}$ is constant and Equation 14–3 can be written $v = k\sqrt{PV}$, and we can say that velocity of the gas molecules is proportional to the square root of pressure times volume. By Equation 13–12 (see page 196), however, we know that PV is directly proportional to the absolute temperature T. We can therefore say that $v = k\sqrt{T}$: that the average velocity of a gas molecule is directly proportional to the square root of the absolute temperature.

If the velocity of 460 m/sec is the average for oxygen molecules at a temperature of 273°K (0°C), what would it be if the temperature were doubled to 546°K (273°C)? The average velocity is then multipled by the $\sqrt{2}$, or approximately 1.4. Oxygen molecules, then, will move at an average velocity of 650 m/sec (0.40 miles per second) at the higher temperature.

But suppose we consider different gases at the same pressure P and volume V. Under these conditions it turns out (as a result of evidence more appropriately considered in a book on chemistry) that the number of particles present (N) is the same in both. We can consider P, V and N to be constant, therefore, and if we write Equation 14–3 as $v = \sqrt{3PV/N}\sqrt{1/m}$, we can simplify this to $v = k\sqrt{1/m}$. We are then able to say that at standard conditions

* Since many properties vary with temperature and pressure, it is usual to give the precise temperature and pressure at which a measurement is carried through. For the sake of standardization, it is common to use 0°C and 1 atmosphere pressure, or to adjust to those values if measurement is made at others. Usually, 0°C and 1 atmosphere pressure are referred to as "standard conditions of temperature and pressure," and this is abbreviated S.T.P.

of temperature and pressure the average velocity of the molecules of a gas is proportional to $\sqrt{1/m}$. Using words, we can say it is inversely proportional to the square root of the mass of individual molecules, *i.e.*, the "molecular weight."

This is sometimes called Graham's law because it was first specifically stated by the Scottish chemist Thomas Graham (1805–1869) in 1829. He noted that the rate of diffusion of a gas (which turns out to depend on the velocity of its molecules) is inversely proportional to the square root of its density (and density in gases depends on molecular weight).

A proper consideration of molecular weights is better taken up in a book on chemistry, but we can say that a molecule of hydrogen has a molecular weight 1/16 that of a molecule of oxygen. Since hydrogen molecules are 16 times less massive than oxygen molecules, they move with a velocity $\sqrt{16}$, or four times more rapidly. If at 273°K (0°C) oxygen molecules move at 460 m/sec (0.28 miles per second), then hydrogen molecules at that same temperature move at 1840 m/sec (1.12 miles per second). At 546°K (273°C) the velocity of both oxygen and hydrogen molecules is multiplied by 1.4, and the latter moves at 2600 m/sec (1.58 miles per second).

The velocity of sound through a gas depends in part upon the rapidity with which gas molecules can swing back and forth to form regions of compression and expansion (see page 165). As the molecules move more quickly with rising temperature, the velocity of sound does also. In different gases, furthermore, the velocity of sound is inversely proportional to the square root of the molecular weight, because that is the way molecular velocity varies.

Air is 4/5 nitrogen (molecular weight, 28) and 1/5 oxygen (molecular weight, 32), so the "average molecular weight" of air is 29. The molecular weight of hydrogen is 2. At a given temperature, the hydrogen molecule moves $\sqrt{29/2}$, or 3.8 times as quickly as the average molecule in air. Since at 20°C sound travels through air at the velocity of 344 m/sec, it would travel through hydrogen at that temperature at the velocity of about 1300 m/sec.

A column of air in an organ pipe, if disturbed, will vibrate at a natural frequency that depends on such things as the size of the column and the velocity of the air molecules. An organ pipe of a given size at a given temperature will therefore produce a note of a given pitch. If the pipe is filled with hydrogen, the molecules of which will move more rapidly than those of air, the same pipe at the same temperature will produce a sound of a much higher pitch. (A man who fills his lungs with hydrogen—not a recommended experiment—will find himself speaking, temporarily at least, in a shrill treble.)

Since both the pitch of the organ pipe and the velocity of sound depend on the velocity of the molecules of a gas, one can be calculated from the other. Indeed, in about 1800, the German physicist Ernst F. F. Chladni (1756–1827), sometimes called the "father of acoustics," calculated the velocity of sound in various gases (something rather difficult to measure directly) by noting the pitch of organ pipes filled with the gas (something quite easy to do).

To be sure, the actual velocities of individual molecules cover a broad range and some molecules of a particular gas move very rapidly. Even at 0°C there would be a very small fraction of the molecules in oxygen gas moving, at least temporarily, at velocities of 7 miles per second, which is some 25 times the average velocity.

It so happens that 7 miles per second is the escape velocity on the earth's surface (see page 63), and a molecule moving this quickly would be expected to leave the earth permanently. For this reason, it might seem that oxygen should constantly be "leaking" out of the atmosphere. So it is, but this should be no cause for panic. For one thing, there are very few oxygen molecules that travel at 25 times the average velocity. Of those that do, all but a vanishingly small number strike other molecules and lose the unusually high velocity, long before they can reach the upper regions of the atmosphere. Such leakage of oxygen as takes place, then, is so slow as to assure the earth its oxygen supply for billions of years to come.

In the case of hydrogen molecules, however, with four times the average velocity of oxygen molecules, a larger fraction can be expected to attain the escape velocity of 7 miles per second (since this is only about six times the average velocity of the hydrogen molecule). Here the leakage is indeed serious and the earth could not hold hydrogen in its atmosphere over the geologic eras—nor has it. There is good reason to think that the earth's atmosphere might have been rich in hydrogen to begin with, but it is all gone now.

The moon, with its much smaller escape velocity (see page 63), could not even hold any oxygen or nitrogen, if it had ever had any; in fact, it lacks an atmosphere altogether. Jupiter and the other outer planets, with larger velocities of escape, and with temperatures lower than those of the earth and the moon, can hold even hydrogen easily. The outer planets therefore have large hydrogen-filled atmospheres.

Real Gases

Boyle's law has always been accepted as useful through the three centuries during which scientists have been aware of its existence. Through the first two of those centuries, it was also (wrongly) considered exact. That Boyle's law, while useful, is only an approximation of the actual situation was first made clear by the French physicist Henri Victor Regnault (1810–1878), who in the 1850's measured the exact volumes of different gases under different pressures and found that the product of the two (PV) was not quite constant after all, even if the temperature were kept carefully constant. Under a pressure of 1000 atmospheres, the product could be twice as high as at 1 atmosphere pressure. Even when he worked with pressures that were only moderately high, he frequently found deviations of up to five percent. Furthermore, there were differences from gas to gas. Up to pressures of 100 atmospheres, hydrogen, nitrogen and oxygen deviated comparatively little from Boyle's law, while carbon dioxide deviated a good deal.

Yet Boyle's law can be derived from the kinetic theory of gases. Is the kinetic theory wrong then? No, not necessarily. However, in deriving Boyle's law from the kinetic theory of gases, it simplifies matters to make two assumptions that are not exactly true for real gases. For instance, it can be assumed that there are no attractive forces among the molecules of a gas, so the motion of one molecule can be considered completely without reference to the others. This is almost correct but not quite, for there are very weak attractive forces among the molecules of gases.

Another assumption is that the molecules are extremely small compared to the empty space separating them—so extremely small that their volume can be taken to be zero. Again, this is almost correct but not quite. The volume of the molecules is indeed very small, but it is not zero.

Now suppose we don't accept the simplifications but consider instead that when a molecule is about to strike the wall of a vessel there is a net pull backward from all the feeble intermolecular forces exerted upon the about-to-collide molecule by the other molecules. (This is a kind of gaseous surface tension, like the more familiar liquid surface tension described on page 126.) Because of this backward pull, the molecule does not strike the surface with full force and its contribution to the pressure is less than one would expect from kinetic theory if no intermolecular forces existed. To bring the pressure of the individual molecule up to the no-intermolecular-force ideal, we must add a small extra quantity of pressure (P_x). The ideal pressure (P_i) is then the actual measured pressure plus this extra quantity ($P + P_x$).

The more molecules present in the gas close to the colliding molecule (the more distant molecules contribute so little to the attractive force that they can be ignored), the greater the backward pull; the more the actual pressure (P) falls short of the ideal pressure (P_i), the greater the value P_x we must add to P in the case of this one colliding molecule. The quantity of nearby molecules is proportional to the density of the gas (D).

But pressure depends upon the total number of molecules striking the walls in a given time. The value for P_x also depends on that number. But that number in turn depends upon the density of the gas. Thus, P_x depends on the density of the gas first in connection with each per colliding molecule, then in connection with the number of colliding molecules per unit time. The total value of P_x depends upon its size per colliding molecule multiplied by the number of colliding molecules per unit time, or upon a factor proportional to density multiplied by another factor proportional to density. The total value is then proportional to the square of the density, D^2. If, on this occasion, we use a for the proportionality constant, we can say that $P_x = aD^2$.

For a given quantity of gas, density is inversely proportional to volume. The denser a gas, the less volume is taken up by a given quantity. If P_x is directly proportional to the square of the density, then, it must be inversely proportional to the square of the volume —that is, $P_x = a/V^2$. Since earlier I said that the ideal pressure was $P + P_x$, we can now write that as $P + a/V^2$.

Next, what about the matter of the finite volume of the molecules? If more and more pressure is put upon a gas, Boyle's law requires that the volume decrease steadily and get closer and closer to zero. The ideal volume (V_i) available for contraction is, if Boyle's law held perfectly, equal to all the volume (V) of the gas. But if a gas is actually put under great pressure, the molecules eventually make virtual contact. After that, there is practically no further shrinkage of volume with increase of pressure. The ideal volume available for contraction is the volume of the gas minus the volume of the molecules themselves. In other words, $V_i = V - b$, where b represents the volume of the molecules.

The ideal gas equation (Equation 13–12, see page 196), based on the assumption of no intermolecular forces and no molecular volume, should really be expressed in terms of ideal pressure and ideal volume: $P_iV_i = RT$. If the expressions containing the actual pressure and volume are used for these ideal values, we get:

$$(P + a/V^2)(V - b) = RT \qquad \text{(Equation 14–4)}$$

This is the *van der Waals equation,* since it was first worked out by the Dutch physicist Johannes Diderik van der Waals (1837–1923) in 1873. The feeble attractive forces between gas molecules that help make this modification necessary are called *van der Waals forces*. The values for a and b in the van der Waals equation are usually quite small and differ from gas to gas, for the various gas molecules have their own characteristic volumes and exert forces of characteristic size among themselves.

The intermolecular forces in gas, while small under ordinary conditions, can be made to bring about important changes in gaseous properties. The attractive force among gas molecules increases as the molecules aproach one another, and the molecules approach more and more closely as the volume of a given quantity of gas decreases with increasing pressure. Where the attractive force is comparatively large to begin with, increased pressure can raise the force to a level higher than that which can be overcome by the kinetic energy of the gas molecules. The molecules will no longer be able to pull apart, but will cling together, and the substance will become a liquid. Gases such as sulfur dioxide, ammonia, chlorine and carbon dioxide can in this way be liquefied by pressure alone and, at this high pressure, be maintained as liquids at room temperature. (If there were no intermolecular forces, liquefaction could not take place under any circumstances. All substances would be gaseous under all conditions.)

Where the intermolecular attractive force is particularly weak, however, it is possible that even when the gas molecules are forced close enough together to touch, the attractive force will still not have increased to the point where it can keep the molecules together against the molecular motion representing their kinetic energy. For that reason, gases such as oxygen, nitrogen, hydrogen, helium, neon or carbon monoxide cannot be liquefied at room temperature under any pressure, no matter how high. During the early nineteenth century, gases of this sort therefore received the name of "permanent gases."

However, one might increase the attractive force and decrease the kinetic energy as well. If the former is brought about by increasing the pressure, the latter can be brought about by decreasing the temperature. If the temperature is brought low enough, the kinetic energy is decreased sufficiently for the attractive forces among the molecules of the so-called permanent gases to suffice to bring about liquefaction. The temperature at which such liquefaction becomes just barely possible is called the *critical temperature.* Above that critical temperature a substance can exist only as a gas. The existence of the critical temperature was first discovered by the Irish physicist Thomas Andrews (1813–1885) in 1869.

The critical temperature for oxygen is 154°K (—119°C), and it was only after oxygen was brought to a lower temperature that it became possible to liquefy it. Hydrogen, with still weaker intermolecular forces, must be lowered to a temperature of 33°K (— 240°C) before the kinetic energy of the molecules is low enough to be neutralized by those forces. The record in this respect is held by helium (not isolated on earth until 1898). Helium is the nearest approach, among real gases, to the gaseous ideal. Its critical temperature is 5°K (—268°C).

On the other hand, there are substances with intermolecular forces so great that they remain liquid at room temperature even under atmospheric pressure. (These intermolecular forces are more than mere van der Waals forces, and they will not be discussed in this book.) Water is the most common example of a substance liquid at ordinary temperatures and pressures. At a temperature of 373°K (100°C) and 1 atmosphere pressure, the intermolecular forces are overcome, thanks to the heightened kinetic energy, and water turns into its gaseous form: steam, or water vapor.* At temperatures over 100°C, water can be kept in liquid form by increasing the pressure. This means that the boiling point rises with increased pressure, a fact taken advantage of in pressure cookers. The critical temperature for water is 647°K (374°C) and it is only at temperatures above that, that liquid water cannot exist under any conditions.

Even in liquids, the attractive forces between molecules are not large enough to prevent the individual moleclules from slipping and sliding about. If, however, the temperature is lowered still further, a point is reached where the energies of the individual molecules are insufficiently large to give it even that much

* A gas that exists as such only at elevated temperatures is usually referred to as a "vapor."

freedom. The intermolecular forces are strong enough to keep the molecules firmly in place. They may vibrate back and forth, but the average position remains fixed and the substance is a solid. If the temperature of a solid is raised, the vibrations become more energetic, and at a certain temperature (depending on the size of the intermolecular forces involved) they become large enough to counter those forces to the extent of allowing the molecules to slide about; the solid has then melted, or liquefied. The melting point is only slightly affected by pressure.

The intermolecular forces of hydrogen are so weak that solid hydrogen melts at a temperature of only 14°K (—259°C), and liquid hydrogen boils (under atmospheric pressure) at a temperature of only 20°K (—253°C). Helium does better still. Its particles consist of individual atoms and the interatomic forces are so weak that even the irreducible bit of kinetic energy still present at absolute zero is enough to keep it liquid. Solid helium cannot exist at any temperature, however low, except under pressures greater than atmospheric. The boiling point of helium under a pressure of one atmosphere is 4°K (—269°C).

On the other hand, some substances possess intermolecular or interatomic forces so strong that they remain solids at ordinary temperatures and even considerably higher. The metal tungsten does not melt until a temperature of 3370°C is reached and does not boil, under atmospheric pressure, until a temperature of 5900°C is reached.

Specific Heat

So far in our discussion of heat in this chapter and the preceding one the emphasis has been on temperature, and we must avoid confusing the two. The terms "heat" and "temperature" are by no means identical. It is all too easy to assume that if one sample of water has a higher temperature than another, it is hotter and therefore has more heat. The final conclusion, however, is not necessarily true.

A thimbleful of water at 90°C is much hotter than a bathtub-full of water at 50°C, but there is more total heat in the bathtub of water. If both are allowed to stand, the thimbleful of water will have cooled to room temperature in an interval during which the bath-tub-full of water would scarcely have cooled at all. The thimbleful loses its heat more quickly because, for one thing, it has far less heat, all told, to lose.

To specify, the heat content of a system is the total internal energy* of the molecules making it up, while temperature is the measure of the average translational kinetic energy of the individual molecules. In other words, heat represents a total quantity and temperature a quantity per molecule.

The difference can be made plainer, perhaps, by an analogy. Consider one liter of water poured into a tall thin cylinder so that it forms a column one meter high. Into a much broader cylinder, five liters of water are poured, and this water stands only 0.1 meters high. The water in the narrow cylinder exerts the greater pressure on the bottom of the container, but the water in the broader cylinder, exerting one-tenth the pressure, is nevertheless five times greater in volume. Volume is a total quantity, while pressure is a quantity per area. Therefore, temperature is to heat, as pressure is to volume.

It may seem that such a distinction between heat and temperature is unnecessary labor. After all, if one heats water, for instance, heat pours into it and the temperature goes up; the two rise together and why can't you use one as the measure of the other? Unfortunately, this parallel behavior of heat and temperature can be counted on only when you deal with a given quantity of a particular substance, and even then only over certain limited temperature ranges. We can see this if we compare the heat contents of two different subjects at identical temperature.

To do this, we need a unit of measurement for heat. Earlier in the book I mentioned such a unit, the calorie, in passing. Now let's go into such matters in a bit more detail.

Suppose we add heat to water, thus raising its temperature. Experiments will show that the amount of heat required to raise the temperature of water by a fixed number of degrees varies with the mass of the water receiving the heat.

We can assume, for instance, that 100 grams of boiling water contain a fixed amount of heat. If 100 grams of boiling water are poured into 5 kilograms (5000 grams) of cold water, the temperature of the cold water will rise about two Celsius degrees. If, on the other hand, the 100 grams of boiling water is poured into 10 kilograms of cold water, the temperature of the cold water will go up only one Celsius degree.

Again, the quantity of heat required to raise the temperature of a fixed mass of water varies with the number of Celsius degrees by which the temperature is raised. It takes twice as large a volume of boiling water to raise a particular quantity of cold water by 10

* The "internal energy" of a substance consists of the kinetic energy of its constituent particles plus the energy involved in the intermolecular attractions.

Celsius degrees than by 5 Celsius degrees. The unit of heat must therefore be defined in terms of a unit mass and a unit rise in temperature; as for instance, the quantity of heat required to raise the temperature of one gram of water by 1 Celsius degree. Actually, refined measurements show that the quantity of heat required to raise the temperature of one gram of water by 1 Celsius degree varies slightly according to the original temperature of the water, so the original temperature must also be included in the definition. We can say then:

One *calorie* is the quantity of heat required to raise the temperature of one gram of water from 14.5°C to 15.5°C.

We might also say that:

One thousand calories, or a *kilocalorie,* is the quantity of heat required to raise the temperature of a kilogram (1000 grams) of water from 14.5°C to 15.5°C.

Suppose now that a gram of aluminum is placed in boiling water for enough time to make certain that it has assumed the temperature of boiling water (100°C). Plunge the hot aluminum quickly into 100 grams of water at 0°C. The aluminum cools off and its heat is added to the water, raising its temperature from 0°C to about 0.22°C.

To raise the temperature of 100 grams of water by 0.22 Celsius degrees takes 100 times 0.22 or about 22 calories. The gram of aluminum, in cooling from 100°C to 0.22°C, has liberated some 22 calories. By the law of conservation of energy, we would expect that if this cooling liberated 22 calories, then adding 22 calories to the cold aluminum would bring it back up to 100°C. Roughly speaking, then, we can say that it takes 22 calories to raise the temperature of a gram of aluminum 100 Celsius degrees, and 0.22 calories to raise it 1 Celsius degree. This represents the *specific heat* of aluminum, where the specific heat of a substance is defined as the quantity of heat required to raise the temperature of 1 gram of that substance by 1 Celsius degree.

By this type of experiment one can find that the specific heat of iron is 0.11, that of copper 0.093, that of silver 0.056, and that of lead 0.03. If one calorie of heat is added to a gram of aluminum at 0°C, that amount of heat will be enough to heat it 1/0.22, or 4.5 Celsius degrees—that is, to a temperature of 4.5°C. The same amount of heat under the same conditions would raise the temperature of a gram of iron to 9°C, of copper to 11°C, of silver to 18°C, and of lead to 33°C.

Here you can see that the distinction between heat and

temperature is indeed a useful one, since the same quantity of heat may be added to a fixed mass of each of a number of different substances and each will attain a different temperature. Temperature in itself is consequently no measure at all of total heat content. (To return to our volume-pressure analogy, this is like pouring equal volumes of water into cyindrical vessels of different diameters. The volumes may be the same; however, the final pressures will vary, and pressure is no measure at all of total volume.)

The conception of specific heat was first advanced by the Scottish chemist Joseph Black (1728–1799) in 1760.

Part of the reason for this variation of specific heat from substance to substance lies in the different masses of the atoms making up each. The lead atom is about 7.7 times the mass of the aluminum atom, the silver atom is 4 times the mass of the aluminum atom, the copper atom 2.3 times the mass of the aluminum atom, and the iron atom 2.1 times the mass of the aluminum atom.

Because of this, a given mass of lead, say 1 gram, contains only 1/7.7 times as many atoms as the same mass of aluminum. In adding heat to 1 gram of lead, you are therefore engaged in setting fewer atoms into motion and less heat is required to increase the kinetic energy of the individual atoms by enough to account for a 1 Celsius degree temperature rise. For this reason, the specific heat of lead, 0.03, is about 1/7.7 that of aluminum, 0.22. Similarly, the specific heat of silver is about 1/4 that of aluminum; the specific heat of copper is about 1/2.3 that of aluminum; and the specific heat of iron is about 1/2.1 that of aluminum.

The general rule is that for most elements the specific heat multiplied by the relative mass of its atoms yields a number that is approximately the same for all. Where the relative mass of the atoms of the different elements (the *atomic weight*) is chosen in such a way that the hydrogen atom, which is the lightest, has a weight of a trifle over 1, then the product of specific heat and atomic weight comes to about six calories for most elements.

This is known as the law of Dulong and Petit, after the French physicists Pierre Louis Dulong (1785–1838) and Alexis Thérèse Petit (1791–1820), who first advanced it in 1819.

Latent Heat

It might occur to you that temperature is very close to being a measure of heat content if only we count by atoms or molecules instead of by grams. This would be so if the law of Dulong and Petit held for all substances under all conditions, but it does not. It holds only for the solid elements and for these only in certain temperature ranges. In fact, it is possible to show cases in which heat content can change a great deal without any change in temperature at all, and that should at once put to rest any notion of using temperature as a measure of heat content.

Suppose 100 grams of liquid water at 0°C is added to 100 grams of liquid water at 100°C. After stirring, the final temperature of the mixture would be 50°C.

Next, suppose that 100 grams of ice at 0°C is added to 100 grams of liquid water at 100°C. After allowing the ice to melt and stirring the mixture (assuming that while we wait there is no over-all loss of heat to the outside world or gain of heat from it—a matter which can be arranged by insulating the system), we find that the temperature of the mixture is only 10°C.

Why should this be? Clearly the liquid water at 0°C had more heat to contribute to the final mixture than the ice at 0°C, and yet both liquid water and ice were at the same temperature. It seems reasonable to suppose that a quantity of the heat in the hot water was consumed, in the second case, in simply melting the ice; so much the less was therefore available for raising the temperature of the mixture.

Indeed, if we heat a mixture of ice and water, we find that no matter how much heat is transferred to the mixture, the temperature remains at 0°C until the last of the ice is melted. Only after all the ice is melted is heat converted into kinetic energy, and only then can the temperature of the water begin to rise. Experiment shows that 80 calories of heat must be absorbed from the outside world in order that 1 gram of ice might be melted, and that no temperature rise takes place in the process. The ice at 0°C is converted to water at 0°C.

But if the heat gained by the ice is not converted into molecular kinetic energy, what does happen to it? If the law of conservation of energy is valid, we know it cannot simply disappear.

The water molecules in ice are bound together by strong attractive forces that keep the substance a rigid solid. In order to convert the ice to liquid water (in which the molecules, as in all liquids, are free of mutual bonds to the extent of being able to slip and slide over, under, and beside each other) those forces must be countered. As the ice melts, the energy of heat is consumed in countering those intermolecular forces. The water molecules contain more energy than the ice molecules at the same temperature, not in the form of a more rapid motion or vibration, but in the form of an ability to resist the attractive forces tending to pull them rigidly together.

The law of conservation of energy requires that the energy change in freezing be the reverse of the energy change in melting. If liquid water at 0° is allowed to lose heat to the outside world, the capacity to resist the attractive forces is lost, little by little. More and more of the molecules lock rigidly into place, and the water freezes. The amount of heat lost to the outside world in this process of freezing is 80 calories for each gram of ice formed.

In short, 1 gram of ice at 0°C, absorbing 80 calories, melts to 1 gram of water at 0°C; and 1 gram of water at 0°C, giving off 80 calories, freezes to 1 gram of ice at 0°C.

The heat consumed in melting ice (or any solid, for that matter) is converted into a sort of potential energy of molecules. Just as a rock at the top of a cliff has, by virtue of its position with respect to gravitational attraction, more energy than a similar rock at the bottom of the cliff, so do freely moving molecules in liquids, by virtue of their position with respect to intermolecular attraction, possess more energy than similar molecules bound rigidly in solids.

It is the kinetic and potential energies of the molecules that, together, make up the internal energy that represents the heat content. It is the kinetic energy only that is measured by the temperature. By changing the potential energy only, as in melting or freezing, the total heat content is changed without changing the temperature.

The discoverer of the fact that heat melted ice without raising its temperature was Joseph Black, who first pointed out the significance of specific heat (see page 215). He referred to the heat consumed in melting as *latent heat.* "Latent" refers to something that is present in essence, but not in such a fashion as to be apparent or visible. This is just about synonymous with "potential," so the connection between "latent heat" and "potential energy" is clear.

Actually, the heat required to melt a gram of ice is its *latent heat of fusion* ("fusion" being synonymous with "melting"). The qualifying phrase "of fusion" is necessary, for another type of latent heat arises in connection with boiling or vaporization. In converting a gram of liquid water at 100°C to a gram of steam at 100°C, what remains of the intermolecular attractions must be completely neutralized. Only then are the molecules capable of displaying the typical properties of gases—that is, the virtually independent motion. In the earlier process of melting, only a minor portion of the intermolecular attractive force was countered, and the major portion remains to be dealt with. For this reason, the *latent heat of vaporization* of a particular substance is generally considerably higher than the latent heat of fusion for that same substance. Thus, the latent heat of vaporization of water—the amount of heat required to convert 1 gram of water at 100°C to 1 gram of steam at 100°C—is 539 calories. For water, the latent heat of vaporization is almost seven times as high as the latent heat of fusion.

The energy content of steam is thus surprisingly high. A hundred grams of water at 100°C can be made to yield 10,000 calories as it cools to the freezing point. A hundred grams of steam at 100°C, however, can be made to give up 53,900 calories merely by condensing it to water. The water produced can then give up another 10,000 calories if it is cooled to the freezing point. It is for this reason that steam engines are so useful and a "hot-water engine" would never do as a substitute. (It is also no accident that James Watt, the perfecter of the steam engine, was a student of Joseph Black.)

The latent heat of vaporization can be put to an important use. Suppose that a gas such as ammonia is placed under pressure in a closed container. If the pressure is made high enough, it will liquefy the gas (see page 210). As the ammonia liquefies, it gives up a certain amount of heat to the outside world. This heat would tend to raise the temperature of the immediate surroundings and of the ammonia itself. However, if the container of ammonia is immersed in running water, the heat evolved is carried off by that water and the liquid ammonia is no warmer than the gas had been.

If the container of ammonia is now removed from the water, and the pressure is lowered so that the liquid ammonia is free to boil again and become a gas, it must absorb an amount of heat equivalent to what it had given up before. It absorbs this heat from the nearest source—itself and its immediate neighbors. Some of the kinetic energy of its own molecules is converted into the potential energy of the gaseous state, and the temperature of the ammonia drops precipitously.

If a gas like ammonia is made part of a mechanical device that alternately compresses it and allows it to evaporate, a heat-pump will have been set up in which heat is pumped from the ammonia and anything in its near neighborhood out (by way of running water, for instance) to the world at large. If such a heat-pump is placed within an insulated box, we have a refrigerator.

The lowering of temperature with vaporization is made use of by our own bodies. The activity of the sweat glands keeps us covered with a thin film of moisture which, as it evaporates, withdraws heat from our body and keeps us cool. Water has the highest latent heat of vaporization of any common substance, so the fact that our perspiration is almost pure water means that little of it need be used, and we are ordinarily unaware of it. In hot weather the process must be accelerated, and if humid conditions cut down the rate of evaporation, perspiration will accumulate in visible quantities. We all know the feeling of discomfort that follows upon this partial breakdown of our own private refrigeration device.

CHAPTER **15**

Thermodynamics

The Flow of Heat

In the previous chapter, I spoke of mixing hot and cold water and said that in the process an intermediate temperature was reached. It is easy to see that this is achieved by the physical intermingling of the molecules of the hot water (which possess a high average kinetic energy) with the molecules of the cold water (which possess a low one). The molecules of the mixture, taken as a whole, are bound to have an average kinetic energy of an intermediate value.

Gases, too, can blunt the extremes of temperature in this fashion. Warm air masses will mingle with cold air masses (and such mingling of air masses is the fount and origin of our weather), and the temperature of the earth's surface is kept at an intermediate value as a result. It might seem that the mixture of warm and cold on earth is not very efficient when one compares the frozen floes of the polar regions with the steaming jungles of the tropics. It could, however, be worse. Our moon is at the same average distance from the sun as the earth itself is, but unlike the earth it lacks an atmosphere. As a result, portions of its sunlit side grow hotter than even the earth's tropics do, and portions of its darkened side grow colder by far than an Antarctic winter.

The transfer of heat by currents of gas or liquid is known as *convection* (from Latin words meaning "to carry together").

Such actual movement of matter is not necessary for transfer of heat, however. If one end of a long metal rod is heated, the heat will eventually make itself felt at the other end of the rod. It is not to be supposed that there are currents of moving matter within the solid metal of the rod. What happens, instead, is something like this. As the end of the rod grows hot, the atoms of that portion of the metal gain kinetic energy. As long as the rod remains solid, the average position of each atom remains fixed, but each can and does vibrate about that position. As the atoms gain energy, the vibrations become more rapid, and the movements extend further from the equilibrium position. The atoms in the hottest portion of the rod, vibrating most energetically, jostle neighboring atoms, and those atoms, as a result of the impacts, vibrate more energetically themselves. In this way, kinetic energy jostles itself from atom to atom and, gradually, from one end of the rod to the other. This transfer of heat through the main body of a solid is *conduction* (from Latin words meaning "to lead together").

The fact that atoms and molecules of solids vibrate with greater amplitude as temperature rises means that each atom or molecule takes up more room. It is not surprising then that the volume of a solid, or a liquid for that matter, will increase with rising temperature and decrease with falling temperature (see page 182), even though the molecules remain in virtual contact throughout the temperature range up to the boiling point.

(This is not the only factor involved in the volume change that solids and liquids undergo with temperature. There is also the matter of the nature of the molecular arrangement. The molecular arrangement for a particular substance is usually more compact in the solid state than in the liquid state, so there is generally a sudden drop in volume—and consequent rise in density—as a substance freezes. Water is exceptional in this respect. Its molecular arrangement is less compact in the solid state than in the liquid. As a result, ice is less dense than liquid water and will float in it rather than sink to the bottom.)

Both convection and conduction are explainable in mechanical terms. In both cases, there are actual impingements of energetic atoms or molecules upon less energetic atoms or molcules, and energy is therefore transferred by direct contact. Heat can, however, be transmitted without direct contact at all. A hot object encased in a vacuum will make its heat felt at a distance, even though there is no matter surrounding it to carry this heat either by convection or by conduction. The sun is separated from us by almost 93,000,000 miles of vacuum better than any we can yet make in the laboratory, and yet its heat reaches us and is evident. Such heat seems to stream out of the hot object in all directions, like the conventional rays drawn about the sun by cartoonists. The word "ray" is "radius" in Latin, and the transference of heat across a vacuum is called *radiation.* The detailed discussion of radiation will be left for the second volume of this book.

Interest in the laws governing the movement of heat by any or all these methods grew sharp in the first part of the nineteenth century because of the growing importance of James Watt's steam engine, which depended in its workings on heat flow. In the steam engine, heat is transferred from burning fuel to water, converting the latter to steam. The heat of the steam then flows into the cold water bathing the condenser, and the steam, now minus its heat, is converted into water again. This heat flow that turned water to steam and back again somehow made available energy that could be converted into the kinetic energy of a piston, which, in turn, could be used to do work.

The study of the movement of heat (with particular attention, at first, to the workings of the steam engine) makes up that branch of physics called *thermodynamics* (from Latin words meaning "motion of heat"). Of course, all consideration of heat flow must assume, to begin with, that none of the heat will vanish into nothing or arise out of nothing. This is the law of conservation of energy, and so important is this generalization, in connection with thermodynamics in particular, that it is frequently called the *first law of thermodynamics.*

The first law of thermodynamics, however, merely states that the total energy content of a closed system is constant; it does not predict the manner in which the energy in such a system may shift from place to place. But even a little experience shows that some of the facts about such energy shifts seem to fall into a pattern.

For instance, suppose a closed system (that is, one that exchanges no energy with the outside world—giving off none and taking up none) consists of a quantity of ice placed in hot water. We can be quite certain that the ice will melt and the water will cool. The total energy has not changed; however, some of it has shifted from the hot water into the ice, and all the experience of mankind tells us that this shift is inevitable. Similarly, a red-hot stone will gradually cool, while the air in its neighborhood will gradually warm.

Such a flow of heat from a hot object to a cool object will continue until the temperature of different portions of the closed system are equal, and this is true whether heat is transferred by convection, conduction or radiation.

Faced with such facts about heat flow, the early workers in thermodynamics found matters most easily visualized if they thought of heat as a kind of fluid, and indeed this fluid even received a name—*caloric,* from a Latin word for "heat."

The flow of heat can be pictured by uses of fluid flow as an analogy. Imagine two vessels connected by a stopcock, with the water level high on the left side and low on the right. Naturally, water pressure is higher on the left than on the right, so there is a net pressure from left to right. If the stopcock is open, water will flow from left to right and continue flowing until the levels are equal on both sides. The high level will fall; the low level will rise; and the final level on both sides will be intermediate in height. Although the total water volume of the system has not changed, there has been a change in the distribution of water within the system leading to an equalization of pressure.

By changing a few key words, we can have the previous sentence read: "Although the total heat of the system has not changed, there has been a change in the distribution of heat within the system leading to an equalization of temperature." (Once again, as on page 213, we have an analogy between volume/pressure and heat/temperature.)

If we think of temperature as a kind of driving force directing the flow of heat, just as water pressure directs the flow of water, then it seems very natural, even inevitable, that heat should flow from a region of high temperature to one of low, without regard to the total heat content in each region.

Consider a gram of boiling water, for instance, and compare it with a kilogram of ice water. To freeze the kilogram of ice water, some 80,000 calories of heat must be withdrawn from it. To reduce the temperature of the gram of boiling water to the freezing point—and then freeze it—would require the withdrawal of 100 plus 80 calories; only 180 altogether. Any further cooling of the kilogram of ice obtained in the first case, as compared with the gram of ice obtained in the second, requires the withdrawal of a thousand times as much heat per Celsius degree from the former as from the latter. It is plain then that despite the difference in temperatures the total heat in the kilogram of ice water is much higher than the total heat in the gram of boiling water.

Nevertheless, if the gram of boiling water is added to the kilogram of ice water, heat flows from the boiling water into the ice water. It is not the difference in total heat content that determines the direction of heat flow. Rather, it is the difference in temperature. Again, our analogy—if in the connected vessels referred to above, the left were of narrow diameter and the right of wide diameter, water would flow from the region of smaller volume to that of greater volume. Not difference of total volume but difference of pressure would dictate the direction of water flow.

The rate at which water flowed from one portion of the system to another would depend on the size of the difference in pressure. When the stopcock is first opened, the water flows quickly, but as the difference in pressure on the two sides of the stopcock decreases, so would the rate of flow. The rate of flow becomes very small as the difference in pressure becomes small; it sinks to zero once the water "finds its level" and the differences in pressure disappear.

The flow of heat by conduction can, apparently, be pictured analogously. The rate of flow of heat from a hot region to a cold one depends in part on the difference in temperature between the two. It is conventional to calculate the quantity of heat that would flow in one second through a one-centimeter cube, where one face of the cube was 1 Celsius degree cooler than the face on the opposite side. This quantity of heat is the *coefficient of conductivity,* and it is measured in calories per centimeter per second per degree Celsius (cal/cm-sec-°C).

Even given a particular difference of water pressure, water flow might yet vary depending on whether it flowed through a wide orifice, a narrow orifice, a series of narrow orifices, a sponge, loosely-packed cotton, well-packed sand, and so on. The same is true for heat, and even where a given temperature difference is involved, heat will flow more rapidly through one substance than through another. In other words, the coefficient of conductivity varies from substance to substance.

Substances for which it is high are said to be good conductors of heat; those for which it is low are said to be poor conductors. In general, metals are good conductors of heat and nonmetals poor ones. The best conductor of heat is copper, with a coefficient of conductivity equal to 1.04 cal/cm-sec-°C. In comparison, water has a coefficient of conductivity of 0.0015 cal/cm-sec-°C, and some kinds of wood have coefficients as low as 0.00009 cal/cm-sec-°C.

It is for this reason that cold metal feels so much colder than cold wood. The metal and wood may be at equal temperatures, but heat leaves the hand much more quickly when it is in contact with the metal than with the wood. The temperature of the portion of the hand making contact with the substance drops much more rapidly in the first case. Analogously, it is safe to lift a kettle of boiling water by its wooden or plastic hand-grip, for the heat from the metal (which it is wiser not to touch) enters the wood or plastic slowly enough for loss by radiation to keep pace.

A system, completely surrounded by material of low heat conductivity, loses heat slowly to the outside world, or gains heat slowly, even though the temperature difference within and without is a great one. The system is made an island, so to speak, of a particular temperature in the midst of an outer sea of a different temperature. It is therefore insulated (from a Latin word for "island"), and a material of low heat conductivity is therefore a *heat insulator.*

Gases have low coefficients of conductivity; air, therefore, is a good heat insulator. Woolen blankets and clothes trap a layer of air in the tiny interstices between fibers; heat therefore travels from our body into the cold outer environment very slowly, and so we have a sensation of warmth that we would not otherwise have. Wool and air are not warm in themselves, but give the effect of warmness by helping us conserve our own body heat. Air alone would do equally well, if it could be relied on to remain still. The warmed air near our bodies is, however, constantly being replaced by cool air as a result of the ubiquitous air currents. Heat is carried away by convection, and a windy day feels colder than a still day at the same temperature.

All substances have coefficients of conductivity greater than zero, and there is no substance, therefore, that can qualify as a perfect insulator of heat. Suppose, though, we take the phrase "no substance" literally and surround a system with a vacuum. We would then have a better insulator than anything we could find in the realm of matter. A perfect vacuum possesses a coefficient of conductivity equal to zero, and cannot bring about heat loss through convection either. Even a vacuum is not a perfect insulator, however, for it will still serve as a pathway for the loss of heat by radiation.

Loss by radiation, however, is a slower process than loss by either conduction or convection. Consequently, some bottles are constructed with a double wall within which a vacuum is formed. Furthermore, the walls can be silvered so that any heat radiating across the vacuum, in either direction, is reflected almost entirely. In the end, passage of heat through such a vacuum flask, or "thermos bottle," is exceedingly slow. Hot coffee placed in such a flask remains hot for an extended period of time, and cold milk remains cold.

Such devices were first constructed by the Scottish chemist James Dewar (1842–1923) in 1892. He used them to store extremely frigid substances, such as liquid oxygen, under conditions that would cut down the entry of heat from outside and thus minimize evaporation. In the laboratory, these are still called "Dewar flasks" in his honor.

The Second Law of Thermodynamics

We might therefore summarize the discussion in the preceding section by saying that it is the experience of mankind that in any closed system heat will spontaneously flow from a hot region to a cold region. It seems fair to consider this *the second law of thermodynamics.*

This view of heat as a kind of fluid reached its peak in the 1820's. A rigorous mathematical analysis of heat flow according to this view was advanced in 1822 by Fourier, the devisor of harmonic analysis. This view was put to further use by another French physicist, Nicolas Léonard Sadi Carnot (1796–1832).

In 1824, Carnot analyzed the workings of a steam engine in terms that we may consider analogous to those that might be applied to a waterfall. The energy of a waterfall can be made to turn a water wheel, the motion of which can then be used to run all the devices attached to the wheel. In this way, energy of falling water is converted into work.

For a given volume of water, the amount of energy that can be converted to work depends on the distance through which the water drops—that is, upon the height of the pool of water at the bottom of the falls subtracted from the height of the cliff over which the water tumbles.

We could measure these two heights from any agreed-upon reference. Taking the level of the pool at the bottom of the falls as our standard, we could say that its height (h_1) was 0. Then, if the height of the cliff (h_2) was 10 meters higher, its height would be +10 meters. The distance fallen by the water would be h_2-h_1 —that is, 10 — 0, or 10 meters.

We could also let sea level be the standard. In that case, h_1

might be + 1727 meters, and h_2 would then be + 1737 meters; h_2–h_1 would be 1737–1727, or still 10 meters. The most strictly rational zero point for height (at least on earth) would be the earth's center. In that case, the values of h_1 and h_2 might be 6,367,212 meters and 6,367,222 meters, respectively, and h_2–h_1 would still be 10 meters. Indeed, we could let the top of the cliff be our zero point. If h_2 is 0, then h_1, representing the water level of the pool, ten meters lower than the cliff height, would have the value –10 meters. In that case, h_2–h_1 would be 0–(–10), or *still* 10 meters.

I have belabored this point in order to make it perfectly clear that it is not the absolute values of h_1 and h_2 that count in deciding the amount of work we can extract from the energy of falling water, but only the difference between them.

Furthermore, if we continue to consider the waterfall, a clear distinction can be drawn between the total energy content of the water and the available energy content. The water drops to the bottom of the waterfall and forms part of a quiet pool there. The pool by itself is not capable of turning a water wheel, yet it contains much potential energy. If a hole were dug, the water in that pool would drop further and some of its energy could be converted to work, provided that a water wheel was placed at the bottom of the hole. Ideally, a hole could be dug to the center of the earth, and then all the potential energy of the water (at least with respect to the earth) could be used. However, in actual practice no hole is dug, and only the energy of the falling water of the actual waterfall is used. That energy is available. The further potential energy of the water, counting down to the center of the earth, is present but unavailable.

We can apply this sort of reasoning to the flow of heat. In the steam engine (or in any heat engine—for example, one that might use mercury vapor instead of steam) heat flows from a hot region, the steam cylinder, to a cold region, the condenser. The heat flows from the high temperature to the low temperature, as water flows from a greater height to a lesser one. It is not the value of either the high or the low temperature which dictates the amount of energy that can be converted to work, but rather the temperature difference. It is fair, then, to represent the *available energy* in terms of the temperature difference within the heat engine. We can express this most conveniently in terms of absolute temperature (see page 193), a concept not yet fully worked out at the time of Carnot's premature death from cholera at the age of 36. If we consider the hot region of the heat engine to be at a temperature T_2 and the cold region to be at T_1, then the available energy can be represented as T_2-T_1.

The cold region of the steam engine still contains heat, of course. If the condenser is at a temperature of 25°C, the water it contains (formed from the condensed steam) can, in principle, be cooled further and frozen, then cooled still further down to absolute zero; in the same way, water can be allowed to drop, in principle, to the earth's center. The *total energy* of the system would be represented by the difference between the temperature of the hot region and absolute zero—that is T_2-0, or simply T_2.

The maximum efficiency (E) of such a heat engine would be the ratio of the available energy to the total energy. If, under the conditions of the heat engine, all the energy of a system could be converted, in principle, to work, then the efficiency would be 1.0; if half the total energy could be converted into work, E would equal 0.5, and so on. Expressing available energy and total energy in terms of temperature differences, we can say then that:

$$E = \frac{T_2-T_1}{T_2} \qquad \text{(Equation 15–1)}$$

Thus, suppose that steam at a temperature of 150°C (423°K) is condensed to water at 50°C (323°K). The maximum efficiency would then be (423–323)/423, or 0.236. Less than a quarter of the total heat in the steam would be available for conversion into work.

What's more, even this value is reached only if the heat engine is mechanically perfect: if there are no losses of energy through friction; none through radiation of heat to the outside world, and so on. In actual practice, heat engines are considerably less efficient than the maximum predicted by Equation 15–1. What equation 15–1 does, however, is to set a maximum beyond which even mechanical perfection cannot pass.

Equation 15–1 is derived on the assumption that heat flows only from a hot region to a cold, never vice versa. It, too, is therefore an expression of the second law of thermodynamics (see page 226). The second law can therefore be viewed as setting a new kind of limitation on the utilization of energy.

The first law of thermodynamics (the law of conservation of energy) makes it plain that one cannot extract more energy from a system than the total energy present in the first place. The second law of thermodynamics maintains that it is impossible to extract more work from a system than the quantity of available energy present, and that the available energy present is invariably less than the total energy present unless a temperature of absolute zero can be attained.*

The second law of thermodynamics points out an important fact. In order to extract work from a heat engine, there must be a temperature difference. Suppose the hot region and the cold region were at the same temperature, both T_2. Equation 15–1 would then become $(T_2–T_2)/T_2$, or 0. There would be no available energy. (In the same way, no work could be done by a waterfall cascading down a height of 0 meters).

If this were not so, it would be conceivable that a ship traveling over the ocean could suck in water, make use of some of its energy content and then expel that water (cooler now than it was before) back into the ocean. All the ships in the world, and indeed all of man's other devices, could be run at the cost of a trifling fraction of the enormous quantity of energy in the ocean. The ocean would cool slightly in the process, and the atmosphere would warm, but the heat would flow back from air to water and all would be well.

If the second law of thermodynamics as expressed by Equation 15–1 is valid, however, this is impossible. To extract heat from the ocean, you would need a reservoir colder than the ocean and a refrigerating device to keep it colder than the ocean. The energy expended on refrigeration would be greater than the energy extracted from the ocean (assuming the refrigeration device to be mechanically imperfect, as it must be) and nothing would be gained. In fact, energy will have been lost. Virtually all "perpetual motion machines" worked up by hopeful inventors violate the second law of thermodynamics in one way or another. Patent offices will not even consider applications for such devices unless working models are supplied, and there seems little chance that a working model of such a device can ever be constructed.

Entropy

In the hands of Carnot, the second law of thermodynamics was of only limited application. He dealt only with heat engines and specifically omitted from consideration engines that worked by other means (by human or animal agency, for instance, or by the power of wind). Indeed, in Carnot's time, even the first law of

* It has been said that the first law of thermodynamics states, "You can't win," and that the second law of thermodynamics adds, "And you can't break even, either."

thermodynamics was not yet thoroughly understood in its broadest sense.

In the 1840's, however, when Joule had demonstrated the interconversion of heat and a variety of other kinds of energy, and Helmholtz had specifically declared the law of conservation of energy to be of universal generality (see page 100), it seemed that the second law, dictating the direction of flow of heat, might also be made universally applicable. In heat engines, a temperature difference was required before energy could be converted to work, but not all work-producing devices were heat engines. It was possible to obtain work out of some systems in which there was only one level of temperature.

Thus, work can be obtained from electric batteries where no temperature differences are involved. Here, however, there are differences in electrical potential (a matter which is not discussed in this book) that represent available energy. Again, chemical reactions can be made to do work though the final products of the reaction might be at the same temperature as the original reagents. The difference in chemical potential would represent the available energy in that case.

To make the second law of thermodynamics fully general, it must be seen to apply to electrical energy, to chemical energy, indeed to all forms of energy, and not to heat alone. In whatever form energy exists, work can only be obtained if the energy is present in a state of greater intensity in one portion of the system and lesser intensity in another portion. (In the case of heat, the intensity is measured as temperature; in other forms of energy, it is measured in other ways.) It is the difference in intensity that measures the available energy. What is left of the total energy content after the available energy is subtracted is the unavailable energy.

In 1850, the German physicist Rudolf Julius Emanuel Clausius (1822–1888) saw the true generality of Carnot's findings and announced it, specifically, as the second law of thermodynamics. (For this reason, Clausius is usually given the credit for being its discoverer.)

Now let's consider the second law again. In a heat engine, the temperature difference between the hot region and the cold region is the measure of the available energy. However, the second law states that in a closed system heat must flow from a hot region to a cold. With time, therefore, this temperature difference must decrease, for as the heat flows in the only direction it can flow, the hot region cools down and the cold region warms up. Consequently, the available energy decreases with time. Since the total energy remains constant, the unavailable energy must increase as the available energy decreases.

Of course, we might remove the restriction of a closed system so that we can allow heat to enter the hot region from outside and keep it from cooling down. We can also pump heat out of the cold region and keep it from warming up. (This is done in actual steam engines, where burning fuel keeps the steam chamber continually hot, and running cold water keeps the condenser continually cold.) It takes energy to pump heat into the hot region and out of the cold region, however. We are increasing the total energy of the system merely to keep the available energy constant. As total energy goes up while available energy remains constant, the unavailable energy goes up, too.

In short, no matter how we argue matters in the case of a heat engine, unavailable energy increases with time. We might make this increase a very slow one, if we insulate the system well enough to minimize heat flow from hot to cold. If we had a perfect insulator, we might even conceive of a situation in which the unavailable energy did not increase.

What applies to heat engines ought also apply to all work-producing devices. We might say then that the unavailable energy in any system can remain unchanged under ideal conditions, but always increases with time under actual conditions.

Clausius invented the word *entropy* (a word of uncertain derivation) to serve as a measure of the unavailability of energy. He showed that entropy could be expressed as heat divided by temperature. The units of entropy therefore are calories per degree Celsius. We can say then that the entropy of a system can remain unchanged under ideal conditions, but always increases with time under actual conditions. And this, too, is an expression of the second law of thermodynamics.

You must remember that the laws of thermodynamics apply to closed systems only. If we consider an open system, it is only too simple to find examples of apparent decreases in entropy.

In a refrigerator, for instance, heat is constantly being pumped from the cold objects within to the warm atmosphere outside in apparent defiance of the second law. A warm object, placed within the refrigerator, cools down; therefore, the available energy (represented by the temperature difference between the air outside and the object within the refrigerator) increases.

Where forms of energy other than heat are concerned, analogous "violations" of the second law of thermodynamics can be demonstrated. A man can walk uphill, increasing the available energy as measured by the difference in potential energy between himself and the bottom of the valley. Iron ore can be refined to pure iron and a spent storage battery can be charged—the former representing an "uphill movement" in chemical energy, the latter an "uphill movement" in electrical energy.

In every case cited, the system is not closed; energy is flowing into the system from outside. In order to make the second law of thermodynamics valid, the source of this outside energy must be included in the system so that it is "outside" no more.

Thus, material within the refrigerator does not spontaneously cool down (and remember that the original expression of the second law, see page 226, speaks only of a spontaneous flow of heat). Instead, the cooling takes place only because a motor is working within the refrigerator. Although the entropy of the refrigerator's interior is decreasing, that of the motor is increasing. Furthermore, the motor's increase is greater than the interior's decrease, so the net change in entropy over the entire system—the refrigerator's interior plus its motor—is an increase.

In the same way, the entropy decrease involved in converting iron ore to iron is smaller than the entropy increase involved in the burning coke and in the other reactions that bring about the refining of iron. The entropy increase in the electric generator supplying the electricity for the charging of the storage battery is greater than the entropy decrease of the storage battery itself as it is charged. The entropy decrease involved in a man walking uphill is less than the entropy increase involved in the reactions within his tissues which make the chemical energy of foodstuffs available for the effort involved in walking uphill.

This is true also of various large-scale, planet-wide processes that seem to involve a decrease in entropy. Examples of such entropy-decreasing phenomena are the uneven heating of the atmosphere, which gives rise to wind and weather; the lifting of uncounted tons of water miles high against the pull of gravity, which gives rise to rain and rivers; the conversion by green plants of carbon dioxide in the atmosphere to complicated organic compounds, which is the basis of the earth's never-ending food supply and of its coal and oil as well. It is because of these phenomena that the available energy on earth remains at approximately the same level through all its history; these phenomena also explain why we are in no danger of running out of available energy in the forseeable future.

Yet all these phenomena must not be considered in isolation, for all take place at the expense of the solar energy reaching the earth. It is solar energy that unevenly heats the atmosphere, that evaporates water, and that serves as the driving force for the photosynthetic activity of green plants. In the course of its radiation of heat and light, the sun undergoes a vast increase in entropy* —one that is much vaster than the relatively puny decreases of entropy in earth-bound phenomena.

In other words, if we include within our system all the activities that affect the system, then it turns out that the net change in entropy is *always* an increase. When we detect an entropy decrease, it is invariably the case that we are studying part of a system and not an entire one.

In actual practice we can never be sure that we are dealing with a closed system. No matter how we insulate, there are always influences from outside—energy gains and energy losses from and to the outside. All processes on the earth are affected by solar energy, and even if we consider the earth and sun together as one large system, there are gravitational and radiational influences from other planets and even other stars. Indeed, we cannot be certain that we are dealing with a truly closed system unless we take for our system nothing less than the entire universe.

In terms of the universe we can (as Clausius did) express the laws of thermodynamics with utmost generality. The first law of thermodynamics would be: *The total energy of the universe is constant.* The second law of thermodynamics would be: *The total entropy of the universe is continually increasing.*

Now suppose the universe is finite in size. It can then contain only a finite amount of energy. If the entropy of the universe (which is the measure of its unavailable energy content) is continually increasing, then eventually the unavailable energy will reach a point where it is equal to the total energy. Since the unavailable energy cannot rise beyond that point, the entropy of the universe will have reached a maximum.

In this condition of maximum entropy, no available energy remains, no processes involving energy transfer are possible, no work can be done. The universe has "run down."

* We might proceed to wonder how the sun was formed, for this formation must have involved a vast entropy decrease in order to make it possible for the sun to continue radiating, at the expense of a continual large entropy increase, for so many billions of years. However, to trace matters back beyond the sun would be more suitable in a book devoted to astronomy.

Disorder

Observations and experiments on heat in the first half of the nineteenth century assumed heat to be a fluid. From the very start of the century, however, evidence indicating that heat was not a fluid, but a form of motion, had begun to mount.

In 1798, for instance, Benjamin Thompson, Count Rumford (1753–1814), a Tory exile from the United States, was boring cannon in the service of the Elector of Bavaria. He noted that great quantities of heat were formed. Neither the cannon being bored nor the boring instrument used was at more than room temperature to begin with, and yet the heat developed by the act of boring was sufficient to bring water to a boil after a time; and the longer the boring was continued the more water could be boiled. It almost appeared as though the quantity of heat contained within the cannon and borer was infinite.

If heat were a fluid, and a form of matter, then to suppose it were formed in the act of boring raised a difficulty. Already, the French chemist Antoine Laurent Lavoisier (1743–1794) had established the law of conservation of matter, according to which matter could be neither created nor destroyed; and there was an increasing tendency among scientists to believe this generalization to be valid. If heat were being formed, then it must be something other than matter. To Rumford, the most straightforward possibility was that the motion of the boring instrument against the metal of the cannon was transformed into the motion of small parts of both borer and metal, and that it was this internal motion that was heat.

This notion was largely disregarded during the following decades. The assumption that small parts of an object might be moving invisibly seemed in 1800 to be just as difficult to accept as the assumption that matter was being created, perhaps even more difficult. A decade after Rumford's experimenting, however, the atomic theory was advanced and began to increase in popularity. By the internal movements of matter, one now meant the motions or vibrations of the atoms and molecules making it up, and the assumption of such motion became continually more acceptible. In the 1840's Joule's experiments in converting work to heat (see page 99) extended Rumford's observations and made the victory of the atomic motion view of heat inevitable. Finally, in the 1860's, the kinetic theory of gases and the concept of heat as a form of motion on the atomic scale were established rigorously by Maxwell and Boltzmann (see page 200).

This did not mean that the laws of thermodynamics, established in the first place on the basis of a fluid theory of heat, turned out to be false. Not at all! The laws were based on observed phenomena, and they remained valid. What had to be changed were the theories that explained why they were valid. The fluid theory of heat, to be sure, explained these phenomena very neatly,* but the atomic motion theory could be made to explain everything the fluid theory of heat could, and proved just as firm a foundation for the observation-based laws of thermodynamics.

To be sure, the view of heat as atomic motion is somewhat more difficult to picture and explain than the view of heat as a fluid. In the latter case, we can think of such familiar objects as waterfalls; in the former, the best we can do is imagine a set of perfectly elastic billiard balls bouncing about eternally in a closed chamber. One might suppose that of two theories one ought to accept the simpler, as Ockham's razor (see page 5) recommends. However, Ockham's razor is applied properly only when two or more theories explain all relevant facts with equal ease. This is not so in the present case.

If we confine ourselves to heat flow only, then it is easier to picture heat as a fluid than as atomic motion. However, if we are to explain the effect of heat on gas pressure and gas volume, if we are to explain specific heat, latent heat, and a host of other phenomena, it becomes very difficult to use the fluid theory. On the other hand, the atomic motion theory not only can explain heat flow but also all the other heat-involved phenomena.

Suppose, for instance, you have a hot body and a cold body in contact. The molecules in the hot body are, on the average, moving or vibrating more rapidly than the molecules in the cold body. To be sure, the molecules in both bodies possess a range of velocities, and there may be some molecules in the cold body that are moving more quickly than some molecules in the hot body, but this is an exceptional situation. When a molecule from the hot body (an "H molecule") collides with one from the cold body (a "C molecule") the chances are very good that it will be the H molecule that will be moving the more quickly of the two. Another way of putting it is that if a great number of H molecules

* In fact, it was just because it explained them so neatly that the fluid theory lasted as long as it did in the face of mounting evidence against it. It was distressing to have to give up something so convenient.

collide with a great number of C molecules, there will be a few cases where the C molecule is moving more rapidly than the H molecule with which it collides, but a vast preponderance of cases where it is the H molecule that is the more rapid of the two.

Now when two moving objects collide and rebound, the velocities of both may change in any of a large number of ways. These changes may be grouped into one of two classes. In the first class, the slower object may lose velocity in the process of collision while the faster object may gain velocity. The result would be that the slower object would finish by moving still more slowly, and the faster object would finish by moving still more quickly. In the second class, the slower object may gain velocity in the process of collision while the faster object may lose velocity. In the first class of collisions, the velocities become more extreme, in the second class more moderate.

There are many more ways in which a collision can belong to the second class than to the first. This means that over a large number of collisions in which velocity redistributes itself in a purely random manner, there will be many more collisions resulting in more moderate velocities than in more extreme velocities. Random collisions will bring about an "averaging out" of velocities.*

When a hot body and a cold body are in contact, a large number of H molecules collide with a large number of C molecules; the result is that after rebounding, the H molecules are moving less quickly on the whole, and the C molecules are moving more quickly. This means that the H molecules have become cooler and the C molecules warmer. There has been a flow of heat from the H molecules to the C molecules. The temperature of the portion of the hot body in contact drops, and that of the portion of the cold body in contact rises.

Such collisions continue not only at the boundary at which the hot and cold bodies meet, but also within the substance of each. In the hot body, for instance, H molecules that have been cooled

* This does not mean that all velocities will ultimately be exactly equal if only there are enough collisions. If two objects collide at equal velocities, it becomes very probable that there will be a gain in velocity of one at the expense of the other. Too much "averaging out" becomes very unlikely, therefore. Instead, "averaging out" proceeds only to a certain point and stops. At a particular temperature, the "averaging out" produces a range of velocities such as that predicted by the Maxwell-Boltzmann equations. A smaller and more limited range is extended to that point by collisions; a wider and more extended range is contracted to that point by collisions.

off by collisions with C molecules collide with neighboring molecules that have not been cooled off; here, too, there is a general moderation of velocities.

The result of these random collisions and random alterations of velocity throughout the entire system is that, eventually, the average velocities of the molecules in any portion of the system will be the same as in any other portion; this average will be a value that will lie between the two original extremes. (Hot and cold mix to produce lukewarm, so to speak.) Once the velocities are the same, on the average, throughout the system, collisions may continue to alter velocities, so a particular molecule may be moving quickly at one moment and slowly at another; however, the average will no longer change. The entire system having reached an intermediate equilibrium temperature, heat flow will cease.

In both the fluid theory of heat and the atomic motion theory, heat can be expected to flow spontaneously from a hot area to a cold area and this, after all, is a statement of the second law of thermodynamics. Yet there is a crucial difference between the two theories with respect to such heat flow.

In the fluid theory, the flow of heat is absolute. It is capable of going "downhill" only, and an "uphill" movement is inconceivable. In the atomic motion theory, however, the flow of heat is a statistical matter and is not absolute. The random changes of velocity as a result of random collisions will result, as a matter of extremely high probability but *not certainty,* in the flow of heat from hot to cold. It is extremely unlikely, but *not inconceivable,* that in every collision, the faster molecule may gain velocity at the expense of the slower one, so heat will flow "uphill" from cold to hot.

Maxwell tried to dramatize this possibility by visualizing a scientific fantasy. Imagine two gas-filled vessels, H and C, connected by a stopcock. The H vessel is the hotter, and its molecules move the more rapidly on the average.

But it is only on the average that H molecules move more rapidly than C molecules. Some H molecules happen to move slowly, and some C molecules happen to move rapidly. Suppose that an intelligent atom-sized creature is in control of the stopcock (this creature is usually referred to as "Maxwell's Demon"). When one of the minority of slow H molecules approaches, Maxwell's Demon opens the stopcock and lets it into the C chamber. When one of the minority of fast C molecules approaches, Maxwell's Demon opens the stopcock and lets it into the H chamber. At other times, the Demon keeps the stopcock closed. In this way, there is a slow but steady drizzle of low-velocity molecules into C and an equally slow and equally steady drizzle of high-velocity molecules into H. The average velocity of the molecules in C drops, while that in H rises—and heat flows uphill from cold to hot.

The chance of such "uphill" flows of heat (or of any other form of energy) is so fantastically small in the ordinary affairs of life that it is quite safe to ignore it. However, the shift from a condition of "certainty" to a condition of "probability" is of crucial importance. As scientists probed deeper and deeper into the subatomic world during the twentieth century, statistical analysis of events and their consequences became more and more important and the improbable (but not impossible) gains a perceptible chance of taking place, while more and more of those cause/effect combinations we usually assume to be certain have been shown to be only very, very, very probable. In short, Maxwell's statistical interpretation of heat flow marks one of the first steps in the transition from the "classical physics" of the nineteenth century (with which this volume is concerned) to the "modern physics" of the twentieth century.

And how can entropy be interpreted in the light of the atomic motion view of heat? Entropy, according to the second law of thermodynamics, always increases. Well, then, what is it that always increases as a result of molecular collisions? In a manner of speaking, moderation does. If in a system, to begin with, an accumulation of heat is concentrated in one portion and there is a deficit in another, molecular collisions increase moderation and spread the heat more evenly throughout the system. In the end, when temperature equilibrium is reached, heat is spread out as evenly as possible.

Entropy can therefore be interpreted as a measure of the evenness with which energy is distributed. This can be applied to any form of energy and not merely to heat. When an electric battery discharges, its electrical energy is more and more evenly distributed over its substance and over the material involved in the electrical flow of current. In the course of a spontaneous chemical reaction, chemical energies are more evenly distributed over the molecules involved.

What's more, the evenness of energy distribution is "most even," so to speak, when it is distributed as random motion among molecules. The conversion of any form of non-heat energy to heat represents a gain in the evenness with which energy is distributed and is, therefore, a gain in entropy.

It is for this reason that any process involving a transfer of energy is bound to produce heat as a side-product. A body in motion will produce heat as a result of friction or air resistance, and some of its kinetic energy will be spread out over the molecules with which it has come in contact. In converting electrical energy to light or to motion, heat is also produced, as we know if we touch an electric light bulb or an electric motor.

This means, in reverse, that if heat were completely converted into some form of non-heat energy, then there would automatically be a decrease in entropy. But a decrease in entropy in a closed system is so extremely unlikely that the possibility of its occurrence under ordinary conditions can be ignored. Some heat, to be sure, can be converted into other forms of energy, but only at the expense of further increasing the entropy of the remaining heat in the system. In the steam engine, for instance, the conversion of the heat energy of the steam into the kinetic energy of the pistons is a piece of decreasing entropy that is at the expense of the (still greater) increasing entropy of the burning fuel that produces the steam.

The increasing evenness with which energy is spread out can be interpreted as increasing "disorder." We interpret order as a quality characterized by a differentiation of the parts of a system: a separating of things into categories; a filing of cards in alphabetical order; a listing of things in terms of increasing quantities. To spread things out with perfect evenness is to disregard all these differentiations. A particular category of objects is evenly spread out among all the other categories, and that is maximum disorder.

For this reason, when we shuffle a neatly stacked deck of cards into random order, we can speak of an increase in entropy. And, in general, all spontaneous processes do indeed seem (in line with the second law of thermodynamics) to bring about an increase of disorder. Unless a special effort is made to reverse the order of things (increasing our own entropy), neat rooms will tend to become messed up, shining objects will tend to become dirty, things remembered will tend to become forgotten, and so on.

We thus find there is an odd and rather paradoxical symmetry to this book. We began with the Greek philosophers making the first systematic attempt to establish the generalizations underlying the order of the universe. They were sure that such an order, basically simple and comprehensible, existed. As a result of the continuing line of thought to which they gave rise, such generalizations were indeed discovered. And of these, the most powerful of all the generalizations yet discovered—the first two laws of thermodynamics—succeed in demonstrating that the order of the universe is, first and foremost, a perpetually increasing disorder.

PART TWO

Light, Magnetism, and Electricity

CHAPTER 16

Mechanism

The Newtonian View

In the first volume of this book, I dealt with energy in three forms: motion (kinetic energy), sound, and heat. As it turned out, sound and heat are forms of kinetic energy after all. In the case of sound, the atoms and molecules making up the air, or any other medium through which sound travels, move back and forth in an orderly manner. In this way, waves of compression and rarefaction spread out at a fixed velocity (see page I–156).* Heat, on the other hand, is associated with the random movement of the atoms and molecules making up any substance. The greater the average velocity of such movement, the greater the intensity of heat (see page I–234).

By the mid-nineteenth century the Scottish physicist James Clerk Maxwell (1831–1879) and the Austrian physicist Ludwig Boltzmann (1844–1906) had worked out, in strict detail, the interpretation of heat as random molecular movement (the "kinetic theory of heat"). It then became more tempting than ever to suspect that all phenomena in the universe could be analyzed as being based on matter in motion.

* When it is necessary to refer to a passage in the first volume, I will precede the page reference by "I." When the reference is to a page in this volume, it will be given without qualification. In other words, I will say "see page I–123" for a reference to the first volume but "see page 123" for a reference to this one.

According to this view, one might picture the universe as consisting of a vast number of parts; each part, if moving, affecting those neighboring parts with which it makes contact. This is exactly what we see, for instance, in a machine like an ordinary clock. One part of the clock affects another by the force of an expanding spring; by moving, interlocking gears; by levers; in short, by physical interconnections of all kinds. In other machines, such interconnections might consist of endless belts, pulleys, jets of water, and so on. On the submicroscopic scale it is atoms and molecules that are in motion, and these interact by pushing each other when they collide. On the cosmic scale, it is the planets and stars that are in motion, and these interact with each other through gravitational influence.

From the vast universe down to the tiniest components thereof, all might be looked on as obeying the same laws of mechanics by physical interaction as do the familiar machines of everyday life. This is the philosophy of mechanism, or the mechanistic interpretation of the universe. (Gravitational influence does not quite fit this view, as I shall point out shortly.)

The interactions of matter in motion obey, first of all, the three laws of motion (see page I–23ff.) propounded by Isaac Newton (1642–1727) in 1687, and the law of universal gravitation that he also propounded. The mechanistic view of the universe may therefore be spoken of, fairly enough, as the "Newtonian view of the universe."

The entire first volume of this book is devoted to the Newtonian view. It carries matters to the mid-nineteenth century, when this view had overcome all obstacles and had gained strength until it seemed, indeed, triumphant and unshakable.

In the first half of the nineteenth century, for instance, it had been found that Uranus traveled in its orbit in a way that could not be quite accounted for by Newton's law of universal gravitation. The discrepancy between Uranus's actual position in the 1840's and the one it was expected to have was tiny; nevertheless the mere existence of that discrepancy threatened to destroy the Newtonian fabric.

Two young astronomers, the Englishman John Couch Adams (1819–1892) and the Frenchman Urbain Jean Joseph Leverrier (1811–1877), felt that the Newtonian view could not be wrong. The discrepancy had to be due to the existence of an unknown planet whose gravitational influence on Uranus was not being allowed for. Independently they calculated where such a planet had to be located to account for the observed discrepancy in

Uranus's motions, and reached about the same conclusion. In 1846 the postulated planet was searched for and found.

After such a victory, who could doubt the usefulness of the Newtonian view of the universe?

And yet, by the end of the century, the Newtonian view had been found to be merely an approximation. The universe was more complicated than it seemed. Broader and subtler explanations for its workings had to be found.

Action at a Distance

The beginnings of the collapse were already clearly in view during the very mid-nineteenth-century peak of Newtonianism. At least, the beginnings are clearly to be seen by us, a century later, with the advantage of hindsight. The serpent in the Newtonian Eden was something called "action at a distance."

If we consider matter in motion in the ordinary world about us, trying to penetrate neither up into the cosmically vast nor down into the submicroscopically small, it would seem that bodies interact by making contact. If you want to lift a boulder you must touch it with your arms or use a lever, one end of which touches the boulder while the other end touches your arms.

To be sure, if you set a ball to rolling along the ground, it continues moving even after your arm no longer touches it; and this presented difficulties to the philosophers of ancient and medieval times. The Newtonian first law of motion removed the difficulty by assuming that only *changes* in velocity required the presence of a force (see page I–24). If the rolling ball is to increase its velocity, it must be struck by a mallet, a foot, some object; it must make contact with something material. (Even rocket exhaust, driving backward and pushing the ball forward by Newton's third law of motion, makes material contact with the ball.) Again, the rolling ball can be slowed by the friction of the ground it rolls on and touches, by the resistance of the air it rolls through and touches, or by the interposition of a blocking piece of matter that it must touch.

Material contact can be carried from one place to another by matter in motion. I can stand at one end of the room and knock over a milk bottle at the other end by throwing a ball at it. I exert a force on the ball while making contact with it; then the ball exerts a force on the bottle while making contact with it. We have two contacts connected by motion. If the milk bottle is balanced precariously enough, I can knock it over by blowing at it. In that case, I throw air molecules at it, rather than a ball, but the principle is the same.

Is it possible, then, for two bodies to interact without physical contact at all? In other words, can two bodies interact across a vacuum without any material bodies crossing that vacuum? Such action at a distance is very difficult to imagine; it is easy to feel it to be a manifest impossibility.

The ancient Greek philosopher Aristotle (384–322 B.C.), for instance, divined the nature of sound partly through a refusal to accept the possibility of action at a distance. Aristotle felt that one heard sounds across a gap of air because the vibrating object struck the neighboring portion of air, and that this portion of the air passed on the strike to the next portion, the process continuing until finally the ear was struck by the portion of the air next to itself. This is, roughly speaking, what does happen when sound waves travel through air or any other conducting medium. On the basis of such an interpretation, Aristotle maintained that sound could not travel through a vacuum. In his day mankind had no means of forming a vacuum, but two thousand years later, when it became possible to produce fairly good vacuums, Aristotle was found to be correct.

It might follow, by similar arguments, that all interactions that seem to be at a distance really consist of a series of subtle contacts, and that no interaction of any kind can take place across a vacuum. Until the seventeenth century it was strongly believed that a vacuum did not exist in nature but was merely a philosophical abstraction, so there seemed no way of testing this argument.

In the 1640's, however, it became clear that the atmosphere could not be infinitely high (see page I–146). Indeed, it was possibly no more than a few dozen miles high, whereas the moon was a quarter of a million miles away, and other astronomical bodies were much farther still. Any interactions between the various astronomical bodies must therefore take place across vast stretches of vacuum.

One such interaction was at once obvious, for light reaches us from the sun, which we now know is 93,000,000 miles away.* This light can affect the retina of the eye. It can affect the chemical reactions proceeding in plant tissue; converted to heat, it can evaporate water and produce rain, warm air, and winds. Indeed, sunlight is the source of virtually all energy used by man. There is

* Our best telescopes can detect light that has traversed some 35,000,000,000,000,000,000 miles of vacuum.

thus a great deal of interaction, by light, between the sun and the earth across the vast vacuum.

Then, once Newton announced the law of universal gravitation in 1687, a second type of interaction was added, for each heavenly body was now understood to exert a gravitational force on all other bodies in the universe across endless stretches of vacuum. Where two bodies are relatively close, as are the earth and the moon or the earth and the sun, the gravitational force is large indeed, and the two bodies are forced into a curved path about their common center of gravity. If one body is much larger than the other, this common center of gravity is virtually at the center of the larger body, which the smaller then circles.

On the earth itself, two additional ways of transmitting force across a vacuum were known. A magnet could draw iron to itself, and an electrically charged body could draw almost any light material to itself. One magnet could either attract or repel another; one electric charge could either attract or repel another. These attractions and repulsions could all be exerted freely across the best vacuum that could be produced.

In the mid-nineteenth century, then, four ways of transmitting force across a vacuum, and hence four possible varieties of action at a distance, were known: light, gravity, electricity, and magnetism. And yet the notion of action at a distance was as unacceptable to nineteenth-century physicists as it had been to the philosophers of ancient Greece.

There were two possible ways out of the dilemma; two ways of avoiding action at a distance.

First, perhaps a vacuum was not really a vacuum. Quite clearly a good vacuum contained so little ordinary matter that this matter could be ignored. But perhaps ordinary matter was not the only form of substance that could exist.

Aristotle had suggested that the substance of the universe, outside the earth itself, was made up of something he called *ether*. The ether was retained in modern science even when virtually all other portions of Aristotelian physics had been found wanting and had been discarded. It was retained, however, in more sophisticated fashion. It made up the fabric of space, filling all that was considered vacuum and, moreover, permeating into the innermost recesses of all ordinary matter.

Newton had refused to commit himself as to how gravitation was transmitted from body to body across the void. "I make no hypotheses," he had said austerely. His followers, however, pictured gravitation as making its way through the ether much as sound makes its way through air. The gravitational effect of a body would be expressed as a distortion of that part of the ether with which it made contact; this distortion would right itself and, in the process, distort a neighboring portion of the ether. The traveling distortion would eventually reach another body and affect it. We can think of that traveling distortion as an "ether wave."

The second way out of the dilemma of action at a distance was to assume that forces that made themselves felt across a vacuum were actually crossing in the form of tiny projectiles. The projectiles might well be far too small to see, but they were there. Light, for instance, might consist of speeding particles that crossed the vacuum. In passing from the sun to the earth, they would make contact first with the sun and then with the earth, and there would be no true action at a distance at all, any more than in the case of a ball being thrown at a bottle.

For two centuries after Newton, physicists vacillated between these two points of view: waves and particles. The former required an ether, the latter did not. This volume will be devoted, in good part, to the details of this vacillation between the two views. In the eighteenth century, the particle view was dominant; in the nineteenth, the wave view. Then, as the twentieth century opened, a curious thing happened—the two views melted into each other and became one!

To explain how this happened, let's begin with the first entity known to be capable of crossing a vacuum—light.

CHAPTER 17

Light

Transmission

Surely light broke in on man's consciousness as soon as he had any consciousness at all. The origins of the word itself are buried deep in the mists of the beginnings of the Indo-European languages. The importance of light was recognized by the earliest thinkers. In the Bible itself, God's first command in constructing an ordered universe was "Let there be light!"

Light travels in straight lines. This, indeed, is the assumption each of us makes from babyhood. We are serenely sure that if we are looking at an object that object exists in the direction in which we are looking. (This is strictly true only if we are not looking at a mirror or through a glass prism, but it is not difficult to learn to make the necessary exceptions to the general rule.)

This straight-line motion of light, its *rectilinear propagation,* is the basic assumption of *optics* (from a Greek word meaning "sight"), the study of the physics of light. Where the behavior of light is analyzed by allowing straight lines to represent the path of light and where these lines are studied by the methods of geometry, we have *geometric optics.* It is with geometric optics that this chapter and the next are concerned.

Consider a source of light such as a candle flame. Assuming that no material object blocks your vision at any point, the flame can be seen with equal ease from any direction. Light, therefore, can be visualized as streaming out from its source in all directions. The sun, for instance, can be drawn (in two dimensions) as a circle with lines, representing light, extending outward from all parts of the circumference.

Such lines about the drawing of the sun resemble spokes of a wheel emerging from the hub. The Latin word for the spoke of a wheel is *radius* (which gives us the word for the straight line extending from the center of a circle to its circumference). For this reason, the sun (or any light source) is said to *radiate* light, and light is spoken of as a *radiation.* A very thin portion of such a light radiation, one that resembles a line in its straightness and ultimate thinness, is a *light ray,* again from *radius.*

Sunlight shining through a hole in a curtain will form a pillar of light extending from the hole to the opposite wall where the intersection of the pillar with the wall will form a circle of bright illumination. If the air of the room is normally dusty, the pillar of light will be outlined in glittering dust motes. The straight lines bounding the pillar of light will be visible evidence of the rectilinear propagation of light. Such a pillar of light is a *light beam* (from the resemblance of its shape to the trunk of a tree; the German word for tree is "Baum," and a similar word, of course, is found in Anglo-Saxon). A light beam may be viewed as a collection of an infinite number of infinitesimally thin light rays.

Light sources vary in brightness. More light is given off by a hundred-watt light bulb than by a candle, and incomparably more light still is given off by the sun. To measure the quantity of light given off by a light source, physicists can agree to use some particular light source as standard. The obvious early choice for the standard was a candle made of a specified material (sperm wax was best) prepared in a particular way and molded to set specifica-

Variation of light intensity with distance

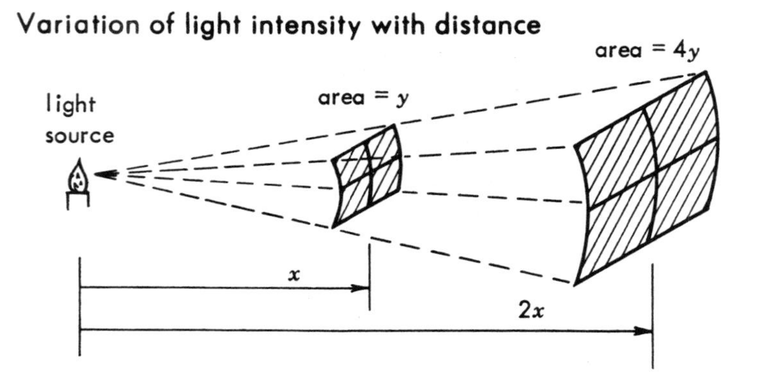

tions. The light emitted by this candle horizontally could then be said to equal 1 *candlepower.* Electric light bulbs of set form have now replaced the candle, especially in the United States, but we still speak of the *international candle,* a measure of light quantity about equal to the older candlepower.

The brightness of a light source varies in some fashion with the distance from which it is viewed: the greater the distance, the dimmer it seems. A book held near a candle may be read easily; held farther away it becomes first difficult and then impossible to read.

This is not surprising. Suppose a fixed amount of light is emerging from the candle flame. As it spreads out in all directions, that fixed amount must be stretched over a larger and larger area. We can imagine the edge of the illumination to be forming a sphere with the light source as center. The sphere's surface grows larger and larger as the light radiates outward.

From plane geometry we know that the surface of a sphere has an area proportional to the square of the length of its radius. If the distance from the light source (the radius of the imaginary sphere we are considering) is doubled, the surface over which the light is spread is increased two times two, or 4 times. If the distance is tripled, the surface is increased 9 times. The total quantity of light over the entire surface may remain the same, but the intensity of light—that is, the amount of light falling on a particular area of surface—must decrease. More, it must decrease as the square of the distance from the light source. Doubling the distance from the light source decreases the light intensity to 1/4 the original; tripling the distance decreases it to 1/9.

Suppose we use the square foot as the unit of surface area and imagine that square foot bent into the shape of a segment of a spherical surface so that all parts of it are equidistant from the centrally located light source. If such a square foot is just one foot distant from a light source delivering 1 candle of light, then the intensity of illumination received by the surface is 1 *foot-candle.* If the surface is removed to a distance of two feet, the intensity of its illumination is 1/4 foot-candle, and so on.

Since light intensity is defined as the quantity of light per unit area, we can also express it as so many candles per square foot. For this purpose, however, a unit of light quantity smaller than the candle is commonly used. This is the *lumen* (from a Latin word for "light"). Thus if one square foot at a certain distance from a light source receives 1 lumen of light, two square feet at that same distance will receive 2 lumens of light, and half a square foot will receive 1/2 lumen. In each case, though, the light intensity will be 1 lumen/foot.^{2} The lumen is so defined that an intensity of 1 lumen/foot^2 equals 1 foot-candle.

Imagine a light source of 1 candle at the center of a hollow sphere with a radius of one foot. The light intensity on each portion of the interior surface of the sphere is 1 foot-candle, or 1 lumen/foot.^{2} Each square foot of the interior surface is therefore receiving 1 lumen of illumination. The area of the surface of the sphere is equal to $4\pi r^2$ square feet. Since the value of r, the radius of the sphere, is set at 1 foot, the number of square feet of surface equals 4π. The quantity π (the Greek letter *pi*) is equal to about 3.14, so we can say that there are about 12.56 square feet on that spherical surface. The light (which we have set at 1 candle) is therefore delivering a total of 12.56 lumens, so we can say that 1 candle equals 12.56 lumens.

Light is transmitted, completely and without impediment, only through a vacuum. All forms of matter will, to some extent at least, absorb light. Most forms do so to such an extent that in ordinary thicknesses they absorb all the light that falls on them and are *opaque* (from a Latin word meaning "dark").

If an opaque object is brought between a light source and an illuminated surface, light will pass by the edges of the object but not through it. On the side of the object opposite the light source there will therefore be a volume of darkness called a *shadow*. Where this volume intersects the illuminated surface there will be a non-illuminated patch; it is this two-dimensional intersection of the shadow that we usually refer to by the word.

The moon casts a shadow. Half its surface is exposed to the direct illumination of the sun; the other half is so situated that the opaque substance of the moon itself blocks the sunlight. We see only the illuminated side of the moon, and because this illuminated side is presented to us at an angle that varies from 0° to 360° during a month, we watch the moon go through a cycle of phases in that month.

Furthermore, the moon's shadow not only affects its own surface, but stretches out into space for over two hundred thousand miles. If the sun were a "point source"—that is, if all the light came from a single glowing point—the shadow would stretch out indefinitely. However, the sun is seen as an area of light, and as one recedes from the moon its apparent size decreases until it can no longer cover all the area of the much larger sun. At that point, it no longer casts a complete shadow, and the complete shadow (or *umbra,* from a Latin word for "shadow") narrows to a point. The umbra is just long enough to reach the earth's surface, however, and on occasion, when the moon interposes itself exactly between earth and sun, a *solar eclipse* takes place over a small area of the earth's surface.

The earth casts a shadow, too, and half its surface is in that shadow. Since the earth rotates in twenty-four hours, each of us experiences this shadow ("night") during each 24-hour passage. (This is not always true for polar areas, for reasons better discussed in a book on astronomy.) The moon can pass through the earth's shadow, which is much longer and wider than that of the moon, and we can then observe a *lunar eclipse.*

Opaque materials are not absolutely opaque. If made thin enough, some light will pass through. Fine gold leaf, for instance, will be traversed by light even though gold itself is certainly opaque.

Some forms of matter absorb so little light (per unit thickness) that the thicknesses we ordinarily encounter do not seriously interfere with the transmission of light. Such forms of matter are *transparent* (from Latin words meaning "to be seen across"). Air itself is the best example of transparent matter. It is so transparent that we are scarcely aware of its existence, since we see objects through it as if there were no obstacle at all. Almost all gases are transparent. Numerous liquids, notably water, are also transparent.

It is among solids that transparency is very much the exception. Quartz is one of the few naturally occurring solids that display the property, and the astonished Greeks considered it a form of warm ice. The word "crystal," first applied to quartz, is from their word for "ice," and the word "crystalline" has as one of its meanings "transparent."

Transparency becomes less pronounced when thicker and thicker sections of ordinarily transparent substances are considered. A small quantity of water is certainly transparent, and the pebbles at the bottom of a clear pool can be seen distinctly. However, as a diver sinks beneath the surface of the sea, the light that can reach him grows feebler and feebler, and below about 450 feet almost no light can penetrate. Thicknesses of water greater than that are as opaque as if they represented the same thickness of rock, and the depths of the sea cannot be seen through the "transparent" water that overlays it.

Air absorbs light to a lesser extent than water does and is therefore more transparent. Even though we are at the bottom of an ocean of air many miles high, sunlight has no trouble penetrating to us, and we in turn have no trouble seeing the much feebler light of the stars.* Nevertheless some absorption exists: it is estimated, for instance, that 30 percent of the light reaching us from space is absorbed by that atmosphere. (Some forms of radiation other than visible light are absorbed with much greater efficiency by the atmosphere, and the thickness of air that blankets us suffices to make the air opaque to these radiations.)

Light is a form of energy, and while it can easily be changed into other forms of energy, it cannot be destroyed. While absorption by an opaque material (or a sufficient thickness of a transparent material) seems to destroy it, actually it is converted into heat.

Reflection

The statement that light always travels in a straight line is completely true only under certain circumstances, as when light travels through a uniform medium—through a vacuum, for instance, or through air that is at equal temperature and density throughout. If the medium changes—as when light traveling through air strikes an opaque body—the straight-line rule no longer holds strictly. Such light as is not absorbed by the body changes direction abruptly, as a billiard ball will when it strikes the edge of a pool table.

This bouncing back of light from an opaque body is called *reflection* (from Latin words meaning "to bend back").

The reflection of light seems to follow closely the rules that govern the bouncing of a billiard ball. Imagine a flat surface capable of reflecting light. A line perpendicular to that surface is called the *normal,* from the Latin name for a carpenter's square used to draw perpendiculars.† A ray of light moving along the normal strikes the reflecting surface head-on and doubles back in its tracks. A speeding billiard ball would do the same.

If the ray of light were traveling obliquely with respect to the reflecting surface, it would strike at an angle to the normal. The

* To be sure, if the atmosphere were compressed to the density of water, it would be only some 33 feet thick; and that thickness of water would retain considerable transparency, too.

† Straightforward behavior that is "square" and "on the beam," like a perpendicular, accurately drawn by a carpenter's square, is also "normal." Other types of behavior are "abnormal" or represent "enormities." In fact, the word "normal" has become so familiar in its sense of natural, commonplace, conforming behavior that its original meaning of "a line perpendicular to a plane, or to another line" has almost been forgotten.

light ray moving toward the surface is the *incident ray,* and its angle to the normal is the *angle of incidence.* The *reflected ray* would return on the other side of the normal, making a new angle, the *angle of reflection.* The incident ray, reflected ray, and normal are all in the same plane—that is, a flat sheet could be made to pass through all three simultaneously without its flatness being distorted.

Experiments with rays of light and reflecting surfaces in dusty air, which illuminates the light rays and makes them visible, will show that the angle of incidence (i) always equals the angle or reflection (r). This can be expressed, simply:

$$i = r \qquad \text{(Equation 2–1)}$$

Actually, it is rare to find a truly flat surface. Most surfaces have small unevennesses even when they appear flat. A beam of light, made up of parallel rays, would not display the same angle of incidence throughout. One ray might strike the surface at a spot where the angle of incidence is 0°; another might strike very close by where the surface has nevertheless curved until it is at an angle of 10° to the light; elsewhere it is 10° in the other direction, or 20°, and so on. The result is that an incident beam of light with rays parallel will be broken up on reflection, with the reflected rays

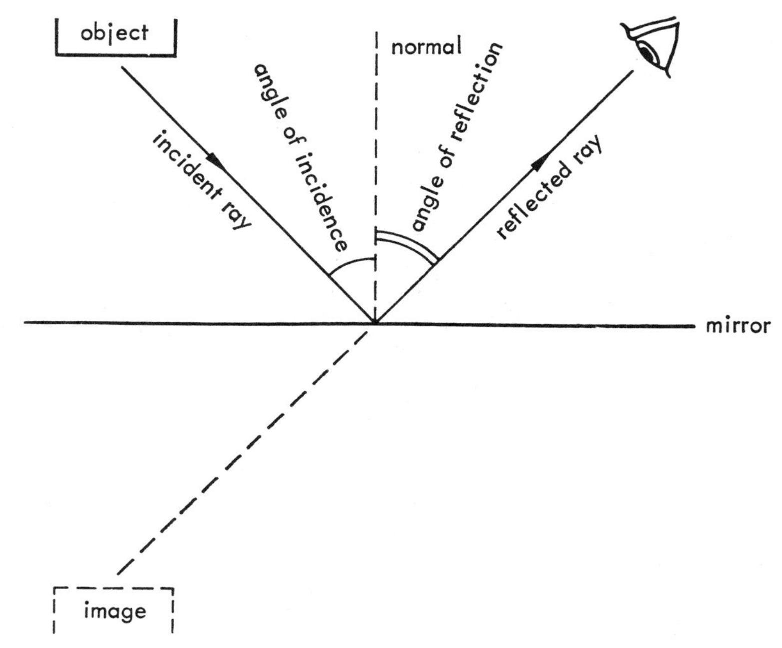

traveling in all directions over a wide arc. This is *diffuse reflection.*

Almost all reflection we come across is of this type. A surface that reflects light diffusely can be seen equally well from different angles, since at each of the various angles numerous rays of light are traveling from the object to the eye.

If a surface is quite flat, a good portion of the parallel rays of incident light will be reflected at the same angle. In such a case, although you can see the reflecting object from various angles, you will see far more light if you orient yourself at the proper angle to receive the main reflection. At that point you will see a "highlight."

If a surface is extremely flat, virtually all the parallel rays of an incident beam of light will be reflected still parallel. As a result, your eyes will interpret the reflected beam as they would the original.

For instance, the rays of light reflected diffusely from a person's face make a pattern that the eyes transmit and the brain interprets as that person's face. If those rays strike an extremely flat surface, are reflected without mutual distortion, and then strike your eyes, you will still interpret the light as representing that person's face.

Your eyes cannot, however, tell the history of the light that reaches them. They cannot, without independent information, tell whether the light has been reflected or not. Since you are used from earliest life to interpreting light as traveling in straight, uninterrupted lines, you do so now, too. The person's face as seen by reflected light is seen as if it were behind the surface of reflection, where it would be if the light had come straight at you without interruption, instead of striking the mirror and being reflected to you.

The face that you see in a mirror is an *image*. Because it does not really exist in the place you seem to see it (look behind the mirror and it is not there) it is a *virtual image.* (It possesses the "virtues" or properties of an object without that object actually being there.) It is, however, at the same distance behind the mirror that the reflected object is before it, and therefore seems to be the same size as the reflected object.

In primitive times virtually the only surface flat enough to reflect an image was a sheet of water. Such images are imperfect because the water is rarely quite undisturbed, and even when it is, so much light is transmitted by the water and so little reflected that the image is dim and obscure. Under such circumstances a primitive man might not realize that it was his own face staring back at him.

(Consider the Greek myth of Narcissus, who fell hopelessly in love with his own reflection in the water and drowned trying to reach it.)

A polished metal surface will reflect much more light, and metal surfaces were used throughout ancient and medieval times as mirrors. Such surfaces, however, are easily scratched and marred. About the seventeenth century the glass-metal combination became common. Here a thin layer of metal is spattered onto a sheet of flat glass. If we look at the glass side, we see a bright reflection from the metal surface covering the other side. The glass serves to protect the metal surface from damage. This is a *mirror* (from a Latin word meaning "to look at with astonishment," which well expresses primitive feelings about images of one's self) or *looking glass.* A Latin word for mirror is *speculum,* and for that reason the phrase for the undisturbed reflection from an extremely flat surface is *specular reflection.*

An image as seen in a mirror is not identical with the object reflected.

Suppose you are facing a friend. His right side is to your left; his left side is to your right. If you want to shake hands, right hand with right hand, your hands make a diagonal line between your bodies. If you both part your hair on the left side, you see his part on the side opposite that of your own.

Now imagine your friend moving behind you but a little to one side so that you can both be seen in the mirror before you. Ignore your own image and consider your friend's only. You are now facing, not your friend, but the image of your friend, and there is a change. His right side is on your right and his left side is on your left. Now the parts in your hair are on the same side, and if you hold out your right hand while your friend holds out his, your outstretched hand and that of the image will be on the same side.

In short, the image reverses right and left; an image with such a reversal is a *mirror image.* A mirror image does not, however, reverse up and down. If your friend is standing upright, his image will be upright, too.

Curved Mirrors

The ordinary mirror with which we are familiar is a *plane mirror*—that is, it is perfectly flat. A reflecting surface, however, need not be flat to exhibit specular reflection. It can be curved, as long as it is smooth. Parallel rays of light reflected from a curved surface are no longer parallel, but neither are they reflected in random directions. The reflection is orderly and the rays of light may *converge* (from Latin words meaning "to lean together") or *diverge* ("to lean apart").

The simplest curvature is that of a section of a sphere. If you are looking at the outside of the section, so that it forms a hill toward you with the center closest to you, it is a *convex surface* (from Latin words meaning "drawn together"). If you are looking at the inside of the spherical section, you are looking into a hollow with the center farthest from you. That is a *concave surface* ("with a hollow").

A spherical segment of glass, properly silvered, is a *spherical mirror.* If it is silvered on its convex surface so that you see it as a mirror if you look into its concave surface, it is, of course, a concave spherical mirror. The center of the sphere of which the curved mirror is part is the *center of curvature.* A line connecting the center of curvature with the midpoint of the mirror is the *principal axis* of the mirror.

Suppose a beam of light, parallel to the principal axis, falls upon the concave reflecting surface. The ray that happens to lie on the principal axis itself strikes perpendicularly and is reflected back upon itself. With a ray of light that strikes near the principal axis but not on it, the mirror has curved in such a way that the ray makes a small angle with the normal. It is reflected on the other side of the normal in a fashion that bends it slightly toward the principal axis. If the ray of light strikes farther from the principal axis, the mirror has bent through a larger angle and reflects the ray more sharply toward the principal axis. Since the mirror is a spherical segment and curves equally in all directions from the principal axis, this is true of rays of light striking either right or left of the principal axis, either above or below it. Reflec-

Concave spherical mirror

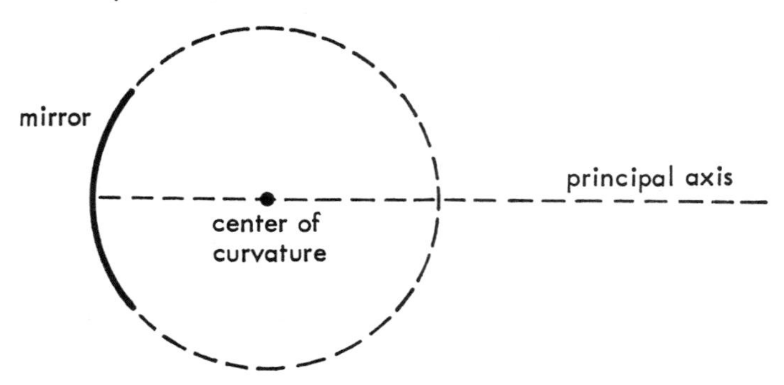

tions from every part of the mirror point toward the principal axis; the reflected rays converge.

If only those rays that strike fairly close to the midpoint of the mirror are considered, it is found that they converge in such a way as to meet in a restricted region—approximately at a point, in fact. This point is called a *focus* (from a Latin word for "hearth," which is where one would expect a concentration of light). The focus falls on the principal axis, halfway between the midpoint of the mirror and the center of curvature.

Actually, the reflected rays do not all meet exactly at the focus. This becomes obvious if we consider rays that fall on the spherical mirror quite a distance from the principal axis. The reflections of these rays miss the focus by a considerable distance. This is called *spherical aberration* (from the Latin, "to wander away"). These distant rays fall between the focus and the mirror itself and are therefore reflected through too great an angle. The mirror, in other words, has curved too sharply to bring all the rays to a focus.

To avoid this, we need a curved mirror that curves somewhat less sharply than a spherical segment does. The necessary curve is that of a *paraboloid of revolution.*

A spherical section, if it is continued, closes in upon itself and finally forms a sphere. A paraboloid of revolution looks like a spherical segment if only a small piece about the midpoint is taken. If it is continued and made larger, it does not close in upon itself. It curves more and more gently till its walls are almost straight, forming a long cylinder that becomes wider only very slowly. A mirror formed of a section (about the midpoint) of such a paraboloid of revolution is called a *parabolic mirror.*

If a beam of light parallel to the principal axis of such a parabolic mirror falls upon its concave surface, the rays do indeed converge upon a focus, and without aberration.

To produce such a beam of light, consisting of parallel rays, we must, strictly speaking, think of a point source of light on the principal axis an infinite distance from the mirror. If the point source is a finite distance away, then the rays striking the mirror from that point source are not truly parallel, but diverge slightly. Each ray strikes the mirror surface at an angle to the normal which is slightly smaller than it would be if the rays were truly parallel, and in consequence is reflected through a smaller angle. The rays therefore converge farther away from the mirror than at the focus. If the distance of the point source is large compared to the distance of the focus (which is only a matter of a few inches for the average parabolic mirror), then the rays converge at a point very near the focus—near enough so that the difference can be ignored.

If the light source is moved closer and closer to the mirror, the reflected rays converge farther and farther from the mirror. When the light source is at twice the distance of the focus from the mirror, eventually the reflected rays converge at the light source itself. If the light source is moved still closer, the reflected rays converge at a point beyond the light source.

Finally, if the light source is located at the focus itself, the reflected rays no longer converge at all, but are parallel. (We might say that the point of convergence has moved an infinite distance away from the mirror.) The automobile headlight works in this fashion. Its inner surface is a parabolic mirror, and the small incandescent bulb is at its focus. Consequently, such a headlight casts a fairly straight beam of light forward.

Let us call the distance of the light source from the mirror D_o, and the distance of the point of convergence of the reflected rays from the mirror, D_i. The distance of the focus from the mirror we can call f. The following relationship then holds true:

$$\frac{1}{D_o} + \frac{1}{D_i} = \frac{1}{f} \qquad \text{(Equation 2–2)}$$

We can check this for the cases we have already discussed. Suppose that the light source is at a very great distance (practically infinite). In that case, D_o is extremely large and $1/D_o$ is extremely small. In fact, $1/D_o$ can be considered zero. In that case, Equation

Parabolic mirror

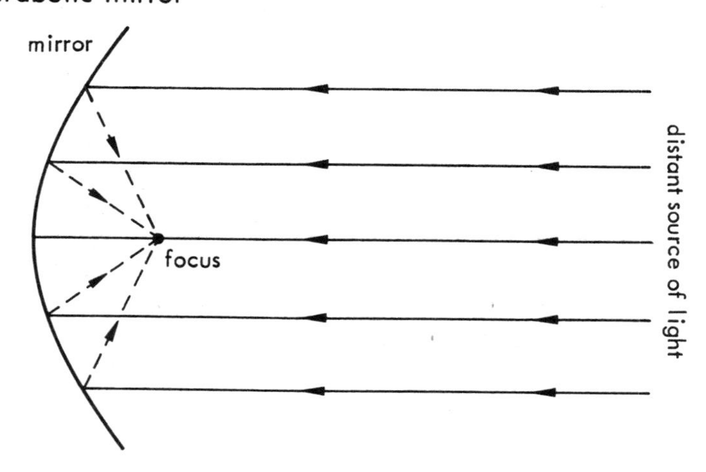

2–2 becomes $1/D_i = 1/f$, and $D_i = f$, which means that the reflected rays of light converge at the focus.

If the light source is on the principal axis but at twice the distance from the mirror that the focus is, then $D_o = 2f$, and Equation 2–2 becomes: $1/2f + 1/D_i = 1/f$. If we solve this equation for D_i, we find that $D_i = f$. In other words, the reflected rays in this case converge upon the location of the light source itself.

And what if the light source is located at the focus? In that case $D_o = f$. Equation 2–2 becomes $1/f + 1/D_i = 1/f$, from which you can see at once that $1/D_i = 0$. But if $1/D_i = 0$, then D_i must be infinitely large. The distance from the mirror at which the reflected rays converge is infinite, and therefore the rays do not converge at all but are parallel.

In the previous section I have been considering the source of light to be a point. Actually, of course, it is not really a point. Suppose the source of light is a candle flame which, naturally, covers an area. Some of the flame is slightly above the principal axis, some slightly below, some to one side, and some to another. The rays of light that originate somewhat above the principal axis are reflected to a point somewhat below the true point of convergence (that is, what would have been the true point if the candle flame had been a point source of light); those that originate below the principal axis are reflected to a point above the point of convergence; those that originate to the right are reflected to the left; those that originate to the left are reflected to the right. If we take any particular ray, the greater the distance from the principal axis it originates, the greater the distance from the point of convergence, but on the opposite side.

The result is that in the area where the reflected rays of light converge, one obtains an image in which not only left and right are interchanged (as in a plane mirror), but also up and down. An upside-down image is formed; indeed, if you look into the shiny bowl of a spoon, you will see your face upside down.

The image produced by such a concave mirror has another important difference from that produced by a plane mirror. The image produced by the plane mirror, as was stated earlier, is not actually behind the mirror where it seems to be, so it is a virtual image. In the case of a concave mirror, the image is formed in front of the mirror by means of converging light rays. The image is really there and can be touched; therefore it is a *real image*.

To be sure, when you actually touch a real image you don't seem to be touching anything, because you are used to considering touch only in connection with matter. A parabolic mirror does not converge matter; it converges light and you cannot touch light in the ordinary sense. However, you can sense light when it is absorbed by the skin and turned to heat; and in that respect, by feeling heat, you are "touching" the image.

A finger held six feet from a candle flame absorbs some heat from the radiation that falls directly upon it. The finger, however, intercepts but a small fraction of the total radiation of the candle, and the heating effect is insignificant. A concave mirror would intercept more of the candle's radiation and converge it to a small volume of space. The finger placed at the area of convergence would feel more heat in that area than elsewhere in the neighborhood. The increase in heat concentration may still be too little to feel, but if the concave mirror is used to concentrate the rays of the sun instead, you will certainly feel it. Large parabolic mirrors have been built that intercept solar radiation over a sizable area and converge it all. Temperatures as high as 7000°C have been reached at the focus of such *solar furnaces*. There is a real image that can be felt with a vengeance.

A mirror of changing curvature can produce odd and humorous distortions in the image, as anyone attending amusement parks knows. However, a proper image from a clean mirror of undistorted shape can seem completely legitimate, particularly if the boundaries of the mirror are masked so that the onlooker has no reason to feel that a mirror is there at all. The casual viewer mistakes image for reality and this is the basis for some of the tricks of the magician. Naturally, a real image is even more tantalizing than a virtual image. At the Boston Museum of Science, a real image is projected in such a way as to make coins seem to tumble about in an upside-down goblet in defiance of gravity. Onlookers (adults as well as children) never tire of placing their hands where the coins seem to be. Not all their insubstantiality can convince the eyes that the coins are not there.

Suppose the light source is moved still closer to the mirror than the distance of the focus. In that case, the reflected rays are neither convergent nor parallel; they actually diverge. Such diverging rays, spraying outward from a surface, may be considered as converging if you follow them backward. Indeed, if you follow them (in imagination) through the mirror's surface and into the space behind, they will converge to a point. At that point you will see an image. Because it appears behind the mirror where the light really doesn't penetrate, it is a virtual image, as in the case of a plane mirror; and, as in the case of a plane mirror, the image is now right-side up.

Equation 2–2 can be made to apply to this situation. If the light source is closer to the mirror than the focus is, then D_o is smaller than f, and $1/D_o$ must therefore be larger than $1/f$. (If this is not at once clear to you, recall that 2 is smaller than 4, and that 1/2 is therefore larger than 1/4.)

If we solve Equation 2–2 for $1/D_i$ we find that:

$$\frac{1}{D_i} = \frac{1}{f} - \frac{1}{D_o} \qquad \text{(Equation 2–3)}$$

Since in the case under consideration, $1/D_o$ is larger than $1/f$, $1/D_i$ must be a negative number. From this it follows that D_i itself must be a negative number.

This makes sense. In the previous cases under discussion, distances have all extended forward from the mirror. In the present case the point at which the reflected rays converge, and where the image exists, lies behind the mirror, and its distance should, reasonably, be indicated by a negative value.

Nor need Equation 2–2 be applied only to concave mirrors; it is more general than that.

Consider a plane mirror again. A beam of parallel rays striking it along its principal axis (any line normal to the plane mirror can be considered a principal axis) is reflected back along the principal axis as parallel as ever. The rays do not converge and therefore the distance of the focus from the mirror is infinitely great. But if f is infinitely great, then $1/f$ must equal zero and, for a plane mirror, Equation 2–2 becomes:

$$\frac{1}{D_o} + \frac{1}{D_i} = 0 \qquad \text{(Equation 2–4)}$$

If Equation 2–4 is solved for D_i, it turns out that $D_i = -D_o$. Because D_o (the distance of the object being reflected) must always be positive since it must always lie before the mirror in order to be reflected at all, D_i must always be negative. In a plane mirror, therefore, the image must always lie behind the mirror and be a virtual one. Since, except for sign, D_i and $-D_o$ are equal, the image is as far behind the mirror as the object being reflected is in front of the mirror.

What, now, if we have a *convex mirror*—that is, a curving mirror which is silvered on the concave side so that we look into and see a reflection from the convex side? A parallel sheaf of light rays striking such a mirror is reflected away from the principal axis (except for the one ray that strikes right along the principal axis). Again, if the diverging reflected rays are continued backward (in imagination) through the mirror and beyond, they will converge to a focus.

The focus of a convex mirror, lying as it does behind the mirror, is a *virtual focus,* and its distance from the mirror is negative. For a convex mirror then, we must speak of $-f$ and, therefore, of $-1/f$. Again, since the reflected rays diverge, no real image will be formed in front of the mirror; only a virtual image (right-side up) behind the mirror. Therefore, we must speak of $-D_i$ and $-1/D_i$. For a convex mirror, Equation 2–2 becomes:

$$\frac{1}{D_o} - \frac{1}{D_i} = -\frac{1}{f} \qquad \text{(Equation 2–5)}$$

or:

$$\frac{1}{D_o} = \frac{1}{D_i} - \frac{1}{f} \qquad \text{(Equation 2–6)}$$

Since the object being reflected must always be in front of the mirror, D_o and, therefore, $1/D_o$ must be positive. It follows then that $1/D_i - 1/f$ must be positive, and for that to be true, $1/D_i$

Real and virtual images

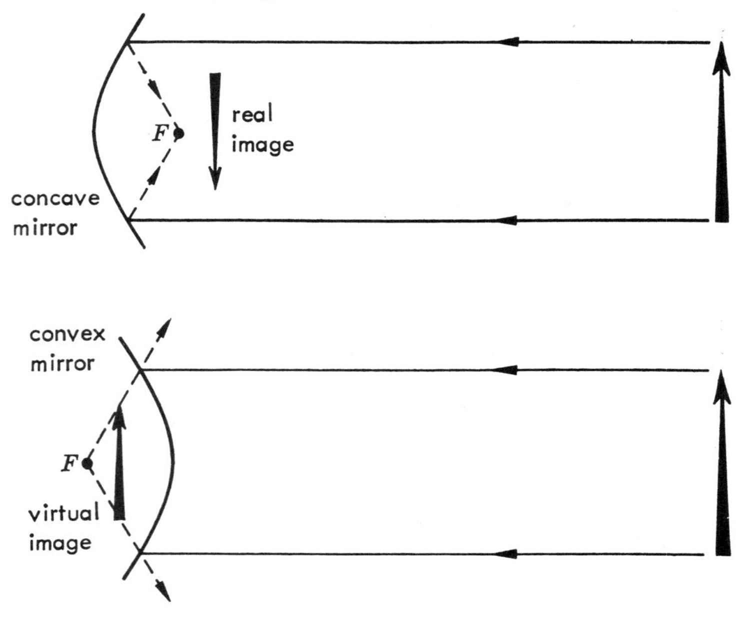

must be greater than $1/f$. But this leads us one step farther and tells us that D_i itself must be smaller than f. In other words, the apparent distance of all the virtual images reflected by a convex mirror must be less than that of the focus, however distant from the mirror the object being reflected is. For this reason, all objects reflected in a convex mirror seem compressed into a tiny space, and small convex mirrors at one corner of a large crowded room can give a panoramic view (albeit distorted) of the entire room.

The size of the image (S_i) is related to the size of the object being reflected (S_o), as the respective distances are related, regardless of whether those distances extend before or behind the mirror. In other words:

$$\frac{S_i}{S_o} = \frac{D_i}{D_o} \qquad \text{(Equation 2–7)}$$

In a plane mirror, where the distance of the image from the mirror is equal to the distance from the mirror of the object being reflected, the sizes of the object and image are likewise equal. An image is neither diminished nor enlarged in a plane mirror. In a convex mirror, where all images must be closer to the mirror than the focus, however distant the objects being reflected are, all the images are small as well. The more distant the object being reflected, the closer and, therefore, the smaller the image.

In a concave mirror, however, when the object being reflected lies between the focus and the center of curvature, the image is beyond the center of curvature. In such a case, since the image is farther from the mirror than is the object being reflected, the image is larger than the object. The closer the object is brought to the focus, the larger the image appears. Of course, the larger the image is, the dimmer it is, for a given amount of light is spread out over a larger and larger volume.

Refraction

Light need not be reflected in order to deviate from straight-line motion. Light, in passing from one transparent medium into another, say from air into water, will generally not be reflected but will continue traveling forward and nevertheless may change direction.

This was undoubtedly first noticed by primitive men when a rod, placed in water in such a way that part remained in the air above, seemed bent at the point where it entered the water. If, however, it was withdrawn, it proved as straight and rigid as ever. Again, it is possible to place an object at the bottom of an empty cup and look at the cup from such an angle that the object is just hidden by the rim. If water is now placed in the cup, the object at the bottom becomes visible though neither it nor the eye has moved. As long ago as the time of the ancient Greeks, it was realized that to explain this, one had to assume that light changed its direction of travel in passing from one transparent medium to another.

Imagine a flat slab of clear glass, perfectly transparent, and imagine a ray of light falling upon it along the normal line—that is, striking the glass at precisely right angles to its flat surface. The light, if one investigates the situation, is found to continue through the glass, its direction unchanged.

Suppose, though, that the light approaches the glass obliquely, forming the angle i with the normal. One might suspect that the light would simply continue moving through the glass, making the same angle i with the normal within the glass. This, however, is not what happens. The ray of light is bent at the point where air meets glass (the air-glass *interface*). Moreover, it is bent toward the normal in such a way that the new angle it makes with the normal inside the glass (r) is smaller than i, the angle of incidence.

This change in direction of a light ray passing from one transparent medium to another is called *refraction* (from Latin words meaning "to break back"). The angle r is, of course, the *angle of refraction.*

If the angle of incidence is made larger or smaller, the angle of refraction also becomes larger or smaller. For every value of i, however, where light passes from air into glass, r remains smaller.

The ancient physicists thought that the angle of refraction was directly proportional to the angle of incidence, and that therefore doubling i would always result in a doubling of r. This is nearly so where the angles involved are small, but as the angles grow larger, this early "law" fails.

Thus, suppose a light ray makes an angle of 30° to the normal as it strikes the air-glass interface and the angle of refraction that results after the light passes into the glass is 19.5°. If the angle of incidence is doubled and made 60°, the angle of refraction becomes 35.3°. The angle of refraction increases, but it does not quite double.

The correct relationship between i and r was worked out first in 1621 by the Dutch physicist Willebrord Snell (1591–1626). He did not publish his finding, and the French philosopher René

Descartes (1596–1650) discovered the law independently in 1637, publishing it in the form (rather simpler than Snell's) that we now use.

The Snell-Descartes *law of refraction* states that whenever light passes from one transparent medium into another, the ratio of the sine of the angle of incidence to the sine of the angle of refraction is constant.* The sine of angle x is usually abbreviated as sin x, so the Snell-Descartes law can be expressed:

$$\frac{\sin i}{\sin r} = n \qquad \text{(Equation 2–8)}$$

When a light ray passes (obliquely) from a vacuum into some transparent substance, the constant, n, is the *index of refraction* of that substance.

If light enters from a vacuum into a sample of gas at 0°C and 1 atmosphere pressure (these conditions of temperature and pressure are usually referred to as *standard temperature and pressure,* a phrase often abbreviated *STP*), there is only a very slight

* The sine of an angle can best be visualized as follows: Imagine the angle to be one of the acute angles of a right triangle. The sine of the angle is then equal to the ratio of the length of the side of the triangle opposite itself, to the length of the hypoteneuse of the triangle. Tables can be found in many texts and handbooks that give the values of the sines of the various angles. Thus, one can easily find that the sine of 10° 17′ is 0.17852, while the sine of 52° 48′ is 0.79653.

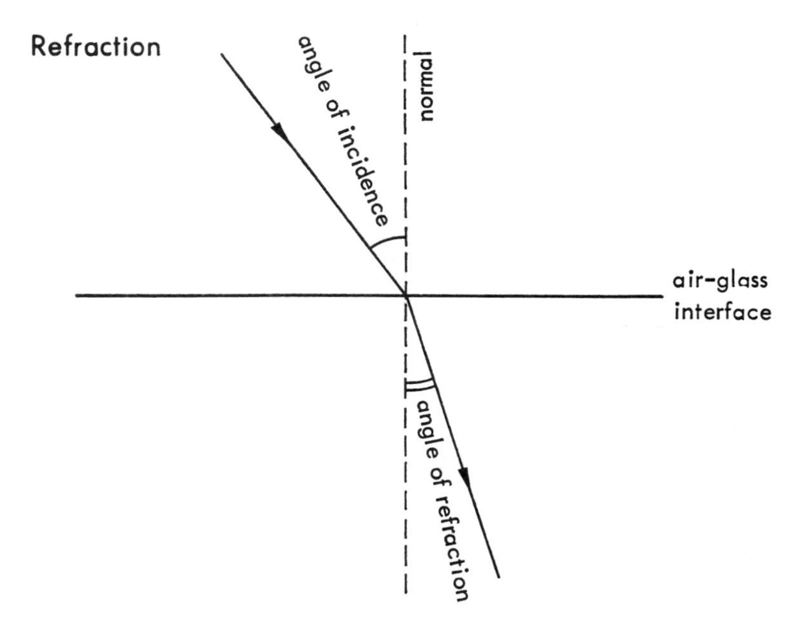

refraction. This means that the angle of refraction is only very slightly smaller than the angle of incidence, and sin r is, in consequence, only very slightly smaller than sin i. Where this is true, we can see from Equation 2–8 that the value of n must be only very slightly greater than 1.

In fact, for hydrogen at STP, the index of refraction is 1.00013, and for air at STP it is 1.00029. There is very little error, therefore, in determining the index of refraction where light passes from air into some transparent substance rather than from vacuum into that transparent substance.

For liquids and solids, the situation is quite different. Water has an index of refraction of 1.33, while the index of refraction of glass varies from 1.5 to 2.0, depending on its exact chemical makeup. For an unusually high value, there is diamond, which has an index of refraction of 2.42. A ray of light entering diamond from air with an angle of incidence of 60°, passes into diamond with an angle of refraction of only 21.1°.

The greater the index of refraction of a material, the greater is its *optical density*. Thus, diamond is optically denser than glass, which is optically denser than water, which is optically denser than air. When a light ray travels from an optically less dense material into an optically more dense one, the direction of the light ray is bent toward the normal. This happens when light travels from air into water, for instance, or from water into diamond. A light ray traveling from an optically more dense material to an optically less dense one is bent away from the normal. One effect cancels the other. Thus if light passes from air into glass, striking at angle i and entering at angle r, and then passes from glass into air, striking at angle r, it will emerge at angle i.

Suppose, for instance, that a ray of light strikes a sheet of glass with an angle of incidence of 60°. The angle of refraction is 35.3°. After traveling through the thickness of the glass, the light ray reaches the other glass-air interface, which in the usual sheet of glass is precisely parallel to the first. As a result, any line which is normal to one interface is also normal to the other. At the second interface, the light is passing from glass into air, so it bends away from the normal. Since now it strikes at 35.3°, it emerges at 60°. The light having emerged from the glass sheet is now traveling in the same direction in which it had been traveling when it had entered; the refractive effect at one interface had been cancelled at the other, and the slight displacement of light rays that result goes unnoticed. (It is for this reason that looking obliquely through a window that is reasonably free of imperfections in the glass does not confuse us. Objects seen through a window are indeed in the direction they seem to be.)

Suppose we rearrange Equation 2–8 in order to solve for sin r. The result is:

$$\sin r = \frac{\sin i}{n} \qquad \text{(Equation 2–9)}$$

If the angle of incidence is 0°, then sin i is equal to 0, and sin r is equal to $0/n$, or 0. The angle of incidence can be increased up to 90°, at which time the ray of light is perpendicular to the normal and just skims along parallel to the interface. If the angle of incidence has its maximum value of 90°, sin i is equal to 1, and the value of sin r is $1/n$. In other words, as i goes through its extreme variation from 0° to 90°, sin r goes through an extreme variation from 0 to $1/n$. In the case of water, where n equals 1.33, the extreme variation for sin r is from 0 to 0.75.

The angle that has a sine of 0 is 0°, and the angle that has a sine of 0.75 is (referring to a table of sines) 48.6°. Therefore as the angle of incidence for light passing from air into water varies from 0° to 90°, the angle of refraction varies from 0° to 48.6°. The angle of refraction cannot be higher than 48.6°, no matter what the angle of incidence.

But what if we reverse the situation and imagine light emerging from water into air? The relationship of the angles is reversed. Now the light is refracted away from the normal. As the light (in passing from water into air) forms an angle of incidence varying from 0° to 48.6°, the angle of refraction (formed by the light emerging into the air) varies from 0° to 90°.

Yet a skin diver under water with a flashlight may easily direct a beam of light so that it makes an angle to the normal (under water) of more than 48.6°. It should emerge at an angle of more than 90°, which means that it really does not emerge at all, since an angle of more than 90° to the normal will direct it under water again. The light, in other words, in passing from water to air will, if it strikes the interface at more than the *critical angle* of 48.6°, be reflected entirely. This is *total reflection.*

As you can see from Equation 2–9, the greater the index of refraction (n) of a substance, the smaller the critical angle. For ordinary glass the critical angle is about 42°, and for diamond, 24.5°. Light can be led through transparent plastic tubes around curves and corners if the rays from the light source, shining in at one end, always strike the plastic-air interface at angles greater than the critical angle for that plastic.

The index of refraction of air itself, while very small, can introduce noticeable effects where great thicknesses are involved. If a heavenly body is directly overhead, its light passes from the vacuum of space into the gas of our atmosphere with an incident angle of 0° and there is no refraction. An object that is not overhead has an angle of incidence greater than 0°, and its light is bent slightly toward the normal. Our eye, following the light backward without making allowance for any bending, sees the light source as somewhat higher in the sky than it actually is.

The lower in the sky a light source is, the greater the angle of incidence and the greater its difference from the angle of refraction. The greater, therefore, the discrepancy between its apparent and its real position. By the time objects at the horizon are involved, the eye sees an object higher than it really is by more than the width of the sun. Consequently, when the sun is actually just below the horizon, the refraction of the atmosphere allows us to see it as just above the horizon. Furthermore, the lowermost part of the sun, being lowest, undergoes the most refraction and is raised the more. As a result, the setting sun seems oval and flattened at the bottom.

Nor is the refractive curve of light as it enters our atmosphere from space a sharp one. The air is not uniformly dense but increases in density as one approaches earth's surface. Its index of refraction increases as its density does. Consequently, as light passes from space to our eye it bends more and more, following what amounts to a smooth curve (rather than the straight line we take so for granted).

The index of refraction of the air varies with temperature, too, and when a layer of air near the ground is heated, and overlaid with cooler air, light will curve in such a way as to make distant objects visible. The temperature conditions of the air may even cause objects on the ground to appear upside down in the air. The *mirages* that have resulted in this way (often in deserts where temperature differences between layers of air may be more extreme than elsewhere) have fooled victims all through history. In modern times such effects may make newspaper headlines, as when a person mistakes the headlights of a distant automobile reaching him through a long, gentle refractive curve, and reports "flying saucers" to be speeding their way through the sky.

CHAPTER 18

Lenses

Focus by Transmission

When the two edges of a piece of glass are not parallel, the normal to one edge will not be parallel to the normal to the other edge. Under such conditions, refraction at the far edge will not merely reverse the refraction at the near edge, and a ray of light passing through the glass will not emerge in the same direction it had on entrance. This is the case, for instance, when light passes through a triangle of glass, or *prism.**

Imagine you are observing a ray of light touching the air-glass interface of such a prism, oriented apex-upward. If the ray of light meets the normal at an angle from below, it crosses into the glass above the normal but makes a smaller angle with it because the glass is optically more dense than the air. When the ray of light reaches the glass-air interface at the second side of the prism, it makes an angle with a new normal altogether, touching the interface above this normal. As it emerges into air, it must bend away from the new normal, because air is optically less dense than glass.

The result is that the ray of light bends twice in the same

* Any solid with parallel lateral edges and with a polygonal cross section when cut at an angle to those edges is a prism. Where the cross section is a triangle (*triangular prism*), we have the solid that is usually referred to simply as a prism, though it is actually only one example of an infinite class.

direction, first on entering the glass and then on leaving it. On leaving the glass it is traveling in a direction different from that in which it had entered. Light always passes through a prism in such a way that it bends away from the apex and toward the base.

Suppose you had two prisms set together, base to base, and a parallel beam of light is striking this double prism in a direction parallel to the mutual base line. The upper half of the beam, striking the upper prism, would be bent downward toward its base. The lower half of the beam, striking the lower prism, would be bent upward toward its base. The two halves of the beam of light, entering the double-prism parallel, would converge and cross on the other side.

The cross section of a double prism has interfaces consisting of two straight lines before and two straight lines behind, so its overall shape is that of a parallelogram (something like the "diamond" on the ace of diamonds in the deck of cards). In such a double prism, the normals to every point on the upper half are parallel because the interface is straight. Therefore all the rays in the light beam striking it make equal angles to the normal and are refracted through equal angles. The same is true for the lower half of the double prism, though there all the rays are bent upward rather than downward. The two half-beams emerge on the other side of the double prism as sheafs of parallel rays of light and cross each other over a broad front.

But what if the double-prism interfaces are smoothed out into a pair of spherical segments? The resulting figure would still be thin at the top and bottom and thickest in the middle, but now the normal to the surface would vary in direction at every point.

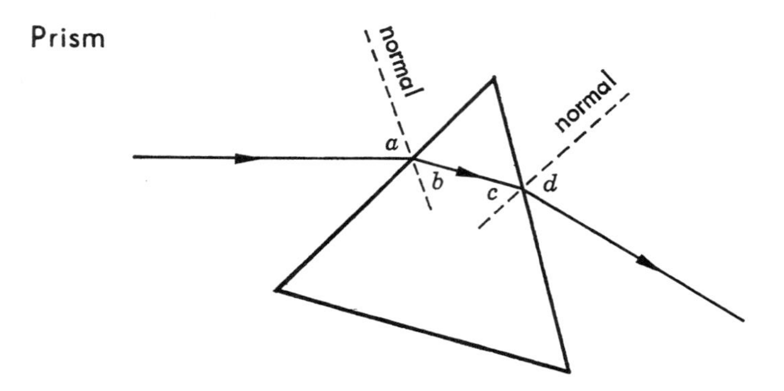

angle *a* greater than angle *b*
angle *d* greater than angle *c*

If the solid is held with its points in an up-down direction, then the normal would be horizontal at the center and would point more and more upward as one traveled toward the upper apex; it would point more and more downward as one traveled toward the lower apex.

Suppose now that a parallel sheaf of light rays strikes such a solid so that the ray striking the central thickest portion travels along the normal. It is not refracted but emerges from the other side unchanged in direction. Light rays striking a little above make a small angle with the upward-tipping normal and are refracted slightly downward. Light rays striking still higher make a somewhat larger angle with the further-tipping normal and are refracted downward more sharply, and so on. Below the center, the light rays are refracted upward more and more sharply as the distance from the center increases. The overall result is that the light rays converge on the other side of the lens, meeting at a focus.

A smoothed-out double prism of the type just described has the shape of a lentil seed and is therefore called a *lens* (from the Latin word for such a seed). By extension, any piece of glass or other transparent material with at least one curved surface is called a lens.

Both surfaces are convex in the particular type of lens that resembles a smoothed-out double prism. Such a lens is therefore a *biconvex lens*. This is the kind that resembles a lentil seed; it is the most familiar kind and, in fact, is what the average man will at once picture if asked to think of a lens.

It is not necessary for the two surfaces of a lens to be evenly curved. One surface might be less convex than the other, or it might even be flat. In the latter case, the lens is *plano-convex*. One of the surfaces can be concave for that matter (*concavo-convex*) so that a cross section of the lens looks something like a crescent moon. Such a lens may be called a *meniscus* (from a Latin word meaning "little moon"). Whatever the comparative shapes of the surfaces of the lens, light rays will be made to converge on passing through it, if the thickness of the lens is least at the edge and increases to maximum at the center. All such lenses can be lumped together as *convex lenses* or *converging lenses*.

The behavior of a convex lens neatly fits that of a convex mirror (see page 22). The light reflected from a convex mirror diverges, but if we imagine that the lines of the diverging rays are carried forward through the mirror, they will come to a focus on the other side; it is there that the virtual (upright) image is formed. In the case of a convex lens, light actually passes through and converges to a real focus where a real (inverted) image is formed. Because the image is real, light is concentrated and the ability of a lens to concentrate sunlight and start fires is well known.

The thicker the central bulge in a converging lens in relation to its diameter, the more sharply the rays of light are converged and the closer the focus is located to the lens itself—that is, the shorter the *focal length* (the distance from the focus to the center of the lens). A lens with a short focal length, which more drastically bends the light rays out of their original direction, is naturally considered a more powerful lens.

The strength of a lens is measured in *diopters* (from Greek words meaning "to see through"), which are obtained by taking the reciprocal of the focal length in meters. If the focal length is 1 meter, the lens has a power of 1/1 or 1 diopter. A focal length of 50 centimeters, or 0.5 meters, implies a power of 1/0.5, or 2 diopters. The larger the diopter value, the more powerful the lens.

A lens can be concave on both sides (a *biconcave lens*) so that it is thickest at the edges and thinnest in the center. It may be plane on one side (*plano-concave*) or even convex (*convexo-concave*). As long as it is thinnest in the center, it may be considered a *concave lens.* Since a parallel sheaf of light rays passing through any concave lens diverges after emerging on the other side, such lenses may also be called *diverging lenses.*

Here again, the properties of a concave lens and a concave mirror fit neatly. Light rays reflected from a concave mirror converge to a focus. If we imagine that the converging rays are carried through the mirror, they will diverge on the other side. In a concave lens, the light actually does pass through and diverge.

In the case of a concave lens, since the light passes through and diverges, it forms no image. However, the diverging light rays can be carried backward in imagination to form a virtual image on the forward side, where a concave mirror would have formed a real one.

The power of a diverging lens is arrived at in a manner similar to that in which the power of a converging lens is dealt with. However, in the case of a diverging lens, a virtual focus is involved, and the focal length therefore has a negative value. A diverging lens would be said to have a power of −2 diopters, for instance.

Spectacles

There is a lens-shaped object within the human eye, just behind the pupil, which is called the *crystalline lens* (not because it contains crystals, but because, in the older sense of the word "crystalline," it is transparent). It is a biconvex lens, and therefore a converging lens, about a third of an inch in diameter. The foremost portion of the eye, the transparent *cornea,* is also a converging lens, with twice the converging power of the crystalline lens itself.

The cornea and crystalline lenses converge the light rays to a focus upon the light-sensitive inner coating (*retina*) of the rear of the eyeball. An inverted image forms on the retina, and the pattern of light and dark is imprinted there. Each light-sensitive cell in the retina's center (where the image of what we are looking at is formed) is connected to an individual nerve fiber, so the pattern is carried without loss of detail to the brain. The brain makes allowance for the inversion of the image, and we see right-side-up.

The image formed by a converging lens cannot, however, always be counted upon to fall upon the focus (which, strictly speaking, is the point at which a sheaf of parallel rays of light are made to converge). Where the light source is far away, the rays are indeed parallel or virtually so, and all is well. As the light source is brought nearer to the lens, however, the light rays are more and more perceptibly divergent, and they then converge beyond the focus—that is, at a distance greater than the focal length.

The relationship between the distances of the object serving as the light source (D_o), of the image (D_i), and of the focus (f) can be expressed by means of Equation 2–2 (see p. 18). In the previous chapter, this equation was used in connection with mirrors, but it will serve for lenses, too. In fact, it is so commonly used for lenses rather than for mirrors that it is usually called the *lens formula.* (In both lenses and mirrors a virtual focus yields a negative value for f and $1/f$, and a virtual image, a negative value for D_i and $1/D_i$. On the other hand, D_o and $1/D_o$ are always positive.)

Let us rearrange the lens formula and write it as follows:

$$\frac{1}{D_i} = \frac{1}{f} - \frac{1}{D_o} \qquad \text{(Equation 3–1)}$$

If the object is at an infinite distance, $1/D_o = 0$, and $1/D_i = 1/f$, which means that $D_i = f$. The image, therefore, is formed at the focus. But let us suppose that the focal length of the cornea-lens combination of the eye is about 1.65 centimeters (which it is) and that we are looking at an object 50 meters (or 5000 centimeters) away. In that case, $1/D_i = 1/1.65 - 1/5000$, and $D_i = 1.6502$. The image forms 0.0002 centimeters beyond the focus, a discrepancy that is small enough to be unnoticeable. Thus, a distance of 50 meters is infinite as far as the eye is concerned.

But what if the object were 30 centimeters away—reading distance? Then $1/D_i = 1/1.65 - 1/30$, and $D_i = 1.68$. The image would form about 0.03 centimeters behind the focus, and on the scale of the eye that would be a serious discrepancy. The light would reach the retina (at the focal length) before the light rays had focused. The image would not yet be sharp, and vision would be fuzzy.

To prevent this, the crystalline lens changes shape through the action of a small muscle. It is made to thicken and become a more powerful light-converger. The focal length shortens. The image, still forming beyond the new and now shorter focal length, forms on the retina. This process is called *accommodation*.

As an object comes nearer and nearer the eye, the crystalline lens must bulge more and more to refract the light sufficiently to form the image on the retina. Eventually it can do no more, and the distance at which accommodation reaches its limit is the *near-point*. Objects closer to the eye than the near-point will seem fuzzy because their image cannot be made to form on the retina.

The ability to accommodate declines with age, and the near-point then recedes. A young child with normal vision may be able to focus on objects as close to the eye as 10 centimeters; a young adult on objects 25 centimeters away; while an old man may not be able to see clearly anything closer than 40 centimeters. In other words, as one grows older one starts holding the telephone book farther away. This recession with age of the near-point is called *presbyopia* (from Greek words meaning "old man's vision").

It may happen that a person's eyeball is deeper than the focal length of the cornea-lens combination. In such a case, the images of objects at a distance form at the focus, which is well in front of the too-deep retina. By the time the light rays reach the retina, they have diverged a bit and vision is fuzzy. As objects come closer, the image is formed at distances greater than the focal length, and these eventually do fall on the retina. Such people can clearly see near objects but not distant ones; they are *nearsighted.* More formally, this is called *myopia,* for reasons to be explained on page 39.

The opposite condition results when an eyeball is too shallow. The focal length is greater than the depth of the eyeball, and the light rays reaching the retina from objects at a great distance have not yet quite converged. The crystalline lens accommodates and bends the light more powerfully so that distant objects can be seen clearly after all. As an object approaches more closely, however, the power of lens accommodation quickly reaches its limit, and near objects can only be seen fuzzily. For such a person, the near-point is abnormally far away, and although he can see distant objects with normal clarity, he cannot see near objects clearly. He is *farsighted* and suffers from *hyperopia* ("vision beyond").

It is easy to produce a new overall focal length by placing one lens just in front of another. One need, then, only add the diopters of the two lenses to find the total refracting power of the two together, and therefore the focal length of the two together.

Imagine a lens with a refracting power of 50 diopters. Its focal length would be 1/50 of a meter, or 2 centimeters. If a second converging lens of 10 diopters were placed in front of it, the refracting power of the lens combination would be 60 diopters, and the new focal length would be 1/60 of a meter, or 1 2/3 centimeters. On the other hand, a diverging lens with a refracting power of —10 diopters would increase the focal length, for the two lenses together would now be 40 diopters and the focal length would be 1/40 meter of 2 1/2 centimeters.

This can be done for the eye in particular and was done as early as the thirteenth-century by such men as the English scholar Roger Bacon (1214?–1294). The results are the familiar *eye-glasses* or *spectacles,* and these represent the one great practical application of lenses that was introduced during the Middle Ages.

The power of the cornea-lens combination of the eye is about 60 diopters, and the lenses used in spectacles have powers ranging from —5 to +5 diopters. For farsighted people with too-shallow eyeballs, the diopters must be increased so that focal length is decreased. To increase the diopters, a lens with positive diopters (that is, a converging lens) must be placed before the eye. The reverse is the case for nearsighted individuals. Here the eyeball is too deep, and so the focal length of the eye must be lengthened by reducing the diopters. A lens with negative diopters (that is, a diverging lens) must be placed before the eye.

For both farsighted and nearsighted individuals, the spectacle lens is usually a meniscus. For the former, however, the meniscus is thickest at the center; for the latter it is thinnest at the center.

As old age comes on, the additional complication of presbyopia may make it necessary to apply two different corrections, one for near vision and one for far vision. One solution is to have two different types of spectacles and alternate them as needed. In his old age, it occurred to the American scholar Benjamin Franklin (1706–1790), when he grew weary of constantly switching glasses, that two lenses of different diopters, and therefore of different focal lengths, could be combined in the same frame. The upper portion might be occupied by a lens correcting for far vision, the lower by one correcting for near vision. Such *bifocals* (and occasionally even *trifocals*) are now routinely produced.

For a lens to focus reasonably well, its curvature must be the same in all directions. In this way, the rays that strike toward the top, bottom, and side of the lens are all equally converged toward the center, and all meet at a true focus.

Suppose the lens curves less sharply from left to right than from top to bottom. The light rays at left and right would then not come to a focus at a point where the light rays from top and bottom would. At that point, instead of a dot of light, there would be a horizontal line of light. If one moves farther back, to a spot where the laggard rays from right and left have finally focused, the rays from top and bottom have passed beyond focus and are diverging again. Now there is a vertical line of light. At no point is there an actual dot of light. This situation is common with respect to the eyeball, and the condition is called *astigmatism** (from Greek words meaning "no point"). This, too, can be corrected by using spectacles that have lenses with uneven curvatures that balance the uneven curvature of the eye, bending the light more in those directions where the eye itself bends it less.

The usual lenses are ground to the shape of segments of spheres, since the spherical shape is the easiest to produce. Such a shape, even if perfectly even in curvature in all directions, still does not converge all the rays of light to an exact point, any more than a spherical mirror reflects all the rays to an exact point. There is spherical aberration (see page 17) here as well as in the case of mirrors.

* The term "astigmatism" as applied to lenses is not quite the same in meaning as when it is applied to the eye. In lenses, it is produced when the source of light is not on the principal axis of the lens. Light in that case strikes the lens obliquely and is not brought to perfect focus but to a line of light.

The extent of this aberration increases with the relative thickness of the lens and with distance from the center of the lens. For this reason, the lens formula (Equation 3–1) holds well only for thin lenses. Near the center of the lens, the spherical aberration is quite small and can usually be ignored. The human eye is fitted with an iris that can alter the size of the pupil. In bright light, the size of the pupil is reduced to a diameter of 1.5 millimeters. The light that enters is still sufficient for all purposes, and spherical aberration is reduced to almost nothing. In bright light, therefore, one sees quite clearly. In dim light, of course, it is necessary to allow as much light to enter the eye as possible, so the pupil expands to a diameter of as much as eight or nine millimeters. More of the lens is used, however, and spherical aberration increases. In dim light, therefore, there is increased fuzziness of sight.

There are other types of aberration (including "chromatic aberration," see page 55), but the usual way of correcting such aberrations in elaborate optical instruments is to make use of two lenses in combination (or a mirror and a lens) so that the aberration of one will just cancel the aberration of the other. By a clever device of this sort, in 1930 a Russian-German optician, Bernard Schmidt (1879–1935) invented an instrument that could without distortion take photographs over wide sections of the sky because every portion of its mirror had had its aberrations canceled out by an irregularly shaped lens called a "corrector plate." (Such an instrument is called a Schmidt camera or a Schmidt telescope.)

Cameras

Images can be formed outside the eye, of course, as well as inside. Consider a single point in space and an object, some distance away, from which light is either being emitted or reflected. From every part of the object a light ray can be drawn to the point and beyond. A ray starting from the right would cross over to the left once it had passed the point, and vice versa. A ray starting from the top would cross over to the bottom once it had passed the point, and vice versa.

Suppose the rays of light, having passed the point, are allowed to fall upon a dark surface. Light rays from a brightly emanating (or reflecting) portion of the light source would yield bright illumination; light rays originating from a dimly lit portion would yield dim illumination. The result would be a real and inverted image of the light source.

Actually, under ordinary conditions we cannot consider a single point in space, since there are also a vast number of neighboring points through which rays from every portion of the light source can be drawn. There are, therefore, a vast number of inverted images that will appear on the surface, all overlapping, and the image is blurred out into a general illumination; in effect, no images are formed.

But suppose one uses a closed box with a hole on the side facing the light source, and suppose one imagines the hole made smaller and smaller. As the hole is made smaller, the number of overlapping images is continually being reduced. Eventually an image with fuzzy outlines can be made out on the surface opposite the hole, and if the hole is made quite small, the image will be sharp. The image will remain sharp no matter what the distance between the hole and the surface on which it falls, for there is no question of focusing since the image is formed of straight-line rays of light that are unrefracted. The farther the surface from the hole the larger the image, since the rays continue to diverge with increasing distance from the hole. However, because the same amount of light must be spread over a larger and larger area, the image grows dimmer as it grows larger.

On a large scale, this can be done in a dark room with the windows thickly curtained except for one small hole. On the opposite wall an image of whatever is outside the hole will appear —a landscape, a person, a building—upside down, of course.

The sun shining through such a hole will form a circle that is actually the image of the sun, and not of the hole. If the hole were triangular in shape but not too small, there would be a triangular spot of light on the wall, but this triangle would be made up of circles, each one of which would be a separate image of the sun. As the hole grows smaller, so does the triangle, until it is smaller than an individual circular image of the sun. At that point, the image will appear a circle despite the triangularity of the hole.

The leaves of a tree form a series of small (though shifting) openings through which sunlight streams. The dappled light on the ground then shows itself as small superimposed circles, rather than reproducing the actual irregular spaces between the leaves. During a solar eclipse, the sun is no longer round but is bitten into and, eventually, shows a crescent shape. When this takes place, the superimposed circles of light under the tree becomes superimposed crescents. The effect is quite startling.

Image-formation in dark rooms began in early modern times,

and such Italian scholars as Giambattista della Porta (1538?–1615) and Leonardo da Vinci (1452–1519) made use of it. The device is called a *camera obscura*, which is a Latin phrase meaning "dark room." Eventually other devices for producing images within a darkened interior were used, and the first part of the phrase, "camera," came to be applied to all such image-forming devices. The original camera obscura, with its very small opening, is now commonly called a *pinhole camera*.

The chief difficulty with a pinhole camera is that to increase the sharpness of the image one must keep the hole as small as possible. This means that the total amount of light passing through the hole is small, and the image is dim. To widen the opening and allow more light to enter, and yet avoid the superimpositions that would immediately destroy the image, one must insert a converging lens in the opening. This will concentrate the light from a large area into a focus, increasing the brightness of the image many times over without loss of sharpness. In 1599, della Porta described such a device and invented the camera as we now know it.

Once a camera is outfitted with a lens, the image will no longer form sharply at any distance, but only at the point where the light rays converge. For cameras of fixed dimensions, sharp images may be formed only of relatively distant objects, if the back of the camera is at the focal length. For relatively close objects, the light rays converge at a point beyond the focal length (see page 34) and the lens must be brought forward by means of an accordionlike extension (in old-fashioned cameras) or by means of a screw attachment (in newer ones). This increases the distance between the lens and the back of the camera, and is the mechanical analog of the eye's power of accommodation.

In an effort to make out objects in the middle distance, people who are nearsighted quickly learn that if they squint their eyes they can see more clearly. This is because the eye is then made to approach more closely to the pinhole camera arrangement, and a clear image depends less on the depth of the eyeball. (Hence "myopia" is the term used for nearsightedness, for this comes from a Greek phrase that means "shut-vision" with reference to the continual squinting.) Of course, the difficulty is that less light then enters the eye, so sharper focus is attained at the expense of brightness. Furthermore, the muscles of the eyelids tire of the perpetual task of keeping them somewhat but not altogether closed; the result is a headache. (Actually, it is "eye-muscle strain" and not "eyestrain" that causes the discomfort.)

The lensed camera came of age when methods were discovered for making a permanent record of the image. The image is formed upon a surface containing chemicals that are affected by light.* A number of men contributed to this, including the French physicist Joseph Nicéphore Niepce (1765–1833), the French artist Louis Jacques Mandé Daguerre (1789–1851), and the English inventor William Henry Fox Talbot (1800–1877). By the mid-nineteenth century, the camera as producer and preserver of images was a practical device, and *photography* ("writing by light") became of infinite use in every phase of scientific work.

To get bright images, as much light as possible must be squeezed together. This requires a lens of large diameter and short focus. The larger the diameter, the more light is gathered together and converged into the image. The reason for the short focus depends on the fact (already discussed in connection with mirrors on page 23 and a point applicable in the case of lenses as well) that the closer the image to the lens, the smaller it is. The smaller the image into which a given quantity of light is focused, the brighter it is. To measure the brightness of the image that a lens can produce, we must therefore consider both factors and take the ratio of the focal length (f) to the diameter (D). This ratio, f/D, is called the *f-number.* As one decreases f or increases D (or both), the f-number decreases. The lower the f-number, the brighter the image.

The image, as originally formed on chemically-coated film, is dark in spots where intense illumination has struck (for the effect of light is to produce black particles of metallic silver) and light where little illumination has struck. The image therefore appears in reverse—light where we see dark and dark where we see light. This is a *negative.* If light is projected through such a negative onto a paper coated with light-sensitive chemicals, a negative of the negative is obtained. The reversal is reversed, and the original light-dark arrangement is obtained. This *positive* is the final picture.

The positive may be printed on transparent film. In that case, a small but intense light source may be focused upon it by a lens and mirror combination, and the image projected forward onto a screen. The rays diverge after leaving the *projector,* and the image on the screen can be greatly enlarged, as compared with the original positive. This can be used for home showing of photo-

* The details of the process are more appropriate for a book on chemistry, and will not be considered here.

graphs, and has been used, far more importantly, as a means of mass entertainment.

The possibility for this arises from the fact that when the cells of the retina react to a particular pattern of light and dark, it takes them a perceptible fraction of a second to recover and be ready for another pattern. If, in a dark room, you wave a long splint, smoldering at the far end, you will not see a distinct point of light changing position, but a connected curve of light out of which you can form circles and ovals.

Imagine, then, a series of photographs taken very rapidly of moving objects. Each photograph would show the objects in slightly different positions. In 1889, the American inventor Thomas Alva Edison (1847–1931) took such photographs on a continuous strip of film with perforations along the side. Such perforations could be threaded onto a sprocket wheel, which, when turning, would pull the film along at a constant velocity. If a projector light could be made to flash on and off rapidly, it would flash onto the screen a quick image of each passing picture. The eye would then see one picture after another, each just slightly different from the one before. Because the eye would experience its lag period in reaction, it would still be seeing one picture when the next appeared on the screen. In this way, an illusion of continual change, or motion, is produced. Thus, *motion pictures* were introduced.

Magnification

Anyone who has handled a converging lens knows that objects viewed through it appear larger. It is very likely that this was known in ancient times, since a round glass bowl filled with water would produce such an effect.

To understand this, we must realize that we do not sense the actual size of an object directly, but merely judge that size from a variety of indirect sensations, including the angle made by light reaching the eye from extreme ends of the object.

Suppose, for instance, that a rod 4 centimeters long is held horizontally 25 centimeters in front of the eyes. The light reaching the eye from the ends of the rod makes a total angle of about 9.14°. In other words, if we looked directly at one end of the rod, then turned to look directly at the other end, we would have turned through an angle of 9.14°. This is the *visual angle,* or the *angular diameter* of an object.

If the rod were only 2 centimeters long, the visual angle would be 4.58°; if it were 8 centimeters long, it would be 18.18°. The visual angle is not exactly proportional to the size, but for small values it is almost exactly proportional. We learn this proportionality through experience and automatically estimate relative size by the value of the visual angle.

However, the angular size of any object is also a function of its distance. Consider the eight-centimeter rod that at 25 centimeters would exhibit a visual angle of 18.18°. At 50 centimeters its visual angle would be 9.14°; at 100 centimeters, 4.58°. In other words, as we also know from experience, an object looks smaller and smaller as it recedes from the eye. A large object far distant from the eye would look smaller than a small object close to the eye. Thus, an eight-centimeter rod 100 centimeters from the eye would produce a smaller visual angle than a four-centimeter rod 25 centimeters from the eye, and the former would therefore appear smaller in size.

It is not likely that we would be fooled by this. We learn at an early age to take distance, as well as visual angle, into account in assessing the real size of an object. In looking first at the distant eight-centimeter rod then at the closeby four-centimeter rod, we must alter the accommodation of the crystalline lens, and we must also alter the amount by which our eyes have to converge in order for both to focus on the same object (the closer the object, the greater the degree of convergence). We may not be specifically aware that our lenses are accommodating and our eyes converging; however, we have learned to interpret the sensations properly, and we can tell that the four-centimeter rod is closer. Making allowance for that as well as for the visual angle, we can usually tell without trouble that the rod that looks smaller is actually larger. We even convince ourselves that it *looks* larger.

Alterations in the accommodation of the lens and the convergence of the eyes are of use only for relatively nearby objects. For distant objects, we judge distance by comparison with neighboring objects whose real size we happen to know. Thus a distant sequoia tree may not look unusually large to us until we happen to notice a tiny man at its foot. We then realize how distant it must be, and its real size is made apparent. It begins to look large.

If there are no neighboring objects of known size with which to compare a distant object, we have only the visual angle, and that by itself tells us nothing. For instance, the moon, high in the sky, presents a visual angle of roughly 0.5°. If we try to judge the real diameter of the moon from this, we are lost. We might decide that the moon looked "about a foot across." However, an object a foot across will produce a visual angle of 0.5° if it is not quite sixty feet away. This is certainly a gross underestimate of the actual distance of the moon, yet many people seem to assume, unconsciously, that that is the distance.

When the moon is near the horizon, it is seen beyond the houses and trees, and we know at once that it must be more than sixty feet away. It might be, let us say, a mile away. To produce a visual angle of 0.5° from a distance of a mile, the moon would have to be 88 feet across. This (unconscious) alteration in our estimate of the moon's distance also alters our (unconscious) estimate of its real size. The moon, as all of us have noticed, seems much larger at the horizon than when it is high in the sky.* This optical illusion has puzzled men ever since the time of the Greeks, but the present opinion of men who have studied the problem is that it is entirely a matter of false judgment of distance.

A converging lens offers us a method for altering the visual angle without altering the actual distance of an object. Consider light rays traveling from an object to the eye and making a certain visual angle. If, on the way, they pass through a converging lens, the light rays are converged and make a larger visual angle. The eye cannot sense that light rays have been converged en route; it judges the light rays as though they came in straight lines from an object larger than the real object. Only by sensing the object as enlarged, can the eye account for the unusually large visual angle. Another way of putting it is that the eye sees not the object

* Actual measurement of the moon's apparent diameter shows that at the horizon it is actually a tiny bit smaller than at zenith, for at the horizon the radius of the earth must be added to its distance. It is therefore 2 percent farther from our eyes at the horizon, and its visual angle is 2 percent smaller.

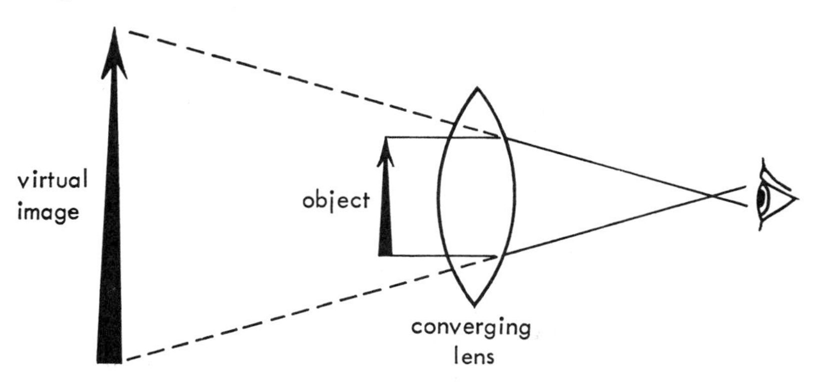

but an enlarged virtual image (hence right-side-up) of the object on the other side of the converging lens. The ratio of the size of the image to the size of the object itself is the *magnification* of the lens.

The magnification can be expressed in terms of the focal length (f) of the lens, provided we turn once more to the lens equation (Equation 2–2 on page 24, or 3–1 on page 33). Since the image is a virtual image, its distance (D_i) receives a negative sign, while the distance of the object itself (D_o) remains positive, as always. The equation can be written then:

$$\frac{1}{D_o} - \frac{1}{D_i} = \frac{1}{f} \qquad \text{(Equation 3–2)}$$

The magnification, as I have said, is the ratio of the size of the image to the size of the object, but this size ratio can be judged in two ways. It can be interpreted as the ratio of the visual angles, if both object and image are at the same distance; or as the ratio of the distances, if both object and image produce the same visual angle. Let us take the latter interpretation and solve Equation 3–2 for the ratio of the distance of the image to that of the object (D_i/D_o). It turns out that:

$$\frac{D_i}{D_o} = \frac{f}{f - D_o} = m \qquad \text{(Equation 3–3)}$$

where m is the magnification.

If the lens is held in contact with the object, which may be a printed page, for instance, D_o is virtually zero and $f - D_o = f$. The magnification m is then equal to f/f, or 1, and the print is not magnified. If the lens is lifted, D_o increases, which means that $f - D_o$ must decrease and, as you can see from Equation 3–3, m must, in consequence, increase. The print seems larger and larger as the lens is lifted. When the distance of the lens from the printed page is equal to the focal length, $f - D_o$ becomes equal to $f - f$, or 0. Magnification is then $f/0$ and becomes infinite. However, no lens is perfect, and if the object is magnified infinitely, so are all the imperfections. As a result, all that can be seen is a blur. Maximum practical magnification comes when the distance of the object is just a little short of the focal length.

If the object is at a distance greater than the focal length, $f - D_o$ becomes negative and therefore m becomes negative. As D_o continues to increase in size, m remains negative, but its absolute value (its value with the negative sign disregarded) becomes smaller. This means that the image becomes inverted and decreases in size again as the object distance becomes greater than the focal length and continues to increase.

It also follows from Equation 3–3 that, for a given distance of the object (D_0), the magnification increases with decrease in the focal length of the lens (provided the focal distance remains greater than the distance of the object). To see this, let us suppose that $D_0 = 1$ and that f takes up successive values of 5, 4, 3 and 2. Since the magnification (m) is equal to $f/(f - D_0)$, it equals, successively, 5/4, 4/3, 3/2 and 2/1; or 1.2, 1.33, 1.5 and 2.0. This is another reason for considering a converging lens to grow more powerful as its focal length decreases (see page 40). After all, its magnifying power increases as its focal length decreases.

All this is reversed for diverging lenses. Here the rays of light converging on their trip to the eye from opposite ends of an object are diverged somewhat by the lens and made to reach the eye at a smaller visual angle. For that reason, objects seem smaller when viewed through a diverging lens.

In this way, you can quickly tell whether a person is nearsighted or farsighted by an extremely simple test with his glasses. A nearsighted man must wear diverging lenses (see page 35), so print looks smaller if those lenses are held a few inches above the printed page. A farsighted man must wear converging lenses, and those will make the print appear larger.

Microscopes and Telescopes

The cells of the retina either "fire" as light strikes them, or do not fire as light does not strike them. As a result, the image that is produced upon them is, so to speak, a combination of light and dark spots. This resembles the appearance of a newspaper halftone reproduction, though the "spots" on the retina are much finer than those on a newspaper photograph.

When an object is considerably larger than the spots that make it up, the object is seen clearly. If it is not much larger, it is seen fuzzily. Thus, if you look at a newspaper photograph with the unaided eye, you will seem to see a clearly delineated face. If you look at it under a magnifying lens, the portion you see in the lens will not be much larger than the magnified dots, and things will not be clear at all. You will not make out "detail."

In the same way, there is a limit to the amount of detail you can see in any object with the unaided eye. If you try to make out finer and finer details within the object, those details begin to be no larger (in the image on your retina) than the dots making up the image. The retinal image becomes too coarse for the purpose.

Light from two dots separated by an angular distance of less than a certain crucial amount activates the same retinal cell or possibly adjacent ones. The two dots are then seen as only a single dot. It is only when light from two dots activates two retinal cells separated by at least one unactivated cell that the two dots can actually be seen as two dots. At 25 centimeters (the usual distance for most comfortable seeing) two dots must be separated by at least 0.01 centimeters to be seen as two dots; the minimum visual angle required is therefore something like 0.006°.

The *resolving power* of the human eye (its ability to see two closely-spaced dots as two dots and, in general, its ability to make out fine detail) is actually very good and is much better than that of the eyes of other species of animals. Nevertheless, beyond the resolving power of the human eye there is a world of detail that would be lost to our knowledge forever, were it not for lenses.

Suppose two dots, separated by a visual angle of 0.001°, were placed under a lens with a magnification of 6. The visual angle formed by those two dots would be increased to 0.006°, and they could be seen as two dots. Without the lens, they could be seen as only one dot. In general, an enlarging lens not only makes the object larger in appearance, it makes more detail visible to the eye.

To take advantage of this, one must use good lenses, that have smoothly ground surfaces and are free of bubbles and imperfections. A lens that is not well constructed will not keep the refracted light rays in good order, and the image, though enlarged, will be fuzzy. Fine detail will be blurred out and lost.

It was not until the seventeenth century that lenses accurate enough to keep at least some of the fine detail were formed. A Dutch merchant, Anton van Leeuwenhoek (1632–1723), used small pieces of glass (it is easier to have a flawless small piece of glass than a flawless large one) and polished them so accurately and lovingly that he could get magnifications of more than 200 without loss of detail. With the use of such lenses, he was able to see blood capillaries, red blood corpuscles, and spermatozoa. Most important of all he could study the detail of independently living animals (protozoa) too small to be made out with the naked eye.

Such strongly magnifying lenses are *microscopes* (from Greek words meaning "to see the small"). A microscope made out of a single lens, as Leeuwenhoek's were, is a *simple microscope*.

There is a limit to the magnifying power of a single lens, however well-ground it may be. To increase the magnifying power, one must decrease the focal length, and Leeuwenhoek was already using minute focal lengths in his tiny lenses. It would be impractical to expect much further improvement in this respect.

However, suppose the light from an object is allowed to pass through a converging lens and form a real image on the other side. As in the case of concave mirrors (see page 23), this real image may be much larger than the object itself, if the object is quite near the focus. (The image would then be much dimmer because the same amount of light would be spread over a greater area. For this reason, the light illuminating the object must be quite intense in the first place, in order to remain bright enough despite this dimming effect.)

Since the image is a real image, it can be treated optically as though it were the object itself. A second converging lens can be used that will further magnify the already-magnified image. By the use of two or more lenses in this way, we can easily obtain a final magnification that will be greater than the best one can do with a single lens. Microscopes using more than one lens are called *compound microscopes.*

The first compound microscopes are supposed to have been built a century before Leeuwenhoek by a Dutch spectacle maker, Zacharias Janssen, in 1590. Because of the imperfect lenses used, it took quite a while for these to be anything more than playthings. By the end of Leeuwenhoek's life, however, compound microscopes were beginning to surpass anything his simple lenses could do.

The *telescope* (from Greek words meaning "to see the distant") also makes use of lenses. Light from an object such as the moon, let us say, is passed through a converging lens and allowed to form a real image on the other side. This image is then magnified by another lens. The magnified image is larger and shows more detail than the moon itself does when viewed by the naked eye.

A telescope can be used on terrestrial objects, too. Here, since the real image formed through the converging lens is inverted, and it would be disconcerting to see a distant prospect with the ground above and the sky below, two lenses are used to form the image, the second lens inverting the inverted image and turning it right-side-up again. This new right-side-up image can then be magnified, and we have a *field glass* for use on landscapes. Small field glasses, designed in pairs to be looked through with both eyes at once, are *opera glasses.*

Astronomical telescopes do not make use of the additional lens, since each lens introduces imperfections and problems, and the fewer lenses the better. An upside-down star or moon does not disconcert an astronomer, and he is willing to let the image remain that way.

The telescope is supposed to have been invented by an apprentice-boy in the shop of the Dutch spectacle maker Hans Lippershey in about 1608.* The next year, the Italian scientist Galileo Galilei (1564–1642), concerning whom I had occasion to speak at length in the first volume of this book, hearing rumors of this new device, experimented with lenses until he had built a telescope. His instrument was exceedingly poor in comparison with modern ones; it only enlarged about thirtyfold. However, in turning it on the sky, he opened virgin territory, and wherever he looked, he saw what no man had ever seen before.

The greater detail visible on the image of the moon made it possible for him to see lunar mountains and craters. He saw spots on the sun and enlarged both Jupiter and Venus into actual globes. He could see that Venus showed phases like the moon (as was required by the Copernican theory) and that Jupiter was circled by four satellites.

The lens of a telescope also serves as a light-collector. All the light that falls upon the lens is concentrated into the image. If the lens is larger than the pupil of the eye (and in a telescope it is bound to be), more light will be concentrated into the image within the telescope than is concentrated into the image within the eye. A star that is too dim to be made out by the unaided eye becomes bright enough therefore to be seen in a telescope. When Galileo turned his telescope on the starry sky, he found a multiplicity of stars that were plainly visible with the telescope and that vanished when he took the instrument from his eye.

Naturally, the larger the lens, the more light it can collect, and the dimmer the stars it can make out. The Yerkes telescope of today (a distant descendant of the first Galilean telescope) has a collecting lens 40 inches in diameter, as compared with the pupil's diameter of no more than 1/3 of an inch. The ratio of the diameters is thus 120 to 1. The light collected depends on the area of the lens, and this is proportional to the square of the diameter. The light-collecting power of the Yerkes telescope is therefore 14,400 times as great as that of the human eye, and stars correspondingly dimmer can be made out by it.

Furthermore, if the light from the telescope is focused on photographic film, rather than on the retina of the eye, there is a further advantage. Light striking the film has a cumulative effect

* The boy was playing with lenses when he should have been working, and you can draw your own moral from that.

(which it does not have on the eye). A star too dim to be seen, even through the telescope, will slowly affect the chemicals on the film and, after an appropriate time exposure, can be photographed even if it cannot be seen.

In theory, lenses can be made larger and larger, and the universe probed more and more deeply. However, practical considerations interfere. The larger the lens, the more difficult and tedious it is to grind it exactly smooth and the more difficult it is to keep it from bending out of shape under its own weight (since it can only be supported about the rim). In addition, the larger the lens, the thicker it must be, and since no lens is perfectly transparent, the thicker it is, the more light it absorbs. After a certain point, it is impractical to build larger lenses. The telescope at the Yerkes Observatory in Wisconsin has a 40-inch lens and is the largest telescope of its sort in the world. It was built in 1897 and nothing larger has been built since. Nor is any likely to be built.

CHAPTER 19

Color

Spectra

So far, I have spoken of light as though all light were the same except that one beam might differ from another in brightness. Actually, there is another distinguishing characteristic, familiar to us all, and that is *color*. We know that there is such a thing as red light, blue light, green light, and so on through a very large number of tints and shades.

The tendency in early times was to consider the white light of the sun as the simplest form of light—as "pure" light. (Indeed, white is still the symbol of purity, and the young bride walks to the altar in a white wedding gown for that reason.) Color, it was felt, was the result of adding impurity to the light. If it traveled through red glass, or were reflected from a blue surface, it would pick up redness or blueness and gain a property it could not have of itself.

From that point of view it would be very puzzling if one were to find the pure white light of the sun displaying colors without the intervention of colored matter at any time. The one such phenomenon known to men of all ages is the *rainbow,* the arc of varicolored light that sometimes appears in the sky, when the sun emerges after a rainshower. The rainbow was startling enough to attract a host of mythological explanations; a common one was that it was a bridge connecting heaven and earth. The first motion toward a rationalistic explanation was that of the Roman philosopher Lucius Annaeus Seneca (4 B.C.?–65 A.D.), who pointed out that the rainbow was rather similar to the play of colors often seen at the edge of a piece of glass.

By the seventeenth century, physicists began to suspect that the rainbow, as well as the colors at the edge of glass, were somehow produced by the refraction of light. The French mathematician René Descartes worked out a detailed mathematical treatment of the refraction and total reflection of light by spheres of water. In this way he could account nicely for the position of the rainbow with relation to the sun, thanks to the refraction of sunlight by tiny droplets of water remaining suspended in air after the rain, but he could not account for the color.

It was left for the English scientist Isaac Newton—whose work takes up so much of Volume I of this book—to make the crucial advance. In 1666, he allowed a shaft of sunlight to enter a darkened room and to fall on a prism. The beam of light, refracted through the prism, was allowed to strike a white surface. There it appeared, not as a spot of white sunlight, but as an extended band of colors that faded into one another in the same order (red, orange, yellow, green, blue, violet) they do in a rainbow. It was a colored image and it received the name of *spectrum,* from the Latin word for "image."

If the light of the spectrum was formed on a surface with a hole in it, a hole so arranged that only one of the colors could pass through, and if that one beam of colored light was allowed to pass through a second prism, the color would be somewhat spread out but no new colors would appear.

Newton's contribution was not that he produced these colors, for that had been done before, but that he suggested a new ex-

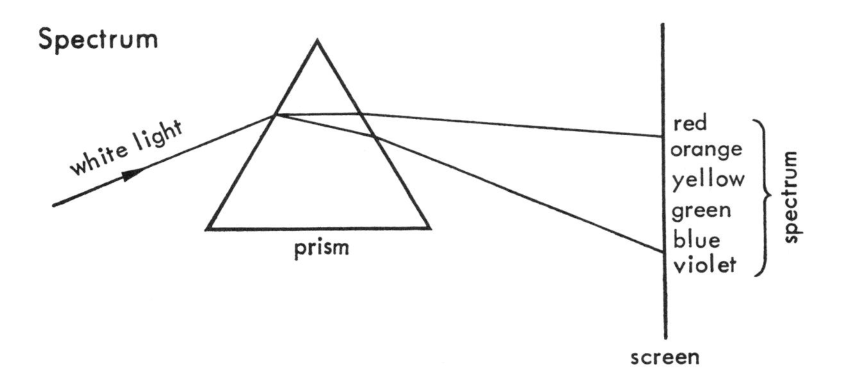

planation for them. The only ingredients that produced the spectrum were the pure white light of sunlight and the pure colorless glass of the prism. Newton therefore stated that despite the long-settled opinion of mankind white light was not pure, but a complex mixture of all the colors of the rainbow. It appeared white only because the combination so stimulated the retina as to be interpreted by the brain as "white."

As a point in favor of the validity of this view, there is Newton's reversal of spectrum formation. Newton allowed the light of the colored spectrum to fall upon a second prism held upside down with respect to the first. The light was refracted in the opposite direction and the situation was reversed. Where previously a round beam of white light had been spread out into a thick line of different colors, now that line was compressed back into a circle of white light.

Apparently, white light is composed of a vast assemblage of different varieties of light, each with its own characteristic way of being refracted. The group of varieties of light that were least refracted gave rise to the sensation of red; the next group, slightly more refracted, to the sensation of orange; and so on to the most refrangible varieties, which seemed violet.

White light, because of this difference in refrangibility of its components, always breaks into color on passing obliquely from one medium into another of different index of refraction. However, if the second medium is bounded by parallel interfaces (as an ordinary sheet of glass is) the effect produced on entering is canceled on leaving. For that reason white light entering a sheet of glass is white once more on leaving. When the edges of a transparent medium are not parallel, as in a prism, at the edge of a sheet of glass, or in the case of round water droplets, the color production process is not canceled, and a spectrum or a rainbow results.

This means that in determining the index of refraction of a transparent substance, the use of white light introduces an uncertainty, since a whole sheaf of indices of refraction is produced by the various colors present in the light. For that reason it is customary to make use of a particular colored light* in determining indices of refraction. One device frequently used is a "sodium lamp," a device in which light is emitted by the heated sodium vapor within the bulb. This light is yellow in color and is refracted by an amount that varies only over a very small range.

* The particular color of light can be defined by its "wavelength." This will be discussed in the next chapter.

By this view of light, it is easy to explain colored objects. It is not necessary to suppose that objects must be either fully transparent (transmitting all colors of light) or fully opaque (transmitting none of them). Some substances may well be opaque to some colors and transparent to others. Red glass, for instance, may possess a chemical admixture that will absorb colors other than red and allow red to pass through. In that case, white light passing through red glass becomes red in color, not because it has gained an impurity from the glass, but only because it has lost to the glass all its components but red. In the same way, an object may reflect some colors and absorb others, appearing colored to the eye for that reason.

It must not be supposed, however, that all yellow objects reflect only yellow light, or that all blue glasses transmit only blue light. It is important to distinguish between physical color and physiological color. A physical color can be identified by the amount of refraction it undergoes in passing from one substance into another. The physiological color is what our brain interprets it to be. The physiological mechanism in the retina of the eye works in such a fashion that a physical orange will give rise to the sensation of orange; therefore it will be a physiological orange as well. However, the retina may be activated in the same way by a mixture of colors that do not include physical orange—for instance, by a mixture of red and yellow. This mixture will then be physiological orange, too.

Light that is colored by transmission through colored glass or reflection from a colored surface need not actually contain the physical colors that correspond to the physiological ones we see. We can determine the physical colors present by passing the light through a prism; for the physiological colors our sight is sufficient, provided, of course, our color vision is normal.

In 1807, the English scientist Thomas Young (1773–1829) pointed out that red, green and blue could, in proper combination, give rise to the sensation of any other color. This was later amplified by the German physiologist Hermann Ludwig Ferdinand von Helmholtz (1821–1894) and is therefore called the Young-Helmholtz theory of color vision.

Many physiologists think that this ability of red, green and blue to create the entire spectrum is a reflection of the situation in the retina of the eye—that is, that there may be three types of color-sensitive retinal cells, one reacting most strongly to red, one to green, and one to blue. The extent to which a particular color of the spectrum or a particular mixture of colors activates each of the three gives rise, therefore, to the color sensation of which we become aware. Light that activates all three equally might then be interpreted as "white"; light that activates the three in one fixed ratio might be "yellow," in another "violet," and so on.

This is made use of in color photography. In one process, the film consists of triple layers, one of which contains a red-absorbing dye so that it is particularly sensitive to red light, another a dye sensitive to blue light, and the third a dye sensitive to green light. The light at each point of the image affects the three in a particular ratio and upon development produces at every point of the image a dye combination in a particular ratio of intensity. The dye combination affects the three pigments of our retina accordingly, and we see the color in the photograph as we would have seen it in the object itself.

Again, a colored print may be obtained by combining dots of a few different colors. Any color can be reproduced by varying the ratio of the colored dots represented. Under a magnifying glass the individual dots may be large enough to be seen in their true color, but if the individual dots cannot be resolved by the unaided eye, neighboring dots will affect the same retinal area and produce a combination of effects that result in the sensation of a color not actually in the dots themselves.

A similar situation is to be found on the screen of a color television set. The screen is covered by an array of dots, some of which react to light by shining blue, some by shining green, and some by shining red. Each particular portion of the TV picture scanned and transmitted by the camera activates these dots in a particular brightness ratio, and we sense that ratio as the same color that was present in the original object.*

Reflecting Telescopes

The fact that white light is a mixture of colors explained what had been observed as an annoying imperfection of the telescope. A parallel beam of light rays passing through a converging lens is brought to a focus on the other side of the lens. The exact position of this focus depends upon the extent to which the light is refracted on passing through the lens, and this at once introduces a complication, since white light consists of a mixture of colors, each

* This three-color system is not universally accepted as actually describing the method by which the human eye detects color, but it can certainly be used to produce color photographs and color television.

with its own refractivity, and it is almost always white light we are passing through the lenses of telescopes and microscopes.

The red component of white light is least refracted on passing through the lens and comes to a focus at a particular point. The orange component being refracted to a somewhat greater extent comes to a focus at a point somewhat closer to the lens than the red light. The yellow light is focused closer still, and next follow green, blue and, closest of all, violet. This means that if the eye is so placed at a telescope eyepiece that the red component of the light from a heavenly body is focused upon the retina, the remaining light will be past its focal point and will be broader and fuzzier. The image of the heavenly body will be circled with a bluish ring. If the eye is placed so that the violet end of the spectrum is focused, the remaining light will not yet have reached its focus and there will be an orange rim. The best that can be done is to focus the eye somewhere in the center and endure the colored rims, which are in this way minimized but not abolished.

This is called *chromatic aberration,* "chromatic" coming from a Greek word for color. It would not exist if light were taken from only a small region of the spectrum (such light would be *monochromatic* or "one color"), but a telescope or microscope must take what it gets—usually not monochromatic light.

Newton felt that chromatic aberration was an absolutely unavoidable error of lenses and that no telescope that depended on images formed by the refraction of light through a lens (hence, a *refracting telescope*) would ever be cleared of it. He set about correcting the situation by substituting a mirror for a lens. As was pointed out earlier in the book, a real image is formed by a concave mirror reflecting light, as well as by a convex lens transmitting light. Furthermore, whereas different colors of light are refracted through lenses by different amounts, all are reflected from mirrors in precisely the same way. Therefore, mirrors do not give rise to chromatic aberration.

In 1668, Newton devised a telescope making use of such a mirror. It was the first practical *reflecting telescope.* It was only six inches long and one inch wide, but it was as good as Galileo's first telescope. Shortly thereafter, Newton built larger and better reflecting telescopes.

In addition to the lack of chromatic aberration, reflecting telescopes have additional advantages over refracting telescopes. A lens must be made of flawless glass with two curved surfaces, front and back, ground to as near perfection as possible, if the faint light of the stars is to be transmitted without loss and focused with precision. However, a mirror reflects light, and for this only the reflecting surface need be perfect. In a telescopic mirror it is the forward end that is covered with a thin, reflecting metallic film (not the rear end as in ordinary mirrors), so the glass behind the forward reflecting surface may be flawed and imperfect. It has nothing to do with the light; it is merely the supporting material for the metallized surface in front. Since it is far easier to get a large piece of slightly flawed glass than a large piece of perfect glass, it is easier to make a large telescopic mirror than a large telescopic lens—particularly since only one surface need be perfectly ground in a mirror, rather than two as in the case of the lens.

Again, light must pass through a lens and some is necessarily absorbed. The larger and thicker the lens, the greater the absorption. On the other hand no matter how large a mirror may be, the light is merely reflected from the surface, and virtually none is lost by absorption. Then too a lens can only be supported about the rim, since all other parts must be open to unobstructed passage of light; it becomes difficult to support a large, thick lens about the rim, for the center sags and this introduces distortion. The mirror on the other hand can be supported at as many points as may be desired.

The result is that all the large telescopes in the world are reflectors. The largest currently in operation is the 200-inch reflector, which went into operation in 1948 at Mount Palomar, California. Then there are the 120-inch reflector at Mount Hamilton and the 100-inch reflector at Mount Wilson, both in California. The Soviet Union has recently put a 103-inch reflector into use in the Crimea and has a 236-inch reflector under construction.

Compare this with the 40-inch refractor at Yerkes Observatory in Wisconsin, which has been the largest refractor in use since 1897 and is likely to remain so.

Nevertheless, even the reflectors have by and large reached their practical limit of size. The gathering and concentration of light implies a gathering and concentration of the imperfections of the environment—the haze in the air, the scattered light from distant cities, the temperature variations that introduce rapid variations in the refractivity of air and set the images of the stars to dancing and blurring.

For the next stage in optical telescopy, we may have to await the day (perhaps not very far off) when an astronomical observatory can be set up on the moon, where there is no air to absorb, refract, and scatter the dim light of the stars, and where an astronomer (given the means for survival in a harsh environment) may well consider himself figuratively, as well as literally, in heaven.

But Newton was wrong in thinking that chromatic aberration in lenses was unavoidable. It did not occur to him to test prisms made of different varieties of glass in order to see whether there were the same differences in refraction of the colors of light in all of them. What's more, he ignored the reports of those who did happen to test the different varieties (even Homer nods!).

The difference in degree of refraction for light at the red end and at the violet end of the spectrum determines the degree to which a spectrum is spread out at a given distance from the prism. This is the *dispersion* of the spectrum. The dispersion varies with different types of glass. Thus, flint glass (which contains lead compounds (has a dispersion twice as great as crown glass (which does not contain lead compounds).

One can therefore form a converging lens of crown glass and add to it a less powerful diverging lens of flint glass. The diverging lens of flint glass will only neutralize part of the convergent effect of the crown glass lens, but it will balance all the dispersion. The result will be a combined lens not quite as convergent as the crown glass alone, but one that does not produce a spectrum or suffer from chromatic aberration. It is an *achromatic lens* (from Greek words meaning "no color"). The English optician John Dollond (1706–1761) produced the first achromatic refracting telescope in 1758. While it did not remove all the disabilities of refractors, it did make moderately large refractors practical.

The development of achromatic lenses was of particularly great importance to microscopy. There it was not practical to try to substitute mirrors for lenses and reflection for refraction. For that reason, microscopists had to bear with detail-destroying chromatic aberration long after telescopists had been able to escape.

Through the efforts of the English optician Joseph Jackson Lister (1786–1869) and the Italian astronomer Giovanni Batista Amici (1786?–1863), microscopes with achromatic lenses were finally developed in the early nineteenth century. It was only thereafter that the smaller microorganisms could be seen clearly and that the science of bacteriology could really begin to flourish.

Spectral Lines

Actually, we must not think of sunlight as being composed of a few different colors, as though it were a mixture of seven pigments. Sunlight is a mixture of a vast number of components separated by very slight differences in refractivity. For example, the red portion of the spectrum is not a uniform red but shades imperceptibly into orange.

In the rainbow and in simple spectra such as those that Newton formed, the light seems to be continuous, as though all the infinite possible refractivities were present in sunlight. This is an illusion, however.

If a beam of light passes through a small hole in a blind, let us say, and then through a prism, a large number of circular images are formed, each imprinted in a variety of light of particular refractivity. These overlap and blend into a spectrum. If light of a certain refractivity were missing, neighboring images in either direction would overlap the spot where the missing refractivity ought to have been, and no gap would be visible.

The situation would be improved if the beam of light were made to pass through a narrow slit. The spectrum would then consist of a myriad of images of the slit, each overlapping its neighbor only very slightly. In 1802, the English chemist William Hyde Wollaston (1766–1828) did see a few dark lines in the spectrum, representing missing slit-images. He, however, felt they represented the boundary lines between colors and did not follow through.

Between 1814 and 1824, however, a German optician, Joseph von Fraunhofer (1787–1826), working with particularly fine prisms, noticed hundreds of such dark lines in the spectrum. He labeled the most prominent ones with letters from A to G and carefully mapped the relative position of all he could find. These *spectral lines* are, in his honor, sometimes called *Fraunhofer lines.*

Fraunhofer noticed that the pattern of lines in sunlight and in the light of reflected sunlight (from the moon or from Venus, for instance) was always the same. The light of stars, however, would show a radically different pattern. He studied the dim light of heavenly objects other than the sun by placing a prism at the eyepiece of a telescope, and this was the first use of a *spectroscope.*

Fraunhofer's work was largely disregarded in his lifetime, but a generation later the German physicist Gustav Robert Kirchhoff (1824–1887) put the spectroscope to use as a chemical tool and founded the science of *spectroscopy.*

It was known to chemists that the vapors of different elements heated to incandescence produced lights of different colors. Sodium vapor gave out a strongly yellow light; potassium vapor a dim violet light; mercury a sickly greenish light, and so on. Kirchhoff passed such light through a spectroscope and found that the various elements produced light of only a few refractive varieties. There would only be a few images of the slit, spread widely apart, and this would be an *emission spectrum*. The exact position of each line could be measured against a carefully ruled background, and it could then be shown that each element always produced lines of the same color in the same place, even when it was in chemical combination with other elements. Furthermore, no two elements produced lines in precisely the same place.

The emission line spectrum could be used as a set of "finger-prints" for the elements, therefore. Thus, in 1859, Kirchhoff and his older collaborator, the German chemist Robert Wilhelm Bunsen (1811–1899), while heating a certain mineral to incandescence and studying the emission spectrum of the vapors evolved, discovered lines that did not correspond to those produced by any known element. Kirchhoff and Bunsen therefore postulated the existence of a new element, which they called *cesium* (from the Latin word for "sky blue," because of the sky-blue color of the brightest of the new lines they had observed). The next year they made a similar discovery and announced *rubidium* (from a Latin word for "dark red"). The existence of both metals was quickly confirmed by older chemical techniques.

Kirchhoff observed the reverse of an emission spectrum. Glowing solids emit light of all colors, forming a *continuous spectrum*. If the light of a carbon-arc, for instance, representing such a continuous spectrum is allowed to pass through sodium vapor that is at a temperature cooler than that of the arc, the sodium vapor will absorb some of the light. It will, however, absorb light only of particular varieties—precisely those varieties that the sodium vapor would itself emit if it were glowing. Thus, sodium vapor when glowing and emitting light produces two closely spaced yellow lines that make up virtually the whole of its spectrum. When cool sodium vapor absorbs light out of a continuous spectrum, two dark lines are found to cross the spectrum in just the position of

Portion of solar spectrum in yellow region

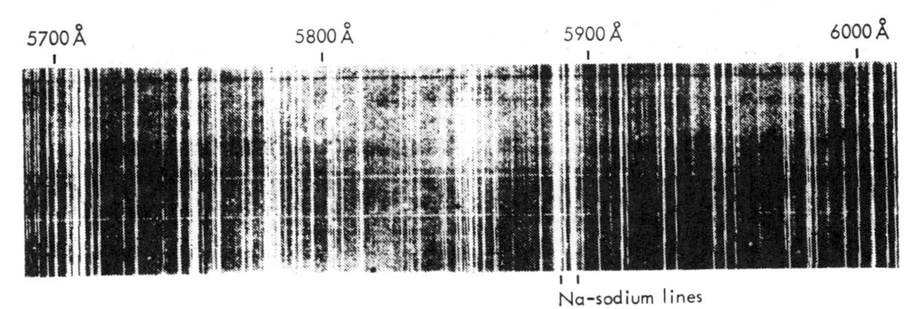

the two bright lines of the sodium emission spectrum. The dark lines represent the sodium *absorption spectrum.*

The dark lines in the solar spectrum seem to be an absorption spectrum. The blazing body of the sun is sufficiently complex in chemical nature to produce what is essentially a continuous spectrum. As the light passes through the somewhat cooler atmosphere, it is partially absorbed. Those parts that would be most strongly absorbed, and that would show up as dark lines in the spectrum, would correspond to the emission spectra of the elements most common in the solar atmosphere. Thus, there are prominent sodium absorption lines in the solar spectrum (Fraunhofer labeled them the "D line"), and this is strong evidence that sodium exists in the solar atmosphere.

In this way, a variety of elements were located in the sun. Indeed, one element, helium, was located in the sun a generation before it was found to exist on the earth. Even the composition of the distant stars could now be determined. While the details of the spectroscopic investigation of the heavens are more appropriately given in a book on astronomy, it might be well to say, in summary, that it was clearly shown that the heavenly bodies are made up of the same chemical elements that make up the earth—though not necessarily in the same proportions.

It also pointed up the dangers of setting limits to human endeavor. The French philosopher Auguste Comte (1798–1857), in an attempt to give an example of an absolute limit set upon man's knowledge, said that it would be forever impossible for man to learn of what material the stars were composed. Had he lived but a few years longer, he would have seen his absolute limit easily breached.

Diffraction

The discovery that white light was actually a mixture of many colors opened new and serious problems for physicists. As long as light could be taken to be an undifferentiated and pure phenomenon, geometrical optics was sufficient. Lines could be drawn representing light rays, and the phenomena of reflection and refraction could be analyzed without any consideration of what the nature of light might be. That question might be left for philosophers.

With light a mixture of colors, it became necessary to seek some explanation for the manner in which light of one color differed from light of another color. For that, the question of the nature of light in general had to be considered, and thus was born *physical optics.*

As was pointed out at the start of the book, there are two ways, in general, of avoiding the problem of action at a distance. One is to suppose particles streaming across a space that might then be considered as empty; the other is to suppose waves being propagated through a space that is not really empty. Both types of explanation were advanced for light in the latter half of the seventeenth century.

The more direct of the two alternatives is the particle theory, which Newton himself supported. To begin with, this at once explains the rectilinear propagation of light. Suppose luminous objects are constantly firing tiny particles outward in all directions. If these particles are considered to be massless, a luminous body does not lose weight merely by virtue of being luminous, and light itself would not be affected by gravity. Light, when traveling in an unobstructed path, if unaffected by gravitational force, must travel in a straight path at a constant velocity, as required by Newton's first law of motion (see page I–24). The particles of light would be stopped and absorbed by opaque barriers, and speeding past the edge of the barrier, would cast a sharp boundary between the illuminated area beyond and the barrier-shaded area.

To Newton, the alternative of a wave theory seemed untenable. The wave forms that were familiar to scientists at the time were water waves and sound waves (see page I–156), and these do not necessarily travel in straight lines or cast sharp shadows. Sound waves curve about obstacles, as we know whenever we hear a sound around a corner; and water waves visibly bend about an obstacle such as a floating log of wood. It seemed reasonable to suppose that this behavior was characteristic of wave forms in general.

And yet the particle theory had its difficulties, too. Beams of light could cross at any angle without affecting each other in direction or color, which meant that the light particles did not seem to collide and rebound as ordinary particles would be expected to do. Furthermore, despite ingenious suggestions, there was no satisfactory explanation as to why some light particles gave rise to red sensations, some to green sensations, and so on. The particles had to differ among themselves, of course, but how?

Some of Newton's contemporaries, therefore, took up the wave theory Newton had discarded. The most vigorous supporter of the wave theory in the seventeenth century was the Dutch physicist Christiaan Huygens (1629–1695). He had no real evidence in favor of waves, but he bent his efforts to show that waves could be so treated as to fit the facts of geometric optics. In 1678, he suggested that when a wave front occupies a certain line, each point on the front acts as a source of circular waves, expanding outward indefinitely. These waves melt together, so to speak, and a line can be drawn tangent to the infinite number of little circles centering about each point on the original wave front. This tangent is a picture of the new wave front, which serves as the starting region for another infinite set of circular waves to which another overall tangent can be drawn, and so on.

If waves are analyzed in this fashion, through use of what is now called *Huygen's principle,* it can be shown that a wave front will travel forward in a straight line (at least if only its middle portion is considered) to be reflected with an angle of reflection equal to the angle of incidence, and so on. Furthermore, the waves themselves would have no mass and would be in virtually infinite supply, after the fashion of water waves and sound waves. Being nonmaterial, these light waves would not affect each other upon crossing (and indeed water waves and sound waves can cross each other without interference).

It seemed, then, that there was much to be said for and against each theory. One must therefore look for places where the two theories differ as to the nature of the phenomena they predict. Through an observation of conditions under which such phenomena should exist, one or the other theory (or conceivably both) may be eliminated. (This is the method generally used wherever theories conflict or overlap.)

For instance, Huygens' wave theory could explain refraction under certain conditions. Suppose a straight wave front of light strikes the plane surface of glass obliquely. One end of the wave front strikes the glass first, but suppose its progress is slowed as it enters the glass. In that case, when the next section of the front hits the glass, it has gained on the first section, for the second has been traveling through air, while the first has been traveling, more slowly, through glass. As each section of the wave front strikes, it is slowed and gained upon by the portion of the wave front that has not yet struck. The entire wave front is in this way refracted and, in entering the glass, makes a smaller angle with the normal. On emerging from the glass, the first section to emerge speeds up again and gains on those portions that have not yet emerged. The emerging light takes on its original direction again.

An analogy can be drawn between this and a line of marching soldiers leaving a paved highway obliquely and entering a plowed field. The soldiers leaving the highway are, naturally, slowed down; those first to enter the field are slowed down first, and the whole line of soldiers (if they make no effort to correct for the change in footing) must alter the direction of march toward the direction of the normal to the highway-field interface.

Thus, the wave theory can explain refraction by supposing that the velocity of light is less in glass than in air. By making a further assumption, it can also explain spectrum formation. If light is a wave form, it must have a *wavelength* (the length from the crest of one wave to that of another, see page I–152). Suppose, then, that this wavelength varies with color, being longest at the red end of the spectrum and shortest at the violet end. It would seem reasonable then to suppose that short wavelengths are slowed more sharply on entering glass from air than are long wavelengths. (Again as an analogy, a marching soldier with a short stride would sink into the plowed field more times in covering a certain distance than would another soldier with a long stride. The short-striding soldier would then be slowed down more, and the marching line of soldiers—if no effort were made to correct matters—would break up into groups marching in slightly different directions depending on the length of their stride.)

In short, red light would be least refracted, orange next, and so on. In this way, light passing through a prism would be expected to form a spectrum.

Newton could explain refraction by his particle theory, too, but he was forced to assume that the velocity of the particles of light increased in passing from a medium of low optical density to one of high optical density. Here, then, was a clear-cut difference in the two theories. One had only to measure the velocity of light in different media and note the manner in which that velocity changed; one could then decide between the Newtonian particles and the Huygensian waves. The only catch was that it was not until nearly two centuries after the time of Newton and Huygens that such a measurement could be made (see page 74).

However, there was a second difference in the predictions of the theories. Newton's light particles traveled in straight lines in all portions of a light beam, so the beam might be expected to cast absolutely sharp shadows. Not so, Huygens' waves. Each point in the wave front served as a focus for waves in all directions, but through most of the wave front, a wave to the right from one point was canceled by a wave to the left from the neighboring point on the right, and so on. After all cancellations were taken into account, only the forward motion was left. There was an exception, however, at the ends of the wave front. At the right end, a rightward wave was not canceled because there was no rightward neighbor to send out a leftward wave. At the left end, a leftward wave was not canceled. A beam of light, therefore, had to "leak" sideways if it was a wave form. In particular, if a beam of light passed through a gap in an opaque barrier, the light at the boundary of the beam, just skimming the edge of the gap, ought to leak sideways so that the illuminated portion of a surface farther on ought to be wider than one would expect from strictly straight-line travel.

This phenomenon of a wave form bending sideways at either end of a wave front is called *diffraction,* and this is, in fact, easily observed in water waves and sound waves. Since light, on passing through a gap in a barrier, did not seem to exhibit diffraction, the particle theory seemed to win the nod. Unfortunately, what was not clearly understood in Newton's time was that the smaller the wavelength of any wave form, the smaller the diffraction effect. Therefore, if one but made still another assumption—that the wavelength of light waves was very small—the diffraction effect would be expected to be very hard to observe, and a decision might still be suspended.

As a matter of fact, the diffraction of light *was* observed in the seventeenth century. In 1665, an Italian physicist, Francesco Maria Grimaldi (1618?–1663), passed light through two apertures and showed that the final band of light on the surface that was illuminated was a trifle wider than it ought to have been if light had traveled through the two apertures in an absolutely straight fashion. In other words, diffraction had taken place.

What was even more important was that the boundaries of the illuminated region showed color effects, with the outermost portions of the boundary red and the innermost violet. This, too, it was eventually understood, fit the wave theory, for if red light had the longest wavelengths it would be most diffracted, while violet light, with the shortest wavelengths, would be least diffracted.

Indeed, this principle came to be used to form spectra. If fine parallel lines are scored on glass, each will represent an opaque region separated by a transparent region. There will be a series of gaps, at the edges of which diffraction can take place. In fact, if the gaps are very narrow, the glass will consist entirely of gap edges, so to speak. If the scoring is very straight and the gaps are very narrow, the diffraction at each edge will take place in the same fashion, and the diffraction at any one edge will reinforce the diffraction at all the others. In this way, a spectrum as good as, or better than, any that can be formed by a prism will be produced. Lines can be scored more finely on polished metal than on glass. In such a case, each line is an opaque region separated by a reflecting region, and this will also form a spectrum (though ordinary reflection from unbroken surfaces will not).

The spectra formed by such *diffraction gratings* are reversed in comparison to spectra formed by refraction. Where violet is most refracted and red least, violet is least diffracted and red most. Consequently, if the spectrum in one case is "red-left-violet-right," it is "red-right-violet-left" in the other. More exactly, in the case of refraction spectra, red is nearest the original line in which light was traveling, while violet is nearest the original line in the case of diffraction spectra.

(Nowadays, diffraction gratings are used much more commonly than prisms in forming spectra. The first to make important use of diffraction gratings for this purpose was Fraunhofer, the man who first made thorough observations of spectral lines.)

Newton was aware of Grimaldi's experiments and even repeated them, particularly noting the colored edges. However, the phenomenon seemed so minor to him that he did not feel he could suspend the particle theory because of it, and so he disregarded its significance. More dramatic evidences of diffraction, and the ability to measure the velocity of light in different media, still remained far in the future. What it amounted to, then, was that physicists of the seventeenth century had to choose between two personalities rather than between two sets of physical evidence. Newton's great prestige carried the day, and for a hundred years afterward, throughout the eighteenth century, light was considered by almost all physicists to be indisputably particulate in nature.

CHAPTER 20

Light Waves

Interference

The eighteenth century confidence in the existence of light particles came to grief at the very opening of the nineteenth century. In 1801, Young (of the Young-Helmholtz theory of color vision) conducted an experiment that revived the wave theory most forcefully.

Young let light from a slit fall upon a surface containing two closely adjacent slits. Each slit served as the source of a cone of light, and the two cones overlapped before falling on a screen.

If light is composed of particles, the region of overlapping should receive particles from both slits. With the particle concentration therefore doubled, the overlapping region should be uniformly brighter than the regions on the outskirts beyond the overlapping, where light from only one cone would be received. This proved not to be so. Instead, the overlapping region consisted of stripes—bright bands and dim bands alternating.

For the particle theory of light, this was a stopper. On the basis of the wave theory, however, there was no problem. At some points on the screen, light from both the first and second cone would consist of wave forms that were *in phase* (that is, crest matching crest, trough matching trough—see page I–151). The two light beams would reinforce each other at those points so that there would be a resultant wave form of twice the amplitude and, therefore, a region of doubled brightness. At other points on the screen, the two light beams would be *out of phase* (with crest matching trough and trough matching crest). The two beams would then cancel, at least in part, and the resultant wave form would have a much smaller amplitude than either component; where canceling was perfect, there would be no wave at all. A region of dimness would result.

In short, whereas one particle of the type Newton envisaged for light could not interfere with and cancel another particle, a wave form can and does easily interfere with and cancel another wave form. *Interference patterns* can easily be demonstrated in water waves, and interference is responsible for the phenomenon of beats; for instance (see page I–150), in the case of sound waves. Young was able to show that the wave theory would account for just such an interference as was observed.

Furthermore, from the spacing of the interference bands of light and dark, Young could calculate the wavelength of light. If the ray of light from one cone is to reinforce the ray of light from the second cone, both rays must be in phase, and that means the distances from the point of reinforcement on the screen to one slit and to the other must differ by an integral number of wavelengths. By choosing the interference bands requiring the smallest difference in distances, Young could calculate the length of a single wavelength and found it to be of the order of a fifty-thousandth of an inch, certainly small enough to account for the difficulty of observing diffraction effects (see page 64). It was possible to show, furthermore, that the wavelengths of red light are about twice the wavelengths of violet light, which fit the requirements of wave theory if spectrum formation is to be explained.

In the metric system, it has proved convenient to measure the wavelengths of light in *millimicrons* (mμ),* where a millimicron is a billionth of a meter (10^{-9}m) or a ten-millionth of a centimeter (10^{-7}cm). Using this unit, the spectrum extends from 760 mμ for the red light of longest wavelength to 380 mμ for the violet light of shortest wavelength. The position of any spectral line can be located in terms of its wavelength.

One of those who made particularly good measurements of the wavelengths of spectral lines was the Swedish astronomer and physicist Anders Jonas Ångström (1814–1874), who did his work

* The Greek letter "mu" (μ) commonly symbolizes a "micrometer" or "micron" which is one-millionth of a meter.

in the mid-nineteenth century. He made use of a unit that was one-tenth of a millimicron. This is called an *angstrom unit* (A) in his honor. Thus, the range of the spectrum is from 7600 A to 3800 A.

The wavelength ranges for the different colors may be given roughly (for the colors blend into one another and there are no sharp divisions) as: red, 7600–6300 A; orange, 6300–5900 A; yellow, 5900–5600 A; green, 5600–4900 A; blue, 4900–4500 A; violet 4500–3800 A.

Incandescent sodium vapor gives off a bright line in the yellow, while sodium absorption produces a dark line in the same spot. This line, originally considered to be single and given the letter D by Fraunhofer has, by the use of better spectroscopes, been resolved into two very closely spaced lines, D_1 and D_2. The former is at wavelength 5896 A, the latter at 5890 A. Similarly, Fraunhofer's C line (in the red) and F line (in the blue) are both produced by hydrogen absorption and have wavelengths of 6563 A and 4861 A respectively. (Indeed, it was Angström who first showed from his study of spectral lines that hydrogen occurred in the sun.) In similar fashion, all the spectral lines produced by any element, through absorption or emission, can be located accurately.

The wave theory of light was not accepted at once despite the conclusiveness (in hindsight) of Young's experiment. However, all through the nineteenth century, additional evidence in favor of light waves turned up, and additional phenomena that, by particle theory, would have remained puzzling, found ready and elegant explanations through wave theory. Consider, for instance, the color of the sky. . . .

Light, in meeting an obstacle in its otherwise unobstructed path, undergoes a fate that depends on the size of the obstacle. If the obstacle is greater than 1000 mμ in diameter, light is absorbed and there is an end to a light ray, at least in the form of light. If the obstacle is smaller than 1 mμ in diameter, the light ray is likely to pass on undisturbed. If, however, the obstacle lies between 1 mμ and 1000 mμ in diameter, it will be set to vibrating as it absorbs the light and may then emit a light ray equal in frequency (and therefore in wavelength) to the original, but traveling in a different direction. This is *light scattering*.

The tiny water or ice particles in clouds are of a size to scatter light in this fashion; therefore a cloud-covered sky is uniformly white (or, if the clouds are thick enough to absorb a considerable fraction of the light altogether, uniformly gray).

The dust normally present in the atmosphere also scatters light. Shadows are therefore not absolutely black but, although darker by far than areas in direct sunlight, receive enough scattered light to make it possible to read newspapers in the shade of a building or even indoors on a cloudy day.

After the sun has set, it is shining past the bulge of the earth upon the upper atmosphere. The light scattered downward keeps the earth in a slowly dimming *twilight*. It is only after the sun has sunk 18° below the horizon that full night can be said to have begun. In the morning, sunrise is preceded by a second twilight period which is *dawn*.

As particles grow smaller, a pronounced difference becomes noticeable in the amount of scattering with wavelength. The light of short wavelength is scattered to a greater extent than the light of long wavelength. Thus, if sunlight shines down upon a cloud of tobacco smoke, it is the short-wave light that is more efficiently scattered, and the tobacco smoke therefore seems bluish.

The British physicist John Tyndall (1820–1893) studied this phenomenon. He found that light passing through pure water or a solution of small-molecule substances such as salt or sugar, underwent no scattering. The light beam, traveling only forward, cannot be seen from the side and the liquid is *optically clear*. If the solution contains particles large enough to scatter light, however (examples are the molecules of proteins or small conglomerates of ordinarily insoluble materials such as gold or iron oxide) some of the light is emitted sideways, and the beam can then be seen from the side. This is the *Tyndall effect*.

The English physicist John William Strutt, Lord Rayleigh (1842–1919), went into the matter in greater detail in 1871. He worked out an equation that showed how the amount of light scattered by molecules of gas varied with a number of factors, including the wavelength of the light. He showed the amount of scattering was inversely proportionate to the fourth power of the wavelength. Since the red end of the spectrum had twice the wavelength of the violet end, the red end was scattered less (and the violet end scattered more) by a factor of 2^4, or 16.

Over short distances, the scattering by particles as small as gas molecules of the atmosphere is insignificant. If, however, the miles of atmosphere stretching overhead are considered, scattering mounts up and, as Rayleigh showed, must be confined almost entirely to the violet end of the spectrum. Enough light is so scattered to drown out the feeble light of the stars (which are, of course, present in the sky by day as well as by night). Furthermore, the scattered light that illuminates the sky, heavily represented in the short-wave region, is blue in color; the sun itself, with that small quantity of shortwave light substracted from its color, is a trifle redder than it would be if the atmosphere were absent.

This effect is accentuated when the sun is on the horizon, for it then shines through a greater thickness of air as its light comes obliquely through the atmosphere. Enough light is scattered from even the midportions of the spectrum to lend the sky a faintly greenish hue, while the sun itself, with a considerable proportion of its light scattered, takes on a ruddy color indeed. This, reflected from broken clouds, can produce a most beautiful effect. Since the evening sky, after the day's activities, is dustier than is the morning sky, and since the dust contributes to scattering, sunsets tend to be more spectacular than sunrises. After gigantic volcano eruptions (notably that of Krakatoa, which blew up—literally—in 1883) uncounted tons of fine dust are hurled into the upper atmosphere, and sunsets remain particularly beautiful for months afterward.

On the moon (which lacks an atmosphere) the sky is black even when the sun is present in the sky. Shadows are pitch black on the moon, and the terminator (the line between the sunlit and shadowed portion of the body) is sharp, because there is neither dawn nor twilight. The earth, as seen from space, would also possess a terminator, but a fuzzy one that gradually shaded from light to dark. Furthermore, its globe would have a distinctly bluish appearance, thanks to light scattering by its atmosphere.

The Velocity of Light

In time, even the question of the velocity of light in various media was settled in favor of Huygens' view as one climax to two centuries of work on the problem. The first effort to measure the velocity of light was made by Galileo about a half-century before the wave-particle controversy began.

Galileo placed himself on a hilltop and an assistant on another about a mile away. It was his intention to flash a lantern in the night and have his assistant flash a lantern in return as soon as he spied Galileo's light. The time lapse between Galileo's exposure of light and his sighting of the return signal would then, supposedly, represent the time it took for light to travel from Galileo to the assistant and back. About that same period, this principle was also used successfully in determining the velocity of sound (see page I–164).

Galileo found a perceptible delay between the light emission and return; however, it was obvious to him that this was due not to the time taken by light to travel but to that taken by the human nervous system to react to a sensation, for the delay was no longer when the two men were a mile apart than when they were six feet apart.

Consequently, all Galileo could show by his experiment was that light traveled far more rapidly than sound. In fact, it remained possible that light traveled with infinite speed, as indeed many scholars had surmised.

It was not until the 1670's that definite evidence was presented to the effect that the velocity of light, while very great, was, nevertheless, finite. The Danish astronomer Olaus Roemer (1644–1710) was then making meticulous observations of Jupiter's satellites (which had been discovered by Galileo in 1610). The orbits of those satellites had been carefully worked out, and the moments at which each satellite ought to pass behind Jupiter and be eclipsed to the sight of an observer on earth could, in theory, be calculated with precision. Roemer found, however, that the eclipses took place off schedule, several minutes too late at some times and several minutes too early at others.

On further investigation, he discovered that at the times that earth and Jupiter were on the same side of the sun, the eclipses

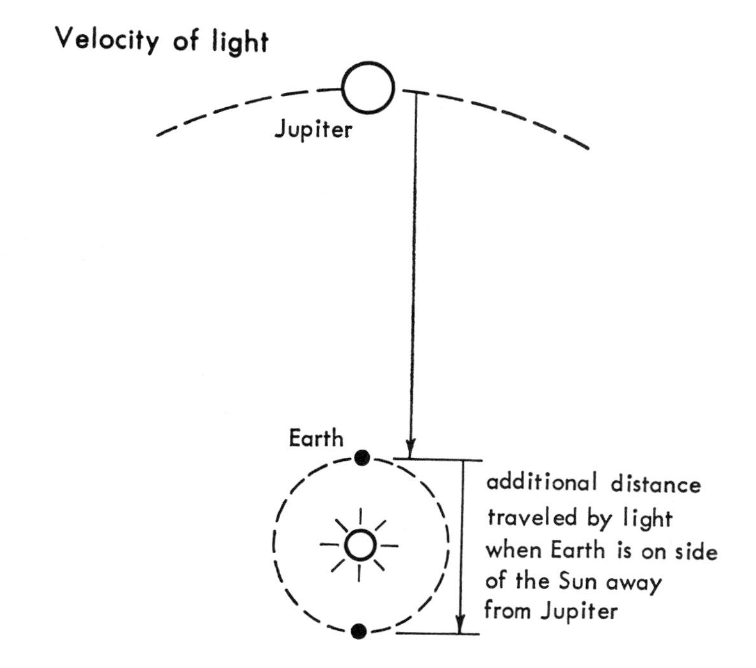

were ahead of schedule; when the two planets were on the opposite sides of the sun, they were behind schedule.

Imagine a beam of light connecting a Jovian satellite with the earth—that is, a beam by means of which we see the satellite. At the moment of eclipse, the beam is cut off, and we no longer see the satellite. At least that would be the situation if light traveled with an infinite velocity. As soon as the beam was cut off, it would, under those conditions, cease to exist all along its path of travel, however long that path might be. It would not matter whether earth was on the same side of the sun as Jupiter was, or on the opposite side.

If, however, light traveled at a finite velocity, then once the beam was cut off by Jupiter, light would continue to travel onward toward earth; the earth observer would therefore continue to see the satellite until such time as the "broken end" of the beam of light reached him. Then, and only then, would the satellite disappear in eclipse. There would be a finite time between the actual eclipse and the eclipse as seen. The greater the distance between Jupiter and the earth, the greater this time lapse.

If the distance separating Jupiter and the earth were always the same, this time lapse would be constant and might, therefore, be ignored. But the distance between Jupiter and the earth is not constant. When earth and Jupiter are on the same side of the sun, they are separated by as little as 400,000,000 miles. When they are on opposite sides, they can be separated by that distance plus the full width of the earth's orbit, or a total of about 580,000,000 miles. If at closest approach the eclipse is, say, eight minutes ahead of schedule, and at furthest distance, eight minutes behind schedule, then it would take light roughly 16 minutes to cross the diameter of the earth's orbit. Knowing the diameter of the earth's orbit, one could then easily calculate the velocity of light and, in 1676, Roemer did so. In the light of modern knowledge, the value he obtained was rather low. However, he succeeded in showing that light traveled at velocities of the order of a hundred fifty thousand miles a second.

Roemer's work was not accepted wholeheartedly, but in 1728 the English astronomer James Bradley (1693–1762) used the phenomenon of the *aberration of light* to perform a similar calculation. Suppose that the light of a star from near the North Celestial Pole is descending vertically upon the earth. The earth, however, is moving in its orbit at right angles to that direction and is therefore moving to meet the beam of light. A telescope must be angled slightly to catch that beam of light, just as an umbrella must be angled slightly to catch the raindrops if you are walking in a storm in which the rain is falling vertically.

The telescope must be angled in a continually changing direction as the earth moves about its curved orbit; therefore, the star seems to move in a very tiny ellipse in the sky. The size of the ellipse depends on the ratio of the velocity of the earth's motion to that of light's motion. (There would be no aberration if the earth were standing still or if light traveled at infinite velocity.) Since the earth's velocity around the sun is known to be 18.5 miles per second, the velocity of light could easily be calculated. Bradley showed that the velocity of light was nearly 190,000 miles a second.

It was not until 1849, however, that the question of the velocity of light was brought down from the heavens and measured upon the earth. The experimenter who did this was the French physicist Armand Hippolyte Louis Fizeau (1819–1896), who returned to Galileo's principle but tried to eliminate the matter of human reaction time.

He did this by allowing the light from one hilltop to be returned from another hilltop, not by a human being, but by a mirror. Furthermore, the light being emitted had to pass between the cogs of a turning wheel; consequently, the light was "chopped up" into a series of fragments—a dotted line of light, so to speak.

Consider the behavior of such an interrupted beam. Light travels so rapidly that if the wheel were turning at an ordinary rate of speed, each bit of light emerging between the cogs of the wheel would streak to the mirror, be reflected, and streak back before the wheel had time to move very far. The light would return through the same gap in the cogs by which it had left. A person viewing the returning light through an eyepiece would see a series of light pulses at such short intervals that he would seem to see one continuous blur of light. Furthermore, the light would be quite bright, because almost all the light that was emitted would be returned.

Of course, the last bit of one of the fragments of light, the bit that had just slipped between the cogs as one cog was about to cut off the beam, would find the cog completely in the way when it returned, and it would be absorbed. Consequently, the reflected light would lose just a tiny bit of its intensity and would not be quite as bright as it would be if there were no cogged wheel in the way at all.

If the cogged wheel were made to rotate faster and faster, a larger and larger fraction of the light would be intercepted by the cog on its return, and the reflected light, as seen through the eyepiece, would grow dimmer and dimmer. Eventually, this dimness would reach a minimum, because all the light that emerged while the gap was passing would return while the cog was passing. But if the wheel were rotated still faster, then some of the light would begin slipping through the next gap, and the light would begin to brighten again. At a certain point, all the light passing through one gap would return through the next, and there would be light at maximum brightness again.

By measuring the speed of rotation of the cogged wheel at the time of both minimum and maximum brightness, and knowing the distance from the light source to the mirror, one could calculate the speed of light. Fizeau's results were not as accurate as Bradley's, for instance, but Fizeau had brought the measurement to earth and had not involved any heavenly bodies.

Fizeau had a co-worker, the French physicist Jean Bernard Léon Foucault (1819–1868), who introduced an improvement that further eliminated human error. In Fizeau's device, it was still necessary to choose the points at which the brightness of the light seemed at a minimum or at a maximum. This required human judgment, which was unreliable. Foucault introduced a second mirror in place of the toothed wheel. The second mirror was set to turning. The turning mirror sent light to the fixed mirror only when it was turned in the proper direction. By the time the light had been reflected from the fixed mirror, the turning mirror had moved slightly. The returning light was reflected, therefore, not back to the fixed mirror again, but at a slight angle. With little trouble, this slight angle could be measured off a scale. From that, from the rate at which the mirror turned and from the distance between the two mirrors, the velocity of light could be measured with considerable accuracy, and was.

What's more, Foucault was able to make the same measurement when light was made to travel through water rather than through air. This he did in 1850 and found that the velocity of light in water was distinctly less than that in air. This was precisely in accord with Huygens' prediction of nearly two centuries before and counter to Newton's prediction. To physicists, this seemed to be the last straw, and there was no important resistance to the wave theory of light after that.

The velocity of light passing through any transparent medium is equal to its velocity in a vacuum divided by the index of refraction (n) of the medium. The velocity of light in a vacuum is customarily represented as c, which stands for *celeritas,* a Latin word for "velocity." We might say then:

$$v = \frac{c}{n} \qquad \text{(Equation 5–1)}$$

If we accept the approximate value of 186,000 miles per second for c, then since the index of refraction of water is 1.33, the velocity of light in water is 186,000/1.33, or 140,000 miles per second. Similarly, the velocity of light in glass with an index of refraction of 1.5 is 124,000 miles per second, while through diamond, with its index of refraction of 2.42, the velocity of light is 77,000 miles per second.

No substance with an index of refraction of less than 1 has been discovered, nor, on the basis of present knowledge, can any such substance exist. This is another way of saying that light travels more rapidly in a vacuum than in any material medium.

Since Foucault's time, many added refinements have been brought to the technique of measuring the velocity of light. In 1923, the American physicist Albert Abraham Michelson (1852–1931) made use of a refined version of Foucault's setup and separated his mirrors by a distance of 22 miles, estimating that distance to an accuracy of within an inch. Still later, in 1931, he decided to remove the trifling interference of air (which has an index of refraction slightly greater than 1 and which carries haze and dust besides) by evacuating a tube a mile long and arranging combinations of mirrors in such a way as to allow the light beam to move back and forth till it had traveled ten miles in a vacuum, all told.

Michelson's last measurement had pinned the velocity down to within ten miles per second of what must be the correct value, but that did not satisfy physicists. In 1905 (as we shall have occasion to see later in the volume, see page 102ff), the velocity of light in a vacuum was revealed to be one of the fundamental constants of the universe, so there could be no resting while it remained possible to determine that velocity with a little more accuracy than had been possible hitherto. Consequently, new and more refined methods for measuring the velocity of light have been brought into use since World War II, and in 1963, the National Bureau of Standards adopted the following value for c: 186,281.7 miles per second.

To be precisely accurate, they adopted the value in metric units, and here, by a curious coincidence, the velocity of light comes out to an almost even value: 299,792.8 kilometers per second.

As you see, this is just a trifle short of 300,000 kilometers per second, or 30,000,000,000 centimeters per second. This latter value can be given as 3×10^{10} cm/sec.

At this velocity, light can travel from the moon to the earth in 1 1/4 seconds, and from the sun to the earth in eight minutes. In one year, light travels 9,450,000,000,000 kilometers, or 5,900,000,000,000 miles, and this distance is called a *light-year.*

The light-year has become a convenient unit for use in astronomy since all objects outside our solar system are separated from us by distances so vast that no smaller unit will do. Our nearest neighbors among the stars, the members of the Alpha Centauri system, are 4.3 light years away, while the diameter of our Galaxy as a whole is some 100,000 light-years.

The Doppler-Fizeau Effect

With light viewed as a wave motion, it was reasonable to predict that it would exhibit properties analogous to those shown by other wave motions. The Austrian physicist Johann Christian Doppler (1803–1853) had pointed out that the pitch of sound waves varied with the motion of the source relative to the listener. If a sound-source were approaching the listener, the sound waves would be crowded together, and more waves would impinge upon the ear per second. This would be equivalent to a raised frequency, so the sound would be heard as being of higher pitch than it would have been heard if the source were fixed relative to the listener. By the same reasoning, a receding sound source emits a sound of lower pitch, and the train whistle, as the train passes, suddenly shifts from treble to base (see page I–174).

In 1842, Doppler pointed out that this *Doppler effect* ought to apply to light waves, too. In the case of an approaching light source, the waves ought to be crowded together and become of higher frequency, so the light would become bluer. In the case of a receding light source, light waves would be pulled apart and become lower in frequency, so the light would become redder.*

Doppler felt that all stars radiated white light, with the light more or less evenly distributed across the spectrum. Reddish

* Please note that this change does not affect the velocity of light. The wavelength may decrease or increase according to the relative motion of the light source and the observer; however, whether the waves are long, short or in-between, they all move at the same velocity.

stars, he felt, might be red because they were receding from us, while bluish stars were approaching us. This suggestion, however, was easily shown to be mistaken, for the fallacy lies in the assumption that the light we see is all the light there is. . . .

So intimately is light bound up with vision that one naturally assumes that if one sees nothing, no light is present. However, light might be present in the form of wavelengths to which the retina of the eye is insensitive. Thus, in 1800, the British astronomer William Herschel (1738–1822) was checking the manner in which different portions of the spectrum affected the thermometer. To his surprise, he found that the temperature rise was highest at a point somewhat below the red end of the spectrum—a point where the eye could see nothing.

When the wave theory was established, the explanation proved simple. There were light waves with wavelengths longer than 7600 A. Such wavelengths do not affect the eye, and so they are not seen; nevertheless, they are real. Light of such long wavelengths can be absorbed and converted to heat; they can therefore be detected in that fashion. They could be put through the ordinary paces of reflection, refraction, and so on, provided that detection was carried through by appropriate heat-absorbing instruments and not by eye. These light waves, as received from the sun, could even be spread out into a spectrum with wavelengths varying from 7600 A (the border of the visible region) up to some 30,000 A.

This portion of the light was referred to as "heat rays" on occasion, because they were detected as heat. A better name, however, and one universally used now, is *infrared radiation* ("below the red").

The other end of the visible spectrum is also not a true end. Light affects certain chemicals and, for instance, will bring about the breakdown of silver chloride, a white compound, and produce black specks of metallic silver. Silver chloride therefore quickly grays when exposed to sunlight (and it is this phenomenon that serves as the basis for photography). For reasons not understood in 1800, but which were eventually explained in 1900 (see page 132), the light toward the violet end of the spectrum is more efficient in bringing about silver chloride darkening than is light at the red end. In 1801, the German physicist Johann Wilhelm Ritter (1776–1810) found that silver chloride was darkened at a point beyond the violet end of the spectrum in a place where no light was visible. What's more, it was darkened more efficiently there than at any place in the visible spectrum.

Thus, there was seen to be a "chemical ray" region of the spectrum, one more properly called *ultraviolet radiation* ("beyond the violet"), where the wavelength was smaller than 3600 A. Even the early studies carried the region down to 2000 A, and in the twentieth century, far shorter wavelengths were encountered.

By mid-nineteenth century, then, it was perfectly well realized that the spectrum of the sun, and presumably of the other stars, extended from far in the ultraviolet to far in the infrared. A relatively small portion in the middle of the spectrum (in which, however, solar radiation happens to be at peak intensity), distinguished only by the fact that the wavelengths of this region stimulated the retina of the eye, was what all through history had been called "light." Now it had to be referred to as *visible light*. What, before 1800, would have been a tautology, had now become a useful phrase, for there was much invisible light on either side of the visible spectrum.

It can now be seen why Doppler's suggestion was erroneous. The amount of the Doppler shift in any wave form depends on the velocity of the wave form as compared with the velocity of relative motion between wave source and observer. Stars within our Galaxy move (relative to us) at velocities that are only in the tens of kilometers per second, while the velocity of light is 300,000 kilometers per second. Consequently, the Doppler effect on light would be small indeed. There would be only a tiny shift toward either the red or the blue—far from enough to account for the visible redness or blueness of the light of certain stars. (This difference in color arises from other causes, see page 128.)

Furthermore, if there is a tiny shift toward the violet, some of the violet at the extreme end does, to be sure, disappear into the ultraviolet, but this is balanced by a shift of some of the infrared into the red. The net result is that the color of the star does not change overall. The reverse happens if there is a shift to the red, with infrared gaining and ultraviolet losing, but with the overall visible color unchanged.

Fizeau pointed this out in 1848 but added that if one fixed one's attention on a particular wavelength, marked out by the presence of a spectral line, one might then detect its shift either toward the red or toward the violet. This turned out to be so, and in consequence the Doppler effect with respect to light is sometimes called the *Doppler-Fizeau effect*.

Important astronomical discoveries were made by noting changes in the position of prominent spectral lines in the spectra of heavenly bodies, as compared with the position of those same spectral lines produced in the laboratory, where no relative motion is involved. It could be shown by spectral studies alone, for instance, that the sun rotated, for one side of the rotating sun is receding and the other advancing; the position of the spectral lines in the light from one side or the other therefore reflected this. Again, light from Saturn's rings showed that the outer rim was moving so much more slowly than the inner rim that the rings could not be rotating as a single piece and must consist of separate fragments.

In 1868, the English astronomer William Huggins (1824–1910) studied the lines in the spectrum of the star Sirius and was able to show that Sirius was receding from us at a speed of some 40 kilometers per second (a value reduced by later investigations). Since then, thousands of stars have had their *radial velocities* (velocities toward or away from us) measured, and most such velocities fall in the range of 10 to 40 kilometers per second. These velocities are toward us in some cases and away from us in others.

In the twentieth century, such measurements were made on the light from galaxies outside our own. Here it quickly turned out that there was a virtually universal recession. With the exception of one or two galaxies nearest us, there was an invariable shift in spectral lines toward the red—an effect which became famous as the *red shift*. Furthermore, the dimmer (and, therefore, presumably the farther) the galaxy, the greater the red shift. This correlation of distance with velocity of recession would be expected if the galaxies were, one and all, moving farther and farther from each other, as though the whole universe were expanding; this, indeed, is the hypothesis usually accepted to explain the red shift.

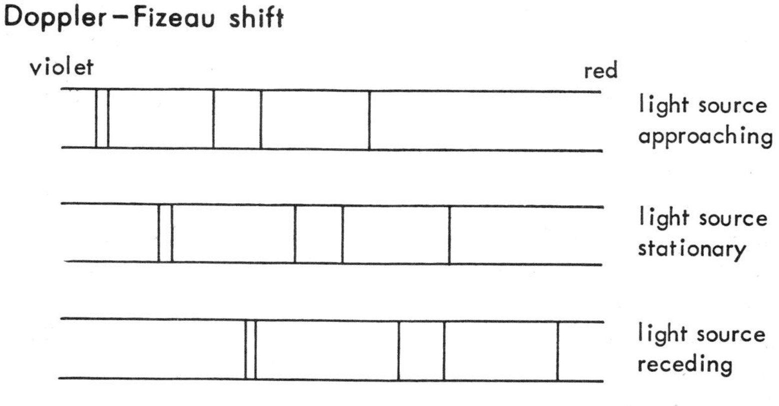

As the red shift increases with the distance of the galaxies, the velocity of recession, relative to ourselves, increases, too. For the very distant galaxies, these velocities become considerable fractions of the velocity of light. Velocities equal to half the velocity of light have been measured among these receding galaxies. Under such circumstances, there is a massive shift of light into the infrared, a greater shift than can be replaced from the ultraviolet radiation present in the light of these galaxies. The total visible light of such distant galaxies dims for that reason and sets a limit to how much of the universe we might see by visible light, no matter how great our telescopes.

Polarized Light

To say that light consists of waves is not enough, for there are two important classes of waves with important differences in properties. Thus, water waves are *tranverse waves,* undulating up and down at right angles to the direction in which the wave as a whole is traveling. Sound waves are *longitudinal waves,* undulating back and forth in the same direction in which the wave as a whole is traveling (see page I–156). Which variety represents light waves?

Until the second decade of the nineteenth century, the scientific minority who felt light to be a wave form believed it to be a longitudinal wave form. Huygens thought this, for instance. However, there remained a seventeenth century experiment on light that had never been satisfactorily explained by either Newton's particles of light or Huygens' longitudinal waves of light, and this eventually forced a change of mind.

The experiment was first reported in 1669 by a Dutch physician, Erasmus Bartholinus (1625–1698). He discovered that a crystal of Iceland spar (a transparent form of calcium carbonate) produced a double image. If a crystal was placed on a surface bearing a black dot, for instance, two dots were seen through the crystal. If the crystal was rotated in contact with the surface, one of the dots remained motionless while the other rotated about it. Apparently, light passing through the crystal split up into two rays that were refracted by different amounts. This phenomenon was therefore called *double refraction.* The ray that produced the motionless dot, Bartholinus dubbed the *ordinary ray;* the other, the *extraordinary ray.*

Both Huygens and Newton considered this experiment but

could come to no clear conclusion. Apparently, if light was to be refracted in two different ways, its constituents, whether particles or longitudinal waves, must differ among themselves. But how?

Newton made some vague speculations to the effect that light particles might differ among themselves as the poles of a magnet did (see page 141). He did not follow this up, but the thought was not forgotten.

In 1808, a French army engineer, Étienne Louis Malus (1775–1812), was experimenting with some doubly refracting crystals. He pointed one of them at sunlight reflected from a window some distance outside his room and found that instead of seeing the shining spot of reflected sunlight double (as he expected) he saw it single. He decided that in reflecting light the window reflected only one of the "poles" of light of which Newton had spoken. The reflected light he therefore called *polarized light*. It was a poor name that did not represent the actual facts, but it has been kept and will undoubtedly continue to be kept.

When the wave theory of light sprang back into prominence with Young's experiment, it soon enough became clear that if light were only considered transverse waves, rather than longitudinal waves, polarized light could easily be explained. By 1817, Young had come to that conclusion, and it was further taken up by a French physicist, Augustin Jean Fresnel (1788–1827). In 1814, Fresnel had independently discovered interference patterns, and he went on to deal with transverse waves in a detailed mathematical analysis.

To see how transverse waves will explain polarization, imagine a ray of light moving away from you with the light waves undulating in planes at right angles to that line of motion, as is required of transverse waves. Say the light waves are moving up and down. They might also, however, move right and left and still be at right angles to the line of motion. They might even be moving diagonally at any angle and still be at right angles to the line of motion. When the component waves of light are undulating in all possible directions at right angles to the line of motion, and are evenly distributed through those planes, we have *unpolarized light*.

Let's concentrate on two forms of undulation, up-down and left-right. All undulations taking up diagonal positions can be divided into an up-down component and a left-right component (just as forces can be divided into components at right angles to each other, see page I–40). Therefore, for simplicity's sake we can consider unpolarized light as consisting of an up-down component and a left-right component only, the two present in equal intensities.

It is possible that the up-down component may be able to slip through a transparent medium where the left-right component might not. Thus, to use an analogy, suppose you held a rope that passed through a gap in a picket fence. If you made up-down waves in the rope, they would pass through the gap unhindered. If you made left-right waves, those waves would collide with the pickets on either side of the gap and be damped out.

The manner in which light passes through a transparent substance, then, depends on the manner in which the atoms making up the substance are arranged—how the gaps between the atomic pickets are oriented, in other words. In most cases, the arrangement is such that light waves in any orientation can pass through with equal ease. Light enters unpolarized and emerges unpolarized. In the case of Iceland spar, this is not so; only up-down light waves and left-right light waves can pass through, and one of these passes with greater difficulty, is slowed up further, and therefore is refracted more. The result is that at the other end of the crystals two rays emerge—one made up of up-down undulations only and one made up of left-right undulations only. Each of these is a ray of polarized light. Because the undulations of the light waves in each of these rays exist in one plane only, such light may more specifically be called *plane-polarized light*.

In 1828, the British physicist William Nicol (1768?–1851) produced a device that took advantage of the different directions in which these plane-polarized light rays traveled inside the crystal of Iceland spar. He began with a rhombohedral crystal of the substance (one with every face a parallelogram) and cut it diago-

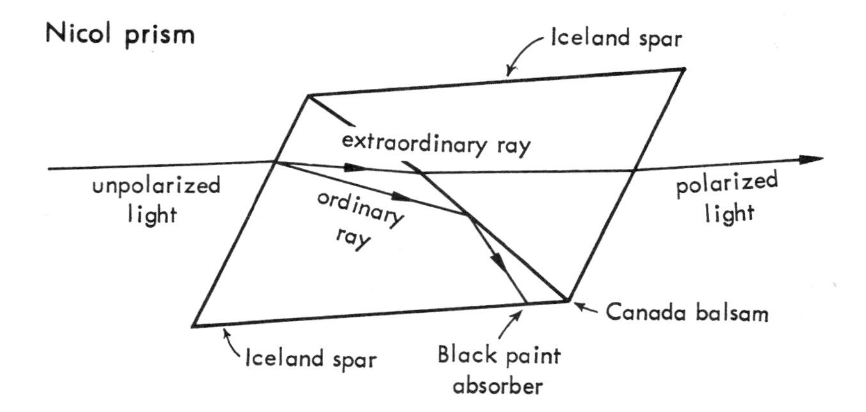

nally. The two halves were cemented together again by way of a layer of Canada balsam (a resin from a tree called the balsam fir). Light entering the crystal would be split up into two plane-polarized rays traveling in slightly different directions. One ray would strike the Canada balsam at an angle such that total reflection would take place. The reflected ray would then strike a painted section of the prism and be absorbed. The other ray, striking the Canada balsam at a slightly different angle, would be transmitted, pass into the other half of the crystal and out into the open air again.

The light emerging from such a *Nicol prism,* then, would consist of a single plane-polarized ray, representing about half the original light intensity.

Suppose the light passing through a Nicol prism is made to pass through a second Nicol prism. If the second prism is aligned in the same fashion as the first, the light will pass through the second unhindered. (That is like a rope with up-down waves passing through two picket fences, one behind the other. Neither fence gets in the way.)

But suppose the second Nicol prism is rotated through a small angle. The polarized light emerging from the first prism cannot get through the second prism in full intensity. There is some loss (as there would be in the up-down waves of the rope if the slats of wood in the second picket fence were tipped a little into the diagonal).

The amount of light that would get through the second prism would decrease as the angle through which that prism was rotated increased. Once the second prism was rotated through 90°, no light at all could get through.

The second prism can thus be used to determine the exact plane in which the light issuing from the first prism is polarized. By twisting the second prism and noticing the alignment at which the light one sees through it is at maximum brightness, one finds the plane of polarization. If one sees no light at all, the alignment of the second prism is at right angles to the plane of polarization. Since it is difficult to judge exactly where maximum or minimum brightness is, the second prism may be manufactured in such a way as to consist of two prisms set at a slight angle to each other. If one is aligned properly, the other is slightly off. Looking through an eyepiece, then, one would see one half distinctly brighter than the other. By adjusting the alignment so that both halves are equally bright, one locates the plane of polarization.

The first prism in such an instrument, the one that produces the polarized light, is the *polarizer*. The second, which determines the plane of polarization, is the *analyzer*. The instrument as a whole is a *polariscope*.

Even before the Nicol prism was invented, it was discovered by the French physicist Jean Baptiste Biot (1774–1862), in 1815, that polarized light, traveling through solutions of certain substances, or through certain transparent crystals, would have its plane of polarization shifted.

Suppose, for instance, that between the two prisms of a polariscope is a cylindrical vessel containing air and that the prisms are aligned in the same direction. If water is poured into the tube, nothing happens; the two halves of the field as seen in the eyepiece remain equally bright. The plane of polarization of the light has not been altered by passing through water. If instead of pure water a sugar solution had been placed in the tube, the two halves seen in the eyepiece would become unequally bright. The analyzer would have to be turned through some definite angle to make them equally bright again. That angle would represent the amount by which the plane of polarized light had been rotated by the sugar solution.

The size of this angle depends on various factors: the concentration of the solution and the nature of the substance dissolved; the distance through which light travels within the solution; the wavelength of the light; and the temperature of the solution. If one standardizes these factors, and either observes or calculates what the angle of rotation would be for light of a wavelength equal to that produced by a sodium vapor lamp, traveling through one decimeter of a solution containing 1 gram per cubic centimeter at a temperature of 20°C, then one obtains the *specific rotation*.

The value of the specific rotation is characteristic for every transparent system. For many this value is 0°—that is, the plane of polarized light is not rotated at all. Such systems are *optically inactive*. Systems that do rotate the plane of polarized light are *optically active*.

Some optically active systems rotate the plane of polarized light in a clockwise direction. This is taken as a right-handed turn, and such systems are *dextrorotatory*. Others turn the light in a counterclockwise direction and are *levorotatory*.

In 1848, the French chemist Louis Pasteur (1822–1895) was able to show that the optical activity of transparent crystals was dependent on the fact that such crystals were asymmetric. Further, if such asymmetric crystals could be fashioned into two mirror-image forms, one would be dextrorotatory, and the other, levorotatory. The fact that certain solutions also exhibited optical activity argued that asymmetry must be present in the very molecules of these substances. In 1874, the Dutch physical chemist Jacobus Hendricus van't Hoff (1852–1911) presented a theory of molecular structure that accounted for such asymmetry in optically active substances. A discussion of this, however, belongs more properly in a book on chemistry, and I will go no further into the subject here.

Nicol prisms are not the only means by which beams of plane-polarized light can be formed. There are some types of crystal that do not merely split the light into two plane-polarized beams but absorb one and transmit the other. Crystals of iodo-quinine sulfate will do this. Unfortunately, it is impossible to manufacture large useful crystals of this material since such crystals are fragile and disintegrate at the least disturbance.

In the mid-1930's, however, it occurred to a Harvard undergraduate, Edwin Herbert Land (1909–) that single large crystals were not necessary. Tiny crystals, all oriented in the same direction, would serve the purpose. To keep them so oriented, and to keep them from further disintegration, they could be embedded in a sheet of transparent, flexible plastic. Land quit school in 1936 to go into business and produced what is now known as *Polaroid*. It can serve all the functions of Nicol prisms more economically and conveniently (though not with quite the same precision).

As Malus had discovered, beams of polarized light can also be produced by reflection, at some appropriate angle, from material such as glass; the exact size of the angle depends on the index of refraction of the material. "Sunglasses" made of Polaroid can block most of this reflected polarized light and cut down glare.

Thus, the nineteenth century saw light established not merely as a wave form but as a transverse wave form; if this solved many problems, it also raised a few.

CHAPTER 21

The Ether

Absolute Motion

If light is a wave form, then it seemed to most scientists, up to the beginning of the twentieth century, that something must be waving. In the case of water waves, for instance, water molecules move up and down; in the case of sound waves, the atoms or molecules of the transmitting medium move to and fro. Something, it seemed, would therefore have to be present in a vacuum; something that would move either up and down, or to and fro, in order to produce light waves.

This something, whatever it was, did not interfere with the motions of the heavenly bodies in any detectable way, so it seemed reasonable to suppose it to be nothing more than an extremely rarefied gas. This extremely rarefied gas (or whatever it was that filled the vacuum of space) was called *ether,* from a word first used by Aristotle to describe the substance making up the heavens and the heavenly bodies (see page I–6). Ether might also be the substance through which the force of gravitation was transmitted, and this might be identical with the ether that did or didn't transmit light. In order to specify the particular ether that transmitted light (in case more than one variety existed) the phrase *luminiferous ether* ("light-carrying ether") grew popular in the nineteenth century.

In connection with the ether, the difference in properties

between transverse and longitudinal waves becomes important. Longitudinal waves can be conducted by material in any state: solid, liquid or gaseous. Transverse waves, however, can only be conducted through solids, or, in a gravitational field, along liquid surfaces (see page I–158). Transverse waves cannot be conducted through the body of a liquid or gas. It was for this reason that early proponents of the wave theory of light, assuming the ether to be a gas, also assumed light to consist of longitudinal waves that could pass through a gas rather than transverse waves that could not.

When the question of polarization, however, seemed to establish the fact that light consisted of transverse waves, the concept of the ether had to be drastically revised. The ether had to be a solid to carry transverse light waves; it had to be a substance in which all parts were fixed firmly in place.

If that were so, then if a portion of the ether were distorted at right angles to the motion of a light beam (as seemed to be required if light were a transverse wave phenomena), the forces tending to hold that portion in place would snap it back. That portion would overshoot the mark, snap back from the other direction, overshoot the mark again, and so on. (This is what happens in the case of water waves, where gravity supplies the force necessary for snap-back, and in sound waves, where intermolecular forces do the job.)

The up and down movement of the ether thus forms the light wave. Moreover, the rate at which a transverse wave travels through a medium depends on the size of the force that snaps back the distorted region. The greater the force, the faster the snap-back, the more rapid the progression of the wave. With light traveling at over 186,000 miles per second, the snap-back must be rapid indeed, and the force holding each portion of the ether in place was calculated to be considerably stronger than steel.

The luminiferous ether, therefore, must at one and the same time be an extremely tenuous gas, and possess a rigidity greater than that of steel. Such a combination of properties is hard to visualize,* but during the mid-nineteenth century, physicists la-

* Such a combination is against "common sense," but this must never be allowed to stand in the way of the acceptance of a hypothesis. We experience only a very limited portion of the universe and are sensitive to only a very limited range of phenomena. It is therefore dangerous to suppose that what seems familiar to us is and must be true of all the universe in all its aspects. Thus, it is only "common sense" to suppose that the earth is flat and motionless, and this argument was strenuously used to oppose the notion that the earth was spherical and in motion.

bored hard to work out the consequences of such a rigid-gas and to establish its existence. They did this for two reasons. First, there seemed no alternative, if light consisted of transverse waves. Secondly, the ether was needed as a reference point against which to measure motion. This second reason is extremely important, for without such a reference point, the very idea of motion becomes vague, and all of the nineteenth century development of physics becomes shaky.

To explain why that should be, let us suppose that you are on a train capable of moving at a uniform velocity along a perfectly straight set of rails with vibrationless motion. Ordinarily, you could tell whether your train were actually in motion by the presence of vibration, or by inertial effects when the train speeds up, slows down, or rounds a curve. However, with the train moving uniformly and vibrationlessly, all this is eliminated and the ordinary methods for noting that you are in motion are useless.

Now imagine that there is one window in the train through which you can see another train on the next track. There is a window in that other train, and someone is looking out at you through it. Speaking to you by sign language, he asks, "Is my train moving?" You look at it, see clearly that it is motionless, and answer, "No, it is standing still." So he gets out and is killed at once, for it turns out that both trains are moving in the same direction at 70 miles per hour with respect to the earth's surface.

Since both trains were moving in the same direction at the same speed, they did not change position with respect to each other, and each seemed motionless to an observer on the other. If there had been a window on the other side of each train, one could have looked out at the scenery and noted it moving rapidly toward the rear of the train. Since we automatically assume that the scenery does not move, the obvious conclusion would be that the train is actually in motion even though it does not seem to be.

Again, suppose that in observing the other train, you noted that it was moving backward at two miles an hour. You signal this information to the man in the other train. He signals back a violent negative. He is standing still, he insists, but you are moving forward at two miles an hour. Which one of you is correct?

To decide that, check on the scenery. It may then turn out that Train A is motionless while Train B is actually moving backward at two miles an hour. Or Train B may be motionless while Train A is moving forward at two miles an hour. Or Train A may be moving forward at one mile an hour while Train B is moving backward at one mile an hour. Or both trains may be moving forward: Train A at 70 miles an hour and Train B at 68 miles an hour. There are an infinite number of possible motions, with respect to the earth's surface, that can give rise to the observed motion of Train A and Train B relative to each other.

Through long custom, people on trains tend to downgrade the importance of the relative motion of one train to another. They consider it is the motion with respect to the earth's surface that is the "real" motion.

But is it? Suppose a person on a train, speeding smoothly along a straight section of track at 70 miles an hour, drops a coin. He sees the coin fall in a straight line to the floor of the train. A person standing by the wayside, watching the train pass and able to watch the coin as it falls, would see that it was subjected to two kinds of motion. It falls downward at an accelerating velocity because of gravitational force and it shares in the forward motion of the train, too. The net effect of the two motions is to cause the coin to move in a parabola (see page I–39).

We conclude that the coin moves in a straight line relative to the train and in a parabola relative to the earth's surface. Now which is the "real" motion? The parabola? The person on the train who is dropping the coin may be ready to believe that although he seems to himself to be standing still, he is "really" moving at a velocity of 70 miles an hour. He may not be equally ready to believe that a coin that he sees moving in a straight line is "really" moving in a parabola.

This is a very important point in the philosophy of science. Newton's first law of motion (page I–24) states that an object not subjected to external forces will move in a straight line at constant speed. However, what seems a straight line to one observer does not necessarily seem a straight line to another observer. In that case, what meaning does Newton's first law have? What is straight-line motion, anyway?

Throughout ancient and medieval times, almost all scholars believed that the earth was affixed to the center of the universe and never budged from that point. The earth, then, was truly motionless. It was (so it was believed) in a state of *absolute rest*. All motion could be measured relative to such a point at absolute rest, and then we would have *absolute motion*. This absolute motion would be the "true" motion upon which all observers could agree. Any observed motion that was not equivalent to the absolute motion was the result of the absolute motion of the observer.

There was some question, of course, as to whether the earth

were truly motionless, even in ancient times. The stars seemed to be moving around the earth in 24 hours at a constant speed. Was the earth standing still and the celestial sphere turning, or was the celestial sphere standing still and the earth turning? The problem was like that of the two trains moving relative to each other, with the "real motion" unverifiable until one turned to look at the scenery. In the case of the earth and the celestial sphere, there was no scenery to turn to and no quick decision, therefore, upon which everyone could agree.

Most people decided it was the celestial sphere that turned, because it was easier to believe that than to believe that the vast earth was turning without our being able to feel that we were moving. (We still speak of the sun, moon, planets and stars as "rising" and "setting.") In modern times, however, for a variety of reasons better discussed in a book on astronomy, it has become far more convenient to suppose that the earth is rotating rather than standing still.

In such a case, while the earth as a whole is not at absolute rest, the axis may be. However, by the beginning of modern times, more and more astronomers were coming to believe that even the earth's axis was not motionless. The earth, all of it, circled madly about the sun along with the other planets. No part of it was any more at rest than was any train careening along its surface. The train might have a fixed motion relative to the earth's surface, but that was not the train's "true" motion.

For a couple of centuries after the motion of the earth had come to be accepted, there was still some excuse to believe that the sun might be the center of the universe. The sun visibly rotated, for sunspots on its surface circled its globe in a steady period of about 27 days. However, the sun's axis might still represent that sought-for state of absolute rest.

Unfortunately, it became clearer and clearer, as the nineteenth century approached, that the sun was but a star among stars, and that it was moving among the stars. In fact, we now know that just as the earth moves around the sun in the period of one year, the sun moves about the center of our Galaxy in a period of 200,000,000 years. And, of course, the Galaxy itself is but a galaxy among galaxies and must be moving relative to the others.

By mid-nineteenth century, there was strong reason to suppose that no material object anywhere in the universe represented a state of absolute rest, and that absolute motion could not therefore be measured relative to any material object. This might have raised a serious heart-chilling doubt as to the universal validity of Newton's laws of motion, on which all of nineteenth century physics was based. However, a material object was not needed to establish absolute motion.

It seemed to nineteenth century physicists that if space were filled with ether, it was fair to suppose that this ether served only to transmit forces such as gravity and waves such as those of light, and was not itself affected, overall, by forces. In that case, it could not be set into motion. It might vibrate back and forth, as in transmitting light waves, but it would not have an overall motion. The ether, then, might be considered as being at absolute rest. All motion became absolute motion if measured relative to the ether. This ether-filled space, identical to all observers, aloof, unchanging, unmoving, crossed by bodies and forces without being affected by them, a passive container for matter and energy, is *absolute space.*

In Newton's time and for two centuries afterward, there was no way of actually measuring the motion of any material body relative to the ether. Nevertheless, that didn't matter. In principle, absolute motion was taken to exist, whether it was practical to measure it or not, and the laws of motion were assumed to hold for such absolute motion and, therefore, must surely hold for all relative motions (which were merely one absolute motion added to another absolute motion).

The Michelson-Morley Experiment

In the 1880's, however, it appeared to Michelson (the latter-day measurer of the velocity of light) that a method of determining absolute motion could be worked out.

Light consists of waves of ether, according to the view of the time, and if the ether moved, it should carry its own vibrations (light) with it. If the ether were moving away from us, it should carry light away from us and therefore delay light in its motion toward us—reduce the velocity of light, in other words. If the ether were moving away from us at half the velocity of light, then light would lose half its velocity relative to ourselves and therefore take twice as long to get to us from some fixed point. Similarly, if the ether were moving toward us, light would reach us more quickly than otherwise.

To be sure, physicists were assuming that the ether itself was not moving under any circumstances. However, the earth must, it seemed, inevitably be moving relative to the ether. In that case, if the earth is taken as motionless, then the ether would seem to be moving relative to us, fixed as we are to the earth. There would seem to be what came to be called an "ether wind."

If there were no ether wind at all, if the earth were at absolute rest, then light would travel at the same velocity in all directions. To be sure, it actually seems to do just this, but surely that is only because the ether wind is moving at a very small velocity compared to the velocity of light; therefore, light undergoes only minute percentage changes in its velocity with shift in direction. In view of the difficulty of measuring light's velocity with any accuracy in the first place, it would not be surprising that small differences in velocity with shifting direction would go unnoticed.

Michelson, however, in 1881, invented a device that was perhaps delicate enough to do the job.

In this device, light of a particular wavelength falls upon a glass plate at an angle of 45°. The rear surface of the glass plate is "half-silvered." That is, the surface has been coated with enough silver to reflect half the light and allow the remaining half to be transmitted. The transmitted light emerges, traveling in the same direction it had been traveling in originally, while the reflected light moves off at right angles to that direction. Both light beams are reflected by a mirror and travel back to the half-silvered glass plate. Some of the originally reflected beam now passes through, while some of the originally transmitted beam is now reflected. In this way, the two beams join again.*

In effect a single beam of light has been split in two; the two halves have been sent in directions at right angles to each other, have returned, and have been made to join in a combined beam again.

The two beams, joining, set up interference fringes, as did the two beams in Young's experiment. One of the mirrors can be adjusted so that the length of the journey of the beam of light to that particular mirror and back can be varied. As the mirror is adjusted, the interference fringes move. From the number of

* The transmitted light travels through the glass plate once on its way outward. When returning it strikes the silver and is reflected without entering the glass. The reflected light travels through the glass plate once in reaching the silver, travels through a second time after reflection, and a third time on its return trip. For this reason, a second glass plate, identical with the first, is placed in the path of the transmitted light. It must travel through it both going and returning, and now each beam of light has traveled through the same total thickness of glass. Michelson had to be very careful to make sure that both beams of light received identical treatment in every possible way.

fringes that pass the line of sight when the mirror is moved a certain distance, the wavelength of the light can be determined. The greater the number of fringes passing the line of sight, the shorter the wavelength.

Michelson determined wavelengths of light with his instrument, which he called the *interferometer* ("to measure by interference"), so precisely that he suggested the wavelength of some particular spectral line be established as the fundamental unit of length. At the time, this fundamental unit had just been established as the *International Prototype Meter.* This was the distance between two fine marks on a bar of platinum-iridium alloy kept at Sèvres, a suburb of Paris.

In 1960, Michelson's suggestion was finally accepted and the fundamental unit of length became a natural phenomenon rather than a man-made object. The orange-red spectral line of a variety of the rare gas krypton was taken as the standard. The meter is now set officially equal to 1,650,763.73 wavelengths of this light.

But Michelson was after bigger game than the determination of the wavelengths of spectral lines. He considered the fact that the beam of light in the interferometer was split into two halves that traveled at right angles to each other. Suppose one of these two light rays happened to be going with the ether wind. Its velocity would be c (the velocity of light with respect to the ether) plus v (the velocity of the light source with respect to the

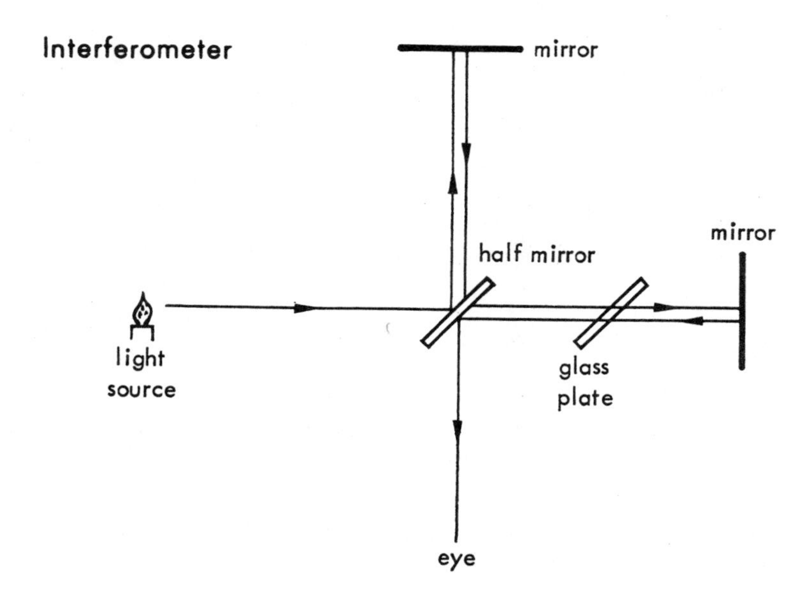

ether). If the distance of the reflecting mirror from the half-silvered prism is taken as d, then the time it would take the light to pass from the half-silvered prism to the reflecting mirror would be $d/(c+v)$. After reflection, the light would move over the distance d in precisely the opposite direction. Now it would be moving into the ether wind, and it would be slowed down, its overall velocity being $c-v$. The time taken for its return would be $d/(c-v)$. The total time (t_1) taken by that beam of light to go and return is therefore:

$$t_1 = \frac{d}{c+v} + \frac{d}{c-v} = \frac{2dc}{c^2 - v^2} \qquad \text{(Equation 6–1)}$$

Meanwhile, however, the second half of the beam is going at right angles to the first; it also returns at right angles to the first. It is going neither with the ether wind nor against it. It is going "crosswind" both ways.

The time taken by the light beam to go and return crosswind (t_2) can be calculated with the help of plane geometry* and turns out to be:

$$t_2 = \frac{2d}{\sqrt{c^2 - v^2}} \qquad \text{(Equation 6–2)}$$

If we divide Equation 6–1 by Equation 6–2, we will determine the ratio of the time taken to cover the ground with-and-against the ether wind and the time taken to cover the same distance crosswind. We would have:

$$\frac{t_1}{t_2} = \frac{2dc}{c^2 - v^2} \div \frac{2d}{\sqrt{c^2 - v^2}} = \frac{c\sqrt{c^2 - v^2}}{c^2 - v^2} \qquad \text{(Equation 6–3)}$$

The expression at the extreme right of Equation 6–3 is of the form, $a\sqrt{x}/x$, and if both numerator and denominator of such an expression is divided by $\sqrt{x}$, the equivalent expression $a/\sqrt{x}$ is obtained. Equation 6–3 can therefore be simplified to:

$$\frac{t_1}{t_2} = \frac{c}{\sqrt{c^2 - v^2}} \qquad \text{(Equation 6–4)}$$

Further simplification can be obtained if both numerator and denominator are multipled by $\sqrt{1/c^2}$. (The multiplication of the numerator and denominator of a fraction by the same quantity

* Anyone curious concerning the details of this calculation may try pages 807–810 of my book *The New Intelligent Man's Guide to Science,* Basic Books, 1965.

does not, of course, alter the value of the expression as a whole.)

The numerator of Equation 6–4 then becomes $c\sqrt{1/c^2}$ or c/c or 1. The denominator becomes $\sqrt{c^2 - v^2}\ \sqrt{1/c^2}$ or $\sqrt{c^2(1/c^2) - v^2(1/c^2)}$ or $\sqrt{1 - v^2/c^2}$. Equation 6–4 can therefore be expressed as:

$$\frac{t_1}{t_2} = \frac{1}{\sqrt{1 - \frac{v^2}{c^2}}} \qquad \text{(Equation 6–5)}$$

If the light source is at rest with respect to the ether, $v = 0$ and $t_1/t_2 = 1$. In that case the time taken by the beam of light going with-and-against the ether wind is the same as the time taken by the beam of light to go crosswind. (Indeed, the time is the same for light beams going in any direction.) If the movable mirror is adjusted so that the two beams of light travel exactly the same distance, they will return exactly in step and there will be no interference fringes. Furthermore there will be no interference fringes if the instrument is then turned so as to have the light beams travel in changed directions.

However, if the light source is moving with respect to the ether, then v is greater than 0, $1 - v^2/c^2$ is less than 1, and t_1/t_2 is greater than 1. The light traveling with-and-against the ether would then take longer to cover a fixed distance than the light traveling crosswind would. To be sure, the ratio is not very much greater than 1 for any reasonable velocity relative to the ether. Even if the light source were moving at one-tenth the velocity of light (so that v was equal to the tremendous figure of 30,000 kilometers per second), the ratio would be only 1.005. At ordinary velocities, the ratio would be very small indeed.

Nevertheless, the difference in time would be enough to throw the wavelengths of the two beams of light out of step and set up interference fringes. Naturally, you could not know in advance which direction would be with-and-against the ether wind and which would be crosswind, but that would not matter. The instrument could be pointed in some direction at random, and the movable mirror could be adjusted so as to remove the interference fringes. If the instrument were turned now, the light beams would change direction and be affected differently by the ether wind so that interference fringes would appear.

From the spacing of the fringes one could determine the velocity of the light source relative to the ether. Since the light source was firmly attached to the earth, this was equivalent to finding the velocity of the earth relative to the ether—that is, the absolute motion of the earth. Once that was done, all bodies, as long as their motions relative to the earth were known, would have absolute motions that were known.

Michelson obtained the help of an American chemist, Edward Williams Morley (1838–1923), and in 1886 he tried this experiment. Michelson had tried it alone, before, but never under conditions that he found satisfactory. Now he and Morley dug down to bedrock to anchor the interferometer, and balanced the instrument with fantastic precautions against error.

Over and over again, they repeated the experiment and always the results were the same—negative! Once they adjusted the device to remove interference fringes, those fringes did not show up to any significant extent when the interferometer was reoriented. One might have thought that they just happened to be unlucky enough to try the experiment at a time when the earth happened to be motionless with respect to the ether. However, the earth travels in an ellipse about the sun and changes the direction of its motion every moment. If it were at rest with respect to the ether on one day, it could not be at rest the next.

Michelson and Morley made thousands of observations over many months, and in July, 1887, finally announced their conclusion. There was no ether wind!

I have gone into detail concerning this experiment because of the shocking nature of the result. To say there was no ether wind meant there was very likely no way of determining absolute motion. In fact, the very concept of absolute motion suddenly seemed to have no meaning. And if that was so, what would become of Newton's laws of motion and of the whole picture of the universe as based upon those laws?

Physicists would have been relieved to find that the *Michelson-Morley experiment* was wrong and that there was an ether wind after all. However, the experiment has been repeated over and over again since 1887. In 1960, devices far more accurate than even the interferometer were used for the purpose, and the result was always the same. There is no ether wind. This fact simply had to be accepted, and the view of the universe changed accordingly.

The FitzGerald Contraction

Naturally, attempts were made to explain the results of the Michelson-Morley experiment in terms of the ether. The most successful attempt was that of the Irish physicist George Francis FitzGerald (1851–1901), who in 1893 proposed that all objects grew shorter in the direction of their absolute motion, being shortened, so to speak, by the pressure of the ether wind. Distances between two bodies moving in unison would likewise shorten in the direction of the motion, since the two bodies would be pushed together by the ether wind. The amount of this "fore-shortening" would increase with the velocity of the absolute motion, of course, as the pressure of the ether wind rose.

FitzGerald suggested that at any given velocity, the length (L) of an object or of the distance between objects would have a fixed ratio to the length (L_0) of that same object or distance at rest; and L_0 may be termed the *rest-length*. This ratio would be expressed by the quantity $\sqrt{1 - v^2/c^2}$, where c is the velocity of light in a vacuum, and v is the velocity of the body, both relative to the ether. In other words:

$$L = L_0\sqrt{1 - v^2/c^2} \qquad \text{(Equation 6–6)}$$

The FitzGerald ratio is equal to the denominator of the expression in Equation 6–5, which represents the ratio of the distances traveled by the two beams of light in the interferometer. Mutiplied by FitzGerald's ratio, the value in Equation 6–5 becomes 1. The distance covered by the beam of light moving with-and-against the ether wind is now decreased by foreshortening to just exactly the extent that would allow the beam to cover the distance in the same time as was required by the beam traveling crosswind. In other words, the existence of the ether wind would make one of the beams take a longer time, but the existence of the *Fitz-Gerald contraction* produced by the same ether wind allows the beam to complete its journey in the same time as one would expect if there were no ether wind.

The two effects of the ether wind cancel perfectly, and this has reminded a thousand physicists of a passage from a poem in Lewis Carroll's *Through the Looking-Glass.*

"But I was thinking of a plan
  To dye one's whisker's green,
And always use so large a fan
  That they could not be seen."

Carroll's book was written in 1872, so it could not have deliberately referred to the FitzGerald contraction, but the reference is a perfect one, just the same.

The contraction is extremely small at ordinary velocities. The earth moves in its orbit about the sun at 30 kilometers per second (relative to the sun), which by earthly standards is a great velocity. If v is set equal to 30 and this is inserted into the FitzGerald ratio, we have $\sqrt{1 - (30)^2/(300{,}000)^2}$, which is equal to 0.999995. The foreshortened diameter of the earth in the direction of its motion would then be 0.999995 of its diameter perpendicular to that direction (assuming the earth to be a perfect sphere). The amount of the foreshortening would be 62.5 meters.

If the earth's diameter could be measured in all directions, and the direction in which the diameter were abnormally short could be located, then the direction of the earth's motion relative to the ether could be determined. Furthermore, from the size of the abnormal decrease in diameter, the absolute velocity of the earth relative to the ether could be worked out.

But there is a difficulty. This difficulty lies not in the smallness of the foreshortening, because no matter how small it is all might be well if it could be detected in principle. It cannot be detected, however, as long as we remain on earth. While on earth, all the instruments we could conceivably use to measure the earth's diameter would share in the earth's motion and in its foreshortening. The foreshortened diameter would be measured with foreshortened instruments, and no foreshortening would be detected.

We could do better if we could get off the earth and, without sharing in the earth's motion, measure its diameter in all directions (very accurately) as it speeds past. This is not exactly practical but it is something that is conceivable in principle.

To make such a thing practical, we must find something that moves very rapidly and in whose motion we do not ourselves share. Such objects would seem to be the speeding subatomic particles* that have motions relative to the surface of the earth of anywhere from 10,000 kilometers per second to nearly the speed of light.

The FitzGerald contraction becomes very significant at such super-velocities. The velocity might be high enough, for instance, for the length of the moving body to be foreshortened to only half its rest-length. In that case $\sqrt{1 - v^2/c^2} = 1/2$, and if we solve for v, we find that it equals $\sqrt{3c^2/4}$. Since $c = 300{,}000$ kilometers per second, $\sqrt{3c^2/4} = 260{,}000$ kilometers per second. At this ferocious velocity, seven-eighths that of light, an object is foreshortened to half its rest-length, and some subatomic particles move more rapidly (relative to the earth's surface) than this.

* These particles, much smaller than atoms, make up the atomic structure. They will be discussed in some detail in the third volume of this book.

At still more rapid velocities, foreshortening becomes even more marked. Suppose the velocity of a body becomes equal to the velocity of light. Under those conditions v is equal to c, and FitzGerald's ratio becomes $\sqrt{1 - c^2/c^2}$, which equals 0. This means that by Equation 6–6 the length of the moving body (L) is equal to its rest-length (L_0) multiplied by zero. In other words, at the velocity of light, all bodies, whatever their length at rest, have foreshortened completely and have become pancakes of ultimate thinness.

But then what happens if the velocity of light is exceeded? In that case, v becomes greater than c, the expression v^2/c^2 becomes greater than 1, and the expression $1 - v^2/c^2$ becomes a negative number. FitzGerald's ratio is the square root of a negative number, and this is what mathematicians call an "imaginary number." A length represented by an imaginary number has mathematical interest, but no one has been able to work out the physical meaning of such a length.

This was the first indication that the velocity of light might have some important general significance in the universe—as something that might, in some fashion, represent a maximum velocity. To be sure, no subatomic particle has ever been observed to move at a velocity greater than that of light in a vacuum, although velocities of better than 0.99 times that of light in a vacuum have been observed. At such velocities, the subatomic particles ought to be wafer-thin in the direction of their motion, but alas, they are so small that it is completely impractical to try to measure their length as they speed past, and one cannot tell whether they are foreshortened or not. If, however, the length of the speeding subatomic particles will not do as a practical test of the validity of the FitzGerald contraction, another property will. . . .

The FitzGerald contraction was put into neat mathematical form, and extended, by the Dutch physicist Hendrik Antoon Lorentz (1853–1928) so that the phenomenon is sometimes referred to as the *Lorentz-FitzGerald contraction.*

Lorentz went on to show that if the FitzGerald contraction is applied to subatomic particles carrying an electric charge, one could deduce that the mass of a body must increase with motion in just the same proportion as its length decreases. In short, if its mass while moving is m and its *rest-mass* is m_0, then:

$$m = \frac{m_0}{\sqrt{1 - v^2/c^2}} \qquad \text{(Equation 6–7)}$$

Again, the gain in mass is very small at ordinary velocities. At a velocity of 260,000 kilometers per second, the mass of the moving body is twice the rest-mass, and above that velocity it increases ever more rapidly. When the velocity of a moving body is equal to that of light, $v = c$ and Equation 6–7 becomes $m = m_0/0$. This means that the mass of the moving body becomes larger than any value that can be assigned to it. (This is usually expressed by saying that the mass of the moving body becomes infinite.) Once again, velocities greater than light produced masses expressed by imaginary numbers, for which there seems no physical interpretation. The key importance of the velocity of light in a vacuum is again emphasized.

But the very rapidly moving charged subatomic particles possessing velocities up to 0.99 times that of light increase markedly in mass; and whereas the length of speeding subatomic particles cannot be measured as they fly past, their mass can be measured easily.

The mass of such particles can be obtained by measuring their inertia—that is, the force required to impose a given acceleration upon them. In fact, it is this quantity of inertia that Newton used as a definition of mass in his second law of motion (see page I–30.)

Charged particles can be made to curve in a magnetic field. This is an acceleration imposed upon them by the magnetic force, and the radius of curvature is the measure of the inertia of the particle and therefore of its mass.

From the curvature of the path of a particle moving at low velocity, one can calculate the mass of the particle and then predict what curvature it will undergo when it passes through the same magnetic field at higher velocities, provided its mass remains constant. Actual measurement of the curvatures for particles moving at higher velocities showed that such curvatures were less marked than was expected. Furthermore, the higher the velocity, the more the actual curvature fell short of what was expected. This could be interpreted as an increase in mass with velocity, and when this was done the relationship followed the Lorentz equation exactly.

The fan had slipped and the green whiskers could be seen. The Lorentz equation fit the observed facts. Since it was based on the FitzGerald equation, the phenomenon of foreshortening also fit the facts, and this explained the negative results of the Michelson-Morley experiment.

CHAPTER 22

Relativity

The Special Theory

If the gain in mass of a speeding charged particle is the result of its motion relative to the ether, then a new method of measuring such motion might offer itself. Suppose some charged particles are measured as they speed along in one direction, others as they speed in another direction, and so on. If all directions are taken into account, some particles are bound to be moving with the ether wind, while others, speeding in the opposite direction, are moving against it. Those moving against the ether wind (one might suspect) will have a more rapid motion relative to the ether and will gain more mass than will those moving at the same velocity (relative to ourselves) with the ether wind. By the changes in gain of mass as direction is changed, the velocity of the ether wind, and therefore the absolute motion of the earth, can be determined.

However, this method also fails, exactly as the Michelson-Morley experiment failed. The gain in mass with motion is the same no matter in which direction the particles move. What's more, all experiments designed to measure absolute motion have failed.

In 1905, in fact, a young German-born, Swiss physicist, Albert Einstein (1879–1955), had already decided that it might be useless to search for methods of determining absolute motion. Suppose that one took the bull by the horns and simply decided that it was impossible to measure absolute motion by any conceivable method* and considered the consequences.

That, then, was Einstein's first assumption: that all motion must be considered relative to some object or some system of objects arbitrarily taken as being at rest; and that any object or system of objects (any *frame of reference*, that is) can be taken, with equal validity, as being at rest. There is no object, in other words, that is more "really" at rest than any other.

Since in this view all motion is taken as relative motion only, Einstein was advancing what came to be called the *theory of relativity*. In his first paper on the subject, in 1905, Einstein considered only the special case of motion at constant velocity; therefore, this portion of his views is his *special theory of relativity*.

Einstein then made a second assumption: that the velocity of light in a vacuum, as measured, would always turn out to be the same, whatever the motion of the light source relative to the observer. (Notice that I speak of the velocity "as measured.")

This measured constancy of the velocity of light seems to be in violation of the "facts" about motion that had been accepted since the days of Galileo and Newton.

Suppose a person throws a ball past us and we measure the horizontal velocity of the ball relative to ourselves as x feet per minute. If the person is on a platform that is moving in the opposite direction at y feet per minute and throws the ball with the same force, its horizontal velocity relative to ourselves ought to be $x - y$ feet per minute. If the platform were moving in the same direction he threw the ball, the horizontal velocity of the ball relative to ourselves ought to be $x + y$ feet per minute.

This actually seems to be the situation as observed and measured in real life. Ought it not therefore be the same if the person were "throwing" light out of a flashlight instead of a ball out of his fist?

In order to make Einstein's second assumption hold true,

* This is not the same as saying flatly that there is no absolute motion. All that scientists know of the physical universe is based, directly or indirectly, on observation and measurement. If there is some phenomenon that can neither be observed nor measured under any conceivable circumstances, then, as far as the world of experimental science is concerned, it can be treated as though it does not exist. Whether it "really" exists, though it can't be either observed or measured, even in principle, is a question that may amuse philosophers and theologians but is completely irrelevant to scientists.

we must suppose that this situation does not hold for light at all and, in fact, that it does not hold for the ball either.

Suppose that the effect of the moving platform on the speed of the ball is not quite as great as we suspect and that when the motion of the platform is added to that of the ball, the overall velocity of the ball is a vanishingly small amount smaller than $x + y$. Again, when the motion of the platform is subtracted from that of the ball the overall velocity of the ball is a vanishingly small amount greater than $x - y$. Suppose, too, that this discrepancy increases as x and y increase, but that at all velocities of material bodies, which we were capable of observing before 1900, the discrepancy remained far too small to measure. Consequently, it would be very natural for us to come to the conclusion that the velocity was exactly $x + y$ or exactly $x - y$, and that this would remain true for all speeds.

But if one could observe very great velocities, velocities of the order of thousands of kilometers per second, the discrepancy would become great enough to notice. If one added the velocity y to the velocity x, the combined velocity would then be noticeably less than $x + y$ and might be hardly any greater than x alone. Similarly, if y were subtracted from x, the combined velocity might be considerably larger than $x - y$ and hardly less than x alone. Finally, at the speed of light, the effect of the movement of the source of the moving body will have declined to zero so that $x + y = x$ and $x - y = x$, regardless of how great y is. And that is another way of expressing Einstein's second assumption.

In order to save that assumption, in fact, it is necessary to add velocities in such a way that the sum never exceeds the velocity of light. Suppose, for instance, a platform is moving forward (with respect to ourselves) at 290,000 kilometers a second, or only 10,000 kilometers a second less than the velocity of light in a vacuum. Suppose, further, that a ball is thrown forward from the platform at a velocity, relative to the platform, of 290,000 kilometers a second. The velocity of the ball relative to ourselves ought to be 290,000 + 290,000 kilometers a second in that forward direction, but at those velocities the effect of the moving platform has so decreased that the overall velocity is, in point of fact, only 295,000 kilometers per second and is still less than the velocity of light.

Indeed, this can be expressed mathematically. If two velocities (V_1 and V_2) are added, then the new velocity (V) according to Newton would be $V = V_1 + V_2$. According to Einstein, the new velocity would be:

$$V = \frac{V_1 + V_2}{1 + \frac{V_1 V_2}{C^2}}$$

Where C is equal to the velocity of light in a vacuum. If V_1 is equal to C, then Einstein's equation would become:

$$V = \frac{C + V_2}{1 + \frac{CV_2}{C^2}} = (C + V_2)\left(\frac{C}{C + V_2}\right) = C$$

In other words, if one velocity were equal to the speed of light, adding another velocity to it, even again up to the speed of light, would leave the total velocity merely at the speed of light.

To put it briefly, it is possible to deduce from Einstein's assumption of the constant measured velocity of light that the velocity of any moving body will always be measured as less than the velocity of light.*

It seems strange and uncomfortable to accept so unusual a set of circumstances just to save Einstein's assumption of the measured constancy of the velocity of light. Nevertheless, whenever it has been possible to measure the velocity of light, that velocity has always been placed at one constant value, and whenever it has been possible to measure the velocity of speeding bodies, that velocity has always been less than the velocity of light. In short, no physicist has yet detected any phenomenon that can be taken as violating either Einstein's assumption of relativity of motion or his assumption of measured constancy of light; and they have looked assiduously, you may be sure.

Einstein could also deduce from his assumptions the existence of the Lorentz-FitzGerald contraction as well as the Lorentz gain of mass with motion. Furthermore, he showed that it was not only electrically charged particles that gained mass with motion, but uncharged particles as well. In fact, all objects gained mass with motion.

It might seem that there is scarcely any reason to crow over the special theory so far. What is the difference between starting with the assumption of the Lorentz-FitzGerald contraction and deducing from it the measured constancy of the velocity of light,

* This is often expressed as "a body cannot move faster than light" but that is not quite right. It is only the *measured velocity* that is less than the measured velocity of light. It is quite conceivable that there are objects in the universe that are traveling at velocities (relative to ourselves) that are greater than the velocity of light, but we could not see such bodies or sense them in any way and therefore could not measure their velocities.

or starting with the assumption of the measured constancy of the velocity of light and deducing from it the Lorentz-FitzGerald contraction?

If that were all, there would be no significant difference, indeed. However, Einstein combined his assumption concerning the measured constancy of the velocity of light with his first assumption that all motion is relative.

This meant that foreshortening or mass-gain was not a "real" phenomenon but only a change in measurement. The amount by which length was decreased or mass increased was not something that could be absolutely determined but differed from observer to observer.

To consider what this means, imagine two identical spaceships moving in opposite directions in a non-collision course, each spaceship possessing equipment that will enable it to measure the length and mass of the other spaceship as it passes by.

Spaceship X watches Spaceship Y flash by (in a particular direction) at 260,000 kilometers per second, and at this velocity Spaceship Y is measured as being only half its rest-length and fully twice its rest-mass. Spaceship X, which to the people on board seems to be motionless, is to them, naturally, exactly at rest-length and rest-mass.

But the people on Spaceship Y have no sensation of moving (any more than we have the sensation of speeding through space on our voyage around the sun). The people on Spaceship Y feel themselves to be motionless and find themselves to be at rest-length and rest-mass. What they see is Spaceship X flashing by (in the opposite direction) at 260,000 kilometers per second. To them it is Spaceship X that is measured as being only half its rest-length and fully twice its rest-mass.

If the observers could communicate while in motion, they could have a glorious argument. Each could say, "I am at rest and you are moving. I am normal length, you are foreshortened. I am normal mass, you are heavy."

Well, which one is really "right"?

The answer is neither and both. It is not a question, you see, of what has "really" happened to length and mass or of which ship is "really" foreshortened or over-massive. The question is only of measurement. (It is—to make a trivial analogy—like measuring the side of a rectangle that is four meters by two meters and then arguing about whether the length of the rectangle is "really" four meters or "really" two meters. It depends on the side you are measuring.)

But suppose you attempt to perform some kind of test that will, perhaps, reach beyond the measurement to the "reality." Suppose, for instance, you brought the two ships together and compared them directly to see which was shorter and heavier. This cannot actually be done within the bounds of Einstein's special theory since that deals only with uniform motion. To bring the ships together means that at least one of them must turn round and come back and thus undergo non-uniform, or accelerated, motion. Even if we did this, however, and imagined the two ships side by side and at rest relative to each other, after having passed each other at super-velocities, we could make no decision as to "realities." Being at rest with respect to each other, each would measure the other as being normal in length and mass. If there had been a "real" change in length or mass in either ship in the past, there would remain no record of that change.

Despite everything, it is difficult to stop worrying about "reality." It is heartening, then, to remember that there have been times when we have abandoned a spurious "reality" and have not only survived but have been immeasurably the better for it.

Thus, a child is pretty certain he knows what "up" and "down" is. His head points "up," his feet point "down" (if he is standing in the normal fashion); he jumps "up," he falls "down." Furthermore, he discovers soon enough that everyone around him agrees as to which direction is "up" and which "down."

If a child with such convictions is shown a picture of the earth's globe, with the United States above and Australia below, and with little Americans standing head-up and little Australians standing head-down, his first thought may well be, "But that's impossible. The little Australians would fall off."

Of course, once the effect of the gravitational force is understood (and this was understood as long ago as Aristotle, at least as far as the earth itself was concerned; see page I–5ff.) then there is no longer fear that anyone would fall off any part of the earth. However, you might still be questioning the nature of "up" and "down." You might call up an Australian on long-distance telephone and say, "I am standing head-up, so you must be standing head-down." He would reply, "No, no. I am clearly standing head-up, so it must be you who is standing head-down."

Do you see, then, how meaningless it is to ask now who is right and who is "really" standing head-up? Both are right and both are wrong. Each is standing head-up in his own frame of reference, and each is standing head-down in the other's frame of reference.

Most people are so used to this that they no longer see a "relative up" and a "relative down" as being in violation of "common sense." In fact, it is the concept of the "absolute up" and the "absolute down" that seems a violation now. If anyone seriously argued that the Australians walked about suspended by their feet, he would be laughed at for his ignorance.

Once the tenets of the relativistic universe are accepted (at as early an age as possible) it, too, ceases to go against common sense.

Mass-Energy Equivalence

During the nineteenth century, chemists were increasingly convinced that mass could neither be created nor destroyed (the *law of conservation of mass*). To Lorentz and Einstein, however, mass was created as velocity increased, and it was destroyed as velocity decreased. To be sure, the changes in mass are vanishingly small at all ordinary velocities, but they are there. Where, then, does created mass come from, and where does destroyed mass go?

Let's begin by considering a body of a given mass (m) subjected to a given force (f). Under such conditions the body undergoes an acceleration (a), and from Newton's second law of motion (see page I–30) one can state that $a=f/m$. The presence of an acceleration means that the velocity of the body is increasing, but in the old Newtonian view this did not affect the mass of the body, which remained constant. If the force is also viewed as remaining constant, then f/m was constant and a, the acceleration, was also constant. As a result of such a constant acceleration, the velocity of the body (in the Newtonian view) would increase indefinitely and would reach any value you care to name—if you wait long enough.

In the Einsteinian universe, however, an observer measuring the velocity of an object under a continuing constant force can never observe it to exceed the velocity of light in a vacuum. Consequently, though its velocity increases under the influence of a constant force, that velocity increases more and more slowly, and as the velocity approaches that of light, it increases exceedingly slowly. In short, the acceleration of a body under the influence of a constant force decreases as the velocity increases and becomes zero when the velocity reaches that of light.

But, again from Newton's second law of motion, the mass of a body is equal to the force exerted upon it divided by the acceleration produced by that force—that is, $m = f/a$. If the force is constant and the acceleration decreases with velocity, then a decreases with velocity while f does not; consequently, f/a increases with velocity. And this means, since $m = f/a$, that mass increases with velocity. (Thus, the increase of mass with velocity can be deduced from Einstein's assumption of the measured constancy of the velocity of light in a vacuum.)

When a body is subjected to a force, it gains kinetic energy, which is equal to one half its mass times the square of its velocity ($e_k = 1/2mv^2$; see page I–95). In the Newtonian view this increase in kinetic energy results only from the increase in velocity, for mass is considered unchanging. In the Einsteinian view the increase in kinetic energy is the result of an increase in both velocity and mass.

Where mass is not involved in energy changes (as in the Newtonian view) it is natural to think of mass as something apart from energy and to think that, on the one hand, there is a law of conservation of energy, and on the other, a law of conservation of mass, and that the two are independent.

Where mass changes and is thus intimately involved in energy changes (as in the Einsteinian view), it is natural to think of mass and energy as different aspects of the same thing, so a law of conservation of energy would include mass. (To make that perfectly clear, in view of our previous convictions, we sometimes speak of the *law of conservation of mass-energy*, but the word "mass" is not really needed.)

Motion does not create mass in any real sense; mass is merely one aspect of a general increase in kinetic energy gained from the force that is maintained by the expenditure of energy elsewhere in the system.

But now suppose the law of conservation of energy (including mass) remains valid in the relativistic universe (and so far it seems to have done so). According to this law, although energy can be neither created nor destroyed, it can be changed from one form to another. This would seem to mean that a certain quantity of mass could be converted into a certain quantity of other forms of energy such as heat, for instance; and that a certain quantity of a form of energy such as heat might, conceivably, be converted into a certain quantity of mass. And this, indeed, Einstein insisted upon.

This equivalence of mass and energy announced by Einstein in his 1905 paper was of great use to the physicists of the time. The discovery of radioactivity nine years earlier (something I will discuss in volume III) had revealed a situation in which energy seemed to be created endlessly out of nowhere. Once the special theory of relativity pointed the way, scientists searched for disappearing mass and found it.

It may seem surprising that no one noticed the interchange of mass and energy until Einstein pointed it out theoretically. The reason for that rests with the nature of the equivalence—in the determination of exactly how much energy is equivalent to how much mass.

To determine that, let's consider the reciprocal of the FitzGerald ratio, which is $1/\sqrt{1 - v^2/c^2}$. This can also be written, according to algebraic convention, as $(1 - v^2/c^2)^{-1/2}$. An expression written in this last fashion can be said to belong to a family of the type $(1 - b)^{-a}$. By the binomial theorem (a mathematical relationship first worked out by Newton himself), the expression $(1 - b)^{-a}$ can be expanded into an endless series of terms that begins as follows: $1 + ab + 1/2(a^2 + a)\ b^2 + \ldots$.

To apply this expansion to the reciprocal of the FitzGerald ratio, we must set $a = 1/2$ and $b = v^2/c^2$. The FitzGerald ratio then becomes: $1 + v^2/2c^2 + 3v^4/8c^4 \ldots$.

Since c, the velocity of light, may be considered to have a constant value, the second and third terms (and, indeed, all the subsequent terms of this infinite series) grow larger as v increases. But v reaches a maximum when the velocity of a moving body attains the velocity of light (at least, we can measure no higher velocity). Therefore the various terms are then at their maximum value, and at $v = c$ the series becomes $1 + 1/2 + 3/8 \ldots$ As you see, the second term is, at most, less than the first, while the third term is, at most, less than the second, and so on.

The decrease is even sharper at lower velocities, and successive terms become rapidly more and more insignificant. When $v = c/2$ (150,000 kilometers per second), the series is $1 + 1/8 + 3/128 \ldots$ When $v = c/4$ (75,000 kilometers per second), the series is $1 + 1/32 + 3/2048 \ldots$

In a decreasing series of this sort it is possible to show that the tail end of the series (even though it includes an infinite number of terms) reaches a finite and small total. We can therefore eliminate all but the first few terms of the series, and consider those first few a good approximation of the whole series.

At ordinary velocities, for instance, all the terms of the series except the first (which is always 1) become such small fractions that they can be ignored completely. In that case, the reciprocal of the FitzGerald ratio can be considered as equal to 1 with a high decree of accuracy (which is why changes in length and mass with motion went unnoticed until the twentieth century). To make it still more accurate, especially at very high velocities, we can include the first two terms of the series. That is accurate enough for all reasonable purposes, and we need not worry at all about the third term or any beyond it.

We can say, then, with sufficient accuracy that:

$$\frac{1}{\sqrt{1 - v^2/c^2}} = 1 + v^2/2c^2 \qquad \text{(Equation 7–1)}$$

Now let us return to the Lorentz mass relationship (Equation 6–7), which states that the mass (m_1) of a body in motion is equal to its rest-mass (m_0) divided by the FitzGerald ratio. This is equivalent to saying that m_1 is equal to m_0 multiplied by the reciprocal of the FitzGerald ratio; therefore, using the new expression for that reciprocal given in Equation 7–1, we can write the mass relationship as follows:

$$\begin{aligned} m_1 &= m_0(1 + v^2/2c^2) \qquad \text{(Equation 7–2)} \\ &= m_0 + m_0v^2/2c^2 \end{aligned}$$

The increase in mass as a result of motion is $m_1 - m_0$, and we can call this difference simply m. If we solve Equation 7–2 for $m_1 - m_0$ (that is, for m), we find that:

$$m = m_0v^2/2c^2 = \tfrac{1}{2}m_0v^2/c^2 \qquad \text{(Equation 7–3)}$$

The expression $\frac{1}{2}m_0v^2$, found in the right-hand portion of Equation 7–3, happens to be the value of the kinetic energy of the moving body (kinetic energy is equal to $\frac{1}{2}mv^2$, see page I–95), if it possesses its rest-mass. Actually, it possesses a slightly higher mass due to its motion, but except for extremely high velocities, the actual mass is only very slightly higher than the rest-mass—so little higher in fact that we can let $\frac{1}{2}m_0v^2$ equal its kinetic energy and be confident of a high degree of accuracy. If we let this kinetic energy be represented as e, then Equation 7–3 becomes:

$$m = e/c^2 \qquad \text{(Equation 7–4)}$$

Remember that m represents the gain of mass with motion. Since very rapid motion, representing a very high value of e, (the kinetic energy) produces only a very small increase in mass, we see quite plainly that a great deal of ordinary energy is equivalent to a tiny quantity of mass. Equation 7–4, which by clearing fractions can be written as the much more familiar:

$$e = mc^2 \qquad \text{(Equation 7–5)}$$

can be used to calculate this equivalence.

In the cgs system (see page I–32), where all the units are in terms of centimeters, grams and seconds, the value of c (the velocity of light in a vacuum) is 30,000,000,000 centimeters per second. The value of c^2 is therefore 900,000,000,000,000,000,000 cm^2/sec^2. If we set the value of m at 1 gram, then mc^2 is equal to 900,000,000,000,000,000,000 gm-cm^2/sec^2; or, since 1 gm-cm^2/sec^2 is defined as an "erg," 1 gram of mass is equal to 900,000,000,000,000,000,000, ergs of energy.

One kilocalorie is equal to 41,860,000,000 ergs. This means that 1 gram of mass is equivalent to 21,500,000,000 kilocalories. The combustion of a gallon of gasoline liberates about 32,000 kilocalories. The mass equivalence of this amount of energy is 32,000/21,500,000,000 or 1/670,000 of a gram. This means that in the combustion of a full gallon of gasoline, the evolution of energy in the form of heat, light, the mechanical motion of pistons, and so on, involves the total loss to the system of 1/670,000 of a gram of mass. It is small wonder that chemists and physicists did not notice such small mass changes until they were told to look for it.

On the other hand, if whole grams of mass could be converted wholesale into energy, the vast concentration of energy produced would have tremendous effects. In Volume III the steps by which it was learned how to do this will be outlined. The results are the *nuclear bombs* that now threaten all mankind with destruction and the *nuclear reactors* that offer it new hope for the future.

Furthermore, Equation 7–5 offered the first satisfactory explanation of the source of energy of the sun and other stars. In order for the sun to radiate the vast energies it does, it must lose 4,600,000 tons of mass each second. This is a vast quantity by human standards but is insignificant to the sun. At this rate it can continue to radiate in essentially unchanged fashion for billions of years.

The *Einstein equation,* $e = mc^2$, as you see, is derived entirely from the assumption of the constant measured velocity of light, and the mere existence of nuclear bombs is fearful evidence of the validity of the special theory of relativity. It is no wonder that of all equations in physics, $e = mc^2$ has most nearly become a household word among the general population of non-physicists.

Relative Time

Einstein deduced a further conclusion from his assumptions and went beyond the Lorentz-FitzGerald dealings with length and mass to take up the question of time as well.

The passing of time is invariably measured through some steady periodic motion: the turning of the earth, the dripping of water, the beating of a pendulum, the oscillation of a pendulum, even the vibration of an atom within a molecule. However, the changes in length and mass with increasing velocity, must inevitably result in a slowing of the period of all periodic motion. Time therefore must be measured as proceeding more slowly as velocity relative to the observer is increased.

Again, the FitzGerald ratio is involved. That is, the time lapse (t) observed on a body moving at a given velocity relative to the time lapse at rest (t_0) is as follows:

$$t = t_0\sqrt{1 - v^2/c^2} \qquad \text{(Equation 7–6)}$$

At a velocity of 260,000 kilometers per second past an observer, t would equal $t_0/2$. In other words, it would take one hour of the observer's time for half an hour to seem to pass on the moving object. Thus, if an observer's clock said 1:00 and a clock on the moving object also said 1:00, then one hour later the observer's clock would say 2:00, but the clock on the moving object would say only 1:30.

At a velocity equal to that of light, t would equal 0. It would then take forever for the clock on the moving object to show any time lapse at all to the observer. As far as the observer would be able to note, the clock on the moving object would always read 1:00; time would stand still on the object. This slowing of time with motion is called *time dilatation.*

Strange as this state of affairs seems, it has been checked in the case of certain short-lived subatomic particles. When moving slowly, they break down in a certain fixed time. When moving very rapidly, they endure considerably longer before decaying. The natural conclusion is that we observe a slowing of time for the speedily moving particles. They still decay in, say, a millionth of a second, but for us that millionth of a second seems to stretch out because of the rapid motion of the particle.

As in the case of length and mass, this change in time is a change in measurement only (as long as the conditions of the special theory are adhered to), and this varies with the observer.

Suppose, for instance, we return to Spaceship X and Spaceship Y as they flash past each other. The men on Spaceship X, watching Spaceship Y flash by at 260,000 kilometers a second and observing a pendulum clock on board Spaceship Y, would see the clock beat out its seconds at two-second intervals. Everything on Spaceship Y would take twice the usual time to transpire (or so it would seem to the observer on Spaceship X). The very atoms would move at only half their usual speed.

The people on Spaceship Y would be unaware of this, of course. Considering themselves at rest, they would insist that it was Spaceship X that was experiencing slowed time. (Indeed, if the spaceships had flashed by each other in such a way that each measured the velocity of the other as equaling the velocity of light, each would insist that it had observed time on the other ship having come to a full halt.)

This question of time is trickier, however, than that of length and mass. If the spaceships are brought together after the flash-by, and placed at mutual rest, length and mass are now "normal" and no record is left of any previous changes, so there need be no worries about "reality."

But, as for time, consider . . . Once at mutual rest, the clocks are again proceeding at the "normal" rate and time lapses are equal on both ships. Yet, previous changes in time lapses *have* left a record. If one clock has been slowed and has, in the past, registered only half an hour while the other clock was registering an hour, the first clock would now be half an hour slow! Each ship would firmly claim that the clock on the other ship had been registering time at a slower-than-normal rate, and each would expect the other's clock to be slow.

Would this be so? Would either clock be slow? And if so, which?

This is the *clock paradox,* which has become famous among physicists.

There is no clock paradox if the conditions of the special theory are adhered to strictly—that is, if both ships continue eternally in uniform motion. In that case, they can never be brought together again, and the difference in measurement remains one that can never be checked against "reality."

In order to bring the ships together, at least one of them must slow down, execute a turn, speed up and overtake the other.

In all this it undergoes non-uniform velocity, or acceleration, and we are promptly outside the special theory.

Einstein worked on problems of this sort for ten years after having enunciated his special theory and, in 1915, published his *general theory of relativity,* in which the consequences of non-uniform or accelerated motion are taken up. This is a much more subtle and difficult aspect of relativity than is the special theory, and not all theoretical physicists entirely agree on the consequences of the general theory.

Suppose we consider our spaceships as alone in the universe. Spaceship Y executes the slowdown, turn, and speedup that brings it side by side with Spaceship X. But, by the principle of relativity, the men on Spaceship Y have every right to consider themselves at rest. If they consider themselves at rest, then it is Spaceship X that (so it seems to them) slows down, turns, and then backs up to them. Whatever effect the men on Spaceship X observe on Spaceship Y, the men on board Spaceship Y will observe on Spaceship X. Thus, it might be that when the two ships are finally side by side, the two clocks will somehow tell the same time.

Actually, though, this will not occur, for the two spaceships are not alone in the universe. The universe is filled with a vast amount of matter, and the presence of this amount of matter spoils the symmetry of the situation of Spaceships X and Y.

Thus, if Spaceship Y executes its turn, Spaceship X observes it make that turn. But as Spaceship X considers itself at rest, it continues to see the rest of the universe (the stars and galaxies) slip past it at a constant, uniform velocity reflecting its own constant, uniform velocity. In other words, Spaceship X sees *only* Spaceship Y and nothing else undergo non-uniform motion.

On the other hand if Spaceship Y considers itself at rest, it observes that not only does Spaceship X seem to undergo an acceleration but also all the rest of the universe.

To put it another way, Spaceship Y and Spaceship X both undergo non-uniform motion relative to each other, but the universe as a whole undergoes non-uniform motion only relative to Spaceship Y. The two ships, naturally enough, are influenced differently by this tremendous difference in their histories, and when they are brought together, it is Spaceship Y (which has undergone non-uniform motion relative to the universe as a whole) that carries the slowed clock. There is no paradox here, for the crews on both ships must have observed the non-uniform motion of the universe relative to Spaceship Y, and so both agree on the difference in histories and cannot seek refuge in a "my-frame-of-reference-is-as-good-as-yours" argument.

Now suppose a space traveler leaves earth and, after a while, is traveling away from us at a speed nearly that of light. If we could observe him as he traveled, we would see his time pass at only perhaps one-hundredth the rate ours does. If he observed us, he would see our time pass at only one-hundredth the rate his does. If the space traveler wanted to return, however, he would have to turn, and he would experience non-uniform motion relative to the universe as a whole. In other words, in turning, he would observe the entire universe turning about him, if he insisted on considering himself at rest. The effect of this is to make the time lapse less for him as far as both he and the stay-at-home earthmen are involved. The round trip may have seemed to him to have lasted only a year, but on the earth a hundred years would have passed. If the space traveler had a twin brother, left behind on earth, that brother would long since have died of old age, while the traveler himself would scarcely have aged. (This is called the *twin paradox.*) It is important, however, to realize that the space traveler has not discovered a fountain of youth. He may have aged only a year in an earth-century, but he would only have lived a year in that earth-century. Moreover, no matter what his velocity, time would never appear, either to him or to the earth-bound observers, to go backward. He would never grow younger.

The variation of the rate at which times passes as velocity changes destroys our concept of absoluteness of time. Because of this, it becomes impossible to locate an event in time in such a way that all observers can agree. In addition, no event can be located in time until some evidence of the event reaches the observer, and that evidence can only travel at the velocity of light.

As a simple example, consider the space traveler returning to earth, after having experienced a time lapse of one year, and finding that his twin brother had died fifty years earlier by earth time. To the traveler this may seem impossible, since fifty years earlier (to him) his twin brother had not even been born yet.

In fact, in the mathematical treatment of the theory of relativity, it does not make sense to deal with space alone or time alone. Rather the equations deal with a fusion of the two (usually called space-time). To locate a point in space-time, one must express a value for each of three spatial dimensions, plus a value for time; time being treated somewhat (but not exactly) like the ordinary three dimensions. It is in that sense that time is sometimes spoken of as the "fourth dimension."

It is sometimes argued that the existence of relative time makes it possible to measure a velocity of more than that of light. Suppose, for instance, a spaceship travels from earth to a planet ten light-years distant and does this at so great a velocity that time dilatation makes it seem to the crewmen that only one year has passed in the course of the voyage.

Since the ship has traveled in one year a distance traversed by light in ten years, has not the ship traveled at ten times the velocity of light?

The answer is it has not. If the crewmen were to argue that they had, they would be measuring the time lapse of one year against their own frame of reference, and the distance of the planet from earth (ten light-years) by earth's frame of reference. They must ask instead: What is the distance of the destination-planet from earth in the frame of reference of the ship?

In the ship's frame of reference, the ship is, of course, motionless, while the universe, including earth and the destination-planet, slips backward past it at an enormous velocity. The entire universe is foreshortened, as one would expect from the FitzGerald contraction (see page 97), and the distance from earth to the destination-planet is much less than ten light-years. The distance is less than one light-year, in fact, so the ship can traverse that distance in one year without having exceeded the velocity of light.

Again, although the ship took only one year to get to its destination, this did not mean they beat light there, even though a light beam, released from earth simultaneously with the ship, would have taken ten years to cover a ten-light-year distance. That ten-year time lapse would be true only in the earth's frame of reference. For the light beam's own frame of reference, since it travels at the velocity of light, the rate of passage of time would decline to zero, and the light beam would get to Alpha Centauri (or to any spot in the universe, however distant) in no time at all.

Nor can one use this to argue that in the light beam's own frame of reference its velocity is then infinite; for in the light beam's own frame of reference, the total thickness of the universe, in the direction of its travel, is foreshortened to zero, and of course it would take no time for light to cross a zero-thickness universe even if its velocity is the finite one of 300,000 kilometers per second.

The General Theory

One of the basic assumptions in the special theory was that it was impossible to measure absolute motion; that any observer had the privilege of considering himself at rest; and that all frames of reference were equally valid.

Yet when we consider non-uniform motion (outside the realm of special theory) the possibility arises that this is not so.

Suppose, for instance, that two spaceships are moving side by side at uniform velocity. The crewmen on each ship can consider both themselves and the other ship to be at rest. Then, suddenly, Spaceship Y begins to move forward with reference to Spaceship X.

The crewmen on Spaceship X could maintain that they were still at rest while Spaceship Y had begun to move forward at an accelerating velocity. The crewmen on Spaceship Y, however, could maintain that, on the contrary, they were at rest while Spaceship X had begun to move backward at an accelerating velocity. Is there any way, now, to decide between these conflicting observations?

In the case of such non-uniform motion, perhaps. Thus, if Spaceship Y were "really" accelerating forward, the men within it would feel an inertial pressure backward (as you are pressed back into your seat when you step on your car's gas-pedal.) On the other hand, if Spaceship X is accelerating backward, the men within it would feel an inertial pressure forward (as you lurch toward the windshield when you step on your car's brake). Consequently, the crews of the spaceships could decide which ship was "really" moving by taking note of which set of crewmen felt inertial pressures.

From this, one could perhaps determine absolute motion from the nature and size of inertial effects. Einstein, in his general theory of relativity, worked out what properties the universe must possess to prevent the determination of absolute motion in the case of non-uniform motion.

The Newtonian view of mass had dealt, really, with two kinds of mass. By Newton's second law of motion, mass was defined through the inertia associated with a body. This is "inertial mass." Mass may also be defined by the strength of the gravitational field to which it gives rise. This is "gravitational mass." Ever since Newton, it had been supposed that the two masses were really completely identical, but there had seemed no way of proving it. Einstein did not try to prove it; he merely assumed that inertial mass and gravitational mass were identical and went on from there.

It was then possible to argue that both gravitation and inertial effects were not the property of individual bodies alone, but of the interaction of the mass of those bodies with all the remaining mass in the universe.

If a spaceship begins to accelerate in a forward direction, the crewmen feel an inertial pressure impelling them to the rear. But suppose the crewmen in the spaceship insist on regarding themselves as at rest. They must then interpret their observations of the universe as indicating that all the stars and galaxies outside the ship are moving backward at an accelerating velocity. The accelerating motion backward of the distant bodies of the universe drags the crewmen back, too, producing an inertial effect upon them, exactly as would have happened if the universe had been considered at rest and the ship as accelerating forward.

In short, inertial effects cannot be used to prove that the ship is "really" accelerating. The same effect would be observed if the ship were at rest and the universe were accelerating. Only *relative* non-uniform motion is demonstrated by such inertial effects: either a non-uniform motion of the ship with reference to the universe or a non-uniform motion of the universe with reference to the ship. There is no way of demonstrating which of these two alternatives is the "real" one.

We might also ask if the earth is "really" rotating. Through most of man's history, the earth was assumed motionless because it seemed motionless. After much intellectual travail, its rotation was demonstrated to the satisfaction of scientists generally and to those non-scientists who followed the arguments or were willing to accept the word of authority. But is it "really" rotating?

One argument in favor of the rotation of the earth rests on the existence of the planet's equatorial bulge. This is explained as the result of a centrifugal effect that must surely arise from a rotation. If the earth did not rotate, there would be no centrifugal effect and it would not bulge. The existence of the bulge, therefore, is often taken as proof of a "real" rotation of the earth.

This argument might hold, perhaps, if the earth were alone in the universe, but it is not. If the earth is considered motionless, for argument's sake, one must also think of the enormous mass of the universe revolving rapidly about the earth. The effect of this enormous revolving mass is to pull out the earth's equatorial bulge—just as the centrifugal effect would if the earth rotated and the rest of the universe were motionless. One could always explain all effects of rotation equally well in either frame of reference.

You might also argue that if the earth were motionless and the rest of the universe revolved about it, the distant stars, in order to travel completely around their gigantic orbits about earth in a mere 24 hours, must move at many, many times the velocity of light. From this one might conclude that the rotation of the universe about the earth is impossible and that therefore the earth is "really" rotating. However, if the universe is considered as rotating about the earth and if the distant stars are traveling, in consequence, at great velocities, the FitzGerald contraction will reduce the distances they must cover to the point where their velocity will be measured as less than that of light.

Of course, one might raise the argument that it simply isn't reasonable to suppose the entire universe is revolving about the earth—that one must naturally prefer to believe that it is the earth's rotation that produces the apparent revolution of the universe. Similarly it is much more sensible to believe that a spaceship is accelerating forward rather than to suppose an entire universe is accelerating backward past one motionless ship.

This is true enough, and it is so much more reasonable to assume a rotating earth (or a moving ship) that astronomers will continue to assume it, regardless of the tenets of relativity. However, the theory of relativity does not argue that one frame of reference may not be simpler or more useful than another—merely that one frame of reference is not more *valid* than another.

Consider that at times it is the motionlessness of the earth that is assumed because that makes for greater simplicity. A pitcher throwing a baseball never takes into account the fact that the earth is rotating. Since he, the ball, and the waiting batter are all sharing whatever velocity the earth possesses, it is easier for the pitcher to assume that the earth is motionless and to judge the force and direction of his throw on that basis. For him the motionless-earth frame of reference is more useful than the rotating-earth frame of reference—yet that does not make the motionless-earth frame of reference more valid.

Gravitation

In his general theory, Einstein also took a new look at gravitation. To Newton it had seemed that if the earth revolved about the sun there must be a force of mutual attraction between the earth and the sun. Einstein showed that one could explain the revolution of the earth about the sun in terms of the geometry of space.

Consider an analogy. A putter is addressing a golf ball toward the cup over a level green. The golf ball strikes the edge of the cup and dips in. It is, however, going too fast so that it spins about the vertical side of the cup (bobsled fashion) and emerges at the other end, rolling in a new direction. It has partly circled the center of the cup, yet no one would suppose that there was a force of attraction between the golf ball and the center of the cup.

Let us imagine a perfectly level, frictionless, putting green of infinite extent. A ball struck by the golf club will continue on forever in a perfectly straight line.

But what if the putting green is uneven; if there are bumps and hollows in it? A ball rising partly up the side of a bump will curve off in a direction away from the center of the bump. A ball dropping down the side of a hollow will curve toward the center of the hollow. If the bumps and hollows are, for some reason, invisible and undetectable, we might be puzzled at the occasional deviations of the balls from straight-line motion. We might suppose the existence of hidden forces of attraction or repulsion pulling or pushing the ball this way and that.

Suppose one imagined a cone-shaped hollow with steep sides on such a green. A ball can be visualized as taking up a closed "orbit" circling around and around the sides like a bobsled speeding endlessly along a circular bank. If friction existed, the circling ball would lose kinetic energy and, little by little, sink to the bottom of the cone. In the absence of friction, it would maintain its orbit.

It is not difficult to form an analogous picture of the Einsteinian version of gravity. Space-time would be a four-dimensional analogy of a flat putting green, if it were empty of matter. Matter, however, produces "hollows"; the more massive the matter, the deeper the "hollow." The earth moves about the sun as though it were circling the sun's hollow. If there were friction in space, it would slowly sink to the bottom of the "hollow" (that is, spiral into the sun). Without friction, it maintains its orbit indefinitely. The elliptical orbit of the earth indicates that the orbit about the "hollow" is not perfectly level with the flatness of the four-dimensional putting green. (The orbit would be a circle, if it were.) A slight tilt of the orbit produces a slight ellipticity, while a more marked tilt produces greater ellipticity.

It is these "hollows" produced by the presence of matter that give rise to the notion of *curved space.*

The consequences of the special theory of relativity—mass increase with motion and the equivalence of mass and energy, for instance—were easily demonstrated. The validity of the general theory was much more difficult to prove. Einstein's picture of gravitation produces results so nearly like those of Newton's picture that it is tempting to consider the two equivalent and then accept the one that is simpler and more "common sense," and that, of course, is the Newtonian picture.

However, there remained some areas where the consequences of the Einsteinian picture were indeed somewhat different from those of the Newtonian picture. By studying those consequences, one might choose between the two on some basis more satisfying than that of mere simplicity. The first such area involved the planet Mercury.

The various bodies in the solar system move, in the Newtonian view, in response to the gravitational forces to which they are subjected. Each body is subjected to the gravitational forces of every other body in the universe, so the exact and complete solution of the motions of any body is not to be expected. However, within the solar system, the effect of the gravitational field of the sun is overwhelming. While the gravitational fields of a few other bodies quite close to the body whose motion is being analyzed are also significant, they are minor. If these are taken into account, the motion of a planet of the solar system can be explained with a degree of accuracy that satisfies everybody. If residual disagreements between the predicted motion and the actual motion remain, the assumption is that some gravitational effect has been ignored.

The presence of a discrepancy in the motion of Uranus, for instance, led to a search for an ignored gravitational effect, and the discovery, in the mid-nineteenth century, of the planet Neptune.

At the time of Neptune's discovery, a discrepancy in the motion of Mercury, the planet nearest the sun, was also being studied. Like the other planets, Mercury travels in an ellipse about the sun, with the sun at one of the foci of the ellipse. This means that the planet is not always at the same distance from the sun. There is a spot in its orbit where it is closest to the sun, the *perihelion,* and a spot at the opposite end of the orbit where it is farthest from the sun, the *aphelion.* The line connecting the two is the *major axis.* Mercury does not repeat its orbit exactly, but moves in such a way that the orbit is actually a rosette, with the major axis of the ellipse slowly revolving.

This can be explained by the gravitational effect of nearby planets on Mercury, but it cannot all be explained. After all known gravitational effects are accounted for, the actual rate at which the major axis (and its two extreme points, the perihelion and aphelion) turned was slightly greater than it ought to have been—greater by 43.03 seconds of arc per century. This meant that the major axis of Mercury's orbit made a complete turn, and an unexplained one, in 3,000,000 years.

Leverrier, one of the men who had discovered Neptune, suggested that an undiscovered planet might exist between Mercury and the sun, and that the gravitational effect of this planet on Mercury could account for that additional motion of the perihelion. However, the planet was never found, and even if it existed (or if a belt of planetoids of equivalent mass existed near the sun) there then would also be gravitational effects on Venus, and these have never been detected.

The situation remained puzzling for some seventy years until Einstein in 1915 showed that the general theory of relativity altered the view of gravity by just enough to introduce an additional factor that would account for the unexplained portion of the motion of Mercury's perihelion. (There would be similar but much smaller effects on the planets farther from the sun—too small to detect with certainty.)

Einstein also predicted that light beams would be affected by gravity, a point that was not allowed for in the Newtonian view. The light of stars passing very close to the sun, for instance, would be affected by the geometry of space and would bend inward toward the center of the sun. Our eyes would follow the ray of light backward along the new direction and would see the star located farther from the center of the sun than it really was. The effect was very small. Even if light just grazed the sun, the shift in a star's position would be only 1.75 seconds of arc, and if the light passed farther from the sun, the shift in the star's position would be even less.

Of course, the light of stars near the sun cannot ordinarily be observed. For a few minutes during the course of a total eclipse, however, they can be. At the time the general theory was published, World War I was in progress and nothing could be done. In 1919, however, the war was over and a total eclipse was to be visible from the island of Principe in the Gulf of Guinea off West Africa. Under British auspices an elaborate expedition was sent to the island for the specific purpose of testing the general theory. The positions of the stars in the neighborhood of the sun were measured and compared with their positions a half year later when the sun was in the opposite end of the sky. The results confirmed the general theory.

Finally, Einstein's theory predicted that light would lose energy if it rose against gravity and would gain energy if it "fell," just as an ordinary object would. In the case of a moving object such as a ball, this loss of energy would be reflected as a loss of velocity. However, light could only move at one velocity; therefore the loss of energy would have to be reflected in a declining frequency and increasing wavelength. Thus, light leaving a star would undergo a slight "red shift" as it lost energy. The effect was so small, however, that it could not be measured.

However, stars had just been discovered (*white dwarfs*) which were incredibly dense and which had gravitational fields thousands of times as intense as those of ordinary stars. Light leaving such a star should lose enough energy to show a pronounced red shift of its spectral lines. In 1925, the American astronomer Walter Sydney Adams (1876–1956) was able to take the spectrum of the white dwarf companion of the star Sirius, and to confirm this prediction.

The general theory of relativity had thus won three victories in three contests over the old view of gravitation. All, however, were astronomical victories. It was not until 1960 that the general theory was brought into the laboratory.

The key to this laboratory demonstration was discovered in 1958 by the German physicist Rudolf Ludwig Mössbauer (1929–), who showed that under certain conditions a crystal could be made to produce a beam of gamma rays* of identical wavelengths. Gamma rays of such wavelengths can be absorbed by a crystal similar to that which produced it. If the gamma rays are of even slighly different wavelength, they will not be absorbed. This is called the *Mössbauer effect*.

Now, then, if such a beam of gamma rays is emitted downward so as to "fall" with gravity, it gains energy and its wavelength becomes shorter—if the general theory of relativity is correct. In falling just a few hundred feet, it should gain enough energy for the decrease in wavelength of the gamma rays, though very minute, to become sufficiently large to prevent the crystal from absorbing the beam.

Furthermore, if the crystal emitting the gamma ray is moved upward while the emission is proceeding, the wavelength of the

* Gamma rays are a form of light-like radiation that will be discussed in Volume III of this book.

gamma ray is increased through the Doppler-Fizeau effect. The velocity at which the crystal is moved upward can be adjusted so as to just neutralize the effect of gravitation on the falling gamma ray. The gamma ray will then be absorbed by the absorbing crystal. Experiments conducted in 1960 corroborated the general theory of relativity with great accuracy, and this was the most impressive demonstration of its validity yet.

It is not surprising, then, that the relativistic view of the universe is now generally accepted (at least until further notice) among physicists of the world.

CHAPTER 23

Quanta

Black-Body Radiation

The theory of relativity does not flatly state that an ether does not exist. It does, however, remove the need for one, and if it is not needed, why bother with it?

Thus, the ether was not needed to serve as an absolute standard for motion since relativity began by assuming that such an absolute standard did not exist and went on to demonstrate that it was not needed. Again, the ether is not needed as a medium to transmit the force of gravity and prevent "action at a distance." If gravity is a matter of the geometry of space-time and is not a transmitted force, the possibility of action at a distance does not arise.

This still leaves one possible use for the ether—that of serving as a medium for transmitting light waves across a vacuum. A second paper written by Einstein in 1905 (in addition to his paper on special relativity) wiped out that possibility, too. Einstein's work on relativity had evolved out of the paradox concerning light that was turned up by the Michelson-Morley experiment (see page 99ff.). Einstein's second paper arose out of a different paradox, also concerning light, that had also arisen in the last decades of the nineteenth century. (It was for this second paper that he later received the Nobel Prize.)

This second paradox began with Kirchhoff's work on spectroscopy (see page 58). He showed that a substance that absorbed certain frequencies of light better than others would also emit those frequencies better than others once it was heated to incandescence.

Supppose, then, one imagined a substance capable of absorbing all the light, of all frequencies, that fell upon it. Such a body would reflect no light of any frequency and would therefore appear perfectly black. It is natural to call such a substance a *black body* for that reason. If a black body is brought to incandescence, its emission should then be as perfect as its absorption, by Kirchhoff's rule. It should emit light in all frequencies, since it absorbs in all frequencies. Furthermore, since it absorbs light at each frequency more efficiently than a non-black body would, it must radiate more efficiently at each frequency, too.

Kirchhoff's work served to increase the interest of physicists in the quantitative aspects of radiation, and in the manner in which such radiation varied with temperature. It was common knowledge that the total energy radiated by a body increased as the temperature increased, but this was made quantitative in 1879 by the Austrian physicist Josef Stefan (1835–1893). He showed that the total energy radiated by a body increased as the fourth power of the absolute temperature. (The absolute temperature, symbolized as °K, is equal to the centigrade temperature, °C, plus 273°; see page I–193.)

Consider a body, for instance, that is maintained at room temperature, 300°K, and is then radiating a certain amount of energy. If the temperature is raised to 600°K, which is that of melting lead, the absolute temperature has been doubled and the total amount of energy radiated is increased by 2^4, or 16 times. If the same body is raised to a temperature of 6000°K, which is that of the surface of the sun, it is at an absolute temperature twenty times as high as it was at room temperature, and it radiates 20^4, or 160,000, times as much energy.

In 1884, Boltzmann (who helped work out the kinetic theory of gases) gave this finding a firm mathematical foundation and showed that it applied, strictly, to black bodies only and that non-black bodies always radiate less heat than *Stefan's law* would require. Because of his contribution, the relationship is sometimes called the *Stefan-Boltzmann law.*

But it is not only the total quantity of energy that alters with rising temperature. The nature of the light waves emitted also changes, as is, in fact, the common experience of mankind. For objects at the temperature of a steam radiator, for instance (less than 400°K), the radiation emitted is in the low frequency infrared. Your skin absorbs the infrared and you feel the radiation as heat, but you see nothing. A radiator in a dark room is invisible.

As the temperature of an object goes up, it not only radiates more heat, but the frequency of the radiation changes somewhat, too. By the time a temperature of 950°K is reached, enough radiation, of a frequency high enough to affect the retina, is emitted for the body to appear a dull red in color. As the temperature goes higher still, the red brightens and eventually turns first orange and then yellow as more and more of still higher frequencies of light are emitted. At a temperature of 2000°K, an object, although glowing brightly, is still emitting radiation that is largely in the infrared. It is only when the temperature reaches 6000°K, the temperature of the surface of the sun, that the emitted radiation is chiefly in the visible light region of the spectrum. (Indeed, it is probably because the sun's surface is at that particular temperature, that our eyes have evolved in such a fashion as to be sensitive to that particular portion of the spectrum.)

Toward the end of the nineteenth century, physicists attempted to determine quantitatively the distribution of radiation among light of different frequencies at different temperatures. To do this accurately a black body was needed, for only then could one be sure that at each frequency all the light possible (for that temperature) was being radiated. For a non-black body, certain frequencies were very likely to be radiated in a deficient manner; the exact position of these frequencies being dependent on the chemical nature of the radiating body.

Since no actual body absorbs all the light falling upon it, no actual body is a true black body, and this seemed to interpose a serious obstacle in the path of this type of research. In the 1890's, however, a German physicist, Wilhelm Wien (1864–1928), thought of an ingenious way of circumventing this difficulty.

Imagine a furnace with a hole in it. Any light of any wavelength entering that hole would strike a rough inner wall and be mostly absorbed. What was not absorbed would be scattered in diffuse reflections that would strike other walls and be absorbed there. At each contact with a wall, additional absorption would take place, and only a vanishingly small fraction of the light would manage to survive long enough to be reflected out the hole again. That hole, therefore, would act as a perfect absorber (within reason) and would, therefore, represent a black body. If the furnace were raised to a certain temperature and maintained there, then the radiation emitted from that hole is *black-body radiation* and its frequency distribution can be studied.

In 1895, Wien made such studies and found that at a given temperature, the energy radiated at given frequencies, increased as the frequency was raised, reached a peak, and then began to decrease as the frequency was raised still further.

If Wien raised the temperature, he found that more energy was radiated at every frequency, and that a peak was reached again. The new peak, however, was at a higher frequency than the first one. In fact, as he continued to raise the temperature, the frequency peak of radiation moved continuously in the direction of higher and higher frequencies. The value of the peak frequency (ν_{max}) varied directly with the absolute temperature (T), so *Wien's law* can be expressed as follows:

$$\nu_{max} = kT \qquad \text{(Equation 8–1)}$$

where k is a proportionality constant.

Both Stefan's law and Wien's law are of importance in astronomy. From the nature of a star's spectrum, one can obtain a measure of its surface temperature. From this one can obtain a notion of the rate at which it is radiating energy and, therefore, of its lifetime. The hotter a star, the more short-lived it may be expected to be.

Wien's law explains the colors of the stars as a function of temperature (rather than as a matter of approach or recession as Doppler had thought—see page 76). Reddish stars are comparatively cool, with surface temperatures of 2000–3000°K. Orange stars have surface temperatures of 3000–5000°K, and yellow stars (like our sun) of 5000–8000°K. There are also white stars with surface temperatures of 8000–12,000°K and bluish stars that are hotter still.

Planck's Constant

At this point the paradox arose, for there remained a puzzle as to just why black-body radiation should be distributed in the manner observed by Wien. In the 1890's, physicists assumed that a radiating body could choose at random a frequency to radiate in. There are many more small gradations of high-frequency radiation than of low-frequency radiation (just as there are many more large integers than small ones), and if radiation could choose any frequency at random, many more high frequencies would be chosen than low ones.

Lord Rayleigh worked out an equation based on the assumption that all frequencies could be radiated with equal probability. He found that the amount of energy radiated over a particular range of frequencies should vary as the fourth power of the frequency. Sixteen times as much energy should be radiated in the form of violet light as in the form of red light, and far more still should be radiated in the ultraviolet. In fact, by Rayleigh's formula, virtually all the energy of a radiating body should be radiated very rapidly in the far ultraviolet. Some people referred to this as the "violet catastrophe."

The point about the violet catastrophe, however, was that it did not happen. To be sure, at very low frequencies the Rayleigh equation held, and the amount of radiation climbed rapidly as the frequency of the radiation increased. But soon the amount of radiation began to fall short of the prediction. It reached a peak at some intermediate frequency, a peak that was considerably below what the Rayleigh equation predicted for that frequency, and then, at higher frequencies still, the amount of radiation rapidly decreased, though the Rayleigh formula predicted a still-continuing increase.

On the other hand, Wien worked up an equation designed to express what was actually observed at high frequencies. Unfortunately, it did not account for the distribution of radiation at low frequencies.

In 1899, a German physicist, Max Karl Ernst Ludwig Planck (1858–1947), began to consider the problem. Rayleigh's analysis, it seemed to Planck, was mathematically and logically correct, provided his assumptions were accepted; and since Rayleigh's equation did not fit the facts, it was necessary to question the assumptions. What if all frequencies were not, after all, radiated with equal probability? Since the equal-probability assumption required that more and more light of higher and higher frequency be radiated, whereas the reverse was observed, Planck proposed that the probability of radiation decreased as frequency increased.

Thus, there would be two effects governing the distribution of black-body radiation. First was the undeniable fact that there were more high frequencies than low frequencies so that there would be a tendency to radiate more high-frequency light than low-frequency light. Second, since the probability of radiation decreased as frequency went up, there would be a tendency to radiate less in the high-frequency range.

At very low frequencies, where the probability of radiation is quite high, the first effect is dominant and radiation increases as frequency rises, in accordance with the Rayleigh formula. However, as frequency continues to rise, the second effect becomes more and more important. The greater number of high frequencies is more than balanced by the lesser probability of radiating at such high frequency. The amount of radiation begins to climb more slowly as frequency continues to rise, reaches a peak, and then begins to decline.

Suppose the temperature is raised. This will not change the first effect, for the fact that there are more high frequencies than low frequencies is unalterable. However, what if a rise in temperature increased the probability that high-frequency light could be radiated? The second effect would therefore be weakened. In that case, radiation (at a higher temperature) could continue to increase, with higher frequencies, for a longer time before it was overtaken and repressed by the weakened second effect. The peak radiation consequently would move into higher and higher frequencies as the temperature went up. This was exactly what Wien had observed.

But how account for the fact that the probability of radiation decreased as frequency increased? Planck made the assumption that energy did not flow continuously (something physicists had always taken for granted) but was given off in discrete quantities. In other words, Planck imagined that there were "atoms of energy" and that a radiating body could give off one atom of energy or two atoms of energy, but never one and a half atoms of energy or, indeed, anything but an integral number of such entities. Furthermore, Planck went on to suppose, the energy content of such an atom of energy must vary directly with the frequency of the light in which it was radiated.

Planck called these atoms of energy, *quanta* (singular, *quantum*) from a Latin word meaning "how much?" since the size of the quanta was a crucial question.

Consider the consequences of this *quantum theory*. Violet light, with twice the frequency of red light, would have to radiate in quanta twice the size of those of red light. Nor could a quantum of violet light be radiated until enough energy had been accumulated to make up a full quantum, for less than a full quantum could not, by Planck's assumptions, be radiated. The probability, however, was that before the energy required to make up a full quantum of violet light was accumulated, some of it would have been bled off to form the half-sized quantum of red light. The higher the frequency of light, the less the probability that enough energy would accumulate to form a complete quantum without being bled off to form quanta of lesser energy content and lower frequency. This would explain why the "violet catastrophe" did not happen and why, in actual fact, light was radiated chiefly at low frequencies and more slowly than one might expect, too.

As the temperature rose, the general amount of energy available for radiation would increase as the fourth power of the absolute temperature. Under this increasing flood of radiation, it would become more and more likely that quanta of high-frequency light might have time to be formed. Thus, as Planck assumed, the probability of radiation in the high frequencies would increase, and the radiation peak would advance into higher frequencies. At temperatures of 6000°K, the peak would be in the visible light region, though the still larger quanta of ultraviolet would be formed in minor quantities even then.

If the energy content (e) of a quantum of radiation is proportional to the frequency of that radiation (ν), we can say that:

$$e = h\nu \qquad \text{(Equation 8–2)}$$

where h is a proportionality constant, commonly called *Planck's constant*. If we solve Equation 8–2 for h, we find that $h = e/\nu$. Since the units of e in the cgs system are "ergs" and those of ν are "1/seconds," the units of h are "ergs" divided by "1/seconds," or "erg-seconds." Energy multiplied by time is considered by physicists to be *action*. Planck's constant, therefore, may be said to be measured in units of action.

Planck derived an equation containing h that, he found, would describe the distribution of black-body radiation, as actually observed, over the entire range of frequencies. At least, it did this if h were given an appropriate, very small value. The best currently-accepted value of h is 0.00000000000000000000000000066256 erg-seconds or 6.6256×10^{-27} erg-seconds.

To see what this means, consider that orange light of wavelength 6000 A has a frequency of 50,000,000,000,000,000, or 5×10^{16} cycles per second. If this is multiplied by Planck's constant, we find that the energy content of a quantum of this orange light is $5 \times 10^{16} \times 6.6256 \times 10^{-27}$, or about 3.3×10^{-10} ergs. This is only about a third of a billionth of an erg, and an erg itself is but a small unit of energy.

It is no wonder, then, that individual quanta of radiant energy were not casually observed before the days of Planck.

Planck's quantum theory, announced in 1900, proved to be

a watershed in the history of physics. All physical theory that did not take quanta into account, but assumed energy to be continuous, is sometimes lumped together as *classical physics,* whereas physical theory that does take quanta into effect is *modern physics,* with 1900 the convenient dividing point.

Yet Planck's theory, when first announced, created little stir. Planck himself did nothing with it at first but explain the distribution of black-body radiation, and physicists were not ready to accept so radical a change of view of energy just to achieve that one victory. Planck himself was dubious and at times tried to draw his quantum theory as close as possible to classical notions by supposing that energy was in quanta form only when radiated and that it might be absorbed continuously.

And yet (with the wisdom of hindsight) we can see that quanta help explain a number of facts about absorption of light that classical physics could not. In Planck's time, it was well known that violet light was much more efficient than red light in bringing about chemical reactions, and that ultraviolet light was more efficient still. Photography was an excellent example of this, for photographic film of the type used in the nineteenth century was very sensitive to the violet end of the spectrum and rather insensitive to the red end. In fact, ultraviolet light had been discovered a century before Planck through its pronounced effect on silver nitrate (see page 77). Was it not reasonable to suppose that the large quanta of ultraviolet light could produce chemical reactions with greater ease than the small quanta of red light? And could not one say that this would only explain the facts if it were assumed that energy was absorbed only in whole quanta?

This argument was not used to establish the quantum theory in connection with absorption, however. Instead, Einstein made use of a very similar argument in connection with a much more recently-discovered and an even more dramatic phenomenon.

The Photoelectric Effect

In the last two decades of the nineteenth century, it had been discovered that some metals behave as though they were giving off electricity under the influence of light. At that time, physicists were beginning to understand that electricity was associated with the movement of subatomic particles called *electrons* and that the effect of light was to bring about the ejection of electrons from metal surfaces. This is the *photoelectric effect.*

On closer study, the photoelectric effect became a prime

puzzle. It seemed fair to assume that under ordinary conditions the electrons were bound to the structure of the metal and that a certain amount of energy was required to break this bond and set the electrons free. Furthermore, it seemed that as light was made more and more intense, more and more energy could be transferred to the metal surface. Not only would the electrons then be set free, but considerable kinetic energy would be available to them, so they would dart off at great velocities. The more intense the light, the greater the velocities. Nor did it seem that the frequency of the light ought to have anything to do with it; only the total energy carried by the light, whatever its intensity.

So it seemed, but that is not what happened.

The German physicist Philipp Lenard (1862–1947), after careful studies in 1902, found that for each surface that showed the photoelectric effect, there was a limiting *threshold frequency* above which, and only above which, the effect was to be observed.

Let us suppose, for instance, that this threshold frequency for a particular surface is 500 quadrillion cycles per second (the frequency of orange light of wavelength 6000 A). If light of lower frequency, such as red light of 420 quadrillion cycles per second, is allowed to fall upon the surface, nothing happens. No electrons are ejected. It doesn't matter how bright and intense the red light is and how much energy it carries; no electrons are ejected.

As soon, however, as the light frequency rises to 500 quadrillion cycles per second, electrons begin to be ejected, but with virtually no kinetic energy. It is as though the energy they have received from the light is just sufficient to break the bond holding them to the surface, but not sufficient to supply them with any kinetic energy in addition. Lenard found that increasing the intensity of the light at this threshold frequency did nothing to supply the electrons with additional kinetic energy. As a result of the increased intensity, more electrons were emitted from the surface, the number being in proportion to the energy of the orange light, but all of them lacked kinetic energy.

If the frequency were increased still further and if violet light of 1000 quadrillion cycles per second were used, electrons would be emitted with considerable kinetic energy. The number emitted would vary with the total energy of the light, but again all would have the same kinetic energy.

In other words, a feeble violet light would bring about the emission of a few high-energy electrons; an intense orange light would bring about the emission of many low-energy electrons; and an extremely intense red light would bring about the emission of no electrons at all.

The physical theories of the nineteenth century could not account for this, but in 1905 Einstein advanced an explanation that made use of Planck's quantum theory, which was now five years old but still very much neglected.

Einstein assumed that light was not only radiated in quanta, as Planck had maintained, but that it was absorbed in quanta also. When light fell upon a surface, the electrons bound to the surface absorbed the energy one quantum at a time. If the energy of that quantum was sufficient to overcome the forces holding it to the surface, it was set free—otherwise not.

If course, an electron might conceivably gain enough energy to break loose after absorbing a second quantum even if the first quantum had been insufficient. This, however, is an unlikely phenomenon. The chances are enormous that before it can absorb a second quantum, it will have radiated the first one away. Consequently, one quantum would have to do the job by itself; if not, merely multiplying the number of quantums (which remain individually insufficient) would not do the job. To use an analogy, if a man is not strong enough to lift a boulder single-handed, it doesn't matter if one million men, each as strong as the first, try one after the other to lift it single-handed. The boulder will not budge.

The size of the quantum, however, increases as frequency increases. At the threshold frequency, the quantum is just large enough to overcome the electron bond to a particular surface. As the frequency (and the energy content of the quantum) increase further, more and more energy will be left over, after breaking the electron bond, to be applied as kinetic energy.

For each substance, there will be a different and characteristic threshold energy depending on how strongly the electrons are bound to their substance. For a metal like cesium, where electrons are bound very weakly, the threshold frequency is in the infrared. Even the small quanta of infrared supply sufficient energy to break that weak bond. For a metal like silver, where electrons are held more strongly, the threshold frequency is in the ultraviolet.

Einstein suggested, then, the following relationship:

$$\frac{1}{2}mv^2 = h\nu - w \qquad \text{(Equation 8–2)}$$

where $\frac{1}{2}mv^2$ is the kinetic energy of the emitted electron; $h\nu$ (Planck's constant times frequency) the energy content of the quanta being absorbed by the surface; and w the energy required to break the electron free. At the threshold frequency, electrons would barely be released and would possess no kinetic energy. For that reason, Equation 8–2 would become $0 = h\nu - w$; and this would mean that $h\nu = w$. In other words, w would represent the energy of the light quanta at threshold frequency.

Einstein's explanation of the photoelectric effect was so elegant, and fit the observations so well, that the quantum theory sprang suddenly into prominence. It had been evolved, originally, to explain the facts of radiation, and now, without modification, it was suddenly found to explain the photoelectric effect, a completely different phenomenon. This was most impressive.

It became even more impresssive when in 1916 the American physicist Robert Andrews Millikan (1868–1953) carried out careful experiments in which he measured the energy of the electrons emitted by light of different frequency and found that the energies he measured fit Einstein's equation closely. Furthermore, by measuring the energy of the electrons ($\frac{1}{2}mv^2$), the frequency of the light he used (ν), and the threshold frequency for the surface he was using (w), he was able to calculate the value of h (Planck's constant) from Equation 8–2. He obtained a value very close to that which Planck had obtained from his radiation equation.

Since 1916, then, the quantum theory has been universally accepted by physicists. It is now the general assumption that energy can be radiated or absorbed only in whole numbers of quanta and, indeed, that energy in all its forms is "quantized"—that is, can only be considered as behaving as though it were made up of indivisible quanta. This concept has offered the most useful views of atomic structure so far, as we shall see in Volume III of this book.

Photons

Einstein carried the notion of energy quanta to its logical conclusion. A quantum seemed to be analogous to an "atom of energy" or a "particle of energy," so why not consider such particles to be particles? Light, then, would consist of particles which were eventually called *photons* (from the Greek for "light").

This notion came as a shock to physicists. The wave theory of light had been established just a hundred years before and for a full century had been winning victory after victory, until Newton's particle theory had been ground into what had seemed complete oblivion. If light consisted of particles after all, what was to be done with all the evidence that pointed incontrovertibly to waves? What was to be done with interference experiments, polarization experiments, and so on?

The answer is that nothing has to be done to them. It is simply wrong to think that an object must be *either* a particle *or* a wave. You might just as well argue that *either* we are head-down and an Australian head-up, *or* we are head-up and an Australian head-down. A photon is *both* a particle *and* a wave, depending on the point of view. (Some physicists, half-jokingly, speak of "wavicles.") In fact, one can be more general than that (as I shall further explain in Volume III of this book) and insist that all the fundamental units of the universe are *both* particles *and* waves.

It is hard for a statement like that to sink in, for the almost inevitable response is: "But how can an object be both a particle and a wave at the same time?"

The trouble here is that we automatically try to think of unfamiliar objects in terms of familiar ones; we describe new phenomena by saying something such as "An atom is like a billiard ball" or "Light waves are like water waves." But this really means only that certain prominent properties of atoms or light waves resemble certain prominent properties of billiard balls and water waves. Not all properties correspond: an atom isn't as large as a billiard ball; a light wave isn't as wet as a water wave.

A billiard ball has both particle and wave properties. However, the particle properties are so prominent and the wave properties so obscure and undetectable that we think of a billiard ball as a particle only. The water wave is also both wave and particle, but here it is the wave properties that are prominent and the particle properties that are obscure. In fact, all ordinary objects are extremely unbalanced in that respect, so we have come to assume that an object must be either a particle or a wave.

The photons of which light is made up happen to be in better balance in this respect, with both wave properties and particle properties quite prominent. There is nothing in ordinary experience among particles and waves to which this can be compared. However, just because we happen to be at a loss for a familiar analogy, we need not think that a wave-particle is "against common sense" or "paradoxical" or, worse still, that "scientists cannot make up their minds."

We may see this more clearly if we consider an indirect analogy. Imagine a cone constructed of some rigid solid such as steel. If you hold such a cone point-upward, level with the eye, you will see its boundary to be a triangle. Holding it in that orientation (point-up), you will be able to pass it through a closely-fitting triangular opening in a sheet of steel, but not through a circular opening of the same area.

Next imagine the cone held point toward you and at eye-level. Now you see its boundary to be that of a circle. In that orientation it will pass through a closely-fitting circular opening in a sheet of steel, but not through a triangular opening of the same area.

If two observers, who were familiar with two-dimensional plane geometry but not with three-dimensional solid geometry, were conducting such experiments, one might hotly insist that the cone was triangular since it could pass through a triangular hole that just fit; the other might insist, just as hotly, that it was a circle, since it could pass through a circular hole that just fit. They might argue thus throughout all eternity and come to no conclusion.

If the two observers were told that both were partly wrong and both partly right and that the object in question had *both* triangular *and* circular properties, the first reaction (based on two-dimensional experience) might be an outraged, "How can an object be both a circle and a triangle?"

However, it is not that a cone *is* a circle and a triangle, but that it has both circular and triangular cross sections, which means that some of its properties are like those of circles and some are like those of triangles.

In the same way, photons are in some aspects wave-like and in others particle-like. The wave-like properties so beautifully demonstrated through the nineteenth century were the result of experiments that served to catch light in its wave-aspect (like orienting the cone properly in order to show it to be a triangle).

The particle-like properties were not so easily demonstrated. In 1901, to be sure, the Russian physicist Peter Nikolaevich Lebedev (1866–1911) demonstrated the fact that light could exert a very slight pressure. A mirror suspended in a vacuum by a thin fiber would react to this pressure by turning, and twisting the fiber. From the slight twist that resulted when a light beam shone on the mirror, the pressure could be measured.

Under some conditions, Lebedev pointed out, radiation pressure could become more important than gravitation. The frozen gases making up the surface of a comet evaporate as the comet approaches the sun, and the dust particles ordinarily held in place by the frozen gas are liberated. These particles are subjected to the comet's insignificant gravitational force and also to the pressure of the sun's tremendous radiation. The unusually large radiation pressure is greater than the unusually weak gravitation, and the dust particles are swept away (in part) by the radiation that is streaming outward in all directions from the sun.

It is this that causes a comet's tail, consisting as it does of light reflected from these dust particles, to be always on the side away from the sun. Thus, a comet receding from the sun is preceded by its tail. This orientation of the comet's tail caused the German astronomer Johannes Kepler (1571–1630) to speculate on the existence of radiation pressure three centuries before that existence could be demonstrated in the laboratory.

The existence of radiation pressure might ordinarily serve as an example of the particle properties of light, since we tend to think of such pressure as resulting from the bombardment of particles as in the case of gas pressure (see page I–199). However, in 1873, Maxwell (who had also worked on the kinetic theory of gases) had shown that there were good theoretical arguments in favor of the fact that light waves might, as waves and not as particles, exert radiation pressure.

A more clear-cut example of particle-like properties was advanced in 1922 by the American physicist Arthur Holly Compton (1892–1962). He found that in penetrating matter an X-ray (a very high-frequency form of light, to be discussed in some detail in Volume III of this book) sometimes struck electrons and not only exerted pressure in doing so, but was itself deflected! In being deflected, the frequency decreased slightly, which meant that the X ray had lost energy. The electron, on the other hand, recoiled in such a direction as to account for the deflection of the X ray, and gained energy just equal to that lost by the X ray. This deflection and energy-transfer was quite analogous to what would have happened if an electron had hit an electron or, for that matter, if a billiard ball had hit a billiard ball. It could not be readily explained by the wave theory. The *Compton effect* clearly demonstrated that an X ray photon could act as a particle. There were good reasons for supposing that the more energetic a photon, the more prominent its particle properties were compared to its wave properties. Therefore, the Compton effect was more easily demonstrated for an X ray photon than for the much less energetic photons of visible light, but the result was taken to hold for all photons. The particle-wave nature of photons has not been questioned since.

Whereas some experiments illuminate the wave properties of

light and some demonstrate its particle properties, no experiment has ever been designed which shows light behaving as both a wave and a particle simultaneously. (In the same way, a cone may be oriented so as to pass through a triangle, or so as to pass through a circle, but not in such a fashion as to pass through both.) The Danish physicist Niels Bohr (1865–1962) maintained that to design an experiment showing light to behave as both a wave and particle simultaneously not only *has not* been done but *cannot* be done in principle. This is called the *principle of complementarity.*

This is not really frustrating to scientists, though it sounds so. We are used to determining the overall shape of a solid object by studying it first from one side and then from another, and then combining, in imagination, the information so gathered. It does not usually occur to us to sigh at the fact that we cannot see an object from all sides simultaneously, or to imagine that only by such an all-sides-at-once view could we truly understand the object's shape. In fact, could we see all sides simultaneously, we might well be confused rather than enlightened, as when we first see a Picasso portrait intended to show a woman both full-face and profile at the same time.

If light is considered as having the properties of both a particle and a wave, there is certainly no need of a luminiferous ether, any more than we need an ether for gravitation or as a standard for absolute motion.

However much light may seem to be a wave form, in its transmission across a vacuum, it is the particle properties that are prominent. The photons stream across endless reaches of vacuum just as Newton had once envisaged his own less sophisticated particles to be doing.

Consequently, once relativity and quantum theory both came into general acceptance—say, by 1920—physicists ceased to be concerned with the ether.

Yet even if we consider light to consist of photons, it remains true that the photons have a wave aspect—that something is still waving. What is it that is waving, and is it anything material at all?

To answer that, let's take up the remaining two phenomena that, since ancient times, have been examples of what seemed to be action at a distance. It will take several chapters to do so, but the answer will eventually be reached.

CHAPTER 24

Magnetism

Magnetic Poles

Forces of attraction between bodies have undoubtedly been observed since prehistoric times, but (according to tradition, at least) the first of the ancient Greeks to study the attractive forces systematically was Thales (640?–546 B.C.).

One attractive force in particular seemed to involve iron and iron ore. Certain naturally occurring types of iron ore ("loadstone") were found to attract iron and, as nearly as the ancients could tell, nothing else. Thales lived in the town of Miletus (on the Aegean coast of what is now Turkey) and the sample of iron ore that he studied purportedly came from the neighborhood of the nearby town of Magnesia. Thales called it "ho magnetes lithos" ("the Magnesian rock") and such iron-attracting materials are now called *magnets,* in consequence, while the phenomenon itself is *magnetism.*

Thales discovered that amber (a fossilized resin called "elektron" by the Greeks), when rubbed, also exhibited an attractive force. This was different from the magnetic force, for whereas magnetism seemed limited to iron, rubbed amber would attract any light object: fluff, feather, bits of dried leaves. In later centuries, objects other than amber were found to display this property when rubbed, and in 1600 the English physician and physicist William Gilbert (1540–1603) suggested that all such objects be called "electrics" (from the Greek word for amber). From this, eventually, the word *electricity* came to be applied to the phenomenon.

Magnetism, while the more restricted force, seemed under the experimental conditions that prevailed in ancient and medieval times to be far the stronger. It was magnetism, therefore, that was the more thoroughly investigated in the two thousand years following Thales.

It was learned, for instance, that the property of magnetism could be transferred. If a sliver of steel is stroked with the naturally occurring magnetic iron ore, it becomes a magnet in its own right and can attract pieces of iron though previously it had not been able to do so.

Furthermore, if such a magnetized needle was placed on a piece of cork and set to floating on water, or if it was pivoted on an axis so that it might freely turn, it was discovered that the needle did not take any position at random, but oriented itself in a specific direction. That direction closely approximates the north-south line. Then, too, if one marks one end of the magnetized needle in some way, as by a notch or a small droplet of paint, it becomes quickly apparent that it is always the same end that points north, while the other end always points south.

Because the ends of the magnetized needle pointed, so it seemed, to the poles of the earth, it became customary to speak of the end that pointed north as the *north pole of the magnet,* and of the other as the *south pole of the magnet.*

It was bound to occur to men that if the north pole of a magnetized needle could really be relied upon to pivot in such a way as always to point north, an unexcelled method of finding direction was at hand. Until then, the position of the North Star by night and the position of the sun by day had been used, but neither would serve except in reasonably clear weather.

The Chinese were supposed to have made use of the magnetized needle as a direction finder in making their way across the trackless vastness of Central Asia. However, the first uses of the needle in ocean voyages are recorded among Europeans of the twelfth century. The needle was eventually mounted on a card on which the various directions were marked off about the rim of a circle. Since the directions encompassed the rim of the card, the magnetized needle came to be called a *compass.*

There is no doubt that the compass is one of those simple inventions that change the world. Men could cross wide tracts of ocean without a compass (some two thousand years ago the Polynesians colonized the islands that dotted the Pacific Ocean without the help of one), but a compass certainly helps. It is probably no accident that it was only after the invention of the compass that European seamen flung themselves boldly out into the Atlantic Ocean, and the "Age of Exploration" began.

The poles of a magnet are distinguished by being the points at which iron is attracted most strongly. If a magnetized needle is dipped into iron filings and then lifted free, the filings will cluster most thickly about the ends. In this sense, a magnet of whatever shape has poles that can be located in this manner. Nor do poles occur singly. Whenever a north pole can be located, a south pole can be located, too, and vice versa.

Nor is it difficult to tell which pole is the north and which the south, even without directly making a compass of the magnet. Suppose that two magnetized needles have been allowed to orient themselves north-south and that the north pole of each is identified. If the north pole of one magnet is brought near the south pole of the second magnet, the two poles will exhibit a mutual attraction, and if allowed to touch, will remain touching. It will take force to separate them.

On the other hand, if the north pole of one magnet is brought near the north pole of the other, there will be a mutual repulsion. The same is true if the south pole of one is brought near the south pole of the other. If the magnets are free to pivot about, they will veer away and spontaneously reorient themselves so that the north pole of one faces the south pole of the other. If north pole is forced against north pole or south pole against south pole, there will be a separation as soon as the magnets are released. It takes force to keep them in contact.

We might summarize this by saying: *Like poles repel; unlike poles attract.*

Once the north pole of a particular magnet has been identified, then, it can be used to identify the poles of any other magnet. Any pole to which it is attracted is a south pole. Any pole by which it is repelled is a north pole. This was first made clear as long ago as 1269 by one of the few experimentalists of the middle ages, the Frenchman Peter Peregrinus.

(In view of this, it might have been better if the north poles of magnets, attracted as they are in the direction of the North Pole, had been called south poles. However, it is too late to do anything about that now.)

It is easy to see that the force exerted by a magnetic pole varies inversely with difference. If one allows a north pole to approach a south pole, one can feel the force of attraction grow stronger. Similarly, if one pushes a north pole near another north pole, one can feel the force of repulsion grow stronger. The smaller the distance, the greater the force.

Of course, we cannot speak of a north pole by itself and a south pole by itself. Every north pole is accompanied by its south pole. Therefore, if a north pole of Magnet A is attracting the south pole of Magnet B, the south pole of Magnet A must be simultaneously repelling the south pole of Magnet B. This tends to complicate the situation.

If one uses long, thin magnets, however, this source of complication is minimized. The north pole of Magnet A is close to the south pole of Magnet B, while the south pole of Magnet A (at the other end of a long piece of metal) is considerably farther away. The south pole's confusing repulsive force is weakened because of this extra distance and may be the more safely ignored.

In 1785, the French physicist Charles Augustin de Coulomb (1736–1806) measured the force between magnetic poles at varying distances, using a delicate torsion balance for the purpose. Thus, if one magnetic needle is suspended by a thin fiber, the attraction (or repulsion) of another magnet upon one of the poles of the suspended needle will force that suspended needle to turn slightly. In doing so, it will twist the fiber by which it is suspended. The fiber resists further twisting by an amount that depends on how much it has already been twisted. A given force will always produce a given amount of twist, and from that amount of twist the size of an unknown force can be calculated. (Fifteen years later, Cavendish used such a balance to measure weak gravitational forces, see page I–50; a century later still, Lebedev used one to detect radiation pressure—see page 137.)

On making his measurements, Coulomb found that magnetic force varied inversely as the square of the distance, as in the case of gravitation. That is, the strength of the magnetic force fell to one-fourth its value when the distance was increased twofold, and the force increased to nine times its value when the distance was decreased to one-third its previous value. This held true whether the force was one of attraction or repulsion.

This can be expressed mathematically as follows: If the magnetic force between the poles is called F, the strength of the two poles, m and m', and the distance between the two poles, d, then:

$$F = \frac{mm'}{d^2} \qquad \text{(Equation 9–1)}$$

If the distance is measured in centimeters, then the force is determined in dynes (where 1 dyne is defined as 1 gram-centimeter per second per second, see page I–32). Suppose, then, that two poles of equal intensity are separated by a distance of 1 centimeter and that the force of magnetic attraction equals 1 dyne. It turns out then that $m = m'$, and therefore, $mm' = m^2$. Then, since both F and d have been set equal to 1, it follows from Equation 9–1 that under those conditions $m^2 = 1$ and, therefore, $m = 1$.

Consequently, one speaks of *unit poles* as representing poles of such strength that on being separated by 1 centimeter they exert a magnetic force (of either attraction or repulsion) of 1 dyne. In Equation 9–1, where F is measured in dynes and d in centimeters, m and m' are measured in unit poles.

If a magnetic pole of 5 unit poles exerts a force of 10 dynes on a unit pole at a certain point, the intensity of the magnetic force is 2 dynes per unit pole. One dyne per unit pole is defined as 1 *oersted* (in honor of the Danish physicist Hans Christian Oersted, whose contribution to the study of magnetism will be discussed on page 201). The oersted is a measure of magnetic force per unit pole or *magnetic intensity*, which is usually symbolized as H. We can say, then, that $H = F/m$, or:

$$F = mH \qquad \text{(Equation 9–2)}$$

where F is magnetic force measured in dynes, m is magnetic strength in unit poles, and H is magnetic intensity in oersteds.

Magnetic Domains

In the existence of both north and south poles, and in the consequent existence of magnetic repulsion as well as magnetic attraction, there is a key difference between magnetism and gravitation. The gravitational force is entirely one of attraction, and no corresponding force of gravitational repulsion has yet been discovered.

For that reason, gravitational force is always at its ideal maximum without the existence of any neutralizing effects. A body possessing the mass of the earth will exert a fixed gravitational attraction whatever its temperature or chemical constitution.

On the other hand, magnetic attraction can always be neutralized to a greater or lesser extent by magnetic repulsion, so magnetic effects will occur only in certain kinds of matter and then in widely varying strengths.

One might suppose (and, as we shall see in Volume III of this book, the supposition is correct) that magnetism is widespread in nature and that magnetic forces exist in all matter. Matter might then be considered to consist of submicroscopic magnets. A point in favor of this view (at least in the case of iron and steel) is the fact, discovered early, that if a long magnetic needle is broken in two, both halves are magnets. The broken end opposite the original north pole becomes a south pole; the broken end opposite the original south pole becomes a north pole. This is repeated when each half is broken again and again. It is easy to imagine the original needle broken into submicroscopic lengths, each of which is a tiny magnet, each with a north pole and south pole.

These submicroscopic magnets, in most substances and under most conditions, would be oriented randomly, so there is little or no concentration of north poles (or south poles) in any one direction and therefore little or no detectable overall magnetic force. In some naturally occurring substances, however, there would be a tendency for the submicroscopic magnets to line up, at least to a certain extent, along the north-south line. There would then be a concentration of north poles in one direction and of south poles in the other; enough of a concentration to give rise to detectable magnetic force.

If, let us say, the north pole of such a magnet is brought near iron, the submicroscopic magnets in the iron are oriented in such a way that the south poles face the magnet and the north poles face away. The iron and the magnet then attract each other. If it is the south pole of the magnet that is brought near the iron, then the submicroscopic magnets in the iron are oriented in the opposite fashion, and again there is attraction. Either pole of a magnet will, for that reason, attract iron. While the iron is near the magnet or in contact with it so that its own magnetic components are oriented, it is itself a magnet. The process whereby iron is made a magnet by the nearness of another magnet is *magnetic induction*. Thus, a paper clip suspended from a magnet will itself attract a second paper clip which will attract a third, and so on. If the magnet is removed, all the paper clips fall apart.

Ordinarily, the submicroscopic magnets in iron are oriented with comparative ease under the influence of a magnet and are disoriented with equal ease when the magnet is removed. Iron usually forms a *temporary magnet*. If a sliver of steel is subjected to the action of a magnet, however, the submicroscopic magnets within the steel are oriented only with considerably greater difficulty. Once the magnet is removed from the steel, however, the disorientation is equally difficult—difficult enough not to take place under ordinary conditions, in fact; therefore, steel generally remains a *permanent magnet*.

Nor is it iron only that is composed of submicroscopic magnets, for it is not only iron that is attracted to a magnet. Other metals, such as cobalt and nickel (which are chemically related to iron) and gadolinium (which is not) are attracted by a magnet. So are a number of metal alloys, some of which contain iron and some of which do not. Thus, Alnico, which as the name implies is made up of aluminum, nickel and cobalt (plus a little copper), can be used to make more powerful magnets than those of steel. On the other hand, stainless steel, which is nearly three-fourths iron, is not affected by a magnet.

Nor need the magnetic substance be a metal. Loadstone itself is a variety of iron oxide, an earthy rather than a metallic substance. Since World War II, a whole new class of magnetic substances have been studied. These are the *ferrites*, which are mixed oxides of iron and of other metals such as cobalt or manganese.

A material that displays, or that can be made to display, a strong magnetic force of the sort we are accustomed to in an ordinary magnet is said to be *ferromagnetic*. (This is from the Latin "ferrum" meaning "iron," since iron is the best-known example of such a substance.) Nickel, cobalt, Alnico, and, of course, iron and steel are examples of ferromagnetic substances.

The question arises, though, why some materials are ferromagnetic and some are not. If magnetic forces are a common property of all matter (as they are), why cannot the submicroscopic magnets of pure copper or pure aluminum, for instance, be stroked into alignment by an already-existing magnet? Apparently, this alignment cannot be imposed from outside unless the substance itself cooperates, so to speak.

In ferromagnetic substances (but only under appropriate conditions even in those) there is already a great deal of alignment existing in a state of nature. The submicroscopic magnets tend to orient themselves in parallel fashion by the billions of billions, producing concentrations of north and south poles here and there within the iron. The regions over which magnetic forces are thus concentrated are called *magnetic domains*.

Iron and other ferromagnetic substances are made up of such magnetic domains, each of which is actually on the border of visibility. A finely divided powder of magnetic iron oxide spread over iron will tend to collect on the boundaries between adjacent domains and make them visible to the eye.

Despite the existence of these domains, iron as ordinarily produced is not magnetic. That is because the domains themselves are oriented in random fashion so that the magnetic force of one is neutralized by those of its neighbors. Therefore, stroking with an ordinary magnet does not orient the submicroscopic magnets themselves (this is beyond its power); it merely orients the domains. Thus, the ferromagnetic material has itself done almost all the work of alignment to begin with, and man proceeds to add one final touch of alignment, insignificant in comparison to that which is already done, in order to produce a magnet.

If a ferromagnetic substance is ground into particles smaller than the individual domains making it up, each particle will tend to consist of a single domain, or of part of one. The submicroscopic magnets within each will be completely aligned. If such a powder is suspended in liquid plastic, the domains can be aligned by the influence of a magnet easily and thoroughly by bodily rotation of the particles against the small resistance of the liquid (rather than against the much greater resistance of the iron itself in the solid state). By allowing the plastic to solidify while the system is still under the influence of the magnet, the domains will be permanently aligned, and particularly strong magnets will have been formed. Furthermore, such magnets can be prepared in any shape and can be easily machined into other shapes.

Anything which tends to disrupt the alignment of the domains will weaken or destroy the magnetic force of even a "permanent" magnet. Two magnets laid parallel, north pole to north pole and south pole to south pole, will, through magnetic repulsion, slowly turn the domains away, ruining the alignment and weaking the magnetic force. (That is why magnets should always be stacked north-to-south.) From a more mechanical standpoint, if a magnet is beaten with a hammer, the vibration will disrupt alignment and weaken the magnetic force.

In particular, increased atomic vibration, caused by a rise in temperature (see page I–203), will disrupt the domains. In fact, for every ferromagnetic substance there is a characteristic temperature above which the alignment of the domains is completely disrupted and above which, therefore, the substance will show no ferromagnetic properties.

This was first demonstrated by the French physicist Pierre Curie (1859–1906) in 1895, and is therefore called the *Curie point*. The Curie point is usually below the melting point of a substance, so liquids are generally not ferromagnetic. The Curie point for iron, for instance, is 760°C whereas its melting point is 1539°C. For cobalt the Curie point is considerably higher, 1130°C, while for gadolinium it is considerably lower, 16°C. Gadolinium is only ferromagnetic at temperatures below room temperature. The Curie point may be located at very low temperatures indeed. For the metal dysprosium, its value is about –188°C (85°K), so it is only at liquid air temperatures that dysprosium forms domains and becomes ferromagnetic.

In some substances, the submicroscopic magnets spontaneously align themselves—but not with north poles pointing all in the same direction. Rather, the magnets are aligned in parallel fashion but with north poles pointing in one direction in half the cases and in the other direction in the remainder. Such substances are *antiferromagnetic,* and because the magnetic forces of one type of alignment are neutralized by those of the other, the overall magnetic force is zero. It may be, however, that the structure of the substance is such that the magnets with north poles pointing in one direction are distinctly stronger than those with north poles pointing in the other. In this case, there is a considerable residual magnetic force, and such substances are called *ferrimagnetic.* (Note the difference in the vowel.)

The ferrites are examples of ferrimagnetic materials. Naturally, a ferrimagnetic material cannot be as strongly magnetic as a ferromagnetic material would be, since in the latter all the domains are, ideally, pointing in the same direction, while in the former a certain amount of neutralization takes place. Thus, ferrites display only about a third the maximum strength that a steel magnet would display.

The Earth as a Magnet

The manner in which a compass needle pointed north-and-south was a tantalizing fact to early physicists. Some speculated that a huge iron mountain existed in the far north and that the magnetized needle was attracted to it. In 1600, the English physicist William Gilbert (1540–1603) reported systematic experiments that led to a more tenable solution.

A compass needle, as ordinarily pivoted, can rotate only about a vertical axis and is constrained to remain perfectly horizontal. What if it is pivoted about a horizontal axis and can, if conditions permit, point upward or downward? A needle so pivoted (in the northern hemisphere) does indeed dip its north pole several degrees below the horizon and toward the ground. This is called *magnetic dip*.

Gilbert shaped a loadstone into a sphere and used that to stand for the earth. He located its poles and decided its south pole, which attracted the north pole of a compass needle, would be equivalent to the earth's Arctic region, while the other would be equivalent to the Antarctic.

The north pole of a compass needle placed in the vicinity of this spherical loadstone pointed "north" as might be expected. In the loadstone's "northern hemisphere," however, the north pole of a compass needle, properly pivoted, also showed magnetic dip, turning toward the body of the loadstone. Above the south pole in the loadstone's "Arctic region," the north pole of the compass needle pointed directly downward. In the loadstone's "southern hemisphere," the north pole of the compass needle angled up away from the body of the loadstone and above its "Antarctic region" pointed straight upward.

Gilbert felt that the behavior of the compass needle with respect to the earth (both in its north-south orientation and in its magnetic dip) was strictly analogous to its behavior with respect to the loadstone. He drew the conclusion that the earth was itself a spherical magnet with its poles in the Arctic and Antarctic. The compass needle located the north by the same force that attracted it to the pole of any other magnet. (It is this natural magnetism of the earth that slowly orients the domains in certain types of iron oxide and creates the magnetized loadstone—from which all magnetic studies previous to the nineteenth century have stemmed.)

It might easily be assumed that the earth's magnetic poles are located at its geographic poles, but this is not so. If they were, the compass needle would point to more or less true north, and it does not. In Gilbert's time (1580), for instance, the compass needle in London pointed 11° east of north. The angle by which the needle deviates from true north is the *magnetic declination*. It varies from place to place on the earth, and in any given place varies from year to year. The magnetic declination in London is now 8° west of north, and since Gilbert's time, the declination has been as great as 25° west of north. In moving from an eastward declination in the sixteenth century to a westward one now, there had to be a time when declination was, temporarily, zero, and when the compass needle, in London, pointed due north. This was true in 1657.

The variation in declination with change in position was first noted by Christopher Columbus (1451–1506) in his voyage of discovery in 1492. The compass needle, which had pointed distinctly east of north in Spain, shifted westward as he crossed the Atlantic Ocean, pointed due north when he reached midocean, and distinctly west of north thereafter. He kept this secret from his crew, for they needed only this clear example of what seemed the subversion of natural law to panic and mutiny.

The existence of magnetic declination and its variation from spot to spot on the earth's surface would be explained if the magnetic poles were located some distance from the geographic poles. This is actually so. The south pole of the earth-magnet (which attracts the north pole of the compass needle) is located in the far north and is called, because of its position, the *north magnetic pole*. It is now located just off Canada's Arctic shore, some 1200 miles from the geographic north pole. The *south magnetic pole* (the north pole of the earth-magnet) is located on the shores of Antarctica, west of Ross Sea, some 1200 miles from the geographic south pole.

The two magnetic poles are not quite on opposite sides of the earth, so the line connecting them (the *magnetic axis*) is not only at an angle of about 18° to the geographic axis, but also does not quite pass through the earth's center.

The fact that magnetic declination alters with time seems to

Earth's magnetic poles

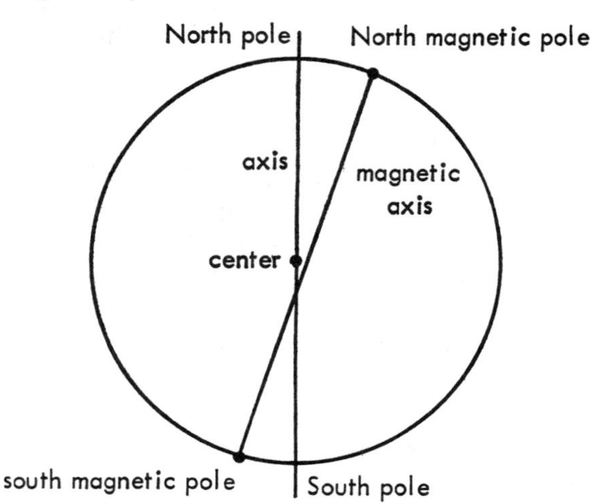

indicate that the magnetic poles change their position, and indeed the position of the north magnetic pole has shifted by several degrees from the time, one century ago, when it was first located.

For all its size, the earth is but a weak magnet. Thus, in even a small horseshoe magnet, the magnetic intensity between the poles can be as high as 1000 oersteds, yet the intensity of the earth's magnetism is only about 3/4 oersted even near the magnetic poles, where the intensity is highest. It sinks to 1/4 oersted at points equidistant from the magnetic poles (the *magnetic equator*).

Lines can be drawn on the face of the earth through points showing equal declination. These are called *isogonic lines* (from Greek words meaning "equal angles"). Ideally, they might be considered lines of "magnetic longitude." However, unlike geographic longitude, they are by no means arcs of great circles but curve irregularly in accordance with local magnetic properties of the earth's structure. And, of course, they change with time and must be constantly redrawn.

If it is agreed that the earth is a magnet, it yet remains to determine why it is a magnet. By the latter half of the nineteenth century, it came to seem more and more likely from several converging lines of evidence that the earth had a core of nickel-iron making up one third of its mass. Nothing seemed more simple than to suppose that this core was, for some reason, magnetized. However, it also seemed more and more likely that temperatures at the earth's core were high enough to keep that nickel-iron mass liquid and well above the Curie point. Therefore, the core cannot be an ordinary magnet, and the earth's magnetism must have a more subtle origin. This is a subject to which I will return.

Magnetic Fields

Magnetic force may fall off with distance according to an inverse square law (see Equation 9–1) as gravitational force does, but there are important differences. As far as we know, gravitational force between two bodies is not in the least affected by the nature of the medium lying between them. In other words, your weight is the same whether you stand on bare ground or insert an iron plate, a wooden plank, a foam rubber mattress or any other substance between yourself and the bare ground. For that matter, the sun's attraction of the earth is not altered when the 2000-mile thickness of the moon slips between the two bodies.

The force between magnetic poles does, however, alter with the nature of the medium between them, and Equation 9–1 holds strictly only where there is a vacuum between the poles. To explain this, we must consider the researches of the English scientist Michael Faraday (1791–1867).

In 1831, he noticed something that had been noticed over five centuries earlier by Peter Peregrinus and, undoubtedly, by numerous other men who had toyed with magnets through the centuries. . . . Begin with a sheet of paper placed over a bar magnet. If iron filings are sprinkled on the paper and the paper is then jarred so that the filings move about and take up some preferred orientation, those filings seem to follow lines that curve from one pole of the magnet to the other, crowding together near the poles and spreading apart at greater distances. Each line begins at one pole and ends at the other, and no two lines cross. (Of course, some of the lines seem incomplete because they run off the paper or because at great distances from the poles they are too weak to be followed accurately by the iron filings. Nevertheless it is reasonable to suppose that all the lines, however far they may have to sweep out into space and however weak they get, are nevertheless continuous from pole to pole.)

The shape of the lines depends on the shape of the magnet

Magnetic lines of force

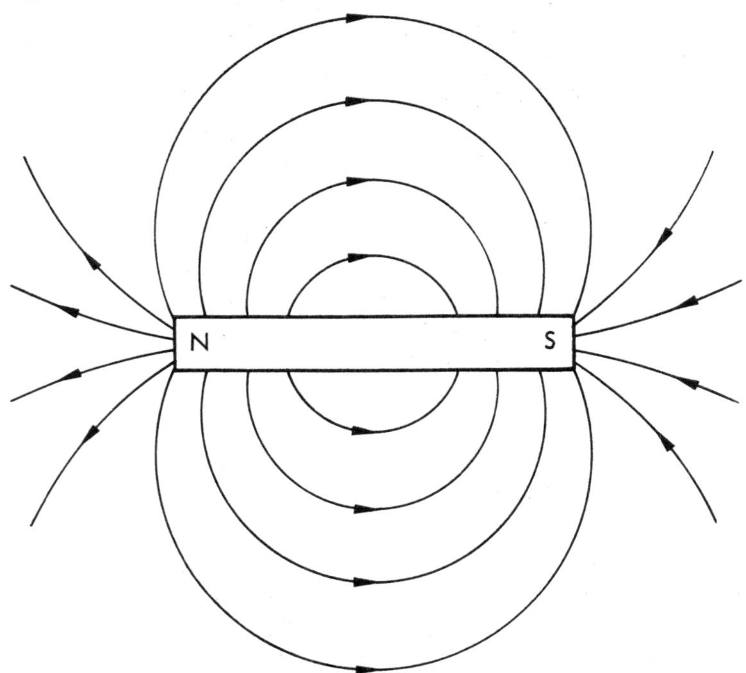

and the mutual relationship of the poles. In a horseshoe magnet, the lines cluster about the two poles and in the space between them are straight. The same is true if a north pole of one bar magnet is brought near the south pole of another. On the other hand, if the north pole of one bar magnet is brought near the north pole of another bar magnet, the lines of force curve away, those from one pole seeming to avoid touching those of the other.

Faraday called these *magnetic lines of force* and believed them to have real existence. He felt they were made of some elastic material that was stretched when extended between two unlike poles and that exerted a force tending to shorten itself, as an extended rubber band would do. It was this shortening tendency that accounted for magnetic attraction, according to Faraday.

The lines of force about a magnet of any shape, or about any system of magnets, can be mapped without the use of iron filings. A compass needle always orients itself in such a way as to lie along one of these lines. Therefore, by plotting the direction of the compass needle at various points in space, one can succeed in mapping the lines. In this way, the shape of the lines of force about the earth-magnet can be determined.

Faraday's view of the material existence of lines of force did not survive long. By mid-nineteenth century, the ether concept had grown strong in connection with light (see page 88ff.), and the magnetic lines of force came to be viewed as distortions of the ether.

With the disappearance of the ether concept at the start of the twentieth century, a further step had to be taken. Once again, it became a matter of the geometry of space itself. Suppose, for instance, you dropped a pencil into a cylindrical hollow. It would automatically orient itself parallel to the axis of the cylinder. If the cylinder were a tube that extended for miles, curving gently this way or that, a pencil dropped into it at any point would orient itself parallel to the axis of the tube, whatever the direction happened to be in that place. In fact, if you could not see the tube, but only the pencil, you could easily map the curves and sinuosities of the tube by noting the position taken up by the pencil at various points. The same is true of the compass needle and the magnetic lines of force.

Each magnetic pole affects the geometry of all of space, and this altered geometry (as compared with what the geometry would be in the absence of the magnetic pole) is called a *magnetic field.* The intensity of the magnetic field (the extent to which its geometry differs from ordinary nonmagnetic geometry of space) drops off as the square of the distance from the pole and soon becomes too small to detect. Nevertheless, the magnetic field of every magnetic pole in existence fills all of space, and the situation is made bearably simple only by the fact that in the immediate neighborhood of any given pole its field predominates over all others to such an extent that it may be considered in isolation. (This is true, of course, of gravitational fields as well.)

The concept of a magnetic field removes the necessity of supposing the magnetic force to represent action-at-a-distance. It is not that a magnet attracts iron over a distance but that a magnet gives rise to a field that influences a piece of iron within itself. The field (that is the space geometry it represents) touches both magnet and iron and no action-at-a-distance need be assumed.

Despite the fact that magnetic lines of force have no material existence, it is often convenient to picture them in literal fashion and to use them to explain the behavior of objects within a magnetic field. (In doing so, we are using a "model"—that is, a representation of the universe which is not real, but which aids thinking. Scientists use many models that are extremely helpful. The danger is that there is always the temptation to assume, carelessly, that the models are real, so they may be carried beyond their scope of validity. There may also arise an unconscious resistance to any changes required by increasing knowledge that cannot be made to fit the model.)

We can define the lines of force between two magnetic poles in the cgs system (using measurements in centimeters and dynes) in such a way that one line of force can be set equal to 1 *maxwell* (in honor of Maxwell, who did so much in connection with both gases and light). In the mks system, where the same measurements are made in meters and newtons, a line of force is set equal to 1 *weber* (in honor of the German physicist Wilhelm Eduard Weber [1804–1891]). The weber is much the larger unit, 1 weber being equal to 100,000,000 maxwells. Maxwells and webers are units of *magnetic flux,* a measurement you can imagine as representing the number of lines of force passing through any given area drawn perpendicularly to those lines.

What counts in measuring the strength of a magnetic field is the number of lines of force passing through an area of unit size. That is the *magnetic flux density*. The flux density measures how closely the lines of force are crowded together; the more closely they crowd, the higher the flux density and the stronger the magnetic field at that point. In the cgs system, the unit area is a square centimeter so that the unit of flux density is 1 maxwell per square centimeter. This is called 1 *gauss,* in honor of the German mathematician Karl Friedrich Gauss (1777–1855).* In the mks system, the unit area is a square meter and the unit of flux density is therefore 1 weber per square meter, a unit which has no special name. Since there are 10,000 square centimeters in a square meter and 100,000,000 maxwells in a weber, 1 weber per square meter is equal to 10,000 gausses.

Imagine a magnetic north pole and south pole separated by a vacuum. Lines of force run from pole to pole and the flux density at any given point between them would have a certain value, depending on the strength of the magnet. Is some material substance were placed between the poles then, even though the strength of the magnet were left unchanged, the flux density would change. The ratio of the flux density through the substance to that through a vacuum is called the *relative magnetic permeability*. Since this is a ratio, it is a pure number and has no units.

The permeability of a vacuum is set at 1 and for most material substances, the permeability is very nearly 1. Nevertheless, refined measurements show that it is never exactly 1, but is sometimes a little larger than 1 and sometimes a little smaller. Those substances with permeability a little larger than 1 are said to be *paramagnetic,* while those with permeability a little smaller than 1 are *diamagnetic*.

In a paramagnetic substance, with a permeability higher than 1, the flux density is higher than it would be in a vacuum. The lines of force crowd into the paramagnetic substance, so to speak, seeming to prefer it to the surrounding vacuum (or air). A paramagnetic substance therefore tends to orient itself with its longest axis parallel to the lines of force so that those lines of force can move through its preferred substance over a maximum distance. Again, since the flux density increases as one approaches a pole, there is a tendency for the paramagnetic substance to approach the pole (that is, to be attracted to it) in order that as many lines of force as possible can pass through it.

On the other hand, a diamagnetic substance, with a permeability of less than 1, has a flux density less than that of a vacuum (or air). The lines of force seem to avoid it and to crowd into the surrounding vacuum. Therefore a diamagnetic substance tends

* Most electrical and magnetic units are named in honor of scientists noted for their work in the field. Gauss and Weber established the first logical system of units in electricity and magnetism. Maxwell's contribution will be discussed on page 236 ff.

to orient itself with its longest axis perpendicular to the lines of force so that these lines of force need pass through the substance over only a minimum distance. Furthermore, the diamagnetic substance tends to move away from the pole (that is, be repelled by it) into a region of lower flux density so that as few lines as possible need pass through it.

Both effects are quite small and become noticeable only when very strong magnetic fields are used. The first to record such effects was Faraday, who found in 1845 that glass, sulfur and rubber were slightly repelled by magnetic poles and were therefore diamagnetic. The most diamagnetic substance known, at ordinary temperatures, is the element bismuth. (At extremely low temperatures, near that of absolute zero, the permeability of some substances drops to zero and diamagnetism is then at a maximum.)

Paramagnetism is considerably more common, and for a few substances permeability may be very high, even in the thousands. These high-permeability substances are those previously referred to as ferromagnetic. Here the attraction of the magnet and the orientation of iron filings parallel to the lines of force is so marked that it is easily noted.

Permeability (symbolized by the Greek letter "mu," μ) must be included in Coulomb's equation (Equation 9–2) to cover those cases where the poles are separated by more than vacuum:

$$F = \frac{mm'}{\mu d^2} \qquad \text{(Equation 9–3)}$$

Since μ is in the denominator, an inverse relationship is indicated. A diamagnetic substance with a permeability of less than 1 increases the magnetic force between poles, while a paramagnetic substance decreases the force. The latter effect is particularly marked when iron or steel, with permeabilities in the hundreds and even thousands, is between the poles. A bar of iron over both poles of a horseshoe magnet cuts down the magnetic force outside itself to such an extent that it almost acts as a magnetic insulator.

CHAPTER 25

Electrostatics

Electric Charge

Gilbert, who originated the earth-magnet idea, also studied the attractive forces produced by rubbing amber. He pivoted a light metal arrow so delicately that it would turn under the application of a tiny force. He could, in that way, detect very weak attractive forces and proceeded to find substances other than amber which, on rubbing, would produce such forces. Beginning in 1570, he found that a number of gems such as diamond, sapphire, amethyst, opal, carbuncle, jet, and even ordinary rock crystal produced such attractive forces when rubbed. He called these substances "electrics." A substance showing such an attractive force was said to be *electrified* or to have gained an *electric charge*.

A number of substances, on the other hand, including the metals in particular, could not be electrified and hence were "non-electrics."

Eventually, electricity came to be considered a fluid. When a substance like amber was electrified, it was considered to have gained electric fluid which then remained there stationary. Such a charge was called *static electricity*, from a Latin word meaning "to be stationary," and the study of the properties of electricity under such conditions is called *electrostatics*.

Before electric forces could be studied easily, the fluid had to be concentrated in sizable quantities—in greater quantity than could be squeezed into small bits of precious and semiprecious materials. Some "electric" that was cheap and available in sizable quantities had to be found.

In the 1660's, the German physicist Otto von Guericke (1602–1686) found such a material in sulfur. He prepared a sphere of sulfur, larger than a man's head, arranged so that it could be turned by a crank. A hand placed on it as it turned gradually electrified it to a hitherto unprecedented extent. Guericke had constructed the first *electrical friction machine.*

Using it, Guericke discovered several similarities between electrostatic forces and magnetic forces. He found, for instance, that there was electrostatic repulsion as well as electrostatic attraction, just as there is both repulsion and attraction in the case of magnets. Again, a substance brought near the electrified sulfur itself exhibited temporary electrification, just as a piece of iron held near a magnet becomes itself temporarily magnetized. Thus there is *electrostatic induction* as well as magnetic induction.

In 1729, an English electrician, Stephen Gray (1696–1736), electrified long glass tubes and found that corks placed into the ends of the tubes, as well as ivory balls stuck into the corks by long sticks, became electrified when the glass itself was rubbed. The electric fluid, which came into being at the point of rubbing, must obviously spread throughout the substance, through the cork and the stick into the ivory, for instance. This was the first clear indication that electricity need not be entirely static but might move.

While the electric fluid, once it was formed within an "electric" by rubbing, might spread outward into every part of the substance, it would not pass bodily through it, entering at one point, for instance, and leaving at another. It was otherwise in the case of "nonelectrics," where such bodily passage did take place. Indeed, the flow of electric fluid took place extremely readily through substances like metals; so readily that a charged electric lost its charge altogether—was *discharged*—if it were brought into contact with metal that was in turn in contact with the ground. The fluid passed from the electric, via the metal, into the capacious body of the earth, where it spread out so thinly that it could no longer be detected.

That seemed to explain why metals could not be electrified by rubbing. The electric fluid, as quickly as it was formed, passed through the metal into almost anything else the metal touched. Gray placed metals on blocks of resin (which did not allow a ready passage to the electric fluid). Under such circumstances, pieces of metal, if carefully rubbed, were indeed electrified, for the fluid formed in the metal, unable to pass readily through the resin, was trapped, so to speak, in the metal. In short, as it eventually turned out, electric forces were universally present in matter, just as magnetic forces were.

As a result of Gray's work, matter came to be divided into two classes. One class, of which the metals—particularly gold, silver, copper and aluminum—were the best examples, allowed the passage of the electric fluid with great readiness. These are *electrical conductors*. The other group, such as amber, glass, sulfur, and rubber—just those materials that are easily electrified by rubbing—presents enormous resistance to the flow of electric fluid. These are *electrical insulators* (from a Latin word for "island," because such a substance can be used to wall off electrified objects, preventing the fluid from leaving and therefore making the objects an island of electricity, so to speak.)

Ideas concerning electrostatic attraction and repulsion were sharpened in 1733 by the French chemist Charles François Du Fay (1698–1739). He electrified small pieces of cork by touching them with an already electrified glass rod, so some of the electric fluid passed from the glass into the cork. Although the glass rod had attracted the cork while the latter was uncharged, rod and cork repelled each other once the cork was charged. Moreover, the two bits of cork, once both were charged from the glass, repelled each other.

The same thing happened if two pieces of cork were electrified by being touched with an already electrified rod of resin. However, a glass-electrified piece of cork attracted a resin-electrified piece of cork.

It seemed to Du Fay, then, that there were two types of electric fluid, and he called these "vitreous electricity" (from a Latin word for "glass") and "resinous electricity." Here, too, as in the case of the north and south poles of magnets, likes repelled and unlikes attracted.

This theory was opposed by Benjamin Franklin. In the 1740's, he conducted experiments that showed quite clearly that a charge of "vitreous electricity" could neutralize a charge of "resinous electricity," leaving no charge at all behind. The two types of electricity were therefore not merely different; they were opposites.

To explain this, Franklin suggested that there was only one electrical fluid, and all bodies possessed it in some normal amount.

When this fluid was present in its normal amount, the body was uncharged and showed no electrical effects. In some cases, as a result of rubbing, part of the electrical fluid was removed from the material being rubbed; in other cases it was added to the material. Where the body ended with an excess of the fluid, Franklin suggested, it might be considered *positively charged,* and where it ended with a deficit, it would be *negatively charged.* A positively-charged body would attract a negatively-charged body as the electric fluid strove (so to speak) to even out its distribution, and on contact, the electric fluid would flow from its place of excess to the place of deficiency. Both bodies would end with a normal concentration of the fluid, and both bodies would therefore be discharged.

On the other hand, two positively-charged bodies would repel each other, for the excess fluid in one body would have no tendency to add to the equal excess in the other—rather the reverse. Similarly, two negatively-charged bodies would repel each other.

Electrostatic induction was also easily explained in these terms. If a positively-charged object was brought near an uncharged body, the excess of fluid in the first would repel the fluid in the second and drive it toward the farthest portion of the uncharged body, leaving the nearest portion of that body negatively charged and the farthest portion positively charged. (The uncharged body would still remain uncharged, on the whole, for the negative charge on one portion would just balance the positive charge of the other.)

There would now be an attraction between the positively-charged body and the negatively-charged portion of the uncharged body. There would also be a repulsion between the positively-charged body and the positively-charged portion of the uncharged body. However, since the positively-charged portion of the uncharged body is farther from the positively-charged body than the negatively-charged portion is, the force of repulsion is weaker than the force of attraction, and there is a net attractive force.

The same thing happens if a negatively-charged body approaches an uncharged body. Here the electrical fluid in the uncharged body is drawn toward the negatively-charged body. The uncharged body has a positively-charged portion near the negatively-charged body (resulting in a strong attraction) and a negatively-charged portion farther from the negatively-charged body (resulting in a weaker repulsion). Again there is a net attractive force. In this way, it can be explained why electrically charged bodies of either variety attract uncharged bodies with equal facility.

Franklin visualized a positive charge and negative charge as being analogous to the magnetic north and south poles, as far as attraction and repulsion was concerned. There is an important difference, however. The magnetism of the earth offered a standard method of differentiating between the magnetic poles, depending upon whether a particular pole pointed north or south. No such easy way of differentiating a positive charge from a negative charge existed.

A positive charge, according to Franklin, resulted from an excess of electric fluid, but since there is no absolute difference in behavior between "vitreous electricity" and "resinous electricity," how could one tell which electric charge represents a fluid excess and which a fluid deficit? The two forms differ only with reference to each other.

Franklin was forced to guess, realizing full well that his chances of being right were only one in two—an even chance. He decided that glass, when rubbed, gained electric fluid and was positively charged; on the other hand, when resin was rubbed, it lost electric fluid and was negatively charged. Once this was decided upon, all electric charges could be determined to be either positive or negative, depending on whether they were attracted or repelled by a charge that was already determined to be either positive or negative.

Ever since Franklin's day, electricians have considered the flow of electric fluid to be from the point of greatest positive concentration to the point of greatest negative concentration, the process being pictured as analogous to water flowing downhill. The tendency is always to even out the unevenness of charge distribution, lowering regions of excess and raising regions of deficit.

Franklin's point of view implies that electric charge can neither be created nor destroyed. If a positive charge is produced by the influx of electric fluid, that fluid must have come from somewhere else, and a deficit must exist at the point from which it came. The deficit produced at its point of origin must exactly equal the excess produced at its point of final rest. Thus, if glass is rubbed with silk and if glass gains a positive charge, the silk gains an equal negative charge. The net electric charge in glass-plus-silk was zero before rubbing and zero after.

This view has been well substantiated since the days of Franklin, and we can speak of the *law of conservation of electric charge.*

We can say that net electric charge can neither be created nor destroyed; the total net electric charge of the universe is constant. We must remember that we are speaking of *net* electric charge. The neutralization of a quantity of positive electric charge by an equal quantity of negative electric charge is not the destruction of electric charge. The sum of $+x$ and $-x$ is 0, and in such neutralization it is not the net charge that has changed, only the distribution of charge. The same is true if an uncharged system is changed into one in which one part of the system contains a positive charge and another part an equal negative charge. This situation is exactly analogous to that involved in the law of conservation of momentum (see page I–67).

Electrons

Actually, both Du Fay's two-fluid theory of electricity and Franklin's one-fluid theory have turned out to possess elements of truth. Once the internal structure of the atom came to be understood, beginning in the 1890's (a subject that will be taken up in detail in Volume III of this book), it was found that subatomic particles existed, and that some of these possessed an electric charge while others did not.*

Of the subatomic particles that possess an electric charge, the most common are the *proton* and the *electron,* which possess charges of opposite nature. In a sense, then, the proton and the electron represent the two fluids of Du Fay. On the other hand, the proton, under the conditions of electrostatic experiments, is a completely immobile particle, while the electron, which is much the lighter of the two, is easily shifted from one body to another. In that sense, the electron represents the single electric fluid of Franklin.

In an uncharged body, the number of electrons is equal to the number of protons and there is no net charge. The body is filled with electric charge of both kinds, but the two balance. As a result of rubbing, electrons shift. One body gains an excess of electrons; the other is left with a deficit.

There is one sad point to be made, however. The electrons move in the direction opposed to that which Franklin had guessed for the electric fluid. Franklin had lost his even-money bet. Where

* Exactly what an electric charge *is,* we cannot say. However, we can say how a substance with an electric charge acts and how we may measure the extent of this action .and, therefore, the size of the electric charge. This is an *operational definition* of electric charge, and it is enough to satisfy scientists, at least for the time being.

he thought an excess of electric fluid existed, there existed instead a deficit of electrons, and vice versa. For this reason, it was necessary to consider the electric charge of the electron to be negative; an excess of electrons would then produce the negative charge required by Franklin's deficiency of fluid, while a deficit of electrons would produce the positive charge of Franklin's excess of fluid. Since the electron is considered as having a negative charge, the proton must have a positive charge.

(Electrical engineers still consider the "electric fluid" to flow from positive to negative, although physicists recognize that electrons flow from negative to positive. For all practical purposes, it doesn't matter which direction of flow is chosen as long as the direction is kept the same at all times and there are no changes in convention in mid-stream.)

Coulomb, who measured the manner in which the force between magnetic poles was related to distance, did the same for the force between electrically charged bodies. Here his task was made somewhat easier because of an important difference between magnetism and electricity. Magnetic poles do not exist in isolation. Any body possessing a magnetic north pole must also possess a magnetic south pole. In measuring magnetic forces between poles, therefore, both attractive and repulsive forces exist, and they complicate the measurements. In the case of electricity, however, charges can be isolated. A body can carry a negative charge only or a positive charge only. For that reason, attractions can be measured without the accompaniment of complicating repulsions, and vice versa.

Coulomb found that the electric force, like the magnetic force, varied inversely as the square of the distance. In fact, the equation he used to express variation of electrical force with distance was quite analogous to the one he found for magnetic forces (see Equation 9–1, page 144).

If the electric charge on two bodies is q and q', and the distance between them is d, then F, the force between them (a force of attraction if the charges are opposite or of repulsion if they are the same) can be expressed:

$$F = \frac{qq'}{d^2} \qquad \text{(Equation 10–1)}$$

provided the charges are separated by a vacuum.

In the cgs system, distances are measured in centimeters and forces in dynes. If we imagine, then, two equal charges separated by a distance of 1 centimeter and exerting a force of 1 dyne upon each other, each charge may be said to be 1 *electrostatic unit* (usually abbreviated *esu*) in magnitude. The esu is, therefore, the cgs unit of electric charge.

The smallest possible charge on any body is that upon an electron. Measurements have shown that to be equal to -4.8×10^{-10} esu, where the minus sign indicates a negative charge.* This means that a body carrying a negative charge of 1 esu contains an excess of approximately 2 billion electrons, while a body carrying a positive charge of 1 esu contains a deficit of approximately 2 billion electrons.

Another commonly used unit of charge—in the mks system—is the *coulomb*, named in honor of the physicist. A coulomb is equal to 3 billion esu. A body carrying a negative charge of 1 coulomb therefore contains an excess of approximately 6 billion billion electrons, while one carrying a positive charge of 1 coulomb contains a deficit of that many.

Imagine two electrons one centimeter apart. Since each has a charge of -4.8×10^{-10} esu, the total force (of repulsion, in this case) between them, using Equation 10–1, is $(-4.8 \times 10^{-10})^2$ or 2.25×10^{-19} dynes.

The two electrons also exert a gravitational force of attraction on each other. The mass of each electron is now known to be equal to 9.1×10^{-28} grams. The force of gravitational attraction between them is equal to Gmm'/d^2, where G is the gravitational constant which equals 6.67×10^{-8} dyne-cm^2/gm^2 (see page I–52). The gravitational force between the electrons is therefore equal to $(9.1 \times 10^{-28})^2$ multiplied by 6.67×10^{-8}, or 5.5×10^{-62} dynes.

We can now compare the strength of the electrical force and that of the gravitational force by dividing 2.25×10^{-19} by 5.5×10^{-62}. The quotient is 4×10^{42}, which means that the electrical force (or the comparable magnetic force in the case of magnets) is some four million trillion trillion trillion times as strong as the gravitational force. It is fair to say, in fact, that gravitational force is by far the weakest force known in nature.

The fact that gravitation is an overwhelming force on a cosmic scale is entirely due to the fact that we are then dealing with the huge masses of stars and planets. Even so, if we stop

* The proton has the same charge exactly, but a positive one. All subatomic particles have been found to have a charge exactly equal to that of an electron, or exactly equal to that of a proton, or to have no charge at all. Both negative and positive charges seem to come in packages of just one size and no other. Why that should be is a matter that is, as yet, unsolved.

to think that we, with our own puny muscles, can easily lift objects upward against the gravitational attraction of all the earth or, for that matter, that a small toy magnet can do the same, it must be borne in upon us that gravitational forces are unimaginably weak. And, in fact, when we deal with bodies of ordinary size, we completely neglect any gravitational forces between them.

Electrically charged objects serve as centers of *electric fields,* which are analogous to magnetic fields.There are *electric lines of force,* just as there are magnetic ones.

As in the case of magnetic lines of force, electric lines of force may pass through a material substance more readily, or less readily, than they would pass through an equivalent volume of a vacuum. The ratio of the flux density of electric lines of force through a medium to that through a vacuum is the *relative permittivity.* (This term is analogous to relative permeability in the case of magnetism.)

In general, insulators have a relative permittivity greater than 1; in some cases, much greater. The relative permittivity of air is 1.00054, while that of rubber is about 3, and that of mica about 7. For water, the value is 78. Where the relative permittivity is greater than 1, the electric lines of force crowd into the material and more pass through it than would pass through an equivalent volume of vacuum. For this reason, insulators are often spoken of as *dielectrics* (the prefix being from a Greek word meaning "through," since the lines of force pass through them). The relative permittivity is therefore frequently spoken of as the *dielectric constant.*

Coulomb's equation for the force between two charged particles might more generally be written, then:

$$F = \frac{qq'}{\kappa d^2} \qquad \text{(Equation 10–2)}$$

where the particles are separated by a medium with a dielectric constant of κ (the Greek letter "kappa").

Electric forces between charged particles decrease, then, if a dielectric is placed between; they decrease more as the dielectric constant is increased. The constituent particles of a substance like common table salt, for instance, are held together by electric attractions. In water, with its unusually high dielectric constant, these forces are correspondingly decreased, and this is one reason why salt dissolves readily in water (its particles fall apart, so to speak) and why water is, in general, such a good solvent.

Electromotive Force

If we rub a glass rod with a piece of silk, electrons shift from glass to silk; therefore, the glass becomes positively charged and the silk negatively charged. With each electron shift, the positive charge on the glass and the negative charge on the silk become higher, and it becomes increasingly difficult to move further electrons. To drag more negatively-charged electrons from the already positively-charged glass means pulling the electrons away against the attraction of the oppositely-charged glass. To add those electrons to the already negatively-charged silk means to push it on against the repulsion of like-charged bodies. As one proceeds to pile up positive charge on glass and negative charge on silk, the attraction and repulsion becomes larger and larger until, by mere hand-rubbing, no further transfer of electrons can be carried through.

This situation is quite analogous to that which arises in connection with gravitational forces when we are digging a hole. As one throws up earth to the rim of the hole, the level of earth around the rim rises while the level of earth within the hole sinks. The distance from the hole bottom to the rim top increases, and it becomes more and more of an effort to transfer additional earth from bottom to top. Eventually, the digger can no longer throw the shovelful of earth high enough to reach the height of the rim, and he has then dug the hole as far as he can.

This points up the value of using the familiar situations involving gravity as an analogy to the less familiar situations involving electric forces. Let us then, for a moment, continue to think of earth's gravitational field.

We can consider that a given body has a certain potential energy depending on its position with relation to earth's gravitational field (see page I–96). The higher a body (that is, the greater its distance from the earth's center), the greater its potential energy. To lift a body against earth's gravity, we must therefore add to its potential energy and withdraw that energy from someplace else (from our own muscles, perhaps). The quantity of energy that must be added, however, does not depend on the absolute value of the original potential energy of the body or of its final potential energy, but merely upon the difference in potential energy between the two states. We can call this difference in potential energy the *gravitational potential difference.*

Thus, an object on the 80th floor of a skyscraper has a greater potential energy than one on the 10th floor of the same skyscraper. All points on the 80th floor have the same potential energy, and all points on the 10th floor have the same potential energy. Both floors represent *equipotential surfaces.* To slide an object from one point on the 10th floor to another point on the 10th floor (ignoring friction) takes no energy since the gravitational potential difference is zero. The same is true in sliding an object from one point on the 80th floor to another point on the 80th floor. Though the absolute value of the potential energy on the 80th floor is greater, the gravitational potential difference is still zero.

Similarly, it is no harder to lift a body from the 80th to the 82nd floor than from the 10 to the 12th floor. (Actually, the gravitational force is a trifle weaker on the 80th floor than on the 10th, but the difference is so minute that it may be ignored.) It is the two-story difference that counts and that is the same in both cases. We can measure the difference in height (which is all that counts) by the amount of energy we must invest to raise a body of unit mass through that difference. In the mks system, the joule is the unit of energy (see page I–90) and the kilogram is the unit of mass. Therefore the unit of gravitational potential difference is a joule per kilogram.

There is an exact analogy between this and the situation in an electric field. Just as one adds energy to move one mass away from another mass, so must one add energy to move a negatively-charged body away from a positively-charged body, or vice versa. (One must also add energy to move a negatively-charged body toward another negatively-charged body or a positively-charged body toward another positively-charged body. For this there is no exact analogy in the gravitational system, since there is no such thing as gravitational repulsion.) The separation of unlike charged bodies or the approach of like charged bodies represents an increase in electric potential energy, and once the charged bodies have changed position with respect to each other, the difference in the electric potential energy is the *electric potential difference.* (The concept of a change in potential energy is so much more commonly used in electrical work than in other branches of physics that when the term *potential difference* is used without qualification, it may invariably be taken to refer to an electric potential difference rather than, say, to a gravitational one.)

Again the electric potential difference can be measured in terms of the energy that must be added to a unit charge to move it a given distance. In the mks system, the unit of charge is the coulomb so that the unit of electric potential difference is the joule per coulomb. This unit is used so often that a special name has been given to it, the *volt,* in honor of the Italian physicist Alessandro Volta (1745–1827), whose work will be described on page 178. As a result, the electric potential difference is sometimes referred to as the "voltage."

Let us return to the gravitational analogy again and consider an object resting on a flat surface. It has no tendency to move spontaneously to another portion of the flat surface, for there is a gravitational potential difference of zero between one point on the surface and another. On the other hand, if the object is suspended a meter above the surface and is released, it will spontaneously fall, moving from the point of higher potential energy to that of lower potential energy. It is the gravitational potential difference that brings about the spontaneous motion.

Similarly, an electric charge will have no spontaneous tendency to move from one point in an electric field to another point at the same potential energy level. If an electric potential difference exists, however, the electric charge will have a spontaneous tendency to move from the point of higher energy to that of lower. Since it is the electric potential difference that brings about spontaneous motion of electric charge, that potential difference can be spoken of as an *electromotive force* (a force that "moves electricity"), and this phrase is usually abbreviated as *emf.* Instead of speaking of a potential difference of so many volts, one frequently speaks of an emf of so many volts.

To create a potential difference, or an emf, in the first place, one must—in one way or another—bring about a separation of unlike charges or a crowding together of like charges. Thus, in rubbing a glass rod, one removes negatively-charged electrons from an increasingly positively-charged rod and adds negatively-charged electrons to an increasingly negatively-charged piece of silk.

It is sometimes possible to create an emf by squeezing certain crystals. A crystal is often made up of both positively- and negatively-charged particles arranged in orderly fashion in such a way that all the positively-charged particles and all the negatively-charged particles are grouped about the same central point. If two opposite faces of a crystal are placed under pressure, the crystal can be slightly flattened and distorted, and the charged particles making up the crystals are pushed together and spread out sideways. In most cases, both types of particles change position in identical fashion and remain distributed about the same central point. In some cases, however, the change is such that the average position of the negatively-charged particles shifts slightly with respect to the average position of the positively-charged particles. This means there is, in effect, a separation of positive and negative charges and a potential difference is therefore created between the two faces of the crystal.

This phenomenon was first discovered by Pierre Curie (who discovered the Curie point, see page 148) and his brother, Jacques, in 1880. They called the phenomenon, *piezoelectricity* ("electricity through pressure").

The situation can also be reversed. If a crystal capable of displaying piezoelectricity is placed within an electric field so that a potential difference exists across the crystal, the crystal alters its shape correspondingly. If the potential difference is applied and taken away, over and over again, the crystal can be made to vibrate and produce sound waves. If the crystal is of the proper size and shape, sound waves of such high frequency can be produced as to be in the ultrasonic range (see page I–180). Such interconversions of sound and electric potential are useful in today's record players.

Condensers

In working with electricity, it is sometimes convenient to try to place as much charge within a body as possible, with as little effort as possible. Suppose you have a metal plate, insulated in such a way that any electric charge added to it would remain. If you touch the plate with a negatively-charged rod, electrons will flow into the metal plate and give it a negative charge.

You can continue this process as long as you can maintain a potential difference between rod and plate—that is, as long as you can keep the rod, by protracted rubbing, more negatively charged than the plate. Eventually, however, you will increase the negative charge of the plate to such a level that no amount of rubbing will make the rod more negatively charged than that. The potential difference between rod and plate will then be zero, and a charge will no longer spontaneously move.

Suppose, however, you next bring a second metal plate, one that is positively charged, down over the first and parallel to it, but not touching. The electrons in the first plate are pulled toward the positively-charged second plate and crowd into the surface facing the positive plate. (The electrons crowding into that surface are now closer together than they were before, when they had been spread out evenly. They are "condensed" so to speak, and so this device of two flat plates held parallel and a short distance apart, may be called a *condenser*.)

With the electrons in the negative plate crowding into the surface facing the positive plate, the opposite surface has fewer electrons and a lower potential. There is once again a potential difference between the negatively-charged rod and that surface of the first plate which is away from the second plate. Electrons can once more pass from the rod into the plate, and the total charge on the plate can be built up considerably higher than would have been possible in the absence of the second plate.

Similarly, the positive charge on the second plate can be built up higher because of the presence of the negatively-charged first plate. Because the plates lend each other a greater capacity for charge, a condenser may also be called a *capacitor*.

The more highly charged the two plates (one positive and one negative), the greater the potential difference between them; this is the same as saying that the higher a mountain peak and the lower a valley, the greater the distance there is to fall. There is thus a direct relationship between the quantity of charge and the potential difference. If we imagine a vacuum between the plates, we can expect the ratio between charge and potential difference to be a constant, and we can express this as follows:

$$\frac{q}{v} = c \qquad \text{(Equation 10–3)}$$

where q is the charge in coulombs, and v is the potential difference in volts. The constant c is the *capacitance*, for which the units are coulombs per volt. One coulomb per volt is referred to as a *farad*, in honor of Michael Faraday.

Thus, a condenser (or capacitor) with a capacitance of 1 farad, will pile up a charge of 1 coulomb on either plate, one negative and one positive for every volt of potential difference between the plates. Actually, condensers with this large a capacitance are not generally met with. It is common, therefore, to use a *microfarad* (a millionth of a farad) or even a *micromicrofarad* (a millionth of a millionth of a farad) as units of capacitance.

Suppose, now, a dielectric (see page 165) is placed between the plates of a condenser. A dielectric decreases the force of attraction between given positive and negative charges (see page

165) and therefore lessens the amount of work required to separate these charges. But, as was explained on page 166, the potential difference is the measure of the work required to separate unlike charges. This means that the potential difference across the condenser, once the dielectric is placed between the plates, is v/κ, where κ is the dielectric constant.

If we call the capacitance of the condenser with the dielectric, c', then, in view of this:

$$c' = \frac{q}{v/\kappa} = \frac{\kappa q}{v} = \kappa\left(\frac{q}{v}\right) \qquad \text{(Equation 10–4)}$$

Combining Equations 10–3 and 10–4:

$$c' = \kappa c \qquad \text{(Equation 10–5)}$$

We see, then, that placing a dielectric between the plates of a condenser multiplies the capacitance of the condenser by the dielectric constant. The dielectric constant of air is only 1.0006 (where that of vacuum is taken as 1), so separation of the plates by air can be accepted as an approximation of separation by vacuum. The dielectric constant of glass is about 5, however, so if the plates of a condenser are separated by glass, its capacitance increases fivefold over the value for plates separated by air. For a given potential difference, a glass-separated condenser will pile up five times the charge an air-separated condenser will.

The capacitance can be further increased by reducing the distance between the plates or by increasing the area of the plates or both. If the distance between the plates is decreased, the potential difference decreases (as gravitational potential difference would decrease if two objects were one story apart instead of two stories apart). If this is so, then v in Equation 10–3 decreases while q remains unchanged and c necessarily increases. Again, if the plates were increased in area, there would be room for more electric charge to crowd in, so to speak. Consequently, q would increase in Equation 10–3 and so therefore would c.

A condenser with large plates can be unwieldy, but the same effect can be attained if one stacks a number of small condensers, and connects all the positive plates by a conducting material such as a metal rod, and all the negative plates by another metal rod. In that way, any charge added to one of the plates would spread out through all the plates of the same type, and the many small pairs of plates would act like one large pair. The condensers are said, in this case, to be connected *in series*.

In such a condenser, one set of plates can be fixed, while the other set is pivoted. By turning a knob connected to the rod about which the other set is pivoted, one can turn the negative plates, let us say, more and more into line with the positive. Essentially, only those portions of the plates which directly face each other have much condenser action. Consequently, as the pivoting set of plates moves more and more into line, the capacitance increases steadily. If the plates are turned out of line, the capacitance decreases. We have here a *variable condenser*.

An electrically charged object can be discharged if, for instance, a finger is placed to it and if the man attached to the finger is standing, without insulation, on the essentially uncharged ground—that is, if the man is *grounded*. If the object is negatively charged, electrons will flow from it through the man and into the earth until the negative charge is dissipated. If the object is positively charged, electrons will flow from the earth through the man and into the object until the positive charge is neutralized. In either case, there is a flow of charge through the body.

Since the sensations of a living body are mediated by the flow of tiny amounts of charge through the nerves, it is not surprising that the flow of charge that results from discharging a charged object can be sensed. If the flow of charge is a small one, the sensation may be no more than a tingle. If it is a large one, the sensation may be a strong pain like that produced by a sudden blow. One then speaks of an *electric shock*. (As in the case of a physical blow, a strong enough electric shock can kill.) Since condensers can pile up large charges of electricity, the shock received from such a condenser is much larger than that received by discharging an ordinary electrified rod of similar size.

This unpleasant property of condensers was discovered accidentally in 1745, when the first condensers were more or less casually brought into existence. This original condenser evolved into a glass jar, coated inside and outside with metal foil. It was corked and a metal rod pierced the cork. A metal chain, suspended from the rod, touched the metal foil inside the glass jar.

Suppose the metal foil outside the glass is grounded. If the metal rod sticking up from the cork is touched with a negatively-charged rod, electrons will enter the metal and spread downward into the internal foil coating. The negative charge on that internal foil repels electrons on the outer foil and forces them down the conductor connecting it to the ground, where those electrons spread into the general body of the planet and can be forgotten. If this is repeated over and over, a large negative charge is built up on the internal foil and a large positive charge on the external foil; a much larger charge (thanks to the fact that the layers of foil act as a glass-separated condenser) than the early experimenters could possibly have expected.

The first men to produce condensers of this sort (the German experimenter Ewald George von Kleist in 1745 and the Dutch physicist Pieter van Musschenbroek [1692–1761] in 1746) were surprised and even horrified when they discharged the devices and found themselves numbed and stunned by the shock. Von Kleist abandoned such experiments at once, and Van Musschenbroek proceeded further only with the greatest caution. Since Van Musschenbroek did his work at the University of Leyden, in the Netherlands, the condenser came to be called a *Leyden jar.*

Through the second half of the eighteenth century, the Leyden jar was used for important electrical experiments. A charge could be collected and then given off in such unprecedented quantities that it could be used to shock hundreds of people who were all holding hands, kill small animals, and so on. These experiments were not important in themselves but served to dramatize

Leyden jar

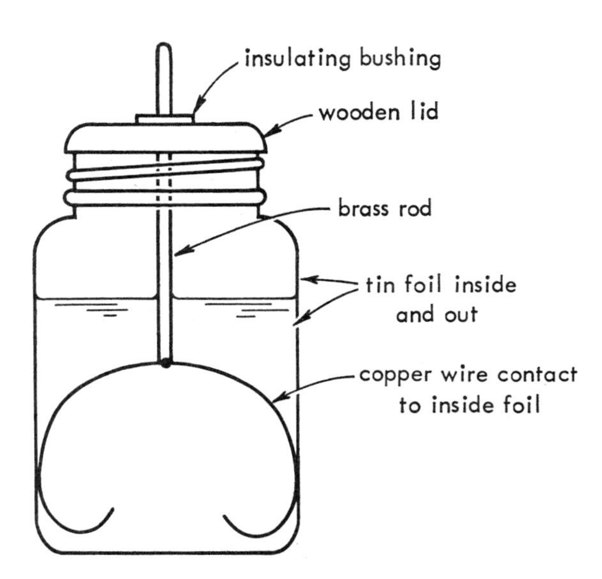

electrical phenomena and to rouse the interest of the scientific community (and of the general public, too).

In particular, the Leyden jar dramatized the matter of discharge through air. Dry air is an insulator, but insulation is never perfect, and if the charge on any object is great enough, it will force itself across a substance that is ordinarily an insulator. (Thus, you can imagine a weight resting upon a wooden plank suspended from its ends, and several feet above the ground. The wooden plank acts as an "insulator" in the sense that the weight cannot move downward despite its tendency to do so as a result of the gravitational potential difference between itself and the ground. If the weight is made heavier and heavier, a point will be reached where the plank breaks, and the weight drops. The "insulator" has been broken down, and the weight is "discharged," to use electrical terminology.)

When an electric charge forces itself across an ordinarily insulating gap of air, the air is heated by the electrical energy to the point where it glows. The discharge is therefore accompanied by a spark. The heated air expands and then, losing its heat to the surrounding atmosphere, contracts again. This sets up sound-wave vibrations, so the discharge is not only accompanied by a spark but also by a crackle. Such sparks and crackling were noted even by Guericke in his work with his charged ball of sulfur. With the Leyden jar and its much greater charge accumulation, sparks and crackling became most dramatic, and discharge would take place over longer gaps of air.

Franklin, who experimented industriously with Leyden jars, could not help but see the similarity between such a discharge and the thunder and lightning accompaniments to rainstorms. The Leyden jar seemed to produce miniature bolts of lightning and tiny peals of thunder, and contrariwise, earth and clouds during a thunderstorm seemed to be the plates of a gigantic Leyden jar. Franklin thought of a way of demonstrating that this was more than poetic fancy.

In June, 1752, he flew a kite during a thunderstorm. He tied a pointed metal rod to the wooden framework of the kite and attached a length of twine to it. This he attached to the cord that held the kite. He also attached an iron key to the end of the twine. To avoid electrocution, he remained under a shed during the storm and held the cord of the kite not directly, but by means of a dry length of insulating silk string.

The kite vanished into one of the clouds and Franklin noted the fibers of the kite cord standing apart as though all were charged and repelling each other. Presumably, the key had also gained a charge. Cautiously, Franklin brought the knuckle of his hand near the key; a spark leaped out, the same kind of spark, accompanied by the same crackle, one would expect of a Leyden jar. Franklin then brought out a Leyden jar he had with him and charged it with electricity from the clouds. The result was precisely the same as though he had charged it from an electrical friction machine. Franklin had thus, beyond doubt, showed that there was electricity in the high heavens just as there was on the ground, that lightning was a giant electrical discharge, and that thunder was the giant crackle that accompanied it.

He went further. Franklin experimented with the manner of discharge where bodies of different shapes were involved. Thus, if a metal sphere were brought near a charged body, there would be a discharge, let us say, across a one-inch gap of air. If a metal needle were brought near the same body charged to the same extent, discharge would take place across an airgap of six to eight inches. This could only be taken to mean that it was easier to discharge a charged body by means of a pointed object than a blunt object. Furthermore, the discharge by way of the point of a needle took place with such ease that it was not accompanied by noticeable sparks and crackles. (Nevertheless the fact of discharge could be detected easily enough, since the charged body suddenly lost the ability to repel a small similarly charged cork ball hanging in the vicinity.)

It occurred to Franklin that this phenomenon could be made use of on a large scale in connection with thunderstorms. If a long pointed metal rod were raised above the roof of a building, it would discharge the charge-laden thunderclouds more efficiently and quietly than would the building itself. The clouds would be discharged before they had built up enough charge to close the gap violently by means of a lightning bolt between themselves and the house. If conductors were attached to such a *lightning rod,* the charge drawn from the clouds could be conducted harmlessly to the earth and the house, in this manner, protected from lightning.

The lightning rod worked very well indeed, and over the next couple of decades, structures throughout America and Europe came under the protecting blanket of Franklin's invention. Franklin was the first great scientist produced by the New World and through this invention in particular he became famous among the scientists of Europe (a fact that had important political consequences when Franklin was sent on a mission to France during the American Revolution, a quarter-century after he flew his kite).

With the invention of the lightning rod, the study of electrostatics reached a climax. By the end of the eighteenth century, a new aspect of electricity came to the fore, and electrostatics receded into the background.

CHAPTER 26

Electric Currents

Continuous Electron Flow

Charge can move from one point to another (something which can also be described as the flowing of an electric current), as has been understood from the time of Gray in the early eighteenth century (see page 158). However, before 1800 only momentary flows of this sort were encountered. Charge could be transferred from a Leyden jar, for instance, to the human body, but after one quick spark, the transfer was done. A much huger charge transfer is that of lightning, yet "as quick as lightning" is a folk saying.

In order to arrange for a continuous transfer of charge, or a continuous flow of current from point A to point B, it is necessary to produce a new supply of charge at point A as fast as it is moved away, and consume it at point B as fast as it is brought there.

Methods for doing so developed out of the observations first made in 1791 by the Italian physician and physicist Luigi Galvani (1737–1798). Galvani was interested in muscle action and in electrical experiments as well. He kept a Leyden jar and found that sparks from it would cause the thigh muscles of dissected frogs to contract, even though there was no life in them. Others had observed this, but Galvani discovered something new: when a metal scalpel touched the muscle at a time when a spark was drawn from a nearby Leyden jar, the muscle twitched even though the spark made no direct contact.

Suspecting this might be caused by induced electric charge in the scalpel, Galvani exposed frogs' thigh muscles to the electrified atmosphere of a thunderstorm, suspending them by brass hooks on an iron railing. He obtained his contractions but found that a thunderstorm was not, after all, necessary. All that was necessary was that the muscle be touched simultaneously by two different metals, whether any electric spark was in the vicinity or not, and whether there was a thunderstorm or not.*

Two dissimilar metals in simultaneous contact with a muscle could not only produce muscle contractions, but they could do so a number of times. It seemed certain that electricity was somehow involved and that whatever produced the electric charge was not put out of action after discharge and muscle contraction; instead the charge could be spontaneously regenerated again and again. Galvani made the assumption that the source of the electricity was in muscle and spoke of "animal electricity."

Others, nevertheless, suspected that the origin of the electric charge might lie in the junction of the two metals rather than in muscle, and outstanding among this group was the Italian physicist Alessandro Volta (1745–1827). In 1800, he studied combinations of dissimilar metals, connected not by muscle tissue but by simple solutions that by no stretch of the imagination could be considered to have any connection with a "life force."

He used chains of dissimilar metals, rightly suspecting that he could get better results from a number of sources combined than from a single one. He first used a series of bowls half full of salt water (each taking the place of a frog muscle) and connected them by bridges of metal strips, each composed of copper and zinc soldered together. The copper end was dipped into one bowl and the zinc end into another. Each bowl contained a copper end of one bridge on one side and the zinc end of another bridge on the other side.

Such a "crown of cups," as Volta called it, could be used as a source of electricity, which was thus clearly shown to originate in the metals and not in animal tissue. What's more, the electricity

* As a result of these experiments, which gained much public attention, a person reacting by muscle contractions to a shock of electricity (or to any unexpected sensation or even to a sudden strong emotion) is said to be "galvanized."

was produced continuously and could be drawn off as a continuous flow.

To avoid quantities of fluid that could slosh and spill, Volta tried another device. He prepared small discs of copper or silver (coins did very well) and other discs of zinc. He then piled them up: silver, zinc, silver, zinc, silver, zinc, and so on. Between each silver-zinc pair, he placed cardboard discs that had been moistened with salt water and that served the purpose of Galvani's frog muscle or Volta's own bowl of salt water. If the top of such a "voltaic pile" was touched with a metal wire, a spark could then be drawn out of the bottom, assuming that the bottom was touched with the other end of the same wire. In fact, if top and bottom were connected, a continuous current would flow through the wire.

The reason for this was not thoroughly understood for another century, but it rested on the fact that atoms of all matter include as part of their internal structure negatively-charged electrons and positively-charged protons. The electric charge produced by a continuously operating voltaic pile is therefore not truly created but is present constantly in matter. For a pile to work it is only necessary that it serve, in some manner, to separate the already-existing negative and positive charges.

Such separation is most simply described where two different metals alone are involved. Imagine two metals, say zinc and copper, in contact. Each metal contains electrons, bound by forces of greater or lesser extent to the atoms of the metal. The forces binding the electrons to zinc atoms are somewhat weaker than those binding electrons to copper. At the boundary, then, electrons tend to slip across from zinc to copper. The copper, with its stronger grip, wrests the electrons, so to speak, from the zinc.

This does not continue for long, for as the electrons enter the copper, that metal gains a negative charge while the zinc comes to have a positive charge. Further transfer of electrons away from the attraction of the positively-charged zinc and into the repelling force of the negatively-charged copper quickly becomes impossible, so an equilibrium is reached while the charge on each metal is still extremely minute. Still, the charge is large enough to be detected, and because unlike charges have been separated, a *contact potential difference* has been set up between the two metals.

If the temperature is changed, the force attracting electrons to atoms is also changed, but generally by different amounts for different metals. Imagine a long strip of zinc and a long strip of copper in contact at the two ends only (a *thermocouple*) and each end kept at a different temperature. There is a contact potential difference at each end, but the two have different values. The copper may be able to seize more electrons at end A than at end B, because at the temperature of end A, its electron-binding force has been strengthened to a greater extent than has that of the zinc.

Since the electron concentration in the copper at end A is greater than in the copper at end B, electrons flow through the copper from A to B. At end B there are now present too many electrons for the copper to retain at its particular temperature. Some of the electrons slip over to the zinc, therefore. Meanwhile, at end A, with some of the electrons lost, the copper can gain still more from the zinc.

The process continues indefinitely, with electrons traveling from end A to end B through the copper and then back from end B to end A through the zinc, and this continues as long as the temperature difference between the two ends is maintained. Such *thermoelectricity* was first observed in 1821 by the German physicist Thomas Johann Seebeck (1770–1831).

Possible practical applications of the phenomenon are not hard to see. The amount of current that flows through the thermocouple varies directly with the size of the temperature difference between the two ends; consequently, a thermocouple may be used as a thermometer. Indeed, if high melting metals such as platinum are used, thermocouples can be used to measure temperature in ranges far too high for ordinary themometers. Furthermore, since even very minute electric currents can be easily detected and measured, thermocouples can be used to detect very feeble sources of heat; for example, that arising from the moon or from Venus.*

Chemical Cells

The junction of dissimilar metals by way of a conducting solution brings about a situation analogous to that of a thermocouple, but without the necessity of a temperature difference.

* Since an electric current can be maintained at the cost of a temperature difference alone, we can generate a continuous flow of electricity from burning kerosene (without moving parts). As we shall see in Chapter 12, other methods of forming electricity from burning fuel gained prominence in the course of the nineteenth century, but in the twentieth century, the potentialities of thermoelectricity are being reinvestigated. In particular, there is the possibility that if sunlight is used to maintain the temperature difference, solar energy may be converted to electricity and made use of directly, possibly in wholesale quantities. This would offer a most welcome addition to man's energy sources.

Suppose, for instance, that a strip of zinc is partially inserted into a container of dilute acid. The zinc has a distinct tendency to go into solution. Each zinc atom, as it goes into solution, leaves two electrons behind so that the zinc rod gains a negative charge. The zinc atom minus two of the electrons it normally carries has a positive charge equal to that of the negative charge of the lost electrons. An electrically-charged atom is called an *ion,* so we may summarize matters by saying that the zinc in zinc sulfate produces positively-charged ions that enter the solution, while the zinc remaining behind gains a negative charge.

Imagine also a strip of copper inserted into a solution of copper sulfate. The copper sulfate solution contains positively-charged copper ions. There is no tendency for the copper metal to form more copper ions. Rather the reverse is true. The copper ions tend to return to the rod carrying with them their positive charge. Now suppose that the acid with its zinc strip and the copper sulfate with its copper strip are connected by a porous barrier so that liquid can slowly seep to and fro. We have a zinc strip carrying a small negative charge on one side and a copper strip carrying a small positive charge on the other.

If the two strips are connected by a wire, the surplus electrons in the negatively-charged zinc flow easily through the wire into the copper strip, which suffers a deficit of electrons. As the zinc loses its electron excess and therefore its negative charge, more zinc ions go into solution to produce a new electron excess. Moreover, as the copper gains electrons and loses its positive charge, more positively-charged copper ions can be attracted to the rod.

In short, electrons flow from the zinc to the copper by way of the wire, and then flow back from copper to zinc by way of the solution. The flow continues in its closed path until such time as all the zinc has dissolved or all the copper ions have settled out (or both). In the thermocouple, the electron flow was maintained by a temperature difference; in a voltaic pile, it was maintained by a chemical reaction.

Although the electron flow through the wire is from the zinc to the copper, electricians, following Franklin's wrong guess (see page 161), accept the convention that the flow of current is from the copper (the *positive pole*) to the zinc (the *negative pole*).

A generation after Volta's experiment, Faraday termed the metal rod that served as poles when placed in solutions, *electrodes,* from Greek words meaning "route of the electricity." The positive pole he called an *anode* ("upper route"), the negative pole a *cathode* ("lower route"), since he visualized the electricity as flowing downhill from anode to cathode.

Different chemicals so arranged as to give rise to a steady flow of electricity make up a *chemical cell,* or an *electrical cell,* or an *electrochemical cell.* All three names are used. Very often, as in Volta's original experiments, groups of cells are used. Groups of similar objects are referred to as "batteries," and for that reason groups of cells such as the voltaic pile are referred to as *electric batteries,* or simply *batteries.* (In ordinary conversation, even a single chemical cell may be referred to as a "battery.")

With Volta's discovery it became possible to study steady and long-continued flows of electric current. It was popular at first to call this phenomenon "galvanism" or "galvanic electricity," in honor of Galvani. However, it is more logical to call the study *electrodynamics* ("electricity-in-motion") as opposed to electrostatics. The study of those chemical reactions that give rise to electric currents is, of course, *electrochemistry.*

Electric currents were put to startling use almost at once. Since a flow of electrons is produced as a result of chemical reactions, it is not surprising that the electrons of a current, routed through a mixture of chemical substances, serve to initiate a chemical reaction. What's more, the chemical reactions that may easily be carried through by this method may be just those that prove very difficult to bring about in other ways.

In 1800, only six weeks after Volta's initial report, two English scientists, William Nicholson (1753–1815) and Anthony Carlisle (1768–1840), passed an electric current through water and found that they could break it up into hydrogen and oxygen. This process of bringing about a chemical reaction through an electric current is termed *electrolysis* ("loosening by electricity") because so often, as in the case of water, the reaction serves to break up a molecule into simpler substances.

In 1807 and 1808, the English chemist Humphry Davy (1778–1829), using a battery of unprecedented power, was able to decompose the liquid compounds of certain very active metals. He liberated the free metals themselves and was the first to form such metals as sodium, potassium, calcium, strontium, barium and magnesium—a feat that till then had been beyond the non-electrical abilities of chemists.

Davy's assistant, Faraday, went on to study electrolysis quantitatively and to show that the mass of substance separated by electrolysis was related to the quantity of electricity passing through the system. Faraday's *laws of electrolysis* (which will be taken up in some detail in Volume III) did much to help establish the atomistic view of matter then being introduced by the English chemist John Dalton (1766–1844). In the course of the next century, they helped guide physicists to the discovery of the electron and the working out of the internal structure of the atom.

As a result of Faraday's studies, a coulomb can be defined not merely in terms of total quantity of charge, or of total current (something not very easy to measure accurately), but as the quantity of current bringing about a certain fixed amount of chemical reaction (and that last could be measured quite easily). For instance, a coulomb of electric current passed through a solution of a silver compound will bring about the formation of 1.18 milligrams of metallic silver.

Chemists are particularly interested in a mass of 107.87 grams of silver for this is something they call "a gram-atomic weight of silver." Therefore, they are interested in the number of coulombs of current required to produce 107.87 grams of silver. But 107.87 grams is equal to 107,870 milligrams, and dividing that by 1.18 milligrams (the amount of silver produced by one coulomb), we find that it takes just about 96,500 coulombs to deposit a gram-atomic weight of silver out of a solution of silver compound. For this reason, 96,500 coulombs is referred to as 1 *faraday* of current.

A coulomb of electricity will deposit a fixed quantity of silver (or bring about a fixed amount of any particular chemical reaction) whether current passes through the solution rapidly or slowly. However, the rate at which the silver is deposited (or the reaction carried through) depends on the number of coulombs passing through the solution per unit of time. It would be natural to speak of the rate of flow of current (or of *current intensity*) as so many coulombs per second. One coulomb per second is called 1 *ampere,* in honor of the French physicist André Marie Ampère (1775–1836), whose work will be described on page 202. The ampere, then, is the unit of current intensity.

If, then, a current flowing through a solution of silver compound deposits 1.18 milligrams of metallic silver each second, we can say that 1 ampere of current is flowing through the solution.

Resistance

The rate of flow of electric current between point A and point B depends upon the difference in electric potential between these two points. If a potential difference of 20 volts serves to set up a current intensity of 1 ampere between those two points, a potential difference of 40 volts will produce a current intensity of 2 amperes, and a potential difference of 10 volts, one of 0.5 amperes.

This direct proportionality between potential difference and current intensity is true only if current is passed over a particular wire under particular conditions. If the nature of the path over which the current flows is changed, the relationship of potential difference and current intensity is changed, too.

Lengthening a wire, for instance, will reduce the current intensity produced in it by a given potential difference. If 20 volts will produce a current intensity of 1 ampere in a wire one meter long, the same 20 volts will produce only 0.5 amperes in a wire of the same composition and thickness but two meters long. On the other hand, if the wire is thickened, the current intensity produced by a given potential difference will be increased as the cross-sectional area or, which is the same thing, as the square of the diameter of the wire. If 20 volts will produce a 1-ampere current through a wire one millimeter in thickness, it will produce a 4-ampere current through an equal length of wire two millimeters in thickness.

Then, too, the nature of the substance conducting the electricity counts. If 20 volts produces a current of 3 amperes in a particular copper wire, it would produce a 2-ampere current in a gold wire of the same length and thickness, and a 1-ampere current through a tungsten wire of the same length and thickness. Through a quartz fiber of the same length and thickness, a current intensity of 0.0000000000000000000000003 amperes would be produced—so little that we might as well say none at all.

This sort of thing was investigated by the German physicist Georg Simon Ohm (1787–1854). In 1826, he suggested that the current intensity produced in a given pathway under the influence of a given potential difference depended on the *resistance* of that pathway. Doubling the length of a wire doubled its resistance; doubling its diameter reduced the resistance to a quarter of the original value. Substituting tungsten for copper increased resistance threefold, and so on.

The resistance could be measured as the ratio of the potential difference to the current intensity. If we symbolize potential difference as E (for "electromotive force"), current intensity as I, and resistance as R, then we can say:

$$R = \frac{E}{I} \qquad \text{(Equation 11–1)}$$

This is *Ohm's law*. By transposition of terms, Ohm's law can of course also be written as $I = E/R$ and as $E = IR$.

Resistance is measured, as one might perhaps expect, in *ohms*. That is, if 1 volt of potential difference produces a current intensity of 1 ampere through some conducting pathway, then the resistance of that pathway is 1 ohm. From Equation 11–1, applied to the units of the terms involved, we see that 1 ohm can be defined as 1 volt per ampere.

Sometimes it is convenient to think of the *conductance* of a substance rather than of the resistance. The conductance is the reciprocal of the resistance, and the unit of conductance was (in a rare fit of scientific whimsy) defined as the *mho,* which is "ohm" spelled backward.

A pathway with a resistance of 1 ohm has a conductance of 1/1, or 1 mho. A resistance of 3 ohms implies a conductance of 1/3 mhos; a resistance of 100 ohms implies a conductance of 1/100 mhos, and so on. If we symbolize conductance by *C,* from Equation 11–1, we can say that:

$$C = \frac{1}{R} = \frac{I}{E} \qquad \text{(Equation 11–2)}$$

so that 1 mho is equal to 1 ampere per volt.

For any given substance, resistance depends upon the length and diameter of the conducting pathway (among other things). In general, the resistance varies directly with length (L) and inversely with the cross-sectional area (A) of the pathway. Resistance is, therefore, proportional to L/A. If we introduce a proportionality constant, ρ (the Greek letter "rho"), we can say that:

$$R = \frac{\rho L}{A} \qquad \text{(Equation 11–3)}$$

The proportionality constant ρ is the resistivity, and each substance has a resistivity characteristic for itself. If we solve for resistivity by rearranging Equation 11–3, we find that:

$$\rho = \frac{RA}{L} \qquad \text{(Equation 11–4)}$$

Since in the mks system, the unit of R is the ohm, that of A is the square meter (or meter^2) and that of L is the meter. The unit of ρ, according to Equation 11–4, would be ohm-meter^2 per meter, or *ohm-meter*.

The better the conductor, the lower the resistivity. The best conductor known is the metal silver, which at 0°C has a resistivity of about 0.0000000152, or 1.52×10^{-8} ohm-meters. Copper is close behind with 0.0000000154, while gold and aluminum come next with 0.0000000227 and 0.0000000263 respectively. In general, metals have low resistivities and are therefore excellent conductors. Even Nichrome, an alloy of nickel, iron and chromium, which has an unusually high resistivity for a metal, has a resistivity of merely 0.000001 ohm-meters. This low resistivity of metals comes about because their atomic structure is such that each atom has one or two electrons that are loosely bound. Charge can therefore easily be transferred through the metal by means of these electrons.*

Substances with atomic structures such that all electrons are held tightly in place have very high resistivities. Even tremendous potential differences can force very little current through them. Substances with resistivities of over a million ohm-meters are therefore nonconductors. Maple wood has a resistivity of 300 million ohm-meters; glass a resistivity of about a trillion; sulfur one of a quadrillion; and quartz something like 500 quadrillion.

Between the low-resistivity conductors and the high-resistivity nonconductors, there is a group of substances characterized by moderate resistivities, higher than that of Nichrome but less than that of wood. The best-known examples are the elements germanium and silicon. The resistivity of germanium is 2 ohm-meters at 0°C and that of silicon is 30,000. Substances like germanium and silicon are therefore called *semiconductors.*

Notice that the resistivities given above are for 0°C. The value changes as temperature rises, increasing in the case of metals. One can picture matters this way. The electrons moving through a conductor are bound to find the atoms of the substance as barriers to motion, and some electrical energy is lost in overcoming this barrier. This energy loss is due to the resistance of the medium. If the temperature of the conductor rises, the atoms of the conductor vibrate more rapidly (see page I–203), and the electrons

* The movement of electrons is not quite the same as the flow of electricity. The electrons move at a certain not-very-high velocity, but the force that makes them move is transmitted much more quickly. If you flick a checker into a row of other checkers in contact, the incoming checker strikes and comes to a halt (perhaps rebounding a bit). The checkers it strikes remain in position for the most part, but the final checker at the opposite end of the line goes shooting off. The individual checkers scarcely moved, but the momentum was transmitted along the line of checkers at a speed that depends on the elasticity of the material making up those checkers. Similarly, quite apart from the actual velocity of electrons, the electric force travels through any substance at the speed of light.

find it more difficult to get through; consequently, resistivity rises. (Compare your own situation, for instance, making your way first through a quietly-standing crowd and then through the same crowd when it is milling about.)

If the resistivity is a given value at 0°C (ρ_o), it rises by a fixed small fraction of that amount ($\rho_0 a$) for each degree rise in temperature (t). The increase in resistivity for any given temperature is therefore $\rho_0 at$. The total resistivity at that temperature (ρ_t) is therefore equal to the resistivity at 0°C plus the increase, or:

$$\rho_t = \rho_o + \rho_0 at = \rho_o(1 + at) \qquad \text{(Equation 11–5)}$$

The constant, a, which is the fractional increase in resistivity per degree, is the *temperature coefficient of resistance.*

As long as the temperature coefficient of resistance remains unchanged, the actual resistance of a particular conductor varies with temperature in a very simple way. From the resistance of a high-melting metal wire of given dimensions it is therefore possible to estimate high temperatures.

For semiconductors, the temperature coefficient of resistivity is negative—that is, resistance decreases as temperature goes up. The reason for this is that as temperature goes up the hold of the material on some of its electrons is weakened; more are available for moving and transferring charge. The increase in the number of available electrons more than makes up for the additional resistance offered by the more strongly vibrating atoms, so overall resistance decreases.

If the temperature coefficient of resistance of conductors was truly constant, one might expect that resistance would decrease to zero at temperatures just above absolute zero. However, at low temperatures, resistivity slowly decreases and the rate at which resistance declines as temperature drops slows in such a way that as the twentieth century opened, physicists were certain that a metal's resistance would decline to zero only at a temperature of absolute zero and not a whit before. This seemed to make sense since only at absolute zero would the metallic atoms lose their vibrational energy altogether and offer no resistance at all to the movement of electrons.

However, actual resistance measurements at temperatures close to absolute zero became possible only after the Dutch physicist Heike Kamerlingh-Onnes (1853–1926) managed to liquefy helium in 1908. Of all substances, helium has the lowest liquefaction point, 4.2°K, and it is only in liquid helium that the study of ultra-low temperatures is practical. In 1911, Kamerlingh-Onnes found to his surprise that the resistance of mercury, which was growing less and less in expected fashion as temperature dropped, suddenly declined precipitously to zero at a temperature of 4.16°K.

A number of other metals show this property of *superconductivity* at liquid helium temperatures. There are some alloys, in fact, that become superconductive at nearly liquid hydrogen temperatures. An alloy of niobium and tin remains superconductive up to a temperature of 18.1°K. Others, like titanium, become superconductive only at temperatures below 0.39°K. Although as many as 900 substances have been found to possess superconductive properties in the neighborhood of absolute zero, there remain many substances (including the really good conductors at ordinary temperatures, such as silver, copper, and gold) that have, as yet, shown no superconductivity at even the lowest temperatures tested.

Electric Power

It takes energy to keep an electric current in being against resistance. The amount of energy required varies directly with the total amount of current sent through the resistance. It also varies directly with the intensity of the current. Since, for a given resistance, the current intensity varies directly with potential difference (as is required by Ohm's law), we can say that the energy represented by a particular electric current is equal to the total quantity of charge transported multiplied by the potential difference.

Since energy can be transformed into work, let us symbolize the electric energy as W. This allows us to keep E for potential difference, and we can let Q stand for the total quantity of charge transported. We can then say:

$$W = EQ \qquad \text{(Equation 11–6)}$$

The unit of potential difference is the volt and that of the total charge, the coulomb. If energy equals potential difference multiplied by total charge transferred, the unit of energy must equal volts multiplied by coulombs. However, a volt has been defined (see page 168) as a joule per coulomb. The units of energy must therefore be (joule/coulomb) (coulomb), or joules. The joule is the mks unit of energy, and we can say, then, that when 1 coulomb of electric charge is transported across a resistance under a potential difference of 1 volt, then 1 joule of energy is expended and may be converted into other forms of energy such as work, light, or heat.

It is often more useful to inquire into the rate at which energy is expended (or work performed) rather than into the total energy (or total work). If two systems both consume the same amount of energy or perform the same amount of work, but one does it in a minute and the other in an hour, the difference is clearly significant.

The rate at which energy is expended or work performed is termed *power*. If we consider the energy expended per second, the units of power become *joules per second*. One joule per second is defined as 1 *watt*, in honor of the Scottish engineer James Watt (1736–1819), whose work was dealt with on page I–93.

If 1 watt is equal to 1 joule per second and 1 joule is equal to 1 volt-coulomb (as Equation 11–6 indicates), then 1 watt may be considered as equal to 1 volt-coulomb per second. But 1 coulomb per second is equal to 1 ampere; therefore a volt-coulomb per second is equivalent to a volt-ampere, and we can conclude that 1 watt is equal to 1 volt-ampere.

What this means is that a current, driven by a potential difference of 1 volt and possessing an intensity of 1 ampere, possesses a power of 1 watt. In general, electric power is determined by multiplying the potential difference and the current intensity. If we symbolize the power as P, we can say:

$$P = EI \qquad \text{(Equation 11–7)}$$

An electrical appliance is usually rated in watts; what is indicated, in other words, is its rate of consuming electrical energy. We are most familiar with this in the case of light bulbs. Here the energy expended is used to increase the temperature of the filament within the bulb. The greater the rate of energy expenditure, the higher the temperature reached, and the more intense is the light radiated. It is for this reason that a 100-watt bulb is brighter than a 40-watt bulb—and hotter to the touch, too.

The potential difference of household current is usually in the neighborhood of 120 volts, and this remains constant. From Equation 11–7, we see that $I = P/E$. For a 100-watt bulb running on household current, then, $I = 100/120 = 5/6$. The current intensity in a 100-watt bulb is therefore 5/6 ampere. From this we can tell at once what the resistance (R) of the light bulb must be. Since, by Ohm's law, $R = E/I$, $R = 120$ divided by 5/6, or 144 ohms.

The watt is the unit of power in the mks system, but it is not the most familiar such unit. Quite frequently, one uses the *kilowatt,* which is equal to 1000 watts. Completely outside the mks system of units is the *horsepower,* which, in the United States at any rate, retains its popularity as the measure of the power of internal combustion engines. The horsepower is a larger unit than the watt; 1 horsepower = 746 watts. It follows that 1 kilowatt = 1.34 horsepower.

Since power is energy per time, energy must be power multiplied by time. This relationship (as always) carries over to the units. Since 1 watt = 1 joule/second, 1 joule = 1 watt-second. The *watt-second* is, therefore, a perfectly good mks unit of energy —as good as the joule to which it is equivalent. A larger unit of energy in this class is the *kilowatt-hour.* Since a kilowatt is equal to 1000 watts and an hour is equal to 3600 seconds, a kilowatt-hour is equal to (1000)(3600) watt-seconds, or joules. In other words, 1 kilowatt-hour = 3,600,000 joules. A 100-watt bulb (0.1 kilowatts) burning for 24 hours expends 2.4 kilowatt-hours of energy. The household electric bill is usually based on the number of kilowatt-hours of energy consumed.

From Ohm's law (Equation 11–1) we know that $E = IR$. Combining this with Equation 11–7, we find that:

$$P = I^2R \qquad \text{(Equation 11–8)}$$

In other words, the rate at which energy is expended in maintaining an electric current varies directly with the resistance involved, and also with the square of the current intensity.

There are times when it is desirable to expend as little energy as possible in the mere transportation of current, as in conducting the current from the battery (or other point of origin) to the point where the electrical energy will be converted into some other useful form of energy (say the light bulb, where part of it will be converted into light). In that case, we want the resistance to be as small as possible. For a given length and thickness of wire, the lowest resistance is to be found in copper and silver. Since copper is much the cheaper of the two, it is copper that is commonly used as electrical wiring.

For really long distance transport of electricity, even copper becomes prohibitively expensive, and the third choice, the much cheaper aluminum, can be used. Actually, this is not bad at all, even though the resistivity of aluminum is 1.7 times as high as that of copper. The higher resistivity can be balanced by the fact that aluminum is only one-third as dense as copper, so a length of aluminum wire 1 millimeter in thickness is no heavier than the same length of copper wire 0.6 millimeters in thickness. Resistance decreases with increase in the cross-sectional area of the wire; consequently, the thicker aluminum wire actually possesses less resistance than does the same weight of thinner (and considerably more expensive) copper wire.

On the other hand, it is sometimes desired that as much of the electrical energy as possible be converted into heat, as in electric irons, toasters, stoves, driers, and so on. Here one would like to have the resistance comparatively high (but not so high that a reasonable current intensity cannot be maintained). Use is often made of high-resistance alloys such as Nichrome.

Within a light bulb, very high temperatures are particularly desired, temperatures high enough to bring about the radiation of considerable quantities of visible light (see page 126). There are few electrical conductors capable of withstanding the high temperatures required; one of these is tungsten. Tungsten has a melting point of 3370°C, which is ample for the purpose. However, its resistivity is only 1/20 that of Nichrome. To increase the resistance of the tungsten used, the filament in the light bulb must be both thin and long.

(At the high temperature of the incandescent tungsten filament, tungsten would combine at once with the oxygen of the air and be consumed. For this reason, light bulbs were evacuated in the early days of incandescent lighting. In the vacuum, however, the thin tungsten wires evaporated too rapidly and had a limited lifetime. To combat this, it became customary to fill the bulb with an inert gas, first nitrogen, and later on, argon. These gases did not react with even white-hot tungsten, and the gas pressure minimized evaporation and extended the life of the bulbs.)

Circuits

Suppose that a current is made to flow through a conductor with a resistance (R) of 100 ohms. Having passed through this, it is next led through one with a resistance (R') of 50 ohms, and then through one with a resistance (R'') of 30 ohms. We will speak of each of these items merely as "resistances," and for simplicity's sake we will suppose that the resistance of the conducting materials other than these three items is negligible and may be ignored.

Such resistances are *in series:* the entire current must pass first through one, then through the second, then through the third.

It is clear that the effect is the same as though the current had passed through a single resistance of 100 + 50 + 30, or 180 ohms. Whenever items are connected in series so that a current passes through all, the total resistance is equal to the sum of the separate resistances.

If we are using household current with a potential difference of 120 volts, and if we assume that R, R' and R'' are the only resistances through which current is passing, then we can use Ohm's law to find the current intensity passing through the resistances. The total resistance is 180 ohms, and since $I = E/R$, the current intensity is equal to 120 volts divided by 180 ohms, or 2/3 amperes. All the current passes through all the resistances, and its intensity must be the same throughout.

Ohm's law can be applied to part of a system through which electricity is flowing, as well as to all of the system. For instance, what is the potential difference across the first of our three resistances, R? Its resistance is given as 100 ohms, and we have calculated that the current intensity within it (as well as within all other parts of the system in series) is 2/3 amperes. By Ohm's law, $E = IR$, so the potential drop across the first resistance is 100 ohms multiplied by 2/3 amperes, or 66 2/3 volts. Across the second resistance, R', it would be 50 ohms multiplied by 2/3 amperes, or 33 1/3 volts. Across the third resistance, R'', it is 30 ohms time 2/3 amperes, or 20 volts. The total potential difference is 66 2/3 + 33 1/3 + 20, or 120 volts.* Whenever items are in series, the total potential difference across all is equal to the sum of the potential differences across each separately.

Suppose that a fourth resistance is added to the series—one, let us say, of 60,000,000,000,000 ohms. The other resistances would add so little to this that they could be ignored. The current intensity would be 120 volts divided by 60 trillion ohms, or two trillionths of an ampere—an intensity so small we might just as well say that no current is flowing at all.

If two conductors are separated by a sizable air gap, current does not flow since the air gap has an extremely high resistance. For current to flow through conductors, there must be no significant air gaps. The current must travel along an unbroken path of reasonably conducting materials, all the way from one pole of a chemical cell (or other source of electricity) to the other pole.

* Even ordinary wiring, though it has as low a resistance as can be managed, does have some resistance. The wiring is in series with the objects to which it leads, and there is some potential difference across the wiring itself (*voltage drop*), though only to the extent of a volt or two, perhaps.

The pathway, having left the cell, must circle back to it, so one speaks of *electric circuits.*

If an air gap is inserted anywhere in a circuit made up of objects in series, a high resistance is added and the current virtually ceases. The circuit is "opened" or "broken." If the air gap is removed, the current flows again and the circuit is "closed."* Electric outlets in the wall involve no current flow if left unplugged, because an air gap exists between the two "terminals." This air gap is closed when an appliance is plugged in. An appliance is usually equipped, however, with an air gap within itself, so that even after it is plugged in, current will not flow. It is only when a switch is thrown or a knob is turned, and that second air-gap is also closed, that significant current finally starts to flow.

It may be desirable to form an air gap suddenly. There are conditions under which the current intensity through a particular circuit may rise to undesirable heights. As the current intensity goes up, the rate of energy expenditure and, therefore, the rate at which heat may develop, increases as the square of that intensity (see Equation 11–8). The heat may be sufficient to damage an electrical appliance or to set fire to inflammable material in the neighborhood.

To guard against this it is customary to include in the circuit, at some crucial point in series, a strip of low-melting alloy. A synonym for "melt" is "fuse," so such a low-melting material is a "fusible alloy." The little device containing a strip of such alloy is therefore an *electric fuse.* A rise in current intensity past some limit marked on the fuse (a limit of 15 amperes on one common type of household fuse, for instance) will produce enough heat to melt the alloy and introduce an air gap in the circuit. The current is stopped till the fuse is replaced. Of course, if the fuse is "blown" repeatedly, it is wise to have the circuit checked in order to see what is wrong.

When objects are in series within the circuit, the whole electric current passing through the first, passes through each of the succeeding objects consecutively. It is possible, though, that the current may have alternate routes in going from point A to point B, which may, for instance, be separately connected by the

* Naturally, if the potential difference is made high enough, current will flow with significant intensity through any resistance. The enormous potential differences between clouds, or between clouds and ground, during thunderstorms, is enough to transfer electric charge over tremendous air gaps, and man can duplicate this on a smaller scale ("man-made lightning"). However, ordinary circuits in everyday use run no risks of bursting across even small air gaps.

three different resistances, R, R', and R'', which I mentioned earlier as being of 100, 50 and 30 ohms respectively. The current flows in response to the potential difference between points A and B, and that has to be the same, regardless of the route taken by the current. (Thus, in going from the twelfth floor to the tenth floor of a building, the change in gravitational potential—two floors—is the same whether the descent is by way of the stairs, an elevator, or a rope suspended down the stairwell.)

Since under such circumstances the three resistances are usually shown, in diagrams, in parallel arrangement, they are said to be *in parallel.* We can then say that when objects are placed in parallel within an electric circuit, the potential drop is the same across each object.

It is easy to calculate the current intensity in each resistance under these circumstances, since the potential difference and resistance are known for each. If household current is used, with a potential difference of 120 volts, then that is the potential difference across each of the three resistances in parallel. Since by Ohm's law, $I = E/R$, the current intensity in the first resistance is 120/100, or 1.2 amperes, that in the second is 120/50, or 2.4 amperes, and that in the third is 120/30, or 4 amperes.

As you see, there is an inverse relationship between current intensity and resistance among objects in parallel. Since the reciprocal of resistance is conductance ($C = 1/R$), we can also say that the current intensity in objects in parallel is directly proportional to the conductances of the objects.

Imagine, a long wire stretching from point A to point B and arranged in a nearly closed loop so that points A and B, though separated by several meters of wire, are also separated by 1 centimeter of air gap. The wire and the air gap may be considered as being arranged in parallel. That is, current may flow from A to B through the long wire or across the short air gap. The resistance of the column of air between the points is, however, much greater than that of the long wire, and only a vanishingly small current intensity will be found in the air between the two points. For all practical purposes, all the current flows through the wire.

If the air gap is made smaller, however, the total resistance of the shortening column of air between A and B decreases, and more and more current intensity is to be found there. The passage of current serves to knock electrons out of atoms in the air, thus increasing the ability of air to conduct electricity by means of those electrons and the positively-charged ions the electrons leave behind. As a result, the resistance across the air gap further decreases. At some crucial point, this vicious cycle of current causing ions causing more current causing more ions builds rapidly to the point where current can be transferred through the air in large quantity, producing the spark and crackle that attracted such attention in the case of the Leyden jar (see page 173). Since the current takes the shortcut from point A to B, we speak of a *short circuit*. The wire from A to B, plus any appliances or other objects in series along that wire, is no longer on the route of the current, and the electricity has been shut off.

When a short circuit takes place so that a sizable portion of what had previously been the total circuit is cut out, there is a sudden decline in the total resistance of the circuit. The resistance across the sparking air gap is now very low, probably much lower than that of the wire and its appliances. There is a correspondingly higher current intensity in what remains of the circuit, and consequently, more heat develops. The best that can then happen is that the fuse blows. If there is any delay in this, then the sparks forming across the air gap may well set fire to anything inflammable that happens to be in the vicinity.

To minimize the possibility of short circuits, it is customary to wrap wires in insulation—silk, rubber, plastic, and so on. Not only do these substances have higher resistivities than air, but being solid, they set limits to the close approach of two wires, which remain always separated (even when pressed forcefully together) by the thickness of the insulation. When insulation is worn, however, and faults or cracks appear, short circuits become all too possible.

If we return to our three resistances in parallel, we might ask what the total resistance of the system is. We know the total current intensity in the system, since that is clearly the sum of the current intensities in each part of the system. The total current intensity in the particular case we have been considering would be $1.2 + 2.4 + 4.0 = 7.6$ amperes. The potential difference from A to B over any or all the routes in parallel is 120 volts. Since, by Ohm's law, $R = E/I$, the total resistance from A to B is 120 volts divided by the total current intensity, or 7.6 amperes. The total resistance, then, is 120/7.6, or just a little less than 16 ohms.

Notice that the total resistance is less than that of any one of the three resistances taken separately. To see why this should be, consider the Ohm's law relationship, $R_t = E/I_t$, as applied to a set of objects in parallel. Here R_t represents the total resistance and I_t the total current intensity; E, of course, is the same whether one of the objects or all of them are taken. The current intensity is equal to the sum of the current intensities (I, I' and I'') in the individual items. We can then say that:

$$R_t = \frac{E}{I + I' + I''} \qquad \text{(Equation 11–9)}$$

If we take the reciprocal of each side, then:

$$\frac{1}{R_t} = \frac{I + I' + I''}{E} = \frac{I}{E} + \frac{I'}{E} + \frac{I''}{E} \qquad \text{(Equation 11–10)}$$

By Ohm's law, we would expect I/E to equal R, I'/E to equal R', and I''/E to equal R'', the individual resistances of the items in parallel. Therefore:

$$\frac{1}{R_t} = \frac{1}{R} + \frac{1}{R'} + \frac{1}{R''} \qquad \text{(Equation 11–11)}$$

We are dealing with reciprocals in Equation 11–11. We can say, for instance, that the reciprocal of the total resistance is the sum of the reciprocals of the individual resistances. It so happens that the lower the value of a quantity, the higher the value of its reciprocal, and vice versa. (Since 11 is greater than 3, 1/11 is smaller than 1/3, for example.) Thus, since the reciprocal of the total resistance ($1/R_t$) is the sum of the reciprocals of the separate resistances and therefore larger than any of the reciprocals of the separate resistances, the total resistance itself (R_t) must be smaller than any of the separate resistances themselves.

An important property of arrangements in parallel is this: If there is a break in the circuit somewhere in the parallel arrangement, the electricity ceases only in that branch of the arrangement in which the break occurs. Current continues to flow in the remaining routes from A to B. Because parallel arrangements are very common, a given electric outlet can be used even though all others remain open. Parallel arrangements also explain why a light bulb can burn out (forming an air gap in place of the filament) without causing all other light bulbs to go out.

Batteries

Throughout the first half of the nineteenth century, the chief source of electric current was the chemical cell, and though this has long since yielded pride of place as far as sheer work load is concerned, it remains popular and, indeed, virtually irreplaceable for many special tasks.

A very common type of electric cell used nowadays has as

its negative pole a cup of metallic zinc, and as its positive pole, a rod of carbon embedded in manganese dioxide.* Between the two is a solution of ammonium chloride and zinc chloride in water. Starch is added to the solution in quantities sufficient to form a stiff paste so that the solution will not flow and the cell is made "unspillable." The unspillability is so impressive a characteristic that the device is commonly called a *dry cell.* It is also called a "flashlight battery" because it is so commonly used in flashlights. It may even be called a *Leclanche cell* because the zinc-carbon combination was first devised in 1868 by the French chemist Georges Leclanché (1839–1882), though it was not until twenty years later that it was converted into its "dry" form.

The potential difference between the positive and negative poles of a chemical cell depends upon the nature of the chemical reactions taking place—that is, on the strength of the tendency of the substances making up the poles to gain or lose electrons. In the case of the dry cell, the potential difference is, ideally, 1.5 volts.

The potential difference can be raised if two or more cells are connected in series—that is, if the positive pole of one cell is connected to the negative pole of the next cell in line. In that case, current flowing out of the first cell under the driving force of a potential difference of 1.5 volts enters the second cell and gives that much of a "push" to the current being generated there. The current put out by the second cell is therefore driven by its own potential difference of 1.5 volts plus the potential difference of the cell to which it is connected—3.0 volts altogether. In general, when cells are connected in series so that all the current travels through each one, the total potential difference is equal to the sum of the potential differences of the individual cells.

Cells might also be connected in parallel—that is, all the positive poles would be wired together and all the negative poles would be wired together. The total current does not pass through each cell. Rather, each cell contributes its own share of the current and receives back its own share, so the potential difference of one has no effect on that of another. There are advantages, however, in having 1.5 volts supplied by ten cells rather than by one. For one thing, the total amount of zinc in ten cells is ten times greater than in one, and the ten-cell combination will continue delivering current for ten times as long.

Then, too, there is the question of *internal resistance* of a

* This is the case where the cell is cylindrical in shape. The two poles are arranged somewhat differently where the cell is box-shaped.

battery. When current is flowing, it flows not only through the wires and devices that make up the circuit connecting positive pole to negative pole, it must also flow from pole to pole within the cell by way of the chemical reactions proceeding there. The internal resistance is the resistance to this electric flow within the cell. The greater the current intensity withdrawn from the cell, the greater, too, the current intensity that must flow through the cell. The potential difference required to drive this current through the cell depends on the current intensity, for (by Ohm's law) $E = IR$. R, in this case, is the internal resistance of the cell, and E is the potential difference driving the current from negative to positive pole (using the electrician's convention). This potential difference is the direction opposite to that driving the current from the positive pole to the negative in the *external circuit* outside the cell, so the internal potential difference must be subtracted from the external one. In short, as you draw more and more current intensity from a cell, the potential difference it delivers drops and drops, thanks to internal resistance.

When cells are arranged in series, the internal resistance of the series is the sum of the internal resistance of the individual cells. The potential difference may go up, but ten cells in series will be as sensitive to high current intensities as one cell would be. When cells are arranged in parallel, however, the total internal resistance of the cells in the system is less than is that of any single cell included, just as in the case of ordinary resistances (see page 196). A cell system in parallel can therefore deliver larger current intensities without significant drop in potential difference than a single cell could, although the maximum potential difference is no higher.

Electric cells of various sorts have been an incalculable boon to technological advance and are still extremely useful. Not only flashlights but a variety of devices from children's toys to radios can be powered by electric cells. Chemists like Davy even used them for important scientific advances requiring the delivery of fairly large amounts of electrical power. However, the really large-scale uses of electricity, such as that of powering huge factories and lighting whole cities, simply cannot be done by piling together millions of electric cells. The expense would be prohibitive.

The dry cell, for instance, obtains its energy by converting metallic zinc to zinc ions. Chemically, this is the equivalent of burning zinc—of using zinc as a fuel. When a dry cell is delivering 1 ampere of current, it consumes 1.2 grams of zinc in one hour. In that one hour, the power delivered by the battery would be 1.5 volts times 1 ampere, or 1.5 watts. Therefore 1.5 watt-hours is equivalent to the consumption of 1.2 grams of zinc, and 1 kilowatt-hour (1000 watt-hours) is equivalent to the consumption of 800 grams of zinc. A typical modern American household would, at the rate it consumes electricity, easily consume eight tons of zinc per year if dry cells were its source of supply (to say nothing of the other components involved.) Not only would this be ridiculously expensive, but the world's production of zinc could not maintain an economy in which individual families consumed the metal at this rate. In fact, it is fair to say that our modern electrified world simply could not exist on a foundation of ordinary chemical cells.*

One way of reducing the expense might be to devise some method of reversing the chemical reactions in a cell so that the original pole-substances might be used over again. This is not practical for the dry cell, but chargeable batteries do exist. The most common variety is one in which the negative pole is metallic lead and the positive pole is lead peroxide. These are separated by a fairly strong solution of sulfuric acid.

When such a cell is discharging, and an electric current is drawn off from it (at a potential difference of about 2 volts for an individual cell), the chemical reactions that proceed within it convert both the lead and the lead peroxide into lead sulfate. In the process, the sulfuric acid is consumed, too. If electricity is forced back into the cell (that is, if the negative pole of an electric source, working at a potential difference of more than 2 volts, is connected to the negative pole of the cell and the positive pole of the source is connected to the positive pole of the cell so that the cell is forced to "work backward" by a push that is stronger than its own), the chemical reaction goes into reverse. Lead and lead peroxide are formed once again, and the sulfuric acid solution grows stronger. The cell is "recharged." Such a cell was first devised in 1859 by the French physicist Gaston Planté (1834–1889).

In a superficial sense, it would seem that as the cell is re-

* Attempts are in progress to devise cells based on substances we would consider as more normal varieties of fuel. Such cells would be based on the combination of hydrogen, methane, alcohol, or even coal, with oxygen. In some cases bacterial action is used to bring about the necessary chemical reactions. Such *fuel cells* would be much cheaper than the more familiar chemical cells. By making it possible to draw electricity from burning fuel directly—and with greater efficiency than is now possible by methods to be described in the next two chapters—such cells could revolutionize various sectors of our economy. Fuel cells are only at the experimental stage, however, and were certainly not available during the decades, at the end of the nineteenth century, when the electrification of the industrialized regions of the earth began.

charged, electricity pours into the cell and is stored there. Actually, this is not so. Electricity is not directly stored; instead a chemical reaction is carried through, producing chemicals that can then react to generate electricity. Thus it is chemical energy that is stored, and such chargeable cells are called *storage batteries.* It is these (usually consisting of three to six lead-plus-lead-peroxide cells in series) that are present under the hood of automobiles.

Such a storage battery is heavy (because of the lead), dangerous to handle (because of the sulfuric acid), and expensive. Nevertheless, because it can be recharged over and over, the same battery can be used for years without replacement under conditions where heavy demands are periodically made upon it. Its usefulness, therefore, is not to be denied.

Yet where does the electricity come from that is used to recharge the storage battery? If such electricity is drawn from ordinary non-rechargeable cells, we are back where we started from. Clearly, in order to make it possible to use storage batteries on a large scale, the electricity used to charge it must have a much cheaper and more easily available source. In the automobile, for instance, the storage battery is continually being recharged at the expense of the energy of burning gasoline—which, while not exactly dirt-cheap, is certainly much cheaper and more available than the energy of burning zinc.

To explain how it is that burning gasoline can give rise to electric power, we must begin with a simple but crucial experiment conducted in 1819.

CHAPTER 27

Electromagnetism

Oersted's Experiment

Until the beginning of the nineteenth century, electricity and magnetism had seemed two entirely independent forces. To be sure, both were strong forces, both showed repulsion as well as attraction, and both weakened with distance according to an inverse square law. On the other hand, magnetism seemed to involve only iron plus (weakly) a few other substances, while electricity seemed universal in its effects; magnetism displayed poles in pairs only, while electricity displayed them in separation; and there was no flow of magnetic poles as there was a flow of electric charge. The balance seemed to be more in favor of the differences than the likenesses.

The matter was settled in 1819, however, as a result of a simple experiment conducted in the course of a lecture (and without any expectation of great events to come) by a Danish physicist, Hans Christian Oersted (1777–1851). He had been using a strong battery in his lecture, and he closed by placing a current-carrying wire over a compass in such a way as to have the wire parallel to the north-south alignment of the compass needle. (It is not certain now what point he was trying to make in doing this.)

However, when he put the wire over the needle, the needle turned violently, as though, thanks to the presence of the current, it now wanted to orient itself east-west. Oersted, surprised, carried the matter further by inverting the flow of current—that is, he connected the wire to the electrodes in reverse manner. Now the compass needle turned violently again, but in the opposite sense.

As soon as Oersted announced this, physicists all over Europe began to carry out further experimentation, and it quickly became plain that electricity and magnetism were intimately related, and that one might well speak of *electromagnetism* in referring to the manner in which one of the two forces gave rise to the other.

The French physicist Dominique François Jean Arago (1786–1853) showed almost at once that a wire carrying an electric current attracted not only magnetized needles but ordinary unmagnetized iron filings, just as a straightforward magnet would. A magnetic force, indistinguishable from that of ordinary magnets, originated in the electric current. Indeed, a flow of electric current *was* a magnet.

To show this more dramatically, it was possible to do away with iron, either magnetized or unmagnetized, altogether. If two magnets attracted each other or repelled each other (depending on how their poles were oriented), then the same should be true of two wires, each carrying an electric current.

This was indeed demonstrated in 1820 by the French physicist Ampère, after whom the unit of current intensity was named. Ampère began with two parallel wires, each connected to a separate battery. One wire was fixed, while the other was capable of sliding toward its neighbor or away from it. When the current was traveling in the same direction in both wires, the movable wire slid toward the other, indicating an attraction between the wires. If the current traveled in opposite directions in the two wires, the movable wire slid away, indicating repulsion between the two wires. Furthermore, if Ampère arranged matters so that the movable wire was free to rotate, it did so when the current was in opposite directions, turning through 180° until the two wires were parallel again with the current in each now flowing in the same direction. (This is analogous to the manner in which, if the north pole of one small magnet is brought near the north pole of another, the second magnet will flip so as to present its south pole end to the approaching north pole.)

Again, if a flowing current is a magnet, it should exhibit magnetic lines of force as an ordinary magnet does, and these lines of force should be followed by a compass needle. Since the compass needle tends to align itself in a direction perpendicular to that of the flow of current in the wire (whether the needle is held above or below the wire, or to either side), it would seem that the magnetic lines of force about a current-carrying wire appear in the form of concentric cylinders about the wire. If a cross section is taken perpendicularly through the wire, the lines of force will appear as concentric circles. This can be demonstrated by running a current-carrying wire upward through a small hole in a horizontal piece of cardboard. If iron filings are sprinkled on the cardboard and the cardboard is tapped, the filings will align themselves in a circular arrangement about the wire.

In the case of an ordinary magnet, the lines of force are considered to have a direction—one that travels from a north pole to a south pole. Since the north pole of a compass needle always points to the south pole of a magnet, it always points in the conventionally accepted direction of the lines of force. The direction of the north pole of a compass needle also indicates the direction of the lines of force in the neighborhood of a current-carrying wire, and this turns out to depend on the direction of the current-flow.

Ampère accepted Franklin's convention of current-flow from the positive electrode to the negative electrode. If, using this convention, a wire were held so that the current flowed directly toward you, the lines of force, as explored by a compass needle, would be moving around the wire in counterclockwise circles. If the current is flowing directly away from you, the lines of force would be moving around the wire in clockwise circles.

As an aid to memory, Ampère advanced what has ever since been called the "right-hand screw rule." Imagine yourself holding the current-carrying wire with your right hand; the fingers close about it and the thumb points along the wire in the direction in which the current is flowing. If you do that, then the sweep of the curving fingers, from palm to fingernails, indicates the direction of the magnetic lines of force.

(It is quite possible to make use of the direction of electron flow instead of that of the conventional current. The electron flow is in the direction opposite to that of the current, so that if you use the same device but imagine yourself to be seizing the wire with the left hand rather than the right, and the thumb in the direction of the electron flow, the fingers will still mark out the lines of force.)

Just as a magnet can come in a variety of shapes and by no means need be restricted in form to that of a simple bar, so a current-carrying wire can come in a variety of shapes. For instance, the wire can be twisted into a loop. When that happens, the lines of force outside the loop are pulled apart, while those inside the loop are crowded together. In other words, the magnetic field is stronger inside the loop than outside.

Now suppose that instead of one loop, the wire is twisted into a number of loops, so that it looks like a bedspring. Such a shape is called a *helix* or *solenoid* (the latter word from a Greek expression meaning "pipe-shaped"). In such a solenoid, the lines of force of each loop would reinforce those of its neighbor in such a way that the net result is a set of lines of force that sweep round the exterior of the entire solenoid from one end to the other. They then enter the interior of the solenoid to return to the first end again. The more individual loops or coils in the solenoid, the greater the reinforcement and the more lines of force are crowded into the interior. If the coils are pushed closed together, the reinforcement is more efficient, and again the magnetic flux increases in the interior of the solenoid.

In other words, the flux within the solenoid varies directly with the number of coils (N) and inversely with the length (L). It is proportional then to N/L. The strength of the magnetic field produced by a flowing current depends also on the current intensity. A 2-ampere current will produce twice the magnetic force at a given distance from the wire that a 1-ampere current will. In the case of the solenoid, then, we have the following relationship for a magnetic field that is virtually uniform in strength through the interior:

$$H = \frac{1.25\ NI}{L} \qquad \text{(Equation 12–1)}$$

where H is the strength of the magnetic field in oersteds, I the current intensity in amperes, N the number of turns in the solenoid, and L the length of the solenoid in centimeters.

The relationship between the strength of the magnetic field and the current intensity makes it possible to define the ampere in terms of the magnetic forces set up. If, in two long straight parallel conductors, one meter apart in air, constant and equal currents are flowing so as to produce a mutual force (either attractive or repulsive) of 2×10^{-7} newtons per meter of length, those currents have an intensity of 1 ampere. In this way, the ampere can be defined on the basis of mechanical units only, and all other electrical units can then be defined in terms of the ampere. (It was because Ampère's work made it possible to supply such a mechanical definition of an electrical unit that his name was given to the unit.)

A solenoid behaves as though it were a bar magnet made out of air. This, of course, suggests that in ordinary bar magnets there is a situation that is analogous to the looping electric current of the solenoid. It was not until the twentieth century, however, that the existence of the electron and its relationship to the atom came to be well enough understood to make that thought more than a vague speculation. Only then could ordinary magnetism be pictured as the result of spinning electron charges within atoms. In some cases, electron spins within atoms can be balanced, as some spin clockwise and some counterclockwise in such a fashion that no net magnetic force is to be observed. In other cases, notably in that of iron, the spins are not balanced and the magnetic force can make itself quite evident if the atoms themselves are properly aligned.

Furthermore, this offers the possibility of explaining the earth's magnetism. Even granted that the earth's liquid iron is at a temperature above the Curie point (see page 148) and cannot be an ordinary magnet, it is nevertheless possible that the earth's rotation sets up slow eddies in that liquid core, that electric charge is carried about in those eddies, and that the earth's core behaves as a solenoid rather than as a bar magnet. The effect would be the same.

If this is so, a planet that does not have a liquid core that can carry eddies, or that does not rotate rapidly enough to set eddies into motion, ought not to have much, if any, magnetic field. So far, the facts gathered by contemporary rocket experiments seem to bear this out. The density of the moon is only three-fifths that of the earth, which makes it seem likely that the moon does not have any high-density liquid iron core of significant size; and lunar probes have made it plain that the moon has, indeed, no significant magnetic field.

Venus, on the other hand, is very like the earth in size and density and therefore is quite likely to have a liquid iron core. However, astronomical data gained in the 1960's make it appear likely that Venus rotates slowly indeed, perhaps only once in 200-plus days. And Venus, too, according to observations of the Mariner II Venus-probe, lacks a significant magnetic field.

Jupiter and Saturn, which are much larger than the earth, nevertheless rotate more rapidly and possess magnetic fields far more intense than that of the earth.

The sun itself is fluid throughout, though gaseous rather than liquid, and as the result of its rotation, eddies are undoubtedly set up. It is possible that such eddies account for the magnetic field of the sun, which makes itself most evident in connection with the sunspots. Stars have been located with magnetic fields much more intense than that of the sun, and the galaxies themselves are thought to have galaxy-wide magnetic fields.

Applications of Electromagnetism

The strength of the magnetic field within a solenoid can be increased still further by inserting a bar of iron into it. The high permeability of iron (see page 156) will concentrate the already crowded lines of force even more. The first to attempt this was an English experimenter, William Sturgeon (1783–1850), who in 1823 wrapped eighteen turns of bare copper wire about a U-shaped bar and produced an *electromagnet*. With the current on, Sturgeon's electromagnet could lift twenty times its own weight of iron. With the current off, it was no longer a magnet and would lift nothing.

The electromagnet came into its own, however, with the work of the American physicist Joseph Henry (1797–1878). In 1829, he repeated Sturgeon's work, using insulated wire. Once the wire was insulated it was possible to wind it as closely as possible without fear of short circuits from one wire across to the next. Henry could therefore use hundreds of turns in a short length, thus vastly increasing the ratio N/L (see Equation 12–1) and correspondingly increasing the strength of the magnetic field for a given current intensity. By 1831, he had produced an electromagnet, of no great size, that could lift over a ton of iron.

In fact, electromagnetism made possible the production of magnetic fields of unprecedented strength. A toy horseshoe magnet can produce a magnetic field of a few hundred gauss in strength, a good bar magnet can produce one of 3000 gauss, and an excellent one a field of 10,000 gauss. With an electromagnet, however, fields of 60,000 gauss are easily obtainable.

To go still higher is in theory no problem, since one need only increase the current intensity. Unfortunately this also increases the heat produced (heat increases as the square of the current intensity), so the problem of cooling the coils soon becomes an extremely critical one. In addition, the magnetic forces set up large mechanical strains. By the twentieth century, ingenious designing and the use of strong materials had made possible the production of temporary fields in the hundreds of thousands of gauss by means of briefly pulsing electric currents. There was even the momentary production of fields of a million and a half gauss while the conducting material was exploding.

Such intense magnetic fields require the continuing use of enormous electric currents and lavish cooling setups. They are therefore exorbitantly expensive. The possibility arose of avoiding much of this expense by taking advantage of the phenomenon of superconductivity (see page 188). If certain conductors are cooled to liquid helium temperatures, their resistance drops to zero, so that current flowing through them develops no heat at all no matter how high the current intensity. Furthermore, an electric current set up in a closed circuit at such temperatures continues flowing forever, and a magnetic field set up in association with it also maintains itself forever (or at least as long as the temperature is maintained sufficiently low). In other words, the magnetic field need not be maintained at the expense of a continuous input of current.

If a superconductive material is used as the windings about an iron core and the whole is kept at liquid helium temperatures, then it might seem that by pumping more and more electricity into it, higher and higher magnetic field strengths could be built up without limit, and that when a desired level is reached, current can be shut off and the field left there permanently.

Unfortunately, superconductivity will not quite open the door in that fashion. A superconductive material is perfectly diamagnetic—that is, no magnetic lines of force at all will enter the interior of the superconductive material. The two properties —superconductivity and perfect diamagnetism—are bound together. If more and more current is pumped into a superconductive electromagnet and the magnetic field strength is built higher and higher, the magnetic flux mounts. The lines of force crowd more and more densely together, and at some point (the *critical field strength*) they are forced into the interior of the superconductive material. As soon as the material loses its perfect diamagnetism, it also loses its superconductivity; heat development starts, and the whole process fails. A superconductive magnet cannot be stronger than the critical field strength of the material making up the coils, and unfortunately, this is in the mere hundreds of gauss for most metals. Lead, for instance, will lose its superconductivity at even the lowest possible temperatures in a magnetic field of 600 gauss. Superconductive magnets of lead can therefore be built no stronger than a toy.

Fortunately, it was discovered in the 1950's that much more could be done with alloys than with pure metals. For instance, an alloy of niobium and tin can maintain superconductivity at liquid helium temperatures even while carrying enough current to produce a continuous and relatively cheap magnetic field in excess of 200,000 gauss, while an alloy of vanadium and gallium may do several times as well. The age of high-intensity superconducting electromagnets would seem to be upon us.

The electromagnet is fit for more than feats of brute strength, however. Consider a circuit that includes an electromagnet. There is a key in the circuit that is ordinarily, through some spring action, kept in an open position, so that an air gap is continuously present and no current flows through the circuit. If the key is depressed by hand so that the circuit is closed, the current flows and the electromagnet exerts an attracting force upon a bar of iron near it.

However, suppose this bar of iron is itself part of the circuit and that when it is attracted to the electromagnet it is pulled away from a connection it makes with the remainder of the circuit. The circuit is broken and the current stops. As soon as the current stops, however, the electromagnet is no longer exerting an attracting force, and the iron bar is pulled back to its original position by an attached spring. Since this closes the circuit again, the electromagnet snaps into action and pulls the iron bar toward itself again.

As long as the key remains depressed, this alternation of the iron bar being pulled to the magnet and snapping back to the circuit will continue. In itself this will make a rapid buzzing sound, and the result is, indeed, what is commonly called a "buzzer." If a clapper that strikes a hemisphere of metal is attached to the iron bar, we have an *electric bell.*

But suppose the iron bar is not itself part of the circuit. In that case, when the key is depressed the electromagnet gains attracting power and pulls the bar to itself and keeps it there. Once the key is released (and not until then), the electromagnet loses its attracting force and the bar snaps back.

The iron bar, under these circumstances, snaps back and forth, not in one unvarying rapid pattern of oscillation, but in whatever pattern one cares to impose upon the key as it is depressed and released. The iron bar makes a clicking sound as it strikes the magnet, and the pattern of the hand's movement upon the key is transformed into a pattern of clicks at the electromagnet.

It might occur to anyone that this could be used as the basis of a code. By letting a particular click pattern stand for particular letters of the alphabet, a message could be sent from one place to another at near the speed of light.

The practical catch is that the current intensity that can be pushed through a wire under a given potential difference decreases as the wire lengthens and its total resistance increases. Over really long distances, the current intensity declines to insignificance, unless prohibitive potential differences are involved, and is then not sufficient to produce a magnetic field strong enough to do the work of moving the heavy iron bar.

It was Henry who found a way to solve this problem. He ran current through a long wire until it was feeble indeed, but still strong enough to activate an electromagnet to the point where a lightweight key could be pulled toward it. This lightweight key, in moving toward the electromagnet, closed a second circuit that was powered by a battery reasonably near the key, so that a current was sent through a second, shorter wire. Thanks to the shortness of the second wire and its consequently smaller resistance, this second current was much stronger than the first. However, this second, stronger current mirrored the first, weaker current exactly, for when the original key was depressed by hand, the far-distant key was at once depressed by the electromagnet, and when the original key was released by hand, the far-distant key was at once released by the electromagnet.

Such a device, which passes on the pattern of a current from one circuit to another, is an *electric relay*. The second circuit may be, in its turn, a long one that carries just enough current intensity to activate a third circuit that, in turn, can just activate a far-distant fourth circuit. By using relays and batteries at regular intervals, there is nothing, in principle, to prevent one from sending a particular pattern of clicks around the world. By 1831, Henry was sending signals across a mile of wire.

Henry did not attempt to patent this or to develop it into a practical device. Instead, he helped an American artist, Samuel Finley Breese Morse (1791–1872) do so. By 1844, wires had been strung from Baltimore to Washington, and a pattern of clicks (reproduced as dots for short clicks and dashes for long ones—the "Morse code") was sent over it. The message was a quotation from the Bible's book of Numbers: "What hath God wrought?" This marks the invention of the *telegraph* (from Greek words meaning "writing at a distance"), and the general public was for the first time made aware of how the new electrical science could be applied in a manner that would change man's way of life.

Eventually telegraph lines spanned continents, and by 1866 a cable had been laid across the Atlantic Ocean. Through the cable, the Morse code could pass messages between Great Britain and the United States almost instantly. Laying the cable was a difficult and heartbreaking task, carried through only by the inhuman perseverance of the American financier Cyrus West Field (1819–1892). Its operation was also attended by enormous difficulties, since relays could not be set up at the bottom of the sea as they could be on land. Many problems had to be solved by such men as the British physicist William Thomson, Lord Kelvin (1824–1907), and even so, intercontinental communication did not become quite satisfactory until the invention of the radio (a matter which will be discussed in Volume III of this book). Nevertheless, by 1900 no civilized spot on earth was out of reach of the telegraph, and after thousands of years of civilization, mankind was for the first time capable of forming a single (if not always a mutually friendly or even mutually tolerant) community.

A more direct method of communication also depends in great part on the electromagnet. This is the *telephone* ("to speak at a distance"), invented in 1876 by a Scottish-American speech teacher, Alexander Graham Bell (1847–1922), and shortly afterward improved by Edison.

To put it as simply as possible, the telephone transmitter (into which one speaks) contains carbon granules in a box bounded front and back by a conducting wall. The front wall is a rather thin and, therefore, flexible diaphragm. Through this box an electric current flows. The resistance of the carbon granules depends on how well they make contact with each other. The better the contact, the lower the overall resistance and (since the potential difference remains constant) the greater the current intensity flowing through it.

As one speaks into the transmitter, sound waves set up a complex pattern of compressions and rarefactions in the air (see page I–156). If a region of compression strikes the diaphragm making up the front end of the box of carbon granules, the diaphragm is pushed inward. When a region of rarefaction strikes it, it is pulled outward. It acts precisely as does the eardrum, and in its motion, mimics all the variations in the compression and rarefaction of the sound wave pattern.

When the diaphragm is pushed inward, the carbon granules make better contact, and the current intensity rises in proportion to the extent to which the diaphragm is pushed. Similarly, the current intensity falls as the diaphragm is pulled outward, so that carbon granules fall apart and make poorer contact. Thus, an electric current is set up in which the intensity varies in precise imitation of the compression-rarefaction pattern of the sound wave.

At the other end of the circuit (which may be, thanks to relays and other refinements, thousands of miles away) the current activates an electromagnet in a telephone receiver. The strength of the magnetic field produced varies with the current intensity, so the strength of this field precisely mimics the sound wave pattern impinging upon the far-off transmitter. Near the electromagnet is a thin iron diaphragm that is pulled inward by the magnetic force in proportion to the strength of that force. The diaphragm in the receiver moves in a pattern that mimics the sound wave pattern impinging upon the transmitter many miles away and, in its turn, sets up a precisely similar sound wave pattern in the air adjoining it. The result is that the ear at the receiver hears exactly what the mouth at the transmitter is saying.

Examples of newer applications of electromagnets involve superconductivity (see page 156). A disk which is itself superconductive might rest above a superconducting magnet. The magnetic lines of force will not enter the perfectly diamagnetic superconducting disk, which cannot, for that reason, maintain physical contact with the magnet. There must be room between the two superconducting materials to allow passage, so to speak, for the lines of force. The disk, therefore, is repelled by the magnet and floats above it. Even weights placed upon the disk will not (up to some certain limit) force it down into contact with the magnet. Under conditions worked with in the laboratory, disks capable of bearing weights up to 300 grams per square centimeter have been demonstrated. Without physical contact, the disk can rotate virtually frictionlessly and thus may serve as a frictionless bearing.

Tiny switches can be made by taking advantage of electromagnetism under superconductive conditions. The first such device to be developed (as long ago as 1935) consisted of a thin wire of niobium about a thicker wire of tantalum. Both are superconducting materials, but they can be arranged to have different critical field strengths. A small current can be set up within the tantalum, for instance, and this will maintain itself indefinitely as long as the temperature is kept low. If an even smaller current is sent through the niobium coil about the tantalum, however, the magnetic field set up is sufficient to destroy the superconductivity of the tantalum (while not affecting that of the niobium). The current in the tantalum therefore ceases. In this way, one current can be switched off by another current.

Such a device is called a *cryotron* (from a Greek word meaning "to freeze," in reference to the extremely low temperatures required for superconductivity to evidence itself). Complex combinations of cryotrons have been made use of as versatile switching devices in computers. The advantage of cryotron switches are that they are very fast, very small, and consume very little energy. There is the disadvantage, of course, that they will only work at liquid helium temperatures.

Measurement of Current

The electromagnet introduced a new precision into the study of electricity itself. It made it possible to detect currents by the presence of the magnetic field they created and to estimate the current intensity by the strength of the magnetic field.

In 1820, following hard upon Oersted's announcement of the magnetic field that accompanied a flowing current, the German physicist Johann Salomo Christoph Schweigger (1779–1857) put that field to use as a measuring device. He placed a magnetized needle within a couple of loops of wire. When current flowed in one direction, the needle was deflected to the right; when current flowed in the other direction, the needle was deflected to the left. By placing a scale behind the needle, he could read off the amount of deflection and therefore estimate the current intensity. This was the first *galvanometer* ("to measure galvanic electricity"), a name suggested by Ampère.

Schweigger's original galvanometer had a fixed coil of wire and a movable magnet, but with time it was found more convenient to have a fixed magnet and a movable coil. The device still depends on the deflection of a needle, but now the needle is attached to the coil rather than to the magnet. A particularly practical device of this type was constructed in 1880 by the French physicist Jacques Arsène d'Arsonval (1851–1940) and is known as a *D'Arsonval galvanometer*.

Galvanometers can be made sensitive enough to record extremely feeble current intensities. In 1903, the Dutch physiologist Willem Einthoven (1860–1927) invented a *string galvanometer.* This consisted of a very fine conducting fiber suspended in a magnetic field. Tiny currents flowing through the fiber would cause its deflection, and by means of such an extremely sensitive galvanometer, the small changes in current intensities set up in a contracting muscle could be detected and measured. In this way, the shifting electric pattern involved in the heartbeat could be studied, and an important diagnostic device was added to the armory of modern medicine.

Galvanometers, in fact, tend to be so sensitive that in unmodified form they may safely be used only for comparatively feeble current intensities. To measure the full intensity of an ordinary household current, for instance, the galvanometer must be deliberately short-circuited. Instead of allowing all the current to flow through the moving coil in the galvanometer, a low-resistance conductor is placed across the wires leading in and out of the coil. This low-resistance short circuit is called a shunt. The device was first used in 1843 by the English physicist Charles Wheatstone (1802–1875).

Shunt and coil are in parallel and the current intensity going through each is in inverse proportion to their respective resistances. If the resistances are known, we can calculate what fraction of the current intensity will travel through the coil, and it is that fraction only that will influence the needle deflection. The sensitivity of the deflection can be altered by adding or subtracting resistances to the shunt, thus decreasing or increasing the fraction of the total current intensity passing through the coil.

By adjusting the fraction of the current intensity reaching the coil so that the deflected needle will remain on the scale, household current intensities or, in principle, current of any intensity, can be measured. The dial can be calibrated to read directly in amperes, and a galvanometer so calibrated is called an *ammeter* (a brief form of "ampere-meter").

Suppose a galvanometer is connected across some portion of a circuit, short-circuiting it. If this galvanometer includes a very high resistance, however, a current of very little intensity will flow over the short circuit through the galvanometer—current of an intensity low enough not to affect the remainder of the circuit in any significant way.

This small current intensity will be under the driving force of the same potential difference as will the current flowing in much greater intensity through the regular route of the circuit between the points across which the galvanometer has been placed. The tiny current intensity driven through the high-resistance galvanometer will vary with the potential difference. The scale behind the moving needle can then be calibrated in volts, and the galvanometer becomes a *voltmeter*.

Once current intensity and potential difference through some circuit or portion of a circuit is measured by an ammeter and voltmeter, the resistance of that same circuit or portion of a circuit can be calculated by Ohm's law. However, with the help of a galvanometer, resistance can also be measured directly by balancing the unknown resistance against known resistances.

Suppose a current is flowing through four resistances—R_1, R_2, R_3, and R_4—arranged in a parallelogram. Current enters at A and can flow either through B to D via R_1 and R_2, or through C to D via R_3 and R_4. Suppose that a conducting wire connects B and C and that a galvanometer is included as part of that wire. If the current reaches B at a higher potential than it reaches C, current will flow from B to C and the galvanometer will register in one direction. If the reverse is true and the current reaches C at a higher potential than at B, current will flow from C to B and the galvanometer will register in the other direction. But if the potential at B and the potential at C are exactly equal, current will flow in neither direction and the galvanometer will register zero.

Suppose the galvanometer does register zero. What can we deduce from that? The current flowing from A to B must pass on, intact, from B to D; none is deflected across the galvanometer. Therefore the current intensity from A to B through R_1 must be the same as the current intensity from B to D through R_2. Both intensities can be represented as I_1. By a similar argument, the current intensities passing through R_3 and R_4 are equal and may be symbolized as I_2.

By Ohm's law, potential difference is equal to current intensity times resistance ($E = IR$). The potential difference from A to B is therefore I_1R_1; from B to D is I_1R_2; from A to C is I_2R_3; and from C to D is I_2R_4.

But if the galvanometer reads zero, then the potential difference from A to B is the same as from A to C (or current would flow between B and C and the galvanometer would not read zero), and the potential difference from B to D is the same as from C to D, by the same argument. In terms of current intensities and resistances, we can express the equalities in potential difference thus:

$$I_1R_1 = I_2R_3 \qquad \text{(Equation 12–2)}$$

$$I_1R_2 = I_2R_4 \qquad \text{(Equation 12–3)}$$

If we divide Equation 12–2 by Equation 12–3, we get:

$$\frac{R_1}{R_2} = \frac{R_3}{R_4} \qquad \text{(Equation 12–4)}$$

Now suppose that R_1 is the unknown resistance that we want to measure, while R_2 is a known resistance. As for R_3 and R_4, they are *variable resistances* that can be varied through known steps.

A very simple variable resistance setup can consist of a length of wire stretched over a meter stick, with a sliding contact capable of moving along it. The sliding contact can represent point C in the device described above. The stretch of wire from end to end of the meter stick is AD. That portion stretching from A to C is R_3 and the portion from C to D is R_4. If the wire is uniform, it is fair to assume that the resistances R_3 and R_4 will be in proportion to the length of the wire from A to C and from C to D respectively, and those lengths can be read directly off the meter stick. The absolute values of R_3 and R_4 cannot be determined, but the ratio R_3/R_4 is equal to AC/CD and that is all we need.

As the sliding contact is moved along the wire, the potential difference between A and C increases as the distance between the two points increases. At some point the potential difference between A and C will become equal to that between A and D, and the galvanometer will indicate that point by registering zero. At that point the ratio R_3/R_4 can be determined directly from the meter stick, and the ratio R_1/R_2 can, by Equation 12–4, be taken to have the same value.

The unknown resistance, R_1, can then easily be determined by multiplying the known resistance R_2 by the known ratio R_3/R_4. Wheatstone used this device to measure resistances in 1843, and (athough some researchers had used similar instruments before him) it has been called the *Wheatstone bridge* ever since.

Generators

The electromagnet, however useful, does not in itself solve the problem of finding a cheap source of electricity. If the magnetic field must be set up through the action of a chemical cell, the field will remain as expensive as the electric current that sets it up, and large-scale uses will be out of the question.

However, the manner in which an electromagnet was formed was bound to raise the question of the possibility of the reverse phenomenon. If an electric current produces a magnetic field, might not a magnetic field already in existence be used to set up an electric current?

Michael Faraday thought so, and in 1831 he attempted a crucial experiment (after having tried and failed four times before). In this fifth attempt, he wound a coil of wire around one segment of an iron ring, added a key with which to open and close the circuit, and attached a battery. Now when he pressed the key and closed the circuit, an electric current would flow through the coil and set up a magnetic field. The magnetic lines of force would be concentrated in the highly permeable iron ring in the usual fashion.

Next, he wound another coil of wire about the opposite segment of the iron ring and connected that coil to a galvanometer. When he set up the magnetic field, it might start a current flowing in the second wire and that current, if present, would be detected by the galvanometer.

The experiment did not work as he expected it to. When he closed the circuit there was a momentary surge of current through the second wire, as was indicated by a quick deflection of the galvanometer needle followed by a return to a zero reading. The zero reading was then maintained however long the key remained depressed. There was a magnetic field in existence and it was concentrated in the iron ring, as could easily be demonstrated. However, the mere existence of the magnetic field did not in itself produce a current. Yet when Faraday opened the circuit again, there was a second quick deflection of the galvanometer needle, in a direction opposite to the first.

Faraday decided it was not the existence of magnetic lines of force that produced a current, but the motion of those lines of force across the wire. He pictured matters after this fashion. When the current started in the first coil of wire, the magnetic field sprang into being, the lines of force expanding outward to fill space. As they cut across the wire in the second coil, a current was initiated. Because the lines of force quickly expanded to the full and then stopped cutting across the wire, the current was only momentary. With the circuit closed and the magnetic field stationary, no further electric current in the second coil was to be expected. However, when he opened the first circuit, the magnetic field ceased and the lines of force collapsed inward again, momentarily setting up a current in a direction opposite to the first.

He showed this fact more plainly to himself (and to the audiences to whom he lectured) by inserting a magnet into a coil of wire that was attached to a galvanometer. While the magnet was being inserted, the galvanometer needle kicked in one direction; and while it was being withdrawn, it kicked in the other direction. While the magnet remained at rest within the coil at any stage of its insertion or withdrawal, there was no current in the coil. However, the current was also initiated in the coil if the magnet was held stationary and the coil was moved down over it or lifted off again. It didn't matter, after all, whether the wire moved across the lines of force, or the lines of force moved across the wire.*

Faraday had indeed used magnetism to induce an electric current and had thus discovered *electromagnetic induction.* In the United States, Henry had made a similar discovery at about the same time, but Faraday's work was published first.

The production of an induced current is most easily visualized if one considers the space between the poles of a magnet, where lines of force move across the gap in straight lines from the north pole to the south pole, and imagines a single copper wire moving between those poles. (It makes no difference, by the way, whether the magnet in question is a permanent one or an electromagnet with the current on.)

If the wire were motionless or were moving parallel to the lines of force, there would be no induced current. If the wire moved in a direction that was not parallel to the lines of force, so that it cut across them, then there would be an induced current.

The size of the potential difference driving the induced current would depend upon the number of lines of force cut across per second, and this in turn would depend on a number of factors. First there is the velocity of the moving wire. The more rapidly the wire moves in any given direction not parallel to the lines of force, the greater the number of lines of force cut across per second, and the greater the potential difference driving the induced current.

Again, there is the question of the direction of the motion of the wire. If the wire is moving in a direction perpendicular to the lines of force, then it is cutting across a certain number of lines of force per second. If the wire is moving, at the same speed, in a direction not quite perpendicular to the lines of force, it cuts across fewer of them per unit time, and the potential difference of the induced current is less intense. The greater the angle between the direction of motion and the perpendicular to the lines of force, the smaller the potential difference of the induced current. Finally when the motion is in a direction 90° to the perpendicular, that

* There is a famous story about such a demonstration, which is probably apocryphal. A woman, having watched the demonstration of the coil and the magnet, is supposed to have said, "But, Mr. Faraday, of what use is this?" Faraday is reported to have answered politely, "Madam, of what use is a newborn baby?" Another version has William Ewart Gladstone (then a freshly elected member of Parliament and eventually to be four times Prime Minister) ask the question. Faraday is supposed to have answered, "Sir, in twenty years, you will be taxing it."

motion is actually parallel to the lines of force and there is no induced current at all.

In addition, if the wire is in coils, and each coil cuts across the lines of force, the potential difference driving the induced current is multiplied in intensity by the number of coils per unit length.

The direction of the induced current can be determined by using the right hand, according to a system first suggested by the English electrical engineer John Ambrose Fleming (1849–1945) and therefore called *Fleming's rule*. It is applied without complication when the wire is moving in a direction perpendicular to that of the lines of force. To apply the rule, extend your thumb, index finger and middle finger so that each forms a right angle to the other two—that is, allowing the thumb to point upward, the forefinger forward, and the middle finger leftward. If, then, the forefinger is taken as pointing out the direction of the magnetic lines of force from north pole to south pole, and the thumb as pointing out the direction in which the wire moves, the middle finger will point out the direction (from positive pole to negative pole) of the induced current in the wire.

Two months after his discovery of electromagnetic induction, Faraday took his next step. Since an electric current was produced when magnetic lines of force cut across an electrical conductor, how could he devise a method of cutting such lines continuously?

He set up a thin copper disk that could be turned on a shaft. Its outer rim passed between the poles of a strong magnet as the disk turned. As it passed between those poles, it continuously cut through magnetic lines of force, so that a potential difference was set up in the disk, a difference that maintained itself as long as the disk turned. Two wires ending in sliding contacts were attached to the disk. One contact brushed against the copper wheel as it turned, the other brushed against the shaft. The other ends of the wires were connected to a galvanometer.

Since the electric potential was highest at the rim where the material of the disk moved most rapidly and therefore cut across more lines of force per unit time, a maximum potential difference existed between that rim and the motionless shaft. An electric current flowed through the wires and galvanometer as long as the disk turned. Faraday was generating a current continuously, without benefit of chemical reactions, and had thus built the first *electric generator*.

The importance of the device was tremendous, for, in essence, it converted the energy of motion into electrical energy. A disk could be kept moving by a steam engine, for instance, at the expense of burning coal or oil (much cheaper than burning zinc), or by a turbine that could be turned by running water, so that streams and waterfalls could be made to produce electricity. It took fifty years to work out all the technical details that stood in the way of making the generator truly practical, but by the 1880's cheap electricity in quantity was a reality; and the electric light and indeed the electrification of society in general became possible.

CHAPTER 28

Alternating Current

The Armature

In modern generators, Faraday's copper disk turning between the poles of a magnet is replaced by coils of copper wire wound on an iron drum turning between the poles of an electromagnet. The turning coils make up the *armature*. To see what happens in this case, let's simplify matters as far as possible and consider a single rectangular loop of wire rotating between a north pole on the right and a south pole on the left.

Imagine such a rectangle oriented parallel to the lines of force (moving from right to left) and beginning to turn in such a fashion that the wire on the left side of the rectangle (the L wire) moves upward, across the lines of force, while the wire on the right side of the rectangle (the R wire) moves downward, across the lines of force.

Concentrate, to begin with, on the L wire and use the Fleming right-hand rule. Point your thumb upward, for that is the direction in which the L wire is moving. Point your forefinger left, for that is the direction of the magnet's south pole. Your middle finger points toward you, and that is the direction of the induced current in the L wire.

What about the R wire? Now the thumb must be pointed downward while the forefinger still points left. The middle finger points away from you and that is the direction of the induced current in the R wire. If the induced current is traveling toward you in the L wire and away from you in the R wire, you can see that it is actually going round and round the loop.

Next imagine that the L wire and R wire are both connected to separate "slip rings" (Ring A and Ring B, respectively), each of which is centered about the shaft that serves as an axis around which the loop rotates. The current would tend to flow from Ring B through the R wire into the L wire and back into Ring A. If one end of a circuit is connected to one ring by way of a brushing contact, and the other end of the same circuit to the other ring, the current generated in the turning armature would travel through the entire circuit.

But let's consider the rectangular loop a bit longer. Since the loop is rotating, the L wire and R wire cannot move up and down, respectively, indefinitely. They are, in fact, constantly changing direction. As the L wire moves up, it curves to the right and moves at a smaller angle to the lines of force, so that the intensity of the induced current decreases. Precisely the same happens to the R wire, for as it moves downward, it curves to the left and moves also at a smaller angle to the lines of force.

The current continues to decrease as the loop turns until the loop has completed a right angle turn, so that the L wire is on top and the R wire on bottom. The L wire is now moving right, parallel to the lines of force, while the R wire is moving left, also parallel to the lines of force. The intensity of the induced current has declined to zero. As the loop continues to rotate, the L wire cuts down into the lines of force, while the R wire cuts up into them. The two wires have now changed places, the L wire becoming the R wire, and the R wire becoming the L wire.

The wires, despite this change of place (as far as the direction of the induced current is concerned), are still connected to the same slip rings. This means that as the armature makes one complete rotation, the current flows from Ring B to Ring A for half the time and from Ring A to Ring B for the other half. This repeats itself in the next rotation, and the next, and the next. Current produced in this manner is therefore *alternating current* (usually abbreviated *AC*) and moves backward and forward perpetually. One rotation of the loop produces one back and forth movement of the current—that is, one *cycle*. If the loop rotates sixty times a second we have a 60-cycle alternating current.

Nor is the current steady in intensity during the period when it is moving in one particular direction. During one rotation of the loop, the current intensity begins at zero, when the moving wires (top and bottom) are moving parallel to the lines of force; it rises smoothly to a maximum, when the wires (right and left) are moving perpendicular to the lines of force, and then drops just as smoothly to zero again, when the wires (bottom and top) are moving parallel to the lines of force once more.

As the loop continues turning, the current changes direction, and we can now imagine the flow to be less than zero—that is, we can decide to let the current intensity be measured in positive numbers when its flow is in one direction and in negative numbers when its flow is in the other. Therefore, after the intensity has dropped to zero, it continues smoothly dropping to a minimum when the wires (left and right) are moving perpendicular to the lines of force; and it rises smoothly to zero again when the wires (top and bottom) are moving parallel to the lines of force once more. This completes one rotation, and the cycle begins again.

If, for convenience, we imagine the maximum current intensity to be 1 ampere, then in the first quarter of the rotation of the loop the intensity changes from 0 to +1; in the second quarter from +1 to 0; in the third quarter from 0 to −1; and in the fourth quarter from −1 to 0. If this change in intensity is plotted against time, there is a smoothly rising and falling wave, endlessly repeated, which mathematicians call a *sine curve.*

A generator can easily be modified in such a way as to make it produce a current that flows through a circuit in one direction only. This would be a *direct current* (usually abbreviated *DC*), and it is this type of current that was first dealt with by Volta and which is always produced by chemical cells.

Suppose the two ends of our rectangular loop are attached to "half-rings" which adjoin each other around the shaft serving as axis of rotation but which don't touch. The L wire is attached to one half-ring and the R wire to the other. The brush contact of one end of the circuit touches one half-ring; the brush contact of the other touches the second half-ring.

During the first half of the rotation of the armature, the current flows, let us say, from Half-Ring A to Half-Ring B. During the second half, the current reverses itself and flows from Half-Ring B to Half-Ring A. However, every time the armature goes through a half-rotation, the half-rings change places. If one brush contact is touching the positive half-ring, the negative half-ring turns into place just as it becomes positive, and leaves its place just as it begins to turn negative again. In other words, the first brush contact is touching each half-ring in turn, always when the rings are in the positive portion of their cycle; the other brush contact always touches the half-rings when they are negative. The current may change direction in the armature, but it flows in one constant direction in the attached circuit.

The intensity still rises and falls, to be sure, from 0 to +1 to 0 to +1, and so on. By increasing the number of loops and splitting the rings into smaller segments, these variations in intensity can be minimized, and a reasonably constant direct current can be produced.

The AC generator is simpler in design than the DC generator, but alternating current had to overcome a number of objections before it could be generally accepted. Edison, for instance, was a great proponent of direct current and, during the closing decades of the nineteenth century, fought the use of alternating current bitterly. (The great proponent of alternating current was the American inventor George Westinghouse [1846–1914]).

In considering this competition between the two types of current, it may seem natural at first that DC should be favored. In fact, AC may seem useless on the face of it. After all, a direct current is "getting somewhere" and is therefore useful, while an alternating current "isn't getting anywhere" and therefore can't be useful—or so it might seem.

Yet the "isn't getting anywhere" feeling is wrong. It is perhaps the result of a false analogy with water running through a pipe. We usually want the water to get somewhere—to run out of the pipe, for instance, so that we can use it for drinking, washing, cooling, irrigating, fire fighting, and so on.

But electricity never flows out of a wire in ordinary electrical appliances. It "isn't getting anywhere" under any circumstances. Direct current may go in one direction only, but it goes round and round in a circle and that is no more "getting anywhere" than moving backward and forward in one place.

There are times when DC is indeed necessary. In charging storage batteries, for instance, you want the current to move always in the direction opposite to that in which the current moves when the storage battery is discharging. On the other hand, there are times when it doesn't matter whether current is direct or alternating. For instance, a toaster or an incandescent bulb works as it does simply because current forcing its way through a resistance heats up that portion of the circuit (to a red-heat in the toaster and to a white-heat in the bulb). The heating effect does not depend upon the direction in which the current is flowing, or even on whether it continually changes direction. By analogy, you can grow heated and sweaty if you run a mile in a straight line, or on a small circular track, or backward and forward in a living room. The heating effect doesn't depend on "getting anywhere."

A more serious objection to AC, however, was that the mathematical analysis of its behavior was more complicated than was that of DC circuits. For the proper design of AC circuits, this mathematical analysis first had to be made and understood. Until then, the circuits were continuously plagued by lowered efficiency.

Impedance

A situation in which the current intensity and the potential difference are changing constantly raises important questions—for instance, as to how to make even the simplest calculations involving alternating current. When a formula includes I (current intensity) or E (potential difference), one is entitled to ask what value to insert when an alternating current has no set value for either quantity, but a value that constantly varies from zero to some maximum value (I_{max} and E_{max}), first in one direction and then in the other.

One must then judge these properties of the alternating current by the effects they produce, rather than by their sheer numerical values. It can be shown, for instance, that an alternating current can, in heat production and in other uses, have the same effect as a direct current with definite values of I and E. The values of I and E therefore represent the *effective current intensity* and the *effective potential difference* of the alternating current, and it is these effective values that I and E symbolize in alternating currents. The effective values are related to the maximum values as follows:

$$I = \frac{I_{max}}{\sqrt{2}} = 0.7\ I_{max} \qquad \text{(Equation 13–1)}$$

$$E = \frac{E_{max}}{\sqrt{2}} = 0.7\ E_{max} \qquad \text{(Equation 13–2)}$$

One might suppose that having defined I and E for alternating currents one could proceed at once to resistance, representing that by the ratio of E/I (the current intensity produced by a given potential difference) in accordance with Ohm's law. Here, however, a complication arises. A circuit that under direct current would have a low resistance, as indicated by the fact that a high current intensity would be produced by a given potential difference, would under alternating current have a much greater resistance, as indicated by the low current intensity produced by the same potential difference. Apparently, under alternating current a circuit possesses some resisting factor other than the ordinary resistance of the materials making up the circuit.

To see why this is so, let's go back to Faraday's first experiments on electromagnetic induction (see page 216). There, as the electric current was initiated in one coil, a magnetic field was produced and the expanding lines of force cut across a second coil, inducing a potential difference and, therefore, an electric current in a particular direction in that second coil. When the electric current was stopped in the first coil, the collapsing lines of force of the dying magnetic field cut across the second coil again, inducing a potential difference of reversed sign and, therefore, an electric current in the opposite direction in that second coil.

So far, so good. However, it must be noted that when current starts flowing in a coil, so that magnetic lines of force spread outward, they not only cut across other coils in the neighborhood but also cut across the very coils that initiate the magnetic field. (The lines of force spreading out from one loop in the coil cuts through all its neighbors.) Again, when the current in a coil is cut off, the lines of force of the collapsing magnetic field cut across the very coils in which the current has been cut off. As current starts or stops in the coil, an induced current is set up in that same coil. This is called *self-induction* or *inductance,* and it was discovered by Henry in 1832. (Here Henry announced his discovery just ahead of Faraday, who made the same discovery independently; Faraday, you will remember, just anticipated Henry in connection with electromagnetic induction itself.)

Almost simultaneously with Henry and Faraday, a Russian physicist, Heinrich Friedrich Emil Lenz (1804–1865), studied inductance. He was the first to make the important generalization that induced potential difference set up in a circuit always acts to oppose the change that produced it. This is called *Lenz's law.*

Thus, when a current in a coil is initiated by closing a circuit, one would expect that the current intensity would rise instantly to its expected level. However, as it rises it sets up an induced potential difference, which tends to produce a current in the direction opposed to that which is building up. This opposition by inductance causes the primary current in the circuit to rise to its expected value with comparative slowness.

Again, if the current in a coil is stopped by breaking a circuit, one would expect the current intensity to drop to zero at once. Instead, the breaking of the circuit sets up an induced potential which, again, opposes the change and tends to keep the current flowing. The current intensity therefore drops to zero with comparative slowness. This opposed potential difference produced by self-induction is often referred to as *back-voltage*.

In direct current, this opposing effect of inductance is not terribly important, since it makes itself felt only in the moment of starting and stopping a current, when lines of force are moving outward or inward. As long as the current flows steadily in a single direction, there is no change in the magnetic lines of force, no induced current, and no interference with the primary current itself.

How different for alternating current, however, which is always changing, so that magnetic lines of force are always cutting the coils as they are continually moving either outward or inward. Here an induced potential difference is constantly in being and is constantly opposed to the primary potential difference, reducing its value greatly. Thus, where a given potential difference will drive a high (direct) current intensity through a circuit, under AC conditions it will be largely neutralized by inductance, and will therefore send only a small (alternating) current intensity through the same circuit.

The unit of inductance is the *henry*, in honor of the physicist. When a current intensity in a circuit is changing at the rate of 1 ampere per second and, in the process, induces an opposed potential difference of 1 volt, the circuit is said to have an inductance of 1 henry. By this definition, 1 henry is equal to 1 (volt per ampere) per second, or 1 volt-second per ampere (volt-sec/amp).

The resistance to current flow produced by self-induction depends not only on the value of the inductance itself but also on the frequency of the alternating current, since with increasing frequency, the rate of change in current intensity with time (amperes per second) increases. Therefore, the more cycles per second, the greater the resistance to current flow by a given inductance. Suppose we symbolize inductance as L and the frequency of the alternating current as f. The resistance produced by these two factors is called the *inductive reactance* and is symbolized as X_L. It turns out that:

$$X_L = 2\pi f L \qquad \text{(Equation 13–3)}$$

If L is measured in henrys, that is in volt-seconds per ampere, and f is measured in per-second units, then the units of X_L must be (volt-seconds per ampere) per second. The seconds cancel and the units become simply volts per ampere, which defines the ohm (see page 185). In other words, the unit of inductive reactance is the ohm, as it is of ordinary resistance.

The ordinary resistance (R) and the inductive reactance (X_L) both contribute to the determination of the current intensity placed in an alternating current circuit by a given potential difference. Together they make up the *impedance* (Z). It is not a question of merely adding resistance and inductive reactance, however. Impedance is determined by the following equation:

$$Z = \sqrt{R^2 + X_L{}^2} \qquad \text{(Equation 13–4)}$$

In alternating currents, it is impedance that plays the role of ordinary resistance in direct currents. In other words, the AC equivalent of Ohm's law would be $IZ = E$, or $I = E/Z$, or $Z = I/E$.

Reactance is produced in slightly different fashion by condensers. A condenser in a direct current circuit acts as an air gap and, at all reasonable potential differences, prevents a current from flowing. In an alternating current circuit, however, a condenser does not keep a current from flowing. To be sure, the current does not flow across the air gap, but it surges back and forth, piling up electrons first in one plate of the condenser, then in the other. In the passage back and forth from one plate to the other, the current passes through an appliance—let us say an electric light—which proceeds to glow. The filament reacts to the flow of current through itself and not to the fact that there might be another portion of the circuit somewhere else through which there is no flow of current.

The greater the capacitance of a condenser, the more intense the current that sloshes back and forth, because a greater charge can be piled onto first one plate then the other. Another way of putting this is that the greater the capacitance of a condenser, the smaller the opposition to the current flow, since there is more room for the electrons on the plate and therefore less of a pile-up of negative-negative repulsion to oppose a continued flow.

This opposition to a continued flow is the *capacitive reactance* (X_C), which is inversely proportional to the capacitance (C) of the condenser. The capacitive reactance is also inversely proportional to the frequency (f) of the current, for the more rapidly the current changes direction, the less likely is either plate of the condenser to get an oversupply of electrons during the course of one half-cycle, and the smaller the negative-negative repulsion set up to oppose the current flow. (In other words, raising the frequency lowers the capacitive reactance, though it raises the inductive reactance.) The inverse relationship can be expressed as follows:

$$X_C = \frac{1}{2\pi fC} \qquad \text{(Equation 13–5)}$$

The capacitance (C) is measured in farads—that is in coulombs per volt or, what is equivalent, in ampere-seconds per volt. Since the units of frequency (f) are per-seconds, the units of $2\pi fC$ are ampere-seconds per volt per seconds, or (with seconds canceling) amperes per volt. The units of capacitive reactance (X_C) are the reciprocal of this—that is, volts per ampere, or ohms. Thus capacitive reactance, like inductive reactance, is a form of resistance in the circuit.

Capacitive reactance and inductive reactance both act to reduce the current intensity in an AC circuit under a given potential difference if either one is present singly. However, they do so in opposite manners.

Under simplest circumstances, the current intensity and potential difference of an alternating current both rise and fall in step as they move along the sine curve. Both are zero at the same time; both are at maximum crest or at minimum trough at the same time. An inductive reactance, however, causes the current intensity to *lag;* to reach its maximum (or minimum, or zero point) only a perceptible interval after the potential difference has reached it. A capacitive reactance, on the other hand, causes the current intensity to *lead;* to rise and fall a perceptible period of time ahead of the rise and fall in potential difference. In either case, current intensity and potential difference are out of phase, and energy is lost.

Yet if there is both a capacitive reactance and an inductive reactance in the circuit, the effect of one is to cancel that of the other. The lead of the capacitive reactance must be subtracted from the lag of the inductive reactance. The total impedance can be expressed as follows:

$$Z = \sqrt{R^2 + (X_L - X_C)^2} \qquad \text{(Equation 13–6)}$$

If the circuit is so arranged that the capacitive reactance equals the inductive reactance, then $X_L - X_C = 0$; and $Z =$

$\sqrt{R^2} = R$. The impedance of the alternating current circuit is then no greater than the ordinary resistance of an analogous direct current circuit would be. Such an alternating current circuit is said to be *in resonance*. Notice that the impedance can never be less than the resistance. If the capacitive reactance is greater than the inductive reactance, then $X_L - X_C$ is indeed a negative number, but its square is positive and when that is added to the square of the resistance and the square root of the sum is taken, the final value of Z will be greater than that of R.

This represents only the merest beginnings of the complications of AC circuitry. A good deal of the full treatment was worked out at the beginning of the twentieth century by the German-American electrical engineer Charles Proteus Steinmetz (1865–1923), and it was only thereafter that alternating currents could be properly exploited.

Transformers

Even before Steinmetz had rationalized the use of alternating circuits—and despite the formidable nature of the difficulties which, in the absence of such rationalization, plagued the circuit-designers; and despite, also, the formidable opposition of men such as Edison and Kelvin—the drive for the use of alternating current carried through to victory. The reason for this was that in one respect, that involving the transmission of electric power over long distances, alternating current was supreme over direct current.

The power of an electric current is measured in watts and is equal to the volts of potential difference times the amperes of current intensity. (Strictly speaking, this is true only in the absence of reactance. Where inductive reactance is present, the power is decreased by a particular *power factor*. However, this can be reduced or even eliminated by the addition of a proper capacitive reactance, so this need not bother us.)

This means that different combinations of volts and amperes can represent an electric current of the same power. For instance, a given appliance might carry one ampere at 120 volts, or two amperes at 60 volts, or five amperes at 24 volts, or twelve amperes at 10 volts. In each case, the power would be the same—120 watts.

There are advantages to having a given electric power appear in a high-volt, low-ampere arrangement under some conditions and in a low-volt, high-ampere arrangement under other conditions. In the former case, the low current intensity makes it possible to use relatively thin copper wire in the circuit without fear of undue heating effects. In the latter case, the low potential difference means that there is a smaller chance of breaking down the insulation or producing a short circuit.

And then there is the previously mentioned problem of transmitting electric currents over long distances. Much of the convenience of electricity would be lost if it could only be used in the near neighborhood of a generator. Yet if the current is sent through wires over considerable distances, so much energy is likely to be lost in the form of heat that either we have too little electricity at the other end to bother with, or we must reduce the heat-loss by using wire so thick as to be uneconomical. The heat produced, however, is proportional to the square of the current intensity. Therefore, if we reduce the current intensity to a very low quantity, while simultaneously raising the potential difference to a correspondingly high value (in order to keep the total electric power unchanged), much less electricity would be lost as heat.

Naturally, it is not very likely that this arrangement of tiny current intensity combined with huge potential differences would be suitable for use in ordinary appliances. Consequently, we want a situation in which the same power can be at very high voltages for transmission and at very low voltages for use.

With direct current, it is highly impractical to attempt to change the potential difference of a current—now up, now down—to suit changing needs. In alternating current, however, it is easy to do this by means of a *transformer* (a device that "transforms" the volt-ampere relationship). In essence, it was a transformer that Faraday had invented when in 1831 he made use of an iron ring with two sets of wire coils on it in his attempt to induce an electric current.

Faraday found that when a direct electric current was put through one of the coils (the *primary*), no current was induced in the other coil (the *secondary*), except at the moments when the current was initiated or ended. It was only then that magnetic lines of force swept over the secondary.

Where the current in the primary is an alternating current, however, the current intensity is always either rising or falling; and the intensity of the magnetic field through the iron ring is always either rising or falling. The lines of force expand outward and collapse inward over and over, and as they do so, they cut across the secondary, producing an alternating current that keeps in perfect step with the alternating current in the primary.

The potential difference of the induced current depends on

the number of coils in the secondary as compared with the number in the primary. Thus, if the current in the primary has a potential difference of 120 volts and if the secondary contains ten times as many turns of wire as does the primary, then the induced current will have a potential difference of 1200 volts. This is an example of a *step-up transformer*. If the induced current produced by such a transformer is used to power the primary in another transformer in which the secondary now has only one-tenth the number of coils that the primary has, the new induced current is back at 120 volts. The second transformer is a *step-down transformer*.

The induced current (if we ignore negligible losses in the form of heat) must have the same power as the original current. Otherwise, energy will either be created or destroyed in the process, and this is inadmissable. This means that as the potential difference goes up, the current intensity must go down, and vice versa. If a one-ampere current at 120 volts activates a step-up transformer in which the secondary has a hundred times the number of coils that the primary has, the induced current will have a potential difference of 12,000 volts and a current intensity of 1/100 ampere. In both primary and secondary, the power will be 120 watts.

If alternating current generators are used, there is no difficulty at all in altering voltages by means of transformers. A step-up transformer in particular will serve to raise the potential difference to great heights and the current intensity to trivial values. Such a current can be transmitted over long distances through wires that are not excessively thick, with little heat loss, thanks to the low current intensity. Thanks to the high potential difference, however, the full power of the electric current is nevertheless being transmitted.

When the current arrives at the point where it is to be used, a step-down transformer will convert it to a lower potential difference and a higher current intensity for use in household appliances or industrial machines. A particular appliance or machine may need low potential differences at one point and high potential differences at the other, and each can be supplied by the use of appropriate transformers.

Long-distance transmission through high-voltage alternating current was made practical by the work of the Croatian-American electrical engineer Nikola Tesla (1857–1943). He was backed by Westinghouse, who in 1893 won the right to set up at Niagara

Falls a hydroelectric station (where the power of falling water would spin turbines that would turn armatures and produce electricity) for the production and transmission of alternating current.

Since then, AC current has come into virtually universal use, and this is responsible for the great flexibility and versatility of electricity as a form of useful energy.

Motors

Thanks to the development of the generator, mechanical energy could be converted to electrical energy, and it was possible to have large supplies of electricity arising, indirectly, out of burning coal or falling water. Thanks to the development of alternating current and transformers, this electrical energy could be transported over long distances and conducted into every home or factory.

However, once in the home or factory, what was the electricity to do there? Fortunately, by the time electricity could be produced and transported in quantity, the question of the manner of its consumption had already been answered.

That answer arose out of the reversal of a known effect. This happens frequently in science. If deforming the shape of a crystal produces a potential difference, then applying a potential difference to opposite sides of a crystal will deform its shape. If an electric current creates a magnetic field, then a magnetic field can be made to produce an electric current.

It is not surprising, therefore, that if mechanical energy can be converted into electrical energy when a conductor is made to move and cut across lines of magnetic force, electrical energy can be converted into mechanical energy, causing a conductor to move and cut across lines of magnetic force.

Imagine a copper wire between the poles of a magnet, north pole on the right and south pole on the left. If the copper wire is moved upward, then by Fleming's right-hand rule we know that a current will be induced in the direction toward us.

Suppose, however, that we keep the wire motionless in mid-field, so that no current is induced in it. Suppose that we then send a current through it from a battery, that current moving toward us. The current-carrying wire now sets up a magnetic field of its own. Since the current is coming toward us, the lines of force run in counterclockwise circles about themselves. Above the wire, those counterclockwise lines of force run in the same direction as do the straight lines of force from the north pole to the south pole of the magnet. The two add together, so that the magnetic flux is increased. Below the wire, the counterclockwise lines of force run in the direction opposed to the lines of force of the magnet, so that there is a canceling of effect and the flux density is decreased. With a high-flux density above the wire, and a low-flux density below it, the wire is pushed downward by the natural tendency of the lines of force to "even out." If the current in the wire is moving away from us, so that its lines of force run in clockwise circles, the magnetic flux density will be greater below and the wire will be pushed upward.

To summarize, consider a magnet with lines of force running from right to left:

If a wire without current is moved upward, current toward you is induced.

If a wire without current is moved downward, current away from you is induced.

If a wire contains current flowing toward you, motion downward is induced.

If a wire contains current flowing away from you, motion upward is induced.

In the first two cases, current is generated out of motion and the device is a generator. In the last two cases, motion is manufactured out of current and the device is a *motor.* (It is the same device in either case, actually, but it is run "forward" in one case and "backward" in the other.)

Notice that in the generator, current toward is associated with motion upward, and current away with motion downward; in the motor, current toward is associated with motion downward, and current away with motion upward. Therefore, in determining the relationship of direction of lines of force, direction of current, and direction of motion, one must, in the case of a motor, use some device that is just the opposite of the device used in the case of the generator.

For the generator we used the Fleming right-hand rule, and since we have an opposite limb in the shape of the left hand, we use a *left-hand rule* (with thumb, forefinger, and middle finger held at mutual right angles) to relate the various directions in a motor. As in the right-hand rule, we allow the forefinger to point in the direction of the lines of force, i.e., toward the south pole. The middle finger points in the direction the current is flowing, i.e., toward the negative pole. The thumb will then automatically point in the direction of the motion imposed upon the wire.

Now let us pass on to a loop of wire between the poles of a

magnet. If a mechanical rotation is imposed upon it, an electric current is induced in the loop. Consequently, it is only natural that if an electric current is put through the wire from an outside source, a mechanical rotation will be induced. (Without going into details, such mechanical rotation can be brought about by both direct current and alternating current. Some motors are designed to run on either.)

You can therefore have two essentially identical devices. The first, used as a generator, will convert the heat energy of burning coal into the mechanical energy of a turning armature and convert that in turn into electrical energy. The electrical energy so produced is poured into the second device, used as a motor, and there it is converted into the mechanical energy of a turning armature. Of course, the generator can be large, supplying enough electric energy to run numerous small motors.

Once large generators made it practical to produce large quantities of electricity, and transformers made it practical to transport large quantities of electricity, it was only necessary that this electricity be conducted to the millions of motors in homes and factories.* The necessary motors were waiting to be used in this fashion; they had been waiting for some half a century, for the first practical motor had been devised by Henry in 1831.

Turning wheels had been used to supply mechanical energy as far back as earliest historic times, for rotational motion is not only useful in itself but it can easily be converted into back-and-forth motion by proper mechanical connections. Through most of man's history, wheels had been turned by the muscle of men and animals, by the action of falling water, and by blowing wind. Muscles, however, were weak and easily tired, water fell only in certain regions, and the wind was always uncertain.

After the invention of the steam engine, wheels could be turned by the application of steam power. However the bulky engines had to exist on the spot where the wheels turned and could profitably be set up only in factories or on large contrivances such as locomotives and ships. They were therefore profitably used only for large-scale work. Small-scale steam engines for home use seemed out of the question. Furthermore, it took time to start steam engines, for large quantities of water had first to be brought to a boil.

With the motor, however, an independent wheel became pos-

* Electricity can be put to good use without motors, of course, as in toasters and electric lights, where the heating effect alone is desired and mechanical motion is not necessary.

sible. The generator, as the source of the energy, did not have to be on the premises or anywhere near them. Moreover, electric motors could start at the flick of one switch and stop at the flick of another. Motors were versatile in the extreme, and wheels of any size and power could be turned. Huge motors were designed to move streetcars or industrial machinery, and tiny motors now power typewriters, shavers, and toothbrushes.

Thanks to Faraday and Henry (with assists from Tesla and Steinmetz), the lives of the citizens of the industrial portions of the world have thus come to be composed, in large measure, of a complex heap of electrical gadgetry.

CHAPTER 29

Electromagnetic Radiation

Maxwell's Equations

By mid-nineteenth century, the connection between electricity and magnetism was well established and being put to good use. The generator and the motor had been invented, and both depended on the interrelationship of electricity and magnetism.

Theory lagged behind practice, however. Faraday, for instance, perhaps the greatest electrical innovator of all, was completely innocent of mathematics, and he developed his notion of lines of force in a remarkably unsophisticated way, picturing them almost like rubber bands.*

In the 1860's, Maxwell, a great admirer of Faraday, set about supplying the mathematical analysis of the interrelationship of electricity and magnetism in order to round out Faraday's non-mathematical treatment.

To describe the manner in which an electric current invariably produced a magnetic field, and in which a magnet could be made to produce an electric current, as well as how both electric charges and magnetic poles could set up fields consisting

* This is not meant as a sneer at Faraday, who was certainly one of the greatest scientists of all time. His intuition was that of a first-class genius. Although his views were built up without the aid of a carefully worked out mathematical analysis, they were solid. When the mathematics was finally supplied, the essence of Faraday's notions was shown to be correct.

of lines of force, in 1864 Maxwell devised a set of four comparatively simple equations,* known ever since as *Maxwell's equations*. From these, it proved possible to deduce the nature of the interrelationships of electricity and magnetism under all possible conditions.

In order for the equations to be valid, it seemed impossible to consider an electric field or a magnetic field in isolation. The two were always present together, directed at mutual right angles, so that there was a single *electromagnetic field*.

Furthermore, in considering the implications of his equations, Maxwell found that a changing electric field had to induce a changing magnetic field, which in turn had to induce a changing electric field, and so on; the two leap-frogged, so to speak, and the field progressed outward in all directions. The result was a radiation possessing the properties of a wave-form. In short, Maxwell predicted the existence of *electromagnetic waves* with frequencies equal to that in which the electromagnetic field waxed and waned.

It was even possible for Maxwell to calculate the velocity at which such an electromagnetic wave would have to move. He did this by taking into consideration the ratio of certain corresponding values in the equations describing the force between electric charges and the force between magnetic poles. This ratio turned out to have the value of 300,000 kilometers per second.

But this was equal to the velocity of light, and Maxwell could not accept that as a mere coincidence. Electromagnetic radiation was not a mere phantom of his equations but had a real existence. In fact, light must itself be an electromagnetic radiation.†

Maxwell's equations served several general functions. First, they did for the field-view of the universe what Newton's laws of motion had done for the mechanist-view of the universe.

Indeed, Maxwell's equations were more successful than Newton's laws. The latter were shown to be but approximations that held for low velocities and short distances. They required the

* Unfortunately, they are differential equations, involving the concepts of calculus, and calculus is not being used in this book. For that reason, Maxwell's equations will be spoken of, but will not be brought on stage.

† An explanation of the exact manner in which light waves, with frequencies in the hundreds of trillions, could be produced electromagnetically had to wait half a century, however—until the quantum theory (undreamed of in Maxwell's time) could be applied to the internal structure of the atom (unknown in Maxwell's time). How this was done will be described in Volume III of this book.

modification of Einstein's broader relativistic viewpoint if they were to be made to apply with complete generality. Maxwell's equations, on the other hand, survived all the changes introduced by relativity and the quantum theory; they are as valid in the light of present knowledge as they were when they were first introduced a century ago.

Secondly, Maxwell's equations, in conjunction with the later development of the quantum theory, seem at last to supply us with a satisfactory understanding of the nature of light (a question that has occupied a major portion of this volume and serves as its central question). Earlier (see page 139) I said that even granting the particle-like aspects of light there remained the wave-like aspects, and questioned what these might be. As we see now, the wave-like aspects are the oscillating values of the electromagnetic field. The electric and magnetic components of that field are set at mutual right angles and the whole wave progresses in a direction at right angles to both.

To Maxwell, wedded to the ether hypothesis, it seemed the oscillation of the electromagnetic field consisted of wave-like distortions of the ether. However, Maxwell's equations rose superior even to Maxwell. Though the ether hypothesis passed away, the electromagnetic wave remained, for now it became possible to view the oscillating field as oscillating changes in the geometry of space. This required the presence of no matter. Nothing "had to wave" in order to form light waves.

Of the four different phenomena which, from Newton's time onward, have threatened to involve action-at-a-distance, no less than three, thanks to Maxwell's equations, were shown to be different aspects of a single phenomenon. Electricity, magnetism and light were all included in the electromagnetic field. Only the gravitational force remained outside. Maxwell, recognizing the important differences between gravity and electromagnetism, made no attempt to include the gravitational field in his equations. Since his time, some attempts have been made, notably by Einstein in the latter half of his life. Einstein's conclusions, however, have not been accepted by physicists generally, and the question of a "unified field theory" remains open.

It seemed to Maxwell that the processes that gave rise to electromagnetic radiation could well serve to produce waves of any frequency at all and not merely those of light and its near neighbors, ultraviolet and infrared radiation. He predicted, therefore, that electromagnetic radiation, in all essentials similar to light, could exist at frequencies far below and far above those of light.

Unfortunately, Maxwell did not live to see this prediction verified, for he died of cancer in 1879 at the comparatively early age of 48. Only nine years after that, in 1888, the German physicist Heinrich Rudolf Hertz (1857–1894) discovered electromagnetic radiation of very low frequency—radiation that we now call *radio waves.* This completely bore out Maxwell's prediction and was accepted as evidence for the validity of Maxwell's equations. In 1895, another German physicist, Wilhelm Konrad Röntgen (1845–1923), discovered what turned out to be electromagnetic radiation of very high frequency: radiation we now call *X rays.*

The decades of the 1880's and 1890's also saw a fundamental advance made in the study of electricity. Electric currents were driven through near-vacuums, and electrons, instead of remaining concealed in metal wires or being considered attachments to drifting atoms and groups of atoms in solution, made their appearance as particles in their own right.

The study of the new particles and radiations introduced a virtual revolution in physics and electrical technology—one so intense that it has been referred to as the Second Scientific Revolution (the First, of course, being that initiated by Galileo).

It is with the Second Scientific Revolution that the third volume of this book will deal.

PART THREE

The Electron, Proton, and Neutron

CHAPTER **30**

The Atom

In the first two volumes of this book I dealt with those aspects of physics in which the fine structure of matter could be ignored.

I discussed gravitation, for instance, since any sphere possessing the mass of the earth would exhibit the gravitational effect of the earth regardless of the type of matter of which it was composed. Furthermore, the question of the ultimate structure of the finest particles of matter need not be considered in working out the laws governing the gravitational interaction of bodies.

The same is true of the laws of motion. A brick moves as a unit and we need not be concerned with the composition of the brick in studying its motion. We can study many phases of the electric charge of a pith ball, or the magnetic field of a magnet, and derive useful laws governing electromagnetic effects without probing into the submicroscopic structure of the magnet. Even heat can be considered a subtle fluid pouring from one object to another, and the laws of thermodynamics can be deduced from this sort of concept.

Yet in the course of these two volumes, it has become plain, every once in a while, that a deeper understanding of phenomena than that offered by the everyday world of ordinary objects can be achieved if we burrow down into the ultra-small.

For instance, the properties of gases are best understood if we consider them to be composed of tiny particles (see page

I–200).* Temperature can best be understood if it is considered to represent the average kinetic energy of tiny particles of matter in motion (see page I–205). Energy, as well as matter, seems to make more sense if it is considered as consisting of tiny particles (see page II–130).

In this third volume, therefore, I will go into the fine structure of matter and energy in some detail. I will try to show how physical experimentation revealed a world beyond the direct reach of our senses, and how knowledge of that world has, in turn, lent more meaning to the ordinary world we can sense directly.

Origin of Atomism

The notion of *atomism* (a name we can give to the theory that matter is composed of tiny particles) arose first among the Greeks—not as a result of experiment, but as a result of philosophic deduction.

Atomism is by no means self-evident. If we can trust our senses, most types of matter seem "all one piece." A sheet of paper or a drop of water does not seem to be composed of particles.

This, however, is not conclusive. The sand making up a beach, if viewed from a distance, seems to be all one piece. It is only upon a close view that we can make out the small grains of which the sand actually consists. Perhaps, then, paper or water is made up of particles too small to see.

One way of testing the matter is by considering the divisibility of a substance. If you had a handful of sand and ignored the evidence of your eyes, seeking instead some other criterion of atomism, you might begin by dividing the handful into two portions with your finger, dividing each of those into two still smaller portions, and so on. Eventually, you would find yourself in possession of a portion so small as to consist of a single grain, and this final portion could no longer be divided by finger. We might consider atomism, then, as implying that matter cannot be divided and subdivided indefinitely. At some point, a unit no longer divisible by a method that had sufficed earlier must be obtained.

If this is true of paper or water, however, the ultimate pieces are for too small to see. No limiting indivisible unit in matter generally can be directly sensed. Can the existence of such ultra-small units be deduced by reason alone?

* References to material in the first two volumes will be indicated by phrases such as "(see page I–123)" or "(see page II–123)." References to material in this third volume will be indicated by a simple "(see page 123)."

The opportunity arose in the fifth century B.C. with the paradoxes raised by Zeno of Elea. Zeno pointed out that one could reach conclusions by reason that seemed to contradict the evidence of the senses and that it was necessary, therefore, to search for flaws either in the process of reasoning or in sense-perception. His most famous paradox is called "Achilles and the Tortoise."

Suppose that the Greek hero Achilles, renowned for his fleetness of foot, can run ten times as fast as a tortoise. The tortoise is now given a hundred-yard headstart, and the two race. By the time Achilles covers the hundred yards that separated him from the tortoise, the tortoise has moved forward ten yards. When Achilles makes up that ten yards, the tortoise has moved forward one yard; when Achilles bounds across that one yard, the tortoise has moved forward a tenth of a yard, and so on. By this line of reasoning, it seems clear that Achilles can never catch up to the tortoise, who always remains ahead (though by a smaller and smaller margin). And yet, in any such race we certainly know that Achilles will, in actual fact, overtake and pass the tortoise.

Nowadays, mathematicians understand that the successive margins of the tortoise—100 yards, 10, 1, 0.1, and so on—make up a "converging series." A converging series may have an infinite number of terms, but these will nevertheless come to a definite and finite sum. Thus, the converging series consisting of $100 + 10 + 1 + 0.1 + 0.01$, etc., has the finite sum of $111\frac{1}{9}$. This means that after Achilles has run $111\frac{1}{9}$ yards he will be exactly even with the tortoise and that thereafter he will forge ahead.

The Greeks, however, knew nothing of converging series and had to find other reasons for reconciling Zeno's argument with the facts of life. One way out was to consider that Zeno divided the distance between Achilles and the tortoise into smaller and smaller portions with no indication that any portion was so small that it could no longer be divided into anything smaller.

Perhaps that was not the way the universe worked. Perhaps there were units so small that they could be divided no further. If this notion of limited divisibility was adopted, perhaps Zeno's paradoxes, based on unlimited divisibility, might disappear.

It may have been reasoning like this that led some Greek philosophers to suggest that the universe was made up of tiny particles that were themselves indivisible. The most prominent of these philosophers was Democritus of Abdera, who advanced his theories about 430 B.C. He called these ultimate particles "atomos," from a Greek word meaning "indivisible," and it is from this that our word, *atom,* is derived.

Democritus went on to interpret the universe in atomic terms and came up with a number of suggestions that sound quite modern. However, it all rested on pure reasoning. He could suggest no evidence for the existence of atoms other than "this is the way it must be."

Other Greek philosophers of the time could offer arguments for the nonexistence of atoms on a "this is the way it must be" basis. On the whole, most ancient philosophers went along with non-atomism, and Democritus' views were buried under the weight of adverse opinions. In fact, so little worthy of attention was Democritus considered that his works were infrequently copied, and none of his voluminous writings have survived into modern times. All we know of him are offhand remarks in the works of such philosophers as have survived, just about all of whom are non-atomists and therefore mention Democritus' views only disparagingly.

Nevertheless, his views, however crushed, did not altogether die. Epicurus of Samos (341–270 B.C.), who began teaching in Athens in 306 B.C., incorporated the atomism of Democritus into his philosophic system. Although Epicureanism proved quite influential in the next few centuries, none of the works of Epicurus have survived, either.

Fortunately, however, the works of one Epicurean philosopher have survived. The Roman poet Lucretius (96?–55 B.C.) wrote a long poem *De rerum natura* ("On the Nature of Things") in which he interpreted the universe in Epicurean fashion, making use of an atomistic viewpoint. *One* copy of this poem survived, and when printing was invented in the fifteenth century, it was one of the first of the ancient classics to be printed.

As modern science came to birth, then, atomistic views were present for the plucking. A French philosopher, Pierre Gassendi (1592–1655), adopted the Epicurean views of Lucretius and was influential in spreading the doctrine of atomism.

One of those who came under the influence of Gassendi was the English scientist Robert Boyle (1627–1691), and with him atomism enters a new phase; it is no longer a matter of philosophy and deduction, but rather one of experiment and observation.

The Chemical Elements

Boyle studied air and found that it could be compressed or expanded (see page I–145). In other words, the volume of a gas could be changed without changing its mass. It is difficult to imagine how this could happen if matter were really continuous. Rubber can be stretched, to be sure, but as a rubber band grows longer, it also grows thinner; the volume is not perceptibly altered.

The behavior of air is much more like that of a sponge, which can be compressed in all directions or pulled apart in all directions—its volume considerably changed without a change in mass. In the case of a sponge, the explanation involves the numerous air-filled cavities. The sponge can be compressed because the cavities can be forced closed by squeezing out the air. It can expand once more if air is allowed to re-enter the cavities.

Is it possible, then, that invisible cavities exist in air itself, cavities which can be squeezed closed when air is compressed and made to open wide when it is expanded? This can, in fact, be visualized if it is supposed that air is made up of a myriad of ultra-tiny particles separated by utter emptiness. Compression would involve the movement closer together of these particles; expansion, the movement farther apart. Volume would change while mass (which would depend merely on the number of particles and not on their distance apart) would not. Other properties of gases could likewise be handily explained by atomistic reasoning.

Atomism could surely be transferred from gases to solids and liquids, since the latter can easily be converted, through heat, to gases or vapors. Thus boiling water (or even water standing at ordinary temperatures) is converted into water vapor, a gas that is much less dense than liquid water. The vapor can be condensed into liquid water once more. To explain this, we might suppose that water consists of atoms packed very closely. (From the fact that water, in common with other liquids and solids, cannot be compressed appreciably by the forces that compress gases easily, we might even suppose that the atoms in liquids and solids are in contact.) When a liquid evaporates, the molecules are pulled apart; when a gas is condensed, its molecules are forced together.

Despite reasoning of this sort, the world of science found it difficult to accept atomism. The philosophical difficulty of dealing with objects infinitesimal in size and undetectable by any device then known was too great.

What finally established atomism firmly was the slow gathering of a quantity of chemical evidence in its favor. To describe this, I will begin with the concept of an *element*.

The Greeks were the first to speculate on the nature of the fundamental substance or substances (elements) of which the universe was composed. Their speculations, in the absence of actual chemical experimentation, were really only guesses, but since the Greek thinkers were highly intelligent men, they produced extraordinarily sensible guesses.

Aristotle (384–322 B.C.) summarized Greek labors in this direction by listing four elements in the world itself: earth, water, air, and fire; and a fifth element, making up the incorruptible heavens above: ether (see page I–6). If, for the four earthly elements, we used the closely allied words: solid, liquid, gas, and energy—we would see that the guesses were indeed sensible.

For two thousand years, the Greek notion of the four earthly elements survived. By 1600, however, the notion of experimentation was beginning to preoccupy scientists, thanks very largely to the work of Galileo Galilei (1564–1642). An element, it seemed, should be defined in experimental terms; it should be defined as something that was capable of doing something (or incapable of doing something) rather than of merely being something. It needs what is now called an *operational definition.*

In 1661, Robert Boyle wrote a book called *The Sceptical Chymist* in which he explained his notion of an element. If an element was indeed one of the simple substances out of which the universe was composed, then it should certainly not be capable of being broken down to still simpler substances or of being produced through the union of still simpler substances. As soon as a substance was broken into simpler substances, it was, at once and forever, not an element.

Since earth is easily separated into different substances, earth is not an element. A century after Boyle's time, air and water were separated into simpler substances and were thus shown not to be elements. As for fire, chemists came to realize that it was not matter at all but a form of energy, and therefore it fell outside the world of elements altogether.

For a considerable time after Boyle, chemists could never know for sure whether a given substance was an element, since one could never tell when new experimental techniques might be developed which would make it possible to break down a previously untouchable substance.

As an example, the substance known as "lime" (or, in Latin, *calx*) had to be considered an element throughout the eighteenth century, since nothing chemists could do would break it down to simpler substances. There were reasons, though, for suspecting that it consisted of an unknown metal combined with a gas, oxygen. However, this was not shown to be fact until 1808, when the English chemist Humphry Davy (1778–1829), succeeded in decomposing lime and isolating a new metallic element, *calcium* (from lime's Latin name). For this, however, he had had to make use of a current of electricity, a new discovery then.

For ease in referring to the elements, a *chemical symbol* for each was introduced in 1814 by the Swedish chemist Jöns Jakob Berzelius (1779–1848). Essentially, these symbols consist of the initial letter of the Latin name (usually, but not always, very similar to the English name) plus (again usually, but not always) a second letter from the body of the name. The symbols used are in almost every case so logical that after very little practice their meanings come to offer no difficulty whatever.

During the course of the nineteenth century, chemists grew to understand the nature of elements, and by the early decades of the twentieth century, elements could be defined with remarkable precision. The manner in which this came about will be described later in this book, but meanwhile, I will list (alphabetically), together with the chemical symbols for each, the substances now recognized as elements (see Table I).

The Modern Atomic Theory

Of course, not all substances found in nature are elements. Most substances are composed of two or more elements, not merely mixed but intimately joined in such a fashion that the final substance has properties of its own that are not necessarily similar to those of any of the elements making it up. Such a substance, formed of an intimate union of elements, is called a *compound.*

In the latter part of the eighteenth century, chemists forming their compounds began to study more than the merely qualitative nature of the products formed in their reactions. It was no longer enough merely to note that a gas had bubbled off or that a flocculent material of a certain color had settled to the bottom of a container. Chemists took to measurement—to determining the actual quantity of substances consumed and produced in their reactions.

The most prominent in establishing this new trend was the French chemist Antoine Laurent Lavoisier (1743–1794), who for this and other services is commonly called "the father of modern chemistry." Lavoisier gathered enough data by 1789 to be able to maintain that in any chemical reaction in a closed system (that is, one from which no material substance may escape and into which no material substance may enter) the total mass is left unchanged. This is the *law of conservation of matter,* or the *law of conservation of mass.*

It was an easy step from this to the separate measurement of the mass of each component of a compound. Important work in this respect was done by the French chemist Joseph Louis Proust (1754–1826). He worked, for example, with a certain compound, now called copper carbonate, which is made up of three elements: copper, carbon and oxygen. Proust began with a pure sample of copper carbonate, broke it down into these three elements, and determined the mass of each element separately. He found the elements always present in certain fixed proportions: for every five parts of copper (by weight) there were four parts of oxygen and one part of carbon. This was true for all the samples

TABLE I—*Elements and Their Symbols*

Actinium	Ac	Erbium	Er
Aluminum	Al	Europium	Eu
Americium	Am	Fermium	Fm
Antimony	Sb	Fluorine	F
Argon	Ar	Francium	Fr
Arsenic	As	Gadolinium	Gd
Astatine	At	Gallium	Ga
Barium	Ba	Germanium	Ge
Berkelium	Bk	Gold	Au
Beryllium	Be	Hafnium	Hf
Bismuth	Bi	Helium	He
Boron	B	Holmium	Ho
Bromine	Br	Hydrogen	H
Cadmium	Cd	Indium	In
Calcium	Ca	Iodine	I
Californium	Cf	Iridium	Ir
Carbon	C	Iron	Fe
Cerium	Ce	Krypton	Kr
Cesium	Cs	Lanthanum	La
Chlorine	Cl	Lawrencium	Lw
Chromium	Cr	Lead	Pb
Cobalt	Co	Lithium	Li
Copper	Cu	Lutetium	Lu
Curium	Cm	Magnesium	Mg
Dysprosium	Dy	Manganese	Mn
Einsteinium	Es	Mendelevium	Md

of copper carbonate he tested, no matter how they were prepared. It was as though elements would fit together in certain definite proportions and no other.

Proust found this was true for other compounds that he tested, and he announced his finding in 1797. It is sometimes called *Proust's law,* sometimes *the law of fixed proportions,* and sometimes *the law of definite proportions.*

It is the law of fixed proportions that forced the concept of atomism to arise out of purely chemical considerations. Suppose that copper consists of tiny copper atoms; oxygen, of oxygen atoms; and carbon, of carbon atoms. Suppose further that copper carbonate is formed when a copper atom, an oxygen atom and a carbon atom all join in a tight union. (The truth of the matter is

Mercury	Hg	Samarium	Sm
Molybdenum	Mo	Scandium	Sc
Neodymium	Nd	Selenium	Se
Neon	Ne	Silicon	Si
Neptunium	Np	Silver	Ag
Nickel	Ni	Sodium	Na
Niobium	Nb	Strontium	Sr
Nitrogen	N	Sulfur	S
Nobelium*	No	Tantalum	Ta
Osmium	Os	Technetium	Tc
Oxygen	O	Tellurium	Te
Palladium	Pd	Terbium	Tb
Phosphorus	P	Thallium	Tl
Platinum	Pt	Thorium	Th
Plutonium	Pu	Thulium	Tm
Polonium	Po	Tin	Sn
Potassium	K	Titanium	Ti
Praseodymium	Pr	Tungsten	W
Promethium	Pm	Uranium	U
Protactinium	Pa	Vanadium	V
Radium	Ra	Xenon	Xe
Radon	Rn	Ytterbium	Yb
Rhenium	Re	Yttrium	Y
Rhodium	Rh	Zinc	Zn
Rubidium	Rb	Zirconium	Zr
Ruthenium	Ru		

* Name not yet official

more complicated than this, but right now we are only trying to observe the consequences of an atomistic supposition.) A tight union of atoms, such as that which I am suggesting, is called a *molecule* (from a Latin word meaning "a small mass"). What I am saying, then, is suppose that copper carbonate is made up of molecules, each containing a copper atom, a carbon atom and an oxygen atom.

What, now, if it happened that a copper atom was five times as massive as a carbon atom, and an oxygen atom was four times as massive as a carbon atom? It would then be expected that copper carbonate would have to contain five parts of copper (by weight) to four parts of oxygen to one part of carbon. In order to have 5.1 parts of copper to one part of carbon, or 3.9 parts of oxygen to one part of carbon, we would need to work with fractions of atoms.

But this never happens. Only certain proportions exist within a compound and these cannot be varied through slight amounts in this direction and that. This shows that from Proust's law of fixed proportions we can not only reasonably speak of atoms, but that we must come to the decision that the atoms were indivisible, as Democritus had imagined so many centuries before.

These thoughts occurred, in particular, to an English chemist, John Dalton (1766–1844). Based on the law of fixed proportions and on other generalizations of a similar nature, he advanced the *modern atomic theory* (so called to distinguish it from the ancient one advanced by Democritus) in 1803. Dalton recognized the honor due Democritus, for he carefully kept the ancient philosopher's term "atom."

Dalton could go much further than Democritus, of course. He did not need to confine himself to the statement that atoms existed. From the law of fixed proportions it was quite plain that:

(1) Each element is made up of a number of atoms all with the same fixed mass.

(2) Different elements are distinguished by being made up of atoms of different mass.

(3) Compounds are formed by the union of small numbers of atoms into molecules.*

* As a matter of fact, each of these three statements proved to be wrong, as chemists found when they probed more deeply into the fundamental structure of matter. Nevertheless, for those substances most easily dealt with by early nineteenth-century techniques, they were reasonably correct. Dalton's propositions represent a "first approximation" that served to start investigations in the right direction and made it possible to improve those starting approximations as further data were gathered. In science, it is not all-important to be Right (it

From the law of fixed proportions it is even possible to come to conclusions about the relative mass of the different kinds of atoms. This relative mass is commonly referred to as *atomic weight.*†

For instance, water is made up of hydrogen and oxygen, and in forming water it is found that one part of hydrogen (by weight) combines with eight parts of oxygen. Dalton was convinced that compounds were formed by the union of as few atoms as possible, so he considered a molecule of water to be made up of one atom of hydrogen combined with one atom of oxygen. In that case, it was easy for him to decide that an oxygen atom must be eight times as massive as a hydrogen atom.

This does not tell us what the actual mass of either the oxygen atom or the hydrogen atom is, but it does not represent checkmate by any means. Dalton decided to use the hydrogen atom as a reference because he suspected it to be the lightest atom (and here, as it happens, he proved to be right), and he set its mass arbitrarily equal to 1. On that hydrogen = 1 basis, he could set the mass of the oxygen atom at 8.

But a refinement became necessary at this point. It turned out that at just about the time that Dalton was working out his atomic theory, water was being split up into hydrogen and oxygen by the action of an electric current. When this was done, it was found that for every liter of oxygen evolved, two liters of hydrogen were produced. The ratio (by volume) was two parts of hydrogen to one part of oxygen. It was not long before this was shown to mean that the water molecule was composed of two hydrogen atoms and one oxygen atom (though Dalton himself never accepted this).

The molecule can be represented by a *chemical formula* in which an atom of each element contained is represented by its chemical symbol. Thus, Dalton's conception of the water molecule would be HO. Where more than one atom of a particular element is present in the molecule, the number is indicated by a numerical subscript. Therefore, the molecule of water, as accepted now, would have a formula of H_2O.

---

may even be that there is no way of ever determining what is Right); it is merely necessary to be right enough for the times, and Dalton was every bit of that.

† Weight is not the same as mass (see page I–53), and it would be more scientifically appropriate to speak of "atomic mass" rather than atomic weight. However, in this case, as in many others, an unfortunate word or phrase has entered the scientific literature and has become so popular and well-known as to be impossible to change. Such things must, with a sigh, be lived with.

Changing one's deductions does not change the nature of the experimental observations. Water remains made up of one part of hydrogen (by weight) to eight parts of oxygen. Under the new interpretation of the molecular structure of water, however, the one oxygen atom in the molecule must now be eight times as massive as both hydrogen atoms taken together and sixteen times as massive, therefore, as a single hydrogen atom. Therefore, if we set the atomic weight of hydrogen arbitrarily equal to 1, the atomic weight of oxygen must be equal to 16.

This system can then be used to leapfrog from element to element. For instance, carbon dioxide is produced when three parts of carbon are combined with eight parts of oxygen (by weight). The molecule of carbon dioxide contains one atom of carbon and two atoms of oxygen (CO_2). This means that one atom of carbon is 3/8 as massive as two atoms of oxygen. Since the atomic weight of oxygen is 16, two atoms of oxygen must have a mass of 32. If the carbon atom has a mass 3/8 times 32, its atomic weight is 12.

The molecule of cyanogen (C_2N_2) contains six parts of carbon (by weight) to seven of nitrogen. The two atoms of carbon have a mass of 24; therefore the two atoms of nitrogen have a mass of 24 times 7/6, or 28, and a single atom of nitrogen has an atomic number of 14.

It would seem from this that atomic numbers can be expressed as integers on a hydrogen = 1 basis, and Dalton was indeed convinced that this was true. However, over the next decades, other chemists, notably Berzelius, made more accurate determinations and found that some atomic weights were not integers at all. The atomic weight of chlorine is approximately 35.5, for instance, and the atomic weight of magnesium is 24.3.

Indeed, even some of the atomic weights that seem integers turn out to be not quite integers if very accurate measurements are made. For instance, the proportions of oxygen and hydrogen in water are not exactly 8 to 1 by weight, but rather 7.94 to 1. This means that if we set the atomic weight of hydrogen arbitrarily equal to 1, then the atomic weight of oxygen is 15.88.

But oxygen combines easily with many elements. Of all the elements readily available to the chemists of the early nineteenth century, oxygen combined most readily with other elements. (It was *chemically active.*) Its readiness to combine made oxygen particularly useful in calculating atomic weights, and to have its atomic weight set at some fractional value meant needless complexity of arithmetical computations. Chemists eventually decided, therefore, to set the atomic weight of oxygen exactly equal to 16.0000 and let that serve as standard. The atomic weight of hydrogen would then be 1.008.

This served satisfactorily for nearly a century. By 1920, however, new facts concerning atoms were learned (see Chapter 8) which made the standard of oxygen = 16.0000 inadequate. However, the standard had become so fixed in the literature and in chemical consciousness that it was difficult to change. In 1961, however, a new and better system was adopted (see page 148) which involved a change so slight that it could be tolerated. By the 1961 system, the atomic weight of oxygen, for instance, is 15.9994.

Of the 103 known elements, 83 occur in the earth's crust to an appreciable extent. In Table II these 83 elements are listed in order of increasing atomic weight, and the atomic weight of each, by the 1961 system, is given. The question of the masses of the 20 remaining elements will be considered in another chapter.

The Periodic Table

By the mid-nineteenth century, two definitions of an element were available. One was Boyle's definition (that of a substance that could not be broken down to two or more still simpler substances), and one was Dalton's definition (that of a substance made up entirely of atoms of a given atomic weight). There was no conflict between the two, for the same list of substances qualified as elements by either definition. However, there was an embarrassment of riches—too many elements for comfort. By the 1860's, more than sixty elements were known.

These came in a wide variety of properties: some were gases at ordinary temperatures, a few were liquids, and most were solids; some were nonmetals, some light metals, some heavy metals, and some semimetals; some were very active, some moderately active, and some quite inactive; some were colored and some were not.

All this was rather upsetting. Scientists must take the universe as they find it, of course, but there is a deep-seated faith (no other word will suffice) dating back to Greek times that the universe exhibits order and is basically simple. Whenever any facet seems to grow tangled and complex, scientists can't help searching for some underlying order that may be eluding them.

Attempts were made in the mid-nineteenth century to find such an order among the elements. As the tables of atomic weights

TABLE II—*The Atomic Weights of Elements*

Element	Atomic Weight	Element	Atomic Weight
Hydrogen	1.00797	Ruthenium	101.07
Helium	4.0026	Rhodium	102.905
Lithium	6.939	Palladium	105.4
Beryllium	9.0122	Silver	107.870
Boron	10.811	Cadmium	112.40
Carbon	12.01115	Indium	114.82
Nitrogen	14.0067	Tin	118.69
Oxygen	15.9994	Antimony	121.75
Fluorine	18.9984	Iodine	126.9044
Neon	20.183	Tellurium	127.60
Sodium	22.9898	Xenon	131.30
Magnesium	24.312	Cesium	132.905
Aluminum	26.9815	Barium	137.34
Silicon	28.086	Lanthanum	138.91
Phosphorus	30.9738	Cerium	140.12
Sulfur	32.064	Praseodymium	140.907
Chlorine	35.453	Neodymium	144.24
Potassium	39.102	Samarium	150.35
Argon	39.948	Europium	151.96
Calcium	40.08	Gadolinium	157.25
Scandium	44.956	Terbium	158.924
Titanium	47.90	Dysprosium	162.50
Vanadium	50.942	Holmium	164.930
Chromium	51.996	Erbium	167.26
Manganese	54.9380	Thulium	168.934
Iron	55.847	Ytterbium	173.04
Nickel	58.71	Lutetium	174.97
Cobalt	58.9332	Hafnium	178.49
Copper	63.54	Tantalum	180.948
Zinc	65.37	Tungsten	183.85
Gallium	69.72	Rhenium	186.2
Germanium	72.59	Osmium	190.2
Arsenic	74.9216	Iridium	192.2
Selenium	78.96	Platinum	195.09
Bromine	79.909	Gold	196.967
Krypton	83.80	Mercury	200.59
Rubidium	85.47	Thallium	204.37
Strontium	87.62	Lead	207.19
Yttrium	88.905	Bismuth	208.980
Zirconium	91.22	Thorium	232.038
Niobium	92.906	Uranium	238.03
Molybdenum	95.94		

grew more and more accurate and as the concept of atomic weight became clearer to chemists generally, it began to seem logical to arrange the elements in order of increasing atomic weight (as in Table II) and see what could be done with that.

Several efforts of this sort failed, but one succeeded. The success was scored in 1869 by a Russian chemist, Dmitri Ivanovich Mendeleev (1834–1907). Having listed the elements in order of atomic weight, he then arranged them in a table of rows and columns, in such a fashion that elements of similar properties fell into the same column (or row, depending on how the table was oriented). As one went along the table of elements, properties of a certain kind would turn up after fixed periods. For this reason, Mendeleev's product was a *periodic table.*

Difficulties arose out of the fact that the list of elements, extensive as it was, was still incomplete. In order to arrange the known elements in such a way that those of similar properties fell into the same column, Mendeleev found it necessary to leave gaps. These gaps, he announced in 1871, must contain elements not as yet discovered. He announced the properties of the missing elements in some detail, judging these by comparing them with the elements in the same column, above and below the gap, and taking intermediate values.

Within fifteen years, all three elements predicted by Mendeleev were discovered, and their properties were found to be precisely those he had predicted. As a result, the periodic table was, by the 1880's, accepted as a valid guide to order within the jungle of elements, and it has never been abandoned since. Indeed, later discoveries (see page 64) have served merely to strengthen it and increase its value. Mendeleev's discovery had been merely empirical—that is, the periodic table had been found to work, but no reason for its working was known. The twentieth century was to supply the reason.

Table III is a version of the periodic table, as presently accepted. The elements are arranged in order of atomic weight (with three minor exceptions shortly to be mentioned) and are numbered in order from 1 to 103. The significance of this "atomic number" will be discussed on page 64.

If you compare Table III with Table II, you will find that in order to put the elements into the proper rows, three pairs of elements must be placed out of order. Element 18 (argon) though lower in number than element 19 (potassium) has a higher atomic weight. Again, element 27 (cobalt) has a higher atomic weight than element 28 (nickel), while element 52 (tellurium) has a higher atomic weight than element 53 (iodine). In each case the difference in atomic weight is quite small and nineteenth century chemists tended to ignore these few and minor exceptions to the general rule. The twentieth century, however, was to find these exceptions particularly significant (see page 64).

The periodic table contains a number of closely-knit families of elements, with many similarities among their properties. For instance, elements 2, 10, 18, 36, 54 and 86 (helium, neon, argon, krypton, xenon and radon) are the *inert gases,* so called because of their small tendency to react with other substances. Until 1962, in fact, it was thought that none of them underwent any chemical

TABLE III—*The Periodic Table*

1 Hydrogen (H) 1.008								
3 Lithium (Li) 6.939	4 Beryllium (Be) 9.012							
11 Sodium (Na) 22.990	12 Magnesium (Mg) 24.312							
19 Potassium (K) 39.102	20 Calcium (Ca) 40.08	21 Scandium (Sc) 44.956	22 Titanium (Ti) 47.90	23 Vanadium (V) 50.942	24 Chromium (Cr) 51.996	25 Manganese (Mn) 54.938	26 Iron (Fe) 55.847	27 Cobalt (Co) 58.933
37 Rubidium (Rb) 85.47	38 Strontium (Sr) 87.62	39 Yttrium (Y) 88.905	40 Zirconium (Zr) 91.22	41 Niobium (Nb) 92.906	42 Molybdenum (Mo) 95.94	43* Technetium (Tc) 98.91	44 Ruthenium (Ru) 101.07	45 Rhodium (Rh) 102.905
55 Cesium (Cs) 132.905	56 Barium (Ba) 137.34	57 Lanthanum (La) 138.91	58 Cerium (Ce) 140.12	59 Prasodymium (Pr) 140.907	60 Neodymium (Nd) 144.24	61* Promethium (Pm) 145	62 Samarium (Sm) 150.35	63 Europium (Eu) 151.96
			72 Hafnium (Hf) 178.49	73 Tantalum (Ta) 180.948	74 Tungsten (W) 183.85	75 Rhenium (Re) 186.2	76 Osmium (Os) 190.2	77 Iridium (Ir) 192.2
87* Francium (Fr) 223	88* Radium (Ra) 226.05	89* Actinium (Ac) 227	90* Thorium (Th) 232.038	91* Protactinium (Pa) 231	92* Uranium (U) 238.03	93* Neptunium (Np) 237	94* Plutonium (Pu) 242	95* Americium (Am) 243

reactions at all. Since 1962, it has come to be realized that at least three of them, krypton, xenon and radon, will take part in chemical reactions with fluorine.

Again, elements 9, 17, 35, 53, and 85 (fluorine, chlorine, bromine, iodine and astatine) are the *halogens* (from Greek words meaning "salt-formers"). These are active nonmetals that get their family name from the fact that one of them, chlorine, combines with sodium to form ordinary table salt, while the others combine with sodium to form compounds quite similar to salt.

Elements 3, 11, 19, 37, 55, and 87 (lithium, sodium, potassium, rubidium, cesium and francium) are soft, easily melted, very

---

								2 Helium (He) 4.003
			5 Boron (B) 10.811	6 Carbon (C) 12.011	7 Nitrogen (N) 14.007	8 Oxygen (O) 15.999	9 Fluorine (F) 18.998	10 Neon (Ne) 20.183
			13 Aluminum (Al) 26.982	14 Silicon (Si) 28.086	15 Phosphorus (P) 30.974	16 Sulfur (S) 32.064	17 Chlorine (Cl) 35.453	18 Argon (A) 39.948
28 Nickel (Ni) 58.71	29 Copper (Cu) 63.54	30 Zinc (Zn) 65.37	31 Gallium (Ga) 69.72	32 Germanium (Ge) 72.59	33 Arsenic (As) 74.922	34 Selenium (Se) 78.96	35 Bromine (Br) 79.909	36 Krypton (Kr) 83.80
46 Palladium (Pd) 106.4	47 Silver (Ag) 107.870	48 Cadmium (Cd) 112.40	49 Indium (In) 114.82	50 Tin (Sn) 118.69	51 Antimony (Sb) 121.75	52 Tellurium (Te) 127.60	53 Iodine (I) 126.904	54 Xenon (Xe) 131.30
64 Gadolinium (Gd) 157.25	65 Terbium (Tb) 158.924	66 Dysprosium (Dy) 162.50	67 Holmium (Ho) 164.930	68 Erbium (Er) 167.26	69 Thulium (Tm) 168.934	70 Ytterbium (Yb) 173.04	71 Lutetium (Lu) 174.97	
78 Platinum (Pt) 195.09	79 Gold (Au) 196.967	80 Mercury (Hg) 200.59	81 Thallium (Tl) 204.37	82 Lead (Pb) 207.19	83 Bismuth (Bi) 208.98	84* Polonium (Po) 210	85* Astatine (At) 210	86* Radon (Rn) 222
96* Curium (Cm) 244	97* Berkelium (Bk) 245	98* Californium (Cf) 246	99* Einsteinium (Es) 253	100* Fermium (Fm) 255	101* Mendelevium (Md) 256	102* Nobelium (No) 255	103* Lawrencium (Lw) 257	

active *alkali metals.* The word "alkali" is from an Arabic phrase meaning "ash." It was from the ashes of certain plants that the original "alkalis," soda and potash ("pot-ash") were derived. From these, sodium and potassium, the first alkali metals to be discovered, were obtained by Davy.

Elements 4, 12, 20, 38, 56, and 88 (beryllium, magnesium, calcium, strontium, barium, and radium) are harder, less easily melted, and less active than the alkali metals. They are the *alkaline earth metals.* (An "earth" is an old-fashioned name given to oxides that are insoluble in water and resistant to change under the influence of heat. Two such earths, lime and magnesia, had certain properties resembling those of soda and potash and were therefore called the "alkaline earths." It was from lime and magnesia that Davy obtained calcium and magnesium, the first two alkaline earth metals to be discovered.)

Elements 57 to 71 inclusive form a closely related family of metals that were originally called the *rare earth elements* but have now come to be called the *lanthanides,* from the first element of the group, lanthanum. Elements 89 to 103 inclusive are the *actinides* from actinium, first element of that group.

Other families also exist within the periodic table, but those I have listed are the best known and the most frequently referred to by the family name.

The Reality of Atoms

Once we have the atomic weight, it is easy to see what one means by *molecular weight:* It is the sum of the atomic weights of the atoms making up a molecule. Let us start, for instance, with oxygen, atomic weight 16, and hydrogen, atomic weight 1.*

There is strong chemical evidence to the effect that under ordinary conditions elementary oxygen and hydrogen do not occur as single, separate atoms. Rather, two atoms combine to form a stable molecule, and the gas consists of these two-atom molecules. For this reason, the chemical formulas for gaseous oxygen and gaseous hydrogen are, respectively, O_2 and H_2. If O and H are written, they refer to individual oxygen and hydrogen atoms. You can see, then, that the molecular weight of oxygen is 32 and that of hydrogen is 2.

Again, consider ozone, a form of oxygen in which the mole-

* It is often convenient to make use of approximate atomic weights, rounding off the actual value to the nearest integer, or one decimal place at most. When more than that is needed, more than that will be used.

cules are made up of three atoms apiece (O_3). Its molecular weight is 48. That of water (H_2O) is 18. Then, since the atomic weight of carbon is 12, the molecular weight of carbon dioxide (CO_2) is 44.

It is useful for a chemist to consider a quantity of substance with a mass equal to its molecular weight in grams. In other words, he may deal with 2 grams of hydrogen, 32 grams of oxygen, 18 grams of water, or 44 grams of carbon dioxide. Such a mass is the *gram-molecular weight,* which is often spoken of, in abbreviated form, as a *mole*. We can say that a mole of carbon dioxide has a mass of 44 grams, while a mole of ozone has a mass of 48 grams.

Sometimes elements do exist in the form of single, separate atoms. This is true of the inert gases such as helium and argon, for instance. Solid elements, such as carbon and sodium, are for convenience sake often considered to be made up of single-atom units. There we can speak of a *gram-atomic weight*. Since the atomic weight of helium is 4 and that of sodium is 23, the gram-atomic weight of helium is 4 grams and that of sodium is 23 grams. Often, the abbreviated form "mole" is used to cover both gram-molecular weights and gram-atomic weights.

The convenience of the mole in chemical calculations stems from a point first grasped in 1811 by the Italian chemist Amedeo Avogadro (1776–1856) and is therefore called *Avogadro's hypothesis*. Expressed in modern terms, this states: Equal volumes of all gases contain equal numbers of molecules under conditions of fixed temperature and pressure.

In later years, this was found to be correct, at least as a first approximation.

A mole of hydrogen (2 grams) at ordinary air pressure and at a temperature of 0°C. takes up a volume of approximately 22.4 liters. A mole of oxygen (32 grams) is sixteen times as massive as a mole of hydrogen but is made up of molecules that are individually sixteen times as massive as those of hydrogen. Therefore, a mole of oxygen contains the same number of molecules as does a mole of hydrogen. By Avogadro's hypothesis (taken in reverse), this means that 32 grams of oxygen should take up just as much room (22.4 liters) as 2 grams of hydrogen—and they do. The same line of reasoning also applies to other gases.

In short, if we deal with different gases by the mole, we end up with quantities that differ in mass but are equal in volume! The number of molecules present in a mole of gas (any gas) is called *Avogadro's number*.

The equal volume rule holds only for gases, but Avogadro's number is of more widespread use. A mole of any substance—solid or liquid as well as gaseous—contains Avogadro's number of molecules. (Where a substance is made up of individual atoms, as in the case of helium, Avogadro's number of atoms is contained in a gram-atomic weight rather than in a mole, properly speaking, but that is merely a detail.)

If only chemists had known the exact value of Avogadro's number, they could have at once determined the mass of an individual molecule. This would have lent atoms and molecules an air of actuality. As long as they were merely objects "too small to see" and nothing more, they were bound to be considered as merely convenient (and possibly fictitious) ways of explaining chemical reactions. Give an individual atom or molecule a fixed mass, however, find a fixed number in a glass of water or in an ounce of iron, and the small objects begin to seem real.

Unfortunately, it was not for over a half-century after the introduction of the modern atomic theory that the value of Avogadro's number could be determined even approximately. Till then, all that chemists could say was that Avogadro's number was very large.

The break came in 1865. The Scottish physicist James Clerk Maxwell (1831–1879) and the Austrian physicist Ludwig Boltzmann (1844–1906) had worked out the properties of gases by mathematically analyzing the random movements of the tiny atoms or molecules making up that gas (see page I–200). From the equations derived by Maxwell and Boltzmann, it was possible, by making some reasonable suppositions, to calculate what Avogadro's number might be. This was done by a German chemist, J. Loschmidt, and it turned out to be approximately six hundred billion trillion—a large number, indeed.

A number of more accurate methods have been used in the twentieth century for determining the value of Avogadro's number. These have yielded virtual agreement among themselves and have shown Loschmidt's first attempt to be remarkably good. The value of Avogadro's number currently accepted as most nearly accurate is 602,300,000,000,000,000,000,000 or, in exponential notation, 6.023×10^{23}.

If a mole of oxygen gas weighs 32 grams and contains 6.023×10^{23} oxygen molecules, then the individual oxygen molecule must have a mass of 32 divided by 6.023×10^{23}, or about 5.3×10^{-23} grams. Since an oxygen molecule is made up of two oxygen atoms, each of those has a mass of about 2.65×10^{-23} grams. If the mass of the oxygen atom is known, that of all the other atoms can be calculated from the table of atomic weights.

For instance, since the atomic weight of hydrogen is about 1/16 that of oxygen, the mass of the hydrogen atom must be about 1/16 that of the oxygen atom. As a matter of fact, the mass of the hydrogen atom (the lightest of all atoms) is, to use the figure now accepted as most nearly accurate, 1.67343×10^{-24} grams or, in non-exponential form 0.00000000000000000000000167343 grams.

From Avogadro's number, it is also possible to calculate the diameter of atoms if one assumes that they are spherical in shape and that, in liquids and solids, they are packed together in virtual contact. It then turns out that the diameter of atoms is approximately 10^{-8} centimeters. In ordinary terms, this means that 250,000,000 atoms placed side by side would make a line an inch long.

With atoms so small and so light, it is no wonder that matter seems continuous to our senses and that men like Democritus, who postulated atoms on purely philosophic grounds, found it so difficult to persuade others of the value of their suggestion.

But even the determinations of the mass and size of the atom rest on indirect evidence. In ordinary life, reality is judged by the direct evidence of the senses—especially that of vision. "Seeing is believing," goes the old bromide.

It is, of course, quite possible to argue that seeing is not necessarily believing; that hallucinations and optical illusions are possible; and that it is not always easy to interpret what one sees (as when one "sees" that the earth is flat). It follows, then, that careful and logical reasoning based on a large accumulation of accurate, but indirect, data can be a more reliable guide to useful conclusions than the senses may be.

Nevertheless, human prejudices being what they are (even among scientists), it is rather exciting to know that atoms have been made visible, at least after a fashion. This came about through the invention by the German-American physicist Erwin Wilhelm Mueller (1911–) of specialized forms of powerful microscopes.

The first of these, devised in 1936, was the *field-emission microscope*. This begins with a very fine needle-tip enclosed in a high vacuum. Under an intense electric field, such a needle can be made to shoot out very tiny particles.* If only these particles would travel in perfectly straight, undeviating lines to a screen enclosing the vacuum tube, they would produce a pattern that would depict the actual atomic makeup of the needle-tip. Unfortunately, in even the best vacuums there are gas molecules here and there. The flying particles that strike these molecules are diverted. The result is a fuzzy, out-of-focus picture.

In the 1950's, Mueller made use of heavier particles. He introduced small quantities of helium atoms. When any of these struck the needle-tip, they were modified by the electric field into helium ions (see page 27) which then raced away from the needle-tip in a straight line.

The heavy helium ions are not easily diverted even by collisions with gas molecules, and a much sharper picture is obtained in such a *field-ion microscope*. The atoms in the needle-tip are then pictured as round dots arranged in orderly and well-packed fashion. This device is applicable only to a limited number of high-melting metals, but it has the effect of making atoms visible and therefore "real." Several photographs of the atom patterns revealed in this fashion have already become scientific classics.

* These particles are called electrons and are even smaller than atoms. They will be discussed in detail throughout this book.

CHAPTER **31**

Ions and Radiation

Electrolysis

With 103 different elements now known and, therefore, 103 different kinds of atoms, there is good reason to feel uncomfortable. The periodic table imposes an order upon them, to be sure, but why should that particular order exist?

Why are there so many elements? Why should slight differences in mass between two sets of atoms make so much difference? For instance, argon has an atomic weight of 39.9 and potassium one of 39.1, and yet that small difference makes the first a very inert gas and the second a very active metal.

To obtain an understanding of atomic properties, one might attempt to delve within the atom. One might wonder whether the atoms might not themselves have a structure and whether the atom might not best be understood in terms of this structure.

Something of this sort occurred in 1816, quite early in the game, to an English physician, William Prout (1785–1850). At the time, the atomic theory was very new and the only atomic weights known were a few that had been determined (not very accurately) by Dalton. These atomic weights, based on a hydrogen = 1 standard, were all integers.

To Prout, this seemed more than one could expect of coincidence. If all the atoms had masses that were integral multiples of the mass of the hydrogen atom, then was it not reasonable to suppose that the more massive atoms were made up of hydrogen atoms? If oxygen had an atomic weight of 16, for instance, might not this be because it was made up of 16 hydrogen atoms tightly mashed together?

Prout published this suggestion anonymously, but his authorship became known and his explanation has been called *Prout's hypothesis* ever since.

For a century afterward, numerous chemists made accurate atomic weight determinations for the purpose (in part, at least) of checking on whether or not they were all integral multiples of the atomic weight of hydrogen. They proved not to be. As stated earlier (see page 12) the oxygen atom was not 16 times as massive as the hydrogen atom, judging by atomic weight determinations, but 15.88 times. The atomic weight of nickel is 58.24 times that of hydrogen, and so on.

Over and over again, Prout's hypothesis was disproved and yet, with the opening of the twentieth century, chemists were still uneasy about it. About half the elements had atomic weights that were quite close to integral values. This was still asking a great deal of coincidence. Surely there had to be significance in this fact.

There was, of course, and that significance was discovered in very roundabout fashion through a line of investigation that began with electricity.*

It was in 1807 and 1808 that Humphry Davy had produced a series of elements (sodium, potassium, calcium, magnesium, strontium, and barium) by passing an electric current through molten compounds that contained atoms of these elements in their molecules. The work was carried on with greater detail by the English chemist Michael Faraday (1791–1867), who in his youth had been Davy's assistant and protégé.

Imagine two metal rods connected to a battery, one to the positive pole, the other to the negative pole. These rods are *electrodes* (from Greek words meaning "the path of the electricity"). Faraday called the one attached to the positive pole the *anode* ("upper path") and the one attached to the negative pole the *cathode* ("lower path"). (Electricity at the time was assumed to flow from the positive pole to the negative pole, like water flowing from an upper level to a lower one.)

If the two electrodes are brought together and allowed to touch, electricity will flow through them. However, if they are

* Electricity makes up the subject matter of the second half of Volume II.

separated by an air gap, the circuit is broken and electricity will not flow. If the electrodes are not in contact but are both immersed in the same container of liquid, electricity may or may not flow, depending on the nature of the liquid. Immersed in a dilute solution of sulfuric acid or of sodium chloride, current will flow; immersed in a dilute solution of sugar or in distilled water, current will not flow. The former liquids are conductors of electricity, the latter are nonconductors. Faraday called the liquid conductors *electrolytes,* and the liquid nonconductors, *nonelectrolytes.*

The passage of an electric current through an electrolyte induces chemical changes. Often these changes consist of the decomposition of some of the molecules contained in the solution and in the production of elements (*electrolysis*), as in the case of the metals produced by Davy from their compounds.

The elements, when produced, appear at the electrodes. If they are gases, they bubble off. If they are metals, they remain clinging to the electrode (*electroplating.*)

Elements can appear at either electrode. If electricity passes through water containing a bit of sulfuric acid, hydrogen appears at the cathode and oxygen at the anode. If an electric current passes through molten salt (sodium chloride), metallic sodium appears at the cathode, gaseous chlorine at the anode.

Faraday did not allow himself to speculate too freely about the exact manner in which an element was transported through the body of the solution to one electrode or the other. One might think of drifting atoms, but Faraday was rather lukewarm on the atomic theory (still new at the time of his experiments) and he preferred not to commit himself. He spoke simply of *ions* (from a Greek word meaning "wanderer") passing through the solution, and said nothing about their nature.

Some ions, like those which ended as sodium or hydrogen, are attracted to the cathode; they are *cations* (pronounced in three syllables). Others, like those which end as chlorine or oxygen, are attracted to the anode and are *anions* (again three syllables).

Faraday carefully measured the mass of element produced by the action of the electric current and, in 1832 and 1833, proposed what have since become known as *Faraday's laws of electrolysis.*

The first law of electrolysis states: The mass of element formed by electrolysis is proportional to the quantity of electric current passing through an electrolyte. The unit of quantity of electricity in the meter-kilogram-second (mks) system is the *cou-*

lomb (see page II–164), and one coulomb of electricity will form 0.001118 grams of metallic silver when passed through a solution of a silver compound. By Faraday's first law two coulombs of electricity would produce twice that mass of silver and, in general, x coulombs will produce $0.001118x$ grams of silver.

A gram-atomic weight of silver is equal to 107.87 grams. How many coulombs would be required to deposit that many grams? It is only necessary to set $0.001118x = 107.87$ and solve for x, which turns out to be equal to about 96,500 coulombs. For this reason, the quantity of electricity represented by 96,500 coulombs is set equal to one *faraday*. The faraday may be defined as that quantity of electricity which will liberate one gram-atomic weight of metallic silver from a silver compound.

To understand Faraday's second law of electrolysis, it is first necessary to grasp the meaning of *equivalent weight.*

One gram-atomic weight of chlorine gas (35.5 grams) will combine with one gram-atomic weight of hydrogen (1 gram) to form hydrogen chloride (HCl). The molecule is composed of one atom of each element, and since a gram-atomic weight of hydrogen and a gram-atomic weight of chlorine contain the same number of atoms of those elements, the two quantities of gas match up neatly. (The fact that in the case of both hydrogen and chlorine the atoms happen to be distributed in the form of two-atom molecules does not alter the case.) By the same reasoning, one gram-atomic weight of chlorine will combine with one gram-atomic weight of sodium (23 grams) to form sodium chloride ($NaCl$.)

However, one gram-atomic weight of chlorine will combine with only half a gram-atomic weight of calcium to form calcium chloride ($CaCl_2$) because every calcium atom takes up two chlorine atoms; consequently only half as many calcium atoms as chlorine atoms are needed for the reaction. The gram-atomic weight of calcium is 40 grams, and half a gram-atomic weight is 20 grams. This means that 20 grams represents the equivalent weight of calcium: the weight that is equivalent, that is, to a gram-atomic weight of chlorine or of hydrogen or of sodium in forming compounds. (It is usually the gram-atomic weight of hydrogen which is taken as the standard.)

In the same way one gram-atomic weight of chlorine will combine with half a gram-atomic weight of magnesium to form magnesium chloride ($MgCl_2$) and with a third of a gram-atomic weight of aluminum to form aluminum chloride ($AlCl_3$). The equivalent weight of magnesium is its gram-atomic weight (24 grams) divided by 2, or 12 grams, while that of aluminum is its gram-atomic weight (27 grams) divided by 3, or 9 grams.

Now we can return to Faraday's second law of electrolysis, which can be stated most simply, as follows: One faraday of electricity will form an equivalent weight of an element when passing through a compound of that element.

If a faraday of electricity will form 108 grams of silver, it will also form 23 grams of sodium, 35.5 grams of chlorine, or 1 gram of hydrogen (in each case equal to the gram-atomic weight). It will form 20 grams of calcium or 12 grams of magnesium (in each case equal to half the gram-atomic weight). It will form 9 grams of aluminum (equal to a third the gram-atomic weight).

Particles of Electricity

Faced with these laws of electrolysis, it is extremely tempting to begin wondering whether electricity might not be particulate in nature. Just as matter consists of indivisible units (atoms), so might electricity.

Let us assume this is so, and let us further assume that such units come in two varieties. There is a positive unit that is attracted to the negatively-charged cathode (opposite electric charges attract, see page II–159). It is such a positive unit that can carry atoms of hydrogen and sodium in the direction of the cathode. Similarly there would be a negative unit that is attracted to the positively-charged anode and that can carry atoms of oxygen and chlorine with it. The two units can be symbolized as + and —.

If we imagine a hydrogen atom being transported toward the cathode by a positive electrical unit, we can symbolize the hydrogen atom in transit as H^+. Using Faraday's term, we can call it a *hydrogen ion.* Similarly, we can have a sodium ion (Na^+) or a potassium ion (K^+). All three are examples of *positive ions* (or cations).

A faraday of electricity can be viewed as containing Avogadro's number of electrical units. Allowing one unit per atom, a faraday of electricity would transport Avogadro's number of hydrogen atoms to the cathode. In other words, a faraday of electricity would produce a gram-atomic weight of hydrogen at the electrode. It would also, by similar reasoning, produce a gram-atomic weight of sodium atoms or potassium atoms or silver atoms.

Since a faraday of electricity has never, under any conditions, been found to produce more than a gram-atomic weight of any element, it seems reasonable to conclude that the electric unit we are dealing with is very likely the smallest unit possible—that it is an indivisible unit and that one unit can transport no more than one atom.

Chlorine atoms are transported to the positive electrode, or anode, and therefore must be transported by a negative electric unit. We can symbolize the chlorine atom in transport as Cl^- and call that the *chloride ion.** Since a faraday of electricity produces exactly one gram-atomic weight of chlorine, we must conclude that the negative unit is exactly equal in size to the positive unit.

What of calcium? A faraday of electricity will produce only half a gram-atomic weight of that element. This is most easily explained by assuming that the atom, on its travels toward the cathode, must be transported by two positive units. In that case the supply of units in a faraday of electricity will only transport half the number of atoms of calcium one would expect if one were dealing, say, with sodium. We can write the *calcium ion* as Ca^{++}, therefore. By similar reasoning, we can write the *magnesium ion* as Mg^{++}, the *barium ion* as Ba^{++}, the *aluminum ion* as Al^{+++}, the *oxide ion* as O^{--}, and so on.

The first to maintain, in complete and logical detail, that Faraday's ions were actually atoms carrying a positive or negative electric charge, was the Swedish chemist Svanté August Arrhenius (1859–1927). These views, presented first in 1887, were based not only on Faraday's work but on other chemical evidence as well.

According to Arrhenius, when an electric current passed through molten sodium chloride, the molecule (NaCl) broke up, or dissociated†, not into atoms, but into charged ions, Na^+ and Cl^-, the sodium ions then drifting toward the cathode and the chloride ions toward the anode. (This is the *theory of ionic dissociation.*) At cathode and anode, the ions are discharged and the uncharged atoms are produced; metallic sodium at the cathode, gaseous chlorine at the anode.

The charged atom, Arrhenius maintained (correctly, as it turned out), did not necessarily have properties in any way resembling those of the uncharged atom. Sodium atoms, for

* It is called "chloride ion" rather than "chlorine ion" for reasons involving chemical nomenclature. These are more fittingly discussed in a book on chemistry. For our purposes here, we can take chemical names as we find them.

† It has since turned out that Arrhenius was wrong is assuming that this dissociation into ions took place only under the influence of the electric current. The atoms in sodium chloride exist in ionic form at all times. However, Arrhenius, like Dalton, was right enough for his time.

instance, would react violently with water, but sodium ions, much milder in character, would not. Chlorine atoms would form chlorine molecules and bubble out of solution; chloride ions would not.

It followed further from Arrhenius' analysis that groups of atoms, as well as individual atoms, might carry an electric charge. Thus, ammonium chloride (NH_4Cl) will dissociate to form NH_4^+ and Cl^-, the former being the *ammonium ion*. Again, sodium nitrate ($NaNO_3$) will break up into Na^+ and NO_3^-, the latter being the *nitrate ion*. Other such *compound ions* (those made up of more than one atom) are the *hydroxyl ion* (OH^-), the *sulfate ion* (SO_4^{--}), the *carbonate ion* (CO_3^{--}) and the *phosphate ion* (PO_4^{---}).

So much in the air was this notion of an indivisible unit of electricity that the Irish physicist George Johnstone Stoney (1826–1911) had even given it a name in a paper published in 1881. He called it an *electron*.

Despite the logic of Arrhenius' views (especially as viewed from hindsight), his theory of ionic dissociation was met with great reserve. The notion of an atom as a featureless, structureless, indivisible object dated back to Democritus and had become a firm part of scientific thinking. The thought of such atoms carrying indivisible units of electric charge ("atoms of electricity" so to speak) was hard to take without heavy evidence on its side.

Such evidence was not obtained in completely acceptable form for a decade after Arrhenius, but it was on its way in Arrhenius' time and even before.

The chief difficulty in detecting particles of electricity under ordinary conditions was that even supposing they existed, they would be lost among the ordinary particles of matter in the path of the electric current.

What was clearly needed was the passage (if possible) of an electric current through a good vacuum. Then the particles of electricity (if any) might show up unmasked. The first to actually force a current of electricity through a vacuum was Faraday himself, in 1838. However, the best vacuum he could obtain was not a very good one, and his observations therefore lacked significance.

In 1854, a German glassblower, Heinrich Geissler (1814–1879), devised a better method for producing vacuums than any hitherto obtained. He manufactured *Geissler tubes* containing these good vacuums. The German physicist Julius Plücker (1801–1868) made use of such Geissler tubes into which two electrodes had been sealed.

Plücker forced electricity to cross the vacuum from one electrode to the other and noted that a greenish luminescence coated the cathode when the current flowed. This greenish luminescence seemed precisely the same whatever the metal out of which the cathode was constructed and whatever the nature of the wisps of gas that still remained in the tube after it had been evacuated. Whatever that luminescence might be, then, it was a property of electricity and not of ordinary matter.

Plücker also showed that the luminescence shifted its position when a magnet was brought near. One pole of the magnet shifted it in one direction; the other pole, in the opposite direction. This also seemed to brand the luminescence an electrical phenomenon, since electricity and magnetism are very closely allied (see page II–237).

It soon became obvious that the phenomenon was not merely confined to the near neighborhood of the cathode but that something was traveling all across the space from the cathode to the anode. What's more, this something traveled in straight lines. If the anode were placed to one side, whatever it was that was traveling missed the anode and went on to strike the glass of the tube, creating a spot of green luminescence where it struck.

Two investigators, the German physicist Johann Wilhelm Hittorf (1824–1914) and the English physicist William Crookes (1832–1919), working independently, showed that if in such a tube an object was enclosed in the path of the traveling entity, that object cast a shadow against the luminescence on the glass. Hittorf published his results first—in 1869.

It was clear, then, that physicists were faced with a kind of radiation that traveled in straight lines and cast sharp shadows. The German physicist Eugen Goldstein (1850–1930), committing himself no further than this and taking note of the appar-

Crookes tube

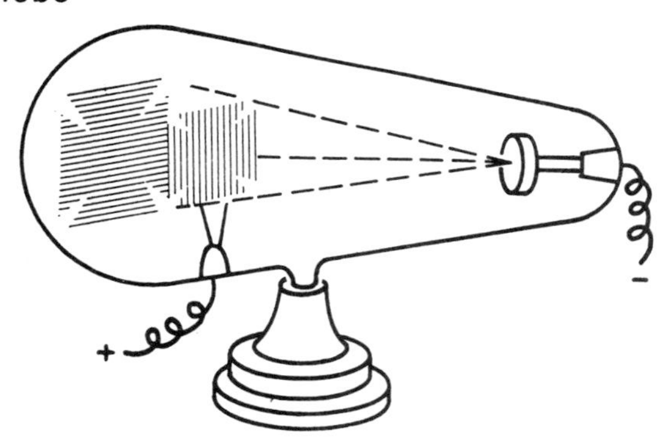

ent origin of the radiation, called it *cathode rays* in 1876. This name was generally adopted.

A controversy then arose as to the nature of the cathode rays. The fact that the rays traveled in straight lines and seemed unaffected by gravity made it appear likely that they were a wave form after the fashion of light. The great argument against this was that the cathode rays were deflected by a magnet, whereas light rays (or any form of radiation resembling light) were not.

The alternative suggestion was that the cathode rays were electrically charged particles, the "atoms of electricity" in fact. They would then naturally be affected by a magnet, and their lack of response to gravitation would be explained by their small mass and rapid motion. The response would be there but would be too small to detect.

The Radiation Spectrum

The controversy over the nature of the cathode rays divided itself almost on national lines, with many German physicists upholding the wave interpretation and many English physicists maintaining the charged-particle suggestion.

This was a natural division, perhaps, for it was in Germany that indisputably new wave forms were discovered in the final decades of the nineteenth century—although the first such discovery was inspired by the theory of an Englishman, James Clerk Maxwell.

Maxwell's analysis of electrical and magnetic phenomena had showed that the two must be so closely and indissolubly related that one could properly speak only of electromagnetism. He went on to show, furthermore, that an oscillating electric charge ought to produce a wave-form type of *electromagnetic radiation* that would travel at the speed of light. It seemed almost inevitable, therefore, that light itself must be an electromagnetic radiation—otherwise the coincidence of its velocity being equal to that of such radiation would be too great for acceptance.

But if Maxwell was correct, there was no reason why man could not deliberately produce an electromagnetic radiation by oscillating an electric current. It could not be oscillated fast enough to produce the tiny wavelengths of light (that would have required about a quadrillion oscillations per second), but Maxwell's theory set no limit on the period of oscillation. A comparatively slow oscillation of, say, a thousand times per second would produce a thousand waves of electromagnetic radiation per second. Since the wave train would travel 300,000 kilometers per second, each individual wave would be 300 kilometers long (vastly longer than the wavelengths of light), but those waves would nevertheless exist.

The attempt to form long wavelength radiation was made in 1887 by a German physicist, Heinrich Rudolf Hertz (1857–1894). He set up an electric circuit that would produce a spark across a small air gap under conditions that would bring about an electrical oscillation of the sort that Maxwell said would produce electromagnetic radiation. To detect the radiation, if any was produced, Hertz used a simple rectangle of wire broken by a small air gap. The electromagnetic radiation crossing the wire receiver would cause an electric current to flow and produce a spark across the air gap.

Such a spark was found and Hertz knew he had detected the electromagnetic radiation predicted by Maxwell—a result that served as strong evidence in favor of the validity of Maxwell's theory. At first, the radiation discovered by Hertz was called "Hertzian waves." However, the noncommittal title of *radio waves* ("waves that radiate") is now usually used.

The discovery of radio waves gave physicists their first notion of the truly broad extent of the electromagnetic spectrum. The wavelength range of visible light is from 380 to 760 millimicrons, representing a single octave of radiation. (A millimicron is equal to a billionth of a meter, and an octave represents a range over which the wavelength doubles.)

It was not until 1800 that this spectrum was broadened beyond the visible. In that year, the German-English astronomer William Herschel (1738–1822) was measuring the effect of the solar spectrum upon a thermometer. He discovered that the temperature-raising effect of the spectrum was most marked at a point somewhat beyond the red, where the eye could see nothing. Herschel correctly concluded that light was present there—light which was incapable of affecting the retina of the eye.

At first, because of the efficient manner in which the glass and mercury of the thermometer absorbed this invisible light, it was referred to as "heat rays." Later, the more noncommittal term, *infrared radiation* ("below the red") was used. With the establishment of the wave theory of light (see page II–66) it was understood that infrared radiation was of longer wavelength than visible light.

Nowadays, the range of infrared radiation is taken as extending from the 760 millimicron limit of visible light to a rather arbitrary upper limit placed at 3,000,000 millimicrons. In expressing the wavelength of infrared radiation, the more convenient unit of length, the micron (equal to 1000 millimicrons), may be used. The range of infrared radiation can then be said to extend from 0.76 microns to 3000 microns, a stretch of about 12 octaves.

Beyond the farthest infrared radiation lie the radio waves. The radiation found in the wavelength region immediately adjacent to the infrared has come in recent years to be known as *microwaves* ("small waves"—small for radio waves, that is). The microwave region extends from 3000 to 300,000 microns. Again we can shift units to millimeters (one millimeter is equal to a thousand microns) and say that the range is from 3 to 300 millimeters, or about 6½ octaves.

Beyond the microwaves are the radio waves, proper. For them there is no definite upper limit. Radio waves of longer and longer wavelength can easily be produced until they become too low in energy to detect by presently available means. (The longer the wavelength of electromagnetic radiation, the lower its energy content, as the quantum theory makes plain, see page II–130). Radio waves up to 30,000,000 millimeters in wavelength have been used in technology, so that we can say that useful radio waves extend over a range of from 300 to 30,000,000 millimeters (0.3 to 30,000 meters), or 16½ octaves.

The electromagnetic spectrum also extends out beyond the violet end of the visible light region. This was first discovered in 1801 by the German physicist Johann Wilhelm Ritter (1776–1810), who was studying the action of light upon silver nitrate. Silver nitrate, white in color, breaks down in the presence of light, liberating black particles of metallic silver and turning visibly gray in consequence. This effect is more marked where shortwave light impinges on the silver nitrate (which is not surprising to us now since shortwave light is known to be the more energetic, so that it naturally initiates an energy-consuming chemical reaction more readily). Ritter discovered that the effect on silver nitrate was even more pronounced when the compound was placed beyond the violet end of the solar spectrum where nothing at all could be seen.

Like Herschel, Ritter concluded that invisible light was present. Because of its effect on silver nitrate it was at first referred to as "chemical rays." This soon gave way, however, to *ultraviolet radiation* ("beyond the violet"), and it came to be understood that such radiation was shorter in wavelength than visible light was.

Nowadays, ultraviolet radiation is taken as covering a range of from 360 millimicrons (the boundary of the visible violet) down to an arbitrary limit of 1 millimicron, a little over eight octaves. Thus, as the 1890's opened, the overall stretch of the electromagnetic spectrum, from ultraviolet radiation to radio waves, represented an extreme range of some 44 octaves, of which only one was visible light.

Even so, the electromagnetic spectrum had not been completely filled in. The next step was taken by the German physicist Wilhelm Konrad Röntgen (1845–1923). He was interested in cathode rays and, in particular, in the luminescence to which they gave rise when they impinged on certain chemicals.

In order to observe the faint luminescence, he darkened the room and enclosed the cathode-ray tube in thin black cardboard. On November 5, 1895, he set the enclosed cathode-ray tube into action, and a flash of light that did not come from the tube caught his eye. He looked up and quite a distance from the tube he noted a sheet of paper that had been coated with barium platinocyanide (a compound that glows under the impact of energetic radiation) shining away. Röntgen would not have been surprised to see it glow if cathode rays were striking it, but the cathode rays were completely shielded off.

Röntgen turned off the tube; the coated paper darkened. He turned it on again; it glowed. He walked into the next room with the coated paper, closed the door, and pulled down the blinds. The paper continued to glow while the tube was in operation.

It seemed to Röntgen that some sort of radiation was emerging from the cathode-ray tube, a radiation produced by the impact of the cathode rays on the solid material with which it collided. The kinetic energy lost by the cathode rays as they were stopped was converted, apparently, into this new form of radiation; a radiation so energetic that it could pass through considerable thicknesses of paper and even through thin layers of metal. Röntgen published his first report on the subject on December 28, 1895.

The radiation is sometimes known as "Röntgen rays" after the discoverer, but Röntgen himself honored its unknown nature by using the mathematical symbol of the unknown and named the radiation *X rays*. The name has clung firmly even though the nature of the radiation is no longer mysterious.

I have mentioned the experiment in some detail because the discovery of X rays is usually taken as initiating the "Second Scientific Revolution" (the first having been initiated by the experiments of Galileo—see page I–9).

In a way, this might be considered as over-dramatic, for Röntgen's experiment did not really represent a sharp break with previous work. It came about in connection with the cathode ray problem, which was occupying many physicists of the time. The new radiation had been observed by Crookes and Hertz even before Röntgen's announcement (although they had not grasped the significance of what they were observing), so that the discovery was inevitable. If Röntgen had not made it, someone else would have done so, perhaps within weeks. Moreover, the existence of X rays was implicit in Maxwell's theory of electromagnetic radiation, and the important discovery in that connection, the one that validated the theory, was that of radio waves, eight years earlier.

Nevertheless, after all this has been allowed for, the X rays caught the fancy of both the scientists and the lay public with unprecedented intensity. Their ability to penetrate matter was fascinating. On January 23, 1896, in a public lecture, Röntgen took an X ray photograph of the hand of the German biologist Rudolf Albert von Kölliker (1817–1905), who volunteered for the purpose. The bones showed up beautifully, for they stopped the X rays where flesh and blood did not. The photographic film behind the soft tissues was fogged by the X rays that reached it, while the portion of the film behind the bone was not. The bones showed up, therefore, as white against gray.

The usefulness to medicine was obvious and X rays were at once applied there and in dentistry. (The dangers of X rays as cancer-producing agents were not understood for a number of years.) A storm of experimentation involving X rays followed, and as a result, discoveries were made which resulted in so quick and rapid an improvement of man's understanding of the universe that it seemed a scientific revolution indeed.

CHAPTER 32

The Electron

The Discovery of the Electron

Since of the new types of radiation, the radio waves were known to be wave forms, and the X rays were strongly suspected of being wave forms (final proof that they were was obtained in 1912, see page 63) it seemed all the more natural to continue to suspect that cathode rays were also wave forms.

For one thing, Hertz showed in 1892 that cathode rays could actually penetrate thin sheets of metal. This seemed quite an unlikely property for particles to possess, whereas a few years later the discovery of X rays made it quite clear that wave forms could possess this property. The German physicist Philipp Lenard (1862–1947), Hertz's assistant, even set up a cathode-ray tube containing a thin metal "window." Cathode rays striking that window passed through and emerged into open air. (Such emerging cathode rays were for a time called "Lenard rays.")

If cathode rays were electrically charged particles, they should be affected not only by a magnetic field but also by an electrostatic field. Hertz passed a beam of cathode rays between two parallel plates, one carrying a positive electric charge and another, a negative one. He detected no deviation in the cathode ray stream and concluded that cathode rays were waves.

That, however, marked the peak of the wave theory. Another experimenter was on the scene, a member of the English group of physicists, Joseph John Thomson (1856–1940). It seemed to him that the experiment involving the electrostatic field would not work unless the cathode rays were passing through a particularly good vacuum. Otherwise the thin wisps of gas present would, by Thomson's reasoning, act to reduce the effect of the electrostatic field upon the cathode rays. In 1897, he therefore repeated Hertz's experiment (Hertz having prematurely died three years earlier), using a cathode-ray tube with a particularly good vacuum. A deflection in the path of the cathode rays was detected.

This observation was the last straw. With cathode rays deflected by both a magnetic field and an electrostatic field, the evidence in favor of particles was too strong to be withstood. From the direction of the deflection, it could be seen that the particles carried a negative charge.

It seemed clear that these cathode ray particles must represent units of electricity, perhaps the indivisible negative unit (see page 29) which some nineteenth century physicists had been postulating. The particles were therefore given Stoney's name of "electron," and it is because of Thomson's crucial experiment that he is usually said to have "discovered the electron" in 1897.

But Thomson did more than merely discover the electron. He went on to determine one of its overwhelmingly important properties.

When an electron passes through a magnetic field, it is deflected by that field and departs from its otherwise straight-line course to take up a curved path. (This is analogous to the fashion in which the moon, when exposed to the gravitational field of the earth, departs from what would otherwise be a straight-line course to take up a curved path.)

The deflection of the electron is the result of the magnetic force exerted upon it. The amount of this force is proportional, first to the strength of the magnetic field (H), then to the size of the electric charge on the electron (e), and finally to the velocity of the electron (v)—for it is the velocity that determines how many magnetic lines of force will be cut by the moving electron. (A stationary electron, or one traveling parallel to those lines of force, would not be affected by the magnetic field.) The force producing the deflection is therefore equal to Hev.

A centrifugal effect must be exhibited by an electron traveling in a curved path. This effect is equal to mv^2/r, where m is the mass of the electron, v its velocity, and r the radius of the curved path it is following.

The electron, in following a particular curved path, must have the magnetic force exactly balanced by the centrifugal effect. If this were not so, it would travel either a tighter curve or a looser one, finding a curve in which the two effects would balance. For the curved path actually followed, we can therefore say that:

$$Hev = \frac{mv^2}{r} \qquad \text{(Equation 3–1)}$$

This can be rearranged and simplified to:

$$\frac{e}{m} = \frac{v}{Hr} \qquad \text{(Equation 3–2)}$$

The strength of the magnetic field is known, and the curvature radius of the beam of cathode ray particles can easily be determined by the shift in the position of the luminescent spot on the wall of the cathode-ray tube. Now if one could only determine the value of v (the velocity of the particles), it would at once be possible to determine the value of e/m (the ratio of the charge of the electron to its mass).

Thomson found the velocity by causing the cathode rays to be under the influence of both an electrostatic field and a magnetic field, but under such conditions that the two deflections were in opposite directions and just balanced. The deflection by the electrostatic field depended upon its strength (F) and upon the charge of the electron (e). It did not depend on the velocity of the electron, for there is attraction between opposite electric charges even if they are stationary relative to each other.

Consequently, when the magnetic and electrostatic fields are adjusted in strength so that the effects on the electrons cancel out:

$$Hev = Fe \qquad \text{(Equation 3–3)}$$

or:

$$v = \frac{F}{H} \qquad \text{(Equation 3–4)}$$

Since the strength of both fields can be measured easily, v can be determined and turns out to be about 30,000 kilometers per second, about one-tenth the velocity of light. This was by far the largest velocity ever measured for material objects up to that time and immediately explained why cathode rays had seemed unaffected by a gravitational field. At that enormous velocity, particles passed from end to end of the cathode-ray tube long before they could show a measurable response to the earth's gravitational field.

With the value of v known, Equation 3–2 makes it at once possible to determine e/m for the electron, and Thomson was amazed to discover that this ratio came out to have a value, far greater than that for any ion (which are also charged particles).

Consider the ions H^+, Na^+ and K^+. All three carry a charge of equal magnitude since a faraday of electricity suffices to produce a gram-atomic weight of each. However, the mass of the potassium ion is 39 times that of the hydrogen atom and the mass of the sodium atom is 23 times that of the hydrogen atom. If e is fixed, then the ratio e/m rises as m decreases. Thus the ratio e/m for H^+ must be 23 times as great as that for Na^+, and 39 times as great as that for K^+.

Indeed, since the hydrogen ion is the least massive ion known, the e/m ratio for it may well be higher than for any other ion that can possibly exist. And yet the e/m ratio for the electron is (using the value now accepted) 1836 times as great as that of the hydrogen ion.

No quantity of electric charge smaller than that on the hydrogen ion had ever been observed, and it seemed reasonable to suppose that the electron carried this smallest-observed charge. If that is so, if e is equal in the case of the electron and the hydrogen ion, and e/m is 1836 times greater in the first case than in the second, it must follow that the difference is to be found in the mass. The mass of the electron must be only 1/1836 that of the hydrogen ion.

Since the mass of the hydrogen atom is known and the mass of the hydrogen ion is only very slightly less, it is easy to calculate the mass of the electron. The best modern determination is 9.1091 $\times 10^{-28}$ grams, or 0.00000000000000000000000000091091 grams.

In one bound, the atoms, which from the time of Democritus on had been assumed to be the smallest particle of matter, were suddenly rendered giants. Here was something much smaller than even the smallest atom; something so small indeed that it could easily be visualized as worming its way through the interstices among the atoms of ordinary matter. That seemed one reasonable explanation for the fact that cathode rays made up of particles could penetrate thin sheets of metal. It also explained why electric currents could be made to flow through copper wires.

Thomson, therefore, had not only discovered the electron, he had also discovered the first of the *subatomic particles,* and opened a new realm of smallness beyond the atom.

The Charge of the Electron

Knowledge of the exact mass of the electron did not automatically provide physicists with an estimate of the exact size of the charge upon the electron. One could only say, at first, that the charge on the electron was exactly equal to the charge on the chloride ion, for instance, or exactly equal (but opposite in sign) to the charge on the hydrogen ion. But then, the exact size of the charge on any ion was not known through the first decade of the twentieth century.

The experiments that determined the size of the electric charge on the electron were conducted by the American physicist Robert Andrews Millikan (1868–1953) in 1911.

Millikan made use of two horizontal plates, separated by about 1.6 centimeters, in a closed vessel containing air at low pressure. The upper plate had a number of fine holes in it and was connected to a battery that could place a positive charge upon it. Millikan sprayed fine drops of a nonvolatile oil into the closed vessel above the plates. Occasionally, one droplet would pass through one of the holes in the upper plate and would appear in the space between the plates. There it could be viewed through a magnifying lens because it was made to gleam like a star through its reflection of a powerful beam of light entering from one side.

Left to itself, the droplet of oil would fall slowly, under the influence of gravity. The rate of this fall in response to gravity, against the resistance of air (which is considerable for so small and light an object as an oil droplet), depends on the mass of the droplet. Making use of an equation first developed by the British physicist George Gabriel Stokes (1819–1903), Millikan could determine the mass of the oil droplets.

Millikan then exposed the container to the action of X rays. This produced ions in the atmosphere within (see page 110). Occasionally, one of these ions attached itself to the droplet. If it were a positive ion, the droplet, with a positive charge suddenly added, would be repelled by the positively-charged plate above, and would rush downward at a rate greater than could be accounted for by the action of gravity alone. If the ion were negative, the droplet would be attracted to the positively-charged plate and might even begin to rise in defiance of gravity.

The change in velocity of the droplet would depend on the intensity of the electric field (which Millikan knew) and the charge on the droplet, which he could now calculate.

Millikan found that the charge on the droplet varied according to the nature of the ion that was adsorbed and on the number of ions that were adsorbed. All the charges were, however, multiples of some minimum unit, and this minimum unit could reasonably be taken as the smallest possible charge on an ion and, therefore, equal to the charge on the electron. Millikan's final determination of this minimum charge was quite close to the value now accepted, which is 4.80298×10^{-10} electrostatic units ("esu," see page II–164), or 0.000000000480298 esu.

As far as we know now, this charge of 4.80298×10^{-10} esu is the only size in which electric charge comes, though it may come in two varieties of that size, positive and negative. Suppose we consider this charge unit as 1, for simplicity's sake. In that case, all objects can be placed in one of three classes:

(1) Objects with a net electric charge of 0. This would include ordinary atoms and molecules.

(2) Objects with a net charge of -1, or some multiple of that. Examples are some negative ions, and, of course, the electron.

(3) Objects with a net charge of $+1$ or some multiple of that. Examples of that are some positive ions.

No one has yet discovered an object with a charge of $+0.5$ or -1.3 or, in fact, with a charge that deviates from an integral value by even the slightest. Such objects may yet be discovered in the future, but the prospects for such an eventuality seem quite small at the moment.

Electronics

It was the existence of electrons, and of subatomic particles generally, that was to bring a new degree of order into the table of elements. Before proceeding in that direction, however, let us consider some of the changes in technology that arose out of the use of streams of electrons in a vacuum. (The study of the behavior of such free electrons and of the techniques for controlling and manipulating them is called *electronics.*)

The flow of electrons across a vacuum was observed under

interesting circumstances in 1883 by the American inventor Thomas Alva Edison (1847–1931). Four years earlier he had devised a practical electric light, and he was still laboring to improve it. The light, at that time, consisted of a carbon filament enclosed in an evacuated bulb. (The vacuum was necessary in order to keep the carbon filament, raised to white-hot temperatures by the current passing through it, from burning to nothing in a flash—as it would if air were present.)

Edison observed on the interior surface of the bulb, a blackening which presumably resulted because some of the carbon vaporized from the hot filament surface and settled on the glass. This weakened the filament and reduced the transparency of the glass, so Edison sought to counter the effect. One of his efforts to do so consisted of sealing a small strip of metal into the bulb near the filament, hoping perhaps that the metal would blacken rather than the glass.

This did not happen, but Edison noticed something else. When he attached this piece of metal (called a *plate* by later workers) to the positive pole of a battery, so that it took on a positive charge with respect to the filament, a current flowed even though there was a gap in the circuit between the filament and the plate. If the plate was given a negative charge, this did not happen. Edison described this phenomenon (the *Edison effect*) and then, since he had no immediate use for the matter, laid it aside.

The Edison effect was no mystery once the cathode rays were understood. The heated filament had a tendency to give off electrons; they "boiled off," so to speak. Ordinarily, this would result in no more than a thin cloud of electrons surrounding the filament.

If, however, a positively-charged plate was placed in the neighborhood, the electrons would be attracted to it. A stream of electrons would pass continuously from the heated filament to the plate, and this is equivalent to a completed electric circuit. If the plate is negatively charged, the electron cloud is repelled, and the circuit is not completed; there is no flow of electricity.

An English electrical engineer, John Ambrose Fleming (1849–1945), who had served as consultant to Edison in the 1880's, remembered the Edison effect twenty years later, in 1904. Suppose the plate was attached to an alternating-current circuit (see page II–221). When the current flowed in one direction, the plate would receive a positive charge; when it flowed in the other, it would receive a negative charge. The nature of the charge would shift some sixty times a second in sixty-cycle alternating current. However, only when the plate was positively charged would the circuit really be complete.

Half the time, then, when the current was flowing in one direction, it would actually flow. The other half the time, when it would ordinarily be expected to flow in the other direction, it would not flow at all, for the circuit would be broken.

The Edison effect made it possible for the circuit to be opened and closed in exact time with the alternation of the current. What would have been an alternating current without the filament-plate combination in the circuit, becomes a direct current with it. The current might flow only intermittently and with fluctuating intensity, to be sure, but it would always flow (when it did flow) in the same direction. The filament-plate combination acted as a *rectifier*.

Fleming called the device a "valve" because it opened and shut the gate to the flow of electricity as an ordinary valve might do for a flow of water. In the United States, the far less significant name of *vacuum tube* has come into use. A better name than either is *diode* ("two electrodes"), since two sealed elements—the filament and the plate—serve as electrodes within the bulb.

Two years later, in 1906, the American inventor Lee De Forest (1873–1961) added a third sealed element to the tube and made it a *triode*. The third element consisted of a network of fine wires placed between the filament and the plate. This network is the *grid*.

The grid serves to make the control of the electron flow much more delicate. In the diode, the current either flows or does not flow; the valve is pretty much either wide open or tight shut. The mere mechanical presence of the grid would have little effect on this, for almost all the electrons would slip through the holes. A very small proportion would strike the wires themselves and be stopped.

Suppose, though, that the grid were part of a separate electrical circuit and that a small negative charge were maintained upon it. Each wire of the grid would then repel the electrons, which would be deflected if they came too close. In addition to the mechanical obstruction of the wire itself, each wire would be thickened, so to speak, by a layer of electrical obstruction. The holes through which the electrons could pass without being turned back would become smaller, so that fewer electrons would reach the plate. If the grid were made slightly more negative, the effect would become more pronounced; it would not take much of a negative charge on the grid to cut off the current completely, despite the positive charge on the plate behind the grid. The ordinary valve action could now be allowed to remain wide open while the grid took over control.

The result would be most important if the grid were part of a circuit in which a very weak and varying current was set up. The negative charge on the grid would vary slightly, in perfect step with the variation in current potential, and this variation would open and shut the valve between the filament and the plate. The very small variation in negative potential on the grid would result in a very large variation on the current getting through the grid. The large current would, however, keep exact step with the weak grid potential, and also imitate its variations exactly. The characteristic of a weak current would be imposed on a strong one, and the triode would act as an *amplifier*.

Inventors now had a method of producing effects by altering the motion of tiny, almost massless electrons, rather than by altering the motion of comparatively large and massive levers and gears. The electrons, with so little mass had equivalently little inertia, so that changes could be enforced upon them in tiny fractions of a second. The proverbial "wink of an eye," fast in comparison to the behavior of mechanical devices, became slowness itself in comparison with the rapid action of electronic instruments.

Radio

Diodes, triodes, and various more complicated descendants were put to work in connection with a device even more dramatic than the electric light that gave them birth.

This dated back to the discovery of radio waves by Hertz, who had produced radio waves at one point and had detected them at another. It was easy to imagine that if radio waves could be produced easily enough and detected sensitively enough, the distance between the point of production and the point of detection might be made miles rather than feet. Consequently, if the radio waves were produced in bursts that imitated the Morse code, for instance, a form of communication would be established. The effect of the telegraph (see page II–209) would be duplicated, with radio waves across space replacing electric currents along wires.

The result might be called "wireless telegraphy" or "radiotelegraphy." Actually, the British call it the former, shortening it to "wireless," while Americans call it the latter, shortening it to *radio*.

An Italian electrical engineer, Guglielmo Marconi (1874–1937), having read a description of Hertz's experiment in 1894, set about making communication by way of radio waves a reality. He made use of Hertz's method of producing the radio waves, and of a device called the "coherer" to detect them. The coherer consisted of a container of loosely-packed metal filings. Ordinarily this conducted little current, but it conducted quite a bit when radio waves fell upon it. In this way, radio waves could be converted into an easily detected electrical current.

Gradually Marconi added devices that facilitated both sending and receiving. In 1895, he sent a signal one mile; in 1896, nine miles; in 1897, twelve miles; and in 1898, eighteen miles. He even established a commercial company for the sending of "Marconigrams."

In all this a seeming paradox appeared. Radio waves, like any other form of electromagnetic radiation, ought to travel in straight lines only, and therefore, like light, should be able to penetrate no farther than the horizon. Beyond the horizon, the bulge of the spherical earth should have interfered.

Marconi noted, however, that radio waves seemed to follow the curve of the earth. He had no explanation for this, but he did not hesitate to make use of the fact. On December 12, 1901, Marconi succeeded in sending a radio wave signal from the southwest tip of England, around the bulge of the earth, to Newfoundland. He had sent a signal across the Atlantic Ocean, and this may be taken as a convenient date for the "invention of radio."

Within the year, an explanation for radio communication around the earth's bulge was offered independently by the British-American electrical engineer Arthur Edwin Kennelly (1861–1939) and the English physicist Oliver Heaviside (1850–1925). In the upper atmosphere, both pointed out, there must be regions rich in electrically charged particles. Such particles, both went on to show, would serve to reflect radio waves, which would then cross the Atlantic Ocean, not in a direct curved path, but in a series of straight-line reflections between heaven and earth.

These regions of charged particles were actually detected in 1924 by the English physicist Edward Victor Appleton (1892–1965). In honor of the original theorists, the region is sometimes referred to as the *Kennelly-Heaviside layer*. The charged particles are, of course, ions, and that portion of the upper atmosphere is therefore called the *ionosphere.*

The use of radio waves to make wireless telegraphy possible was only the beginning. Might they not be used to transmit sounds, and not merely pulses? Suppose radio waves could be made to pull a diaphragm in and out and thus set up sound waves in the air?

At first this thought might seem impractical. Radio waves, though far lower in frequency than light waves, are nevertheless far higher in frequency than sound waves. A typical radio wave might have a frequency of 1,000,000 cycles per second (or 1000 kilocycles per second) and it would not be useful to force a diaphragm to vibrate at that frequency. The sound would be far too high-pitched for the human ear to hear. To produce sounds within the range of human hearing, a diaphragm must be made to vibrate between 20 and 20,000 cycles per second. These are the *audiofrequencies.* To use radio waves of such frequencies would be to involve one's self with radiation so low in energy as to be unusable.

The attack was made differently. The radio wave itself was allowed to be uniform and featureless, and with a frequency far above the audio range. It was a *carrier wave,* which served merely to transport the message that was to be impressed on it. Sounds picked up by a microphone could then be used to set up a current that would alter the intensity of the carrier wave in exact step with the fluctuations of the sound waves as in the case of a telephone mouthpiece (see page II–210). This fluctuating current is then made to alter the energy of the carrier wave, the amplitude of which will rise and fall with the rise and fall of the sound waves.

Amplitude modulation

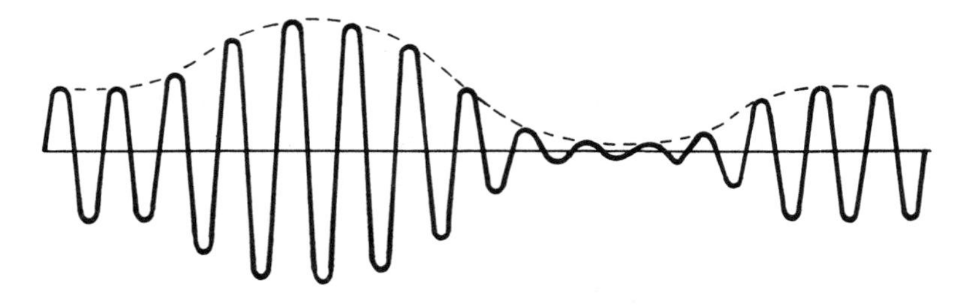

Frequency modulation

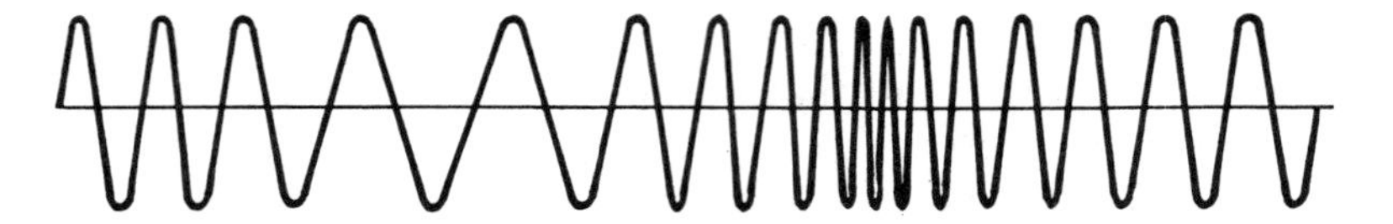

The carrier wave thus regulated is said to be *modulated.* Since the modulation takes the form of alterations in amplitude to match the variability of a sound wave, it is said to be *amplitude modulation,* often abbreviated AM.

When such a modulated radio wave is received, it is first rectified so that only the top half of the wave is allowed through. That half of the wave then acts upon a diaphragm by setting up a fluctuating magnetic force, as in the case of a telephone receiver. The diaphragm cannot react to the rapid fluctuations of the carrier wave itself but only to the much slower variations in its amplitude. In this way, sound waves are reproduced that exactly mimic those that had originally been impressed upon the carrier wave.

In 1906, the Canadian-American physicist Reginald Aubrey Fessenden (1866–1932) first made use of a modulated carrier wave to send out an appropriate message that allowed receivers actually to pick up music. Thus radio meant not only "radiotelegraphy" but also "radiotelephony."

None of this would be truly practical without the use of vacuum tubes for properly manipulating the excessively feeble electric currents set up by radio waves. In fact, so important were these devices to radio that they came to be commonly called *radio tubes.*

Each radio transmitting station makes use of a carrier wave of distinctive frequency. The radio set can be tuned by adjusting a variable condenser (see page II–172) to a point that will allow the set to respond to a particular frequency. In the first two decades of radio, this was not an easy task, and radio enthusiasts had to develop considerable skill at it.

During World War I, however, the American electrical engineer Edwin Howard Armstrong (1890–1954) invented what came to be called a *superheterodyne receiver.* Armstrong had tried to work out a system for detecting airplanes at a distance by picking up the electromagnetic waves sent out by their ignition systems. Those waves were too high in frequency to be received easily; Armstrong therefore arranged to produce a second electromagnetic wave of somewhat different frequency from that which he was trying to detect. The two combined to produce "beats" exactly as sound waves would (see page I–169). The beats were of far lower frequency than either original wave and could easily be detected.

World War I ended before Armstrong could perfect his device, but it was thereafter applied to radio sets in such a way as to make it simple to tune in stations by the turn of a dial. Radio moved into the home in consequence.

In later years, Armstrong tackled another problem involved in radio reception—that of "static." Electromagnetic waves are set up by spark discharges in automobile ignition systems, in the brushes of electric motors, in thermostats, and in all sorts of electrical appliances. (They are also set up in lightning discharges—giant sparks—during thunderstorms.) These waves interfere with the entire range of carrier waves, modulating them in random fashion so that one hears sharp, crackling noises that can be very distracting and cannot be tuned out.

Armstrong devised circuits that modulated not the amplitude of a carrier wave, but its frequency. Such *frequency modulation,* or FM, is not affected by the electromagnetic waves that pulse randomly all about us; consequently, static is largely abolished. In addition, FM allows better reproduction in the extreme portions of the audio-frequency range.

Television and Radar

The cathode-ray tube itself came into direct use in an electronic instrument that was fated to replace radio in the public heart. The beginning here came when physicists learned to take advantage of the low inertia of the electrons in order to move the stream with great rapidity.

Imagine, for instance, a cathode-ray tube with its anode in the form of a hollow cylinder. The electron beam, hurrying in the direction of the anode, would pass through the cylinder to the other end of the tube, which flares out to a flat circular piece of glass coated inside with some fluorescent chemical. Where the electron beam strikes, there would be a brilliant spot of flourescence.

Suppose, though, that on its way to the screen, the electron beam passed between two vertical electrodes. The electron beam will be deflected, naturally, in the direction of the positive electrode. If one electrode carries a strong positive charge to begin with, the electron beam would be strongly deflected in that direction, and the fluorescent spot would appear at the very edge of the screen.

If the positive charge is gradually weakened, the beam's deflection is decreased and the spot moves toward the center of the screen. Eventually, as the positive charge is decreased to zero and as the electrode in question then becomes negative (with the other electrode taking its turn at being positive), the spot passes the center and moves all the way to the other end of the screen. If the maximum positive charge is then placed once more on the first electrode, the beam flashes backward and the spot appears in its original position again.

This can be repeated over and over again, the fluorescent spot drifing across the screen over and over again. This can easily be done quickly enough to cause the spot to become a bright, horizontal line—the eye being unable to see it as a moving spot. (It is a similar effect that allows the eye to see the successive stills of a motion picture film as representing moving objects.)

Next imagine the electron beam also passing between a second pair of plates, a pair oriented horizontally. This second pair, acting alone, could be used to make the electron beam mark out a vertical line.

If both plates work together, however, the results can be most useful. The first plate may have superimposed upon it the change in voltage required to bring about a steady horizontal line. The second pair of plates may be hooked up to an ordinary alternating current so that the charge on the plates oscillates rapidly and evenly. The action of the two taken together would form a sine wave.

If the current passing through the second pair of plates is made to vary in accordance with a particular set of sound waves, the electron beam would trace out a varying curve that would mimic the properties of the sound wave (translating the longitudinal sound wave into an analogous transverse wave, however—see page I–150). For this reason, when the German physicist Karl Ferdinand Braun (1850–1918) introduced such a device, it came to be called a *cathode-ray oscillograph* ("wave-writer").

The cathode-ray oscillograph can do more. Imagine that the second pair of plates increases its voltage in steps, so that after the electron beam marks out a horizontal line, it moves up a bit and marks out another horizontal line, then moves up and marks out still another, and so on. The entire screen may thus be divided into hundreds of lines, but so fast do voltages shift, and so quickly do the electrons shift with them, that the entire screen can be scanned many times per second. To the eye, then, the entire screen will appear lit up, though a close look will show that the lighting consists of horizontal lines separated by narrow dark spaces that represent the step through which the second pair of plates has lifted the electron beam.

This, in essence, is a *television tube*. In order to impress a picture on it, the electron beam must be strengthened and weakened according to some fixed pattern so that the fluorescent spot is made to grow brighter and dimmer, producing the light-dark pattern we would recognize as an image.

The first to produce a practical method for doing this was the Russian-American physicist Vladimir Kosma Zworykin (1889–). In 1938, he invented the *iconoscope* (from Greek words meaning "picture-viewer"). It was a camera of sorts, one in which the rear surface was coated not with photographic film, but with a large number of tiny droplets of an alloy of cesium and silver. Cesium readily gives off electrons when light falls upon it, the intensity of electron emission being proportional to the intensity of the light that falls upon it. When the light-dark pattern of the scene in front of the camera is focussed on the cesium-silver rear surface, an analogous pattern of many-electrons/few-electrons is produced.

This electron pattern can be made to influence the electron beam emitted in the television tube which causes the fluorescent spot on the television screen to brighten and dim in exact analogy to the light-dark pattern being viewed by the iconoscope. The entire picture is reproduced on the screen; and since this is done over and over many times per second, each time in a slightly different pattern (as the scene being viewed changes), the eye seems to make out motion.

The cathode-ray oscillograph is also used in connection with a device that makes use of electromagnetic waves to judge distance, much as sound waves are used in echo location (see page I–180).

Electromagnetic waves move at the precisely known, very high velocity of 300,000 kilometers per second. Imagine a short pulse of electromagnetic waves moving outward, striking some obstacle, and being reflected backward and received at the point from which it had issued forth an instant before. What is needed is a wave form of low enough frequency to penetrate fog, mist and cloud, but of high enough frequency to be reflected efficiently. The ideal range was found to be in the microwave region, with wavelengths of from 0.5 to 100 centimeters.

From the time lapse between the emission of the pulse and the return of the echo, the distance of the reflecting object can be estimated. And, of course, the direction of the reflecting object would be that in which reflection was sharpest.

A number of physicists worked on devices making use of

this principle, but the Scottish physicist Robert Alexander Watson-Watt (1892–) was the first to make it thoroughly practical. By 1935, he had made it possible to follow an airplane by the microwave reflections it sent back. The system was called "radio detection and ranging" (to "get a range" on an object is to determine its distance), and this was abbreviated to "ra. d. a. r." or *radar*.

The microwave pulse sent out in radar can be made to deflect the electron beam of a cathode-ray oscillograph upward, producing a sharp spike in what would otherwise be a horizontal line. The returning echo (much feebler than the original pulse, since only a portion of the pulse strikes the object it is aimed at, and some of the pulse that does strike is scattered in other directions) produces a smaller spike. The electron beam moves sideways with such rapidity that even though the echo arrives only a fraction of a millisecond after the pulse has been sent out, there is still ample space on the fluorescent line between pulse and echo—a space that can be measured and made to yield distance.

Another way in which an electron beam can be made to do this work is to have it start at the center of the screen and move out to the edge along any radius. The radius it chooses is governed by the direction in which the large radar antenna (designed to receive and magnify feeble echoes) is pointing. As the antenna makes a complete circle, the electron beam very rapidly sweeps out a series of radii all around the screen.

Returning echoes make themselves evident not by sharp deviations in the beam itself, but by a brightening of the beam intensity; consequently, an obstructing object, returning echoes, shows up as a bright spot on the screen. If the screen is coated with a

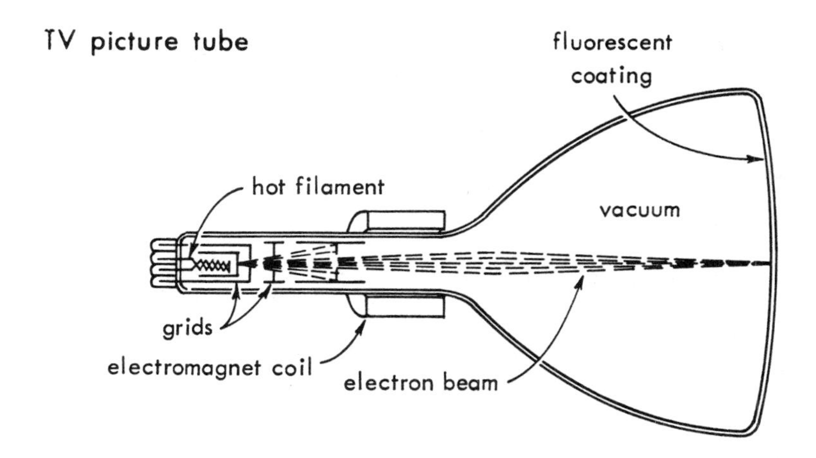

substance whose fluorescence lingers a few seconds, the shape of the object may be roughly scanned out as the beam completes its sweep. From an airplane, the radar screen may even present a rough map of the ground below, since land, water, green leaves and concrete all reflect microwaves at differing intensity.

It is not only man-produced microwaves that can now be detected by the help of electronic instruments. The various heavenly bodies and the phenomena with which they are associated produce among themselves the entire range of the electromagnetic spectrum. Little of that spectrum, however, can penetrate the earth's atmosphere. Among that little, fortunately, is the visible light region in which our sun's radiation happens to be particularly rich.

Another region, however, to which the atmosphere is transparent is that in which the microwaves are found.

In 1931, the American radio engineer Karl Jansky (1905–1950) was engaged in the problem of tracking down causes of static. Having eliminated static caused by known disturbances, he found a new kind of weak static from a source which, at first, he could not identify. It came from overhead and moved steadily from hour to hour. At first it seemed to Jansky that the source moved with the sun. However, it gained slightly on the sun to the extent of four minutes a day. Since this is just the amount by which the vault of the stars gains on the sun, the source must lie somewhere among the stars outside the solar system.

By 1932, Jansky had decided the source was strongest in the direction of the constellation of Sagittarius—in which direction, astronomers had decided, lay the center of the Galaxy.

The center of the Galaxy, hidden from optical view by dust clouds that efficiently absorb all the light, is nevertheless apparent through its microwave emissions which penetrate the dust clouds. *Radio telescopes* were built to receive and focus the very weak signals (especially after World War II when advances in radar technology could be put to this use), and the new science of *radio astronomy* came into its own.

CHAPTER 33

Electrons Within Atoms

The Photoelectric Effect

For a brief while after the discovery of the electron, it might have been tempting to feel that the universe contained at least two sets of ultimate particles without necessary connection with each other. One set consisted of the atoms of matter, these being comparatively massive objects existing in dozens of varieties. The other set consisted of the electrons associated with an electric current which, to all appearances, came in but a single variety.

Yet there was reason for doubting the independence of these two sets of particles. When an electric current was first produced by Volta a century before the discovery of the electron, it was done by combining certain metals and solutions. Since that time any number of chemical cells—devices whereby an electric current originates as a result of some chemical reaction—were devised. The ordinary "flashlight battery" and the storage battery present in every automobile are the best-known examples of these.

If a group of chemicals, each electrically neutral when taken by itself, could give rise to electric current made up of myriads of electrons, then certainly the worlds of atoms and of electrons must have some connection. Furthermore, one had to believe that either the electrons of the current were formed in the process of the atomic joinings and atomic separations that make up a chemical reaction, or that the electrons were present in the chemicals at all times, and were merely released in the course of the reaction.

Both views had their difficulties. If electrons were formed, that meant that mass was created, and that seemed impossible in the light of the law of conservation of mass (see page II–107), a generalization which, during the 1890's, was completely accepted by scientists. On the other hand, if electrons were present in chemicals at all times, why was there ordinarily no evidence either of their existence or, particularly, of the electric charge associated with them?

The dilemma was made the more acute through a phenomenon that was already known to physicists at the time of the discovery of the electron.

When Hertz was experimenting with radio waves during the 1880's, he found that he could elicit a spark from his radio-wave detector more easily if light fell upon the metal points giving out the spark. Light drew electricity from metal, so to speak, and this came to be called the *photoelectric effect.*

In 1888, the German physicist Wilhelm Hallwachs (1859–1922) discovered that light affected the two varieties of electric charge differently. A negatively-charged zinc plate lost its charge if it was exposed to ultraviolet light, the charge being drawn out by the light. On the other hand, a positively-charged zinc plate was not affected by the ultraviolet light.

Once the electron was discovered, a reasonable explanation of the phenomenon at once offered itself. Those were electrons that were ejected from metal through the impact of light. It was these electrons that formed the spark. It was from a negatively-charged zinc plate, containing an excess of electrons, that those particles were easily ejected. Such particles were not ejected from a positively-charged plate, which clearly did not contain an excess of electrons.

In 1899, Thomson tested this notion by measuring the e/m ratio for the particles being ejected from metals under the influence of light; it turned out to be virtually identical with that of cathode-ray particles. They were accepted as electrons from that time.

Again the same problem arose. When light forced electrons out of an electrically-neutral metallic surface, were those electrons formed as they were emitted or did they exist within the metal at all times? By 1905, Einstein had shown that the law of conservation of mass was incomplete in the form that had generally been accepted during the nineteenth century. He showed that energy and mass could be interconverted and that one ought to speak of the law of conservation of mass-energy. Nevertheless, the bookkeeping involved in the interconversion of mass and energy was rigorous, and there was insufficient energy in ordinary light—and even in ultraviolet light—to serve the purpose of manufacturing electrons.

Electrons, then, must exist in the metal at all times, and one could ask another question. Did the electrons exist in the interstices between the atoms, or did they actually occur within the atoms themselves? It was hard to accept the latter view, for that would mean that the atom was not the featureless, ultimately indivisible object that Democritus and Dalton had proposed and that the scientific world had finally accepted.

Yet there were phenomena that seemed to make this anti-Democritean view necessary, perhaps. Philipp Lenard had observed that the energy with which electrons were ejected depended on the frequency of the light, and that light of less than a certain frequency (the *threshold value*) did not eject electrons. The quantum theory (see page II–130), which was beginning to come into acceptance in the first decade of the twentieth century, made it clear that light consisted of photons that increased in energy content as frequency increased.

The threshold value represented quanta of just sufficient energy to break the bonds holding the electrons to matter. The strength of those bonds varies from substance to substance, since electrons are forced out of some metals only by energetic ultraviolet light, whereas they are forced out of other metals by light as un-energetic as the visible red. If electrons are tied to matter, it must be to the atoms they are bound, and with differing bond strengths, so to speak, depending on the nature of the particular atom. It seems only sensible to consider something always present near the atom, always bound to the atom with a characteristic force, to be part of the atom.

Furthermore, once the view is accepted, there are advantages to it. There are many varieties of atom and only one type of electron (since the particles emitted from all metals by the photoelectric effect are of identical properties). Perhaps the troublesome variety of the atoms could be explained in terms of the number of electrons each contained, of their arrangement, of the strength with which they were held, and so on. Perhaps the order enforced empirically upon the elements by the periodic table could now be made more systematic. If so, the indivisible atom of Democritus was well lost.

Indeed, there were some facets of the photoelectric effect that fit in well with the periodic table. For instance, the elements that most readily give up electrons in response to light are the alkali metals. These give up electrons with increasing ease as atomic weight goes up—that is, as one moves down the column in the periodic table. Thus cesium, the naturally-occurring alkali metal with the highest atomic weight,* releases its electrons most easily of all—hence Zworykin's use of the metal in his iconoscope.

Here is an indication of how Mendeleev's periodic table established a kind of order with respect to a property completely undreamed of in Mendeleev's time. This is an example of how a truly useful scientific generalization can be superior to the state of knowledge that brought it forth, and how a great scientist must almost necessarily produce more than he realizes.

The photoelectric effect can be put to good use. A vacuum tube can be devised that does not require a heated filament for the production of electrons—merely a filament (if one chooses the right metal) that can be exposed to light. When light falls upon a cathode capable of showing a photoelectric effect in response to such light, electrons are ejected and a current flows. The current can be used to activate an electromagnet that can open doors or perform other tasks. This is a *photoelectric cell.*

A common version of such a cell places it in one post with a source of light from another post shining constantly into the cell, keeping a current constantly flowing and a door constantly closed against a pull that would otherwise open it. A person walking between the posts intercepts the light beam, the current in the cell ceases, and the door flies open.

The Nuclear Atom

The apparent existence of electrons within the atom raised some important questions.

The atoms were electrically neutral; if negatively-charged electrons existed about or within the atom, there had to be a positive charge somewhere to neutralize the negative charge of the electrons. If so, where was it? Why didn't light ever bring about the ejection of very light positively-charged particles? Why were there only cathode rays, never analogous anode rays?

* Francium, an alkali metal of still higher atomic weight, does not occur in nature in any significant quantity (see pag 130).

Thomson offered an answer to these questions. In 1898, he suggested that the atom was a solid, positively-charged sphere into which just enough electrons were embedded (like raisins in pound cake, so to speak) to bring about an overall electrical neutrality.

This was an attractive suggestion, for it seemed to explain a great deal. Light quanta would jar loose one or more of these electrons, but could scarcely budge the large atom-sphere of positive charge. Again, the heat in a vacuum tube filament would indeed "boil off" electrons, for as atoms vibrated more strongly with rising temperature (in accordance with kinetic theory, see page I–205), the electrons would be jarred loose while the atom itself would be essentially unaffected. This would explain why only negative particles appeared and never positive ones.

Then, too, Thomson's theory explained ions neatly. An atom that lost one or more electrons would retain a net positive charge —the size of the charge depending on the number of electrons lost. A hydrogen ion (H^+) or a sodium ion (Na^+) would be a hydrogen atom or a sodium atom that had lost a single electron. A calcium ion (Ca^{++}) would be a calcium atom minus two electrons, and an aluminum ion (Al^{+++}) would be an aluminum atom minus three electrons.

On the other hand, what if more than the normal quantity of electrons could be jabbed into the positively-charged atom substance? The chloride ion (Cl^-) would be a chlorine atom bearing an extra electron, while a sulfate ion (SO_4^{--}) and a phosphate ion (PO_4^{---}) would represent groups of atoms possessing among themselves two and three extra electrons, respectively.

In this view, the negatively-charged electron is the only subatomic particle, but by means of it ions of both types of electric charge can be explained.

Thomson's theory, although so attractive, nevertheless, had a fatal shortcoming. Lenard had noted that cathode rays could pass through small thicknesses of matter. To be sure, the electrons making up the cathode rays were very small and might be pictured as worming their way between the atoms. If so, they would most likely emerge badly scattered. Instead, cathode rays passed through small thicknesses of matter still traveling in an essentially parallel beam, as though they had passed through atoms without very much interference.

In 1903, therefore, Lenard suggested that the atom was not a solid mass but was rather mostly empty space. The atom, in his view, consisted of tiny electrons and equivalent particles of positive charge, existing in pairs so that the atom as a whole was electrically neutral.

But, in that case, why were there only cathode rays and never anode rays?

The reconciliation of the Thomson and Lenard views fell to the lot of the New Zealand-born physicist Ernest Rutherford (1871–1937). Beginning in 1906, he conducted crucial experiments in which he bombarded thin gold leaf with alpha particles.* Behind the gold leaf was a photographic plate.

The stream of alpha particles passed right through the gold leaf as though it were not there and fogged the photographic plate behind it. The gold leaf was only 1/50,000 of a centimeter thick, but this still meant a thickness of 20,000 atoms. The fact that alpha particles could pass through 20,000 gold atoms as though they weren't there was strongly in favor of Lenard's notion of an empty atom, (an atom, that is, made up of nothing more than a scattering of light particles).

But the truly interesting point was that not all the alpha particles passed through unaffected. The spot of fogging on the plate would, in the absence of the gold leaf, have been sharp; but with the gold leaf in place, the boundary of the fogged spot was rather diffuse, fading out gradually. It was as though some of the alpha particles were, after all, slightly deflected from their path. In fact, Rutherford was able to show that some were deflected more than slightly! About one alpha particle out of every 8000 was deflected through a right angle or even more.

This was amazing. If so many alpha particles went through thousands of atoms untouched or nearly untouched, why should a very few be twisted in their path so badly? The alpha particle is not a light particle such as the electron. It is 7350 times as massive as an electron; four times as massive as a hydrogen atom. If the alpha particle encountered electrons within an atom, it would brush them aside as a man might brush aside a sparrow. For an alpha particle to be set back on its heels, it must at the very least meet something nearly as massive as itself—something, in short, of atom-sized mass. And yet this atom-sized mass was only rarely encountered on the journey of the alpha particle through matter, so it must take up a very small volume.

It was as though one were faced with contacting fluffy balls of foam with a lead pellet at the center of each. If lead pellets were

* Alpha particles are rapid-moving, massive particles obtained from radioactive substances (see page 111) capable of penetrating matter with much greater effectiveness than electrons can.

tossed at such a barrier, most would pass through the foam as though nothing were there, but occasionally a tossed pellet would strike one of the buried pellets and bounce off. From the frequency with which such bouncing took place, you could calculate the comparative size of the foam ball and the central pellet.

To be sure, the alpha particles were not actually bouncing off the massive object within the atom. Instead, from the nature of the scattering, Rutherford could show there was an electrical interaction. The alpha particles are themselves positively charged (each carrying a charge of +2), and the massive object within the atom is also charged (positively, as it turns out), so the alpha particle is repelled by electric forces even if it scores a near miss.

By 1911, Rutherford was ready to describe his picture of the atom. In his view, Thomson's massive positively-charged atom was still there as far as mass was concerned, but it was drastically shrunken in volume. It had shrunk down to an extremely small object in the very center of the atom. This massive central object was the *atomic nucleus,* and what Rutherford was proposing was the *nuclear atom,* a concept that has remained valid ever since and that is more firmly accepted now than ever.

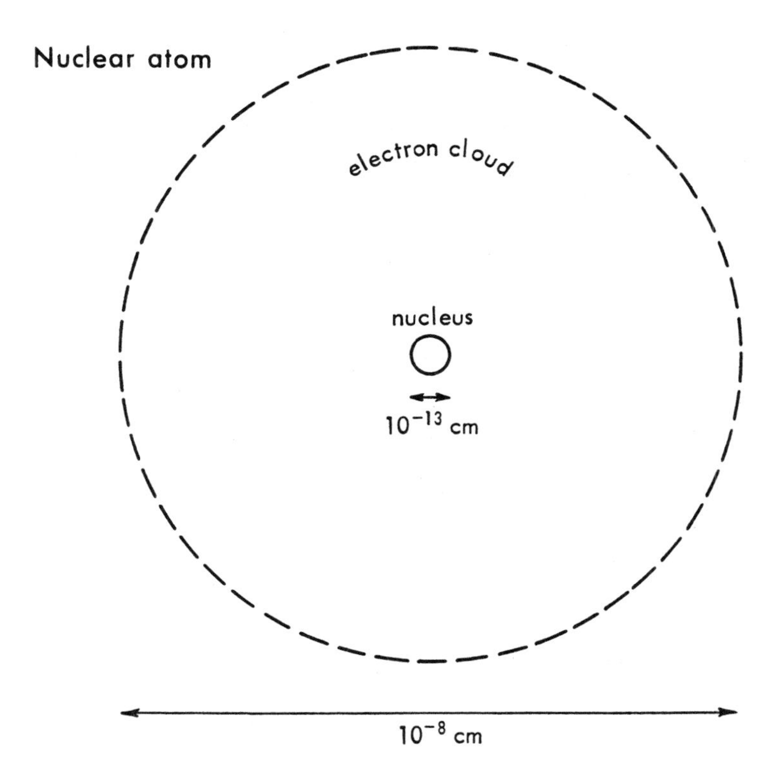

The atomic nucleus, as could be seen from the pattern of deflections of alpha particles, was tiny indeed, not more than 10^{-13} to 10^{-12} centimeters in diameter, or only 1/100,000 to 1/10,000 the diameter of the atom as a whole. The volumes of nucleus and atom are in proportion to the cube of the diameter, so the volume of the nucleus is rather less than one trillionth (1/1,000,000,000,-000) of the atom as a whole.

Yet virtually all the mass of the atom is concentrated in that tiny nucleus. Even the lightest nucleus, that of the hydrogen atom, is 1836 times the mass of an electron, while the nuclei of the really massive atoms are nearly half a million times as massive. Such a nucleus would be much less mobile than electrons would be, and it is not surprising that light ejects negatively-charged electrons and not positively-charged nuclei from metals—that heated filaments emit electrons and not nuclei.

Outside the nucleus, the comparatively vast remainder of the atom is made up of nothing but the ultra-light electrons. These electrons offer little obstacle to speeding cathode ray particles, and virtually no obstacle at all to alpha particles; consequently, Rutherford's nuclear atom is as thoroughly empty as Lenard's model was.

And, of course, the nuclear atom can explain ions in terms of loss or gain of electrons as easily as Thomson's raisin-cake atom could explain them. In short, the nuclear atom proved completely satisfactory; only the details required elaboration.

Characteristic X Rays

Thanks to Rutherford, physicists now saw the atom as a tiny but massive, positively-charged nucleus surrounded by electrons. The nucleus, if it contained virtually all the mass of the atom, as Rutherford maintained, must vary in mass with the atomic weight.

It seemed reasonable to suppose that the greater the mass of the nucleus, the larger the size of the positive charge it carried and the greater the number of negatively-charged electrons necessarily present outside the nucleus to balance that positive charge. If this were so, it would mean that physicists were beginning to probe close to what might prove the crucial difference between the atoms of one element and another. It was not just the difference in mass, which was all that Dalton, and nineteenth century chemists generally, could put their finger on. A possible new difference was emerging, an electrical difference that made itself manifest in two ways: first, in the size of the positive charge on the nucleus and second, in the number of electrons outside the nucleus.

These two aspects of the electrical differences among atoms are closely related, but the nuclear charge is more fundamental than the electron number. Electrons can be removed from atoms by heat or by light, leaving positive ions behind. Additional electrons can be forced onto atoms in chemical reactions, forming negative ions. While these ions have properties that differ radically from those of the neutral atom, they are not completely divorced from the neutral atom; they do not constitute a new element. In other words, the sodium ion is very different from the sodium atom, but one can be changed into the other by recognized nineteenth century chemical or physical procedures. Neither can be changed into either a potassium atom or a potassium ion, at least not by those procedures. Therefore, changes in the electron number in an atom are not necessarily crucial, and it is not by means of the number of electrons within an atom that elements are best distinguished.

On the other hand, the nuclear charge could not be altered by any method known to the chemists and physicists of 1900; no alteration of the number of electrons, one way or the other, would alter that nuclear charge. It was the size of the nuclear charge, then, that best characterized the different varieties of atoms and, therefore, the different elements.

But if all this is so, how can one go about finding the exact size of the nuclear charge of the atoms of a particular element? The answer to that question was arrived at through X rays.

When Röntgen first discovered X rays, he had produced them as a result of the impact of cathode ray particles on the glass at the end of the cathode-ray tube. Speeding electrons can penetrate small thicknesses of matter, but they are slowed down; if the obstructing matter is thick enough, they are stopped completely and absorbed. The deceleration of electrically charged particles will, according to Maxwell's theory of electromagnetism, result in the production of electromagnetic radiation, and this does indeed appear in the form of X rays.

It is to be expected that material made up of massive atoms will more effectively and rapidly decelerate speeding electrons and will produce more intense beams of X rays. For that reason, physicists took to placing metal plates directly opposite the cathode inside cathode-ray tubes. This metal plate, sometimes called the *anticathode* ("opposite the cathode") is subjected to the collision of electrons, and from its surface, powerful beams of

X rays are emitted. Such a cathode-ray tube is usually called an *X-ray tube.*

The X rays produced from the anticathode varied in properties according to the nature of the material making up the anticathode. The first to show this was the English physicist Charles Glover Barkla (1877–1944). In 1911, Barkla showed that among the X rays produced at a given anticathode, certain groups predominated. He could only judge the difference among the X-ray groups produced by their ability to penetrate thicknesses of matter. One group would penetrate a relatively large thickness, another group a lesser thickness, and so on. The greater the thickness penetrated, the "harder" the X rays. It became customary to call the hardest X rays produced at a given anticathode, the *K-series,* the next the *L-series,* then the *M-series,* and so on. These are the *characteristic X rays* for a given element.

The hardness of these sets of characteristic X rays varies with the nature of the metal making up the anticathode. In general, the higher the atomic weight of the metal, the harder the X rays produced. It seemed reasonable to suppose that if the hardness could be measured accurately, interesting information concerning the atomic nuclei could be obtained.

Unfortunately, measuring the hardness of X rays by their penetrability is rather imprecise. Something better was needed. It was strongly suspected, for instance, that X rays were electromagnetic radiation (though when Barkla did his work this had not yet been conclusively demonstrated). If so the shorter the wavelength of a particular beam of X rays, the more energetic it would be and the more penetrating. Measurement of the wavelength of the X rays (or of their frequency) would thus offer a possibly precise method for estimating their hardness.

However, how could their wavelength be measured? In principle, the best method would be to use a diffraction grating (see page II–65). A diffraction grating, made of a series of parallel scratches on an otherwise clearly transmitting surface, can only work under certain circumstances. The distance between scratches must approximate the size of the wavelength being measured. The wavelength of X rays was far shorter than that of ultraviolet radiation, so short, in fact, that it was impractical to expect scratches to be produced with sufficiently close spacing.

A way out of this dilemma occurred in 1912 to the German physicist Max von Laue (1879–1960). Crystals, he realized, were natural diffraction gratings far more finely spaced than any man could make. In crystals, atoms existed in orderly rows and files.

The nuclei of the atoms, which would deflect X rays just as scratches would deflect ordinary light, are about 10^{-8} centimeters apart (this being roughly the diameter of a typical atom), and this might very well be about the size of an X ray wavelength.

Laue used a crystal of zinc sulfide, allowing a beam of X rays to fall upon it and, passing through, to strike a photographic plate. The X rays were indeed diffracted, producing a pattern of dots, instead of a single, centrally located dot. This was the definite proof, at last, that X rays were wave-like in nature.

This approach was carried further that same year by a pair of physicists, the Englishman William Henry Bragg (1862–1942) and his son, the Australian–born William Lawrence Bragg (1890–). They analyzed the manner in which X rays would be reflected by the planes of atoms within a crystal, and showed that this reflection would be most intense at certain angles, the values of which depended upon the distance between the planes of atoms within a crystal and upon the wavelength of the X ray. If the distance between the planes of atoms was known, the wavelength could then be calculated.

It was found that by this method the wavelength could be calculated with a satisfactory degree of precision. X rays, produced by the deceleration of speeding electrons, have been found across the entire range of from 1 millimicron (the arbitrary lower limit of ultraviolet radiation wavelengths, see page 34) down to somewhat less than 0.01 millimicrons, a range of about seven octaves.

Atomic Numbers

With the Bragg technique at hand, it was now possible to turn to Barkla's characteristic X rays and study them carefully and precisely. This was done in 1913 by the English physicist Henry Gwyn-Jeffreys Moseley (1887–1915).

Moseley worked with the K-series of characteristic X rays for about a dozen consecutive elements in the periodic table, from calcium to zinc, and found that the wavelength of the X rays went down (and the frequency therefore went up) as the atomic weight increased. By taking the square root of the frequency, he found that there was a constant increase as one went from one element to the next.

Moseley decided that there was something about the atom which increased by regular steps as one went up the periodic table. It was possible to demonstrate that this "something" was most likely the positive charge on the nucleus. The most straightforward conclusion Moseley could reach was that the simplest atom had a charge of +1 on its nucleus; the next, a charge of +2; the next, a charge of +3, and so on. Moseley called the size of this charge the *atomic number*.

This has turned out to be correct. Hydrogen is now considered to have an atomic number of 1; helium, one of 2, lithium, one of 3, and so on. In the periodic table presented on page 16, the elements are given atomic numbers of from 1 to 103, and the atomic number of every known element has been determined.

The atomic number is far more fundamental to the periodic table than the atomic weight is. Mendeleev had been forced to put some elements out of atomic weight order so that they would fit into their proper families. For instance, cobalt fits the table better if it is placed ahead of nickel. Yet cobalt, with an atomic weight of 58.93, should fall behind nickel, which has an atomic weight of only 58.71.

Moseley, however, found that cobalt, despite its heavier atomic weight, produced X rays that were lower in frequency than those of nickel. Cobalt therefore has the lower atomic number, 27, and the atomic number of nickel is 28. Mendeleev's chemical intuition, working without the guide of X-ray data, had led him aright.

To summarize, there are three pairs of elements in the periodic table (argon-potassium, cobalt-nickel, and tellurium-iodine) which are out of order if increasing atomic weight is taken as the criterion. If increasing atomic number is taken as the criterion instead, not one element in the table is out of order.

The atomic number concept also brought new power to the periodic table in another way. Not only could chemists predict missing elements (as Mendeleev had), but they could now also predict the nonexistence of elements.

As long as atomic weight had been the only guide, one could never be certain that whole new families of undiscovered elements might not exist. In the 1890's, for instance, the family of inert gases—helium, neon, argon, krypton, and xenon—was discovered and fitted into a new column in the periodic table, a column no one had previously suspected of existing. Again, the lanthanides were discovered one by one over the space of a century, and until Moseley's time no chemist could be certain how many remained to be found—thousands, for all one could then tell.

With atomic numbers, such uncertainties were smashed. As long as one could assume the nonexistence of fractional electric charges on the nucleus, one could be sure there were no unknown elements between hydrogen (atomic number 1) and helium (atomic number 2), or between phosphorus (atomic number 15) and sulfur (atomic number 16).

In fact, for the first time chemists could tell how many elements remained to be discovered. The first element in the periodic table was hydrogen (atomic number 1), and there could be no element preceding it. The element with the most massive known atom (in Moseley's time) was uranium (atomic number 92). Between these two limits, all the atomic numbers but seven were filled, and only seven unknown elements therefore remained to be discovered. The seven gaps were those with atomic numbers 43, 61, 72, 75, 85, 87, and 91.

X-ray analysis could also be used to check the identity of possibly newly-discovered elements. For instance, the French chemist Georges Urbain (1872–1938) had in 1911 isolated what he thought was a new element, and he had named it "celtium." When Moseley's work was published, Urbain decided his new element must fit into the gap in the periodic table at number 72 and brought a sample to Moseley for testing. Moseley analyzed the characteristic X rays and found the "new element" to be a mixture of ytterbium and lutetium (elements number 70 and 71) both of which were already known. Painstaking chemical work confirmed this and Urbain, very impressed, labored mightily to popularize the concept of the atomic number.

Within a dozen years, three of the gaps were filled. Protactinium (atomic number 91) was discovered in 1917; hafnium (atomic number 72), in 1923; and rhenium (atomic number 75), in 1925. After that, over a decade passed before the last four gaps (43, 61, 85, and 87) were filled. These last elements will be taken up in due course (see page 175).

Once the nuclear charge of an element was known, something was also known about the number of electrons in the atoms of that element. An element might lose an electron or two, or gain an electron or two, and become an electrically-charged ion, but in the *neutral atom,* the number of electrons had to be precisely enough to neutralize the nuclear charge. If, in the oxygen atom, the nucleus has a charge of $+8$, there must be eight electrons (each with a charge of -1) to balance that. We may say, then, that the number of electrons in a neutral atom is equal to the atomic number of the element. The neutral hydrogen atom possesses 1 electron, the neutral sodium atom possesses 11 electrons and the neutral uranium atom possesses 92 electrons.

Electron Shells

The next general question was this: How are the electrons in an atom arranged? Thomson, in his raisin-cake model of the atom, had suggested that the electrons embedded in the positively-charged substance of the atom were arranged in circles. If there were a large number of electrons, there might well be a number of circles.

After Thomson's model had been abandoned and replaced by Rutherford's nuclear atom, it remained possible that the electrons possessed some regular arrangement outside the nucleus. This notion seemed to be backed by the several series of characteristic X rays produced by various elements. Perhaps each series was produced by a separate group of electrons enclosing the central nucleus. The group nearest the nucleus would be most firmly held and would produce the hardest X rays, the K-series. The next group would produce the L-series, and so on. If the electrons were pictured as arranged spherically about the nucleus (like the shells making up an onion), one could speak of the *K-shell,* the *L-shell,* the *M-shell,* and so on, as one worked outward from the nucleus.

Then, consider the inert gases—helium, neon, argon, kryp-

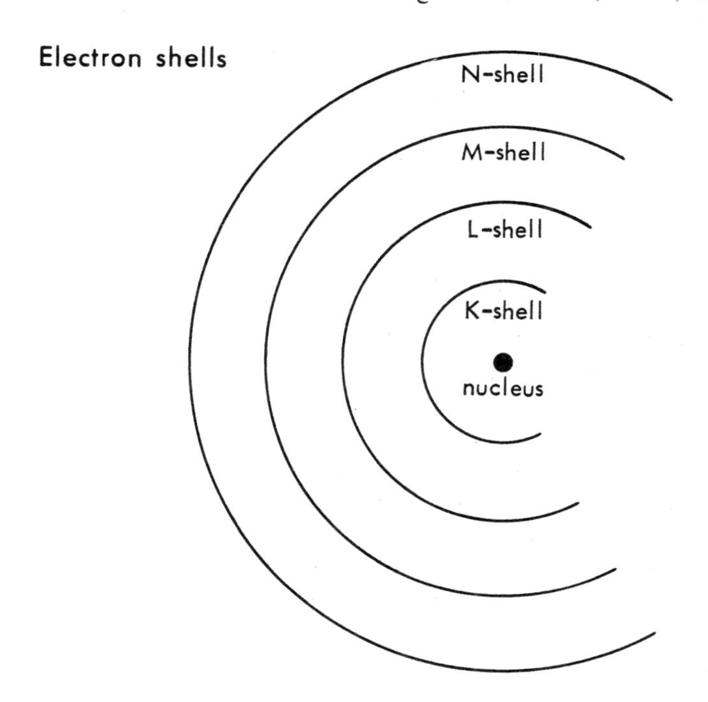

ton, xenon and radon. Of all the elements, they were the least apt to engage in chemical reactions. (Until 1962, it was taken for granted they could not engage in any chemical reactions at all. It was then discovered that krypton, xenon and radon could engage in a very few.) Why is this so?

One reason is that chemical reactions must involve the interactions of the electrons within atoms. For instance, when sodium metal reacts with gaseous chlorine, sodium chloride is formed and this consists of sodium ions and chloride ions. In the reaction of sodium and chlorine, then, the sodium atom loses an electron to become Na^+, and the chlorine atom gains an electron to become Cl^-.

Perhaps, if the inert gases do not easily engage in chemical reactions, this is because their atoms already possess a particularly stable arrangement of electrons and have only the most minor tendency to upset that arrangement by indulging in the loss or gain of electrons.

It seemed logical to suppose that this stable arrangement is represented by the complete filling of a particular shell of electrons.

For instance, helium has an atomic number of 2, and is inert. One can assume that since the neutral helium atom possesses two electrons, it requires but two electrons to fill the innermost shell, or K-shell. The next inert gas is neon, which has an atomic weight of 10 and which, in its neutral state, possesses ten electrons in its atoms. With two electrons filling the K-shell, the remaining eight must suffice to fill the L-shell. The next inert gas is argon, which has an atomic number of 18, and has eighteen electrons per atom. With two electrons in the K-shell and eight in the L-shell, the remaining eight must fill the M-shell. Based on this reasoning, Table IV contains the distribution of the electrons among the shells of the first twenty elements. (Only the first twenty are included because the distribution becomes more complicated—see page 83—for the higher elements.)

Soon after Moseley's work, an attempt was made to rationalize the chemical reactions on the basis of electron distributions inside the atom. A relatively successful attempt was made, independently, by the American chemists Gilbert Newton Lewis (1875–1946) and Irving Langmuir (1881–1957). The essence of their views was that in any chemical reaction an element gained or lost electrons in such a way as to gain an "inert-gas configuration," that being the most stable arrangement.

Thus, sodium, with its electrons divided 2/8/1, had a strong tendency to give up one electron and become sodium ion (Na^+)

with its electrons divided 2/8. The sodium ion has the neon configuration of electrons but, of course, does not actually become neon, for the nuclear charge of the sodium ion (the characteristic property of a particular element) remains +11, while that of neon is +10. The same argument will hold for chlorine. The chlorine atom, with an electron arrangement of 2/8/7, has a strong tendency to gain an electron and form the chloride ion (Cl^-), which has a 2/8/8 arrangement, like that of argon.

The ease with which sodium and chlorine interact can be viewed as the consequence of the manner in which their electron shifting tendencies complement each other. The electron that a sodium atom will so easily give up will be accepted just as easily by a chlorine atom. The oppositely charged ions that result cling together to make up sodium chloride.

In the same way, calcium (2/8/8/2) will easily give up two electrons to form calcium ion (Ca^{++}) with a 2/8/8 configuration, that of argon; while oxygen (2/6) will readily accept two

TABLE IV—*Electron Arrangements*

Element	*Atomic Number*	*Electrons in*			
		K-shell	*L-shell*	*M-shell*	*N-shell*
Hydrogen	1	1	—	—	—
Helium	2	2	—	—	—
Lithium	3	2	1	—	—
Beryllium	4	2	2	—	—
Boron	5	2	3	—	—
Carbon	6	2	4	—	—
Nitrogen	7	2	5	—	—
Oxygen	8	2	6	—	—
Fluorine	9	2	7	—	—
Neon	10	2	8	—	—
Sodium	11	2	8	1	—
Magnesium	12	2	8	2	—
Aluminum	13	2	8	3	—
Silicon	14	2	8	4	—
Phosphorus	15	2	8	5	—
Sulfur	16	2	8	6	—
Chlorine	17	2	8	7	—
Argon	18	2	8	8	—
Potassium	19	2	8	8	1
Calcium	20	2	8	8	2

electrons to form oxide ion (O^{--}) with a 2/8 configuration, that of neon. Thus, calcium oxide (CaO) is formed.

Or calcium may give up one electron to a chlorine atom and a second electron to another chlorine atom to form calcium chloride ($CaCl_2$). In this way, calcium combines with two chlorine atoms, so that a gram-atomic weight of chlorine combines with only half a gram-atomic weight of calcium. The existence of equivalent weights (with the equivalent weight of calcium being only half its gram-atomic weight) can thus be explained electronically.

Any such theory had to explain why it is that two chlorine atoms cling together tightly to form a chlorine molecule. Each chlorine atom has a strong tendency to accept one electron, but virtually no tendency to give one up. The Lewis–Langmuir suggestion was that each of two chlorine atoms might contribute an electron to a "shared pool" of two. These two electrons would be within the outermost electron shell of both atoms (provided they were so close to each other as to be in virtual contact), and each atom would then have the 2/8/8 configuration of argon.

Anything that would pull the chlorine atoms apart would disrupt this stable electron arrangement by making the existence of the shared pool impossible. For this reason, the two-atom chlorine molecule is very stable, and considerable energy is required to decompose it to individual chlorine atoms.

Similar arguments will explain why fluorine, hydrogen, oxygen and nitrogen all form stable two-atom molecules.

The electron configuration of the carbon atom is 2/4. It may contribute one electron to form a shared pool of two electrons with a hydrogen atom, contribute a second electron to a second hydrogen atom, and so on. In the end, there will be four shared pools, of two electrons each, with each of four hydrogen atoms. The carbon atom shares in eight electrons altogether, four of its own and one each from the four hydrogen atoms, to achieve the 2/8 neon configuration. Each hydrogen atom possesses a share in two electrons to achieve the helium configuration. Thus, the molecule of methane (CH_4) is stable.

Indeed, the Lewis-Langmuir picture of electrons being transferred and shared has turned out to be a very useful way of picturing how the molecules of a great many of the simpler chemical compounds are held together.

Furthermore, the Lewis-Langmuir theory made it plain why the periodic table was periodic (something which Mendeleev, of course, had been unable to explain). To begin with, the inert gases all have their electrons arranged in a way that yields them maximum stability. They are all chemically inert, therefore, and form a natural chemical family of very similar elements.

The alkali metals are all located in positions one atomic number higher than the inert gases. Thus lithium (one past helium) has the electron configuration 2/1; sodium (one past neon) is 2/8/1; potassium (one past argon) is 2/8/8/1, and so on. Every alkali metal has but one electron in its outermost shell and has a strong tendency to lose that one. For that reason, all are very active elements, with similar properties, forming a natural family.

The alkaline earth elements form a similar family in which each has two electrons in the outermost shell of the atom. Beryllium is 2/2, magnesium is 2/8/2, calcium is 2/8/8/2, and so on.

Again, the halogens all are to be found one atomic number before the inert gas configuration. Fluorine is 2/7, chlorine is 2/8/7, and so on. All have a strong tendency to accept one electron and they also form a natural family of elements of similar chemical properties.

And thus, by way of electrons and electron shells, the periodic table was rationalized a half-century after its inception.

CHAPTER 34

Electrons and Quanta

Spectral Series

Useful as the Lewis-Langmuir view of the atom is in explaining the structure of many of the simpler chemical compounds, it does not explain everything. It does not, for instance, describe in satisfactory manner the structure of the boron hydrides (compounds of boron and hydrogen) or explain the peculiar properties of the well-known compound, benzene (C_6H_6). Furthermore, it does not adequately explain the behavior of many of the elements with atomic weights beyond that of calcium. The Lewis-Langmuir view does not, for instance, explain why the lanthanides, with atomic numbers of from 57 to 71 inclusive, should be so similar in properties.

One obvious shortcoming in the Lewis-Langmuir view is that it considers the electrons to be stationary particles distributed about the atom in certain fixed positions. Indeed, the eight electrons of the L-shell and the M-shell were usually depicted as being located at the eight corners of a cube, so that simple molecules could be presented in diagrams as being made up of interlocking cubes.

This is a convenient picture from the chemical point of view, but it is unacceptable to physicists and must be replaced by something else if the Lewis-Langmuir view is to be made more useful. After all, if the negatively-charged electron is stationary with respect to the positively-charged nucleus, then electromagnetic theory requires that it fall into the nucleus (just as the earth would fall into the sun if it were stationary with respect to the sun).

Consequently, physicists tended to assume that the electrons were circling the nucleus at great velocity in order not to fall into it. In 1904, a Japanese physicist, Hantaro Nagaoka, specifically suggested that electrons circled in orbits within the atom, just as planets circled in orbits within the solar system.*

There is, however, a fundamental difficulty that had to be faced by all models that involved electrons revolving about a nucleus. A revolving electron undergoes a continual acceleration toward the center and, by Maxwell's electromagnetic theory, such an accelerating charge should be constantly emitting electromagnetic radiation.

Indeed, Nagaoka made that part of his model. The electron in its circular movement about the nucleus acted as a charge oscillating from one end of its orbit to the other, and this should create radiation of corresponding frequency (as in the case of Hertz's spark discharge oscillations, see page 32). If the electron made five hundred trillion revolutions per second in its orbit (which it would do if it traveled at the not-impossible velocity of 150 kilometers per second), it would produce radiation with a frequency of five hundred trillion cycles per second; this would be in the visible light range. Here was an explanation of light as an electromagnetic radiation.

This was so attractive a suggestion that it almost hurts to break it down, but one must. If the revolving electron emits electromagnetic radiation continually, it must lose energy, and kinetic energy (the energy of motion) is all that the electron can lose, as far as we know. Consequently, its motion about the central nucleus must constantly slow and the electron must spiral into the nucleus.†

Since electrons do not, in actual fact, spiral into the nucleus,

* This picture of the atom caught the public fancy, perhaps because it compared the atom with something that was already familiar. Although the solar system model was quickly replaced by more complex and more useful models, it has remained in the minds of many nonphysicists. Innumerable science fiction stories, for instance, have been written in which atoms were considered to be tiny solar systems and in which the electron-planets were supposed to be inhabited; sometimes by creatures very much like earthmen.

† As the earth revolves about the sun, it must, by analogy, constantly radiate "gravitational radiation." However, the force of gravity is so much weaker than the electromagnetic force (see page II–164) that the loss of energy by gravi-

another model must be found. Such a model must account not only for the fact that atoms radiate light (and absorb it, too) but that they only radiate and absorb light of certain characteristic wavelengths (or frequencies). By studying the interrelationships of these characteristic wavelengths, hints may be found as to what that structure might be. Hydrogen would be the element to tackle, for it produces the simplest and most orderly spectrum.

Thus, the most prominent line in the hydrogen spectrum has a wavelength of 656.21 millimicrons. Next to that is one at 486.08 millimicrons; then one at 434.01 millimicrons; then one at 410.12 millimicrons; then one at 396.81 millimicrons, and so on. If the wavelengths of these lines are plotted to scale, they will be seen to be separated by shorter and shorter intervals. Apparently, some order exists here.

In 1885, a German mathematician, Johann Jakob Balmer (1825–1898), tinkered with the series of numbers representing the wavelengths of the lines of the hydrogen spectrum and found a simple formula that expressed the wavelength (λ) of the lines. This was:

$$\lambda = \frac{364.56\, m^2}{m^2 - 4} \qquad \text{(Equation 5–1)}$$

where m can have successive whole-number values starting with 3. If $m = 3$, then λ can be calculated as equal to 656.21 millimicrons, which is the wavelength of the first line. If m is set equal to 4, then to 5, and then to 6, the wavelengths of the second, third, and fourth lines of the hydrogen spectrum turn up in the value calculated for λ. This series of lines came to be called the *Balmer series.*

Eventually as m becomes very high, $m^2 - 4$ becomes very little different from m^2, so the two terms would cancel in Equation 5–1. In that case, λ would become equal to 364.56 millimicrons (*Balmer's constant*), and this would be the limit toward which all the lines in the series would tend.

Some years after Balmer's work, the Swedish physicist Johannes Robert Rydberg (1854–1919) put the formula into a more convenient form. He began by taking the reciprocal of both sides of Equation 5–1, and this gave him:

---

tational radiation is insignificant. It would take many trillions of years for the earth to lose a noticeable amount of kinetic energy in this fashion. The electron, subjected to a much stronger force than gravitation, would have its orbit decay in a very short time indeed.

$$\frac{1}{\lambda} = \frac{m^2 - 4}{364.56\, m^2} \qquad \text{(Equation 5–2)}$$

Multiplying both numerator and denominator of the right-hand side of Equation 5–2 by four:

$$\frac{1}{\lambda} = \frac{4\,(m^2 - 4)}{364.56\,(4m^2)} = \frac{4}{364.56}\left(\frac{m^2 - 4}{4m^2}\right) = 0.0109\left(\frac{m^2 - 4}{4m^2}\right) \qquad \text{(Equation 5–3)}$$

Let's take each part of the extreme right-hand portion of Equation 5–3 separately. The value 0.0109 is obtained through the division of 4 by Balmer's constant, which is 364.56 millimicrons. The units of the quotient are therefore "per millimicron." Rydberg chose to use the unit "per centimeter." There are 10,000,000 millimicrons in a centimeter, so there are ten million times as many of anything per centimeter as per millimicron. If we multiply 0.0109 by ten million we get 109,000. The exact value, as determined by modern measurements, is 109,737.31 per centimeter. This is called the *Rydberg constant* and is symbolized R. In terms of centimeters, then, we can express Equation 5–3 as follows:

$$\frac{1}{\lambda} = 109{,}737.31\left(\frac{m^2 - 4}{4m^2}\right) = R\left(\frac{m^2 - 4}{4m^2}\right) \qquad \text{(Equation 5–4)}$$

The value of λ determined by Equation 5–4 is, of course, expressed in centimeters, so that the wavelength of the principal line comes out to 0.000065621 centimeters.

Now consider that portion of the equation which is written $(m^2 - 4)/4m^2$. This can be written as $m^2/4m^2 - 4/4m^2$, or, reducing to lowest terms, $1/4 - 1/m^2$. To make this look more symmetrical, we can now express 4 as a square also and make it $1/2^2 - 1/m^2$. Now Equation 5–4 becomes:

$$\frac{1}{\lambda} = R\left(\frac{1}{2^2} - \frac{1}{m^2}\right) \qquad \text{(Equation 5–5)}$$

where m can equal any integer from 3 up.

It is possible to imagine similar series such as:

$$\frac{1}{\lambda} = R\left(\frac{1}{1^2} - \frac{1}{m^2}\right) \qquad \text{(Equation 5–6)}$$

$$\frac{1}{\lambda} = R\left(\frac{1}{3^2} - \frac{1}{m^2}\right) \qquad \text{(Equation 5–7)}$$

$$\frac{1}{\lambda} = R\left(\frac{1}{4^2} - \frac{1}{m^2}\right) \qquad \text{(Equation 5–8)}$$

and so on. In Equation 5–6, the values of *m* must be integers greater than 1; in Equation 5–7, integers greater than 3; and in Equation 5–8, integers greater than 4.

The wavelengths given by Equation 5–6 would be shorter than those of the Balmer series and would exist only in the ultraviolet range. This series was actually discovered in 1906 by the American physicist Theodore Lyman (1874–1954) and is consequently known as the *Lyman series.*

The wavelengths given by Equation 5–7 would be longer than those of the Balmer series and would exist only in the infrared range. These were observed in 1908 by the German physicist Friedrich Paschen. The wavelengths given by Equation 5–8 would be still deeper in the infrared, and these were discovered by the American physicist Frederick S. Brackett. Equations 5–7 and 5–8 therefore represent the *Paschen series* and the *Brackett series,* respectively. Other series have also been discovered.

The Bohr Atom

A useful model of the hydrogen atom must therefore not only account for the fact that the circling electron gives off radiation without spiraling into the nucleus, but also that it gives off radiation of highly specific wavelengths, in such a fashion as to make them fit the simple Rydberg equations.

The necessary model was suggested in 1913 by the Danish physicist Niels Bohr (1885–1962). It seemed to him that one ought to apply the then-newly-established quantum theory (see page II–130) to the problem.

If the quantum theory is accepted, then any object which is converting kinetic energy into radiation ought to radiate energy in whole quanta only. This would be true if the earth, for instance, lost energy steadily as it revolved about the sun. The quanta of

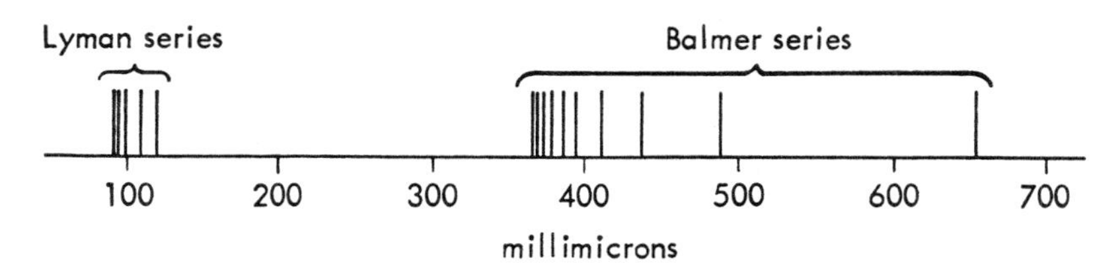

energy radiated by the earth in this fashion would, however, be so incredibly small in comparison to the total kinetic energy of the planet that even the most delicate observations would not suffice to detect any unevenness in the motion of the earth. It would seem to be spiraling gradually and smoothly into the sun.

Not so for electrons. The total kinetic energy of so small a body as an electron is not much larger than the individual quanta of visible light. Therefore, if a quantum of visible light is radiated by the electron revolving about the nucleus, a sizable fraction of its kinetic energy is lost all at once. Instead of spiraling gradually inward toward the nucleus (as one would expect according to the tenets of pre-quantum times—that is, of "classical physics"), the electron would suddenly take on a new orbit closer to the nucleus. On the other hand, if light were absorbed by an electron, it would be absorbed only a whole quantum at a time. With the absorption of a whole quantum, an electron would gain a sizable fraction of the energy it already possessed and it would suddenly take on a new orbit farther from the nucleus.

Bohr suggested that the electron had a certain minimum orbit, one that represented its *ground state*; at which time it was as close to the nucleus as it could be, and possessed minimum energy. Such an electron simply could not radiate energy (though the reason for this was not properly explained for over a decade, see page 105). Outside the ground state were a series of possible orbits extending farther and farther from the nucleus. Into these orbits, the *excited states*, the electron could be lifted by the absorption of an appropriate amount of energy.

Bohr arranged the orbits about the nucleus of the hydrogen atom in such a way as to give the electron a series of particular values for its angular momentum. This momentum had to involve Planck's constant (see page II–131) since it was that constant that dictated the size of quanta. Bohr worked out the following equation:

$$p = \frac{nh}{2\pi} \qquad \text{(Equation 5–9)}$$

In Equation 5–9, p represents the angular momentum of the electron, h is Planck's constant, and π is, of course, the familiar ratio of the circumference of a circle to its diameter. As for n, that is a positive integer that can take any value from 1 upward. By bringing in Planck's constant and making the electron capable of assuming only certain orbits in which n is a whole number, the atom is said to be *quantized*.

The expression $h/2\pi$ is commonly used in calculations involving the quantized atom, and is usually expressed by the single symbol $\hbar$, which is referred to as "$\hbar$ bar." Since the value of h is approximately 6.6256×10^{-27} erg-seconds and that of π is approximately 3.14159, the value of $\hbar$ is approximately 1.0545×10^{-27} erg-seconds.

We can therefore express Equation 5–9 as:

$$p = n\ (1.0545 \times 10^{-27}) \qquad \text{(Equation 5–10)}$$

The symbol n is sometimes referred to as a "quantum number" or, more properly, the *principal quantum number,* for there are others. It can be imagined to represent the various orbits. Where n equals 1, it refers to the ground state; where n equals 2, 3, 4, and so on, it refers to the higher and higher excited states.

If the single electron of the hydrogen atom dropped from orbit 2 to orbit 1, it emits a quantum of fixed size, and this is equivalent to a bit of radiation of fixed frequency. This would show up as a bright spectral line in a fixed position. (If the single electron rose from orbit 1 to orbit 2, this would be through the absorption of a quantum of the same fixed size, and this would produce a dark line against a bright background in the same position.)

If the single electron of the hydrogen atom dropped from orbit 3 to orbit 1, this would represent a greater difference in energy, and light of higher frequency would be emitted. Light of still higher frequency would result in a drop of an electron from orbit 4 to orbit 1, and higher frequency still in a drop from orbit 5 to orbit 1.

The series of possible drops from various orbits to orbit 1 would produce a series of successively higher frequencies (or successively lower wavelengths) that would correspond to those in the Lyman series. A series of possible drops from various outer orbits to orbit 2 would give rise to the Balmer series; from various outer orbits to orbit 3 to the Paschen series, and so on.

In the equations defining the wavelengths of the spectral lines included in the various series (Equations 5–5, 5–6, 5–7, and 5–8), the integer in the denominator of the first fraction on the right-hand side of the equation turns out to be the principal quantum number of the orbit into which the electrons drop (or out of which they rise).

If we consider atoms that are more complicated than hydrogen and contain more electrons, we must remember that they also contain nuclei of higher positive charge. The innermost electrons are held progressively more firmly as that nuclear charge increases.

It takes larger increments of energy to move such electrons away from the nucleus into excited states. Conversely, larger quanta of energy are given off when an electron drops closer to its ground state. Whereas the shortest wavelengths hydrogen can produce are those represented by the Lyman series in the ultraviolet, more complicated atoms can produce radiation in the X-ray region. The X-ray wavelength decreases with increasing atomic number, as Moseley had noticed.

So far, so good. If the lines of the hydrogen spectrum had been simple lines, the Bohr model of the hydrogen atom might have been reasonably satisfactory. However, as spectral analysis was refined, it turned out that each line had a *fine structure;* that is, it consisted of a number of distinct lines lying close together. It was as though an electron dropping down to orbit 2, for instance, might drop into any of a number of very closely spaced orbits.

This threatened the quantum interpretation of the atom, but in 1916 the German physicist Arnold Sommerfeld (1868–1951) offered an explanation. Bohr had pictured the electron orbits as

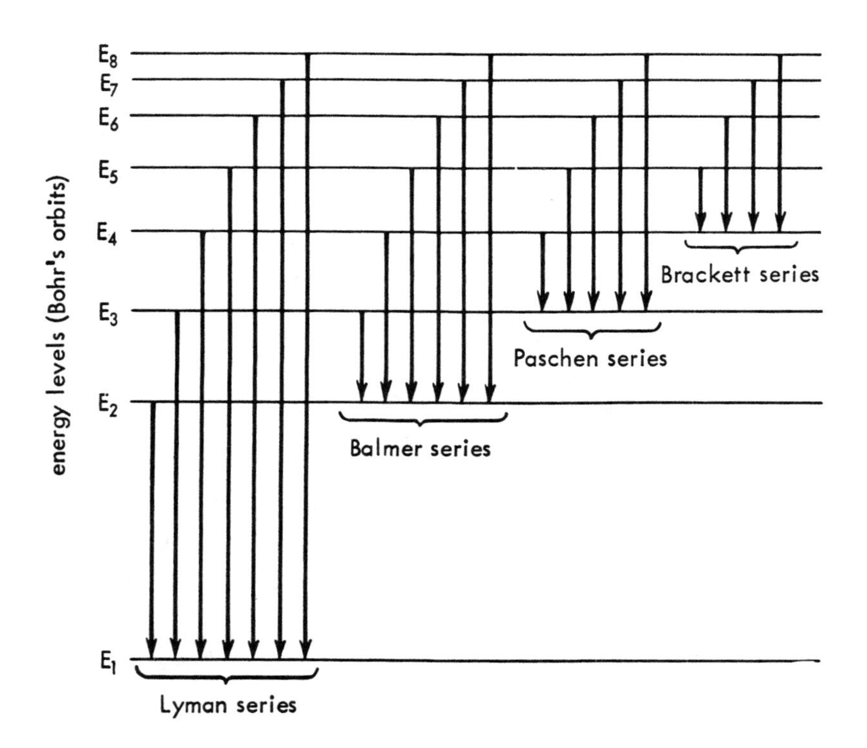

uniformly circular, but Sommerfeld suggested they might also be elliptical. Elliptical orbits of only certain eccentricities could be fitted into the quantum requirements, and for any principal quantum number, a fixed family of orbits—one circular and the rest elliptical—was permissible, the angular momenta among the orbits being slightly different. A drop to each of the various members of the family produced radiation of slightly different frequency.

To take into account the elliptical orbits, Sommerfeld introduced the *orbital quantum number,* for which we can use the symbol L.* The orbital quantum number can have any whole number value from zero up to one less than the value of the principal quantum number. Thus, if $n = 1$, then L can only equal 0; if $n = 2$, then L can equal 0 or 1; if $n = 3$, then L can equal 0, 1, or 2, and so on.

But the spectral lines can be made even more complicated, for in a magnetic field, lines that seem single, split further. In order to account for this, a third number, the *magnetic quantum number* had to be introduced, and this was symbolized as m.

The magnetic quantum number was visualized as extending the family of orbits through three-dimensional space. Not only could an orbit be elliptical rather than circular, but it could also be tilted to the principal orbit by varying amounts. The possible values of m are the same as those for L, except that negative values are also included. Thus if $n = 2$—so that L could be either 0 or 1—m could be either 0, 1, or -1. If $n = 3$—so that L could be either 0, 1, or 2—m could be either 0, 1, 2, -1, or -2, and so on.

Finally, a fourth and last quantum number had to be introduced, the *spin quantum number,* symbolized as s. This was visualized as representing the spin of the electron (analogous to the rotation of the earth about its axis). This spin could either be clockwise or counterclockwise, so that in connection with any value of n there can be only two values of s, $1/2$ and $-1/2$.

Sub-shells

The orbits described by the quantum numbers are all that are available. When an atom contains more than one electron (as is true for all atoms other than those of hydrogen), they must be distributed among these orbits, filling them from the one closest to the nucleus outward.

* Actually, the usual symbol is the lower-case "l," which I am not using here because of the ease with which it is confused with the numeral "1".

But how many electrons may be permitted in each orbit? In 1925, the Austrian physicist Wolfgang Pauli (1900–1958) suggested that in order to account for the various spectral characteristics of the different elements, one must assume that no two electrons in a given atom can have all four quantum numbers identical. This means that in any orbit (circular, elliptical, or tilted) two electrons at most may be present; and of these two, one must spin clockwise and the other must spin counterclockwise. Thus, the presence of two electrons of opposite spin in a given orbit excludes other electrons, and this is called Pauli's *exclusion principle.*

We can now determine the number of electrons that can be included in each of the different orbit-families represented by the principal quantum numbers.

Suppose $n = 1$. In that case, $L = 0$ and $m = 0$. No other combinations are possible, and the only orbit is 1/0/0. This may contain two electrons of opposite spins. The total number of electrons which may be contained in the first orbit-family ($n = 1$) is therefore 2.

Next, suppose that $n = 2$. In that case, L can equal either 0 or 1, and m can equal 0, 1, or -1. To be more specific, if $n = 2$ and $L = 0$, then m must equal 0, too. If $n = 2$ and $L = 1$, then m can equal either 0, 1, or -1. There are therefore four possible orbits for $n = 2$. These are 2/0/0, 2/1/0, 2/1/1, and 2/1/-1. In each one of these orbits two electrons of opposite spins can exist; consequently, the second orbit-family ($n = 2$) can contain eight electrons altogether.

By similar reasoning, the total number of electrons that can be present in the next orbit-family ($n = 3$) turns out to be eighteen. In fact, it can be shown that the maximum number of electrons in any orbit-family is equal to $2n^2$. Therefore for orbit-family $n = 4$, a total of thirty-two electrons may be found; for orbit-family $n = 5$, a total of fifty electrons may be found, and so on.

The orbit-families, represented by the principal quantum number n and deduced from physical data, correspond to the different electron shells deduced from chemical data and made use of in the Lewis-Langmuir model of the atom.

The number of electrons in each orbit-family or electron shell can be divided into *sub-shells,* according to the value of L. Thus, where $n = 1$, L can equal only 0, so that the first electron shell consists of only a single sub-shell, which can hold two electrons.

Where $n = 2$, on the other hand, L can equal either 0 or 1. Where $L = 0$, only one orbit (2/0/0), containing two electrons at most, is possible; but where $L = 1$, three orbits (2/1/0, 2/1/1, and 2/1/−1) containing, at most, a total of six electrons are possible. The eight electrons of the second shell can, therefore, be divided into two sub-shells, one of 2 electrons and one of 6.

In the same way, it can be shown that the eighteen electrons of the third electron shell can be divided into three sub-shells, one capable of holding 2 electrons, one of holding 6, and one of holding 10. In general, the electron shell of principal quantum number n can be divided into n sub-shells, where the first can contain two electrons, and where each one thereafter can contain four more than the one before (6, 10, 14, 18, and so on).

The subgroups are symbolized as *s, p, d, f, g, h,* and *i*. We may therefore say that the first electron shell contains only the 1*s* sub-shell, that the second electron shell contains a 2*s* sub-shell and a 2*p* sub-shell, and so on.

Now let's see how all this applies to the individual elements. The first two present no problem. Hydrogen has one electron and helium two, and both can be accommodated in the single sub-shell of the first electron shell.

	Number of Electrons in 1*s*	Atomic Number
Hydrogen	1	1
Helium	2	2

All elements of atomic number greater than 2 contain two electrons in the first shell and distribute the remainder among the outer shells. The elements immediately after helium make use of the second electron shell, which is made up of sub-shells 2*s* (electron capacity, two) and 2*p* (electron capacity, six).

	Number of Electrons in			Atomic Number
	1*s*	2*s*	2*p*	
Lithium	2	1	—	3
Beryllium	2	2	—	4
Boron	2	2	1	5
Carbon	2	2	2	6
Nitrogen	2	2	3	7
Oxygen	2	2	4	8
Fluorine	2	2	5	9
Neon	2	2	6	10

In neon the second electron shell is full, and elements of higher atomic number must place electrons in the third electron shell. The third shell has three sub-shells, 3*s*, 3*p* and 3*d* with electron capacities of two, six and ten, respectively.

	Number of Electrons in					Atomic Number
	1*s*	2*s* 2*p*	3*s*	3*p*	3*d*	
Sodium	2	8	1	—	—	11
Magnesium	2	8	2	—	—	12
Aluminum	2	8	2	1	—	13
Silicon	2	8	2	2	—	14
Phosphorus	2	8	2	3	—	15
Sulfur	2	8	2	4	—	16
Chlorine	2	8	2	5	—	17
Argon	2	8	2	6	—	18

Notice the similarities in the electron configurations of these atoms and the ones in the previous list. The 2*s*/2*p* configuration of lithium is like the 3*s*/3*p* configuration of sodium. There is the same comparison to be made between beryllium and magnesium, between boron and aluminum, and so on. No wonder the periodic table is as it is.

Argon, which has a 3*s*/3*p* combination of 2/6, just as neon has a 2*s*/2*p* combination of 2/6, is also an inert gas. Yet argon does not have its outermost shell completely filled. There is still room for ten more electrons in the 3*d* sub-shell. The conclusion must be that the inert gas properties are brought on not by a truly completed electron shell but only by the complete filling of the *s* and *p* sub-shells. These two sub-shells always contain a total of eight electrons; consequently, this total of eight in the outermost shell is the hallmark of the inert gas. The one exception is helium. That contains electrons only in the first electron shell, which is made up of the 1*s* sub-shell only. It therefore contains only two electrons in its outermost (and only) electron shell.

You might suppose that the elements immediately after argon would possess electrons in sub-shell 3*d*. This is not so, however. We can understand this by viewing the sub-shells of a particular electron shell as taking up room. As one goes outward from the nucleus, each succeeding electron shell has more sub-shells, and eventually the outer sub-shells of one will begin to overlap the inner sub-shells of the next one out. In this case, 3*d*, the outermost sub-shell of $n = 3$ overlaps $4s$, the innermost sub-shell of $n = 4$ and it is $4s$ that is therefore next to be filled. Thus:

	Number of Electrons in					Atomic Number
	$1s$	$2s$ $2p$	$3s$ $3p$	$3d$	$4s$	
Potassium	2	8	8	—	1	19
Calcium	2	8	8	—	2	20

Potassium, with a single electron in sub-shell $4s$, is like sodium with a single electron in sub-shell $3s$ and lithium, with a single electron in sub-shell $2s$. Calcium similarly resembles magnesium and beryllium.

Transition Elements

If scandium, the element after calcium, possessed an electron in sub-shell $4p$, it would have an s/p combination of $2/1$, and would resemble aluminum in its properties. However, this is not what happens. With sub-shell $4s$ filled in the case of calcium, the additional electrons of the next few elements are added to sub-shell $3d$ as follows:

	Number of Electrons in					Atomic Number
	$1s$	$2s$ $2p$	$3s$ $3p$	$3d$	$4s$	
Scandium	2	8	8	1	2	21
Titanium	2	8	8	2	2	22
Vanadium	2	8	8	3	2	23
Chromium	2	8	8	5	1	24
Manganese	2	8	8	5	2	25
Iron	2	8	8	6	2	26
Cobalt	2	8	8	7	2	27
Nickel	2	8	8	8	2	28
Copper	2	8	8	10	1	29
Zinc	2	8	8	10	2	30

The overlapping of sub-shells $3d$ and $4s$ is not very pronounced, so there is not much difference between a $3d/4s$ arrangement of $5/1$ and $4/2$, or of $10/1$ and $9/2$. In the case of chromium and copper there are reasons for preferring to assign only one electron to $4s$, but this is a mere detail.

What is important in the ten elements from scandium to zinc inclusive is that the difference in electron arrangement concentrates on an inner sub-shell, 3*d*. The outermost sub-shell, 4*s*, is the same (or virtually the same), in all. This series of elements makes up a group of *transition elements,* and the progressive difference among them in chemical properties is not as sharp as among the succession of elements from hydrogen to calcium, where it is the outermost sub-shell in which the difference in electron distribution shows up.

Indeed, the three successive elements—iron, cobalt, and nickel—resemble each other so closely as to form a tight-knit family group of elements.

The Lewis-Langmuir model of the atom makes no allowance for changes in the electron content of inner shells, and it is for this reason that this model does not work well for transition elements (and, as it happens, about three-fifths of all the elements are transition elements).

With zinc, the third electron shell is completely filled and contains a grand total of eighteen electrons. Sub-shell 4*s* is also filled and additional electrons must be added to 4*p* and beyond. Sub-shell 4*p* has a capacity of six electrons:

	Number of Electrons in					Atomic Number
	1*s*	2*s* 2*p*	3*s* 3*p* 3*d*	4*s*	4*p*	
Gallium	2	8	18	2	1	31
Germanium	2	8	18	2	2	32
Arsenic	2	8	18	2	3	33
Selenium	2	8	18	2	4	34
Bromine	2	8	18	2	5	35
Krypton	2	8	18	2	6	36

These six elements have the same *s/p* electron configuration as do the series of elements from aluminum to argon or from boron to neon. Thus gallium resembles aluminum and boron in its properties; germanium resembles carbon and silicon, and so on. Krypton with a 4*s*/4*p* combination of 2/6 is, of course, an inert gas.

The fourth shell has two additional sub-shells: 4*d,* which can hold ten electrons, and 4*f,* which can hold fourteen. Both these sub-shells overlap the innermost sub-shell of the fifth shell, 5*s*:

	Number of Electrons in							Atomic Number
	1*s*	2*s* 2*p*	3*s* 3*p* 3*d*	4*s* 4*p*	4*d*	4*f*	5*s*	
Rubidium	2	8	18	8	—	—	1	37
Strontium	2	8	18	8	—	—	2	38

The next elements possess electrons in sub-shell 4*d* so that a new series of transition elements is produced like those from scandium to zinc inclusive. Thus, we have:

	Number of Electrons in							Atomic Number
	1*s*	2*s* 2*p*	3*s* 3*p* 3*d*	4*s* 4*p*	4*d*	4*f*	5*s*	
Yttrium	2	8	18	8	1	—	2	39
Zirconium	2	8	18	8	2	—	2	40
Niobium	2	8	18	8	4	—	1	41
Molybdenum	2	8	18	8	5	—	1	42
Technetium	2	8	18	8	5	—	2	43
Ruthenium	2	8	18	8	7	—	1	44
Rhodium	2	8	18	8	8	—	1	45
Palladium	2	8	18	8	10	—	—	46
Silver	2	8	18	8	10	—	1	47
Cadmium	2	8	18	8	10	—	2	48

We next return to the 5*p* column and produce half a dozen elements with the *s*/*p* combination similar to that of the series of elements from boron to neon. These are not transition elements.

	Number of Electrons in							Atomic Number
	1*s*	2*s* 2*p*	3*s* 3*p* 3*d*	4*s* 4*p* 4*d*	4*f*	5*s*	5*p*	
Indium	2	8	18	18	—	2	1	49
Tin	2	8	18	18	—	2	2	50
Antimony	2	8	18	18	—	2	3	51
Tellurium	2	8	18	18	—	2	4	52
Iodine	2	8	18	18	—	2	5	53
Xenon	2	8	18	18	—	2	6	54

Xenon is another inert gas.

There remains sub-shell 4*f*, capable of holding fourteen electrons. There are also sub-shells 5*d*, 5*f* and 5*g* capable of holding 10, 14, and 18 electrons, respectively. All of these overlap sub-shell 6*s*, however.

	Number of Electrons in										Atomic
	1*s*	2*s* 2*p*	3*s* 3*p* 3*d*	4*s* 4*p* 4*d*	4*f*	5*s* 5*p*	5*d*	5*f*	5*g*	6*s*	Number
Cesium	2	8	18	18	—	8	—	—	—	1	55
Barium	2	8	18	18	—	8	—	—	—	2	56

With lanthanum, the element beyond barium, electrons start entering sub-shell 4*f* or 5*d*, and this gives us a new sort of transition element. In the ordinary transition elements from scandium to zinc or from yttrium to cadmium, the sub-shell to which electrons were being added was covered only by the one or two electrons in the next higher *s* sub-shell. Here, however, where 4*f* is involved, the electrons being added are covered not only by two electrons in 5*s* but by six electrons in 5*p* and by two electrons in 6*s*. In these elements, electrons are being added to a sub-shell that is deeper within the atom, so to speak, than was true in the case of the transition elements considered earlier. The sub-shell in which the electron difference occurs is more efficiently covered by outer electrons. For this reason, these elements (the lanthanides) resemble each other particularly closely.

	Number of Electrons in										Atomic
	1*s*	2*s* 2*p*	3*s* 3*p* 3*d*	4*s* 4*p* 4*d*	4*f*	5*s* 5*p*	5*d*	5*f*	5*g*	6*s*	Number
Lanthanum	2	8	18	18	—	8	1	—	—	2	57
Cerium	2	8	18	18	1	8	1	—	—	2	58
Praseodymium	2	8	18	18	3	8	—	—	—	2	59
Neodymium	2	8	18	18	4	8	—	—	—	2	60
Promethium	2	8	18	18	5	8	—	—	—	2	61
Samarium	2	8	18	18	6	8	—	—	—	2	62
Europium	2	8	18	18	7	8	—	—	—	2	63
Gadolinium	2	8	18	18	7	8	1	—	—	2	64
Terbium	2	8	18	18	8	8	1	—	—	2	65

Dysprosium	2	8	18	18	9	8	1	—	—	2	66
Holmium	2	8	18	18	10	8	1	—	—	2	67
Erbium	2	8	18	18	11	8	1	—	—	2	68
Thulium	2	8	18	18	13	8	—	—	—	2	69
Ytterbium	2	8	18	18	14	8	—	—	—	2	70
Lutetium	2	8	18	18	14	8	1	—	—	2	71

The elements after lutetium add further electrons to sub-shell 5*d*, which can hold up to ten electrons. This sub-shell is still covered by the two electrons in sub-shell 6*s*, so that we continue after lutetium with a set of ordinary transition elements:

	Number of Electrons in									Atomic
	1*s*	2*s* 2*p*	3*s* 3*p* 3*d*	4*s* 4*p* 4*d* 4*f*	5*s* 5*p*	5*d*	5*f*	5*g*	6*s*	Number
Hafnium	2	8	18	32	8	2	—	—	2	72
Tantalum	2	8	18	32	8	3	—	—	2	73
Tungsten	2	8	18	32	8	4	—	—	2	74
Rhenium	2	8	18	32	8	5	—	—	2	75
Osmium	2	8	18	32	8	6	—	—	2	76
Iridium	2	8	18	32	8	7	—	—	2	77
Platinum	2	8	18	32	8	9	—	—	1	78
Gold	2	8	18	32	8	10	—	—	1	79
Mercury	2	8	18	32	8	10	—	—	2	80

With mercury, sub-shell 5*d* is filled. Sub-shells 5*f* and 5*g* remain untouched, and electrons are next found in sub-shell 6*p*, so that we have a group of elements with the familiar *s/p* arrangement of the boron-to-neon group:

	Number of Electrons in									Atomic
	1*s*	2*s* 2*p*	3*s* 3*p* 3*d*	4*s* 4*p* 4*d* 4*f*	5*s* 5*p* 5*d*	5*f*	5*g*	6*s*	6*p*	Number
Thallium	2	8	18	32	18	—	—	2	1	81
Lead	2	8	18	32	18	—	—	2	2	82
Bismuth	2	8	18	32	18	—	—	2	3	83
Polonium	2	8	18	32	18	—	—	2	4	84
Astatine	2	8	18	32	18	—	—	2	5	85
Radon	2	8	18	32	18	—	—	2	6	86

Radon is an inert gas.

There still remain sub-shells 5*f* and 5*g*, with capacities for 14 and 18 electrons, respectively. There are also sub-shells 6*d*, 6*f*, 6*g* and 6*h*, with electron capacities of 10, 14, 18, and 22 respectively. All of these, however, overlap sub-shell 7*s*.

	Number of Electrons in													Atomic
	1*s*	2*s* 2*p*	3*s* 3*p* 3*d*	4*s* 4*p* 4*d* 4*f*	5*s* 5*p* 5*d*	5*f*	5*g*	6*s* 6*p*	6*d*	6*f*	6*g*	6*h*	7*s*	Number
Francium	2	8	18	32	18	—	—	8	—	—	—	—	1	87
Radium	2	8	18	32	18	—	—	8	—	—	—	—	2	88

There then arises just such a situation as occurs in the case of the lanthanides:

	Number of Electrons in											Atomic Number
	1*s*	2*s* 2*p*	3*s* 3*p* 3*d*	4*s* 4*p* 4*d* 4*f*	5*s* 5*p* 5*d*	5*f*	5*g*	6*s* 6*p*	6*d*	6*f* 6*g* 6*h*	7*s*	
Actinium	2	8	18	32	18	—	—	8	1	—	2	89
Thorium	2	8	18	32	18	—	—	8	2	—	2	90
Protactinium	2	8	18	32	18	2	—	8	1	—	2	91
Uranium	2	8	18	32	18	3	—	8	1	—	2	92
Neptunium	2	8	18	32	18	4	—	8	1	—	2	93
Plutonium	2	8	18	32	18	5	—	8	1	—	2	94
Americium	2	8	18	32	18	7	—	8	—	—	2	95
Curium	2	8	18	32	18	7	—	8	1	—	2	96
Berkelium	2	8	18	32	18	8	—	8	1	—	2	97
Californium	2	8	18	32	18	9	—	8	1	—	2	98
Einsteinium	2	8	18	32	18	10	—	8	1	—	2	99
Fermium	2	8	18	32	18	11	—	8	1	—	2	100
Mendelevium	2	8	18	32	18	12	—	8	1	—	2	101
Nobelium	2	8	18	32	18	13	—	8	1	—	2	102
Lawrencium	2	8	18	32	18	14	—	8	1	—	2	103

The group of elements from actinium to lawrencium are the actinides. When the element with atomic number 104 is studied, it is fully expected that the 104th electron will be added to sub-shell 6*d* and that this element will resemble hafnium in its chemical properties.

If you will now compare the electron arrangements of the various elements as given in this section with the periodic table presented on page 16, you will see how the periodic table reflects similarities in electron arrangements.

CHAPTER 35

Electron Energy Levels

Semiconductors

While the notion of electron shells and sub-shells finally rationalized the periodic table, even down to the until-then puzzling lanthanides, the Bohr model itself, even as modified by Sommerfeld and others, did not stand up in its original form. The attempt to produce a literal picture of the atom as consisting of electron particles moving in orbits that were circular, elliptical and tilted—much more complicated than the solar system, but still with some key points of similarity—grew top-heavy and collapsed.

During the early 1920's, it became more common to think, not of orbits, but of *energy levels*. Electrons moved from one energy level to another and the difference in energy levels determined the size of the quantum (hence the frequency of the radiation) emitted or absorbed.

In 1925, in fact, the German physicist Werner Heisenberg (1901–) worked out a system whereby the energy levels of atoms could be written out as a set of numbers. These could be arranged in rectangular arrays called "matrices," and these matrices could be manipulated according to mathematical principles (*matrix algebra*). The proper manipulation applied to atomic data (*matrix mechanics*) produced values from which spectral lines could be calculated. No actual picture of any sort was required for the atom by this view; it had faded away completely into a mere collection of numbers.

In the case of a single atom, energy levels could be pictured as simple lines at given heights above the base of a schematic drawing. Two electrons of opposite spin could occupy any of the energy levels and could shift from one level to any other that was not fully occupied. The spaces between the lines represented "forbidden gaps" within which no electron could be located. Each element had its own characteristic collection of lines and gaps, of course.

If two atoms of an element are in close proximity, the picture becomes more complicated. The outer electrons of the two atoms are close enough for the energy levels to merge. For each energy level, the electron population is doubled. An energy level cannot hold more than its capacity (two electrons of opposite spin); consequently, what happens is that the energy levels associated with the two atoms shift a bit—one becoming slightly higher than the other. Each can then hold its own electrons.

In a solid, where there are a vast number of atoms existing in near proximity, this happens on a grand scale. An energy level can no longer be depicted as a line but as a dense assemblage of lines at slightly different heights. What was an energy line is now actually an *energy band*. Electrons can rise from band to band, rather than from line to line, and there are forbidden gaps between the bands.

If each of the atoms making up a solid has its outermost energy levels containing all the electrons they can hold, then the result is a filled energy band. In such a case, the electrons are fixed in place. They cannot pass from one atom to the next since the neighboring atom has no room for it. Such a solid is a nonconductor of electricity. Extreme examples are sulfur and quartz.

If the outermost energy levels of the individual atoms contain fewer electrons than capacity, the resulting energy band of the solid is only partly filled, and electrons can move easily from atom to atom by way of unfilled energy levels. The electric impulse can easily travel across a path containing these "free electrons," and the substance is a conductor of electricity. Extreme examples are silver and copper.

Even if the energy band is electron-filled, there is a chance of electrical conduction. Above the filled energy band is another energy band that is empty. The absorption of energy may kick a few electrons up into the higher energy band and there they may move freely. The likelihood of this happening depends on the width of the forbidden gap between the filled energy band and the higher empty one. If the gap is quite wide, the likelihood of an electron leaping across is low, and the substance is an excellent nonconductor.

In some substances (for example, the elements silicon and germanium) the forbidden gap is comparatively narrow, and the likelihood of an electron leap into the higher band becomes appreciable. The result is a *semiconductor.* If the temperature is raised, the tendency of an electron to reach the higher band is increased, for more energy becomes available to kick it upward. For this reason, the resistance of a semiconductor decreases with temperature. (A semiconductor differs in this respect from a metallic conductor, on which the chief effect of heightened temperature is to produce intensified atomic vibrations that interfere with transmission of the electrical impulse and increase the resistance.)

The semiconductors have proven unexpectedly and fabulously useful, where their chemical composition and physical structure are suitably tailored to need.

Consider germanium, for instance. Like carbon, the germanium atom has four electrons in its outermost shell. Each germanium atom can contribute one electron to form a shared pool of two with each of four other germanium atoms. In the end, the germanium atoms will be stacked in such a fashion that each is connected with four others. Under such conditions, all electrons are firmly in place and the substance's semiconducting properties are at a minimum.

In order for this to happen, all the atoms must be stacked perfectly. Imperfections in the crystal means that some atoms are going to be out of place with respect to their neighbors and will not be able to share electrons. It is these few unshared electrons that contribute to the semiconducting properties of germanium.

Those properties are more useful if they arise out of a deliberately added impurity, rather than out of the random imperfections that are almost inevitably found in any germanium crystal. Imagine a perfect germanium crystal formed out of germanium to which a small trace of arsenic had been added as an impurity. Arsenic has five electrons in its outermost shell. When an arsenic atom tries to fit in with the germanium arrangement, it can find room for four of its electrons in the shared pools formed with neighboring germanium atoms. The fifth arsenic electron, however, is at loose ends. It acts as a free electron.

Under the influence of an electric potential applied across

the crystal, the free electrons, negatively charged of course, drift away from the negative electrode and toward the positive. Because it is a negatively-charged particle that is drifting, the result is an *n-type semiconductor,* "n" for negative.

Next, consider a germanium crystal to which a trace of boron has been added. The boron atom has three electrons in its outermost shell. Each of the three can join in a shared pool with electrons of a neighboring germanium atom. But only three germanium atoms can be thus accommodated; the fourth will be left with a "hole" where an electron ought to be.

Under the influence of an electric potential across such a crystal, a negatively-charged electron will be pushed or pulled into the hole, traveling always from the side of the repelling negative electrode toward the attracting positive electrode. But the electron that has filled the hole has left another hole in the place it had earlier occupied. Since the electron came from the direction of the negative electrode, the new hole is now closer to the negative electrode than the old hole had been. The same thing happens over and over again, and the hole drifts steadily toward the negative electrode and away from the positive one. Indeed, the drifting hole behaves as though it were a positively-charged particle, and so such a crystal is a *p-type semiconductor,* "p" for positive.

Solid-State Devices

The drift of electrons through a semiconductor can be governed and manipulated in order to achieve the ends that had earlier been achieved by a vacuum tube. The key difference is that electrons move across solids in the former case and across a vacuum in the latter. For this reason, electronic instruments in which semiconductors play a part are called *solid-state devices.*

Electrons and holes

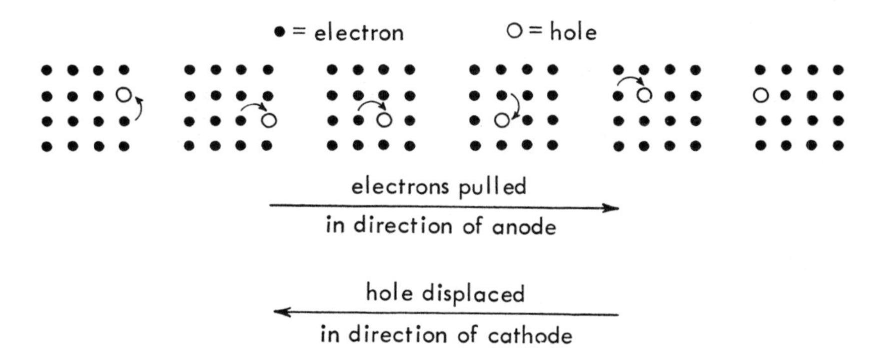

Consider, for instance, a crystal of which one half is n-type and the other half is p-type. Imagine that the n-type end is connected to the negative pole of a battery, while the p-type is attached to the positive pole. If the circuit is closed, the electrons in the n-type end of the crystal are repelled from the negative pole and move toward the junction between the two halves. Meanwhile the holes in the p-type end are repelled from the positive pole and also move toward the junction. There they can meet and neutralize each other, while electrons flood into the n-type end and are withdrawn from the p-type end (making new holes). By the movement of electrons and holes, ever-renewed as long as the circuit is closed, current can flow across the crystal.

But suppose the circuit is arranged in the opposite way, with the n-type end of the crystal attached to the positive pole of the battery and the p-type attached to the negative pole. Now the electrons in the n-type end are attracted to the positive pole and move away from the junction. The holes in the p-type end are attracted to the negative pole and also move away from the junction. First the junction area and then the entire crystal is emptied of both free electrons and of holes; consequently, it becomes a nonconductor and no current flows.

In short, such an n-p crystal acts to allow current to pass in only one direction. If connected to a source of alternating current, it would serve to rectify the current. In fact the n-type end of the crystal is similar in its behavior to the heated filament of a vacuum tube, while the p-type end is similar in its behavior to the plate. The crystal is like a diode with the two parts meeting to form a junction. Such a device is therefore called a *junction diode.*

It is possible to construct a semiconductor analog of a triode also. In an ordinary triode, a third element, the grid, is inserted between the filament and the plate, and the same may be done in solid-state devices. A crystal can be made up of three regions, with both ends n-type and the middle region p-type. There are now two junctions, an n-p and a p-n.

Imagine one n-type end of such a crystal attached to the negative pole of a battery, and the other attached to the positive pole. Electrons at the end attached to the negative pole are repelled from the pole and into the p-type middle. Electrons at the end attached to the positive pole are attracted toward the pole and away from the middle, so that electrons are pulled out of the p-type middle. Electrons flow from one end to the other, with the p-type middle encouraging the first half of the flow and hindering the second half. The rate of flow can be sharply altered, then, by the size of the charge placed on the p-type center.

Such a "junction triode" was first made practical in 1948 by the English-American physicist William Bradford Shockley (1910–) and his American co-workers John Bardeen (1908–) and Walter House Brattain (1902–). Because the device transferred a current across a material that ordinarily had high resistance (a "resistor"), it was called by a shortened version of the phrase "transfer-resistor." It was a *transistor.*

In the early days of radio, before the vacuum tube had been devised, natural crystals had sometimes been found that in proper combination with other materials displayed rectifying action. It was the use of these that gave the name to the old-fashioned "crystal sets."

The development of the vacuum tube had thrown the crystals out of use, but now the specially tailored transistor crystals turned the tables. Transistors had several advantages over the vacuum tubes. They required no vacuum but were solid throughout; as a result they were sturdier than tubes. They made no use of heat (as was required in a vacuum tube, for there only a hot filament would emit electrons), so they required no warm-up period and also lasted longer. Furthermore, they could be made much, much smaller than vacuum tubes.

Since it was the vacuum tube that made the radio so bulky, the use of transistors made it possible to develop radios no larger than a pack of cigarettes. In fact, "to transistorize" came to be synonymous with "to miniaturize." The effect of transistorization was most spectacular in connection with electronic computers requiring thousands of vacuum tubes. The size of computers was drastically reduced with the coming of solid-state devices.

Semiconductors could also be used in the direct conversion of heat into electricity (*thermoelectricity*). The general phenomenon whereby a temperature difference can be made to give rise to an electric current was first observed in 1821 by the German physicist Thomas Johann Seebeck (1770–1831). He found that when part of a circuit made up of two different metals is heated, a magnetic needle would be deflected if placed near the point at which the two metals meet. This is the *Seebeck effect.*

The effect on the magnet came from the magnetic field set up by the electric current that was produced by the appearance of a temperature difference. Unfortunately, Seebeck did not interpret his observations in this fashion (the connection between electricity and magnetism was just being discovered then—see page II–201). He thought it purely a magnetic effect. Consequently, interest in the effect languished, and it is only in recent decades that it has revived.

Consider an n-type semiconductor, one end of which is heated. The hot end is more likely to have its electrons kicked into a higher energy band, wherein they can easily drift away. As a result, electrons tend to move from the hot end toward the cold end, and an electric potential (cold end, negative; hot end, positive) is set up. (This would also happen in an ordinary conductor, but an ordinary conductor has a high concentration of free electrons at the cold end, which would repel incoming electrons from the hot end; consequently, there would only be a small net drift. In a semiconductor, the number of free electrons at the cold end is very small, and the repelling effect is much weaker. For that reason a much larger drift takes place, and a much higher potential difference is set up in a semiconductor than in an ordinary conductor.)

If a p-type semiconductor is used, the holes are more easily filled at the hot end, where the electrons are made more mobile by the greater energy associated with high temperature. New holes are formed further away from the hot end and, in brief, the holes drift from hot end to cold. Again a potential difference is set up but this time with the cold end positive and the hot end negative.

If an n-type and a p-type semiconductor are joined at the hot end, electrons will flow from the p-type cold end to the hot junction and then on to the n-type cold end. If the two ends are connected through a closed circuit, a current will flow and useful work will be done as long as the temperature difference is maintained. A kerosene lamp, then, can be used to power an electric generator without moving parts.

The situation can be reversed. If an electric current is forced through a circuit made up of different materials, a temperature difference is set up. This was first observed in 1834 by a French physicist, Jean Charles Athanase Peltier (1785–1845), and is therefore called the *Peltier effect*. By using p-type and n-type semiconductors joined at one end, the heat of the system can be concentrated at that end and removed from the other, so that the device can function either as a heater or a refrigerator.

Semiconductors can also convert light to electricity. Essentially, a device of this sort, a *solar battery,* consists of an n-type semiconductor overlaid with a thin layer of p-type semiconductor. There is a potential difference between the electron-rich n-type section and the electron-poor p-type section which could set up a current for a very short time until the electrons had flooded into the p-type area and filled the holes.

If the plate is exposed to sunlight, the high-energy quanta of the light act to knock electrons loose in the p-type area, forming more holes as fast as incoming electrons fill them and thus allowing a continuous current of significant proportions to flow through a circuit connected to the battery. Such batteries will hold good for extended periods of time, and some have successfully been used to power artificial satellites for years.

Masers and Lasers

The fact that energy levels are separated by fixed distances and that a change of levels can only be accomplished by the absorption or emission of photons of specific size has given rise in recent years to important new devices.

Consider the ammonia molecule (NH_3), for instance. It possesses two energy levels separated by a gap that is equal in size to the energy content of a photon equivalent to an electromagnetic wave with a frequency of 24 billion cycles per second. The wavelength of such radiation is 1.25 centimeters, which places it in the microwave portion of the electromagnetic spectrum.

This difference in energy levels can be pictured in terms of the architecture of the molecule. The three hydrogen atoms of the ammonia molecule can be viewed as occupying the three vertices of an equilateral triangle, while the single nitrogen atom is some distance above the center of the triangle. With the change in energy level, the nitrogen atom moves through the plane of the triangle to an equivalent position on the other side. The ammonia molecule can be made to vibrate back and forth with a frequency of just 24 billion times a second.

This vibration period is extremely constant, much more so than the period of any man-made vibrating device; much more constant, even, than the movement of astronomical bodies. A vibrating molecule, producing a microwave of highly constant frequency, can therefore be used to control time measuring devices with unprecedented precision. By means of such *atomic clocks,* accuracies in time measurement of one second in 100,000 years are looked forward to as attainable.

But let us leave molecular architecture and consider only energy levels. If a beam of microwaves passes through ammonia gas, a beam containing photons of the proper size, the ammonia molecules will be raised to the higher energy level. In other words, a portion of the microwave beam will be absorbed.

What if an ammonia molecule is already in the higher energy level, however? As early as 1917, Einstein pointed out that if a photon of just the right size struck such an upper-level molecule, the molecule would be nudged back down to the lower level and would emit a photon of exactly the size—and moving in exactly the direction—of the entering photon.

Ammonia exposed to microwave radiation could, therefore, undergo two possible changes: Molecules could be pumped up from lower level to higher, or be nudged down from higher level to lower. Under ordinary conditions, the former process would predominate, for only a very small percentage of the ammonia molecules would at any one instant be at the higher energy level.

Suppose, though, that some method were found to place all or almost all the molecules in the upper energy level. Then it would be the movement from higher level to lower that would predominate. Indeed, something quite interesting would happen. The incoming beam of microwave radiation would supply a photon that would nudge one molecule downward. A second photon would be released, and the two would speed on, striking two molecules, so that two more were released. All four would bring about the release of four more, and so on. The initial photon would let loose a whole avalanche of photons, all of exactly the same size and moving in exactly the same direction.

Physicists both in the United States and the Soviet Union labored to achieve such a situation, but the lion's share of credit for success goes to the American physicist Charles Hard Townes (1915–). In 1953, he devised a method for isolating excited ammonia molecules and subjecting them to stimulation by microwave photons of the exact energy content. A few photons entered, and a flood of photons left. The incoming radiation was thus greatly amplified.

The process was described as "microwave amplification by stimulated emission of radiation." The phrase was initialed as "m. a. s. e. r." and the instrument came to be known by the acronym *maser*, a word which quickly replaced the more dramatic, but too narrow, phrase "atomic clock."

Solid-state masers were soon developed, using paramagnetic atoms or molecules (see page II–155) in a magnetic field. Here an electron can be pictured as occupying two energy levels, depending upon its spin: the lower level, when it is spinning in a direction parallel to the magnetic field; the upper, when it is spinning in the opposite direction. The electrons are slowly pumped upward to the higher level and then made to release all their stored energy in a sudden burst of radiation at a single frequency (*monochromatic radiation*).

The first masers, both gaseous and solid state, were intermittent. That is, they had to be pumped up first and then released. After a burst of radiation, no more could be emitted until the pumping progress had been repeated.

To get round this, it occurred to the Dutch-American physicist Nicholaas Bloembergen (1920–) to make use of a three-level system. This is possible, for instance, if the solid that forms the radiating core of the maser contains atoms of metals such as chromium or iron. It then becomes possible to distribute electrons among three energy levels, a lower, a middle and an upper. In this case, both pumping and stimulated emission can go on simultaneously. Electrons are pumped up from the lowest energy level to the highest. Once at the highest, proper stimulation will cause them to drop down first to the middle level, then to the lower. Photons of different size are required for pumping and for stimulated emission, and the two processes will not interfere. Thus, we end with a continuous maser.

As microwave amplifiers, masers can be used as very sensitive detectors in radio astronomy, where exceedingly feeble microwave beams received from outer space will be greatly intensified with great fidelity to the original radiation characteristics. (Reproduction without loss of original characteristics means to reproduce with little "noise." Masers are extraordinarily "noiseless" in this meaning of the word.)

In principle, the master technique could be applied to electromagnetic waves of any wavelength, notably to those of visible light. Townes pointed out the possibility of such applications to light wavelengths in 1958. Such a light-producing maser might be called an *optical maser*. Or this particular process might be called "light amplification by stimulated emission of radiation," and the acronym *laser* might be used. It is the latter that has grown popular.

The first successful laser was constructed in 1960 by the American physicist Theodore Harold Maiman (1927–). He used a bar of synthetic ruby for the purpose—this being, essentially, aluminum oxide with a bit of chromium oxide added. (It is the chromium oxide that lends the synthetic ruby its red color.) If the ruby bar is exposed to light, the electrons of the chromium atoms are pumped to higher levels and after a short while begin to fall back. The first few photons of light emitted (with a wavelength of 694.3 millimicrons) stimulate the production of other such photons, and the bar suddenly emits a beam of deep red light. Before the end of 1960, continuous lasers were prepared.

The laser made possible light in a completely new form. The light was the most intense and the most narrowly monochromatic that had ever been produced, but it was even more than that.

Light produced in any other fashion, from a wood fire to the sun, consists of relatively short wave packets oriented in all conceivable directions. Ordinary light is made up of countless numbers of these packets.

The light produced by a stimulated laser, however, consists of photons of the same size and moving in the same direction. This means that the wave packets are all of the same frequency, and since they are lined up precisely end to end, so to speak, they melt together. This is *coherent light,* because the wave packets seem to stick together. Physicists had learned to prepare coherent radiation for radiation of long wavelength, like radio waves. (It is a coherent radio wave that acts as a carrier wave in radio.) However, it had never been done for light until 1960.

The laser was so designed, moreover, that the natural tendency of the photons to move in the same direction was accentuated. The two ends of the ruby tube were accurately machined and silvered so as to serve as plane mirrors. The emitted photons flashed back and forth along the rod, knocking out more photons at each pass, until they had built up sufficient intensity to burst through the end which was more lightly silvered. Those that did come through were precisely those that had happened to be emitted in a direction exactly parallel to the long axis of the rod, for only those would move back and forth, striking the mirrored ends over and over. If any photon of proper size happened to enter the rod in a different direction (even a very slightly different direction) and started a train of stimulated photons in that different direction, they would quickly pass out the sides of the rod after only a few reflections at most.

A beam of laser light is made up of coherent waves so firmly parallel that it can travel through long distances without widening to uselessness. Laser beams even reached to the moon, in 1962, having spread out to a diameter of only two miles after having crossed nearly a quarter of a million miles of space.

In the short time since their invention, lasers have proliferated in variety. They can be formed not only out of metallic oxides, but out of fluorides and tungstates, out of semiconductors, and out of columns of gas. Light can be produced in any of a variety of wavelengths in the visible and infrared ranges.

The narrowness of the beam of laser light means that a great deal of energy can be focused into an exceedingly small area, and in that area the temperature reaches extreme levels. The laser can vaporize metal for quick spectral investigation and analysis, and it can punch holes of any desired shape through high-melting substances. By shining laser beams into the eye, surgeons have succeeded in welding loosened retinas so rapidly that surrounding tissues have no time to be affected by the heat.

The possible applications of laser beams are exciting and dramatic, and they will probably come quickly. Rather than speculate about them now, it would be more appropriate to wait for later editions of this book and discuss them in actuality.

Matter-Waves

Bohr's application of quantum theory to atoms had thus proven incalculably fruitful, both in theory and application. Not only was the periodic table rationalized but a whole realm of solid-state devices had grown out of it. Physicists had every reason to be delighted with the results.

And yet, taken by itself, the quantized atom did not solve the problems of the chemist. It left him, for a while, with no clear-cut method for explaining the manner in which atoms clung together to form molecules. Where the Lewis-Langmuir atom, with all its faults and deficiencies, had enabled him to depict interlocking cubes and shared-electron pools, the quantized atom, with its electrons hopping nimbly from energy level to energy level, seemed impossible to handle.

An answer grew out of a second seeming source of confusion, that between particle and wave. The physicists of the early twentieth century had become convinced that light, and electromagnetic radiation generally, while wave form, also displayed particle-like properties. The Compton effect (see page II–138) had been the final convincer that wave-particle duality existed and that an entity could demonstrate both wave-like properties and particle-like properties.

But was this confined to electromagnetic radiation only? If entities commonly viewed as wave forms exhibited particle-like properties that can be detected if properly searched for, what of entities commonly viewed as particles? Would they exhibit wave-like properties that would be detected if properly searched for?

The French physicist Louis Victor de Broglie (1892–) considered this last problem. He made use of some of the relationships developed in treating a photon as a particle and applied them to electrons. In 1923, he announced the relationship:

$$\lambda = \frac{h}{mv} \qquad \text{(Equation 6–1)}$$

where h is Planck's constant (see page II–131), m is the mass of a moving particle, and v its velocity (and the product mv is its momentum). As for λ (the Greek letter "lambda"), that is the wavelength associated with its wave-like properties.

This equation will, in theory, apply to any moving body—to a baseball, a cannonball, a planet. However, as momentum increases, wavelength decreases, and for all ordinary bodies, the associated wavelength is far too small to be detected by any known method. Ordinary bodies can therefore be viewed as particles, without any worry about associated wave properties.

When the mass of an object decreases to that of an electron, however, the associated wavelength is significantly large—as large as that of an X ray. (The wave form associated with an electron is not, however, identical with an X ray in nature, though the wavelength may be the same. The wave forms associated with particles of matter are not electromagnetic in nature; these non-electromagnetic waves may be called *matter-waves.*)

A matter-wave with a wavelength equal to that of an X ray ought to have its wave nature as easily detectable as that of X rays. The wave nature of X rays was detected by the diffraction of X rays by the atom lattices of crystals (see page 63). Might not then the matter-waves associated with electrons be demonstrated in equivalent fashion?

This feat was independently carried out in 1927 by the American physicists Clinton Joseph Davisson (1881–1958) and Lester Halbert Germer (1896–) on the one hand, and by the English physicist George Paget Thomson (1892–) on the other. In later years, the wave properties of other, more massive, particles were also detected, and there is no reasonable doubt now that wave-particle duality is a general phenomenon in nature. All entities that display wave properties must also display particle properties, and vice versa.

The analogy between matter-waves and electromagnetic radiation showed up in the matter of microscopy.

There is a limit to the resolution possible when using a wave form such as light. Objects of a size less than about three-fifths the wavelength of the light being used for the purpose cannot be made out, however perfect the optical portions of the microscope. The light "steps over" the small object, so to speak. This means that even when viewing with the shortest wavelengths of visible light, say 380 millimicrons, an object less than 200 millimicrons in diameter cannot be made out. Viruses, which are smaller than that, cannot therefore be seen by visible light, however one attempts to magnify. Indeed, the greatest useful magnification of an optical microscope is about 2000 times.

Successful attempts have been made to use electromagnetic radiation of wavelength smaller than that of visible light, but greater success was achieved with matter-waves. Electrons, with an associated wavelength about equal to that of X rays, can be used for the purpose. Electrons can be focused sharply by magnetic fields, just as light waves can be focused by lenses. A specimen subjected to focused electrons must be quite thin to allow the electrons to pass through; it must also be encased in a good vacuum, otherwise the electrons will be scattered by air. This limits the nature of objects that can be studied by electron microscopy, but not too drastically.

The electrons, having passed through the specimen, form an image on a fluorescent screen or on a photographic plate. Those portions more opaque to electrons absorb and scatter them more efficiently, and for this reason, a meaningful light-dark pattern is produced.

The first *electron microscope* was prepared in Germany in 1931, the German electrical engineer Ernst August Friedrich Ruske (1906–) being prominent in its development. By 1934, electron microscopes were developed that surpassed optical microscopes in magnifying power, and by 1939 these were being produced commercially. The magnifications made possible by modern electron microscopes are a hundred times those within the range of possibility of the best optical microscopes.

Matter-waves entered the realm of atomic theory, too. The Austrian physicist Erwin Schrödinger (1887–1961) tackled the problem of interpreting the structure of atoms in terms of particle waves, rather than of particles alone.

Schrödinger pictured the electron as a wave form circling the nucleus. It seemed to him that the electron could then exist only in orbits of such size that the wave form occupies it in a whole number of wavelengths. When this happens, the wave form repeats itself as it goes round, falling exactly on itself, so to speak. The electron is then a stable *standing wave*.

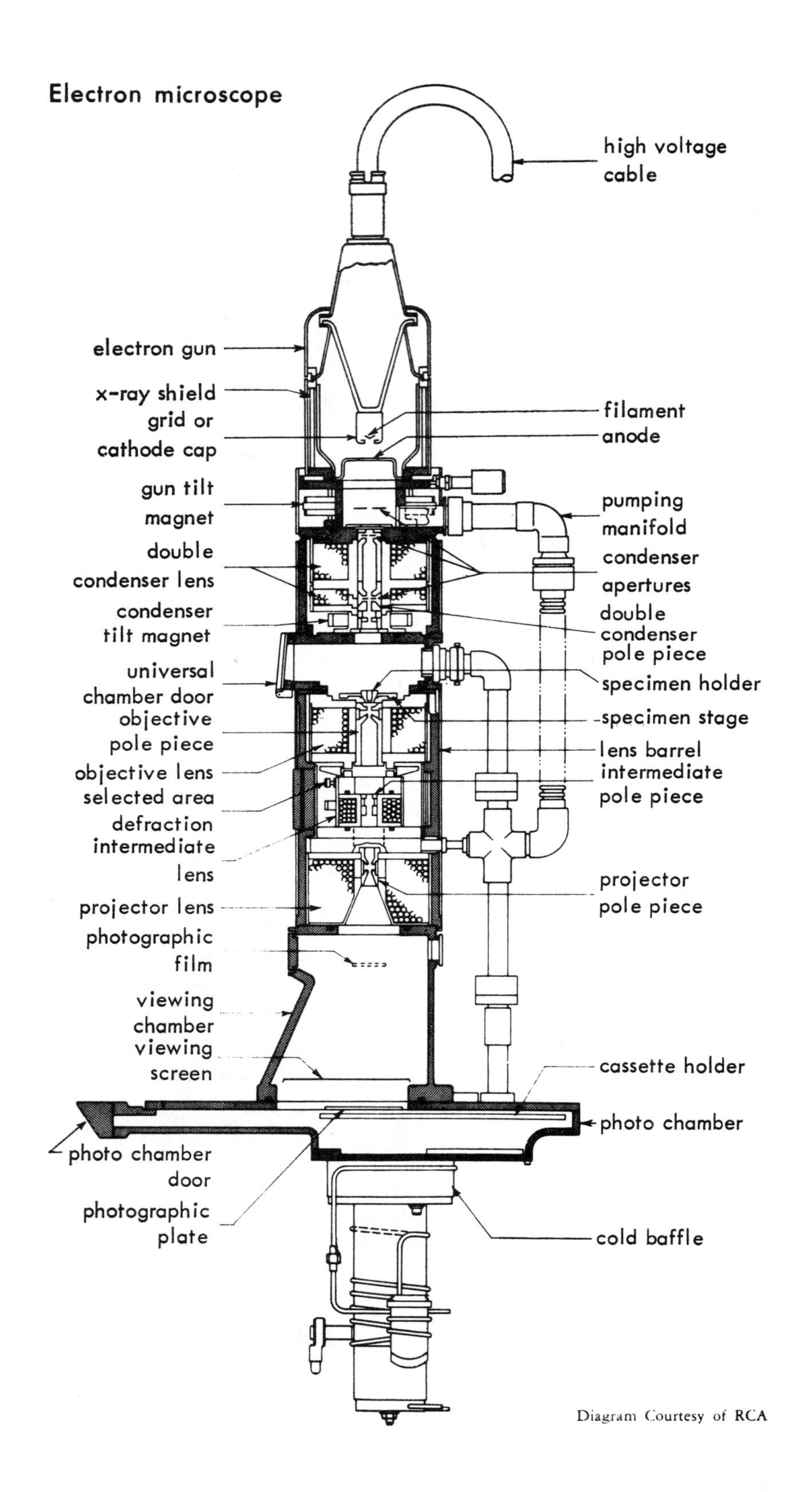
Electron microscope
high voltage cable
electron gun
x-ray shield
grid or cathode cap
filament
anode
gun tilt magnet
pumping manifold
double condenser lens
condenser apertures
condenser tilt magnet
double condenser pole piece
universal chamber door
specimen holder
objective pole piece
specimen stage
objective lens
lens barrel
selected area defraction
intermediate pole piece
intermediate lens
projector lens
projector pole piece
photographic film
viewing chamber
viewing screen
cassette holder
photo chamber
photo chamber door
photographic plate
cold baffle
Diagram Courtesy of RCA

If the electron gains a bit more energy, its wavelength decreases slightly and the orbit no longer contains a whole number of wavelengths. The same is true if the electron loses a bit of energy, so that its wavelength increases somewhat. If it is assumed that the electron cannot possess an amount of energy that will force it to circle a nucleus in a non-integral number of wavelengths, then the electron cannot gain or lose just any amount of energy.

The electron must gain (or lose) just enough energy to decrease (or increase) the wavelength to the point where an integral number of wavelengths can again fit the orbit. Instead of, say, four wavelengths to the orbit, there would be five somewhat shorter wavelengths to the orbit, with a gain of a specific quantity of energy, or three somewhat longer wavelengths, with a loss of a specific quantity of energy. If enough energy is lost and the wavelength increases to the point where a single wavelength fits the orbit, this is ground state and there can be no further loss of energy.

The different energy levels, then, represent different standing waves. Schrödinger analyzed this point of view mathematically in 1926, working out for the purpose what is now called the *Schrödinger wave equation.*

The analysis of the details of atomic behavior on the basis of the Schrödinger model is termed *wave mechanics.* Since the energy can only be absorbed or given off in quanta of given energy content, designed to maintain standing waves, it can also be called *quantum mechanics.*

Quantum mechanics has proved highly satisfactory from the physicist's standpoint. Psychologically, it seemed superior to Heisenberg's matrix mechanics (see page 90), for Schrödinger offered a picture, however hard to grasp, of wave forms, whereas Heisenberg's array of pictureless-numbers lacked something for the image-seeking mind to grasp.

In 1944, the Hungarian-American mathematician John von Neumann (1903–1957) presented a line of argument that seemed to show that quantum mechanics and matrix mechanics were mathematically equivalent—that everything that was demonstrated by one could be equally well demonstrated by the other.*

In principle, it would seem that quantum mechanics offers a complete analysis of the atom and that all facets of chemical be-

* In 1964, however, the English physicist Paul Adrien Maurice Dirac (1902–   ) raised doubts. He suggests the two theories are not mathematically equivalent and that matrix mechanics more accurately fits reality.

havior can be accounted for and predicted by means of it. In actual fact, however, a complete analysis is impractical, even by present-day techniques, because of the sheer difficulty of the mathematics involved. Chemistry is therefore far from being a completely solved science.

Nevertheless, quantum mechanics could be used to explain the manner in which atoms linked together to form molecules. The American chemist Linus Pauling (1901–) showed how two electrons could, in combination, form a more stable wave arrangement than they could separately. The shared-electron pool of the Lewis-Langmuir model of the atom became two wave forms resonating with each other (see page I–176). This *theory of resonance* was expounded fully in Pauling's book *The Nature of the Chemical Bond* published in 1939.

Resonance explains the structure and behavior of molecules far more satisfactorily than the old Lewis-Langmuir model does. It explains just those points that the older model left unexplained —such as the boron hydrides and benzene—and modern chemistry is more and more built about the quantum-mechanical viewpoint.

Another important consequence of the wave nature of the electron (and of particles generally) was pointed out by Heisenberg in 1927. You can see that if a particle is viewed as a wave, it is a rather fuzzier object than it would be if it were viewed as a particle only. Everything in the universe becomes slightly fuzzy, precisely because there is no such thing as a particle without wave-like properties.

A particle (or its center) can be located precisely in space —in principle, at least—but a wave form is somewhat harder to think of as being located at a particular point in space.

Thinking about this, Heisenberg advanced reasons for supposing that it is not possible to determine both the position and momentum of a particle simultaneously and with unlimited accuracy. He pointed out that if an effort is made to determine the position accurately (by any conceivable method, and not merely by those methods which are technically possible at the moment) one automatically alters the velocity of the particle, and therefore its momentum. Therefore, the value of the momentum at the moment at which the position was exactly determined becomes uncertain. Again, if one attempts to determine the momentum accurately, one automatically alters the position, the value of which becomes uncertain. The closer the pinning down of one, the greater the uncertainty in the other.

The concisest expression of this is:

$$(\Delta p)(\Delta x) \approx h \qquad \text{(Equation 6–2)}$$

where Δp represents the uncertainty of position, Δx the uncertainty of momentum, and h is Planck's constant. The symbol $\approx$ signifies "is approximately equal to." This is Heisenberg's *principle of uncertainty*.

Philosophically, this is an upsetting doctrine. Ever since the time of Newton, scientists and many nonscientists had felt that the methods of science, in principle at least, could make measurements that were precise without limits. One needed to take only enough time and trouble, and one could determine the *n*th decimal place. To be told that this was not so, but that there was a permanent wall in the way of total knowledge, a wall built by the inherent nature of the universe itself, was distressing.

Even Einstein found himself reluctant to accept the principle of uncertainty, for it meant that at the subatomic level, the law of cause and effect might not be strictly adhered to. Instead events might take place on the basis of some random effect. After all, an electron might be here or it might be there; if you couldn't tell, you couldn't be sure exactly how strongly a particular force at a particular point might affect it. "I can't believe," said Einstein, "that God would choose to play dice with the world."

Nevertheless, Einstein failed to devise any line of reasoning that would involve the principle of uncertainty in a contradiction. Nor could anyone else, and the principle is now firmly accepted by physicists.

Nor need one be overly downhearted about the loss of certainty. Planck's constant is very small, so for any object that is above the atomic in size, the relative uncertainties of position and momentum are vanishingly small. Only in the subatomic world need the principle be made a part of everyday life, so to speak.

What's more, the existence of uncertainty need not be a source of humiliation for science, either. If a tiny, but crucial, uncertainty is part of the fabric of the universe, it is a tribute to scientists to have discovered the fact. And surely, to know the limits of knowledge is itself an item of knowledge of the first importance.

CHAPTER **36**

Radioactivity

Uranium

So far, the discussion of the internal structure of the atom has been confined to the outer electrons. In a way, it might almost seem that in doing so we were discussing virtually all the atom, for the nucleus has a diameter in the range of from 10^{-13} to 10^{-12} of a centimeter and makes up an insignificant portion of the atom. Indeed, if the atom were visualized as having been expanded to the size of the earth, the nucleus would be a sphere at the center of the planet, about 700 feet in diameter.

Yet the nucleus contains more than 99.9 percent of the mass of the atom, and almost from the first it was recognized (despite its minute size) as having an intricate structure of its own.

The first indication of this dates back to a discovery in 1896 by the French physicist Antoine Henri Becquerel (1852–1908). It was during the first year after Röntgen's discovery of X rays, and Becquerel, like many other physicists, was eagerly investigating the new phenomenon further.

Becquerel's father, himself a famous physicist, had been interested in fluorescent materials: substances which absorbed light of a particular wavelength and then gave off light of a longer wavelength.* Becquerel wondered if among the fluorescent radiation, there might not be X rays.

Becquerel's father had, in particular, worked with the fluorescent compound, potassium uranyl sulfate, $K_2UO_2(SO_4)_2$, the molecule of which, as you can see, contains a uranium atom. Becquerel, finding samples of this compound handy, used it in his experiments. He quickly discovered that after exposure to the sun the fluorescent radiation from the compound would penetrate black paper (opaque to ordinary light) and darken a photographic plate on the other side.

On March 1, 1896, however, he made the startling discovery that the compound would do this even when it had not been exposed to sunlight and when it was not fluorescing. Indeed, the compound constantly and ceaselessly emitted strong and penetrating radiation.

This radiation was not only as penetrating as X rays but, like X rays, possessed the ability to ionize the atmosphere. To demonstrate this, Becquerel made use of a *gold-leaf electroscope*. This device consists of two thin and very light sheets of gold leaf attached to a rod and enclosed in a box designed to protect the gold leaf from disturbing air currents. The rod emerges from the upper end of the box. If an electrically charged object is brought near the rod, the charge enters the gold leaf. Since both sheets of gold

* This is most spectacular when a fluorescent substance absorbs ultraviolet radiation and gives off light in the visible range. It seems, then, to glow eerily, and rather beautifully, in the dark.

Gold-leaf electroscope

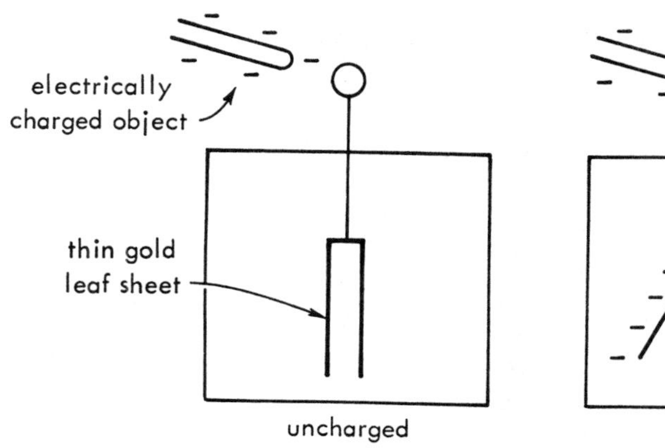

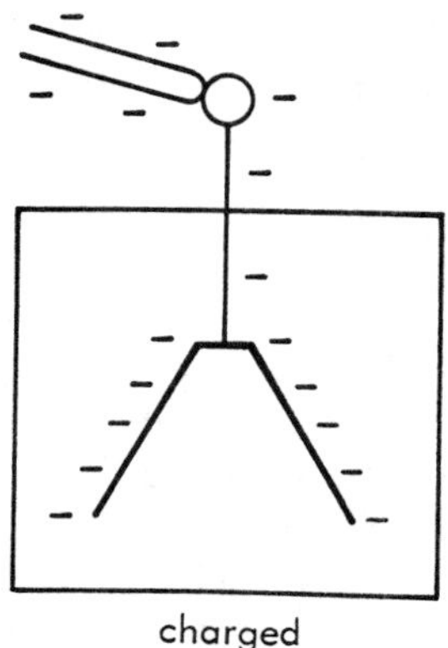

leaf are now similarly charged, they repel each other and stand stiffly apart, like an inverted V.

Left to itself, this situation will persist for an extended period of time. If, however, the air within the box is ionized, the charged particles in the air will gradually neutralize the charge on the gold leaf. The two sheets will then slowly come together as mutual repulsion vanishes. When potassium uranyl sulfate was brought near an electroscope, just this happened, so one could conclude that the compound liberated *ionizing radiation.*

This property of constantly emitting penetrating and ionizing radiation was termed *radioactivity* by the Polish-French physicist Marie Sklodowska Curie (1867–1934) in 1898. Madame Curie went on to show that different uranium compounds were all radioactive and that the intensity of radioactivity was in proportion to the uranium content of the compound. It seemed a fair conclusion that it was the uranium atom itself that was radioactive. Madame Curie was also able to show that the thorium atom was radioactive. (Both elements have particularly complex atoms. Thorium with an atomic number of 90 and uranium with one of 92 were the two most massive atoms known in the 1890's.)

It appeared almost at once that the radiation given off by uranium and by thorium was not homogeneous in its properties. In a magnetic field, part of the radiation was deflected in one direction by a very slight amount; part was deflected in the opposite direction by a considerable amount; and part remained undeflected. Ernest Rutherford (who was later to advance the nuclear model of the atom) gave these three parts of the radiation names taken from the first three letters of the Greek alphabet: *alpha rays, beta rays,* and *gamma rays,* respectively. These radiations also differed in respects other than their response to a magnetic field. Thus, it was the gamma rays which displayed the X ray-like penetrability. Beta rays were much less penetrating, and alpha rays were scarcely penetrating at all.

From the direction and extent of the beta ray deflection, Becquerel recognized that it must contain negatively-charged particles quite like those in cathode rays. He suggested this in 1899, and later investigations corroborated this over and over. Beta rays were shown to be streams of rapidly moving electrons. A speeding electron, emitted by a radioactive substance, is therefore commonly called a *beta particle.*

The gamma rays, undeflected by a magnetic field, were at once suspected of being electromagnetic in nature, with wavelengths even shorter than those of X rays since they were even more penetrating than X rays. This was unmistakably demonstrated in Rutherford's laboratory in 1914, when gamma rays were shown, like X rays, to be diffracted by crystals.

With the advent of the nuclear atom, it came to be realized that these radioactive radiations must originate out of events taking place within the nucleus. For instance, there are no energy level differences among the electrons of atoms which are large enough to produce photons as energetic as those of most gamma rays. Presumably there are *nuclear energy levels* within the nucleus, with differences large enough to produce gamma ray photons.

Yet the division between X rays and gamma rays is not a sharp one. While X rays, as a whole, have the longer wavelengths, some of the more massive atoms can produce X rays that are rather shorter in wavelength than some of the longest wave gamma rays originating from nuclei.

A wavelength of 0.01 millimicrons splits the overlap down the middle, so that as a rough rule of thumb it is possible to consider electromagnetic radiation on the short side of 0.01 millimicrons to be gamma rays and those on the long side to be X rays. The discovery of gamma rays completed the electromagnetic spectrum as we know it today. The stretch of radiation from the shortest gamma rays studied to the longest radio waves covers a range of some sixty octaves.

Alpha Particles

But what of the alpha rays? Their deflection in the direction opposite that of the beta rays showed that they must consist of positively-charged *alpha particles*. The fact that they were only slightly deflected by the same magnetic field that deflected the beta rays considerably made it quite likely that the alpha particles were much more massive than electrons.

This was not an unprecedented situation. Streams of massive particles were encountered a decade before the discovery of radioactivity. In 1886, Goldstein (who named the "cathode rays") had used a perforated cathode in a cathode-ray tube. He found that when an electric potential sent negatively-charged cathode rays streaking out of the cathode toward the anode, another type of radiation passed through the perforations of the cathode and shot off in the other direction. Goldstein called this second radiation "channel rays," because they passed through the channels, or holes, in the cathode.

Since the channel rays move in the direction opposite that of the cathode rays, they must consist of positively-charged particles. In consequence, J. J. Thomson suggested that they be called *positive rays.*

It might be suspected that the positive rays were the positive analog of the cathode ray particles; that here were the equivalents of "anode rays." Actually, this was not so. The German physicist Wilhelm Wien (1864–1928) measured their *e/m* ratio and showed that the low values of that ratio made it quite likely that the positive-ray particles were much more massive than electrons. They were, by and large, as massive as atoms.

Furthermore, the *e/m* ratio of the positive rays varied according to the nature of the substance making up the cathode, or according to the nature of the wisps of gas in the cathode-ray tube. Once Rutherford had developed his nuclear model of the atom, it seemed to make sense to suppose that where cathode rays consisted of electrons knocked out of atoms the positive rays consisted of what remained of atoms after some electrons had been removed. They were, in short, positively-charged atomic nuclei (varying in mass according to the element from which they were derived).*

The positively-charged particle that was found to have the highest *e/m* ratio and, therefore, presumably the lowest mass, was the nucleus of the hydrogen atom. If its charge is taken to be +1, equal to that of the electron but opposite in sign, then its mass had to be 1836 times as great as that of the electron. By 1914, Rutherford had given up hope of finding within the atom a positively-charged particle that was lighter than the hydrogen nucleus, and he suggested that this nucleus might as well be settled upon as the electron's opposite number, despite the difference in mass. (The discovery of the electron's true opposite number had to wait two more decades; see page 222.)

In 1920, Rutherford suggested that the hydrogen nucleus be given the name of *proton* (from a Greek word meaning "first").

* Of late, a dramatic use for such positively-charged particles has been forecast. The atom most easily stripped of at least some of its electrons is cesium, and, moreover, this atom is a comparatively massive one. A stream of cesium ions, accelerated out of a rocket tube by an appropriate electric potential, would by Newton's third law (see page I–34) accelerate the rocket in the opposite direction. The force of a stream of even massive ions is small compared to the vast thrusts of the exhaust of burning fuel, but it can be long-continued. After chemical fuels do the heavy work of getting a rocket through earth's atmosphere and into outer space, an *ion-drive* may then prove the most economical method of slowly building up velocities near that of light. Perhaps only in this way can long space voyages become practical.

This harked back to Prout's hypothesis (see page 24), for what Rutherford was suggesting was that all atomic nuclei were made up, at least to a certain extent, of hydrogen nuclei. Prout's hypothesis was thus reborn in a more sophisticated form. The question of nonintegral atomic weights, which had seemed to destroy the hypothesis in the nineteenth century, was settled in a manner to be discussed later (see page 146).

Now let's return to the alpha particle. In 1906, Rutherford measured its e/m ratio and found that it was equivalent to that of the nucleus of the helium atom. In 1909, he settled this matter by placing radioactive material in a thin-walled tube, which was in turn surrounded by a thick-walled tube. The space between the inner and outer walls was evacuated. The alpha particles could penetrate the thin wall but not the thick one. After entering the space between the walls they picked up electrons and became ordinary atoms; then they could not pass through the thin wall either, but were trapped in the space between. After several days, enough atoms had been collected there to allow of spectroscopic investigation and the atoms proved to be those of helium.

The atomic weight of helium is 4, and the helium nucleus is therefore four times as massive as the hydrogen nucleus. If the e/m ratio of the helium nucleus were like that of the hydrogen nucleus, the helium nucleus would have to have a positive charge four times that of the hydrogen nucleus. However, the e/m ratio of the helium nucleus is only half that of the proton, so that its electric charge is only half the expected amount, or only twice that of the hydrogen nucleus. The alpha particle (as we can fairly term the helium nucleus) has, therefore, a mass of 4 and a charge of $+2$, whereas the proton (or hydrogen nucleus) has a mass of 1 and a charge of $+1$.

It would seem from this that to account for its mass the alpha particle must consist of four protons. Yet it cannot consist of four protons only, for then its charge would be $+4$. There seemed, however, an easy solution to this apparent paradox. Since radioactive substances emitted beta particles (electrons) as well as alpha particles, it seemed quite reasonable to suppose that the nucleus contained electrons as well as protons. The alpha particle, from this point of view, could be made up of four protons and two electrons. The two electrons would add virtually nothing to the mass, which would remain 4, but they would cancel the charge on two of the protons, leaving a net charge of $+2$.

The existence of electrons in the nucleus also seemed satisfactory from another standpoint. The nucleus could not very well consist of protons only, it seemed, for all the protons would be positively charged and there would be a colossally strong repulsion among them when forced into the ultra-narrow confines of an atomic nucleus. The presence of the negatively-charged electrons acted as a kind of "cement" between the protons.

Considerations of this sort gave rise to the *proton-electron model* of the atomic nucleus. Every nucleus, according to this view, was made up of both protons and electrons (except the hydrogen nucleus, which was made up of a single proton requiring, since it was alone, no electron-cement).

The number of protons in each variety of nucleus would be equal to the atomic weight (A),* while the number of electrons was equal to the number required to cancel the charge of enough protons to leave uncanceled only the amount required to account for the atomic number (Z). The number of electrons in a nucleus would, therefore, be equal to $A - Z$. Those protons remaining uncanceled in the nucleus would have their charge canceled by the electrons outside the nucleus, so that in the neutral atom there would be Z "extra-nuclear electrons."

Thus, to give a few examples, the nucleus of the carbon atom, which has an atomic weight of 12 and an atomic number of 6, must be made up of twelve protons and 12 — 6, or six electrons. The nucleus of the arsenic atom with an atomic weight of 75 and an atomic number of 33, must be made up of seventy-five protons and 75 — 33, or forty-two electrons. The nucleus of the uranium atom with an atomic weight of 238 and an atomic number of 92 must be made up of 238 protons and 238 — 92, or 146 electrons. Even the nucleus of the hydrogen atom fits this view, for with an atomic weight of 1 and an atomic number of 1, it must be made up of one proton and 1 — 1, or zero electrons.

Unfortunately, the proton-electron model of the atomic nucleus met with difficulties. For example, there is the question of *nuclear spin*. Each particle in the nucleus contributes its own spin to the overall nuclear spin. The spin of each proton and electron is either $+1/2$ or $-1/2$, and the sum of a number of such values can be either a whole number (positive, negative, or zero) or a half-number such as 1/2, 3/2, 5/2, etc. (either positive or negative).

The nitrogen nucleus, with an atomic weight of 14 and an atomic number of 7, ought, by the proton-electron model, have

* Strictly speaking, this applies only to those elements whose atomic weights are approximately whole numbers. The situation with respect to the other elements will be considered later.

fourteen protons and seven electrons for a total of 21 particles in the nucleus. Therefore, if the spins of 21 particles (each $+1/2$ or $-1/2$) are totaled, regardless of the distribution of negative or positive spins, the sum must be a half-number. Nevertheless, measurements convinced physicists that the spin of the nitrogen nucleus was the equivalent of a whole number. This gave them good reason to suppose that the nitrogen nucleus could not be made up of twenty-one protons and electrons, or indeed of any odd number of protons and electrons. Yet no even number of protons and electrons could produce an atomic weight of 14 and an atomic number of 7.

It seemed more and more necessary, as additional data on nuclear spins were accumulated, for the proton-electron model to be scrapped altogether.

Particle Detection

Yet what alternative was there? One possibility was that the electron within the nucleus ought not be counted as a separate particle. Perhaps, in the close confines of the tiny nucleus, the electron melted together with a proton to form a single "fused particle," with a mass of about that of a proton (since the electron contributes very little mass) and with an electric charge of 0 (since the electron's charge of -1 cancels the proton's charge of $+1$). If this were so then a nitrogen nucleus would contain seven protons plus seven "fused particles" for 14 particles altogether—an even number.

Speculations concerning the possible existence of uncharged particles with the mass of a proton began as early as 1920. For over a decade, however, no signs of such a particle could be found. This did not necessarily mean that it did not exist, for physicists expected an uncharged particle to be elusive.

The usual methods for detecting a subatomic particle took advantage of its ability to ionize atoms and molecules. It was by their ionizing abilities, for instance, that radioactive radiations were detected by the electroscope.

Two devices, in particular, used for the purpose of detecting subatomic particles, grew famous in the early days of research into radioactivity. The prototype of the first of these was constructed in 1913 by the German physicist Hans Geiger (1882–1945), who had worked with Rutherford in the experiments that led to the working out of the nuclear model of the atom. It was greatly improved by Geiger in 1928 in collaboration with the German physicist S. Müller, and is therefore commonly known as the *Geiger-Müller counter*, or the *G-M counter*.

The G-M counter consists, essentially, of a cylindrical glass tube lined with metal, with a thin metal wire running down the center of the tube. Filling the tube is a gas such as argon. The tube is placed under an electric potential, with the central wire the anode and the metal cylinder as cathode. The potential is not quite great enough to cause a spark discharge across the argon.

If a charged subatomic particle comes flying into the G-M counter, it will strike an argon atom and knock one or more electrons loose. The electrons so formed will then speed toward the anode under the lash of the electric potential and will in turn ionize other argon atoms, producing more electrons which will in turn ionize still other argon atoms, and so on. In short, the first particle starts a process that in a small instant of time produces so many ions that the argon becomes capable of conducting a current, and there is then an electric discharge within the tube that momentarily reduces its potential to zero.

The discharge, or pulse, of electric current can be converted into a clicking sound that marks the passage of a single subatomic particle. From the number of clicks per second one can estimate, by ear, how much ionizing radiation is present. (Hence G-M counters are used in uranium prospecting.) The pulses can also be counted accurately by automatic devices.

To do more than merely count subatomic particles, one can use a device invented in 1911 by the Scottish physicist Charles Thomas Rees Wilson (1869–1959). He was primarily interested in cloud formation to begin with and had come to the conclusion that water droplets in clouds were formed about dust particles and could also form about ions. If air were completely free of dust or ions, clouds would not form and the air would become *supersaturated*—that is, it would retain water vapor in quantities greater than it could ordinarily hold.

Wilson placed dust-free air, saturated with water vapor, into a chamber fitted with a piston. If the piston were pulled outward, the air would expand and its temperature would drop. Cold air cannot hold as much water vapor as warm air can, and ordinarily some of the water vapor would have to condense out in drops of liquid as the temperature dropped. In the absence of dust or ions, this could not happen and the cooled air became supersaturated.

If, now, a subatomic particle entered the chamber while the air was supersaturated, it would form ions along its path of travel, and small droplets of water would form about these ions. These droplets would mark out the route taken by the subatomic particle.

The tracks formed in such a *Wilson cloud chamber* are packed with information. Different types of particles can be identified. A massive alpha particle, for instance, forms many ions in its path and continues in a straight line, for it is too massive to be diverted by electrons. It is diverted only when it approaches a nucleus, and then the diversion is likely to be a sharp one. The nucleus, stripped of some electrons, recoils and becomes an ionizing particle itself. The track of an alpha particle is therefore thick and straight, and it usually ends with a fork. From the length of the track one can estimate the original energy of the alpha particle.

A beta particle, which is much lighter, changes its direction of motion more easily and forms fewer ions than an alpha particle does. It leaves a thinner and more wavering track. Gamma rays and X rays knock electrons out of atoms, and these act as ionizing particles that mark out short tracks to either side of the path of the gamma ray or X ray. Such radiation therefore leaves faint fuzzy tracks.

If a cloud chamber is placed between the poles of a magnet, charged particles travel in curved paths, and the water droplets indicate that. From the direction of curve, one can determine whether the charge is negative or positive; and from the sharpness of the curve, one can make deductions as to the e/m ratio.

For a speeding particle to form ions, however, the presence of an electric charge is essential. A positively-charged particle attracts electrons out of the atoms it passes through, and a negatively-charged particle repels them out of the atom. An uncharged

Wilson cloud chamber

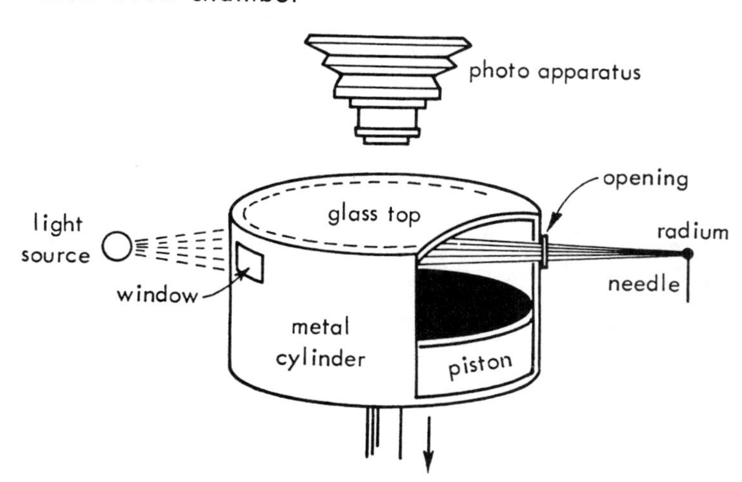

particle would neither attract nor repel electrons and would not form ions. Such an uncharged particle would, therefore, go unmarked by either a G-M counter or a Wilson cloud chamber (or, for that matter, by the more sophisticated devices that have been invented since). If an uncharged particle exists at all, therefore, it would have to be detected indirectly.

It was this which stood in the way of the easy detection of a neutral particle and delayed for over a decade the development of a nuclear model more satisfactory than the proton-electron model.

The Neutron

Beginning in 1930, evidence was obtained to the effect that when beryllium was exposed to alpha rays a radiation was emitted that differed from those that were already known. It was very penetrating, and it was not affected by a magnetic field, which seemed to give it the hallmark of gamma rays. However, the radiation was not gamma ray in nature either, for unlike gamma rays, it was not an ionizing radiation and could not, for instance, be detected by an electroscope.

Indeed, the radiation was not detected directly at all. When a substance such as paraffin was put in the way of the radiation, however, protons were hurled out of the paraffin and these protons gave it away.

In 1932, the English physicist James Chadwick (1891–) explained the phenomenon satisfactorily. Energetic electromagnetic radiation might push electrons out of the way, but not the more massive protons. For a proton to be banged about so cavalierly, another particle had to be involved, and a particle in the mass-range of the proton. Since this particle did not ionize the atmosphere, it had to be uncharged. In short, here was the massive uncharged particle that physicists had been seeking for a decade, and because it was electrically neutral, it was named the *neutron*.

Almost as soon as the neutron was discovered, Heisenberg suggested a *proton-neutron model* of the atomic nucleus. According to this model, the nucleus would be made up of protons and neutrons only. The neutron is just about equal to the proton in mass, so the total number of protons (p) and neutrons (n) would equal the atomic weight (A). On the other hand, only the protons would contribute to the positive charge of the nucleus, so the total number of protons in the nucleus would be equal to the atomic number (Z). In short:

$$p + n = A \qquad \text{(Equation 7–1)}$$

$$p = Z \qquad \text{(Equation 7–2)}$$

The number of neutrons, then, can be obtained by subtracting Equation 7–2 from Equation 7–1:

$$n = A - Z \qquad \text{(Equation 7–3)}$$

Using this new view, it is easy to specify the structure of the nuclei of those atoms that have atomic weights that are approximately whole numbers.

The nucleus of the hydrogen atom ($A = 1$, $Z = 1$) is made up of one proton and zero neutrons; that of the helium atom ($A = 4$, $Z = 2$) is two protons and two neutrons; that of the arsenic atom ($A = 75$, $Z = 33$) is thirty-three protons and forty-two neutrons; and that of the uranium atom ($A = 238$, $Z = 92$) is ninety-two protons and one hundred and forty-six neutrons.

This proton-neutron model of the nucleus quickly showed itself satisfactory in those respects in which the proton-electron model failed. The nitrogen nucleus, for instance ($A = 14$, $Z = 7$) is made up of seven protons and seven neutrons for a total of 14 particles, altogether. The neutron, as well as the proton, has a spin of $+1/2$ or $-1/2$, and 14 such particles (or any even number) must have a net spin represented by a whole number, in agreement with observation.

The proton-neutron model is still accepted at the present writing, and the two particles are lumped together as *nucleons* because of their characteristic appearance in atomic nuclei.

The model does, of course, raise questions. One of these is this: If the nucleus contains protons and neutrons only, and does not contain electrons, then where do the electrons come from that make up the beta rays emitted by radioactive substances? It was, after all, the existence and nature of these beta rays that led to the belief in nuclear electrons in the first place.

The answer to this arises out of the nature of the neutron which, in one particular respect, differs crucially from the proton and the electron. Both the proton and the electron are examples of a *stable particle*. That is, if either a proton or an electron were alone in the universe it would persist, unchanged, indefinitely (at least as far as we know). Not so with the neutron, which is an *unstable particle.*

An isolated neutron will eventually cease to exist, and in its place will be two particles, a proton and an electron. (This is not a complete description of the breakdown, but it will do for now.

The subject will be explored further on page 236.) We can write this change symbolically, using superscripts to indicate charge, as follows:

$$n^{\circ} \longrightarrow p^{+} + e^{-} \qquad \text{(Equation 7–4)}$$

An important point demonstrated by this equation is that electric charge is not created. All experience involving the behavior of subatomic particles indicates that the neutron cannot merely change into a proton, for there would be no way in which an uncharged particle could develop a positive charge (or, for that matter, a negative charge) out of nothing. By forming both a proton and an electron, the net charge of the product remains zero.

This is an example of the *law of conservation of electric charge,* which states that the net charge of a closed system cannot be altered by changes taking place within the system. This was first recognized in the study of electrical phenomena (see page II–161), long before the existence of subatomic particles was suspected.

A neutron existing, not free, but within an atomic nucleus, is often stabilized for reasons to be discussed later (see page 243). Thus, the nucleus of a nitrogen atom is stable despite its neutron content and, left to itself, will continue to be made up of seven protons and seven neutrons indefinitely.*

On the other hand, there are some nuclei within which neutrons retain a certain degree of instability. In such cases, the neutron within the nucleus will, at some point, change into a proton and an electron. The proton is at home in the nucleus and remains there, but the electron comes flying out as a beta particle. Thus, although the beta particle does emerge from the nucleus, this is no indication, after all, that it was a constituent of the nucleus; rather, it was created at the moment of its emergence.

New Radioactive Elements

If, in the course of the emission of a beta particle, a neutron within the nucleus of an atom is converted into a proton, it is clear that the proton-neutron makeup of the nucleus changes and that the nature of the atom itself is altered. Since the number of

* It might occur to you to wonder what holds the seven positively-charged protons together against the powerful electric repulsion of like charges, in the absence of an "electron-cement." This problem will be considered later in the book (see page 243).

protons is increased by one, so is the atomic number, and the atom within which the change has taken place is transformed from one element into another.

In fact, radioactivity is almost invariably a sign of a fundamental change in the nature of the atom displaying the phenomenon. This came to be realized shortly after the discovery of radioactivity, and well before the internal structure of the nucleus was worked out.

As early as 1900, Crookes, one of the cathode-ray pioneers, discovered that when a uranium compound was thoroughly purified it showed virtually no radioactivity. It was his suggestion, therefore, that it was not uranium that was radioactive, but some impurity in the uranium.

However, the next year Becquerel confirmed Crookes' findings but went on to show that as the purified uranium compound remained standing, its radioactivity gradually grew more intense, until it was at the normal level associated with uranium. In 1902, Rutherford and his co-worker, the English physicist Frederick Soddy (1877–1956), showed this to be also true of thorium compounds.

It seemed reasonable to conclude that if the radioactivity was that of an impurity, it was an impurity that was gradually formed from uranium. In other words, the radioactivity of uranium was a symptom of the change of uranium atoms into some other form of atom. This new atom was itself radioactive and changed into a third atom which was also radioactive, and so on. In short, as Rutherford and Soddy pointed out, one ought not speak of a radioactive element, but of a *radioactive series* of elements.

The radioactivity detected in uranium and thorium might not then be so much characteristic of uranium and thorium itself (which might be, and indeed proved to be, only very mildly radioactive) as of the various "daughter elements." The latter were much more strongly radioactive and were always present in the uranium and thorium—except immediately after those elements had been rigorously purified.

The "daughter elements," if slowly formed and rapidly broken down, ought to be present in uranium and thorium minerals only in vanishingly small quantities. Even so, while they would remain immune to discovery by ordinary chemical methods, they could be detected and traced by the radiations they gave off, since these could be detected with great sensitivity and since it was only to be expected that each different element would give off radiations of characteristic type and intensity.

This feat was successfully carried through by Madame Curie, in collaboration with her husband, the French physicist Pierre Curie (1859–1906). In 1898, the Curies began with large quantities of uranium ore and divided it, by standard chemical techniques, into fractions of different properties. They followed the track of intense radioactivity, keeping those fractions that displayed it and discarding those that did not. Before the end of the year, they had discovered two hitherto unknown elements, the first of which they named *polonium* after Madame Curie's native land, and the second, *radium,* after the element's intense radioactivity.

Both elements were, in fact, far more radioactive than either uranium or thorium. In fact the rapidity with which polonium and radium broke down and ceased being polonium and radium was such that no detectable quantity could possibly have survived the five-billion-year history of the earth, even if large quantities had existed in the planet's structure when it was formed. The existence of these elements today was due entirely to their constant formation from uranium and thorium. The latter elements broke down so slowly that a sizable fraction of the original supply remains in existence today, despite a steady diminution over the last five billion years.

How many such short-lived elements might exist as daughter products of uranium and thorium? In the time of the Curies this was uncertain, for there was no telling how much room there might remain in the periodic table. Once Moseley had worked out the concept of atomic numbers, in 1913, the subject grew less mysterious.

As of 1913, all elements with atomic numbers up to and including 83 (bismuth) were nonradioactive. It was thoroughly expected that the as-yet-undiscovered elements in this range (43, 61, 72, and 75) would also be nonradioactive. And, to be sure, when hafnium (72) was discovered in 1923 and rhenium (75) in 1925, both turned out to be nonradioactive. Attention, then, was focused on elements of atomic number higher than 83.

Thorium (atomic number 90) and uranium (atomic number 92) were the first radioactive elements discovered. Those discovered by the Curies also fit into this region, for polonium had an atomic number of 84 and radium one of 88.

Other discoveries followed. In 1899, the French chemist André Louis Debierne (1874–1949) discovered *actinium* (atomic number 89), and in 1900, the German chemist Friedrich Ernst Dorn (1848–1916) discovered *radon* (atomic number 86). In

1917, the German chemist Otto Hahn (1879–   ) and his co-worker, the Austrian physicist Lise Meitner (1878–   ), discovered protactinium (atomic number 91).

By that time (and for a quarter-century afterward) only two gaps remained in that region of the periodic table, gaps corresponding to atomic numbers 85 and 87. Chemists confidently expected the elements with those atomic numbers to prove radioactive when discovered (and this turned out to be so).

And yet, as we shall see in the next chapter, this listing of elements, which seems to fit the periodic table so neatly, actually involved chemists in a problem that began by seeming to shake the very concept of a periodic table and ended by establishing it more firmly and more fruitfully than ever.

CHAPTER 37

Isotopes

Atomic Transformations

The discovery of new elements in radioactive minerals in 1898 and immediately thereafter was, in a way, too successful for comfort. The periodic table had room for exactly nine radioactive elements with atomic numbers of from 84 to 92 inclusive. Room for new elements such as radium and polonium could be found, but how many more were there? If one judged by the number of distinct and characteristic types and the intensities of radiation among the daughter elements of uranium and thorium, then physicists seemed to have discovered dozens of different elements.

Names were applied to each distinct type of radiation: There were, for instance, uranium X_1, uranium X_2, radium A, radium B, and so on through radium G. There was also a list of thoriums from A to D, two mesothoriums, a radiothorium, and so on. But if each type of radiation did indeed belong to a different element, where could one place them all? Once Moseley had worked out the atomic number structure of the periodic table, the problem had grown crucial.

To answer this problem, let's consider the nature of the radioactive radiations and the manner in which they must affect the atom giving them off. (I will make use of the proton-neutron model of the atomic nucleus, though the analysis I will describe was worked out originally on the basis of the proton-electron model.)

Let's begin with an element, Q, and suppose that its nucleus is made up of x protons and y neutrons. Its atomic number, then, is x and its atomic weight is $x + y$. Placing the atomic number as a subscript before the symbol of the element and the atomic weight as a superscript after it, we can write the element as ${}_{x}Q^{x+y}$.

Next let us suppose that an atom of this element gives off an alpha particle (symbolized by the Greek letter α, which is "alpha"). The alpha particle is made up of two protons and two neutrons and therefore has an atomic number of 2 and an atomic weight of 4. It can be written: ${}_{2}\alpha^{4}$.

What is left of the original atom after the departure of an alpha particle must contain $x - 2$ protons and $y - 2$ neutrons. The atomic number is decreased by 2 (producing a new element, *R*), and the atomic weight is decreased by 4. We can write this:

$${}_{x}Q^{x+y} \longrightarrow {}_{x-2}R^{x+y-4} + {}_{2}\alpha^{4} \qquad \text{(Equation 8–1)}$$

If the original atom had given off a beta particle (symbolized as β, the Greek letter "beta") instead, the situation would be different. The emission of a beta particle means that within the nucleus a neutron had been converted into a proton. The nucleus would therefore contain $x + 1$ protons and $y - 1$ neutrons. The atomic number would be increased by one, but the atomic weight would remain unchanged for $x+1+y-1=x+y$.

The beta particle itself can be considered as having an atomic weight of about 0. (Actually, it is 0.00054, which is close enough to zero for our purposes.) Since atomic number is equivalent to the number of units of positive charges present and since the beta particle is an electron with a unit negative charge, we can consider its atomic number as equal to -1. The beta particle can therefore be written as ${}_{-1}\beta^{0}$ and beta particle emission can be represented as:

$${}_{x}Q^{x+y} \longrightarrow {}_{x+1}R^{x+y} + {}_{-1}\beta^{0} \qquad \text{(Equation 8–2)}$$

Notice that in both Equation 8–1 and Equation 8–2, the sum of the atomic numbers on the right-hand side of the equation is equal to that on the left-hand side. This is in accordance with the law of conservation of electric charge. The same is true for atomic weights in accordance with the law of conservation of mass. (The minor deviations involving the conversion of some mass into energy need not concern us just yet.)

A gamma ray can be symbolized as γ, the Greek letter

"gamma." Since it is electromagnetic radiation, it has neither atomic weight nor atomic number and can be written $_0\gamma^0$. We can, therefore, write the following equation:

$$_xQ^{x+y} \longrightarrow {}_xQ^{x+y} + {}_0\gamma^0 \qquad \text{(Equation 8–3)}$$

In short, then, when an atom emits an alpha particle, its atomic number decreases by two and its atomic weight decreases by four. When it emits a beta particle, its atomic number increases by one and its atomic weight is unchanged. When it emits a gamma ray, its atomic number and atomic weight are both unchanged. This is the *group displacement law,* first proposed in its complete form by Soddy in 1913.

Let us apply the group displacement law to the specific case of the uranium atom with an atomic number of 92 and an atomic weight of 238—that is, $_{92}U^{238}$. The feeble radioactivity of highly purified uranium consists of alpha particles. An alpha particle emission reduces the atomic number of uranium to 90, which is that of thorium, and reduces its atomic weight to 234. We can write:

$$_{92}U^{238} \longrightarrow {}_{90}Th^{234} + {}_2\alpha^4 \qquad \text{(Equation 8–4)}$$

The thorium atom that has arisen as a result of this breakdown of the uranium atom is not quite like the thorium atom that occurs in quantity in thorium ores. The latter possesses an atomic number of 90, to be sure, but it has an atomic weight of 232. It is $_{90}Th^{232}$.

Both types of thorium atoms possess the atomic number of 90 and fit into the same place in the periodic table. Soddy pointed this out in 1913 and suggested that atoms differing in atomic weight but not in atomic number, as in the case of $_{90}Th^{234}$ and $_{90}Th^{232}$, be referred to as *isotopes,* from Greek words meaning "same place," because they occupy the same place in the periodic table.

Since such isotopes always share the same atomic number and differ in atomic weight only, chemists concentrate on the latter and usually leave out the subscript in writing the symbol of the isotope. They will write the two thorium isotopes as Th^{234} and Th^{232} or, less compactly, as thorium-234 and thorium-232.

From the chemist's point of view the placing of different isotopes in the same place in the periodic table is justified. Both thorium-234 and thorium-232 have 90 protons in the nucleus and therefore 90 electrons outside the nucleus in the neutral atom. The chemical properties depend on the number and distribution of the electrons, and therefore thorium-234 and thorium-232 have virtually identical chemical properties. This reasoning holds for other sets of isotopes as well.*

But if different isotopes have identical complements of outer electrons, they nevertheless have nuclei of different structures. Since the proton number in the nuclei of different isotopes of an element is fixed, the difference must rest in the neutron number. The thorium-234 atom, for instance, has a nucleus made up of 90 protons and 144 neutrons, while the thorium-232 atom has one made up of 90 protons and 142 neutrons.

In the case of changes involving the atomic nucleus, such as those that mark the phenomenon of radioactivity (as contrasted with chemical changes that involve only the electrons and not the nucleus), the difference in neutron number is important.

Thus, thorium-232 breaks down with exceeding slowness, which is exactly why it is still present in the crust. It emits an alpha particle, so that its atomic number is reduced to 88, which is that of radium. We can write:

* There are some minor differences in chemical property, particularly among the lighter atoms, because one isotope is more massive than another and therefore somewhat more sluggish in taking part in chemical reactions. These differences are small, however, and in ordinary chemical work can be ignored.

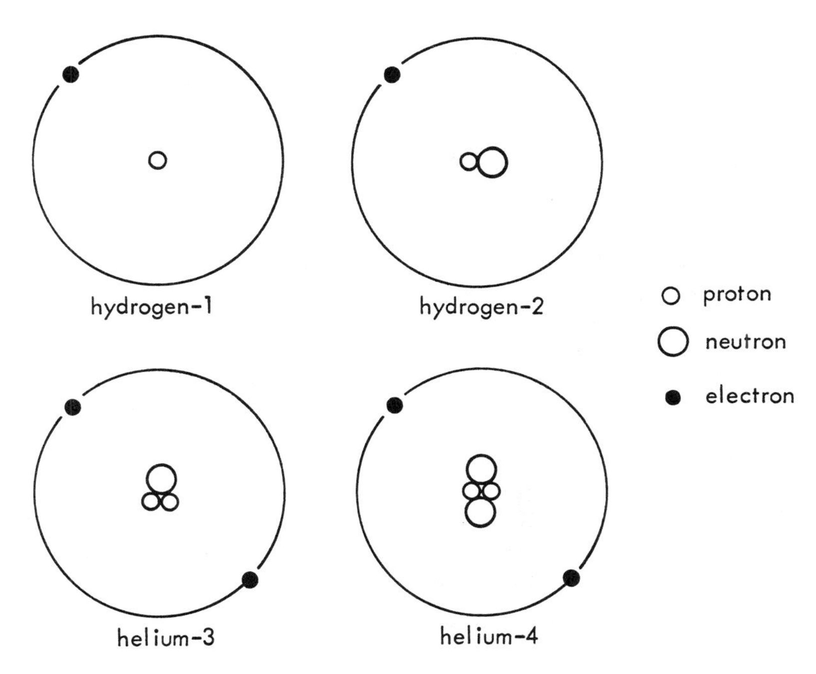

$${}_{90}Th^{232} \longrightarrow {}_{88}Ra^{228} + {}_{2}\alpha^{4} \qquad \text{(Equation 8–5)}$$

Atoms of thorium-234, on the other hand, break down with exceeding rapidity, which is why this isotope does not occur in nature except in vanishingly small quantities in uranium ores. Furthermore, it breaks down with the emission of a beta particle. This raises its atomic number to 91, that of protactinium:

$${}_{90}Th^{234} \longrightarrow {}_{91}Pa^{234} + {}_{-1}\beta^{0} \qquad \text{(Equation 8–6)}$$

Where either an alpha particle or a beta particle is emitted, the new atom may be formed at a nuclear energy level above the ground state. In dropping to the ground state thereafter, a gamma ray is emitted. In some cases this takes an appreciable time and the excited nucleus has a lifetime of its own and different radiation characteristics. To indicate a nucleus in the excited state, an asterisk is added to the symbol. When protactinium-234 is formed, it is in the excited state and:

$${}_{91}Pa^{234*} \longrightarrow {}_{91}Pa^{324} + {}_{0}\gamma^{0} \qquad \text{(Equation 8–7)}$$

Atoms that are identical in atomic weight and atomic number but differ in nuclear energy level are called *nuclear isomers,* a name suggested by Lise Meitner in 1936. The first evidence for nuclear isomerism had been obtained by her long-time partner, Otto Hahn, in connection with protactinium-234, back in 1921.

Radioactive Series

Once the group displacement law was worked out, the trivial names given to the different atoms formed from uranium and thorium could be abandoned. They are still to be found in physics books because of their historical interest, but they will not be used here. The proper isotope names will be used instead. When this is done it turns out that despite the dozens of isotopes formed in the course of the radioactive breakdown of uranium and thorium, all can be made to fit into one or another of the places in the atomic table.

You can see this to be true, for instance, of the different atoms formed from uranium-238, the so-called *uranium series,* which are listed in Table V.

If we consider this series in detail, a number of points arise. First, lead-206 is a stable isotope that does not undergo radioactive breakdown. The series therefore ends there. Nevertheless, there are also included in the series such lead isotopes as lead-214 and lead-210, which are radioactive. Here is a clear indication that isotopes are a phenomenon that are not confined to radioactive atoms alone, but that a particular element may have both stable and radioactive isotopes.

If we leave lead-206 out of consideration and take into account just the radioactive members of the series, only uranium-238 breaks down with exceeding slowness. All the rest break down with comparative rapidity. Consequently, only uranium-238 can endure over the full stretch of the earth's existence. It is the "parent" of the series, and none of the daughter atoms would exist on earth today if uranium-238 did not.

TABLE V—*The Uranium Series*

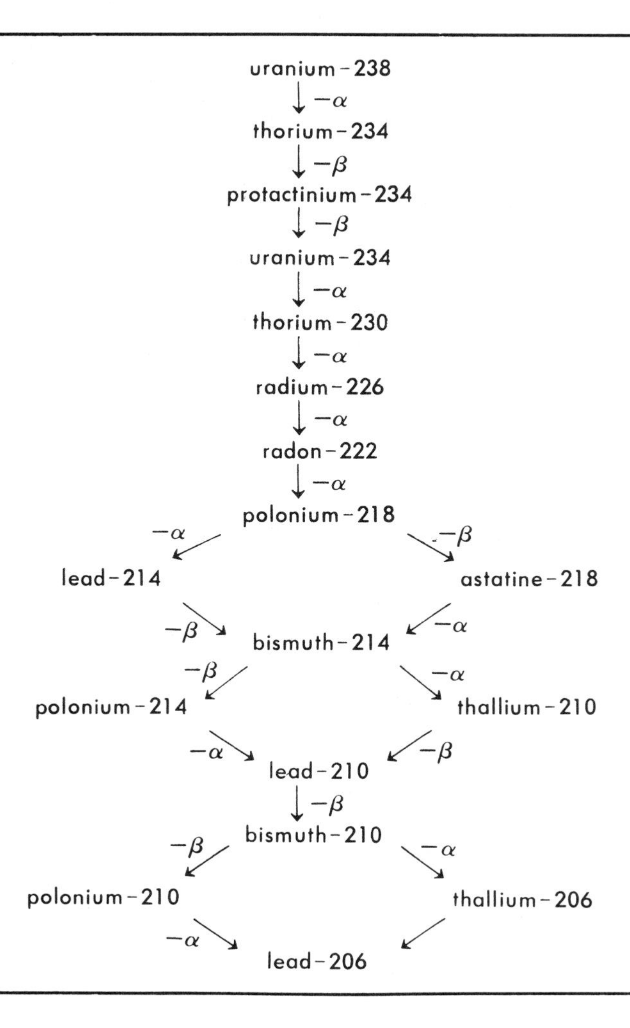

A particular radioactive atom need not always have only one mode of breakdown. Polonium-218, for instance, may give off an alpha particle to form lead-214, or it may give off a beta particle to form astatine-218. This is an example of *branched disintegration.* Often, in these cases, one of the branches is the overwhelming favorite. For instance out of every 10,000 atoms of polonium-218, only two break down to astatine-218, all the rest breaking down to lead-214. (In this case, it is the alpha particle emission that is the favored alternative, in other cases the beta particle emission is favored.)

Astatine (atomic number 85), when formed at all in radioactive breakdowns, is usually formed at the very short end of a branched disintegration. That is why it exists naturally in almost unimaginably small traces and why it evaded discovery so long. The same is true of francium (atomic number 87), which is not formed at all in this particular series.

In a radioactive series, the atomic weight of any atom faces one of two fates. Either its value does not change at all, when a beta particle or a gamma ray is emitted, or it decreases by 4 units,

TABLE VI—*The Thorium Series*

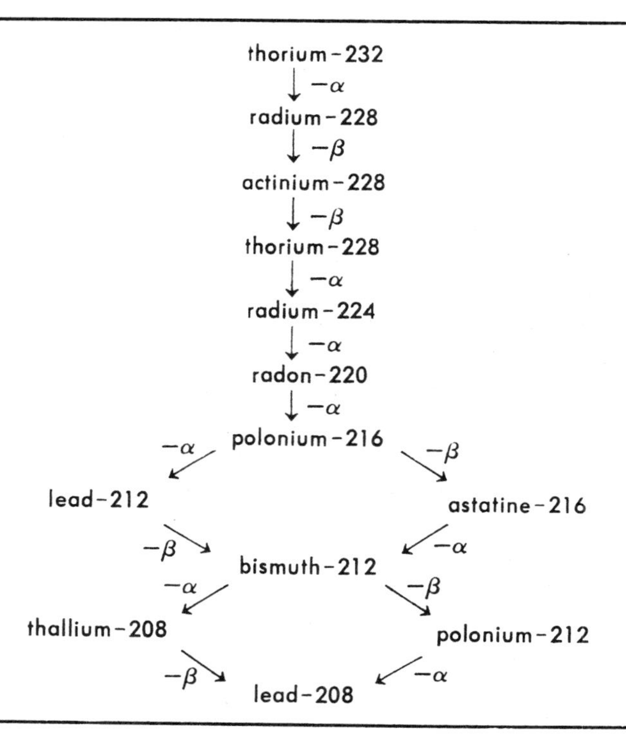

when an alpha particle is emitted. That means that the difference in atomic weights between any two members of the series must be 0 or else a multiple of 4.

The atomic weight of uranium-238 is 238 and this, when divided by 4, gives a quotient of 59 with a remainder of 2. For any number that differs from 238 by a multiple of 4, division by 4 will yield a different quotient but will always leave a remainder of 2. The value of the atomic weight of every member of the uranium series therefore has the form of $4x + 2$, where x can vary from 59 for uranium-238 down to 51 for lead-206. For this reason, one can call the uranium series the *$4x + 2$ series.*

Thorium, the second element to be discovered to be radioactive, is also the parent of a group of daughter atoms, the *thorium series* (see Table VI).

Here, too, the atomic weight of all the atoms in the series differs by multiples of 4. Since thorium-232 has an atomic weight, 232, which is evenly divisible by 4, all the other atomic weights in the series must be evenly divisible by 4, and the series may be referred to as the *$4x + 0$ series.*

It might be thought that since uranium and thorium are the only two radioactive elements that occur in appreciable quantities in the soil, there would be only two radioactive series. However, atoms appeared in radioactive minerals with atomic weights that were neither of the form $4x + 0$ nor $4x + 2$ and which therefore could belong to neither the thorium series nor the uranium series.

At first, it was felt that these formed part of a series originating with actinium-227, an isotope with a $4x + 3$ type of atomic weight, so it was named the *actinium series.* The name persists even though this supposition was sent tumbling by the discovery that actinium-227 broke down far too quickly to permit its existence through the eons of earth's history, so that it could not possibly serve as the parent atom of a series.

When protactinium was discovered, it was found that protactinium-231 (to use present terminology) broke down to form actinium-227, and that, indeed, was the reason the new element received its name (which means "before actinium"). Protactinium-231, however, is also too short-lived to qualify as parent of a series.

In 1935, the Canadian-American physicist Arthur Jeffrey Dempster (1886–1950) discovered that not all uranium atoms were uranium-238. Out of every thousand atoms of uranium isolated from the natural ore, seven were *uranium-235.* These atoms, possessing 92 protons and 143 neutrons in their nuclei, broke down very slowly (though not quite as slowly as uranium-238) and qualified as the parent atom of the actinium series. (For this reason, uranium-235 is sometimes called "actinouranium.") The atomic weights of all the atoms making up the actinium series (see Table VII) are of the $4x + 3$ variety.

It is plain that still a fourth radioactive series should exist, one in which all the atomic weights are of the form $4x + 1$. Isotopes of this form, such as uranium-233, cannot belong to any of the three series already described. No such series was discovered in the 1920's and 1930's, and physicists decided (correctly, as it turned out) that no isotope with an atomic weight of this form was long-lived enough to serve as a parent for such a series in nature.

Each of the three radioactive series ends with a lead isotope with an atomic weight of appropriate form. The uranium series ends with lead-206 ($4x + 2$), the thorium series ends with lead-208 ($4x + 0$), and the actinium series ends with lead-207 ($4x +$

TABLE VII—The Actinium Series

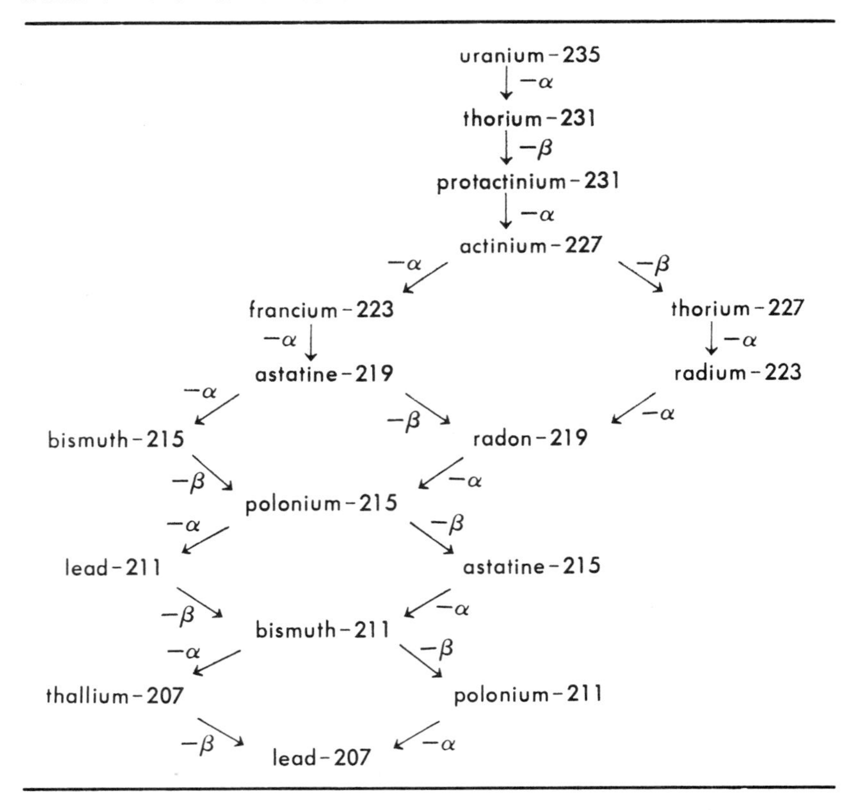

3). All three lead isotopes are stable, which indicates that an element may not only contain both stable and unstable isotopes but more than one stable isotope as well.

Half-Lives

So far I have talked about radioactive isotopes that underwent radioactive breakdown very slowly and others that broke down rapidly, but I have made no attempt to assign actual figures to these qualitative descriptions.

The first attempt to do so was made by Rutherford and Soddy in investigations beginning in 1902. Making use of a short-lived radioactive isotope, they traced the variation of the intensity of radiation with time. They found the intensity fell off with time in what is called an "exponential manner."

This can be true only if individual radioactive atoms break down at a rate that is a fixed fraction of the total number of such atoms present. This fraction, let us say, is 0.02 of the atoms present per second. If we begin with 1,000,000,000,000 atoms; then in the first second 20,000,000,000 atoms will break down. We can't tell which twenty billion will break down, of course. If we were considering a particular atom, we couldn't tell whether that one atom would break down in the first second or after five seconds or after five years.

This is quite analogous to a much more familiar situation. Given a million Americans aged 35, insurance companies (from a thorough study of statistics) can predict with reasonable accuracy how many of them will die in the course of the next year, assuming the year to be a "normal" one. They could not point out individual Americans who will die in that year, or predict the particular year of death of a particular American. They can only make general predictions where large numbers of faceless individuals are concerned. Where insurance men work with millions of human beings, physicists work with trillions of trillions of atoms, and the predictions of the latter are correspondingly more accurate.

Radioactive breakdown involves a fixed breakdown rate as time goes on. Let us suppose that this was true of human deaths. Let us suppose that out of 1,000,000 Americans aged 35 some 0.2 percent—that is 2000—will die in the course of the year. At the end of the year 998,000 men are left. If now 0.2 percent of those die in the next year, 1996 will die, leaving 996,004. In the third year, keeping the rate constant, 1992 will die, leaving 994,012, and so on.

The number of men dying would, in such a case, decrease evenly with the number of men still alive. There would be no particular year in which you could predict that the last man would die, for in any given year only a small percentage of those alive would die. Naturally, this analysis would only be reasonably correct as long as the number of men remained great enough for statistical methods to be reasonably accurate. However, if you started with an extremely large number of men, you would expect some of them to live hundreds of thousands of years.

This does not actually happen because the human death rate does not remain constant as the men grow older. It rises steadily, and very old men have a very high death rate. For that reason, no matter how large a number of 35-year-old men you begin with, all will be dead in less than a century.

In the case of radioactive atoms, the "death rate" through breakdown does not change with time, and while some atoms break down almost at once, other atoms of the same kind may refrain from breaking down for indefinitely long periods of time. One cannot therefore speak of the "lifetime" of a radioactive atom since that "lifetime" can be anything.

It is characteristic of such a "fixed death rate" situation, however, that for a given value of that death rate there is a specific interval of time during which half the original atoms would break down. This specific interval of time, named the *half-life* by Rutherford in 1904, would remain the same, however large or

however small (within statistical reason) the original number of atoms. For given isotopes, such a half-life was found to be virtually independent of environmental conditions such as temperature and pressure. Physicists have found ways of imposing minor changes—a few percent at most—on the half-lives of a few specific types of radioactive atoms, but such cases are quite exceptional.

Let us say that the half-life of a particular isotope is one year. This means that if you begin with two trillion atoms of that isotope, you will have only one trillion left at the end of the year. With the number of atoms present having declined to half, the number of breakdowns also declines to half, and only half a trillion vanish in the next year, leaving half a trillion. After a third year, a quarter-trillion would be left, and so on.

To generalize: Given any number of radioactive atoms, half will break down during the first half-life period, half of those left will break down during the second half-life period, and so on indefinitely—or at least until the total number of atoms involved becomes small enough for statistical methods no longer to apply with reasonable accuracy.

To know the half-life of an isotope, then, is to know in capsule form how many breakdowns will take place in a given quantity in a given time, and therefore, how intensely radioactive the isotope is. You can also trace what the intensity will be at any time in the future and what it was at any time in the past.

The half-lives of radioactive isotopes vary in length from the vanishingly small to the immensely long. In the intermediate range it is possible to determine half-lives directly from observed breakdown rates. For instance, the half-life of radium-226 is 1620 years.

To find half-lives much longer than this, indirect methods may be used. Consider the case of uranium-238, for instance. In any given sample of uranium ore, the uranium atoms are breaking down, but at so small a rate that we can safely assume that over some limited period of time, the number of uranium atoms present is virtually constant. Call that number, N_u. A given fraction (F_u) of the number of uranium atoms present breaks down each second. The total number of uranium atoms breaking down each second is therefore $F_u N_u$.

In the course of its breakdown, uranium-238 forms radium-226. It doesn't do this directly, for there are four other radioactive isotopes between, but this can be shown not to matter. We can legitimately simplify matters for the present by assuming that radium-226 is formed directly from uranium-238. Since $F_u N_u$ uranium-238 atoms are breaking down each second, $F_u N_u$ radium-226 atoms are being formed each second.

As radium-226 is formed, it also begins to break down at a rate that is a fixed fraction of the number of radium-226 atoms present, $F_r N_r$. As the radium-226 atoms form and accumulate from uranium-238, the number of radium-226 atoms breaking down increases until the point is reached where the number breaking down is equal to the number being formed. At that point, the number of radium-226 atoms actually present reaches a constant value, and radium-226 is in *radioactive equilibrium* with uranium-238. At radioactive equilibrium:

$$F_u N_u = F_r N_r \qquad \text{(Equation 8–8)}$$

or, rearranging:

$$\frac{F_u}{F_r} = \frac{N_r}{N_u} \qquad \text{(Equation 8–9)}$$

It can be shown that the fraction of a particular isotope breaking down each second is inversely proportional to the half-life of that particular isotope. The longer the half-life, the smaller the fraction of atoms present breaking down in one second. If the half-life of uranium-238 is symbolized as H_u and that of radium-226 as H_r, we can say:

$$\frac{F_u}{F_r} = \frac{H_r}{H_u} \qquad \text{(Equation 8–10)}$$

Combining Equations 8–9 and 8–10, we have:

$$\frac{H_r}{H_u} = \frac{N_r}{N_u} \qquad \text{(Equation 8–11)}$$

At radioactive equilibrium, in other words, the ratio of the quantity of parent and daughter atoms present is equal to the ratio of their half-lives. In uranium ores, there are 2,800,000 times as many uranium-238 atoms present as radium-226 atoms. The half-life of uranium-238 must be 2,800,000 times as long as that of radium-226, or just about 4,500,000,000 years.

It is not surprising then, that uranium-238 still exists in the earth's crust. If the solar system is five to six billion years old (as is now believed), then there has been time for only little more than half the uranium-238 originally present to have broken down.

The half-life of uranium-235 is shorter than that of uranium-

238; it is only 713,000,000 years. This is still long enough for something like one percent of the original amount present at the time of the origin of the solar system to remain in existence today. However, it is not surprising that only seven out of a thousand uranium atoms now existing are uranium-235.

Any radioactive isotope with a half-life of less than 500,000,000 years would not be present on earth today in more than vanishingly small quantities, unless it were formed out of a longer-lived ancestor. In the $4x + 2$ series, only uranium-238 qualifies for existence, and in the $4x + 3$ series, only uranium-235 does.

The only $4x + 0$ atom with a long enough half-life to exist today and give rise to a radioactive series is, of course, thorium-232. Its half-life is no less than 13,900,000,000 years.

Indirect methods can also be used to determine very short half-lives. For instance, it has been found that among those radioactive isotopes that emit alpha particles, the energy of the particles is inversely proportional to the half-life, in a moderately complicated fashion. Therefore, the half-life can be calculated from the energy of the alpha particles (which can be determined by noting how far they will penetrate a given type of substance). The half-life of polonium-212, for example, is found to be 0.0000003 seconds.

If isotopes of a given element differ among themselves inappreciably in chemical properties, they differ among themselves enormously in nuclear properties such as half-life. Thorium-232, as stated above, has a half-life of nearly fourteen billion years, but thorium-231 (differing in the lack of a single neutron in the nucleus) has a half-life of just about one day!

Stable Isotopes

If we go over the three radioactive series presented in Tables V, VI, and VII, we see that they include radioactive isotopes of elements ordinarily considered stable. There are five such isotopes of bismuth, with atomic weights of 210, 211, 212, 214 and 215; four of thallium, with atomic weights of 206, 207, 208 and 210; and four of lead, with atomic weights of 210, 211, 212 and 214. Each of these must possess at least one stable isotope as well, since each is found in the soil in appreciable quantities and in nonradioactive form. Indeed, the radioactive series include three different isotopes of lead—206, 207 and 208—each one of which is stable.

Nevertheless, all these isotopes, stable as well as unstable, are involved with radioactivity. It is fair to ask if elements that are in no way involved with radioactivity may nevertheless consist of two or more isotopes. If so, the fact would be difficult to establish, since ordinary laboratory methods do not suffice to separate isotopes (save in exceptional cases, see page 143) and since radioactivity cannot be relied on to help.

But suppose the atoms of an element are ionized, as in the formation of positive rays (see page 112). The atoms, each with an electron removed, would all have an identical charge of +1. If the element consisted of two or more isotopes, however, the ions would fall into groups that differ in mass.

Suppose, now, a stream of these positive ions are made to pass through a magnetic field. Their path is bound to curve and the extent of the curvature would depend upon the charge and mass of the individual particles. The charge would be the same in every case, but the mass would not be. The more massive ions curve less sharply than do the less massive ones. If the stream of positive rays is allowed to fall on a photographic plate, they would form one spot if all the ions were alike in mass, but more than one spot if the ions formed groups of different mass. Furthermore, if the groups were unequal in size, the larger group would form a larger, darker spot.

In 1912, J. J. Thomson, the discoverer of the electron, performed an experiment of this sort with neon. Positive rays formed of neon ions made two spots on the plate, corresponding to what would be expected for neon-20 and neon-22. The former spot was some ten times as large as the latter; from this is could be concluded that neon consisted of two stable isotopes, neon-20 and neon-22, in a ratio of about 10 to 1. (Eventually, it was discovered that a third stable isotope, neon-21, existed in very small quantities, and that in every 1000 neon atoms, 909 were neon-22, 88 were neon-22 and 3 were neon-21.)

In 1919, the English physicist Francis William Aston (1877–1945), who had worked with Thomson on this problem, constructed an improved device for analyzing positive rays. In his device, the positive rays consisting of ions of a given mass did not simply form a smear on the photographic plate. They were curved in such a way as to focus at a point, thus allowing for finer resolution. As a result, the beam of ions produced from a given element was spread out into a succession of points (a "mass spectrum" rather than a light spectrum). From the position of the points one could deduce the mass of the individual isotopes, and from their darkness, the frequency (or *relative abunda* with which each occurred in the element. The instrument termed a *mass spectrograph.*

The use of the mass spectrograph made it quite plain that most stable elements consisted of two or more stable isotopes. A complete list of such stable isotopes* is given in Table VIII.

A number of points can be made concerning Table VIII. In the first place, although most of the 81 stable elements consist of two or more stable isotopes (with tin made up of no less than ten), there remain no less than 20 elements consisting of but a single isotope. (Indeed, two elements of atomic number less than 84 possess no stable isotopes at all. These two, with atomic numbers 43 and 61, will be discussed on page 175.)

Properly speaking, one ought not speak of "a single isotope," since isotopes were originally defined as two or more kinds of atom falling in the same place in the periodic table. One might as well speak of "one twin." For that reason one sometimes speaks of *nuclides,* meaning an atom with a characteristic nuclear structure. One can certainly speak of one nuclide. However, the term "isotope" is so well-established that I will continue to speak of "a single isotope" with the assurance that I will be correctly understood.

Not all the 282 nuclides listed in Table VIII are indeed completely stable. Some eighteen of them, it turns out, are radioactive, though always with such extended half-lives that the radioactivity they display is feeble indeed. Some, with half-lives in the quadrillions of years, have radioactivities so weak that they may be ignored for all practical purposes. Seven, however, are perceptibly radioactive, and these are included in Table IX.

* Or isotopes that are so feebly radioactive that they may just about be considered stable.

TABLE VIII—*The Stable Isotopes*

Atomic Number	*Element*	*Isotope Weights*
1	Hydrogen	1, 2
2	Helium	3, 4
3	Lithium	6, 7
4	Beryllium	9
5	Boron	10, 11
6	Carbon	12, 13
7	Nitrogen	14, 15

The Stable Isotopes **(continued)**

Atomic Number	*Element*	*Isotope Weights*
8	Oxygen	16, 17, 18
9	Fluorine	19
10	Neon	20, 21, 22
11	Sodium	23
12	Magnesium	24, 25, 26
13	Aluminum	27
14	Silicon	28, 29, 30
15	Phosphorus	31
16	Sulfur	32, 33, 34, 36
17	Chlorine	35, 37
18	Argon	36, 38, 40
19	Potassium	39, 40, 41
20	Calcium	40, 42, 43, 44, 46, 48
21	Scandium	45
22	Titanium	46, 47, 48, 49, 50
23	Vanadium	50, 51
24	Chromium	50, 52, 53, 54
25	Manganese	55
26	Iron	54, 56, 57, 58
27	Cobalt	59
28	Nickel	58, 60, 61, 62, 64
29	Copper	63, 65
30	Zinc	64, 66, 67, 68, 70
31	Gallium	69, 71
32	Germanium	70, 72, 73, 74, 76
33	Arsenic	75
34	Selenium	74, 76, 77, 78, 80, 82
35	Bromine	79, 81
36	Krypton	78, 80, 82, 83, 84, 86
37	Rubidium	85, 87
38	Strontium	84, 86, 87, 88
39	Yttrium	89
40	Zirconium	90, 91, 92, 94, 96
41	Niobium	93
42	Molybdenum	92, 94, 95, 96, 97, 98
44	Ruthenium	96, 98, 99, 100, 101, 102, 104
45	Rhodium	103
46	Palladium	102, 104, 105, 106, 108, 110
47	Silver	107, 109

The Stable Isotopes (continued)

Atomic Number	*Element*	*Isotope Weights*
48	Cadmium	106, 108, 110, 111, 112, 113, 114, 116
49	Indium	113, 115
50	Tin	112, 114, 115, 116, 117, 118, 119, 120, 122, 124
51	Antimony	121, 123
52	Tellurium	120, 122, 123, 124, 125, 126, 128, 130
53	Iodine	127
54	Xenon	124, 126, 128, 129, 130, 131, 132, 134, 136
55	Cesium	133
56	Barium	130, 132, 134, 135, 136, 137, 138
57	Lanthanum	138, 139
58	Cerium	136, 138, 140, 142
59	Praseodymium	141
60	Neodymium	142, 143, 144, 145, 146, 148, 150
62	Samarium	144, 147, 148, 149, 150, 152, 154
63	Europium	151, 153
64	Gadolinium	152, 154, 155, 156, 157, 158, 160
65	Terbium	159
66	Dysprosium	156, 158, 160, 161, 162, 163, 164
67	Holmium	165
68	Erbium	162, 164, 166, 167, 168, 170
69	Thulium	169
70	Ytterbium	168, 170, 171, 172, 173, 174, 176
71	Lutetium	175, 176
72	Hafnium	174, 176, 177, 178, 179, 180
73	Tantalum	180, 181
74	Tungsten	180, 182, 183, 184, 186
75	Rhenium	185, 187
76	Osmium	184, 186, 187, 188, 189, 190, 192

The Stable Isotopes (**continued**)

Atomic Number	*Element*	*Isotope Weights*
77	Iridium	191, 193
78	Platinum	190, 192, 194, 195, 196, 198
79	Gold	197
80	Mercury	196, 198, 199, 200, 201, 202, 204
81	Thallium	203, 205
82	Lead	204, 206, 207, 208
83	Bismuth	209

It might seem surprising that this radioactivity among the lighter elements was not detected sooner than it was—particularly in the case of potassium-40. Potassium is a very common element, and potassium-40 (alone among the isotopes listed in Table IX) has a shorter half-life and is therefore more intensely radioactive than is either uranium-238 or thorium-232.

The answer to that is twofold. In the first place, potassium-40 makes up only one atom out of every 10,000 in potassium, so that it is not as common as it seems. In the second place, uranium and thorium are the parents of a series of intensely radioactive isotopes. It is the daughter atoms, rather than uranium or thorium themselves, that gave rise to the effects observed by Becquerel and the Curies.

None of the long-lived radioactive isotopes of the lighter elements serve as parents for a radioactive series. In every case they emit beta particles and are, in a single step, converted into a stable isotope of the element one atomic number higher. Thus, rubidium-87 becomes stable strontium-87, lanthanum-138 becomes stable cerium-138, and so on.

Potassium-40 introduces a slight variation. Of all the potassium-40 atoms that break down, some eighty-nine percent do indeed emit a beta particle and become stable calcium-40. The remaining eleven percent absorb an electron into the nucleus. This electron is taken from the innermost extranuclear shell, the K-shell (see page 66), and the process is therefore known as *K-capture*. An electrón taken into the nucleus serves to cancel the positive charge of one proton and produce an additional neutron. The total number of nucleons is not changed and neither, therefore, is the atomic weight. The atomic number, however, decreases by one. By K-capture, potassium-40 (atomic number 19) becomes the stable argon-40 (atomic number 18).

In some ways, the most remarkable of the stable isotopes is hydrogen-2, the nucleus of which is made up of one proton and one neutron, instead of one proton only as in hydrogen-1. The mass ratio between the two stable isotopes of hydrogen is much greater than is true of any two stable isotopes of any other element.

Thus, uranium-238 is 238/235, or 1.013 times the mass of uranium-235. Tin-124, the heaviest of the stable isotopes of that element, is 1.107 times the mass of tin-112, the lightest. Oxygen-18 is 1.125 times the mass of oxygen-16. But compare that with hydrogen-2, which is 2.000 times the mass of hydrogen-1.

This great difference in relative masses between the two hydrogen isotopes means that the two differ considerably more in their physical and chemical properties than isotopes generally do. The boiling point of ordinary hydrogen is 20.38° K, whereas hydrogen made up of hydrogen-2 only ("heavy hydrogen") has a boiling point of 23.50° K.

Again, ordinary water has a density of 1.000 gram per cubic centimeter and a freezing point of 273.1°K (0°C). Water with molecules containing hydrogen-2 only ("heavy water") has a density of 1.108 grams per cubic centimeter and a freezing point of 276.9°K (3.8°C).

So marked are the differences between hydrogen-1 and hydrogen-2 that the latter is given the special name of *deuterium* (from a Greek word for "second"). Its symbol is D and heavy hydrogen can be written D_2, while heavy water can be written D_2O.

In the early days of isotope work, physicists had suspected the existence of deuterium because the atomic weight of hydrogen seemed slightly higher than it ought to be. The single electron, it

TABLE IX—*Lighter Radioactive Nuclides*

Nuclide	*Half-Life (years)*
Potassium-40	1,300,000,000
Rubidium-87	47,000,000,000
Lanthanum-138	110,000,000,000
Samarium-146	106,000,000,000
Lutetium-176	36,000,000,000
Rhenium-187	70,000,000,000
Platinum-190	700,000,000,000

was calculated, would have its energy levels somewhat differently distributed in hydrogen-2 than in hydrogen-1, so faint lines of the former ought to appear near the heavy lines of the latter in the hydrogen spectrum. This was not observed, nor was hydrogen-2 located by mass spectrograph. One reason for this is that hydrogen-2 is quite rare; only one atom out of 7000 in ordinary hydrogen is hydrogen-2.

The American chemist Harold Clayton Urey (1893–) began, in 1931, with four liters of liquid hydrogen and let it slowly evaporate to one cubic centimeter. He reasoned that hydrogen-2 would evaporate more slowly and would be concentrated in the final bit. He was right. When he studied the spectrum of that last residue, he detected the lines of deuterium precisely where calculations had predicted they would be.

CHAPTER 38

Nuclear Chemistry

Mass Number

One might try, out of a spirit of neatness, to divide the atom cleanly between the two major physical sciences, awarding the electrons to the chemist and the nucleus to the physicist.

To attempt such a cleancut division would, however, be a violation of the spirit of science, which is all-one-piece. The structure of the nucleus, however remote that might seem from the world of ordinary chemical reactions, must nevertheless be of keen interest to the chemist, if only because of its effect upon that fundamental chemical datum, the atomic weight.

By the time the nineteenth century had come to its end, the matter of the atomic weight seemed settled. Each element had a characteristic atomic weight, chemists believed, and the only future in that respect was the ever greater precision with which the fourth and fifth decimal places might be determined.

Then came the discovery of isotopes and all that had seemed certain about atomic weights immediately went into discard. The notion, dating from Dalton, that all atoms of a single element possessed identical mass and that the atomic weight expressed this mass was seen to be false. Instead, most elements were made up of two or more varieties of atoms differing in mass. The atomic weight was merely the weighted average of the masses of these isotopes.

If the term "atomic weight" is reserved for this weighted average of the isotope masses as found in their natural distribution within an element, then one ought not to speak of the atomic weight of an individual isotope (as I have been doing so far in this book). It is better to use a different phrase and speak of the relative mass of an individual isotope as its *mass number*.

We can say, therefore, that neon is made up of three isotopes of mass numbers 20, 21, and 22. Neon-20 makes up about nine-tenths of all the neon atoms, while neon-22 makes up most of the remaining tenth. We can neglect neon-21 as occurring in too small a concentration to affect the result materially, and content ourselves by taking the average of ten atoms of neon, nine of which have a mass of 20 and one of which has a mass of 22. We arrive at a result of 20.2, which is roughly the atomic weight of neon.

Again, chlorine is made up of two isotopes of mass numbers 35 and 37, with chlorine-35 making up three-quarters of the whole, and chlorine-37 making up the remaining quarter. If we average the mass of four atoms, three of which have a mass of 35 and one of which has a mass of 37, we end with a result of 35.5, which is also roughly the atomic weight of chlorine.

All the nineteenth century demonstrations that Prout's hypothesis (see page 24) was false—because the atomic weights of the various elements were not necessarily integral multiples of the atomic weight of hydrogen—were shown to be irrelevant. The mass numbers of the various isotopes were, without exception, all found to be very nearly exact multiples of the mass of the hydrogen atom, and Prout's hypothesis was re-established in a more sophisticated form. The various elements were not built up out of hydrogen atoms exactly, but (ignoring the almost massless electrons) they were built up out of nucleons of nearly identical mass, while the hydrogen atom itself is built up out of a single nucleon.

Atomic weights that are nearly whole numbers in value are so because the particular element is made up of a single isotope, as in the case of aluminum, or of two or more isotopes, with one vastly predominant in relative abundance. An example of the latter situation is calcium, which is made up of six stable isotopes with mass numbers of 40, 42, 43, 44, 46, and 48, but with calcium-40 making up ninety-seven percent of the whole. It is because so many of the lighter elements fall into one of these two classes that Prout found reason to advance his hypothesis in the first place.

It is the imbalance of isotopes that causes some elements to be "out of order" in the periodic table. Thus, cobalt, with an atomic number of 27, consists of the single isotope of mass number 59. Its atomic weight, therefore, is about 58.9.* We would expect nickel, with the higher atomic number 28, to have a higher atomic weight as well. Nickel consists of five isotopes with mass numbers 58, 60, 61, 62, and 64; and, not surprisingly, four of those isotopes have mass numbers higher than the single cobalt isotope. However, it is the lightest of the nickel isotopes, nickel-58, which happens to be predominant. There are twice as many nickel-58 atoms than all the other nickel atoms put together. The atomic weight of nickel is thus pulled down to 58.7, which is somewhat less than that of cobalt.

The atomic weight is thus deprived of its fundamental character, and it is not truly characteristic of an element. What made it seem characteristic was the fact that the various isotopes of an element have virtually identical properties. The processes that led to the concentration of the compounds of an element in various places in the earth's structure, or to the isolation of the element in the laboratory, affected all the isotopes alike. Each sample of an element, however produced, would therefore contain the various isotopes in virtually identical proportions and would therefore display the same, apparently characteristic, atomic weight.

Yet there are exceptional cases, and the most dramatic is that of lead. Each of the radioactive series (see Chapter 8) ends in a particular lead isotope. The two series that begin with uranium isotopes as parent atoms produce lead-206 and lead-207, with lead-206 far in the lead since there is so much more uranium-238 than uranium-235. As for the thorium series, that ends in uranium-208.

The atomic weight of ordinary lead, found in nonradioactive ores, is about 207.2. In uranium ores, with lead-206 having been produced steadily over geologic periods, the atomic weight should be distinctly less, while in thorium ores it should be distinctly more. In 1914, the American chemist Theodore William Richards (1868–1928) carried through atomic-weight determinations and found, indeed, that lead obtained from uranium ores had atomic weights that ran as low as 206.1. Lead from thorium ores gave atomic weights as high as 207.9.

Where radioactivity is not involved, such large variations are not to be expected. Still, the atomic weights of some of the lighter elements were found to vary slightly in accordance with the con-

* Why not exactly 59? See page 185.

ditions under which the element was produced. For instance, the relative distribution of oxygen-16 and oxygen-18 in the calcium carbonate ($CaCO_3$) of seashells has been shown to depend on the temperature of the water in which the organism that formed the seashell was living. Delicate measurements of isotope ratios in fossil seashells have therefore been used to determine the temperature of ocean water at different periods of earth's geologic past.

The existence of oxygen isotopes introduced a particular difficulty in connection with atomic weights. By a convention as old as Berzelius, the atomic weights had been determined on a standard that set the atomic weight of oxygen arbitrarily equal to 16.0000. In 1929, however, the American chemist William Francis Giauque (1895–) showed that oxygen consisted of three isotopes—oxygen-16, oxygen-17 and oxygen-18—and that the atomic weight of oxygen had therefore to represent the weighted average of three mass numbers.

To be sure, oxygen-16 was by far the most common of the three, making up 99.759 percent of the whole, so that one wasn't very far off in pretending that oxygen was made up of a single isotope. For a generation after the discovery, therefore, chemists tended to ignore the oxygen isotopes and continue the atomic weights on the old basis. Such atomic weights came to be called *chemical atomic weights.*

Physicists, however, preferred to set the mass of the oxygen-16 isotope at 16.0000 and to determine all other masses on that basis. Their reasoning was that the mass number of an isotope was characteristic and unalterable, whereas the atomic weight of a multi-isotope element would alter with changes in relative abundance of those isotopes from sample to sample.

On the basis of oxygen-16 = 16.0000, a new list of atomic weights, the *physical atomic weights,* was drawn up. The atomic weight of oxygen on this new basis was 16.0044 (oxygen-17 and oxygen-18 pulling up the average), and this is 0.027 percent higher than its chemical atomic weight of 16.0000. This same difference would exist throughout the entire list of elements, and while this difference is small, it acts as an unnecessary source of confusion in refined work.

In 1961, physicists and chemists reached a compromise. It was agreed to determine atomic weights on the basis of allowing the carbon-12 isotope to have a mass of 12.0000. As the physicists desired, this tied the atomic weights to a fixed and characteristic mass number. In addition, an isotope was chosen for the purpose which would produce a set of atomic weights as nearly identical to the old chemical atomic weights as possible. Thus the atomic weight of oxygen under this new system is 15.9994, which is only 0.0037 percent less than the chemical atomic weight. The atomic weights given in Table II of Chapter 1, are on a carbon-12 = 12.0000 basis.

The atomic weight, since it is the weighted average of the mass numbers of the naturally-occurring isotopes, can truly be applied only to those elements that are primordial; that is, that have been in the earth from the beginning, when, presumably, the different isotopes made their appearance at the same time. This includes only 83 elements altogether. There are first the 81 stable elements (with atomic numbers from 1 to 83 inclusive, minus atomic numbers 43 and 61), and then the nearly stable elements thorium and uranium.

The elements that appear in the earth only because they are formed from uranium or thorium appear in the form of isotopes of different mass number, depending on whether they are found in uranium or thorium ore. One cannot form a true average and obtain a real atomic number. It is therefore usual, in the case of these elements (and of other unstable elements to be considered in the next chapter), to use the mass number of the most long-lived known isotope as a kind of substitute atomic weight. In tables of atomic weights, these mass numbers are usually included in brackets, as in Table X. Of the isotopes which appear in this table, those of radon and radium appear naturally in the uranium series, while those of francium, actinium and protactinium appear naturally in the actinium series. Polonium-209 and astatine-210 do not occur naturally at all but have been artificially produced (see page 175).

TABLE X—*"Atomic Weight" of Radioactive Elements*

Element	*Mass Number of Most Long-Lived Isotope*	*Half-Life*
84–Polonium	[209]	103 years
85–Astatine	[210]	8.3 hours
86–Radon	[222]	3.8 days
87–Francium	[223]	22 minutes
88–Radium	[226]	1,620 years
89–Actinium	[227]	21.2 years
91–Protactinium	[231]	32,480 years

Radioactive Dating

The lead isotopes played a role not only in re-orienting the chemical view of atomic weights, but also of the geological view of terrestrial history.

As long ago as 1907, the American physicist Bertram Borden Boltwood (1870–1927) suggested that radioactive series could be used as a method for determining the age of minerals.

Suppose a particular layer of rock containing uranium or thorium was laid down at a certain time in the past as a solid by sedimentation from the sea or by freezing from a volcanic melt. Once such a solid had made its appearance, the uranium or thorium atoms within it would be "trapped." When some broke down and eventually formed lead atoms, those atoms would be trapped, too.

During the entire stretch of time that would have elapsed since the layer had solidified, the uranium or thorium would be breaking down and the lead content would, in consequence, be rising. It would seem, then, that the uranium/lead and thorium/lead ratios within solid rocks would increase steadily with time.

Since Rutherford had already worked out the concept of half-life, it seemed furthermore that this increase would take place at a known rate, so that from the uranium/lead or thorium/lead ratio at any instant of time (the present, for instance), the time lapse since the rock had solidified could be calculated. Since the half-life of both uranium-238 and thorium-232 is so immensely long, time lapses of billions of years could be calculated with assurance.

A possible difficulty rests in the fact that one cannot be sure that the lead in such rocks was produced entirely through uranium or thorium breakdown. An indefinite amount might be primordial and might have been trapped along with uranium in the rock at the moment of its solidification. Such lead would obviously have no connection with the uranium and would seriously confuse the issue.

This difficulty was resolved when the mass spectrograph made it possible to determine the relative abundance of the isotopes in lead found in nonradioactive rocks. Such lead contained four stable isotopes, with mass numbers 204, 206, 207, and 208; and lead-204 made up 1.48 percent (1/67.5) of the whole.

This is fortunate, for lead-204 is not produced as the end product of any radioactive series, and its occurrence is not affected by radioactivity. If in the lead content of radioactive rocks the lead-204 concentration is determined and multiplied by 67.5, then the total quantity of primordial lead can be determined. Any lead over and above this quantity would have been produced by radioactive breakdown.

By using the uranium/lead ratio, and allowing for the presence of lead-204, rocks have been found which have been solid for over 4,000,000,000 years. This is taken as the best evidence yet obtained for the extreme age of the earth.

Uranium and thorium are not, of course, among the more common elements, and rocks containing sufficient uranium and thorium to make reasonably reliable age determinations of this sort are found only in restricted areas. However, use can also be made of the long-lived radioactive isotopes rubidium-87 and potassium-40, each of which is much more widely distributed than is either uranium or thorium. In the case of rubidium-87, one can determine the rubidium/strontium ratios in rocks, since rubidium-87 decays to the stable strontium-87. (Primordial strontium can be estimated by noting the quantity of other stable strontium isotopes present, these others not being formed in radioactive breakdown.) Rubidium-containing minerals that have been solid for nearly 4,000,000,000 years have been located.

Potassium-40 offers an interesting situation. Mostly, it breaks down to calcium-40; but calcium-40 is very common in the earth's crust, and it is impractical to try to distinguish "radiogenic" calcium-40 (that which has arisen through radioactive breakdown) from primordial calcium-40. However, a fixed proportion of the potassium-40 atoms break down by K-capture (see page 142) to form argon-40.

Argon is one of the inert gases found in the atmosphere. All the isotopes of the various inert gases, with the single exception of argon-40, are present in almost vanishingly small quantities. This situation probably reflects a time in earth's early history when its mass was too small or its temperature too high to retain any of the gaseous elements except in the form of solid compounds. Since the inert gases do not form any compounds to speak of, they were all lost.

Argon-40, however, occurs in quantity, making up about one percent of the atmosphere. It seems likely that all this argon-40 was only formed after the earth had attained its present mass and temperature (at which time it could retain the heavier inert gases), and that it had been formed, presumably, from potassium-40. If one calculates the time it would have taken the present quantity of argon-40 to have accumulated from scratch, it would appear that the earth has existed in approximately its present form for 4,000,000,000 years.

Since a variety of methods independently agree, the question of the earth's age is taken as settled—or at least (remember the fate of many previous "settled" points) it is so taken until further notice.

Nuclear Reactions

As long as it was thought that an atom was a structureless, indivisible particle, it seemed an inevitable consequence that its nature could not be altered in the laboratory. However, once the atom was found to consist of numerous subatomic particles in a characteristic arrangement, the thought at once arose that this arrangement might somehow be altered.

The outer electrons of an atom may have their arrangement altered easily enough. Collisions among atoms and molecules with forces to be expected at temperatures attainable in nineteenth century laboratories sufficed for the purpose. It was these electron rearrangements that produced the familiar chemical reactions that were the established province of the chemist.

But what about rearrangements among the particles within the nucleus? These would alter the very fundamental nature of an atom and convert one element into another.

To smash atoms together so hard that the outer cushion of electrons is smashed through and nucleus meets nucleus requires extraordinarily high temperatures. Fortunately, as the twentieth century opened there was an obvious way of bypassing the need for such temperatures. Radioactive elements furnished a supply of subatomic particles at room temperature. One of them, the alpha particle, was a bare atomic nucleus (that of helium). Alpha particles are emitted with enough energy to smash through the electron barrier and, if properly aimed, strike the nucleus of a target atom.

It is impossible, of course, to aim an alpha particle at a given nucleus, but, statistically speaking, if enough alpha particles are fired, some will strike nuclei. It was through such collisions and near-collisions that Rutherford worked out the concept of the nuclear atom and estimated the size of the nucleus (see page 58).

Still, a collision that merely results in a deflection or a bounce alters the nature of neither the target nucleus nor the alpha particle. Something more is needed, and in a series of experiments.

the results of which he described in 1919, Rutherford worked out the necessary evidence that something more is occasionally obtained. He began by placing a source of energetic alpha particles inside a closed cylinder, one end of which was coated with a layer of zinc sulfide.

Now, whenever an alpha particle strikes the zinc sulfide, it gives rise to a tiny flash of luminescence, or scintillation. This arises because the kinetic energy of the particle excites the zinc sulfide molecule which, in returning to its ground state, emits a photon of visible light. (This phenomenon was first observed by Becquerel in 1899 and was later put to use in the preparation of luminescent objects. Tiny quantities of radium compounds mixed with zinc sulfide or some other appropriate substance would produce light flashes that would be clearly visible in the dark. In the 1920's there grew up quite a fad for watches with numerals marked out with such luminescent materials.)

When scintillating zinc sulfide screens were viewed under some magnification in the dark (with eyes well-accustomed to darkness and therefore particularly sensitive to feeble light) individual scintillations could be made out and therefore, surprisingly enough, so could the effect of single alpha particles. By counting the number of flashes over a given area in a given time, one could estimate the total number of disintegrations per second in a known mass of radioactive substance, and from this (for instance) one could calculate the half-life. Rutherford in his experiments was making use of what is now known as a *scintillation counter*.

Modern scintillation counters make use of more efficient scintillators, of phototubes to detect the flashes, and of appropriate electronic circuits to count them.

The number of scintillations produced by a given alpha particle source is reduced if a gas such as oxygen or carbon dioxide is introduced into the tube. Through collision and deflection, the gas slows the alpha particles to the point where some pick up electrons and become ordinary helium atoms. Those that manage to reach the screeen are fewer and less energetic.

If hydrogen is introduced into the tube, however, particularly bright scintillations suddenly appear. These can best be interpreted by supposing that occasionally an alpha particle will strike the nucleus of a hydrogen nucleus (a single proton) squarely and send it hurtling forward, away from what had been its associated electron. In this fashion, the bare proton can be made to move far more rapidly than could the massive nuclei of carbon and oxygen. In fact, the proton moves quickly enough to strike the screen with sufficient force to produce the unusually bright scintillations.

Rutherford found that when nitrogen was introduced into the tube, what looked like proton scintillations appeared. The nitrogen nucleus itself could not be forcibly hurled forward any more than could those of carbon or oxygen, but perhaps a proton had been knocked out of the nitrogen nucleus by the alpha particle.

This was confirmed in 1925 by the English physicist Patrick Maynard Stuart Blackett (1897–), who allowed the alpha particle bombardment of nitrogen to proceed in a Wilson cloud chamber. The alpha particle usually made a straight streak of water droplets without striking any nucleus, disappearing when its energy had been sufficiently nibbled away for the particle to pick up electrons and become an atom. Once in every 50,000 cases or so, however, there was a collision.

The alpha particle streak therefore ended in a fork. One side of the fork was long and thinner than the original track; it was the proton, carrying a smaller charge (+1, rather than the alpha particle's +2) and producing fewer ionizations. The other side of the fork was thick and short. It was the recoiling nitrogen nucleus from which numerous electrons had been stripped, and its high positive charge made it an efficient ionizer. However, it moved rather slowly, quickly picked up electrons once more, and, neutral again, ceased ionizing. There was no sign of the alpha particle after the collision, so it must have joined the nitrogen nucleus.

In the light of all this, it was not difficult to see that Rutherford, in 1919, had produced the first case of a deliberate rearrangement of nuclear structure through human efforts. It was the first man-made *nuclear reaction.* (In a sense, this is a kind of "nuclear chemistry," for the nucleons were being shuffled about in fashion analogous to the shuffling of electrons in ordinary chemistry.)

Suppose we begin with a nitrogen nucleus (seven protons and seven neutrons), add to it an alpha particle (two protons and two neutrons) and subtract the single proton that is knocked out. What is left then is an atom of eight protons and nine neutrons, which is oxygen-17. We can therefore write:

$$_7N^{14} + {_2He^4} \longrightarrow {_1H^1} + {_8O^{17}} \qquad \text{(Equation 9–1)}$$

where the subscripts are atomic numbers and the superscripts mass numbers. The $_2He^4$ represents the helium nucleus, or alpha particle, while the $_1H^1$ is the hydrogen nucleus, or proton. Notice that the atomic numbers add up to 9 on either side of the arrow and the mass numbers add up to 18. Such a balance must be preserved in all nuclear reactions if the laws of conservation of electric charge and of mass are to be preserved.

Physicists have devised briefer methods of writing such nuclear reactions. The atomic number is omitted since the name of the element fixes that number. The alpha particle is symbolized as α, and the proton as p. The nuclear reaction given in Equation 9–1 can then be written as: $N^{14}(\alpha,p)O^{17}$.

According to this system we have the target nucleus at the far left, then, in the parentheses, first the nature of the particle striking the target and then the particle knocked out of the target. Finally, on the extreme right, is the residual nucleus. The usefulness of this system, quite apart from its conciseness, is that it makes it easy to speak of a whole family of (α,p) nuclear reactions. In all such reactions, the residual nucleus is one higher in atomic number and three higher in mass number than the target nucleus.

Other (α,p) reactions were brought about by Rutherford, but there is a limit to what can be done in this direction. Both the alpha particle and the target nucleus are positively charged and repel each other. This repulsion increases with the atomic number of the nucleus, and for nuclei of elements beyond potassium (with a charge of $+19$) the repulsion is so strong that even the most energetic alpha particles produced by radioactive atoms lack the energy required to overcome that repulsion and strike the nucleus.

The search was on, therefore, for methods of obtaining subatomic particles with energies greater than those encountered in radioactivity.

The Electron-Volt

A charged particle can be accelerated by being subjected to the influence of an electric field so oriented as to pull the particle forward. The greater the electric potential to which the particle is subjected, the greater the acceleration, and the greater the energy gain of the particle.

A particle with a unit charge, such as an electron, which is accelerated by a field with an electric potential of one volt, gains an energy of one *electron-volt*. The electron-volt, often abbreviated to *ev*, is equal to 1.6×10^{-12} ergs. For larger units of this sort, we have the *kiloelectron-volt* (*Kev*), which is equal to 1000 electron-volts. Beyond that is the *Mev* (a million electron-volts) and the *Bev* (a billion electron-volts.)* A Bev is equal to 1.6×10^{-3} ergs. This is a small quantity of energy in ordinary terms, but it is simply enormous when we consider that it is packed into a single subatomic particle.

Mass can be expressed in electron-volts and subatomic masses are expressed in this way with increasing frequency. The mass of an electron is 9.1×10^{-28} grams. This can be expressed as its equivalent in energy (as calculated by means of Einstein's mass-energy equivalence equation, $e = mc^2$, see page II–111) and turns out to be 8.2×10^{-7} ergs. This, in turn, equals 510,000 electron-volts, or 0.51 Mev.

The wavelength of electromagnetic radiation can also be expressed in electron-volts. According to the quantum theory, $e = h\nu$, where e is the energy of a quantum of electromagnetic radiation in ergs, h is Planck's constant in erg-seconds, and ν (the Greek letter "nu") is the frequency of the radiation in cycles per second.

The wavelength of that radiation (represented by λ, the Greek letter "lambda") is equal to the distance in centimeters traveled, in a vacuum, by the radiation in one second (c) divided by the number of wavelengths formed in that time—that is, by the frequency of the radiation (ν).

In other words:

$$\lambda = \frac{c}{\nu} \qquad \text{(Equation 9–2)}$$

or:

$$\nu = \frac{c}{\lambda} \qquad \text{(Equation 9–3)}$$

Substituting c/λ for ν in the quantum theory equation $e = h\nu$, we have:

$$e = \frac{hc}{\lambda} \qquad \text{(Equation 9–4)}$$

or:

$$\lambda = \frac{hc}{e} \qquad \text{(Equation 9–5)}$$

* The term "billion" has different meanings in different parts of the world. To an American, for instance, it means 1,000,000,000, but to an Englishman it means 1,000,000,000,000; and what we call a billion, they would call a "thousand million." In Great Britain, then, 1,000,000,000 electron-volts is spoken of as a "giga-electron-volt" and is abbreviated Gev.

The value of h is 6.62×10^{-27} erg-seconds, while that of c is equal to 3.00×10^{10} centimeters per second. Consequently, hc is equal to 1.99×10^{-16} ergs. We can therefore write Equation 9–5 thus:

$$\lambda = \frac{1.99 \times 10^{-16}}{e} \qquad \text{(Equation 9–6)}$$

If we substitute the value of 1.6×10^{-12} ergs (the value of one electron-volt) for e in Equation 9–6, we obtain a value of 1.24×10^{-4} centimeters. In other words, radiation with a wavelength of 1.24 microns (in the infrared range) is made up of photons with an energy of 1 ev.

It follows that one kev is the energy content of radiation with a wavelength one-thousandth as great—that is, of 1.24 millimicrons, or 12.4 angstrom units. This is in the X-ray range. Similarly, one Mev is the energy content of radiation with a wavelength of 0.0124 angstrom units, which is in the gamma-ray range.

Conversely, Equation 9–6 can be used to show that visible light has an energy content varying from 1.6 ev at the red end of the spectrum to 3.2 ev at the violet end. Ordinary chemical reactions are brought about by visible light and ultraviolet light and, in turn, produce such radiation. You can see then that ordinary chemical reactions involve energies of from not more than one to five electron-volts. It is a measure of the increased difficulty of bringing about nuclear reactions that particles with energies in the thousands of electron-volts, and even millions of electron-volts, are required for the purpose.

Particle Accelerators

Devices intended to produce subatomic particles with energies in the kev range and beyond are called *particle accelerators*. Since the energetic particles produced by these accelerators were used to disrupt atom nuclei and induce nuclear reactions, the devices were popularly called "atom-smashers," though this term has rather gone out of fashion.

The first particle accelerator to achieve useful results was one that was adapted to accelerate protons by the English physicist John Douglas Cockcroft (1897–) and his Irish co-worker Ernest Thomas Sinton Walton (1903–), in 1929.

Protons are preferable to alpha particles in that the former carry a smaller positive charge and are therefore subjected to a smaller repulsive force from atomic nuclei. In addition, protons are ionized hydrogen atoms (H^+), while alpha particles are ionized helium atoms (He^{++}); and hydrogen is both far more common and far more easily ionized than is helium.

The Cockcroft-Walton device used an arrangement of condensers to build up potentials to extraordinarily high levels (it was called a *voltage multiplier*) and to accelerate protons to energies of as high as 380 Kev.

In 1931, they were able to use such accelerated protons to bring about the disruption of a lithium nucleus:

$$_3Li^7 + {_1H^1} \longrightarrow {_2He^4} + {_2He^4} \qquad \text{(Equation 9–7)}$$

This was the first completely artificial nuclear reaction, for here even the bombarding particles were artificially produced.

In that same year, 1931, no less than three other important types of particle accelerators were introduced.

The American physicist Robert Jemison Van de Graaf (1901–) built a mechanism shaped like half a dumbbell standing on end. Within it, a moving belt was so arranged as to carry positive electric charge upward and negative electric charge downward, producing a large electrostatic charge on either end. This *electrostatic generator* produced a huge potential difference that could accelerate particles to an energy of 1.5 Mev. Later such devices produced particles of still higher energies—as much as 18 Mev.

A second variety of accelerator was built up of separate tubes. This made it possible to accelerate particles by separate individual potential "kicks," instead of attempting to do it all in one powerful kick. In each tube the particle gained additional energy and took on additional velocity. Since the potential kicks were administered at equal intervals of time, the accelerating particle covered longer and longer distances between kicks, and each successive tube had to be made longer. For this reason, the *linear accelerator,* or *linac,* quickly grew inconveniently long.

The most compact arrangement for building up huge energies was the product of the American physicist Ernest Orlando Lawrence (1901–1958), who sought to save space by having the particles travel in a curved path, rather than in a straight line.

A high-temperature filament at the center of a closed flat circular vessel ionizes low-pressure hydrogen to produce protons. Opposite halves of the vessel are placed under a high potential that accelerates the protons. The poles of a magnet above and below the vessel force the protons to follow a curved path.

Ordinarily, the protons following this curved path would

eventually find themselves moving toward the positively-charged portion of the vessel and begin to slow up. However, the vessel is under an alternating potential, so that cathode and anode flip back and forth rapidly, at a carefully adjusted rate.

Each time the protons turn in such a way as to be moving toward the anode, there is a flip and the protons are moving toward the cathode after all. They are therefore pulled forward and accelerated further. (It is very like a greyhound pursuing an electric rabbit that always remains just ahead.)

As the protons accelerate, they move faster and faster, and it might be thought that they would make their turns about the vessel in less and less time. In such a case, the flip-flop of the electric field, which continues at a constant rate, would fall out of synchronization. The protons would find themselves heading toward the repulsive force of the anode, which would not be replaced by the cathode in time, and the proton would be slowed up. (This would be like the greyhound putting on a burst of speed and catching the electric rabbit.)

Fortunately, as the protons are accelerated, they naturally curve to a lesser degree under the influence of the magnetic field. They move in a larger circle and their greater velocity is just compensated for by the longer distance through which they must travel. They therefore continue to move from one half to the other in a fixed cycle that matches the alternation of the potential, spiraling outward from the center of the container as they do so. Eventually, they spiral out of a prepared exit as a stream of high-energy particles.

Linear particle accelerator

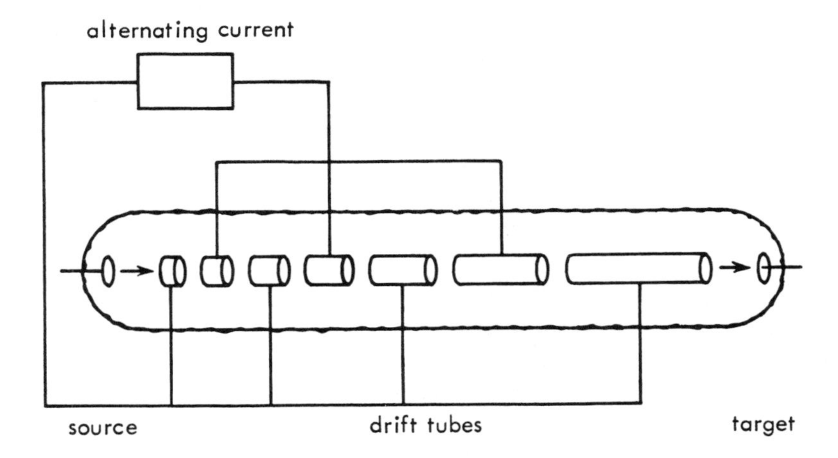

Lawrence called his instrument a *cyclotron,* and even his first model, which was no more than eleven inches in diameter and intended only as a small-scale test of the principle, managed to produce particles of 80 Kev. Over the next ten years larger and larger cyclotrons were built, and particles with more than 10 Mev of energy were produced.

This perfect matching of particle movement and potential alternation works only if the mass of the particle remains unchanged. Under ordinary conditions, it does (just about), but as acceleration proceeds, particles eventually move at velocities that are sizable fractions of that of light. Acceleration begins to involve increasingly minor additions to the particle's velocity (which cannot, in any case, exceed that of light) and increasingly major additions to its mass, in accordance with the special theory of relativity (see page II–102).

As the mass of the particle increases, it takes longer than would otherwise be expected to make its semicircle, and the particle motion falls out of synchronization with the alternation of the potential. This sets a limit to the energies that can be piled on the proton, and this limit was reached by World War II.

In 1945, two men, the American physicist Edwin Mattison McMillan (1907–) and the Russian physicist Vladimir I. Veksler, independently suggested a way of getting around this. They showed how the alternation of the potential could be decreased gradually at just the rate required to keep it synchronized to the motion of the increasingly massive particle. The result is a *synchrocyclotron.*

A synchrocyclotron cannot produce high-energy particles continuously, for the alternation period of the potential that is suitable for particles in the late stages of acceleration is far too slow for particles in the early stages. Therefore the particles had to be produced in separate bursts of 60 to 300 per second, each burst being carried from beginning to end before a new batch could be started. However, the increase in possible energies was well worth the cut in total quantity. The first synchrocyclotron was built in 1946, and within a few years instruments capable of producing particles with energies up to 800 Mev made their appearance.

The problem of relativistic mass increase had appeared even sooner in connection with electron acceleration. Electrons are so light that they must be made to move at extremely high velocities to achieve even moderate energies. If an electron is to attain an energy of even one Mev, it must be made to move at about

270,000 kilometers per second, which is better than 9/10 the speed of light. At that speed, the mass of the electron is 2.5 times what it would be if the particle were at rest. Using the cyclotron principle on the electron is therefore impractical, for the electron would lose synchronization at very low energies.

A solution to this was found even before the principle of the synchrocyclotron was established. In 1940, the American physicist Donald William Kerst devised an accelerator in which the electrons were made to move in a circle through a doughnut-shaped vessel. As they gained velocity, the strength of the magnetic field that made the electrons move circularly was increased. Since the increase in magnetic field intensity (which tended to make the electrons move in a more sharply curved path) was matched with the increase in the electrons' mass (which tended to make them move in a less sharply curved path), the net result was to keep the electrons moving in the same path. At a given moment, a sudden change in the magnetic field hurled a burst of high-energy electrons out of the instrument. Because beta particles are a natural example of high-energy electrons, Kerst called his instrument the *betatron*. Kerst's first instrument produced electrons with an energy of 2.5 Mev, and the largest betatron built since produces electrons with an energy of 340 Mev.

Electrons whirling very rapidly in circular orbits are sharply accelerated toward the center and, as Maxwell's theory of electromagnetic phenomena would require, give off energy in the form of radiation. This sets a limit to how much energy can be pumped into electrons by any device requiring the particles to travel in circles. (This phenomenon is less marked for protons, which for a given energy need not travel so quickly nor be subjected to so great a consequent acceleration.)

For this reason new efforts are now being made to manufacture linear accelerators long enough (and a two-mile-long unit is being planned) to produce electrons with energies up to 20,000 Mev, or 20 Bev.

The synchrocyclotron has one defect, not in theory, but in practice. As the particle spirals outward, it sweeps through curves of greater and greater radius, and the magnet must be large enough to cover the maximum radius. Magnets of the proper enormous size were a bottleneck in construction of larger machines.

There was an advantage then in adjusting the magnetic field to allow protons to travel in circles rather than in spirals. The design was such that "strong-focusing" was introduced, making the proton stream hold together tightly in as narrow a beam as possible. In this way, *proton synchrotrons* and *electron synchrotrons* were built.

By 1952, proton synchrotrons capable of producing particles in the Bev range were built. There is such a device at the University of California; it is appropriately called the *Bevatron* and can produce protons with energies of 6.2 Bev.

In the 1960's, particularly large strong-focusing accelerators were built (one in Geneva, and one at Brookhaven, Long Island) and are capable of producing protons with energies in excess of 30 Bev. Still larger machines are envisaged, but the plans are, of necessity, colossal. Present large accelerators are three city blocks in diameter.

CHAPTER **39**

Artificial Radioactivity

Radioisotopes

The first nuclei produced by artificial transmutation were stable ones that exist in the elements as found in nature. Examples are the oxygen-17 produced by Rutherford and the helium-4 produced by Cockcroft and Walton.

This precedent was shattered in 1934 through the work of the French physicist Frédéric Joliot-Curie (1900–1958) and his wife, Irène (1897–1956)—who were the son-in-law and daughter of Pierre and Marie Curie, the discoverers of radium.

The Joliot-Curies continued Rutherford's work on alpha particle bombardment of nuclei. In bombarding aluminum, they found emissions of protons and neutrons, emissions which ceased when the alpha particle bombardment was interrupted. Another type of radiation* did not cease but fell off in an exponential manner, with a half-life of 2.6 minutes. It seemed quite plain that something in the aluminum that had not originally been radioactive had become radioactive as a result of the bombardment.

The following equation describes what happens when aluminum-27 absorbs an alpha particle and emits a proton:

* The nature of this radiation will be discussed later in the book; see page 224.

$$_{13}Al^{27} + {}_{2}He^{4} \longrightarrow {}_{14}Si^{30} + {}_{1}H^{1} \qquad \text{(Equation 10–1)}$$

or:

$$Al^{27}(\alpha,p)Si^{30}$$

Silicon-30 is a stable isotope, occurring in silicon with a relative abundance of just about 3 percent.

But under the bombardment aluminum also emits neutrons. It follows then that a reaction might be taking place in which the target aluminum nucleus absorbs an alpha particle and emits a neutron, making a net gain of two protons and one neutron. In such an (α,n) reaction, then, the atomic number is increased by two rather than by one, and aluminum is converted to phosphorus rather than to silicon. The equation can be written:

$$_{13}Al^{27} + {}_{2}He^{4} \longrightarrow {}_{15}P^{30} + {}_{0}n^{1} \qquad \text{(Equation 10–2)}$$

or:

$$Al^{27}(\alpha,n)P^{30}$$

But phosphorus as it occurs in nature is made up of a single isotope, phosphorus-31. No other stable phosphorus isotope is known, and it is to be presumed that if any other phosphorus isotope is synthesized in the course of a nuclear reaction, it would be radioactive; and it is because of this radioactivity (combined with a short half-life) that it does not occur in nature.

The Joliot-Curies confirmed the presence of radioactive phosphorus in the aluminum by dissolving the metal and allowing it to undergo reactions that would put any phosphorus present into the form of either a gaseous compound or a solid precipitate. Sure enough, the radioactivity was found in the gas or the precipitate.

Phosphorus-30 was the first isotope to be produced in the laboratory that did not occur on earth naturally, and it is also the first example of *artificial radioactivity.*

It was by no means the last. Over the next generation, nuclear reactions induced in the laboratory produced over a thousand such artificial isotopes. Since every single one of those so produced is radioactive, they are often called *radioisotopes.*

Radioisotopes of every stable element have been formed, sometimes in considerable number. In the case of cesium, for instance, which has a single stable isotope, cesium-133, no less than twenty different radioisotopes have been formed, with mass numbers of from 123 to 148.

None of the radioisotopes so produced have half-lives long

enough to allow them to remain in the earth's crust over the planet's lifetime. Some of the half-lives are long by human standards to be sure (cesium-135 has a half-life of 2,000,000 years), but none are long enough.

One might suspect that at the time the matter making up the solar system was created, all conceivable nuclear arrangements were brought into existence. Those that happened to be stable, and those that were only very slightly radioactive (as, for instance, potassium-40 and uranium-238), survived. And indeed it seems quite likely that all stable or nearly stable isotopes that can exist do exist on earth, and the chances are virtually zero that an unknown stable or nearly stable isotope will ever be discovered.

As for those isotopes that are sufficiently unstable to have half-lives of less than 500,000,000 years, they may have also been formed, but they broke down and disappeared, some rapidly and some less rapidly. It is only the labor of the physicist that now brings them back to life.

The Biochemical Uses of Isotopes

Once physicists began isolating rare isotopes and synthesizing new ones, it became possible to prepare chemical compounds containing them. If the isotopes could be prepared cheaply enough, then the compounds containing them could be used in chemical experiments in quantity.

The first isotope to be used in comparatively large-scale experimentation in this manner was the stable hydrogen-2, which could be prepared in the form of "heavy water" (see page 144).

By carrying out organic chemical reactions in heavy water, it was possible to prepare other compounds with molecules containing one or more atoms of hydrogen-2. If such compounds were allowed to take part in chemical reactions, their ultimate fate could be determined by isolating the products and checking to see which of them contained hydrogen-2. A compound containing a more-than-normal amount of a rare isotope may therefore be said to be a *tagged compound,* and the abnormal atom itself, an *isotopic tracer.*

This technique is particularly important where the tagged compound is one that ordinarily undergoes chemical changes in living tissue, for then it can be followed through the rapid and extraordinarily complicated transformations that take place there. Beginning in 1935, the German-American biochemist Rudolf Schoenheimer (1898–1941) carried on such experiments, making use of fat molecules rich in hydrogen-2. This introduced a veritable revolution in biochemistry, for it quickly became possible to work out details of tissue reactions that might otherwise have remained impenetrable.

Schoenheimer, and others as well, also worked with the heavier isotopes of nitrogen and oxygen. These were nitrogen-15 and oxygen-18, with relative abundances of 0.37 percent and 0.20 percent, respectively. Both are rare enough in nature to serve as effective tracers when used in concentrated fashion.

The production of radioisotopes made possible an even greater sensitivity in the use of isotopic tracers, for radioactive isotopes can generally be detected more easily, more quickly, and in much smaller concentration than can stable isotopes.

Radioactive tracers were used as early as 1913 by the Hungarian physicist Georg von Hevesy (1885–   ). At the time, the only radioactive isotopes that were available were those that were members of the various radioactive series. Hevesy made use of lead-210 in determining the solubility of very slightly soluble lead compounds. (He could determine the fraction of lead-210 that went into solution by measuring the radioactivity of the solution before and after, and it seemed reasonable to assume that this fraction held good for all lead isotopes generally.)

In 1923, Hevesy tagged a lead compound with lead-212 and studied the uptake of lead by plants. This was the first biological application of isotopic tracers. However, lead is not a compound that occurs naturally in living tissue; indeed, lead is an acute poison. The behavior of tissue in the presence of lead is not necessarily normal. The use of radioisotopes of the more biologically useful elements did not become really large-scale until after World War II, when methods for preparing such isotopes in quantity were developed.

One unavoidable shortcoming of the radioisotope technique is that few good radioisotopes are available for those elements most common in tissue. The four elements making up over 90 percent of the soft tissues of the body are carbon, hydrogen, oxygen and nitrogen. In the case of nitrogen, the most long-lived radioisotope known is nitrogen-13, which has a half-life of ten minutes. That means that once nitrogen-13 is formed, it must be incorporated into a suitable compound, made available to the tissues, meet whatever fate it will, and have its products isolated and investigated—all in a matter of half an hour or so. Even after a mere half an hour, the radioactivity is already only 1/8 what it was to start with.

For oxygen the situation is much worse because the most long-lived radioisotope known here is oxygen-15, which has a half-life of only two minutes.

In the case of carbon, the most long-lived radioisotope known before 1940 was carbon-11, which has a half-life of twenty minutes. This was a borderline situation. It left little time for maneuver, but of all the elements in living tissue, carbon was by far the most important; biochemists therefore worked out methods for squeezing information out of experiments using compounds tagged with carbon-11, despite the tight time-limit enforced by the short half-life.

It was not expected that any longer-lived carbon isotope would be discovered. In 1940, however, a new radioisotope of carbon was discovered as the result of the bombardment of carbon itself with *deuterons* (the nuclei of deuterium, H^2.)

A deuteron is made up of a proton and a neutron, and the carbon atoms undergoing the deuteron bombardment give off protons, retaining the neutrons. In a (d,p) reaction, the atomic number remains unchanged, but the mass number increases by one. Carbon is made up of two stable isotopes, carbon-12 and carbon-13. The former is converted to the latter by a (d,p) reaction, but the latter undergoes the following:

$$_6C^{13} + _1H^2 \longrightarrow _6C^{14} + _1H^1 \qquad \text{(Equation 10–3)}$$

or:

$$C^{13}(d,p)C^{14}$$

Carbon-14 is radioactive and has the unexpectedly long half-life of 5770 years. In terms of the duration of any laboratory experiment likely to be conducted with carbon-14, its radioactivity rate can be considered constant. Numerous biological and biochemical experiments were conducted with compounds tagged with carbon-14, and it is undoubtedly the most useful single radioisotope.

In 1946, the American chemist Willard Frank Libby (1908–) pointed out that carbon-14 should exist in nature as a result of nuclear reactions indirectly induced in the nitrogen-14 present in the atmosphere by energetic radiations from outer space.* This reaction is, in essence, the gaining of a neutron and the loss of a proton. In such an (n,p) reaction, there is no net change in the mass number, but a decrease of one in the atomic number. Thus:

* These radiations, called cosmic rays, will be taken up on page 217.

$$_{7}N^{14} + _{0}n^{1} \longrightarrow _{6}C^{14} + _{1}H^{1} \qquad \text{(Equation 10–4)}$$

or:

$$N^{14}(n,p)C^{14}$$

Carbon-14 is continually being formed in this way, and it is also continually breaking down after being formed. There is an equilibrium between the two processes, and the carbon-14 in the atmosphere (occurring in part of its carbon dioxide content) is at a constant, though very low, level.

Libby further pointed out that since plant life constantly absorbed and made use of carbon dioxide, its tissues ought to contain a constant, though very low, concentration of carbon-14, and so ought animal tissues, since animals feed on plants (or on other animals which feed on plants).

The constant concentration of carbon-14 in tissue was only maintained, however, while that tissue was alive, since only then was radioactive carbon being continually incorporated, either by absorption of atmospheric carbon dioxide or by the ingestion of food. Once a creature dies, intake of carbon-14 ceases, and the amount already present begins to decrease in a fixed manner.

Anything that was once part of a living organism can be analyzed for its carbon-14 content, and the time-lapse since life ended can be determined. This method of *radiocarbon dating* has been much used in archaeology. Wood from an old Egyptian tomb was found to be roughly 4800 years old, for instance, while wood from an old Etruscan tomb was about 2730 years old. The age of the Dead Sea Scrolls has been confirmed in this manner.

Wood from ancient trees knocked over by advancing glaciers can be tested, as can driftwood that once lay on the shores of lakes formed from melting glaciers. Scientists were surprised to discover that the last advance of the ice sheets that covered much of North America began but 25,000 years ago and reached its maximum extent about 18,000 years ago. This was not as long ago in the past as had previously been thought. Even as recently as 10,000 years ago, the retreating glaciers made a new partial advance, and it wasn't until 6000 B.C. (when men were already preparing to build their first civilizations) that the glaciers finally disappeared from the Great Lakes regions.

The (d,p) reaction that had led to the discovery of carbon-14 had earlier led to the discovery of the only radioisotope of hydrogen. In 1934, the Australian physicist Marcus Laurence Elwin Oliphant (1901–   ) had bombarded deuterium gas with deuterons. The heavy hydrogen nucleus (H^2) was thus both target and bombarding particle:

$$_1H^2 + {}_1H^2 \longrightarrow {}_1H^3 + {}_1H^1 \qquad \text{(Equation 10–5)}$$

or:

$$H^2(d,p)H^3$$

The hydrogen-3 formed in this way has the unexpectedly long half-life of 12.26 years. It has been named *tritium* (from a Greek word meaning "third") and its nucleus, composed of one proton and two neutrons, is a *triton*. Tritium is also formed naturally in the atmosphere through the action of high-energy radiation, so extremely small quantities are present in ordinary water. In very special cases, the decline in tritium content can be used in dating.

Units of Radioactivity

In using radioisotopes, what counts is not the mass alone but also the breakdown rate, for it is the latter that governs the quantity of particles being emitted per unit mass, and it is those particles which must be detected.

The breakdown rate (R_b) of a radioisotope can be expressed as follows:

$$R_b = \frac{0.693N}{T} \qquad \text{(Equation 10–6)}$$

where N is the total number of radioactive atoms present, and T is the half-life in seconds.

Let's consider a gram of radium. The mass number of the most long-lived radium isotope (and the one almost invariably meant when the unqualified word "radium" is used) is 226. This means that 226 grams of radium contains Avogadro's number of atoms, 6.023×10^{23} (see page 20). One gram of radium therefore contains Avogadro's number divided by 226, or 2.66×10^{21} atoms. The half-life of radium-226 is 1620 years, or 5.11×10^{10} seconds.

Substituting 2.66×10^{21} for N in Equation 10–6, and 5.11×10^{10} for T, we find a value of 3.6×10^{10} for R_b. This means that in a gram of radium, 36,000,000,000 atoms are breaking down each second.

In 1910, it was decided that the number of atomic breakdowns in one gram of radium be taken as a unit called a *curie,* in honor of the discoverers of radium. At the time, the calculation of this figure yielded the value of 37,000,000,000 breakdowns per second. One therefore defines 1 curie as equal to 3.7×10^{10} atomic breakdowns per second. The number of breakdowns per gram of radioisotope is its *specific activity*. Thus the specific activity of radium is 1 curie per gram.

What about other isotopes? The breakdown rate is inversely proportional to the half-life. The longer the half-life, the fewer the atomic breakdowns per second in a given quantity of radioisotope, and vice versa. The breakdown rate is therefore proportional to T_r/T_i where T_r is the half-life of radium-226 and T_i is the half-life of the other isotope.

For a fixed breakdown rate, the actual number of breakdowns in a gram of isotope is inversely proportional to the mass number of the isotope. If the isotope is more massive than radium-226, fewer atoms will be squeezed into one gram, and there will be fewer breakdowns in that one gram. The reverse is also true. The number of breakdowns will be proportional to M_r/M_i, where M_r is the mass number of radium-226 and M_i is the mass number of the isotope.

The specific activity (S_a) of a radioisotope—that is the number of breakdowns per second in one gram, as compared with that in one gram or radium—depends on the half-lives and mass numbers as follows:

$$S_a = \frac{T_r M_r}{T_i M_i} \qquad \text{(Equation 10–7)}$$

Since the half-life of radium-226 is 5.11×10^{10} seconds and its mass number is 226, the numerator of Equation 10–7 is equal to $226(5.11 \times 10^{10})$, or 1.15×10^{13}. Therefore:

$$S_a = \frac{1.15 \times 10^{13}}{T_i M_i} \qquad \text{(Equation 10–8)}$$

Thus, carbon-14, which has a half-life of 5770 years, or 1.82×10^{11} seconds, and a mass number of 14, has 2.55×10^{12} for its value of $T_i M_i$. If 1.15×10^{13} is divided by 2.55×10^{12}, we find that the specific activity of carbon-14 is 4.50 curies per gram. Carbon-14 has a longer half-life than radium-226, and that cuts down its breakdown rate. However, carbon-14 is a much lighter atom than radium-226; consequently, there are many more of the former per gram, and the actual number of breakdowns in that gram is greater than in the case of radium-226, despite the lower breakdown rate.

On the whole, most radioisotopes used in the laboratory have half-lives shorter and mass numbers smaller than that of radium, so that the specific activity is generally very high. Thus, carbon-11 has a half-life of 20.5 minutes, or 1230 seconds, a mass number of 11, and a specific activity of 850,000,000 curies per gram.

To be sure, gram lots of these isotopes are not used. They are generally not available in such quantities in the first place and would be highly dangerous if they were. Besides, such quantities are not needed. Particle detection is so delicate that the curie turns out to be a unit too large for convenience, and one more often speaks of *millicuries* (1/1000 of a curie) or *microcuries* (1/1,000,000 of a curie). Thus a microgram of carbon-11 is equivalent to 850 microcuries.

Even a microcurie represents a breakdown rate of 36,000 per second. Under the best conditions, four breakdowns per second may be detected with reasonable precision. This would represent 1/9000 of a microcurie, or 1.1×10^{-10} curie.

The use of the curie is made inconvenient to some extent by the fact that it represents a large and "uneven" number of atomic breakdowns per second. It has been suggested that the *rutherford* be used instead (named in honor of the discoverer of the nuclear atom). One rutherford is defined as a million atomic breakdowns per second.

This means that 1 curie = 37,000 rutherfords and that 1 rutherford = 270 microcuries.

Neutron Bombardment

As soon as the neutron was discovered, it occurred to physicists to use it as a bombarding particle to bring about nuclear reactions (and it was this, really, which eventually led to the wholesale production of radioisotopes). However, an apparent disadvantage of the neutron in such a role is its lack of charge. This means it cannot be accelerated by the electric fields used by all particle accelerators.

One way out of this dilemma was provided in 1935 by the American physicist John Robert Oppenheimer (1904–), who suggested the use of a deuteron instead. The deuteron is made up of a proton and a neutron in comparatively loose combination. A deuteron, with a charge of +1, can be accelerated. As the energetic deuteron approaches the positively-charged target nucleus, the proton component is repelled, sometimes strongly enough to break the combination. The proton veers off, but the neutron, unaffected by the repulsion, continues on, and if its aim is true, may be absorbed by the nucleus. The result resembles a (*d,p*) reaction of the type shown in Equations 10–3 and 10–5.

However, the inability to accelerate neutrons themselves is by no means a fatal defect. Indeed, it scarcely matters. A neutron, being neither attracted nor repelled by electric charge, can strike a nucleus (if aimed in the correct direction) regardless of how little energy it carries.

During the 1930's, streams of neutrons were produced from atoms subjected to bombardment by alpha particles. An alpha particle source mixed with beryllium served as a particularly useful neutron source.

A neutron may be absorbed by a target nucleus without the immediate emission of some other particle. Instead, the nucleus reaches an excited state as a result of absorbing the kinetic energy of the neutron and simply radiates off that excess energy as a gamma-ray photon. This is a (n,γ) reaction. The energy may not be written explicitly into the equation representing this reaction, thus:

$$_{48}Cd^{114} + {_0n^1} \rightarrow {_{48}Cd^{115}} \qquad \text{(Equation 10–9)}$$

or:

$$Cd^{114}\ (n,\gamma)\ Cd^{115}$$

The neutron, even more surely than the deuteron, can thus be used to produce higher isotopes of a target element.

It often happens that the higher isotope so produced is radioactive and breaks down by emitting a beta particle. This does not affect the mass number but raises the atomic number by one. Cadmium-115, for instance, is a beta-emitter with a half-life of 43 days, and is converted to indium-115.

If cadmium-116 had been bombarded with neutrons and converted to cadmium-117, there would follow a double change. Cadmium-117 is a beta-emitter with a half-life of about three hours and becomes indium-117, which is a beta-emitter with a half-life of about two hours and is converted, by beta particle emission, to the stable tin-117.

In many cases, then, neutron bombardment can produce an element one or two atomic numbers higher than the target element. The efficiency with which this may be achieved depends upon the probability of a (n,γ) reaction taking place. This probability can be dealt with as follows:

Imagine a target material one square centimeter in area and containing *N* atomic nuclei. Suppose it is bombarded by *I* particles per particle and that *A* atomic nuclei are hit per second.

The part of the target actually struck by the particles in one second is therefore A/N.

That part, however, is hit by all I particles lumped together. The part of the target hit by a single particle has to be A/N divided by I. The size of the target hit by a single particle is the *nuclear cross section,* which is symbolized as σ (the Greek letter "sigma"). We can say then that:

$$\sigma = \frac{A}{NI} \qquad \text{(Equation 10–10)}$$

By this analysis it would seem that in order to induce reaction a single bombarding particle must strike a particular area, σ square centimeters in size, centered about a particular target nucleus. The value of the nuclear cross section, as worked out by Equation 10–10, usually comes out in the neighborhood of 10^{-24} square centimeters. For convenience, nuclear physicists have defined 1 *barn* as equal to 10^{-24} square centimeters. (The story is that the name of the unit arose out of a statement that on the subatomic scale hitting an area 10^{-24} square centimeters in size, was like hitting the side of a barn on a familiar everyday scale.)

The value of the nuclear cross section varies with the nature of the target nucleus and with the nature of the bombarding particle. The Italian physicist Enrico Fermi (1901–1954) found in 1935 that neutrons became more efficient in bringing about nuclear reactions after they had been passed through water or paraffin. The nuclear cross section for bombarding neutrons on a given target nucleus increased, in other words, after the neutron's passage through water or paraffin.

In passing through water or paraffin, neutrons collided with light atoms that were particularly stable and therefore had little tendency to absorb an additional neutron. (They had low cross sections for neutron absorption, in other words.) As a result, the neutron bounced off.

When two objects bounce, there is usually a redistribution of kinetic energy between them. If one of the objects is moving and one is at rest, the moving object loses some energy and the object at rest gains some. The division of energy is most likely to be equal if the two bouncing objects are more or less equal in mass.

We can see this on a large scale if we imagine ordinary objects in place of subatomic particles. If a moving billiard ball collides with a ping-pong ball (the case of a neutron striking an electron), the ping-pong ball will bounce away vigorously, but the billiard ball loses little energy and goes on its way as before. On the other hand, if a moving billiard ball collides with a cannonball (the case of a neutron striking the nucleus of a lead atom), the billiard ball merely bounces, retaining its energy, while the cannonball is virtually unaffected. However, if a billiard ball strikes another billiard ball, it is quite likely that the two will end with roughly equal energies.

Consequently, a neutron is most efficiently slowed if it bounces off light nuclei such as those of hydrogen, beryllium or carbon, and it does this when passing through compounds, such as water and paraffin, made up of these light atoms. Such substances act as *moderators*. In the end, neutrons can be slowed until they are moving at no more than the velocity of atmospheric atoms and molecules under the influence of the local temperature (see page I–205). These are *thermal neutrons*.

Why then should nuclear cross sections rise as neutrons are slowed down? To answer this, we must remember that while neutrons have some particle-like properties, they also have wave-like properties. It had already been shown in the 1920's that electrons exhibit the wave-like properties predicted for them by de Broglie (see page 102), but there remained some question as to whether this might not apply only to charged particles. In 1936, it was shown that neutrons passing through crystals were diffracted, and the wave-particle duality was demonstrated for all matter and not for electrically charged matter only.

As a particle slows down, it loses energy. In its wave aspect this lowering of energy is represented by an increase in wavelength. A neutron therefore "spreads out" and grows "fuzzier" as it slows down. The larger, slow neutron is more likely to strike a nucleus than the smaller, fast neutron and is therefore more likely to bring about a nuclear reaction. It is also true that a slow neutron remains in the vicinity of the target nucleus a longer interval of time than a fast neutron, and this, too, encourages reactions.

Synthetic Elements

The development in the 1930's of new methods for bringing about nuclear reactions led to the formation not only of isotopes not found in nature, but of elements not found there.

In the 1930's, there existed just four gaps in the list of elements from atomic numbers 1 to 92 inclusive. These were the elements of atomic numbers 43, 61, 85 and 87.

The first of the four gaps to be filled was that of element number 43. Lawrence, the inventor of the cyclotron, had exposed molybdenum to a stream of accelerated deuterons, and it was possible that element number 43 had been produced in a (*d,n*) reaction:

$${}_{42}Mo^{98} + {}_{1}H^{2} \longrightarrow {}_{43}X^{99} + {}_{0}n^{1} \qquad \text{(Equation 10–11)}$$

or:

$Mo^{98}(d,n)X^{99}$

A sample of the irradiated molybdenum reached the Italian physicist Emilio Segrè (1905–) in 1937. He tested it by chemical methods to see if any part of the new radioactivity would follow the course to be expected of element number 43. It did; and it was amply confirmed that element 43 existed in the molybdenum. Since it was the first element to be discovered as the result of man-made nuclear reactions, it was named *technetium* ("artificial").

Not only was technetium the first man-made element, but it was also the first case of a light element (one with an atomic number of less than 84) that lacked any stable isotope whatever. There are no less than three technetium isotopes with quite long half-lives—technetium-97, 2,600,000 years; technetium-98, 1,500,000 years; and technetium-99, 210,000 years. However, no isotopes are completely stable. Since none of the half-lives of these isotopes are long enough to survive the ages of the earth's existence and since no technetium isotopes are part of a radioactive series, there are no measurable quantities of technetium present in the earth's crust.

In 1939, the French chemist Mlle. M. Perey discovered an isotope of element 87 among the breakdown products of uranium-235. She named it *francium,* after her native land. Element 85 was also later detected in the radioactive series, but as early as 1940 it had been produced artificially by the bombardment of bismuth with alpha particles and was named *astatine* ("unstable"). The reaction was:

$${}_{83}Bi^{209} + {}_{2}He^{4} \longrightarrow {}_{85}At^{211} + {}_{0}n^{1} + {}_{0}n_{1} \qquad \text{(Equation 10–12)}$$

or:

$Bi^{209}(\alpha,2n)At^{211}$

(Segrè had by now come to the United States and was a member of the group that isolated astatine.)

Element number 61 was discovered in 1948 (under circumstances to be described later in the book) by a team working under the American chemist Charles DuBois Coryell (1912–) and was named *promethium*. It was the second case of a light element without stable isotopes. Indeed, its most long-lived isotope, promethium-145, has a half-life of only 18 years.

By 1948, therefore, the periodic table had finally been filled up and its last gap removed. Meanwhile, however, the table had opened at its upper end. Fermi, mindful of the ability of the neutron to raise the atomic number of a target nucleus by one or two, had since 1934 been bombarding uranium with neutrons.

He felt that uranium-239 might be formed from uranium-238. By emitting beta particles this might become an isotope of element 93 and then possibly of element 94. He thought at first that he had actually demonstrated this and referred to the hypothetical element 93 as "Uranium X."

The discovery of uranium fission (see page 192) showed that Fermi had done far more then prepare element 93 and, for a while, element 93 was forgotten. However, when the furor of fisson had died down a bit, the question of element 93 was taken up again. The formation of uranium-239 was not the chief result of the neutron bombardment of uranium, nor the most important, but it was a result. It took place.

This was finally demonstrated in 1940 by the American physicist Edwin Mattison McMillan and his colleague, the American chemist Philip Hauge Abelson (1913–). They traced radioactivity showing a 2.3 day half-life and found it belonged to an isotope with an atomic number of 93 and a mass number of 239. Since uranium had originally been named for the planet Uranus, the new element beyond uranium was named *neptunium*, for Neptune, the planet beyond Uranus.

It seemed quite likely that this isotope, neptunium-239, was a beta particle emitter and decayed to an isotope of element number 94. However, the isotope so produced was apparently so weakly radioactive as to be difficult to detect in small quantities. By the end of the year, however, McMillan and a new assistant, the American chemist Glenn Theodore Seaborg (1912–), bombarded uranium with deuterons and formed neptunium-238:

$$ {}_{92}U^{238} + {}_{1}H^{2} \longrightarrow {}_{93}Np^{238} + {}_{0}n^{1} + {}_{0}n^{1} \quad \text{(Equation 10–13)} $$

or:

$$ U^{238}(d,2n)Np^{238} $$

Neptunium-238 emitted a beta particle and formed an isotope of element 94, one which was indeed radioactive enough to detect. The new element was named *plutonium,* after the planet Pluto, which is beyond Neptune.

Once plutonium was formed in sufficient quantity, it was bombarded with alpha particles, and in 1944 a research team headed by Seaborg formed isotopes of element 95 (*americium,* for America) and 96 (*curium,* for the Curies).

Still higher elements were formed by Seaborg's group. In 1949 and 1950, elements 97 and 98 were formed by the bombardment of americium and curium with alpha particles. Element 97 was named *berkelium* and element 98 was named *californium,* after Berkeley, California, where the work was done.

Elements 99 and 100 were formed in the laboratory in 1954, but two years earlier, in 1952, isotopes of these elements were found in the residue of a hydrogen bomb test explosion (see page 208) at a Pacific atoll. By the time these discoveries were confirmed and the announcement was made, both Einstein and Fermi had died, and in their honor, element 99 was named *einsteinium* and element 100, *fermium.*

In 1955, element 101 was formed by the bombardment of einsteinium with alpha particles, and was named *mendelevium,* in honor of Mendeleev, the discoverer of the periodic table. In 1957, the discovery of element 102 was announced at the Nobel

TABLE XI—*Transuranium Elements*

Atomic Number	*Element*	*Mass Number of Most Long-Lived Isotope*	*Half-Life*
93	Neptunium	[237]	2,140,000 years
94	Plutonium	[242]	37,900 years
95	Americium	[243]	7,650 years
96	Curium	[247]	c. 40,000,000 years
97	Berkelium	[247]	c. 10,000 years
98	Californium	[251]	c. 800 years
99	Einsteinium	[254]	480 days
100	Fermium	[253]	c. 4.5 days
101	Mendelevium	[256]	1.5 hours
102	Nobelium	[253]	c. 10 minutes
103	Lawrencium	[257]	8 seconds

Institute in Stockholm and was named *nobelium,** and in 1961, element 103 was identified and named *lawrencium* in honor of the discoverer of the cyclotron, who had died some years earlier. In 1964, Soviet physicists announced the formation of element 104, but this has not yet been confirmed.

The elements beyond uranium are generally spoken of as the *transuranium elements.* Nearly a hundred isotopes of these elements have been formed. In Table XI, the most long-lived known isotopes of these elements are presented.

The chief theoretical interest in these elements is in the light they have thrown on the higher reaches of the periodic table. Before the discovery of the transuranium elements, thorium had been placed under hafnium in the periodic table; protactinium under tantalum; and uranium under tungsten. There was some chemical evidence in favor of this arrangement.

Working on this basis, when neptunium was discovered it should have fitted under rhenium. However, the chemical properties of neptunium revealed themselves almost at once to be much like uranium, and the other transuranium elements agreed in this respect. It turned out (as Seaborg was first to suggest) that the elements from actinium on formed a new "rare earth" series (see page 18) and should be fitted under the first series from lanthanum on. This is done in the periodic table presented on page 16.

The first series, from lanthanum to lutetium inclusive, is now called the *lanthanides* after the first member. Analogously, the second series, from actinium to lawrencium inclusive, is that of the *actinides.* Lawrencium is the last of the actinides and chemists are quite certain that when element 104 is obtained in quantity sufficient for its chemical properties to be studied it will turn out to resemble hafnium.

While a few of the transuranium isotopes have half-lives that are long in human terms, none are long in geologic terms, and none have survived over the eons of earth's history. (Nevertheless, traces of neptunium and plutonium have been located in uranium ores. They have arisen from the reaction of neutrons—occurring naturally in air as a result of the nuclear reaction induced by high-energy radiation from outside earth—with uranium.)

Neptunium-237 is of particular interest. Its mass number

* Attempts to duplicate the Swedish work failed. Element 102 has been formed by other methods than those described at the Nobel Institute, and the name "nobelium" is not as yet officially accepted.

divided by 4 leaves a remainder of 1, so that it belongs to the $4n + 1$ group of mass numbers, the group for which there is no naturally-occurring radioactive series (see page 132). With a half-life of over two million years, it is the longest-lived member (as far as is known) of this group. It can serve therefore as parent atom for a *neptunium* series. It gives rise to a series of daughter atoms that do not duplicate any of the products of the other three series, see Table XII.

The most distinctive feature of the neptunium series is that it ends with bismuth rather than with lead, as do the other three series. Naturally, since the parent atom of the series could not survive through earth's history, neither could any of the shorter-lived daughter atoms. The entire series is extinct except for the final stable product, bismuth-209.

TABLE XII—*The Neptunium Series*

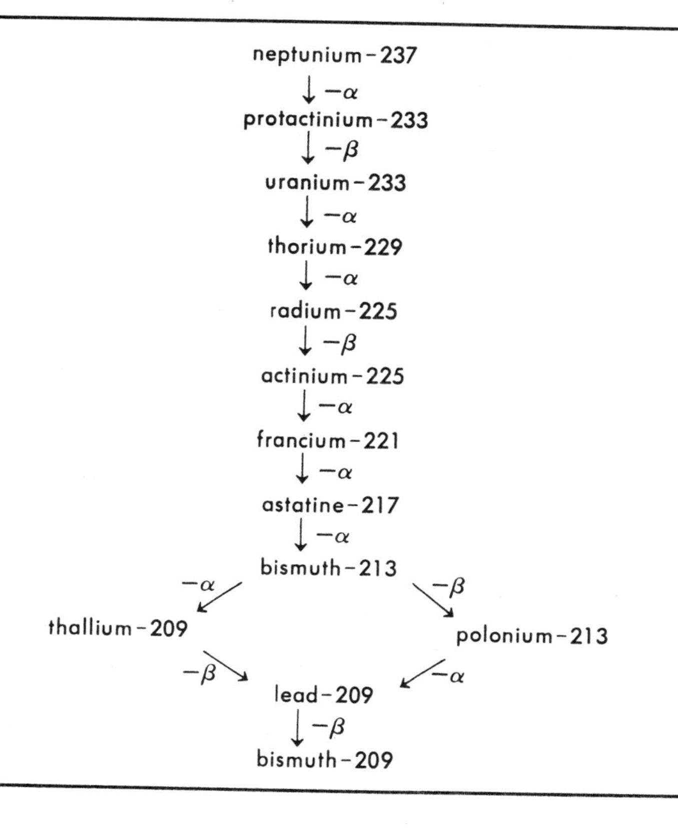

CHAPTER **40**

Nuclear Structure

Nucleons, Even and Odd

With the entire list of isotopes, stable and unstable, spread before us, it is possible to make certain statements about nuclear structure.

To begin with, one can have an atom with a single proton as its nucleus; that is the case with hydrogen-1. No nucleus, however, can contain more than one proton without also containing a neutron. Among the elements with small atoms, stable nuclei tend to be made up of equal or nearly equal numbers of protons and neutrons. Thus, hydrogen-2 contains one of each; helium-4, two of each; carbon-12, six of each; oxygen-16, eight of each: sulfur-32, sixteen of each; and calcium-40, twenty of each.

The trend does not persist. All stable nuclei more massive than calcium-40 contain more neutrons than protons, and the unbalance becomes more marked as the mass number increases. Thus, the most common iron isotope, iron-56, contains 26 protons and 30 neutrons, for a neutron/proton (n/p) ratio of 1.15. The most common silver isotope, silver-107, contains 47 protons and 60 neutrons for an n/p ratio of 1.27. The only stable bismuth isotope, bismuth-209, which has the distinction of being the most massive of the stable isotopes, contains 83 protons and 126 neutrons, for an n/p ratio of 1.52; the most massive naturally-occurring isotope, uranium-238, with 92 protons and 146 neutrons, has an n/p ratio of 1.59.

Apparently, as more and more protons are packed into the nucleus, a larger and larger excess of neutrons is required to keep the nucleus stable. By the time 84 protons, or more, exist in the nucleus, no number of neutrons will suffice for stability. (And, of course, too many neutrons are as bad as too few.)

It seems quite clear that the existence of protons in pairs has a stabilizing effect on the nucleus. Of the nuclei containing more than one nucleon, those with protons in pairs (and therefore possessing an even atomic number) are the more widespread in the universe. Thus, six elements make up about 98 percent of the planet we live on (counting its interior as well as its crust) and these are: iron, oxygen, magnesium, silicon, sulfur and nickel. The atomic numbers are 26, 8, 12, 14, 16, and 28 respectively—all even.

This is reflected also in the ease with which even numbers of protons are stabilized as compared with odd numbers. For elements with atomic numbers over 83, no number of neutrons will suffice to stabilize the nucleus, but for two elements in this group, stability is nearly achieved. They are thorium and uranium, with atomic numbers of 90 and 92, both even. On the other hand, there are only two elements with atomic numbers under 83 that possess no stable isotopes. These are technetium and promethium, with atomic numbers 43 and 61, both odd.

Consider next the number of isotopes per element. There are 21 elements possessing only one naturally-occurring isotope. Of these, two have even atomic numbers: beryllium (atomic number 4) and thorium (atomic number 90). The other 19 all have odd atomic numbers. Then there are 23 elements with only two naturally-occurring isotopes. Again two of these have even atomic numbers: helium (atomic number 2) and uranium (atomic number 92). And again the other 21 all have odd atomic numbers.

Indeed, it would appear that the possession of an odd number of protons in the nucleus makes stabilization so touch-and-go that only one particular number of neutrons, or at most two, will do. Only a single element of odd atomic number possesses more than two naturally-occurring isotopes, and this is potassium (atomic number 19). It has three isotopes: postassium-39, potassium-40, and potassium-41. Of these three, however, potassium-40 is slightly radioactive and quite rare.

On the other hand, all but four of the naturally-occurring elements with even atomic numbers possess more than two naturally-occurring isotopes; indeed, tin (atomic number, 50) possesses ten. It is as though the possession of an even number of protons makes stabilization so easy that it is possible to carry it through with any of a wide variety of neutron numbers.

Neutrons also seem to occur most readily in pairs. Of the six elements earlier referred to as making up 98 percent of the earth, the most common isotopes are iron-56, oxygen-16, magnesium-24, silicon-28, sulfur-32, and nickel-58. The proton-neutron contents are 26-30, 8-8, 12-12, 14-14, 16-16, and 28-30. In every case there are even numbers of both protons and neutrons ("even-even nuclei").

Among the elements of odd atomic number which possess only one naturally-occurring isotope, in every case the single isotope possesses an even number of neutrons ("odd-even nuclei"). Examples are fluorine-19 (9 protons, 10 neutrons), sodium-23 (11 protons, 12 neutrons), phosphorus-31 (15 protons, 16 neutrons) and gold-197 (79 protons, 118 neutrons).

Where elements of odd atomic number possess two naturally-occurring isotopes, in almost every case both have an even number of neutrons. Thus chlorine occurs as chlorine-35 and chlorine-37, with 17 protons and either 18 or 20 neutrons. Copper occurs as copper-63 and copper-65, with 29 protons and either 34 or 36 neutrons. Silver occurs as silver-117 and silver-119, with 47 protons and either 60 or 62 neutrons.

Elements of even atomic number, with three or more naturally-occurring isotopes, usually have a larger number with even numbers of neutrons than of odd (the latter being "even-odd nuclei"). As an example, xenon possesses nine naturally-occurring isotopes, of which seven are "even-even" (xenon-124, 126, 128, 130, 132, 134, and 136). The number of protons in each is 54, while the number of neutrons is 70, 72, 74, 76, 78, 80, and 82 respectively. Only two "even-odd" naturally-occurring xenon isotopes exist. These are xenon-129 and xenon-131, with neutron numbers of 75 and 77.

With one exception no element possesses more than two "even-odd" isotopes. The exception is tin, which contains three of them, tin-115, 117 and 119. Here the number of protons is 50, and the number of neutrons is 65, 67, and 69. (However, tin possesses seven "even-even" isotopes.)

The rarest of all nuclei are the "odd-odd nuclei" which contain odd numbers of both protons and neutrons. Only nine of these are naturally-occurring; of these nine, five are slightly radioactive and only the four simplest are completely stable.

The four stable "odd-odds" are hydrogen-2 (one proton, one neutron); lithium-6 (three protons, three neutrons); boron-10 (five protons, five neutrons); and nitrogen-14 (seven protons, seven neutrons). Of these, three are rare within their own element. Hydrogen-2 makes up only 1 out of 7000 hydrogen atoms; lithium-6 only 2 out of 27 lithium atoms; and boron-10 only 1 out of 5 boron atoms.

Nitrogen-14 is the surprising member of the group. It makes up 996 out of every 1000 nitrogen atoms, far outweighing the only other stable nitrogen isotope, nitrogen-15, an "odd-even" made up of seven protons and eight neutrons.

The alpha particle, made up of a pair of protons and a pair of neutrons, is particularly stable. When radioactive atoms eliminate nucleons, they never do so in units of less than an alpha particle.

The alpha particle is so stable that a nucleus made up of a pair of them (four protons plus four neutrons) is extremely unstable, almost as though the alpha particles are far too self-contained to have any capacity whatever to join together. Such a nucleus would be that of beryllium-8, which has a half-life of something like 3×10^{-16} seconds.

On the other hand, carbon-12, oxygen-16, neon-20, magnesium-24, silicon-28, sulfur-32, and calcium-40, which may be looked upon, after a fashion, as being made up of the union of 3, 4, 5, 6, 7, 8, and 10 alpha particles, respectively, are all particularly stable.

Some of the phenomena of natural radioactivity may be interpreted in the light of what has already been said. Atoms such as those of uranium-238 or thorium-232, in order to achieve stability, must reduce the number of protons in the nucleus to not more than 83.

To do so, alpha particles are ejected, but this eliminates neutrons as well as protons. When an equal number of protons and neutrons are eliminated, where neutrons are already present in excess, the n/p ratio rises. Thus the n/p ratio in uranium-238 (92 protons, 146 neutrons) is 1.59. If a uranium-238 managed to eject five alpha particles, it would lose ten protons and bring its atomic number down to 82 (that of lead) for possible stability. However, it would also have lost ten neutrons for a total decline in mass number of 20, and it would be lead-218. There the n/p ratio (82 protons, 136 neutrons) would be 1.66. So high an n/p ratio is completely incompatible with stability, and indeed lead-218 has never been detected. The most massive known lead isotope is lead-214, with a half-life of less than half an hour.

As the atomic weight is decreased, the n/p ratio must decrease also if stability is to be achieved. To do this, a neutron is converted to a proton and a beta particle is emitted. By a combination of alpha and beta emission, uranium-238 eventually becomes lead-206, with a loss of 10 protons and 22 neutrons and a decline in the n/p ratio from 1.59 to 1.51.

The regularities in proton-neutron combinations show clearly that stable nuclei are not built up in random fashion, but according to some orderly system. It has seemed to some physicists that just as orderliness was introduced into the chemical aspects of the elements by means of a periodic table eventually found to be based on electron shells (see Chapter 5), so order could be brought to the nuclear properties by means of a system of nuclear shells.

Such a system was advanced in 1948 by the Polish-American physicist Maria Goeppert-Mayer (1906–). She pointed out that isotopes containing certain numbers of protons or neutrons were particularly common or particularly stable. These numbers are called *shell numbers,* or, more dramatically, *magic numbers,* and they are 2, 8, 20, 50, 82, and 126.

Thus, helium-4 is made up of two protons and two neutrons; oxygen-16 of eight protons and eight neutrons and calcium-40 of 20 protons and 20 neutrons, and all three are particularly stable isotopes. Again, the element with the greatest number of stable isotopes is tin, whose nuclei contain 50 protons. There are also six naturally-occurring isotopes containing 50 neutrons (among which is rubidium-87, which is very slightly radioactive). There are seven stable isotopes containing 82 neutrons and four (those of lead) containing 82 protons.

Nor is it numbers of isotopes alone that count. Other nuclear properties seem to reach significant maxima or minima at the magic numbers. Thus isotopes containing a magic number of either protons or neutrons seem to have lower cross sections with respect to neutron absorption than do other isotopes of similar complexity.

Mrs. Goeppert-Mayer has attempted to account for these magic numbers by assuming the protons and neutrons to be arranged within the nucleus in *nucleon shells,* which are filled according to an arrangement of nuclear quantum numbers. The magic numbers are those at which key shells are completely filled (analogously to the situation of the inert gases in connection with electron shells).

This "nuclear periodic table" has had some triumphs. It has been used to predict which nuclides could exist in excited states for a significant length of time, forming nuclear isomers (see page 128). However, this model is still a subject of considerable controversy.

Packing Fractions

The stability of a particular nuclide rests not only on *n/p* ratios, but, more fundamentally, on the energy relationship of a particular nuclide with other nuclides of equal nucleon counts.

To see how this can be, let's begin by considering that though the mass number of an isotope is usually given as a whole number, it is not quite a whole number. We speak of oxygen-18, potassium-41 and uranium-235, assuming the mass numbers to be 18, 41, and 235 respectively.

Aston's mass spectrograph (see page 139) made it possible, however, to measure the mass of individual isotopes with great precision. We now know that on the carbon-12 scale the actual mass of the oxygen-18 nucleus is 17.99916, while that of potassium-41 is 40.96184 and that of uranium-235 is 235.0439.

This may seem strange in view of the belief that the nucleus is made up of protons and neutrons only, with each one of those particles possessing unit mass. But do protons and neutrons indeed have a mass of precisely 1? They do not. On the carbon-12 scale, the mass of the proton is 1.007825 and that of the neutron is 1.00865.

But this raises another question. The carbon-12 nucleus is made up of six protons and six neutrons. The 12 nucleons, considered one at a time, have a total mass of 12.098940, yet the mass of these same nucleons combined into a carbon-12 nucleus is 12.00000. There is a *mass defect* of 0.098940; what has happened to it?

In the light of Einstein's equation showing the equivalence of mass and energy (see page II–111), it seems clear that the extra mass has been turned into energy.

Six protons and six neutrons in combining to form a carbon-12 nucleus lose a little less than 1 percent of their total mass and liberate that as energy. If it were desired to break up the carbon-12 nucleus into individual nucleons again, that quantity of energy must be supplied it again. It is the difficulty of collecting and delivering this energy that keeps carbon-12 stable. The energy holding the nucleons within a nucleus together is much greater than the energy holding the atoms of a molecule, or the molecules of a solid mass together. It is also greater than the energy holding electrons within the atom. For that reason, procedures which suffice to melt a solid or decompose a compound, or even ionize an atom, fail utterly in any effort to break up an atomic nucleus.

Yet if nuclei cannot be broken up into individual nucleons

without prohibitive expenditure of energy, less drastic changes are possible; some of these less drastic changes even take place spontaneously.

To begin with, the more energy per particle given off in forming the nucleus by "packing together" individual nucleons, the more stable that nucleus tends to be (all other things being equal). A way of measuring this energy of nuclear formation is to subtract the mass number (A) from the actual mass of the isotope (A_m). This difference, the mass defect, can be divided by the actual mass to give the fractional mass defect. To remove the decimal, the result is customarily multiplied by 10,000 to yield what Aston called the *packing fraction*. If we let P_f represent the packing fraction, then we can say:

$$P_f = \frac{10{,}000\,(A_m - A)}{A_m} \qquad \text{(Equation 11–1)}$$

The lower the packing fraction, the greater the loss of mass in forming the nucleus, and the greater the tendency toward stability.

Among the elements, the packing fraction is highest for hydrogen. The actual mass of the hydrogen-1 nucleus (the bare proton) is 1.007825. If this is substituted for A_m in Equation 11–1, and 1 is substituted for A, then the packing fraction turns out to be 78.25. This is not surprising since a single proton isn't

Aston's packing fraction curve

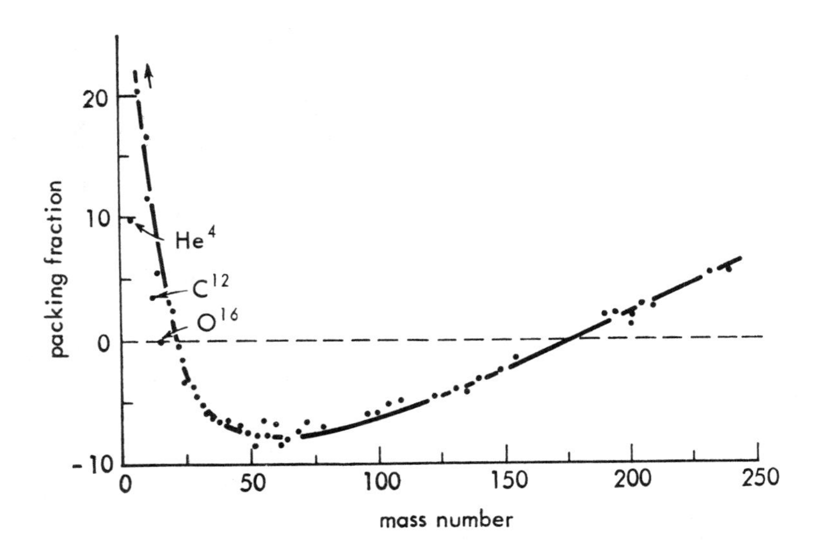

"packed together" at all in forming a nucleus. The packing fraction of a single neutron is even higher, for it is 86.7.

On the other hand, lithium-7, with A_m equal to 7.01601 and A equal to 7, has a packing fraction of 22.9, while carbon-13 with A_m equal to 13.00335 and A equal to 13, has a packing fraction of 2.4.

In general, starting at hydrogen-1, the packing fraction decreases for quite a while; this indicates that the nuclei of proper n/p ratio tend to grow more stable as they grow more complicated. To put it another way, if two very simple nuclei are combined to form a more complicated one, energy is released.

By the time nitrogen-15 is reached, the packing fraction is just about 0, but for still more complex nuclei the packing fraction falls into negative numbers. (This is a consequence of the fact we have chosen to let carbon-12 equal exactly 12. Had we established mass numbers on the basis of iron-56 equal to exactly 56, there would be no negative packing fraction numbers.)

For potassium-41, for instance, the value of A_m is 40.96184 and that of A is 41, so that the packing fraction is −9.3. A minimum is finally reached at iron-56, which has a packing fraction of −11.63. Thereafter, the packing fraction increases again, so that for tin-120, for instance, it is −8.1, and for iridium-191, −2.0. At the extreme end of the list of tables the packing fractions are positive again. For uranium-238, it is +2.1.

This means that the middle-sized atoms such as iron and nickel are the most stable of all. Not only is energy released if very simple atoms are built up to more complicated ones, but also if very complicated elements are broken down to less complicated ones.

All this is reflected in the general composition of the universe. On the whole, estimates of the distribution of elements in the universe, based on astronomic data, indicate that the more complicated an element, the rarer it is. Some 90 percent of the atoms in the universe are hydrogen (the simplest element) and another 9 percent are helium (the next simplest). Nevertheless, because of the particular stability of iron, it is to be expected that iron ought to be more common than other elements of similar complexity. This is indeed true, and our own planet, which is not massive enough to have retained the very simplest atoms, is about 35 percent iron in mass.

Packing fractions are particularly low for carbon-12 and oxygen-16 (which can be looked upon as being made up of alpha particles) and are especially low for helium-4 itself (which *is* the alpha particle). Thus the packing fraction of lithium-6 is 25.2 and that of hydrogen-2 is 70. Since helium-4 is midway between these in mass, one might suppose that its packing fraction would also fall roughly midway between. It is, however, only 6.5, much lower than either. It is not surprising then that helium, carbon and oxygen are among the more common atoms in the universe.

Whether a particular nuclide is stable or not, depends not only on its own packing fraction, but on the packing fraction of nuclides of equal nucleon number. For instance, sodium-24 (11 protons, 13 neutrons) might be suspected of stability if it could be considered by itself. However, magnesium-24 (12 protons, 12 neutrons) has a lower packing fraction. Therefore, if sodium-24 emits a beta particle and changes its nucleon arrangement from 11-13 to 12-12, it loses energy and gains stability. Beta emission is a cheap price to pay for this gain. Whereas it would take prohibitive energy to break up the sodium-24 nucleus altogether, only a slight amount of energy suffices to set off the change involved in a beta particle emission. Consequently, sodium-24 emits beta particles spontaneously and breaks down to magnesium-24, with a half-life of 15 hours.

Two neighboring isotopes of the same mass number cannot possibly both be stable. The one with the higher packing fraction spontaneously changes into the one with the lower. It's like rolling down an "energy hill" and the steeper the hill the shorter the half-life.

Two isotopes of the same mass number, but not neighbors, can both be stable. Thus, zinc-64 (30 protons, 34 neutrons) and nickel-64 (28 protons, 36 neutrons) are both stable, for they are separated by copper-64 (29 protons, 35 neutrons), which has a higher packing fraction than either. Both zinc-64 and nickel-64 may be visualized as occupying "energy hollows" with an "energy hump" (on which copper-64 is perched) lying between. Copper-64 is indeed unstable and can break down in either of two ways. It can emit a beta particle to become zinc-64, or it can emit the opposite of a beta particle (see page 223) to become nickel-64.

Sometimes the "energy hump" is very low, and the isotope existing there is nearly stable. This is the case of potassium-40 (19 protons, 21 neutrons), which lies between two stable isotopes, argon-40 (18 protons, 22 neutrons) and calcium-40 (20 protons, 20 neutrons). Potassium-40 is only slightly radioactive and it, too, breaks down in two fashions, one of which yields calcium-40 and the other argon-40.

Nuclear Energy

As soon as the existence of nuclear energy* was accepted, scientists began to speculate on the possibilities of putting it to use. Some isotopes already exist which, in effect, stand at the top of a very gentle slope and have been slowly rolling down it an atom at a time. These are the isotopes, uranium-238, uranium-235 and thorium-232, of course.

For instance, uranium-238 breaks down in a number of steps to form lead-206. In doing so, it gives off beta particles and gamma rays, the mass of which may be ignored, and also eight alpha particles, the mass of which may not be ignored. Including only the massive items, we can write: $U^{238} \longrightarrow Pb^{206} + 8He^{4}$.

The mass of the uranium-238 nucleus is 238.0506, that of the lead-206 nucleus is 205.9745, and that of the alpha particle is 4.00260. The total mass of the lead-206 nucleus plus eight alpha particles is 237.9953. This means that in the radioactive breakdown of uranium-238 to lead-206, each uranium-238 nucleus loses 238.0506 — 237.9953, or 0.0553 atomic mass units.

This can be escalated to the gram level. If 238 grams of uranium break down completely to lead, then 55.3 milligrams of mass are converted to energy. This is a conversion of 0.225 milligrams of mass for each gram of uranium breaking down completely.

According to Einstein, $e = mc^2$, where e represents energy in ergs, m represents mass in grams, and c represents the velocity of light in centimeters per second. The velocity of light is 3×10^{10} centimeters per second, and squaring that gives us 9×10^{20}. If that is multiplied by 0.225 milligrams (or 2.25×10^{-4} grams) we find that a gram of uranium, breaking down completely, liberates 2.5×10^{17} ergs, or just under 5,000,000 kilocalories.

If one gram of gasoline is burned, some 12 kilocalories of energy are liberated. We see then that the energy delivered through

* Nuclear energy is commonly referred to as "atomic energy," and the phrase is even enshrined in such names as the "Atomic Energy Commission." This is certainly wrong, for the electrons are as much a part of the atom as the nucleus is, and energy derived from chemical reactions, which involve electron transfers, have every right to be referred to as "atomic energy." Nevertheless, names such as "atomic energy" and equivalent misnomers such as "atomic bombs" and "atomic submarines" can never be wiped out for the more accurate "nuclear bombs" and "nuclear submarines." In this book, I shall use the adjective "nuclear" as a matter of principle and not because I expect it will change anything.

the radioactive breakdown of a gram of uranium is 420,000 times as great as that delivered by an equal mass of burning gasoline and is, in fact, equivalent to the explosion of about 5000 tons of TNT. This is a fair comparison of the relative intensities of nuclear energy and chemical energy.

But why, then, had man remained so unaware of this vast energy release by uranium? (He was always aware of the comparatively tiny energy release of, say, a candle flame.) The answer is clear. Uranium releases a great deal of energy in breaking down, but spreads that energy over a vast expanse of time. The gram of uranium delivers half its energy, or 2,500,000 kilocalories, over a period of 4,500,000,000 years. In one second it delivers far less energy than a candle flame does.

To be sure there are isotopes more intensely radioactive than uranium-238. Consider polonium-212, one of the daughter nuclides of thorium-232 and therefore always present in thorium ores. A gram of polonium-212 will decay to lead-208 by emitting one alpha particle. In doing so, it loses only 0.046 milligrams. This is only a sixth the mass lost by uranium-238 all-told, and so polonium-212 liberates only a sixth the energy. It liberates less than 1,000,000 kilocalories, or only about as much energy as would be turned loose by a mere 1000 tons of exploding TNT. However, polonium-212 has a half-life of less than a millionth of a second, and all that energy would be delivered in an instant. The explosion would be shattering. However, there is no way of accumulating a full gram of natural polonium-212. All earth's crust would have to be combed for it, and that still might prove insufficient.

After 1919, it became possible to deal with intensely radioactive nuclides without searching for them in nature; they could be synthesized. By bombardment with alpha particles or with artificially accelerated protons, stable nuclides could be knocked out of their "energy hollows," so to speak, and sent skittering up to some "energy hump" in the form of a radioisotope. The radioisotope would then, either quickly or slowly, slide back down into a hollow. Might not the energy released then be utilized?

Of course it might; and it is, every time a radioisotope is detected by a counter, and every time it is used as a source of bombarding particles. However, the amount of energy expended uselessly, in order to knock only an occasional atom out of its "energy hollow" is much greater than the amount that is gained when the resulting radioisotope rolls back.

How, then, arrange matters so as to turn a profit? For one thing, the return on the original outlay might be increased by making the exploding radioisotope itself do the work of forming more radioisotopes.

Thus, a carbon-12 nucleus, if struck by a neutron under the proper conditions, may absorb it and emit two neutrons. The reaction can be written:

$${}_6C^{12} + {}_0n^1 \longrightarrow {}_6C^{11} + {}_0n^1 + {}_0n^1 \qquad \text{(Equation 11–2)}$$

or:

$$C^{12}(n,2n)C^{11}$$

Suppose then, a carbon-12 nucleus struck by a neutron gave up two neutrons, each of which struck a carbon-12 nucleus to produce a total of four neutrons, each of which . . . Nuclear reactions take place in millionths of seconds or less, so that if the number of breakdowns increases to 2, 4, 8, 16, 32 and so on in steps of millionths of seconds, the entire supply of carbon-12 would have undergone a nuclear reaction. Carbon-12, a very common substance, would deliver as much energy, as quickly, as polonium-212.

It is precisely this sort of thing that happens (on a chemical energy scale) when with a single match we burn down a forest. The match supplies the energy to set a leaf burning; the heat developed by the burning leaf ignites neighboring objects, and so on. Where the product of a reaction serves as the condition for continuing the reaction, chemists speak of a *chain reaction*. What the $(n,2n)$ process offered was the chance of a *nuclear chain reaction.*

However, the $(n,2n)$ process does not work. All the examples so far discovered require fast, energetic neutrons. A fast neutron sent into a carbon-12 target will liberate two neutrons, but slow ones. They are invariably much less energetic than the incident neutron; not energetic enough to initiate a new reaction of the sort that had given rise to them.

It is like trying to burn wet wood. You may set a tiny flame going, but it will not deliver enough heat to dry neighboring sections of wood so that they may burn—and the fire sputters out. Nor is this a bad thing. There are always stray neutrons present in the atmosphere and they are low energy. If they sufficed to start a nuclear chain reaction, then large parts of earth's crust might be subject to almost instantaneous nuclear explosions, and planets as we know them might not exist. The fact that the earth does exist may be evidence that no $(n,2n)$ process involving common atoms offers a practical nuclear chain reaction.

That was the situation, then, up to 1939. Although physicists knew that nuclear energy existed in tremendous quantities, there was no practical method of tapping it. There even seemed reason to believe that no practical method could conceivably exist. Rutherford, for example, was convinced (and said so) that the development of a practical source of large-scale nuclear energy was an idle dream. He died just a few years too soon to see his reasons refuted.

Nuclear Fission

The situation with respect to the utilization of nuclear energy changed radically in the late 1930's. Fermi had been bombarding uranium with thermal neutrons and had felt he had formed element 93. In a way he was right, but he had also induced other nuclear reactions that confused the results and left him puzzled.

Other physicists tackled the problem and were equally puzzled. Up to that point all nuclear reactions studied, whether natural or artificially produced, had involved the emission of small particles, no more massive than an alpha particle. Consequently, physicists tried to associate the various types of radioactivity in the bombarded uranium with atoms only slightly smaller than uranium.

The German physicist Otto Hahn (1879–) and his coworker, the Austrian physicist Lise Meitner (1878–), found in 1938 that when barium compounds were added to the bombarded uranium, a certain type of radioactivity followed the barium through all the chemical manipulations to which it was subjected. Since barium is very like radium from the chemical standpoint (radium is just under barium in the periodic table), Hahn supposed he was dealing with a radium isotope.

However, nothing he could do would separate the barium carrier from the radium he supposed was accompanying it. Even manipulations that would ordinarily separate barium from radium failed. He found himself forced, little by little, to suppose that it was not a radium isotope he was dealing with, but a radioactive barium isotope.

Consider the consequences of this thought. The barium isotopes have an atomic number of 56, which is 32 less than that of uranium. To form a barium isotope, a uranium atom would have to unleash a flood of eight alpha particles, and no such flood of alpha particles was detected in neutron-bombarded uranium. It began to seem necessary to suppose that the uranium nucleus, upon absorbing a neutron, might simply break in half (more or less). This process came to be called *uranium fission* or, more generally, *nuclear fission,* since isotopes other than those of uranium were eventually found to be subject to it.

Such nuclear fission makes sense in that it involves a slide down an "energy hill" even more extensive than that brought about in ordinary radioactive transformations. Where under ordinary conditions uranium is converted to lead, with a lower packing fraction, in the case of fission, uranium is converted to such atoms as barium and krypton, which have still lower packing fractions.

Thus, while a gram of uranium, converted to lead by the ordinary radioactive process, will lose about 1/4 milligram, that same gram of uranium undergoing fission will lose just about 1 milligram in mass. In other words, uranium fission will yield some four times as much energy, gram for gram, as ordinary radioactivity will.

Uranium fission seems to fit in well with a theoretical model of nuclear structure advanced by Bohr. In this model, the nucleus is viewed as analogous to a drop of liquid (it is referred to, in fact, as the *liquid-drop model*). Instead of considering the nucleons as occupying different shells and behaving with relative independence, as in the shell model, the nucleons are considered as jostling each other randomly like molecules in a drop of liquid.

A neutron entering such a nucleus has its energy absorbed and distributed among all the nucleons very quickly, so that no one nucleon retains enough energy to eject itself from the nucleus. The surplus energy could be gotten rid of as a gamma ray, but there is also a distinct possibility that the entire nucleus will be set to oscillating as a liquid drop might under similar conditions. There is then a further possibility that before the energy could be eliminated as a gamma ray, the nucleus might oscillate strongly enough to break in two.

When a uranium nucleus breaks in two in this fashion, it does not always divide in exactly the same way. The packing fraction among nuclei of moderate size does not vary a great deal, and the nucleus may well break at one point in one case and at a slightly different point in another. For this reason, a great variety of radioisotopes are produced, depending on just how the division takes place. They are lumped together as *fission products.* The probabilities are highest that the division will be slightly unequal, with a more massive half in the mass number region of from 135 to 145 and a less massive half in the region of from 90 to 100.

It was among these fission products that isotopes of element number 61 were first isolated in 1948. The new element was named "promethium" because it was snatched out of the nuclear furnace in the same way that fire was supposed to have been snatched from the sun by the Greek demigod Prometheus.

In producing relatively small fission products, the uranium atom is brought to a portion of the list of elements where the n/p ratio is smaller. Fewer neutrons are needed in the nuclei of the fission products than in the original uranium nucleus, and these superfluous neutrons are liberated. In consequence, each uranium atom undergoing fission liberates two or three neutrons.

One might wonder why, if uranium fission liberates more energy than ordinary uranium breakdown does, uranium does not spontaneously undergo fission rather than ordinary breakdown. Apparently, before any change can take place, the nucleus must absorb a small quantity of energy that will carry it over an "energy hump" before it can start sliding down the "energy hill." We have here a sort of "nuclear ignition" that is analogous to the heat of friction that starts a match burning.

The higher the energy hump, the less likely an individual nucleus is to gain sufficient energy to pass over it in the ordinary course of events in which energy is constantly being randomly distributed and redistributed among subatomic particles. Therefore, the higher the energy hump, the fewer the nuclei that will undergo breakdown in any particular time interval, and the longer the half-life.

The energy hump is higher for uranium fission than for ordinary uranium breakdown, and it is therefore the latter which takes place and is detected, even though the former represents the greater overall stabilization.

Still, the energy hump for fission ought very occasionally to be overcome on a purely random basis (and not merely by the deliberate introduction of a neutron), and when that happens, the uranium nucleus ought to undergo fission without neutrons. Such *spontaneous fission* was discovered in 1940 by a pair of Russian physicists, G. N. Flerov and K. A. Petrjak.

Naturally, since fission has the higher energy hump, its half-life is longer. Whereas uranium-238 has an alpha emission half-life of some 4,500,000,000 years, it has a spontaneous fission half-life of some 1,000,000,000,000 years.

For the more massive transuranium isotopes, the spontaneous fission half-life decreases. For curium-242 it is a mere 7,200,000 years, and for californium-250 it is only 15,000 years.

CHAPTER 41

Nuclear Reactors

Uranium-235

When Hahn first came to the conclusion that neutrons were initiating uranium fission, he hesitated to announce his finding since it seemed so "far out" a suggestion. At that time, however, Lise Meitner, his long time co-worker, being Jewish, had to flee Hitler's anti-Semitism and was in Stockholm.* The uncertainties of her own position made the risk of a "far out" scientific suggestion seem less dangerous, and she sent a letter to the scientific periodical *Nature,* discussing the possibility of uranium fission. The letter was dated January 16, 1939.

Niels Bohr learned of this by word of mouth and, on a visit to the United States to attend a conference of physicists, spread the news. The physicists scattered to see if they could confirm the suggestion. They promptly did so, and nuclear fission became the exciting discovery of the year.

To the Hungarian-American physicist Leo Szilard (1898–1964), what seemed most exciting (and unsettling) was the possibility of a nuclear chain reaction. He had been one of those who had considered the possibility before, and he had even

* Until 1938 she had been relatively safe, for she was an Austrian national, but in that year Hitler's Germany had forcibly annexed Austria.

patented a nuclear process that might possibly give rise to one (but didn't).

Uranium fission, however, offered a new approach. It was induced by slow thermal neutrons even more readily than by fast ones. The neutrons produced in the process of fission possessed ample energy for the induction of further nuclear fission. If anything, they had to be slowed down, which was easy.

World War II had started now, and Szilard, a refugee from Hitler's psychopathic tyranny, was fully aware of the terrible danger that faced the world if the Nazis tamed nuclear energy and put it to war use. Together with two other physicists of Hungarian birth, Eugene Paul Wigner (1902–) and Edward Teller (1908–), he set about interesting the American government in pursuing the project of developing methods for obtaining and controlling such a chain reaction.

They chose Albert Einstein as the only man with enough prestige to carry weight with non-scientists in such a matter. Overcoming Einstein's pacifistic scruples with difficulty, they persuaded the gentle physicist to write a letter on the subject to President Franklin D. Roosevelt. In 1941, Roosevelt was persuaded and he agreed to initiate a massive research program to develop a war weapon involving uranium fission. The final order was issued on December 6, the day before Pearl Harbor.

In order to establish a nuclear chain reaction, it is necessary to set up conditions radically different from those prevailing in the earth's crust. In the crust, although uranium is present and stray neutrons are to be found in the atmosphere, no chain reaction exists or, as far as we can tell, ever has existed.

The reason for this is that when an atom of uranium undergoes fission (either spontaneously or through absorption of a neutron), the neutrons liberated are absorbed by surrounding atoms. Most of these surrounding atoms are not uranium and are not themselves nudged into fission. The neutrons from fissioning uranium are thus absorbed and no neutrons are re-emitted, so that the potential chain reaction is effectively quenched. There is enough non-uranium material in even the richest natural concentration of uranium to quench any potential chain reaction at once.

What was necessary, then, if a nuclear chain reaction was to have any chance at all, was to make use of pure uranium, in the form of an oxide or even as the metal itself. In the metal, where almost all atoms would be uranium atoms, any neutron liberated by one uranium atom undergoing fission would stand an excellent chance of being absorbed by another uranium atom and therefore of bringing about another fission—the next link in the chain.

This, in itself, was a stiff requirement. In 1941, uranium had virtually no important uses, so that only small amounts of the metal were produced and those small amounts were not of high purity. Then, even as attempts to prepare large quantities of pure uranium began, an even more stringent qualification made itself evident.

Shortly after the idea of uranium fission had been accepted, Niels Bohr pointed out that on theoretical grounds uranium-235 was much more likely to undergo fission than uranium-238 was. Experiment soon showed Bohr to be right. This meant that ordinary uranium, even if highly purified, was still a poor material with which to set up a nuclear chain reaction, for 993 atoms out of every 1000 in such uranium would be uranium-238 which would absorb neutrons without undergoing fission—thus quenching the chain reaction.

To give a nuclear chain reaction a decent chance, uranium would have to be prepared in which uranium-235 was present in greater than usual amounts. Such a preparation would involve isotope separation, a difficult task—particularly, if it is to be carried through on a large scale.

Different isotopes of a given element have virtually identical chemical properties, and such differences as do exist depend on the fact that one isotope is more massive, and therefore reacts more sluggishly, than another. This difference is most marked in the case of hydrogen where hydrogen-2 is just twice as massive as hydrogen-1. This makes it possible for hydrogen-2 to be separated from hydrogen-1 with relative ease. The difference in mass between uranium-238 and uranium-235, however, is only 1.3 percent.

The best-established method for separating isotopes of small percentage difference in mass is by forcing a gas containing these isotopes as part of their molecules through some porous material (*diffusion*). The molecules must find their way through the pores, and those that contain less massive isotopes do so a bit more rapidly than do those which contain more massive ones.

The first samples of gas to emerge from the porous material are therefore "enriched" with a more than usual percentage of the light isotope, while the last samples to come through are "depleted" because they have smaller than usual percentage of the light isotope. The difference between the two fractions is very small, but the process may be repeated on each fraction. The smaller fractions can be recombined according to a fixed pattern and then separated again. Eventually, if this is continued long enough, the isotopes are nearly completely separated. The smaller the difference in mass between the isotopes, the greater the number of individual diffusions required.

Such a diffusion method requires a gas, of course, and neither uranium itself nor its most common compounds are gaseous. P. H. Abelson, however, suggested the use of uranium hexafluoride (UF_6) which, if not itself a gas at ordinary temperatures, is at least a volatile liquid with a boiling point at 56° C. It can therefore be maintained as a gas with little trouble.

The molecular weight of uranium hexafluoride containing uranium-238 is 352, while that of uranium hexafluoride containing uranium-235 is 349. The difference in molecular weight is only about 0.85 percent and diffusion had to be prolonged indeed to take advantage of so small a percentage mass difference. Giant installations (*diffusion cascades*) were set up at Oak Ridge, Tennessee, for this purpose in the early 1940's. In these, the UF_6 was put through numbers of porous barriers under conditions which automatically separated and recombined fractions in an appropriate manner. Eventually, enriched uranium hexafluoride was turned out at one end and depleted uranium hexafluoride at the other.

The "Atomic Pile"

Even as work on the purification of uranium and the separation of its isotopes proceeded, it was realized that a nuclear chain reaction could not, under even the best of conditions, be set up in a limited volume of uranium. Even uranium-235 atoms will not necessarily always absorb a neutron that comes blundering its way, toward the uranium atom. The neutron may merely bounce off, unabsorbed. It may do this over and over again, and it may be only the hundredth or the thousandth uranium-235 atom that will absorb it.

If in the process of bouncing from atom to atom the neutron manages to make its way out of the uranium and into the open air, it is lost. If enough neutrons do so, the nuclear chain reaction will be quenched. To prevent this, one must see to it that the chances of loss of neutrons to the surrounding environment, before absorption and consequent fission have a chance to take place, are minimized. The simplest way to do this is to increase the size of the uranium core in which fission is to take place. The larger its size the more bounces a neutron must undergo before reaching the edge of the core and the greater the chance of its absorption.

If the core is just large enough to lose so few neutrons that the nuclear chain reaction may just barely keep going, it is said to be at *critical size*. A smaller core, one of "subcritical size," cannot maintain a "self-sustaining nuclear reaction."

The critical size is not an absolute. It depends on the nature of the core, on its shape, and so on. A core of enriched uranium naturally has a smaller critical size than one of ordinary uranium, since the greater the concentration of uranium-235, the fewer the bounces required before absorption and the smaller the chance (at any particular core size) of escape into the air.

Then, too, the critical size can be reduced if slow neutrons, rather than fast neutrons, are used, since uranium-235 has a greater cross section for slow neutrons and fewer bounces will be required in their case. To slow the neutrons, a moderator (see page 174) is required, and very pure graphite serves the purpose well. Such a moderator could also serve as a neutron reflector. If the moderator is built about the uranium core, neutrons emerging from the core and striking the graphite will bounce here and there without being absorbed, and a number will bounce back into the uranium. In this way, the critical size is further reduced.

To control the nuclear chain reaction and prevent the uranium core from exploding, the reverse of a moderator is required. Instead of atoms that bounce the neutrons without absorbing them, as a moderator does, we need atoms that readily absorb the neutrons without either bouncing them or re-emitting them. Cadmium, some of the isotopes of which have a high cross section for neutrons, serves the purpose and "control rods" can be formed out of it.

Toward the end of 1942, the first attempt was made to set up a self-sustaining nuclear reaction. This took place under the guidance of Enrico Fermi (who had emigrated from Italy to the United States in 1938, but who was not yet an American citizen and was therefore technically an "enemy alien") under the stands of a football stadium at the University of Chicago.

At the time, some pure uranium was available in both metallic form and in the form of the oxide. It was not enriched and so the critical size was extraordinarily high. A very large "atomic pile" had to be built. (It was called a "pile" because it was, literally, a pile of bricks of uranium, uranium oxide, and graphite. In addition, "pile" was a neutral term that would not betray the actual nature of the structure if outsiders heard of it. After the war, "atomic pile" continued to be used for a short while and then gave way to a much more appropriate term, *nuclear reactor.*

When this first nuclear reactor was completed, it was 30 feet wide, 32 feet long and 21½ feet high. It weighed 1400 tons, of which 52 tons was uranium. The uranium, uranium oxide, and graphite were arranged in alternate layers with, here and there, holes into which long rods of cadmium could be fitted.

Suppose that in such a reaction a certain number of uranium atoms (n) undergo fission in a fixed unit of time, liberating x neutrons. Of these x neutrons, y do not find their mark but are absorbed by materials other than uranium, or are absorbed by

Oak Ridge Nuclear Reactor

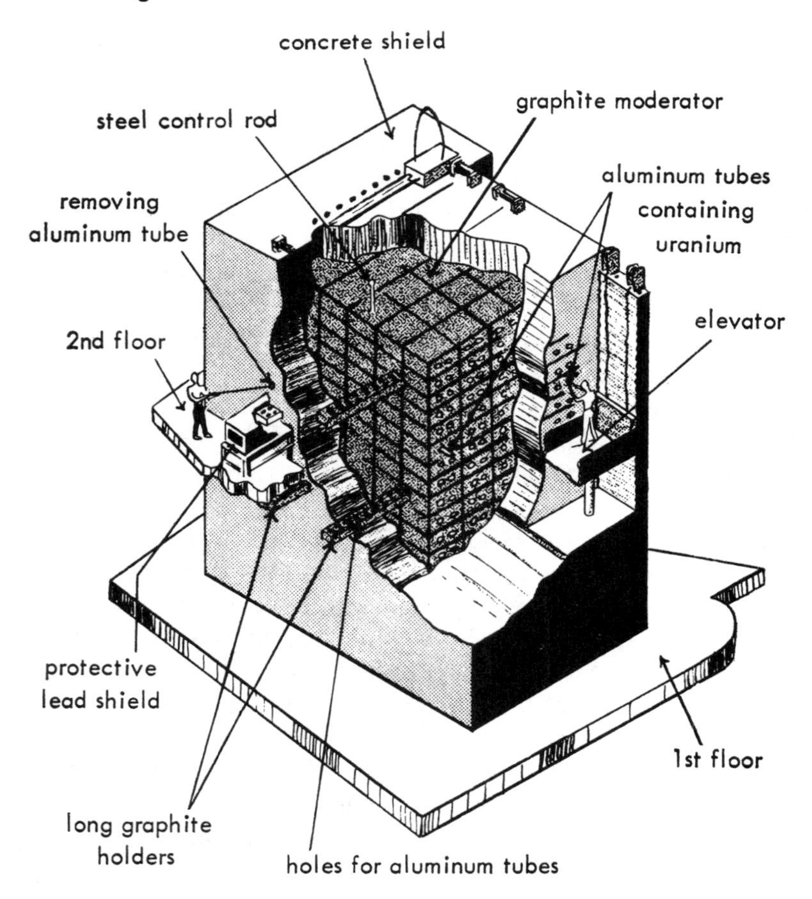

uranium atoms which nevertheless do not undergo fission, or escape out of the reactor altogether. This means that *x-y* neutrons actually strike a uranium-235 atom and bring about fission. The ratio $(x\text{-}y)/n$ is the *multiplication factor.*

If the multiplication factor is less than 1, then at each succeeding link in the chain reaction, fewer atoms undergo fission and fewer neutrons are produced. The nuclear chain reaction is quickly quenched.

If the multiplication factor is greater than 1, then at each link in the chain a larger number of uranium atoms undergo fission and a greater number of neutrons are produced. In a fraction of a second, the intensity of the chain reaction escalates itself into a fearsome explosion.

In the reactor, as constructed at the University of Chicago, the multiplication factor was distinctly less than 1 with the cadmium control rods pushed all the way in. As the rods were slowly pulled out, less and less cadmium remained within the reactor to absorb neutrons, and more and more neutrons were consequently available to spur uranium atoms to fission. The multiplication factor rose.

It might be supposed that as the control rods are removed and the multiplication factor rises nothing happens until the multiplication factor edges the tiniest trifle over 1—at which time the entire pile explodes carrying part of the city of Chicago with it.

Fortunately, this need not happen. Almost all (but not quite all) the neutrons produced in the course of a nuclear chain reaction are produced virtually instantaneously as a uranium atom undergoes fission. These are *prompt neutrons.* About 0.75 percent of the neutrons, however, are produced by fission products and are emitted over a period of several minutes. These are *delayed neutrons.*

If the multiplication factor is above 1.0075, then prompt neutrons alone are sufficient to escalate the reaction and bring about an explosion at once. If the multiplication factor is between 1.0000 and 1.0075, the prompt neutrons cannot do this of themselves but must have the cooperation of the delayed neutrons. This means that for a short while the intensity of the fission reaction increases only slowly. During this period of slow increase there is time to push the cadmium rods inward thus reducing fission intensity. Automatic control of the cadmium rods can keep the multiplication factor between 1.0000 and 1.0075 indefinitely, keeping the nuclear reaction alive but not allowing an explosion. If anything goes wrong with the control system, matters are so arranged that the cadmium rods fall inward of themselves, quenching the reaction. This is a "fail-safe" situation, and a quarter-century of experience shows that nuclear reactors are quite safe when properly designed.

On December 2, 1942, at 3:45 P.M., the cadmium rods in Fermi's "atomic pile" were pulled out just enough to produce a self-sustaining reaction. That day and minute are taken to mark the beginning of the "atomic age." (Had the control rods been pulled out all the way, the multiplication factor would have been 1.0006—safe enough.)

News of this success was announced to Washington by the cautious telegram reading: "The Italian navigator has entered the new world." There came a questioning wire in return: "How were the natives?" The answer was sent off at once: "Very friendly."

The "Atomic Age"

Nuclear reactors have multiplied in number and in efficiency since Fermi's first "pile." Many nations now possess them, and they are used for a variety of purposes.

Neutrons are produced by the uranium atoms undergoing fission in unprecedented amount. They can be used to bombard a variety of targets and to produce radioisotopes in quantities that would be impossible under any other conditions. It is only since World War II, therefore, that radioisotopes have been available in relatively large quantity and at relatively low prices. Consequently, techniques involving such isotopes in biochemical research, in medicine and in industry have multiplied and flourished in the last few decades.

The nuclear reactor can also be used to produce power. The heat produced by the reactor can heat some high-boiling fluid passing through it (liquid sodium, for instance). This in turn can be used to boil water and form steam that will turn a turbine and produce electricity.

In 1954, the first nuclear submarine, the U.S.S. *Nautilus,* was launched by the United States. Its power was obtained entirely from a nuclear reactor, and it was not constrained to rise to the surface at short intervals to recharge batteries. Since it could remain underwater for extended periods of time, it was that much safer from enemy detection and attack.

The first American atomic surface ship was the N.S. *Savannah,* launched in 1959. Its nuclear reactors make use of enriched uranium dioxide as the fuel, and its 21 control rods contain neutron-absorbing boron.

In the mid-1950's, nuclear power stations were designed for the production of electricity for civilian use. The Soviet Union built a small station of this sort in 1954. It had a capacity of 50,000 kilowatts. The British built one of 92,000 kilowatt capacity which they called Calder Hall. The first American nuclear reactor for civilian purposes began operations at Shippingport, Pennsylvania, in 1958.

The greatest problem presented by such power stations (aside from expense, which may be expected to decrease as techniques grow more sophisticated) is the fact that the products of uranium fission are themselves radioactive.

What's more, as these fission products accumulate in the uranium core, they begin to interfere with operations. Some of them are relatively efficient absorbers of neutrons, so that they act to quench the nuclear chain reaction. Every two or three years, therefore, a nuclear reactor must be shut down (even though its fuel is very far from exhausted) and the fission products separated from the core.

The half-life of some of the fission products is 20 years or more, so it may be over a century before a batch of them can be considered no longer dangerously radioactive. For this reason, they must be disposed of with great care. Concentrated solutions can be encased in concrete and sealed in steel tanks, and then buried underground. Methods are also being investigated for fusing such fission products with silicates to form "glasses." This would be completely leak-proof and therefore safer to store.

The fission products themselves still contain energy, and some of them can be used in lightweight *nuclear batteries.* Such batteries are popularly termed SNAP ("Systems for Nuclear Auxiliary Power.") In such batteries, the heat given off by the radioactive breakdown of an isotope is used to raise the temperature of one end of a thermocouple (see page II–180) and produce electricity.

The first SNAP was constructed in 1956, and since then over a dozen varieties have been built. Some have been put to use in powering man-made satellites over long periods. SNAP batteries can be as light as four pounds, can deliver as much as 60 watts, and can have a lifetime of up to ten years.

Not just any radioisotope will do for nuclear batteries. It must have an appropriate half-life so that it will deliver heat neither too rapidly nor too slowly; it must be free of dangerous gamma ray emission, and be relatively cheap to prepare. Only a few radioisotopes meet all necessary qualifications. The one most frequently used is the fission product strontium-90, which in another connection (see page 212) is one of the great new dangers to humanity.

The vision of a world in which uranium fission ekes out the energy supplies stored in coal and oil is dimmed somewhat by the fact that the prime nuclear fuel, uranium-235, is not extremely common. Uranium itself is not one of the rarest elements, but it is widely scattered throughout the earth's crust and concentrated pockets are rare. In addition, uranium-235 makes up only a small percentage of the metal.

Fortunately, uranium-235 is not the only isotope that can be stimulated into fission by neutron bombardment. Another isotope of this sort is plutonium-239. This does not exist in significant quantities in nature, but it can be formed by the neutron bombardment of uranium-238. This forms neptunium-239 first, then plutonium-239.

Once formed, plutonium-239 is easily handled, for it has a half-life of over 24,000 years. Therefore, in human terms its existence is just about permanent. Furthermore, it is not an isotope of uranium but a distinct element, so separating it from uranium is not nearly as difficult a problem as was isolating uranium-235.

During World War II, plutonium-239 was painstakingly gathered together so that its ability to undergo fission might be studied. A self-sustaining nuclear reaction can be maintained in plutonium-239 even under the impact of fast neutrons. Plutonium reactors (*fast reactors*) require no moderators and are therefore more compact than ordinary reactors.

Plutonium-239 can be produced as a by-product of power obtained from uranium-235. The neutrons emerging from a uranium-235 core can be used to bombard a shell of ordinary uranium surrounding the core. Quantities of uranium-238 in the shell are converted to plutonium-239. In the end, the quantity of fissionable material produced in the shell may actually be greater than that consumed in the core. This is a *breeder reactor*.

Such a breeder reactor makes uranium-238 indirectly available as a nuclear fuel and increases the fissionable resources of mankind over a hundredfold.

Another fissionable material is uranium-233, an isotope first discovered by Seaborg and his group in 1942. It is a daughter isotope of the neptunium series and therefore does not occur in nature. However, it has a half-life of 162,000 years, so once it is formed it can be handled without trouble.

When thorium-232 is exposed to neutron bombardment, it becomes thorium-233, which emits a beta particle with a half-life of 22 minutes to become protactinium-233. The latter, in turn, emits a beta particle with a half-life of 27 days to become uranium-233. Thus, if a thorium shell surrounds a nuclear reactor, fissionable uranium-233 can be formed within it and easily separated from the thorium. In this way, the earth's supply of thorium is added to its nuclear fuel potential.

Despite this enumeration of the peaceful aspects of nuclear fission, it must be remembered that the research project set up in 1941 had as its first purpose the development of an explosive weapon. What was wanted was a core that would exceed the multiplication factor by as much as possible. For this purpose, the critical mass must be as small as possible, since such a bomb ought to be transportable. Hence, pure uranium-235 or plutonium-239 ought to be used.

Such a bomb can safely be transported in two halves, since each portion would then be of subcritical size. At the crucial point, one half can be slammed against the other by means of an explosive. The stray neutrons in the air will suffice to build up an immediate nuclear explosion.

By 1945, uranium isotopes and plutonium had been prepared in sufficient quantity to construct three *fission bombs*.* At 5:20 A.M. on July 16, 1945, at Alamogordo, New Mexico, one of them was exploded and was a horrifyingly complete success. The explosion had the force of 20,000 tons (20 kilotons) of TNT.

By that time, World War II was over in Europe, but not the war with Japan. It was decided to use the two remaining nuclear bombs against Japan. On August 6, 1945, one of them was exploded over the town of Hiroshima, and on August 8, the other was exploded over Nagasaki. Japan surrendered and World War II was over.

Nuclear Fusion

Even under the best of circumstances, energy drawn from fissionable fuel has its disadvantages. Taken together, uranium and thorium make up only about 1.2 parts per hundred thousand of the earth's crust. This represents, to be sure, perhaps ten times

* These are popularly termed "atomic bombs" or "A-bombs."

as much potential energy as can be obtained from the earth's total supply of coal, oil, and gas, but only a small part of earth's fissionable fuel supply can be extracted from the crust with reasonable ease. Then, too, even if all could be used, what would we do with the mounting accumulations of fission products—products impossible to keep and dangerous to dispose of?

A bright alternative offers itself at the other end of the packing fraction curve. Energy can be obtained not only by breaking down massive atoms into less massive ones, but by building up simple atoms into less simple ones. The latter situation is termed *nuclear fusion.*

The most obvious case is that in which hydrogen, the simplest atom, is fused to helium, the next simplest. Suppose, for instance, that we consider the following reaction:

$$_1H^2 + {}_1H^2 \longrightarrow {}_2He^4$$

The mass number of hydrogen-2 is 2.01410 and that of two of such nuclei is 4.02820. The mass number of helium-4 (which has an unusually low packing fraction) is 4.00280. The loss of mass is 0.0254 out of a total of 4.0282. The percentage loss of mass is 0.63, whereas in uranium fission it is only 0.056. In other words, on a weight for weight basis, over ten times as much energy is available in nuclear fusion as in nuclear fission.

Nuclear fusion brought itself to human attention in the sky first. In the mid-nineteenth century, when the law of conservation of energy was clearly established, physicists began to question the origin of the vast energies of the sun. The German physicist Hermann Ludwig Ferdinand von Helmholtz (1821–1894) had seen in the force of gravitation the only possible source of the sun's energy and had suggested that a slow contraction powered the solar radiation.

Unfortunately, with gravitation as the source of radiant energy, it seemed as though the earth could not have lasted more than a hundred million years or so. Prior to that, the sun would have had to be large enough to more than fill the earth's orbit; it would have had to be that large if enough contraction were to have taken place to support its radiation for a hundred million years.

Once radioactivity was discovered, however, it was possible to take a new look at the problem. The atomic nucleus was an energy source unknown to Helmholtz and to the men of his generation. It came to seem more and more reasonable to suppose that the sun's radiation was supported by nuclear reactions.

The nature of such a reaction, however, remained a puzzle for some decades. The earliest nuclear reactions known, those of uranium and thorium breakdown (or, for that matter, the later-discovered uranium fission) could not be useful in the sun, since there wasn't enough uranium, or massive atoms generally, in the solar sphere to supply the necessary energy under any circumstances.

Indeed, by atom count the sun was something like 85 percent hydrogen and 10 percent helium. It seemed quite likely, therefore, that if nuclear reactions powered the sun, they would have to be reactions involving hydrogen.

However, hydrogen does not, under conditions on earth, spontaneously participate in nuclear reactions. Conditions on the sun differ most dramatically with respect to temperature (the sun's surface is known to be at a temperature of 6000° C), but it was not at all certain that this difference was significant.

Early experiments with uranium and with other naturally radioactive elements made it appear that the radioactive process, unlike ordinary chemical reactions, was not affected by heat. The half-life of radium was not decreased by extreme cold or increased by extreme heat. Nor could two atoms which, at ordinary temperatures, did not engage in nuclear reactions, be made to do so by extreme heat.

Of course this depends on what is meant by "extreme heat." The temperatures available in the laboratories of the early twentieth century were insufficient to smash atoms together with such force as to break through the electronic "bumpers" and force nucleus against nucleus. Even the temperature of the solar surface was quite insufficient for the purpose.

However, the English astronomer Arthur Stanley Eddington (1882–1944) produced a convincing line of argument to show that if the sun were gaseous throughout, then it could be stable only if its interior temperature was extremely high—at millions of degrees.

At *such* temperature extremes, atomic nuclei could indeed be forced together, and nuclear reactions that would not take place at ordinary temperatures would become "spontaneous." A nuclear reaction that proceeds under the lash of such extreme heat intensities is termed a *thermonuclear reaction* ("thermo-" is from a Greek word for "heat"). Clearly, it must be thermonuclear reactions proceeding somewhere deep in the sun's interior that serve as the source of its radiant energy.

In 1938, the German-American physicist Hans Albrecht

Bethe (1906–) worked through the list of possible thermonuclear reactions involving the light elements, eliminating those that took place too quickly and would explode the sun, and those that took place too slowly and would let the sun's radiation die. The reaction upon which he finally settled began with the most likely candidate, the overwhelmingly present hydrogen.

He postulated that the hydrogen reacted with carbon to build up first nitrogen and then oxygen in a series of reactions. The oxygen atom broke up to helium and carbon. The carbon was thus ready to begin a new cycle and, since it was in the long run unchanged, behaved as a sort of "nuclear catalyst." The net effect of the series of reactions was to convert hydrogen-1 to helium-4. In later years, other sets of reactions involving more direct changes of hydrogen-1 to helium-4 were also proposed.*

The energy released by such nuclear fusion of hydrogen to helium (with or without a carbon catalyst) is quite sufficient to maintain the sun's radiation. The energy is obtained, of course, at the expense of the sun's mass. In order to keep its radiation going at the observed rate, the sun must lose 4,600,000 tons of mass every second. To do that, it must convert 650,000,000 tons of hydrogen-1 to helium-4 every second. However, the sun has so huge a supply of hydrogen-1 that although it has been radiating for five or six billion years there is enough left to power it for billions of years more.

With the development of the fission bomb, scientists had a method of achieving, even if only momentarily, temperatures high enough to bring about nuclear fusion here on earth. The dreadful power of such a weapon (a *fusion bomb*) was such that numerous scientists hesitated to proceed in that direction. Among those who hesitated was Oppenheimer, who in 1954 was to pay for this by suffering a form of political-scientific disgrace when his access to secret information was withdrawn. Prominent among those who condemned Oppenheimer and pushed for the development of the fusion bomb was Edward Teller, whose contribution to the problem was such that he later received the rather unenviable distinction of being called the "father of the hydrogen bomb."

In 1952, the first "thermonuclear device," or "hydrogen bomb," or "H-bomb" (for the fusion bomb is known by all these names) was exploded by the United States in the Marshall Islands. It was not long after that, that the Soviet Union developed its

* The details of these reactions—together with those reactions that involve helium fusion, and so on, during the later stages of a star's life-cycle—are more fittingly discussed in a book on astronomy.

own fusion bomb, and that, later still, Great Britain became the third thermonuclear power. (France and China, which have exploded fission bombs, have not yet developed fusion bombs.)

Where the first fission bomb had the explosive force of 20,000 tons of TNT, fusion bombs with an explosive force of 50,000,000 tons of TNT (50 megatons) and beyond have been exploded.

Radiation Sickness

The fusion bomb escalated the danger to humanity (and to life on earth generally) by several notches. It was not merely that the explosion was far worse than that of the fission bomb, but rather that the long-lived effects of the products of a fusion bomb (even one exploded experimentally in peacetime) were insidious in the extreme. This is owing to the effect of high-energy radiation on living tissue.

Soon after the discovery of X rays, it was discovered that overexposure to such radiation gave rise to skin inflammations and burns that healed very slowly. The same proved to be true of the radiations from radioactive substances. Pierre Curie deliberately exposed himself to such radiations and reported the lingering symptoms that resulted.

The energy of X rays, gamma rays or speeding subatomic particles is sufficient, if absorbed by a molecule, to break chemical bonds with the production of high-energy molecular fragments (*free radicals*). These will, in turn, react with other compounds. A subatomic particle that is absorbed by an atom may alter its nature and, therefore, that of the molecule of which it is part. If the new atom is radioactive and emits a particle, the recoil will rupture the molecule even if it had survived intact till then.

Such chemical changes may well disrupt the intricately interrelated chemical machinery of a cell and upset those systems of reactions that control cellular cooperation. Changes may be induced, for instance, which will allow the unrestrained growth of certain cells at the expense of their neighbors and cancer will result. The skin, which bears the brunt of the onslaught of radiation, and those portions of the body such as the lymphoid tissue and the bone marrow, which produce blood cells, are particularly subject to this. (Even excessive exposure to ultra violet radiation increases the likelihood of the development of skin cancer.)

Leukemia, an unrestrained production of white blood cells (a condition which is slowly, but invariably, fatal) is one of the more likely results of excessive exposure to radiation. Both Marie

Curie and her daughter, Irène Joliot-Curie, died of leukemia, presumably as a result of longtime exposure to the radiation of radioactive substances.

Where radiation exposure is particularly great, enough destruction is wrought among the particularly sensitive tissues to break down cellular chemistry completely and bring about death in a period of weeks or months. Such *radiation sickness* was studied on a large scale for the first time among the survivors of the fission bombing of Hiroshima and Nagasaki.

Even worse than the death, fast or slow, of the individual is the long-term danger that continues over the generations. An altered molecule may not very seriously affect the individual within which it exists, for it will be present in only a few cells; however, it may be transmitted to a child born of that individual, and that child may have the altered molecule in every cell. The child will have undergone a *mutation.*

Mutations may also take place spontaneously, as the result of the natural radiation arising from radioactive substances in the soil, and natural radiation arising in outer space, as well as the result of random imperfections in the reproduction of key molecules. The rate of mutation will increase, however, as the general radiation from the environment rises because of nuclear bombs. Such mutations are generally for the worse and if produced at too great a rate will swamp the human species with a "mutation load" too great for safety.

Attempts have therefore been made to determine what amount of radiation can reasonably be borne by individuals (and by mankind in general) without making the danger acute.

A unit of radiation is the *roentgen,* abbreviated as *r,* named in honor of the discoverer of the X rays. This is defined as the quantity of X rays or gamma rays required to produce a number of ions equivalent to 1 electrostatic unit of charge (see page II–164) in a cubic centimeter of dry air at 0° C and 1 atmosphere pressure. (For this, a little over two billion ions of either sign must be formed.)

This unit applied originally to energetic electromagnetic radiation only. However, energetic particles produce the same sort of symptoms and effects that radiation does, and an effort was made to apply the unit to those particles. A *roentgen equivalent physical,* or *rep,* was spoken of as that quantity of radiation of particles which, on absorption by living tissue, produce the same effect as the absorption of 1 r of X rays or gamma rays.

The same effect is not always produced by a given quantity

of a given radiation on all living species. If one wishes to specify the effect on man, one speaks of a *roentgen equivalent man,* or *rem,* as that quantity of radiation of particles which, on absorption by the tissues of a living man, produces the same effect as the absorption of 1 r of X rays or gamma rays.

Massive particles are particularly dangerous to man. Thus, 1 r of X rays, gamma rays or beta particles can also be expressed as 1 rem. However 1 r of alpha particles must be expressed as 10 to 20 rem. In other words, the absorption of alpha particles is at least ten times as dangerous to man as the absorption of the same amount of ionizing potential in the form of beta particles.

The roentgen and the units derived from it are unsatisfactory in some respects because they measure ion-production, and the quantity of energy required to form ion-pairs in the case of some types of radiation can be a rather complicated quantity to determine. Therefore, the *rad* (short for "radiation") was introduced and has grown popular. This is a direct measure of energy. One rad is equivalent to the absorption of enough radiation in any form to liberate 100 ergs of energy per gram of absorbing material. Under most cases, the rad is just about equal to the roentgen.

Background radiation is the unavoidable radiation arising from outer space, from radioactive substances in the soil, and so on. It is estimated that the average human being receives 0.050 rem per year from radiation from outer space and another 0.050 rem per year from the natural radioactivity of the soil. In addition, there is 0.025 rem per year from the body's own radioactivity in the form of potassium-40 and carbon-14. The total background radiation is thus about 0.125 rem per year, and this must be consistent with life since we are all subjected to it and life generally has been subjected to it from the beginning. Indeed, in parts of the world in which radioactivity is higher than average and in which high altitudes and high latitudes combine to make radiation from space more intense, background radiation of as much as 12 rem per year have been reported.

Obviously, experimentation to see how high a level of background radiation can be tolerated is unthinkable, but experts in the field had estimated that general body exposure to 500 rem per year is tolerable. Those working with radioactive materials are guarded against absorbing more than some safe limit of radiation each week (it is supposed that absorption at a higher rate than 500 rem per year is tolerable if it is for short periods or is localized to parts of the body). For instance, badges may be worn containing strips of film behind various filters that will be penetrated only by the sort of energetic radiation being guarded against. The extent to which the film blackens will measure the extent of exposure.

Exposure to 100 r over a few days will kill most mammals, but it takes several million r to sterilize food completely by killing all the microorganisms. All the nuclear testing so far performed has not brought the radiation level anywhere near such lethal levels and indeed has not even contributed more than a comparatively small fraction of the background radiation that already exists.

However, every little bit hurts, and it was the general disapproval by public opinion that finally forced the thermonuclear powers to agree to a ban on such nuclear bomb testing as would increase the radiation level.

It was the fusion bomb that made the danger intense. The fission fragments produced by fission bombs are spread only locally and present only a limited danger (however horrible within that limit). The far greater force of the fusion bomb, however, lifts the fragments of its fission-trigger high into the stratosphere, where they may circulate for a period of years and then slowly settle over the world generally. It is the danger of this stratospheric *fallout* (a word coined in 1945, after the first nuclear explosions) that presents mankind with its greatest radiation hazard.

That the fallout danger is real was made plain at once. The first large fusion bomb, exploded in the Marshall Islands on March 1, 1954, contaminated 7000 square miles with radiation.

Among the more dangerous fission fragments are strontium-90 and cesium-137. Strontium-90 has a half-life of 28 years, so it remains dangerously radioactive for a century and more. Because of the chemical similarity of strontium to calcium, strontium-90 is concentrated in the calcium-rich milk of mammals feeding on strontium-90 contaminated vegetation. Children drinking such contaminated milk then concentrate the strontium-90 in their calcium-rich bones. The atomic turnover in the bones is relatively slow, and therefore the *biologic half-life* of strontium-90 is long. (That is, it takes a long time for the body to remove half of what it has absorbed, even after it is protected against further absorption.) In the bones, moreover, strontium-90 is in dangerously close contact with sensitive blood-cell forming tissues.

Cesium-137, with a half-life of 30 years, is another dangerous fragment. It remains in the soft tissues and has a shorter biologic half-life. However, it emits energetic gamma rays and, while in the body, can do significant damage.

Fusion Power

Naturally, it is not only for the sake of its destructive potentialities that fusion processes are of interest. If nuclear fusion could be made to proceed at a controlled pace, the energy requirements of mankind would be solved for the foreseeable future.

The advantage of fusion over fission involves first the matter of fuel. Where the fission fuels are comparatively rare metals, uranium and thorium, the fusion fuel is a much more common and readily available element, hydrogen. It would be most convenient if it were hydrogen-1 that were the specific isotope suitable for man-made fusion since that is the most common form of hydrogen. Unfortunately, the temperatures required for hydrogen-1 fusion, at a rate fast enough to be useful, are prohibitively high. Even at the temperatures of the solar interior, hydrogen-1 undergoes fusion slowly. It is only because of the vast quantity of hydrogen-1 available in the sun that the small percentage that does fuse is sufficient to keep the sun radiating as it does. (To be sure, if hydrogen-1 were more readily fusible than it is, the sun—and other stars—would explode.)

Hydrogen-2 (deuterium) can be made to undergo fusion at a lower temperature, and hydrogen-3 at a lower temperature still. However, hydrogen-3 is unstable and would be extremely difficult to collect in reasonable quantities. That leaves hydrogen-2 as the best possible fuel.

Two atoms of deuterium can fuse in one of two ways with equal probability:

$$H^2 + H^2 \longrightarrow He^3 + n^1$$

and:

$$H^2 + H^2 \longrightarrow H^3 + H^1$$

In the latter case, the H^3 formed reacts quickly with another H^2, thus:

$$H^3 + H^2 \longrightarrow He^4 + n^1$$

The overall reaction, then, would be:

$$5\ H^2 \longrightarrow He^3 + He^4 + H^1 + 2\ n^1$$

The energy produced from such a fusion of five deuterium atoms (let's call it a "deuterium quintet") is 24.8 Mev. Since 1 Mev is equivalent to 1.6×10^{-6} ergs, the deuterium quintet, on fusion, yields 4.0×10^{-5} ergs.

A gram-molecular weight of hydrogen-2 contains 6.023×10^{23} atoms. Since a gram-molecular weight of hydrogen-2 is two grams, one gram of hydrogen-2 contains 3.012×10^{23} atoms. Dividing this figure by five, we find that a gram of hydrogen-2 contains 6.023×10^{22} deuterium quintets. The total energy produced by the complete fusion of one gram of hydrogen-2 is therefore 2.4×10^{18} ergs. Since there are 4.186×10^{10} ergs to a kilocalorie, we can say that the complete fusion of one gram of hydrogen-2 produces 5.7×10^{7} kilocalories.

To be sure, only 1 out of every 7000 hydrogen atoms is hydrogen-2. Allowing for the fact that that one atom is twice as massive as the remaining 6999, we can say that one liter of water weighs 1000 grams, that 125 grams of it are hydrogen, and that of that hydrogen 43 milligrams are hydrogen-2. We can therefore say that the complete fusion of the hydrogen-2 contained in a liter of water will yield about 2.5×10^{6} kilocalories.

This means that by the fusion of the hydrogen-2 contained in a liter of ordinary water, we would obtain as much energy as we would get through the combustion of 300 liters of gasoline.

Considering the vastness of the earth's ocean (from all of which hydrogen-2 is easily obtainable) we can see that the earth's supply of hydrogen-2 is something like 50,000 cubic miles. The energy that could be derived from this vast volume of hydrogen-2 is equivalent to the burning of a quantity of gasoline some 450 times the volume of the entire earth.

Obviously, if fusion power could be safely and practically tapped, mankind would have at its disposal an energy supply that should last for many millions of years. And to top off that joyful prospect, the products of the fusion reaction are hydrogen-1, helium-3, and helium-4, all of which are stable and safe, plus some neutrons which could be easily absorbed.

There is one catch to this prospect of paradise. In order to ignite a hydrogen-2 fusion reaction, a temperature of the order of 100,000,000° C must be reached. This is far higher than the temperature of the solar interior, which is only 15,000,000° C, but then the sun has the advantage of keeping its hydrogen under enormous pressures, pressures unattainable on the earth.

Any gas at such a temperature on earth would, if left to itself, simply expand to an excessively thin vapor and cool almost instantaneously. That this does not happen to the sun is due to the sun's mass, which produces a gravitational field capable of holding gases together even at the temperature reached in the solar interior.

Such gravitational fields cannot be produced on earth, of course, and the hot gas must be kept in place some other way. Material confinement would seem to be out of the question, for a hot gas making contact with a cool container would cool off at once—or heat the container itself to a thin gas. A gas cannot be both hot enough for fusion and contained within a solid substance.

Fortunately, another method offers itself. As the temperature rises, all atoms are progressively stripped of their electrons and all that then exists are charged particles, negatively-charged electrons plus positively-charged nuclei. Substances made up of electrically charged atom-fragments, rather than intact atoms, are called *plasma.*

Investigators grew interested in *plasma physics* chiefly as a result of interest in controlled fusion, but, by hindsight, we now see that most of the universe is plasma. The stars are plasma, and here on earth phenomena such as ball lightning are isolated bits of plasma that have achieved temporary stability. Plasma even exists in man-made devices—for instance, within neon light tubes.

Plasma, consisting as it does of charged particles, can be confined by a nonmaterial container, a properly shaped magnetic field. The effort of physicists is now engaged in attempting to design magnetic fields that will keep plasma stably confined for periods long enough to initiate a fusion reaction—and to make the plasma hot enough for the fusion reaction to ignite. It is estimated that at the critical point, using gas, which at ordinary temperatures would be only 1/100 or less the density of the atmosphere, the pressures which would have to be withstood by the magnetic field at the point of fusion ignition would be something like 1500 pounds per square inch, or 100 atmospheres.

The requirements are stringent, and after a decade of research, success still lies frustratingly beyond the fingertips. Temperatures of about 20,000,000° C have been attained. Magnetic fields capable of containing the necessary pressures have been produced. Unfortunately, the combined temperature and pressure can be maintained only for millionths of a second, and it is estimated that at least a tenth of a second duration must be obtained in order for the first man-made controlled fusion reaction to be produced.

There is nothing (as far as we know) but time and effort standing in the way.

CHAPTER 42

Anti-Particles

Cosmic Rays

So far, we have populated our atomic world with nothing more than electrons, protons and neutrons, and yet have managed to explain a great deal. In the early 1930's, these subatomic particles were the only ones known and it was rather hoped they would suffice, for there would then be an agreeable simplicity to the universe. However, some theoreticians were pointing out the necessity for further types of subatomic particles, and the first discoveries of such particles arose out of the vast energies present in radiation bombarding earth from outer space. It is to this radiation that we will turn now.

As the twentieth century opened, physicists were on the watch for new forms of radiation. The coming of radio waves, X rays and the various radioactive radiations had sensitized them to such phenomena, so to speak.*

* The sensitization was too great in one case. In 1903, a reputable French physicist, Prosper Blondlot, reported the existence of a new type of radiation from metal under strain. He and others published many reports on this radiation, which Blondlot termed "N rays," the N standing for Nancy, the French city in which he held his university appointment. There seems no question but that Blondlot was utterly sincere. Nevertheless, the N rays were an illusion, his reports proved worthless, and his scientific career was blasted. The story is important if only to demonstrate that scientists are not infallible and that "scientific evidence" is not necessarily trustworthy.

Nevertheless, the most remarkable discovery of this sort arose out of the attempt to exclude radiation rather than to detect it. The gold-leaf electroscope, which was early used to detect penetrating radiation (see page 109), worked too well. A number of investigators, notably C. T. R. Wilson, of cloud chamber fame, had reported, by 1900, that the electroscope slowly lost its charge even when there were no known radioactive materials in the vicinity. Presumably, most reasoned, the earth's crust was permeated with small quantities of radioactive materials everywhere, so that stray radiation was always present.

Yet other investigators found that even when the electroscope was taken out over stretches of water remote from land, or better yet, when it was shielded by a metal opaque to known radiation and producing no perceptible radiation of its own, the loss of charge on the part of the electroscope was merely diminished. It did not disappear.

Finally, in 1911, the Austrian physicist Victor Franz Hess (1883–1964) took the crucial step of carrying an electroscope up in a balloon, in order that several miles of atmosphere might serve as the shield between the earth's slightly radioactive crust and the charged gold leaf. To his surprise, the rate of discharge of the electroscope did not cease; instead it increased sharply. Later balloon flights confirmed this, and Hess declared that the radiation, whatever it was, did not originate on earth at all, but in outer space.

Robert Millikan (who measured the charge on an electron), took a leading part in the early investigations of this new radiation, and suggested in 1925 that it be named *cosmic rays,* because this radiation seemed to originate in the cosmos generally.

Cosmic rays are more penetrating than either X rays or gamma rays, and Millikan maintained that they were a form of electromagnetic radiation even shorter in wavelength and higher in frequency than gamma rays. Nevertheless, as in the case of X rays and gamma rays, many physicists suspected that the radiation might be particulate in nature.

In this case, since the radiation came from outer space, a method of distinguishing between electromagnetic radiation and particles offered itself. If the cosmic rays were electromagnetic radiation, they would fall on all parts of earth's surface equally, assuming that they originated from all directions. They would not be affected by the earth's magnetic field.

If, on the other hand, they were charged particles, they would be deflected by earth's magnetic lines of force, those particles lower in energy being the more deflected. In this case, cosmic rays would be expected to concentrate toward earth's magnetic poles and to strike earth's surface with least frequency in the vicinity of its magnetic equator.

This *latitude effect* was searched for through the 1920's, particularly by the American physicist A. H. Compton (1892–1962). By the early 1930's, he was able to show that such a latitude effect did exist and that cosmic rays were particulate and not electromagnetic. One might therefore refer to *cosmic particles.*

The Italian physicist Bruno Rossi (1905–) pointed out, in 1930, that if the cosmic rays were particulate in nature, earth's magnetic field ought to deflect them eastward if the particles were positively charged, so that more of them would seem to be coming from the west than from the east. The reverse would be true if the particles were negatively charged.

To detect such an effect, it was insufficient merely to detect the arrival of a cosmic particle; one had to tell the direction from which it had arrived. To do this, use was made of a *coincidence counter,* which had first been devised by the German physicist Walther Bothe (1891–1957). This consisted of two or more G-M counters placed along a common axis. An energetic particle would pass through all of them, provided it came along that axis. The electric circuit was so arranged that only the discharge of all the counters at once (and an energetic particle passes through all the counters with so little a time interval between that it may be considered a simultaneous discharge) will register and be counted. The counters can be oriented in different directions to form a "cosmic-ray telescope."

By placing a cloud chamber among the counters, one can arrange the circuit so that the chamber is automatically expanded when the counters discharge. The ions linger a short interval and are caught by the droplets formed by the expanding cloud chamber. If a camera is also rigged to take photographs automatically as the chamber expands, the cosmic particle ends by taking its own picture.

Using coincidence counters, the American physicist Thomas Hope Johnson (1899–) was able to show, in 1935, that more cosmic particles approached from the west than from the east. Thus it was determined that cosmic particles were positively charged.

An understanding of the actual nature of the cosmic particles was hampered by the fact that many did not survive to reach earth's surface. Instead, they struck one or another of the atomic nuclei present in the atmosphere, inducing nuclear reactions and producing a highly energetic *secondary radiation.* Some of this secondary radiation consists of neutrons which can, in turn, react with nitrogen-14 to produce carbon-14 in an (n,p) reaction. Or it can knock a triton (H^3) out of a nitrogen-14 nucleus, producing carbon-12 in an (n,t) reaction. These tritons are the source of the small quantities of H^3 existing on earth.

Cosmic particles can produce other events that cannot easily be duplicated in the laboratory simply because even now we have no way of producing particles with the energy of the most penetrating of the particles from outer space. Where man-made accelerators can now produce particles with energies of 30 Bev or more, cosmic particles with energies in the billions of Bev have been recorded.

Such super-energetic particles possess these energies partly because they are massive and partly because their velocities are great—nearly as high as the ultimate velocity, that of light in a vacuum. When such extremely rapid particles burst through transparent matter (water, mica, glass), they are scarcely slowed. Light itself, however, is slowed down appreciably in these substances, in inverse proportion to the index of refraction (see page II–25). It may follow, then, that within some forms of matter a charged particle may travel considerably faster than light does in that form of matter (but never faster than light does in a vacuum).

Such a "faster-than-light" particle throws back light radiation in a sort of shock effect, analogous to the manner in which a faster-than-sound bullet throws back a cone of sound waves. This effect was first noticed by the Russian physicist Pavel Alekseyevich Cerenkov (1904–) in 1934, and it is therefore called *Cerenkov radiation.*

The wavelength of the Cerenkov radiation, its brightness, and the angle at which it is emitted can all be used to determine the mass, charge and velocity of the moving particle. Following a suggestion in the late 1940's by the American physicist Ivan Alexander Getting (1912–), *Cerenkov counters* were developed that react to the radiation and thus distinguish very energetic particles from among floods of ordinary ones, and serve also as source for much information about the former.

The late 1940's also saw the beginning of investigations of radiation by high-altitude balloons and by rockets. At elevated altitudes, the *primary radiation*—the original cosmic particles, and not those produced by collisions of those particles with nuclei—could be detected. It turned out that the large majority (roughly 80 percent) of the cosmic particles were very energetic protons and most of the remainder were alpha particles. About 2.5 percent of the particles were still heavier nuclei, ranging up to iron nuclei.

It very much seemed as though the cosmic particles were the basic material of the universe, stripped down to bare nuclei. The proportion of the elements represented was very much like that in typical stars such as our sun.

In fact, the sun is at least one source of cosmic particles. A large solar flare will give rise, shortly afterward, to a burst of cosmic particles falling upon the earth. Nevertheless, even though the sun is one source, it can't be the only one, or even a major one, as otherwise the direction from which cosmic particles arrive would vary markedly with the position of the sun in the sky—which it doesn't. Moreover cosmic particles from the sun are comparatively low in energy.

This raises the question: How do cosmic particles gain their tremendous energies? No nuclear reactions are known which would supply sufficient energy for the more energetic cosmic particles. Even the complete conversion of mass into energy would not turn the trick.

It seems necessary to suppose that the cosmic particles are, at the start, protons and other nuclei of high, but not unusually high, energy. They are then accelerated in some natural accelerator on a cosmic scale. The magnetic fields associated with the sun's spots might accelerate such particles to moderate energies. More energetic particles might have been produced by stars with more intense magnetic fields than our sun, or even by the magnetic field associated with the Galaxy as a whole.

The Galaxy, in this respect, might be looked upon as a gigantic cyclotron in which protons, and atomic nuclei generally, whirl about, gaining energy and moving in a constantly widening spiral. If they do not collide with some material body for a long enough time, they eventually gain enough energy to go shooting out of the Galaxy altogether.

The earth interrupts the flight of these particles (in all directions) at all stages of their energy-gaining lifetime. The most energetic particles may be those that went shooting out of some other galaxy at their energy peak. It may be that some galaxies, with unusually intense magnetic fields, may accelerate cosmic particles to far greater energies than our Galaxy does and may be important sources for these most energetic particles. Such "cosmic galaxies" have not yet been pinpointed.

The Positron

Now let us look at the list of particles known in the early 1930's when the nature of cosmic radiation was first being unraveled. There are the proton, neutron and electron, of course. In addition, there is a massless "particle," the photon, which is associated with electromagnetic radiation.

The photon makes it unnecessary to rely on the notion of action at a distance in connection with electromagnetic phenomena (see page II–139), and that can give rise to speculation concerning that other long-distance phenomenon, gravitation.

Some physicists suggest that gravitational effects also involve the emission and absorption of particles, and the name *graviton* has been given to these particles. Like the photons, these particles are visualized as massless objects that must therefore (like all massless particles) travel at the velocity of light.

Gravitation is an incredibly weak force, however. The electrostatic attraction between a proton and an electron, for instance, is about 10^{40} times as strong as the gravitational attraction between them. The graviton must therefore be correspondingly weaker than the average photon—so weak that it has never been detected and, as nearly as can be told now, is not likely to be detected in the foreseeable future. Nevertheless, to suppose its existence rounds out the picture of the universe and helps make it whole.

We can now list the five particles in Table XIII and include some of the properties determined for them. (Those for the graviton are predicted and not, of course, observed.)

In the 1950's, the custom arose of lumping the light particles

TABLE XIII—*Some Subatomic Particles*

Particle	*Symbol*	*Mass (electron = 1)*	*Spin (photon = 1)*	*Electric Charge*	*Half-Life (seconds)*
graviton	g	0	2	0	stable
photon	γ	0	1	0	stable
electron	e	1	½	− 1	stable
proton	p	1836	½	+ 1	stable
neutron	n	1839	½	0	1013

together as *leptons* (from a Greek word meaning "small") and the heavy particles as *baryons* (from a Greek word meaning "heavy"). Using this classification, the graviton, photon and electron are leptons, while the proton and neutron are baryons.

It would seem quite neat if these three leptons and two baryons represented all that existed in the universe—both matter and energy—and that out of them were built the hundred-odd atoms—out of which, in turn, all the manifestations of the universe from a star to a human brain were constructed.

The first indication that mankind was not to rest in this Eden of simplicity came even before the neutron was discovered. In 1930, the English physicist Paul Adrien Maurice Dirac (1902–), working out a theoretical treatment of the electron, showed that it ought to be able to exist in either of two different energy states. In one of those energy states, it was the ordinary electron; in the other, it carried a positive charge rather than a negative one.

For a while, however, this remained a theoretical suggestion only. In 1932, however, the American physicist Carl David Anderson (1905–) was investigating cosmic particles with cloud chambers divided in two by a lead barrier. A cosmic particle crashing through the lead would lose a considerable portion of its energy, and it seemed to Anderson that the less energetic particle emerging from the barrier would curve more markedly in the presence of a magnetic field and, generally, reveal its properties more clearly. However, some cosmic particles in bursting through the lead smashed into atomic nuclei and sent out secondary radiations.

One of Anderson's photographs showed a particle of startling characteristics to have been ejected from the lead. From the extent of its curvature it seemed to have a mass equal to that of the electron, but it curved in the wrong direction. It was Dirac's positively-charged electron.

Anderson named it, naturally enough, the *positron,* and that is the name by which it is now universally known. The positron, since it is a particle opposed in certain key properties to that of a more familiar particle, belongs to a class now termed *anti-particles*. Were it discovered now it would be called the *anti-electron* and, indeed, it may be referred to in this fashion at times.

The question of symbolism is a little confused. One could use a full symbol, including the charge as subscript and mass as superscript, so that the electron is ${}_{-1}e^0$ while the positron is ${}_{1}e^0$. The disadvantage to this is that it is cumbersome. Most physicists do not feel that they need to be reminded of the size of the charge and the mass (particularly since the mass is not truly 0 but only very close to it). For that reason, it is very common, to symbolize the electron simply as e^- and the positron as e^+. This, too, has its difficulties, for, as it turned out later, there are anti-particles that have the same charge (or lack of charge) as the particles they oppose. For this reason, it is sometimes more convenient to indicate the anti-particle with a bar above the symbol. Thus, an electron would be e and a positron would be $\bar{e}$.

Positrons play a role in radioactivity, and this can best be understood if we once again go over the role played by the electron.

When the number of neutrons in a nuclide is too great for stability, the situation can be corrected by the conversion of a neutron to a proton with the emission of an electron. If we write the symbols in full (so that we can observe the manner in which mass and charge are conserved), we can say:

$${}_0n^1 \longrightarrow {}_1p^1 + {}_{-1}e^0 \qquad \text{(Equation 13–1)}$$

In this process, the atomic number of the nuclide increases by 1 because an additional proton appears, but the mass number remains unchanged, since the proton appears at the expense of a disappearing neutron.

Consider phosphorus, for instance. Its only stable isotope is phosphorus-31 (15 protons, 16 neutrons). If we found ourselves with phosphorus-32 (15 protons, 17 neutrons) and observed it to be radioactive, we would expect that, because of its neutron excess, it would eliminate an electron in the form of a beta particle. Sure enough, it does. It emits a beta particle and becomes the stable isotope sulfur-32 (16 protons, 16 neutrons).

All naturally occurring radioactive isotopes, long-lived or short-lived, possess a neutron excess and, in the process of rearranging the nuclear contents to achieve stability, sooner or later emit electrons (though they may also emit alpha particles).

What if an artificial radioisotope is formed which has a neutron deficit? To achieve stability, a neutron must be gained and this must be at the expense of a proton. This can be done by a direct reversal of Equation 13–1, the absorption of an electron by a proton as in K-capture (see page 142).

$${}_1p^1 + {}_{-1}e^0 \longrightarrow {}_0n^1 \qquad \text{(Equation 13–2)}$$

There is also the possibility, however, of another type of reversal. While a neutron can be converted to a proton with the emission of an electron, a proton can, analogously, be converted to a neutron with the emission of a positron:

$$_1p^1 \longrightarrow {}_0n^1 + {}_1e^0 \qquad \text{(Equation 13–3)}$$

The emission of a positron (or "positive beta particle") has the reverse effect of the emission of an electron. The atomic number of the nuclide is decreased by one, since a proton disappears. Again, the mass number is unchanged since a neutron appears in place of the proton.

As it happened, the very first artificial radioisotope formed, phosphorus-30, suffered from a neutron deficit. Where the stable phosphorus-31 is made up of 15 protons and 16 neutrons, phosphorus-30 is made up of 15 protons and only 15 neutrons. Phosphorus-30, with a half-life of 2.6 minutes, emits a positron and becomes the stable silicon-30 (14 protons, 16 neutrons). In forming phosphorus-30, the Joliot-Curies nearly anticipated Anderson in the discovery of the positron.

A large number of positron-emitters have been prepared among the radioactive isotopes artificially produced in the laboratory. Perhaps the best-known is carbon-11, which before the discovery of carbon-14 was much used as an isotopic tag.

The most important positron-producing process in nature is the hydrogen fusion that proceeds in the sun and in other stars. The overall change of four hydrogen-1 nuclei to a helium-4 nucleus is that of 4 protons to a 2-proton/2-neutron nucleus. Two of the protons, therefore, have been converted to neutrons with the emission of two positrons:

$$_1H^1 + {}_1H^1 + {}_1H^1 + {}_1H^1 \longrightarrow {}_2He^4 + {}_1e^0 + {}_1e^0 \qquad \text{(Equation 13–4)}$$

Matter Annihilation

The electron is a stable particle—that is, left to itself it undergoes no spontaneous change. This is in accord with the law of conservation of electric charge, which states that net charge can neither be created nor destroyed. The electron is the least massive particle known to carry a negative electron charge, and physicists work on the assumption that no smaller negatively-charged particle can conceivably exist. In breaking down, an electron would have to become a less massive particle, and then there is no room, so to speak, for an electric charge—so the electron doesn't break down.

This same argument holds for the positron, which cannot break down, for it is the least massive particle known to carry a positive electric charge, and has nowhere to dispose of it if it does break down. The positron is therefore also considered a stable particle and would, presumably, remain in existence forever if it were alone in the universe.

However, the positron is not alone in the universe. When formed, it exists in a universe in which electrons are present in overwhelming numbers. Under ordinary conditions on earth, it cannot move for more than a millionth of a second or so before it collides with an electron. What happens then?

If we consider a positron and electron together, the net electric charge is zero. The two can therefore merge and cancel each other's charge. In doing so, they apparently also cancel each other's mass in *mutual annihilation*. It is not true annihilation, however, for something is left since the law of conservation of mass-energy remains in force regardless of the situation with respect to electric charge. If the mass of the electron and positron disappear, an equivalent amount of energy must appear.

The total mass of an electron and a positron is 1.822×10^{-27} grams. Making use of Einstein's equation, $e = mc^2$ (see page II–111), we can determine that the energy equivalence of the two particles is 1.64×10^{-6} ergs, or 1.02 Mev.

In this conversion of mass to energy, however, there are other conservation laws that must be observed. The law of conservation of angular momentum (see page I–81) governs the distribution of spin, for instance.

The photon's spin is accepted, by definition, to be either +1 or −1. If an electron and positron can annihilate each other with the formation of a photon with an energy of 1.02 Mev (a gamma ray photon), and if it is assumed, as seems likely, that electron and positron have equal spin, then both must have a spin of 1/2. If both have a spin of +1/2, then a photon of spin +1 is formed, and if both have a spin of −1/2, then a photon of spin −1 is formed.

However, the difficulty here is that the law of conservation of linear momentum (see page I–69) must also be respected. If the positron-electron system has a net momentum of zero with respect to its surroundings, then a single photon could not move after it was produced. Since a photon must move, and at the velocity of light, it follows that the production of a single photon is unlikely.

Instead three photons, each of 0.34 Mev energy (still gamma

rays), must be produced simultaneously, and shoot off toward the apices of an equilateral triangle. If the three photons have spins of +1, +1, and −1 respectively, the net spin is +1, while if the spins are −1, −1, and +1, the net spin is −1. In either case both angular momentum and linear momentum are conserved.

If the electron and positron spin in the same sense (that is, if both have a positive spin or both have a negative spin), then three photons can be produced, but not two. Two photons taken together can have a spin of 0 (+1 plus −1), +2 (+1 plus +1) or −2 (−1 plus −1), whereas the total spin of an electron and positron spinning in the same sense can only be +1 (+1/2 plus +1/2) or −1 (−1/2 plus −1/2). Angular momentum is not conserved.

On the other hand, if an electron and positron spin in opposite senses (+1/2 and −1/2), they can produce two photons (+1 and −1), for the net angular momentum is 0 both before and after; consequently, angular momentum is conserved. The two photons are gamma rays of 0.51 Mev each and dart off in opposite directions to conserve linear momentum.

I have gone into some detail here to show how nuclear physicists use the various conservation laws to decide what events on the subatomic scale can take place and what events cannot. They work on the assumption that any nuclear event that can happen will indeed happen if one waits long enough and looks hard enough. If, therefore, some particular event does not take place despite hard, long search, and nevertheless does not seem to be forbidden by any conservation law, a new conservation law is tentatively introduced. On the other hand, if an event that is forbidden by a conservation law takes place, it is necessary to recognize that this conservation law will hold only in certain circumstances and not in others, or that a deeper and more general conservation law must be sought.

It has been observed that when electrons and positrons annihilate each other, gamma rays of just the energies predicted by theory are formed. This is one of the neatest verifications of Einstein's special theory of relativity and of the mass-energy equivalence that is a part of it.

The reverse of all this is also to be expected. Energy ought to be converted into mass. No amount of energy can form an electron alone, or a positron alone, for in either case where is the electric charge to come from? Neither a net negative nor a net positive charge can be created.

However, an electron and a positron can be created simultaneously. The net charge of such an *electron-positron pair* is still zero. A gamma ray of at least 1.02 Mev energy is required for this, and if a more energetic gamma ray is used, the pair of particles possesses a kinetic energy equal to the energy excess over 1.02 Mev. The observed energy bookkeeping works out perfectly to the credit of Einstein.

Indeed, it is because of the superabundant energy of cosmic particles that energetic positrons are formed, and it is those which, detected by Anderson, marked the discovery of the first antiparticle.

When Dirac first worked out the theoretical reasoning that gave rise to the concept of the anti-particle, he felt that the electron's opposite number was the proton. However, this did not prove to be the case. The proton and the electron are exact opposites in electric charge, but in hardly anything else. The proton is, for instance, 1836 times as massive as the electron. (Why this should be so, and why the mass ratio should be 1836, no more and no less, is one of the more interesting unanswered questions of nuclear physics.)

The electron and the proton attract each other, as would any objects carrying opposing electric charges, but they do not and cannot annihilate each other. At best, the electron is captured by the proton and can approach to a minimum distance, representing the lowest possible energy-state. (If proton-electron annihilation were possible, matter could not exist.)

The electron and positron, which can annihilate each other, may also, at least temporarily, capture each other without annihilation. The "atom" consisting of an electron and positron circling each other (if we accept an ordinary particle view and ignore the wave manifestations) about a mutual center of gravity, is called *positronium*.

Two varieties exist, ortho-positronium, in which the two particles spin in the same sense, and para-positronium, in which they spin in opposite senses. The average existence of the former is a ten-millionth of a second (or a tenth of a microsecond) before annihilation takes place. The latter lasts only a ten-thousandth of a microsecond. The former forms three photons on annihilation, the latter forms two. The Austrian-American physicist Martin Deutsch (1917–) was able to detect positronium in 1951 by the light (if I may permit myself a play on words) of the gamma rays they emitted.

Anti-Baryons

There is nothing in Dirac's theory that cannot be applied to the proton as well as to the electron. If the electron has an antiparticle, then so must the proton. Such an *antiproton* could annihilate a proton and produce photons in pairs or in triplets, just as is true for the positron and electron.

However, since a proton has 1836 times the mass of an electron and the antiproton 1836 times the mass of a positron, the energy produced must be 1836 times that produced in electron/positron annihilation. The total energy produced is 1.02×1836, or 1872 Mev. This can also be expressed as 1.872 Bev. We are, as you see, in the billion electron-volt range.

In reverse, the formation of a proton/antiproton pair requires an input of 1.872 Bev at the very least. In fact, more energy is needed since the pair must be formed as the result of the collision of two highly energetic particles, and by using an excess of energy we would make the chances of antiproton production that much better. Physicists estimated that an energy of 6 Bev would turn the trick comfortably.

This energy is present in the more energetic cosmic particles. However, the more energetic cosmic particles are not common and to expect one of them to form a proton/antiproton pair just at the moment when someone is waiting with an appropriate detecting device is to ask a great deal of coincidence.

As it turned out, the discovery of the antiproton was not made until physicists had developed accelerators capable of producing particles in the billion electron-volt range. Particles in the Bev range could then be concentrated on some target at a time when specialized detecting setups were operating. At the University of California a proton synchroton, appropriately called the "Bevatron," was used for the purpose.

The energetic particles produced by the Bevatron were allowed to fall on a copper block, and a vast number of particles were formed by the colossally energetic collisions that resulted. It was then necessary to sort out any antiprotons that might have formed from among all the other explosion debris. The debris was led through a magnetic field that sorted out the negatively-charged particles. Among these, the antiproton was the most massive and traveled most slowly. The debris was therefore led across two scintillation counters some 40 feet apart, and only when those two counters registered with a time interval exactly equivalent to that of the time it would take an antiproton to cover that distance

(0.051 microseconds) was an antiproton considered to be detected.

This was accomplished in 1956 by Segrè (the discoverer of technetium, who by that time had emigrated to the United States) and the American physicist Owen Chamberlain (1920–).

The antiproton is, as might be expected, the twin of the proton, equal to it in mass but differing in charge. The proton is positively charged, but the antiproton is negatively charged. Proton and antiproton can therefore be symbolized as ${}_{1}p^{1}$ and ${}_{-1}p^{1}$ respectively, or as p^{+} and p^{-}, or as p and $\overline{p}$.

The proton is a stable particle and, if left to itself, will presumably exist forever. There seems no obvious conservation law to account for this stability. Might not a proton break down to a positron of 0.51 Mev energy and yield up the remainder of its vast energy in the form of photons? Would not electric charge be conserved?

The fact that this has never been observed to happen means a new conservation law may well be involved. This is the *law of conservation of baryon number,* which states that in any subatomic event, the net number of baryons must be the same before and after. This has always been observed to be so in all the subatomic events studied, and physicists are convinced the law is valid.

If a proton breaks down to a positron, 1 baryon is changed to 0 baryons. This violates the law of conservation of baryon number, and so it doesn't happen. In fact, the proton is the least massive of all baryons, so it can't break down, and thus its stability is a reflection of a conservation law.

Similarly, an antiproton is stable and can't break down to an electron, for instance. It is an *anti-baryon,* the least massive of all anti-baryons, and the law of conservation of baryon number applies to anti-baryons as well.

In the actual universe, however, the antiproton encounters one of the protons (present in overwhelming numbers) almost at once, and mutual annihilation takes place. The net charge of a proton/antiproton pair is zero, so annihilation is possible without violation of the law of conservation of electric charge. In addition, an antiproton is considered as having a baryon number of -1, while a proton has a baryon number of 1. Consequently, the baryon number of a proton/antiproton pair is zero, and annihilation can take place without violating the conservation of baryon number.

The energy resulting from proton/antiproton annihilation may make itself evident in the formation of other particles, rather than as photons only. It sometimes happens, for instance, that where the proton and antiproton score a near miss, it is the charge only and not the mass that is annihilated. One might suppose that an uncharged particle is formed, but one such particle alone cannot be formed. The baryon number of a proton/antiproton pair is 0, but if a neutron, say, is formed, its baryon number is 1, and baryon number is not conserved. Instead, two particles must be formed, a neutron and an *antineutron,* with baryon numbers of 1 and −1, respectively, for a net baryon number of 0. In this way baryon number is conserved. This sort of "semi-annihilation" was first noted in 1956, shortly after the discovery of the antiproton, and that marked the discovery of the antineutron.

It is fair enough to ask what the difference between the neutron and antineutron might be. In the case of the other two particle/anti-particle pairs, the electric charge offers a handy means of differentiation. The electron is negative, the positron, positive. The proton is positive, the antiproton, negative.

There is, however, another difference as well, for these are all particles which possess spin. A spinning particle, if viewed as a tiny sphere, can be pictured as turning about an axis and possessing two poles. If viewed from above one pole, it would seem to be spinning counterclockwise; if viewed from above the other, it would seem to be spinning clockwise. Let us suppose the particle always to be pictured with the counterclockwise pole on top.

A spinning electric charge sets up a magnetic field with a north magnetic pole and a south magnetic pole. In the proton, viewed with the counterclockwise pole on top, the north magnetic pole is on top and the south magnetic pole is on the bottom. In the antiproton, on the other hand, with the counterclockwise pole still on top, it is the south magnetic pole that is on top and the north magnetic pole that is on bottom. In other words, if a particle and an anti-particle are so oriented as to spin in the same sense, the magnetic field of one is reversed with respect to the other. This is also true of the electron and positron.

Although the neutron has no electric charge, it does have a magnetic field associated with it. This is so because although the neutron has a net charge of zero, it apparently has local regions of charge associated with it. The American physicist Robert Hofstadter (1915–), in experiments from 1951 onward, has probed individual nucleons with beams of high-energy electrons. His results seem to indicate that both protons and neutrons are made up of shells of electric charge, and that they differ only in the net total charge.

Because of the neutron's magnetic field, it is possible to speak of both a neutron and an antineutron, the orientation of the magnetic field of one being opposed to that of the other. Because neither neutron nor antineutron has a charge, the symbol $_0n^1$ can apply equally well to both. The two are therefore invariably symbolized as n and $\bar{n}$ respectively.

The neutron decays, with a half-life of 1013 seconds, to a proton and an electron. A baryon is thus converted to a slightly less massive baryon, so baryon number is conserved. A net charge of 0 produces a net charge of 0, so electric charge is conserved. To be sure an electron is created but there is an added refinement to this reaction that will be discussed in the next chapter and that will bring the electron under the guardianship of a conservation law, too (see page 238).

In the same way, an antineutron can decay, with a half-life of 1013 seconds, to an antiproton and a positron, with conservation of baryon number (—1 before and after) and charge (0 before and after). The two events might be written:

$$n \longrightarrow p + e \qquad \text{(Equation 13–5)}$$

and $$\bar{n} \longrightarrow \bar{p} + \bar{e} \qquad \text{(Equation 13–6)}$$

Antimatter

We have now extended Table XIII by three more particles, the positron, the antiproton, and the antineutron, each the mirror-image, so to speak, of one of the particles in the table. Nor may we expect further mirror-images among these particles, for the photon and graviton cannot contribute anti-particles to the table. From theoretical considerations, each of these massless particles is considered to be its own anti-particle. The "anti-photon" and "anti-graviton" are, in others words, identical with the photon and graviton respectively.

We now have, then, four leptons (including one anti-lepton) and four baryons (including two anti-baryons).

Our universe (or at least that part of it which we can study) is very lopsided as far as particle/anti-particle distribution is concerned. It is composed almost entirely of particles, while anti-particles are extremely rare and, even when they are formed, live for only a fraction of a microsecond.

It is fair to wonder why this should be. Most physicists seem to assume that matter was created out of energy, either continuously—little by little—or all at once—at some long-past time.

We might assume, for instance, that matter is produced in the form of neutrons that then decay to form protons and electrons, and that the universe is built out of all three, plus additional energy in the form of photons and gravitons.

But if a neutron is formed, the law of conservation of baryon number would seem to require that an antineutron be simultaneously created. This antineutron would then break down to form antiprotons and positrons. The net result would be that particles and anti-particles would be formed in equal quantities, and any set of nuclear events that could be imagined to have led to the creation of the universe would yield the same result.

Still, if particles and anti-particles were created simultaneously, they would surely interact in mutual annihilation and return to the energy from which they sprang. Under these conditions, the universe could not be created.

It may be, therefore, that although particles and anti-particles were formed simultaneously, they were formed under such conditions that they separated at once, so that the chance for interaction was lost.

Thus, the effect of gravity on individual subatomic particles is so small that it has never been actually measured. It is possible that whereas particles are very feebly attracted to a gravitational field, anti-particles are very feebly repelled by it. In other words, anti-particles produce "anti-gravity." If particles and anti-particles are formed in vast numbers, the gravitational field of one may strongly repel the gravitational field of the other, so that in the end, two universes, driven violently apart, may be formed. The Austrian-American physicist Maurice Goldhaber (1911–   ) has speculated on just this possibility and refers to the two universes as a "cosmon" and an "anti-cosmon." We live in the cosmon, of course.

In the cosmon, atomic nuclei are made up of protons and neutrons and are surrounded by electrons. In the anti-cosmon, consisting as it does almost entirely of anti-particles, there would be nuclei made up of antiprotons and antineutrons surrounded by positrons. Such "anti-atoms" would make up what is called *antimatter*.

A universe of antimatter, totally unobservable by us perhaps, would be in all resects analogous to ours, consisting of "anti-galaxies" made up of "anti-stars" about which "anti-planets" circled, bearing, perhaps, "anti-life" and including even "anti-intelligent observers" studying their universe just as we study ours. They would note that their universe consisted almost entirely of what we consider anti-particles and that particles would have but a rare and fugitive existence. However, it is safe to bet that they would consider their universe to be made up of particles and matter and ours to be of anti-particles and antimatter—and they would be as justified in supposing so as we are.

Another alternative is to suppose that there is only one universe (ours) within which matter and antimatter are distributed equally but in separate chunks. The only safe way of separating these chunks is to suppose that individual galaxies (or galaxy-clusters) are made up of only one variety of substance, either matter or antimatter, but that both galaxies and anti-galaxies might exist in the universe.

If this is so, observation of the fact would be difficult. The only information we receive from other galaxies rests upon their gravitational influence and on the radiation they emit——that is, upon the gravitons and photons that flow from them to us. And, since gravitons and photons are considered to be their own anti-particles, they are produced with equal ease by both matter and antimatter. In other words, an anti-galaxy emits the same gravitons and photons that a galaxy does, and the two cannot be distinguished in that fashion. (Unless it turns out to be true that matter and antimatter repel each other gravitationally and that there is, after all, such a thing as the anti-graviton. The chances of this seem small.)

It is possible, of course, that a galaxy and an anti-galaxy may occasionally approach each other. If so, the mutual annihilation that results should emit energy of a magnitude much more intense than that produced under ordinary conditions. There are indeed galaxies that release unusually colossal energies, and every once in a while the possibility of antimatter raises its head among speculative scientists.

In 1962, certain unusual objects called "quasi-stellar objects," or *quasars*, were discovered. These radiate with the energy of a hundred ordinary galaxies, though they are only one to ten light years in diameter (as opposed to an ordinary galactic diameter of as much as 100,000 light years).

However, every effort is being made to explain this radiation by processes that do not involve antimatter. Antimatter will be turned to only as a last resort, since it would be so difficult to confirm such a speculation.

CHAPTER 43

Other Particles

The Neutrino

In Chapter 11, disappearance in mass during the course of nuclear reactions was described as balanced by an appearance of energy in accordance with Einstein's equation, $e = mc^2$. This balance also held in the case of the total annihilation of a particle by its anti-particle, or the production of a particle/anti-particle pair from energy.

Nevertheless, although in almost all such cases the mass-energy equivalence was met exactly, there was one notable exception in connection with radioactive radiations.

Alpha radiation behaves in satisfactory fashion. When a parent nucleus breaks down spontaneously to yield a daughter nucleus and an alpha particle, the sum of the mass of the two products does not quite equal the mass of the original nucleus. This difference appears in the form of energy—specifically, as the kinetic energy of the speeding alpha particle. Since the same particles appear as products at every breakdown of a particular parent nucleus, the mass-difference should always be the same, and the kinetic energy of the alpha particles should also always be the same. In other words, the beam of alpha particles should be *monoenergetic*. This was, in essence, found to be the case.

In some instances, to be sure, the beam of alpha particles could be divided into two or more subgroups, each of which was monoenergetic, but with the energy of the subgroups differing among themselves. It was shown without much trouble that this was so because the parent nucleus could exist at various energy levels to begin with. An excited parent nucleus had a bit more energy content than a non-excited one, and the alpha particles produced by the former had correspondingly more kinetic energy. For each different energy level of the parent nucleus, there was a separate subgroup of monoenergetic alpha particles, but in each case, mass-energy equivalence (or, in a broader sense, the law of conservation of energy) was upheld.

It was to be expected that the same considerations would hold for a parent nucleus breaking down to a daughter nucleus and a beta particle. It would seem reasonable to suppose that the beta particles would form a monoenergetic beam, too, or, at worse, a small group of monoenergetic beams.

Instead, as early as 1900, Becquerel indicated that beta particles emerged with a wide spread of kinetic energies. By 1914, the work of James Chadwick demonstrated the "continuous beta particle spectrum" to be undeniable.

The kinetic energy calculated for a beta particle on the basis of mass loss turned out to be a maximum kinetic energy that very few attained. (None surpassed it, however; physicists were not faced with the awesome possibility of energy appearing out of nowhere.)

Most beta particles fell short of the expected kinetic energy by almost any amount up to the maximum. Some possessed virtually no kinetic energy at all. All told, a considerable portion of the energy that should have been present, wasn't present, and through the 1920's this missing energy could not be detected in any form.

Disappearing energy is as insupportable, really, as appearing energy, and though a number of physicists, including, notably, Niels Bohr, were ready to abandon the law of conservation of energy at the subatomic level, other physicists sought desperately for an alternative.

In 1931, an alternative was suggested by Wolfgang Pauli. He proposed that whenever a beta particle was produced, a second particle was also produced, and that the energy that was lacking in the beta particle was present in the second particle.

The situation demanded certain properties of this hypothetical particle. In the emission of beta particles, electric charge was conserved; that is, the net charge of the particles produced after emission was the same as that of the original particle. Pauli's postulated particle therefore had to be uncharged. This made additional sense since, had the particle possessed a charge, it would have produced ions as it sped along and would therefore have been detectable in a cloud chamber, for instance. As a matter of fact, it was not detectable.

In addition, the total energy of Pauli's projected particle was very small—only equal to the missing kinetic energy of the electron. The total energy of the particle had to include its mass, and the possession of so little energy must signify an exceedingly small mass. It quickly became apparent that the new particle had to have a mass of less than 1 percent of the electron and, in all likelihood, was altogether massless.

Enrico Fermi, who interested himself in Pauli's theory at once, thought of calling the new particle a "neutron," but Chadwick, at just about that time, discovered the massive, uncharged particle that came to be known by that name. Fermi therefore employed an Italian diminutive suffix and named the projected particle the *neutrino* ("little neutral one"), and it is by that name that it is known.

An uncharged, massless particle struck physicists as being a "ghost particle," since it could be detected by neither charge nor mass. Its existence would have been rather difficult to swallow, even for the sake of saving the law of conservation of energy, and the neutrino might have been ignored had it not turned out to save three other conservation laws as well.

This came up most clearly in the application of neutrino theory to the breakdown of neutrons. The neutron breaks down, with a half-life of 12 minutes, to a proton and an electron, and the electron can emerge with any of a wide range of kinetic energies. It should therefore follow from Pauli's theory that the neutron

Neutron breakdown

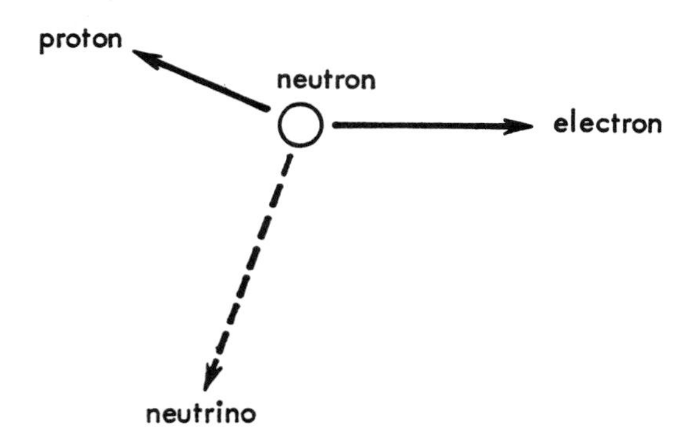

should break down to three particles—a proton, an electron, and a neutrino.

The difference in breaking down to three particles rather than two is significant in connection with the law of conservation of momentum (see page I–69). If a stationary neutron broke down to two particles only, these two would have to be ejected in opposite directions, their lines of travel forming a straight line. Only so could momentum be conserved.

If the same neutron broke down to three particles, then any two of those particles would have to be ejected to one side of an imaginary straight line, yielding a net momentum in a particular direction that would be exactly balanced by the momentum of the third particle shooting off in the opposite direction.

Studies of neutron breakdown indicated clearly enough that the proton and electron, when formed, went shooting off to one side of a straight line and that the existence of a third particle shooting off to the other side was absolutely necessary if momentum were to be conserved.

Once the matter of particle spin was understood, it became clear that the neutrino was useful in connection with the law of conservation of angular momentum (see page 81) as well. The neutron, proton and electron all have spins of either $+1/2$ or $-1/2$. Suppose a neutron broke down to a proton and electron only. The proton and electron together could have a spin of $+1$, 0, or -1 ($+1/2+1/2$, $+1/2-1/2$, or $-1/2-1/2$). In no case could they match the neutron's original spin of $+1/2$ or $-1/2$, and angular momentum would not be conserved.

But suppose the neutrino also had a spin of either $+1/2$ or $-1/2$. In that case, the sum of the spin of all three particles could easily be $+1/2$ or $-1/2$. It could, for instance, be $+1/2+1/2-1/2$, and thus angular momentum would be conserved.

Finally, there is a more subtle conservation law. In the prevous chapter, I made use of the conservation of baryon number (see page 229). A proton and neutron each have a baryon number of $+1$, and an antiproton and antineutron each have a baryon number of -1. In the neutron breakdown, baryon number is conserved, for we begin with a neutron (baryon number $+1$) and end with a proton (baryon number $+1$).

Can we recognize a similar law involving electrons, with an electron possessing a number of $+1$ and a positron a number of -1? The answer is: not if those two particles are the only ones considered. Thus, in neutron breakdown, we begin with no electrons (or positrons), and we end with one electron.

However, suppose we consider an *electron family* that includes not only electrons and positrons, but neutrinos, too. To make matters work out, it will be necessary to have not only a neutrino but an *antineutrino* as well. The difference between the neutrino and antineutrino would involve the direction of the magnetic field associated with the spinning particles, exactly as in the case of the neutron and antineutron (see page 231). The neutrino can be given an electron family number of $+1$ and the antineutrino an electron family number of -1.

With that in mind, let's consider neutron breakdown again. The neutron begins with an electron family number of 0, since it is itself not a member of the family. In breaking down, it produces a proton (electron family number 0) and an electron (electron family number $+1$). If we add to that, not a neutrino, but an antineutrino (electron family number -1), we have preserved the *law of conservation of electron family number*, which is 0 before and after breakdown.

The antineutrino saves the laws of conservation of energy, momentum, and angular momentum just as the neutrino would, and it enables us to retain the law of the conservation of electron family number as well. If we symbolize the neutrino as ν (the Greek letter "nu") and the antineutrino as $\bar{\nu}$, we can write the equation for neutron breakdown as follows:

$$n^{o} \longrightarrow p^{+} + e^{-} + \bar{\nu}^{o} \qquad \text{(Equation 14–1)}$$

On the other hand, in the conversion of a proton to a neutron with the ejection of a positron (see page 224), we have produced a particle with an electron family number of -1. To balance that we must add the production of a neutrino (electron family number $+1$). We can therefore write:

$$p^{+} \longrightarrow n^{o} + e^{+} + \nu^{o} \qquad \text{(Equation 14–2)}$$

Indeed, if we introduce neutrinos or antineutrinos into nuclear reactions, we can, whenever necessary, save the four conservation laws of energy, momentum, angular momentum, and electron family number. With this fourfold benefit, neutrinos and antineutrinos have to be accepted, whether they can be detected or not.

Neutrino Interactions

Despite the tightness of the reasoning from laws of conservation, physicists recognized that great satisfaction would come with the actual detection of the neutrino or antineutrino. To make detection possible, however, a neutrino or antineutrino must interact with some other particle in a recognizable manner.

Thus, a neutron changes to a proton, emitting an antineutrino in the process. Why cannot the reverse hold true and an antineutrino be absorbed by a proton to form a neutron? If so, this antineutrino absorption could leave a recognizable mark.

Unfortunately, the chance of such antineutrino absorption is vanishingly small. A neutron will break down to a proton with a half-life of 12 minutes. This means that in 12 minutes there is an even chance of a particular neutron producing an antineutrino. It follows that if an antineutrino remained in the immediate neighborhood of a proton for 12 minutes, there could be an even chance of the absorption taking place.

However, an antineutrino will not remain in the neighborhood of a proton for 12 minutes or, for that matter, even for a millionth of a second. Massless particles such as the neutrino, the antineutrino, the photon, or the graviton all begin moving at the speed of light at the moment of creation and keep moving at that speed until the moment of absorption. This means that an antineutrino remains in the immediate neighborhood of a proton only for about 10^{-23} seconds, and the chances of interaction in that short interval of time are exceedingly small. They are so small, in fact, that a neutrino or antineutrino would have to travel through some 3500 light-years of solid matter, on the average, before undergoing absorption.

The situation with regard to a photon is completely different. A photon also travels at the speed of light, but when the energy situation is such that a photon must be emitted by an atom, that photon is emitted in only about 10^{-8} seconds. Hence a photon need be in the vicinity of an atom for only about 10^{-8} seconds to stand a good chance of being absorbed. In addition, the photon has a considerably longer wavelength than the neutrino (viewing both as wave forms) and takes a longer time to pass an object, even though both are traveling at the same speed.

Gamma rays will, in fact, penetrate only ten feet of lead before being absorbed. Ordinary light, which is of much longer wavelength than gamma rays and takes even longer to pass a single atom, is even more readily absorbed and rarely penetrates more than a couple of dozen atom-thicknesses into a solid.

All this has an important consequence in astronomy. In the course of the fusion of hydrogen to helium, protons are converted to neutrons, so that neutrinos are formed as well as photons.

Photons carry off about 90 to 95 percent of the energy produced in the sun's core, while neutrinos carry off the remaining 5 to 10 percent.

The photons, once formed, are absorbed and re-emitted over and over again by the matter making up the sun; consequently, it takes something like a million years for the average photon to make its way from the core of the sun, where it is formed, to the surface, where it is radiated out into space. This insulating effect of solar material (thanks to the way in which photons so readily interact with matter) is dramatically demonstrated by the fact that the sun's core is at a temperature of 15,000,000° C, while the surface, only 430,000 miles away, is at a temperature of merely 6000° C.

The neutrinos formed in the core, however, are not re-absorbed by the matter of the sun. They shoot off instantly, at the speed of light, passing through the solar matter as though it were vacuum and taking less than three seconds to reach the solar surface and pass into space. This instant loss of energy has a small cooling effect on the sun's core, but not enough to matter.

A certain number of the solar neutrinos reach the earth and, after doing so, pass right through the planet in 1/125 of a second or less. About ten billion neutrinos pass through every square centimeter of the earth's cross section (passing through us, too). We are steadily bombarded both day and night, for the intervention of the bulk of the earth between ourselves and the sun does not interfere. However, the neutrinos pass through us without interacting, so they do not disturb us in any way.

It is possible that neutrinos and antineutrinos can be formed by methods that do not involve protons and neutrons. For instance, an electron-positron pair may be formed from gamma ray photons. The electron and positron may then react to form a neutrino and antineutrino:

$$e^- + e^+ \longrightarrow \nu^0 + \bar{\nu}^0 \qquad \text{(Equation 14–3)}$$

Energy, charge, momentum and angular momentum are all conserved in this reaction, and so is electron family number. The net electron family number of an electron and positron is zero, and that of a neutrino and antineutrino is also zero.

Such an electron-positron interaction is extremely unlikely even at the temperature of the sun's core, so that it makes no important contribution to the neutrino supply. In the course of a star's evolution, however, the core grows hotter and hotter, and as it does so the probability of conversion of photons to neutrinos via the electron-positron pair increases.

The American physicist Hong-Yee Chiu has calculated that when a temperature of 6,000,000,000° C is reached, the conversion of photons to neutrinos becomes so massive that the major portion of the energy formed in the core of such a star appears as neutrinos. These leave the core at once and withdraw so much energy that the core collapses and the star with it, resulting in a tremendous outburst of energy. This, it is suggested, is the cause of a supernova.

To say that a neutrino is extremely unlikely to interact with another particle is not the same, of course, as saying that it will never interact at all. If a neutrino must travel through an average of 3500 light-years of solid matter to be absorbed, that length of travel remains an average. Some neutrinos may survive for much longer distances, but some may be absorbed long before traversing such a path. There is a finite chance, exceedingly small but not zero, of a neutrino interacting after traveling only a mile or even only a foot.

Evidence for such interactions was sought by the American physicists Clyde L. Cowan, Jr. (1919–) and Frederick Reines (1918–) in experiments beginning in 1953. As the proton target they used large tanks of water (rich in hydrogen atoms and, therefore, in nuclei consisting of single protons), and they placed these in the path of a stream of antineutrinos originating in a fission reactor. (These antineutrinos arise in the course of the rapid conversion of neutrons to protons within the nuclei of fission products.)

If an antineutrino were to join a proton to form a neutron, in the reverse of the interaction of Equation 14–1, an electron would have to be absorbed simultaneously. The necessity of such a double joining makes the reaction less likely than ever. However, the absorption of an electron is equivalent to the emission of a positron, so the expected neutrino interaction with a proton may be described as follows:

$$\bar{\nu}^{0} + p^{+} \longrightarrow e^{+} + n^{0} \qquad \text{(Equation 14–4)}$$

In such a reaction, baryon number is conserved, for a proton (+1) is replaced by a neutron (+1). Electron family number is also conserved, for an antineutrino (−1) is replaced by a positron (−1).

Cowan and Reines calculated that in the water targets they

were using, such an antineutrino/proton interaction ought to take place three times an hour. The trouble was that a large number of other events were also taking place, events originating in cosmic radiation, stray radioactive radiations, and so on. At first these unwanted events took place with many times the frequency of the antineutrino reactions being searched for. With time, this interference was reduced to manageable levels by using heavy shielding that excluded most extraneous subatomic particles and photons but offered no barrier whatever, of course, to antineutrinos.

It remained to identify the antineutrino interaction precisely and certainly. The interaction produces a positron and a neutron. The positron interacts almost at once with an electron, producing two gamma rays of a known energy content, coming off in opposite directions.

The neutron produced by the interaction lasts a few millionths of a second longer before it is absorbed by a cadmium atom (introduced into the water tank in the form of a solution of cadmium chloride precisely for the purpose of absorbing neutrons). The cadmium atom, excited by neutron absorption, releases a gamma ray (or possibly three) of known frequency. It is this combination of events, a double gamma ray of fixed frequency, followed after a fixed interval by a third gamma ray of fixed frequency, that is the identifying mark of the antineutrino. No other particle could produce just the duplicate of these results, at least as far as is known.

In 1956, the antineutrino was finally detected by this characteristic pattern of gamma radiation, and Pauli's original suggestion of a quarter of a century earlier was vindicated.

The Muon

While Pauli was advancing his solution for the problem of the continuous beta particle spectrum, a second problem, just as puzzling, had arisen.

The atomic nucleus contains protons held together in a volume of something like 10^{-40} cubic centimeters. The electromagnetic force of repulsion between such protons, jammed so closely together, is tremendous. As long as it was believed that electrons also existed within the nucleus, it could be supposed that the attraction between protons and electrons (also jammed tightly together) could make up for the inter-proton repulsion. The electrons, then, would serve as a "nuclear cement," and electromagnetic forces would explain the situation within the nucleus as they did the situation between atoms and molecules.

When, however, it became quite clear, in 1932, that the atomic nucleus was made up of protons and neutrons and that electrons did not exist there, the problem was thrown wide open. Since only electromagnetic repulsion could exist within the nucleus, why did not all atomic nuclei explode at once?

The only way of explaining the stability of the nucleus was to suppose that there was a nuclear force of attraction between nucleons, one that was in evidence only at extremely small distances and that was then much stronger than the electromagnetic force.

In the early 1930's, quantum mechanical analysis made it seem that a force, which seemed to act at a distance as the electromagnetic force did, actually acted through the emission and absorption of photons. Electrically-charged particles, exchanging photons, experienced *exchange forces,** a term introduced by Heisenberg in 1932. By analogy, it was decided that the gravitational force had to make itself evident through the emission and absorption of gravitons (see page 221).

Both the electromagnetic force and the gravitational force are long-distance forces, diminishing only as the square of the distance between the objects exerting the force, and making themselves evident even over astronomical distance.

The hypothesized nuclear force, however, had to be extremely short-range and, however strong within the nucleus, it had to be imperceptible outside the nucleus. In fact, in the larger nuclei, the nuclear force must barely reach across the full diameter, and it may be for this reason that nuclear fission takes place as easily as it does in the more massive atoms.

The Japanese physicist Hideki Yukawa (1907–) set himself the task of working out the mechanism of such an unusually strong and unusually short-range force. Without going into the quantum mechanics of the theory, we can present a simplified picture of the reasoning involved.

The principle of uncertainty states that position and momentum cannot be simultaneously determined with complete accuracy. The uncertainty in the determination of one multiplied by the uncertainty in the determination of the other is approximately equal to Planck's constant. It can be shown that time and

* Actually, the word "force" is going out of fashion among physicists. In subatomic physics, particularly, *interaction* is preferred to "force" when describing the consequences of emission and absorption of particles.

energy can be substituted in place of position and momentum. This means that the precise energy content of a system cannot be determined at an exact moment of time. There is always a small time interval during which the energy content is uncertain. The uncertainty in energy content multiplied by the uncertainty in time is again approximately equal to Planck's constant.

During the interval of time in which energy content is uncertain, a proton might, for instance, emit a small particle. It doesn't really have the energy to do this, but for the short instant of time during which its energy cannot be exactly determined, it can violate the law of conservation of energy with impunity—because, so to speak, no one can get there fast enough to enforce it.

By the end of the time period that particle emitted by the proton must be back where it started, and the proton must be obeying energy conservation. The particle, which is emitted and re-absorbed too quickly to be detected, is a *virtual particle*. Reasoning shows it can exist, but no system of measurement can detect it.

During the period of existence of the virtual particle it can move away from the parent proton, but it can only move a limited distance because it must be back when the time-uncertainty period is over. The more massive the particle (and the greater its energy content), the greater the uncertainty represented by this energy and the shorter the time interval permitted its existence, for the two together must yield the same product under all circumstances so that as one uncertainty goes up the other goes down in precise step.

Even if the virtual particle were traveling at the speed of light, it could not move very far from its proton, for Planck's constant is a very small quantity, and the time interval permitted the particle's existence is excessively tiny. Ordinarily, the virtual particle never reaches far enough from the proton to impinge on any other particle. The only exception arises when protons and neutrons are in the close proximity found within the atomic nucleus. Then, one of the particles leaving the proton may be picked up by a neutron before it has a chance to return to the proton. It is this emission and absorption of virtual particles that produces the nuclear force.

In 1935, Yukawa advanced his views that this virtual particle served as the exchange particle of the nuclear force. Unlike the exchange particles of the electromagnetic and gravitational forces, the exchange particle of the nuclear force had to have mass so that its permitted time of existence would be brief enough to make it sufficiently short-range. Yukawa showed that the particle would have to be about 270 times as massive as an electron in order for its permitted time of existence to be short enough to make it as short-range as observation showed the nuclear force must be.

Because such a particle is intermediate in mass between the light electrons and the massive particles of the nucleus, it came to be called a "mesotron," from a Greek word meaning "intermediate," and this was quickly shortened to *meson*.

Yukawa's theory indicated that in the process of exchange, a proton would become a neutron and a neutron would become a proton. In other words the meson, in being emitted by one and absorbed by the other, would have to carry the charge with it. You would expect a positive meson, therefore. In the case of antiprotons and antineutrons you would expect a negative meson as an anti-particle, holding the nucleus of antimatter together.

Then, too, it turned out that exchange forces existed between proton and proton and between neutron and neutron; for this a neutral meson was needed. This neutral meson served as its own anti-particle and served equally well to bind antiproton and antiproton or antineutron and antineutron.

The proton-neutron exchange force is somewhat stronger than the proton-proton exchange force, which means that the *p-n* combination within a nucleus has a lower packing fraction than the *p-p* combination. It therefore takes an energy input to convert a *p-n* to a *p-p* within a nucleus.

The conversion of *n* to *p* yields a small amount of energy (which is why a neutron breaks down spontaneously), but the quantity of energy so obtained is not always sufficiently great to change the *p-n* combination to a *p-p* combination. It is for this reason that in some nuclei the neutron does not change to a proton but stays put; and thus stable nuclei exist.

To check Yukawa's theory, the mesons would actually have to be detected. Within the nucleus, where they are virtual particles only, this cannot be done. However, if enough energy is added to the nucleus, the meson can be formed without violating conservation of energy. It then becomes a real particle and can leave the nucleus.

In 1936, Carl Anderson, who had earlier discovered the positron among the tracks produced by cosmic particles, now found a track which curved less sharply than an electron and more sharply than a proton. It was obviously produced by a particle of intermediate mass, and physicists assumed at first that it was Yukawa's predicted particle.

This proved not to be the case. Anderson's particle was only 207 times the mass of an electron, distinctly less than Yukawa's prediction. It came only in a positive and negative variety, without any sign of a neutral variety; and it was the negative variety, rather than the positive variety, that was the particle. Worst of all, it did not seem to interact with protons or neutrons. If it was Yukawa's exchange particle it should have been absorbed by any nucleon it encountered. Anderson's meson, however, passed through matter almost undisturbed.

It eventually turned out that there were not one but a number of different mesons and that Anderson's meson was not Yukawa's exchange particle. The different types of mesons were given different prefixes (often one or another of the Greek letters) and Anderson's was named a *mu-meson*, a term which is now commonly shortened to *muon*.

As the properties of the muon were studied more and more closely, it turned out that the muon seemed more and more similar to an electron. It was identical in charge, with the negative variety serving as the particle; the positive, as the anti-particle. The muon was the same as the electron in spin and in magnetic properties—in everything but mass and stability.

Indeed, for every interaction involving the electron, there is an analogous interaction involving the muon. The muon, while it lives, can even replace the electron within atoms to form a *mesonic atom*. Angular momentum must be conserved in the process. If we view the electron in the old-fashioned way as a particle circling the nucleus, the muon (moving at the same speed) must circle in an orbit closer to the nucleus. Its greater mass is thus countered by the shortened radius of revolution to keep the angular momentum the same (see page I–82).

Since the muon is 207 times the mass of the electron, it must be at only 1/207 the distance from the nucleus. In very massive atoms, it means that the orbit of the innermost meson must actually be within the nucleus! The fact that it can circle freely within the nucleus shows how small a tendency it has to interact with protons and neutrons.

The difference in mesonic energy levels in such atoms is correspondingly larger than in the electronic energy levels in ordinary atoms. Mesonic atoms emit and absorb X ray photons in place of the visible light photons emitted and absorbed by ordinary atoms.

To be sure, the muon is unstable, decaying in about 2.2 microseconds, and changing to an electron. However, on the subatomic scale 2.2 microseconds is quite a long time, and the muon does not seem too different in this respect from the completely stable electron.

It seems now that the muon is virtually a "heavy electron" and nothing more. But why there should be a heavy electron at all, and why it should be so much heavier, is as yet not known.

The Pion

Though the muon failed to fill the role of Yukawa's particle, there was success elsewhere. In 1947, the English physicist Cecil Frank Powell (1903–) duplicated Anderson's feat and uncovered meson tracks in photographic plates exposed to cosmic radiation in the Bolivian Andes. These new mesons were distinctly more massive than the muon. Their mass, in fact, was equal to 273 electrons, almost exactly the predicted mass for Yukawa's exchange particles.

On investigation they proved to interact strongly with nuclei, as Yukawa's exchange particle would be expected to do. The new meson was a particle when positively charged and an anti-particle when negatively charged, as was to be expected. Eventually, a neutral version of this meson was also found, one that was somewhat lighter than the charged varieties (only 264 times the mass of the electron).

The new meson was named the *pi-meson,* or, as it is now commonly known, the *pion,* and it is the pion that is Yukawa's exchange particle. Both the neutron and proton are now viewed as consisting, essentially, of clouds of pions. This was demonstrated in the 1950's by Robert Hofstadter, who bombarded protons and neutrons with electrons of 600 Mev energy, produced in a linear accelerator. These electrons, in being scattered, actually penetrated the proton passed through the outer portion of the pion cloud.*

Pions are unusual in the nature of their spin. Most of the

* Findings such as this raise the question of just which subatomic objects are *elementary particles*—that is, which are not composed of still smaller and simpler components. For that matter, do elementary particles exist at all? Does the phrase have meaning? Physicists have no good answer to this at the moment.

particles so far discussed—the neutrino, electron, muon, proton and neutron, together with their anti-particles—have spins of 1/2. Particles with such nonintegral spins behave according to *Fermi-Dirac statistics* (a mathematical analysis worked out by Fermi and Dirac) and are in consequence all lumped together as *fermions.* An outstanding property possessed by fermions generally is that of adhering to Pauli's exclusion principle (see page 80).

The photon has a spin of 1 and the graviton a spin of 2. These particles, and all others possessing integral spin, including a number of atomic nuclei, behave according to *Bose-Einstein statistics,* worked out by Einstein and by the Indian physicist Satyenda Nath Bose (1904–). Such particles are *bosons,* and the exclusion principle does not hold for them.

The pions were the first individual particles to be found with a spin of 0, and the first particles, possessing mass, which were bosons.

The ready reaction of a pion with nuclear particles is an example of what is known as the *strong interaction.* This is characterized by extreme rapidity. A pion traveling at almost the speed of light remains within short range of a proton or neutron for only 10^{-23} seconds, yet this is enough time for the strong interaction to take place. It is this strong interaction which is the nuclear force that holds the nucleus together against the repulsion of the electromagnetic interaction.

There are, however, interactions involving subatomic particles that take place in much longer intervals of time—only after a hundred-millionth of a second or more. These are the *weak interactions* which are very short-range, like the strong interactions but are only a trillionth as intense as the strong interactions. The weak interaction is, in fact, only a ten-billionth as intense as the electromagnetic interaction, but it is still tremendously stronger than gravitation, which retains its status as the weakest force in nature.

The pions are the exchange particles of the strong interactions and there should also be exchange particles for the weak interactions. This "weak exchange particle," symbolized by *w,* should be more elusive than either the pion or the photon, though not as elusive as the graviton. It should be one of the boson family of particles and should be more massive than such bosons as the photons, though less massive than such bosons as the pions. It is, for that reason, sometimes referred to as the *intermediate boson.* Some recent reports have indicated its detection, but this is not yet certain.

The proton, antiproton, positive pion and negative pion can all be involved in any of the four types of interaction: strong, weak, electromagnetic, and gravitational. The neutron, antineutron, and neutral pion, being uncharged, cannot be involved in electromagnetic interactions, but can engage in any of the remaining three. The electron, positron, positive muon and negative muon cannot take part in strong interactions, but can be involved in the remaining three.

The neutrino and antineutrino are the most limited in this respect. They do not take part in the strong interaction. Being uncharged, they do not take part in the electromagnetic interaction, and being massless, they do not take part in the gravitational interaction. The neutrino and antineutrino take part only in weak interactions and nothing more. The appearance of a neutrino or antineutrino in the course of the breakdown of a particle is thus a surefire indication that this breakdown is an example of a weak interaction. The breakdown of a neutron is, for instance, a weak interaction.

Once negative pions and positive pions are formed in the free state, they too break down in a weak interaction, with a half-life of 25 billionths of a second. They break down to muons and neutrinos, and if we allow pions to be represented as π (the Greek letter "pi") and muons as μ (the Greek letter "mu"), we can present the breakdown as follows:

$$\pi^+ \longrightarrow \mu^+ + \nu^0 \qquad \text{(Equation 14–5)}$$

$$\pi^- \longrightarrow \mu^- + \bar{\nu}^0 \qquad \text{(Equation 14–6)}$$

At first, physicists suspected that the neutrino produced in the course of pion breakdown might be distinctly more massive than the ordinary neutrino and, in fact, have a mass perhaps 100 times that of the electron. For a while, they called this particle the "neutretto." However, further study scaled the apparent mass downward, until finally it was decided that the small neutral product of pion breakdown was a massless neutrino.

If the muon is considered as only a "heavy electron," it might seem reasonable to include the muons in the electron family and give the negative muon (like the electron) an electron family number of $+1$, and the positive muon (like the positron) an electron family number of -1.

If so, then in Equation 14–5 the production of a positive muon (-1) and a neutrino ($+1$) gives a net electron family number of 0, matching that of the original pion (which, not being a member of the electron family at all, has an electron family number of 0). In the same way, in Equation 14–6, the production of a negative muon (+ 1) necessitates the simultaneous production of an antineutrino (— 1) for, again, a net electron family number of 0.

So far, so good, but difficulty arises when muon breakdown is considered. A muon breaks down to form an electron and two neutrinos. If electron family number is to be conserved, then one of the neutrinos must be an antineutrino. The reaction (in the case of a negative muon) can be written as follows:

$$\mu^- \longrightarrow e^- + \nu^0 + \bar{\nu}^0 \qquad \text{(Equation 14–7)}$$

Beginning with an electron family number of + 1 for the negative muon, one ends with a total electron family number of + 1 by taking the sum of + 1, + 1 and — 1 for the electron, neutrino, and antineutrino respectively. The conservation holds.

However, if this is so, should not the neutrino and antineutrino at least sometimes annihilate each other in a burst of energy, as any particle/anti-particle combination would? Should not then a negative muon break down to form an electron only, at least sometimes, with the remaining mass of the muon appearing as photons?

This is never observed and the suspicion therefore arose that the neutrino and antineutrino produced by muon breakdown were not true opposites. Could it be that the neutrino was produced in association with the muon and that the antineutrino was produced in association with the electron, for instance, and that muons and electrons produced different kinds of neutrinos?

In 1962, this possibility was tested as follows: Very high energy protons were smashed into beryllium atoms in such a way as to produce an intense stream of pions. The pions broke down rapidly to muons and neutrinos, and all then smashed into a wall of armor plate about 13.5 meters thick. All particles but the neutrinos were stopped. The neutrinos passed through easily, and inside a detecting device would every once in a while interact with a neutron to form a proton plus either a negative muon or an electron.

If there were only one kind of neutrino then it ought to produce negative muons and electrons without discriminating between the two:

$$\nu^0 + n^0 \longrightarrow p^+ + e^- \qquad \text{(Equation 14–8)}$$

or $$\nu^0 + n^0 \longrightarrow p^+ + \mu^- \qquad \text{(Equation 14–9)}$$

As you see, electric charge and baryon number is conserved in either case. Electron family number would seem to be conserved in either case, too, for you would begin with a neutrino (electron family number + 1) and end with either an electron or a negative muon, each of which has an electron family number of + 1. In subatomic interactions, anything which can happen does happen, so physicists were sure that if there was only one kind of neutrino then negative muons and electrons would be produced in equal numbers.

They weren't! Only negative muons were produced.

This meant that when pions broke down to form muons and neutrinos, the neutrinos were *muon-neutrinos,* a special variety that could interact only to form muons, never electrons. Similarly, the ordinary neutrinos formed in association with electrons and positrons were *electron-neutrinos,* which could interact only to form electrons or positrons, never muons.

If we symbolize the muon-neutrino as ν_μ and the electron-neutrino as ν_e, we can rewrite Equations 14–1, 14–2, 14–3, 14–4, 14–5, and 14–6 as follows:

$$n^o \longrightarrow p^+ + e^- + \bar{\nu}_e{}^o \qquad \text{(Equation 14–10)}$$

$$p^+ \longrightarrow n^o + e^+ + \nu_e{}^o \qquad \text{(Equation 14–11)}$$

$$e^- + e^+ \longrightarrow \nu_e{}^o + \bar{\nu}_e{}^o \qquad \text{(Equation 14–12)}$$

$$\bar{\nu}_e{}^o + p^+ \longrightarrow e^+ + n^o \qquad \text{(Equation 14–13)}$$

$$\pi^+ \longrightarrow \mu^+ + \nu_\mu{}^o \qquad \text{(Equation 14–14)}$$

$$\pi^- \longrightarrow \mu^- + \bar{\nu}_\mu{}^o \qquad \text{(Equation 14–15)}$$

Equations 14–10, 14–11, 14–12, and 14–13, still exhibit conservation of electron family number. Equations 14–14 and 14–15 now exhibit a *conservation of muon family number,* where the members of the muon family include the negative muon and the muon-neutrino (each with a muon family number of + 1) and the positive muon and the muon-antineutrino (each with a family number of − 1). In Equations 14–14 and 14–15, as you see, there is a net muon family number of 0 both before and after the breakdown of the pion.

The one interaction which deals with both electrons and muons is Equation 14–7. This can now be rewritten:

$$\mu^- \longrightarrow e^- + \nu_\mu{}^o + \bar{\nu}_e{}^o \qquad \text{(Equation 14–16)}$$

This exhibits conservation of muon family number because you begin with a muon family number of + 1 (the negative muon) and end with a muon family number of + 1 (the muon-neutrino). It also exhibits conservation of electron family number because you begin with an electron family number of 0 (no members of the electron family present) and end with an electron family number of + 1 for the electron and — 1 for the electron-anti-neutrino, making a net electron family number of 0 again.

By the same reasoning, the breakdown of the positive muon would be:

$$\mu^{+} \longrightarrow e^{+} + \bar{\nu}_{\mu}{}^{o} + \nu_{e}{}^{o} \qquad \text{(Equation 14–17)}$$

producing a positron, an electron-neutrino, and a muon-antineutrino.

In the breakdown of either the negative muon or the positive muon, you could expect no mutual annihilation among the neutrinos and antineutrinos since they are not true anti-particles. Mutual annihilation would violate conservation of both electron family number and muon family number.

However, whatever difference there may be between electron-neutrinos and muon-neutrinos, when both are massless, chargeless particles with a spin of 1/2 remain a mystery.

The Frontier

Since 1947, a variety of other particles have turned up. With the exception of the muon-neutrino (which was more of a realization than a discovery), all the new particles are quite unstable, quite massive, and subject to strong interactions.

There are, for instance, a group of *K-mesons,* or *kaons,* which are 966.5 electrons in mass and, in this respect, lie roughly midway between protons and pions. Like the pions, the kaons have 0 spin and are bosons. Like the pions, also, there is a positive kaon, which is the particle, and the negative kaon, which is the anti-particle. There is also a neutral kaon, a trifle less massive than the charged ones and somewhat more unstable. However, the neutral kaon is not its own anti-particle as the neutral pion is. There is a neutral kaon and a neutral anti-kaon.

In addition to these, particles more massive than protons or neutrons were discovered. These fell into three groups, distinguished by Greek letters: *lambda particles, sigma particles* and *xi particles.*

There is one lambda particle (neutral), three sigma particles (positive, negative, and neutral), and two xi particles (negative and neutral). Each has its anti-particle. The lambda particle has a mass of 2182 electrons (or 1.18 protons). The three sigma particles are more massive, about 1.27 protons, while the xi particles are still more massive, about 1.40 protons. All of these are lumped together as *hyperons* (from a Greek word meaning "beyond," because they are beyond the proton in mass), and all are fermions.

Just as muons can replace electrons in atomic structure to form mesonic atoms, the lambda particle can replace a particle in the atomic nucleus to produce a short-lived *hypernucleus.*

The 1960's saw the discovery of numerous extremely short-lived particles, with lifetimes as short as 10^{-23} seconds. These are *resonance particles.* It is not certain that these are truly single particles. They may be merely momentary associations of two or more particles.

The proliferation of particles has been an embarrassment to physicists, for it is difficult to reduce them to order. New rules of behavior had to be deduced for them.

For instance, although the hyperons are produced under conditions that make it clear that they are strongly-interacting particles, and though they can break down to strongly-interacting products, they neverthelesss take a strangely long time about it. The lambda particle breakdown can be represented thus, for instance:

$$\Lambda^0 \longrightarrow p^+ + \pi^- \qquad \text{(Equation 14–18)}$$

where Λ (the Greek capital letter "lambda") represents the lambda particle. All the usual conservation laws are obeyed in this interaction. Since the pion has zero spin, for instance, angular momentum is conserved. Since hyperons are considered baryons, baryon number is conserved. (The pion is not a member of any of the family groups that are conserved, and can appear and disappear freely as long as other conservation laws are observed.)

Since Equation 14–18 seems to represent a strong interaction, it should take place in not much longer than 10^{-23} seconds or so; instead, however, it takes place in 2.5×10^{-10} seconds. This is ten trillion times as long as it ought to take—immensely long on the subatomic time scale.

In 1953, an explanation was independently advanced by the American physicist Murray Gell-Mann (1929–) and the Japanese physicist Kazuhiko Nishijima (1926–). They suggested a new quantity that had to be conserved, one which Gell-Mann called *strangeness.*

The various members of the electron family and muon family, as well as the pions, nucleons and their anti-particles, all have a strangeness number of zero. Other particles, with strangeness numbers other than zero, are lumped together as *strange particles*. The kaon has a strangeness number of + 1; the lambda particle and sigma particle, one of — 1; and the xi particle, one of — 2. For the anti-particles the sign of the strangeness number is reversed, of course.

These numbers are not assigned arbitrarily but are deduced from experiment. A particle with strangeness number of + 1 is never formed without the simultaneous production of particle with strangeness number — 1, if the strangeness number was 0 to begin with. Strangeness, if the given strangeness numbers are used, is conserved.

When the lambda particle decays, as in Equation 14–18, a lambda particle (strangeness number — 1) becomes a proton and a negative pion (both with strangeness number 0). Strangeness is not conserved in that reaction and the breakdown ought not to occur.

However, the *law of conservation of strangeness* holds only for the strong interactions. Therefore, the breakdown can occur, provided it occurs by way of a weak interaction which takes a much longer time. Despite appearances, then, Equation 14–18 represents weak interactions and that accounts for the long lifetime of the lambda particle.

An older conservation law that was also found to have its limitations was the *law of conservation of parity*.

Parity is a quantity that is conserved in the same manner that "even" and "odd" are conserved in the realm of numbers. If an even number such as 8 is broken down into the sum of two smaller numbers such as 6 + 2 or 5 + 3, the two smaller numbers are either both even or both odd. If an odd number is so treated, as 7 = 4 + 3, the smaller numbers are always one odd and one even. More complicated transformations will also yield universal rules of this sort.

In 1956, however, it turned out that some kaons decayed to two pions, and some to three pions. Since the pion is assigned odd parity, two pions together are even, while three pions together are odd. This meant that there were even-parity kaons and odd-parity kaons, and the two were given different names.

Yet the two types of kaons were absolutely identical in all properties but the manner of their breakdown. Was it necessary that parity be conserved? Tackling this problem, two Chinese-American physicists, Tsung-Dao Lee (1926–) and Chen Ning Yang (1922–), produced theoretical reasons why parity need not be conserved in weak interactions, even though it was conserved in strong interactions.

There was a possibility of testing this. As long ago as 1927, Eugene Wigner had considered the problem and had shown that the conservation of parity meant a lack of distinction between left and right, or (which is equivalent) between a situation and its mirror-image. This could only be true if all interactions took place symmetrically in space. If electrons were given off by nuclei, for instance, they would have to be given off in all directions equally so that the mirror-image would show no clear distinction from the actual state of affairs. If electrons were given off in one direction predominantly (say to the left), then the mirror-image would show them given off predominantly to the right, and the two states of affairs could be distinguished.

The Chinese-American physicist Madame Chien-Shiung Wu (1913–) tested the Lee-Yang theory, making use of cobalt-60, which gives off electrons in a weak interaction. She cooled the cobalt-60 to nearly absolute zero and subjected it to a magnetic field that lined up all the nuclei with their north magnetic poles in one direction and with their south magnetic poles in the other. At nearly absolute zero, they lacked the energy to move out of alignment.

It turned out then that electrons were not given off equally in all directions at all. They emerged in uniformly greater numbers from the south magnetic pole than from the north. This situation could be distinguished from its mirror-image, and the law of conservation of parity turned out not to hold for weak interactions.

Consequently, it is perfectly possible for a kaon to be odd parity sometimes and even parity other times, when only weak interactions are involved.

An attempt has been made to produce a more general conservation law by combining parity with *charge conjugation,* a quantity which involves the interchange of particles and anti-particles. This combination, usually abbreviated *PC conservation,* means that a shift in parity implies an appropriate change in connection with anti-particles. Thus the antimatter version of cobalt-60 would give off positrons predominantly from the north magnetic pole. Parity and charge conjugation are separately conserved in strong interactions and jointly conserved in weak interactions.

Gell-Mann went on, in 1961 (as did, independently, the Israeli physicist Yuval Ne'eman) to attempt to produce order among the growing dozens of strongly interacting particles by arranging them in a certain manner making use of eight properties conserved in strong interactions. Gell-Mann, making use of an advanced branch of mathematics called "group theory" to guide his arrangements, called the result the *eightfold way.*

In one case, for instance, Gell-Mann dealt with a group of four *delta particles,* new hyperons that came in varieties charged — 1, 0, + 1 and + 2. Above these he could place the somewhat more massive sigma particles (with charges — 1, 0, and + 1), and above these the still more massive xi particles (with charges — 1 and 0).

The regularity could be continued if at the apex of the triangle he was forming he could put an even more massive single particle with a charge of — 1. Since this was the end of the triangle, Gell-Mann named it the *omega-minus particle,* "omega" being the last letter of the Greek alphabet. From the arrangement of the various conserved properties, the omega-minus particle would have to possess particular values for these, including, most unusually, an unprecedented strangeness number of — 3.

In 1964, the omega-minus particle was detected and shown to have all the predicted properties with amazing fidelity, down to the strangeness number of — 3. It was a discovery as significant, perhaps, as those of Mendeleev's missing elements.

Here is the present frontier of physics—the world of subatomic particles which in the last two decades has become a jungle of strange and mystifying events but which, if the proper key is found, may yield a new, subtle, and intensely bright illumination of the physical universe.

Appendix

Since the description of physics herein is presented from the historical viewpoint, most of it does not require any changes. Newton's laws of motion are still his laws of motion and electromagnetism is still electromagnetism.

In the last twenty years, however, some views of subatomic physics have sharpened and deepened, and an overview of those developments should be given.

In the third and final part of the book, for instance, I discussed the strong interaction (page 728). The particles that can interact with each other very rapidly by means of this strong interaction are the mesons (p. 725), the baryons (p. 702) and the hyperons (p. 733).

These now tend to be lumped together as *hadrons*, from the Greek word for "strong," precisely because they are subject to the strong interaction. The least massive of the hadrons is the pion (p. 727). The best known are the proton (p. 402) and neutron (p. 598).

At the very end of the book I refer to the fact that, as physicists experimented with particle accelerators (p. 635 ff) of larger and larger energies, more and more hadrons were detected until their multiplicity became a serious embarrassment. I then explained that Murray Gell-Mann (p. 733) worked out a way of dealing with these hadrons that could organize them into families or groups that seemed to make sense:

Quarks

Gell-Mann's organization made most sense if he assumed that hadrons were made up of still more fundamental particles. These Gell-Mann called *quarks*, from a phrase in James Joyce's *Finnegans*

Wake: "three quarks for Muster Mark."

Gell-Mann needed only two types of quarks to begin with; they are called *up-quarks* and *down-quarks*, or *u-quark* and *d-quark*, for the purpose of telling them apart. The description must not be taken literally, however, for there is nothing that is particularly down or up about these quarks.

In order to make sense out of the hadron families, we must assume that the quarks have electric charges of fractional values (something no one has ever observed, so this was quite a revolutionary notion). Thus, the electric charge of the u-quark is $+2/3$—that is, two-thirds that of a proton or positron—while the electric charge of the d-quark is $-1/3$—$1/3$ that of an antiproton or electron.

The quarks join together to form hadrons in such a way that the total charge is always 0, 1, 2 and so on. There are no fractional charges in the quark combinations. Thus two u-quarks and a d-quark have a total charge of $+1$ and make up a proton, while a u-quark and two d-quarks have a total charge of 0 and make up a neutron.

Quarks in other combinations of three make up the hyperons; in combinations of two, they make up the mesons.

But the u-quarks and d-quarks don't account for all the hadrons. In order to explain some of the hadrons encountered in experiments involving large energies, another pair of quarks were required. They are the *s-quark* and the *c-quark*—*s* stands for "strangeness" (or "sideways") and *c* for "charm." A third, still more energetic, pair is the *t-quark* and the *b-quark*—the *t* and *b* stand for "truth" and "beauty," or "top" and "bottom." (Physicists, just like other human beings, have a sense of whimsy, you see, but such names mustn't be taken seriously.)

As of 1983, evidence for the b-quark has not been uncovered, but physicists are sure it exists.

The world of quarks is rapidly becoming rather complicated. Each pair of quarks is called a *flavor*. So far, three flavors of quarks are known: u-d, s-c, and t-b. What's more, for each quark there is an *anti-quark*; there are three "anti-flavors," so to speak—it is the combination of anti-quarks that makes up antiparticles (p. 702 ff).

Furthermore, each different flavor of quark comes in three varieties that are distinguished by *color*. (There isn't really any color; that's just the way physicists find it convenient to describe them.) The three colors are *red*, *blue*, and *green*.

When quarks get together in threes, one quark must be red, one blue, and one green. The combination is *white*, and therefore lacks color. The colors always disappear in quark combinations, as the fractional charges do. Thus, when two quarks combine to form a meson, one may be red and the other an anti-red. The study of quark combinations is called *quantum chromodynamics*, or *QCD*, the *chromo* coming from the Greek for *color*.

Quarks are held together by the existence of exchange particles (p. 728), which in this case are called *gluons* because they glue the exchange particles together.

Thanks to the gluons, which quarks constantly exchange with each other, the quark combinations are so firm that physicists have not yet been able to pull hadrons apart into the individual quarks that make them up. Some physicists even speculate that it is impossible to produce free quarks.

It is possible, however, to force electrons and positrons together with such force that quarks are formed out of the energy released; they are not detected in the free state but instantly combine to form hadrons and antihadrons. The hadrons move off in one direction and the antihadrons in another. If there is enough energy, gluons move off in a third direction, the whole forming a "three-leaf clover." In 1979, such a three-leaf clover was detected; it was taken as strong evidence for the existence of gluons and quarks.

Leptons

There are particles, generally less massive than the hadrons, that are not subject to the strong interaction but to the weak interaction (p. 728). These are the leptons (p. 702). The most common lepton is the electron (p. 372), which is associated with a neutrino (p. 716).

The electron and neutrino are now considered the two particles that make up the least energetic level of leptons, as the u-quark and d-quark make up the least energetic level of quarks.

The muon and its associated neutrino (pp. 729-732) form a second, more energetic, level. The muon is distinguished from the electron only by the former's greater mass, which is 207 times that of the electron. The two levels of neutrino (which may be termed the "electron-neutrino" and the "muon-neutrino") are not distinguished by any property we know of, but nevertheless behave in different ways.

In recent years a still more energetic lepton, the "tau-electron," or "tauon," has been detected. It is about 3,600 times as massive as an ordinary electron and therefore some eighteen times as massive as a muon. (It is actually twice as massive as a proton.) Associated with it is, presumably, a tauon-neutrino, though that has not yet been detected as of 1983.

Thus, we have three levels of leptons, as we have three levels of quarks and, for each level, an anti-level as well, for such particles as antielectrons, or positrons (p. 702), and antineutrinos (p. 718 ff). Leptons, unlike quarks, do not have color designations.

Just as quarks interact by the exchange of gluons, leptons interact by the exchange of intermediate bosons (p. 728), which are also called *W particles*—"W" stands for "weak."

Present theories suggest that there are three different W particles and that each is about 80 times as massive as a proton, or nearly 150,000 times as massive as an electron. To demonstrate the existence of such particles, enough energy must be disposed of to be equivalent to such a large mass, and concentrating that quantity of energy into a small volume is very difficult. Nevertheless, in 1983, experiments with a large particle accelerator in Geneva, Switzerland, were reported. These clearly demonstrated the existence of two of the three W particles.

Neutrinos are particularly troublesome because they are so difficult to detect; not because of the presence of too much mass, but for the opposite reason. They have neither mass nor electric charge and virtually do not interact at all with other particles (p. 716 ff).

Physicists have set up devices in deep mines that can detect neutrinos that might be coming from the sun. (The reason for having detectors deep in the earth is that other forms of radiation cannot penetrate that far.) Throughout the 1970s, the number of neutrinos detected have never been more than one third the number that present-day theories of what is happening inside the sun would lead us to expect.

When people began to realize there were different kinds of neutrinos, it occurred to them that an explanation for this mystery of the missing neutrinos might perhaps arise if these neutrinos kept "oscillating," that is, changing back and forth among the different types.

The sun should be emitting only electron-neutrinos and the devices on earth should detect only electron-neutrinos. En route to earth, however, it might be that electron-neutrinos might turn into muon-neutrinos and tauon-neutrinos as well. As they shift back and forth, only one-third of them would be detected on earth as electron-neutrinos.

In 1980, Frederick Reines (p. 721) presented experiments that seemed to show that this actually took place (though other experiments then and since seem to cast doubt on Reines's results).

If Reines were correct, though, not only would the mystery of the missing neutrinos be solved, but a couple of other mysteries as well.

Thus, the oscillations would take place only if the neutrinos had a tiny bit of mass, say 1/10,000 that of an electron. If they had that small a mass, they would be subject to a very weak gravitational pull and in response would tend to form clusters. Because of the weakness of the pull, the clusters would be about the size of groups of galaxies—and this might explain why galaxies exist.

Although each neutrino would have such an extremely tiny mass, there are so many neutrinos in the universe that it would turn out that 99 percent of the total mass of the universe would be in the form of neutrinos. Astronomers have some reason to think that there may be considerably more mass in the universe than can be accounted for by the stars, so neutrinos might be the ideal way of explaining the "mystery of the missing mass."

But it is too soon to tell whether or not this explanation can be relied on.

Grand Unified Theories

In addition to the strong interaction and the weak interaction, there are the electromagnetic interaction and the gravitational interaction. These are thought to be the only four ways in which particles can interact, and through these four interactions, everything that is scientifically observable in the universe can, in principle, be explained.

Yet each one of the four interactions seems to behave in a different way, to affect different particles, to have different strengths. It seems to physicists that there ought to be some mathematical treatment that could serve to describe all four; that, in effect, all four are just different aspects of a single interaction.

In the 1970s, two American physicists, Steven Weinberg (1933–) and Sheldon L. Glashow (1932–) succeeded in working out a mathematical treatment that showed that the weak interaction and the electromagnetic interaction were related. Their work won them the 1979 Nobel Prize in physics. Later this treatment was extended to cover the strong interaction as well; the result is now referred to as the "Grand Unified Theory" (or GUT). So far, no one has been able to make the theory cover the gravitational interaction.

Physicists now tend to believe that the universe may have begun as a very tiny object at an enormously high temperature and that it expanded very rapidly, cooling down as it did so. At the beginning, during the period of enormously high temperatures, there was only one type of interaction, but as the temperature dropped, the four types that now exist "froze out."

If all this is so, physicists think that there should be a very minute chance that protons are not absolutely stable, but that every once in a while one can break down into leptons. A proton should exist, on the average, 10^{31} years (10,000,000,000,000,000,000,000,000,-000,000 years) before breaking down. This is so enormously long a time that it might seem hopeless to check whether such events actually take place. There are so many protons in the universe, however, that a few of them are breaking down every second, even in samples of matter small enough to handle in the laboratory.

Physicists are attempting to set up experiments to detect this occasional breakdown. As of 1983, they are not yet quite ready, but should the breakdown eventually be detected, it would represent powerful evidence in favor of GUT.

If GUT is correct, it would also mean that when particles formed at the very beginning of the universe, particles and antiparticles were not necessarily formed in equal quantities. Particles might have been formed in an excess of one out of every billion pairs. The billion pairs would have combined with each other to form photons, but that extra one would have been left over. Out of those extras our present universe would have been built up, and that would account for the fact that the universe seems to be made up only of particles, and not of antiparticles.

It would seem, then, with quarks, oscillating neutrinos, and Grand Unified Theories all having entered the consciousness of physicists since this book was first published, that exciting days are ahead.

Suggested Other Reading

Cajori, Florian, *A History of Physics*. New York: Dover, 1929.

Communications Research Machines, Inc., *Concepts in Physics*. Del Mar, California: CRM Books, 1973.

Feather, Norman, *The Physics of Mass, Length and Time*. Edinburgh, Scotland: Edinburgh University Press, 1959.

Feather, Norman, *The Physics of Vibrations and Waves*. Edinburgh: Edinburgh University Press, 1961.

Feinberg, Gerald, *What Is the World Made Of?* Garden City, N.Y.: Anchor Press/Doubleday, 1977.

Feynman, Richard P.; Leighton, Robert B.; and Sands, Matthew, *The Feynman Lectures on Physics* (Volumes I, II, and III). Reading, Massachusetts: Addison-Wesley, 1963.

Ford, Kenneth W., *The World of Elementary Particles*. New York: Blaisdell, 1963.

Friedlander, Gerhart; Kennedy, Joseph W.; and Miller, Julian Malcolm, *Nuclear and Radiochemistry*. New York: Wiley, 1964.

Gardner, Martin, *Relativity for the Million*. New York: Macmillan, 1962.

Hoffman, Banesh, *The Strange Story of the Quantum*. New York: Dover, 1959.

Kaplan, Irving, *Nuclear Physics*. Reading, Massachusetts: Addison-Wesley, 1963.

Lengyel, Bela A., *Lasers*. New York: Wiley, 1962.

Lilley, Sam, *Discovering Relativity for Yourself*. Cambridge, England: Cambridge University Press, 1981.

Miller, Franklin, Jr., *College Physics*. San Diego, California: Harcourt Brace Jovanovich, 1959.

Rothman, Milton A., *The Laws of Physics*. New York: Basic Books, 1963.

Taylor, Lloyd W., *Physics* (vols. 1 and II). New York: Dover, 1941.

Weinberg, Steven, *The Discovery of Subatomic Particles*. New York: Scientific American Library, 1982.

Wolf, Fred Alan, *Taking the Quantum Leap*. San Francisco: Harper & Row, 1981.

Index

Abelson, Philip H., 656, 678
Aberration, light, 312–313
chromatic, 295; spherical, 257, 276–277
A-bomb, 685n
Absolute motion, 329–331
Absolute rest, 329–331
Absolute scale, 193
Absolute space, 331
Absolute temperature, 193–195
Absolute zero, 193, 204, 212
Absorption spectrum, 299–300
Acceleration, 13ff.
angular, 79–80; definition of, 15–16; direction of, 29n; free fall and, 20; force and, 30; mass and, 30–31; negative, 28; units of, 16, 21; vectors and, 28–29
Accelerators, particle, 637–642
Accommodation, 274
Achromatic lens, 297
Achromatic microscope, 297
Acoustics, 178
Actinides, 498, 658
electron structure of, 568
Actinium, 602, 611
Actinium series, 611, 612
Actinouranium, 612
Action, 371
Action at a distance, 243–246
Adams, John C., 242
Adams, Walter S., 363
Adhesion, 127–130
Air, 251–252
index of refraction of, 266, 268; light scattering by, 308–310
Air pressure, 138–139
Air resistance, 11–12
Airplanes, 142–143, 165–166
Alkali metals, 498
Alkaline earth metals, 498
Alnico, 386
Alpha Centauri, 316
Alpha particles, 538, 591, 669
atomic changes and, 605; cosmic rays and, 700; energy of, 714–715; half-life determination and, 617; nuclear reactions and, 632–635; stability of, 663; structure of, 593; tracks of, 597
Alpha rays, 590
Alternating current, 460ff.
Aluminum, 214–215, 426, 643–644
electric wiring and, 430–431
Amber, 380–381, 397
Americium, 657
Amici, Giovanni B., 297
Ammeter, 453
Ammonia, 210, 218, 577
Ammonium chloride, 437
Amontons, Guillaume, 191, 199
Ampère, André M., 423, 442–443, 452
Ampere, 423
Amplifier, 524
Amplitude, 150
Amplitude modulation, 527
Anderson, Carl D., 702, 725
Andrews, Thomas, 211
Angle, critical, 267
Angle, visual, 281–283
Angle of refraction, 264
Angstrom, Anders J., 307–308
Angstrom unit, 308
Angular acceleration, 79–80
Angular momentum, 81–83
Angular velocity, 73
Anions, 505
Anisotropic substances, 190n
Annihilation, mutual, 705, 709
Anode, 421, 504
Antibaryons, 708–711
Anticathode, 541
Anticosmon, 712
Antielectron, 702
Antiferromagnetism, 388
Antigravity, 712
Antilepton, 711
Antimatter, 712–713
Antineutrino, 718ff.
detection of, 721–722

Antineutron, 710
  breakdown of, 711
Antinodes, 153
Antiparticles, 702ff.
Antiproton, 708–713
Appleton, Edward V., 525
Arago, Dominique F. J., 442
Archimedes, 9, 88, 123–124
Archimedes' principle, 124
Archytas, 154
Argon, 431, 495
Argon-40, 623, 631–632
Aristotle, 4, 6–7, 135–136, 154–155, 244–245, 326, 486
Armature, 460–464
Armstrong, Edwin H., 527–528
Arrhenius, Svante A., 508–509
Artificial radioactivity, 644
Assumptions, 4–5
Astatine, 610, 655
Astatine-210, 629
Astigmatism, 276
Aston, Francis W., 618, 665–666
Astrophysics, 3
Atmosphere, 207
  density of, 146; pressure of, 138
Atmosphere (unit), 138
Atom(s), 146, 483
  electrons and, 533ff., 547–550, 560ff.; mass of, 500–501; neutral, 545; quantum theory and, 555ff.; radioactive changes in, 604–606; structure of, 503–504, 537ff.; visualization of, 501–502
Atomic bomb, 685n
Atomic clocks, 577
Atomic energy, 669n
Atomic nucleus, 539–540
  charge of, 540ff.; energy levels of, 591; masses of, 665–667; neutrons in, 598–600; structure of, 588ff., 660–664, 722ff.
Atomic number, 544–545
Atomic pile, 678–682
Atomic theory, 490
Atomic wave structure, 583–585
Atomic weight, 215, 491
  chemical, 628; isotopes and, 625–629; non-integral, 504; physical, 628; standards of, 628–629
Atomism, 135ff., 482ff.
  gases and, 143ff.
Attraction, magnetic, 382ff.
Audiofrequencies, 526
Available energy, 227–229
Avogadro, Amedeo, 499
Avogadro's hypothesis, 499
Avogadro's number, 499–501
Axis, magnetic, 390
Axis, principal, 256

Background radiation, 691
Back-voltage, 466
Bacon, Roger, 275
Balance, 57, 85–86
Balloons, 141
Balmer, Johann J., 553
Balmer series, 553, 557–558
Balmer's constant, 553
Bar, 139
Bardeen, John, 575
Barium, 426
Barkla, Charles G., 542
Barn, 653
Barometer, 138
Bartholinus, Erasmus, 320
Baryon, 702
Baryon number, conservation of, 709
Bats, 180
Batteries, nuclear, 683–684
Battery, electric, 422
Battery, storage, 439–440
Beam, light, 248
Bearing, frictionless, 451
Beats, 169–170
Beats, radio wave, 527
Becquerel, Antoine H., 588–590, 601, 633, 715
Bel, 160
Bell, Alexander G., 160, 450
Bell, electric, 448
Berkelium, 657
Bernoulli, Daniel, 134, 197–198
Bernoulli's principle, 133–134, 142–143
Beryllium, 658
Berzelius, Jöns J., 487
Beta particles, 590
  atomic changes and, 605; energy of, 715–716; nuclear structure and, 600–601; positive, 704; tracks of, 597
Beta rays, 590
Betatron, 641
Bethe, Hans A., 687–688
Bev, 636
Bevatron, 708
Bifocals, 276
Billion, 636n
Bimetallic strip, 184
Biologic half-life, 692
Biot, Jean B., 324
Bismuth, 396, 659
Black body, 366
Black-body radiation, 368–371
Black, Joseph, 215, 217–218

Blackett, Patrick M. S., 634
Bloembergen, Nicholaas, 579
Blondlot, Prosper, 696n
Boethius, Anicius M. S., 155
Bohr, Niels, 379, 555–556, 673, 675, 677, 715
Boltwood, Bertram B., 630
Boltzmann, Ludwig, 200, 235, 241, 366, 500
Bose, Satyenda Nath, 728
Bose-Einstein statistics, 728
Bosons, 728
Bothe, Walther, 698
Boyle, Robert, 145, 191, 484, 486
Boyle's law, 145, 191, 208–210
  kinetic theory of gases and, 199
Brackett, Frederick S., 555
Brackett series, 555, 557–558
Bradley, James, 312–313
Bragg, William H., 543
Bragg, William L., 543
Branched disintegration, 610
Brattain, Walter H., 575
Braun, Karl F., 529
Breeder reactors, 684
British thermal unit, 99
Broglie, Louis V. de, 582
Bunsen, Robert W., 299
Buoyancy, 122–125

Cadmium, 652
Cadmium chloride, 722
Calcium 422, 487
  isotopes of, 626
Calcium-40, 622, 631
Calcium carbonate, 628
Californium, 657
Californium-250, 674
Caloric, 223
Calorie, 99, 214
Camera, 277–281
  pinhole, 279; Schmidt, 277
Camera obscura, 279
Candlepower, 249
Capacitance, 410
Capacitive reactance, 467–468
Capacitor, 410
Capillary action, 129–130
Carbon, 437, 489–490
  atomic weight of, 492
Carbon-11, 647, 651
Carbon-12, 665, 671
  atomic weight standards and, 628–629;
    packing fraction and, 667
Carbon-13, 647, 667
Carbon-14, 647–648, 650
Carbon dioxide, 210, 499
Carlisle, Anthony, 422
Carnot, Nicolas L. S., 226–227, 229
Carrier wave, 526
Carroll, Lewis, 337
Cathode, 421, 504
Cathode rays, 510–511, 514, 528–532
  particles of, 517
Cathode-ray oscillograph, 529
Cation, 505
Cavendish, Henry, 50–51
Celestial mechanics, 88
Cell, electrical, 422
  dry, 437; fuel, 439n; Leclanche, 437
Celsius, Anders, 187
Celsius scale, 187–188, 193
Celtium, 545
Center of gravity, 77
Center of mass, 76
Centigrade scale, 187–188
Centimeter, 21
Centipoise, 133
Centrifugal force, 55
Centripetal force, 55
Cerenkov, Pavel A., 699
Cerenkov counters, 699
Cerenkov radiation, 699
Cesium, 299, 530, 536, 592n
Cesium-133, 644
Cesium-135, 645
Cesium-137, 692
cgs system, 32
Chadwick, James, 598, 715
Chain reaction, 691
Chamberlain, Owen, 709
Channel rays, 591
Characteristic X-rays, 542–545
Charge, electric, 397ff.
Charge conjugation, 735
Charles, Jacques, A. C., 192
Charles' law, 192, 195
Chemical atomic weight, 628
Chemical cell, 422
Chemical rays, 318
Chemical reactions, 547–549
Chemistry, 147n
Chladni, Ernst F. F., 207
Chlorine, 210, 626
Chord, 172
Chromatic aberration, 295
Chromium, 426
Circuits, electric, 431–436
Circular motion, 104–105
Classical physics, 372
Clausius, Rudolf J. E., 230
Clock paradox, 353
Clockwise, 73
Cloud chamber, Wilson, 596–597

Cobalt, 386, 495, 625, 627
  Curie point of, 388
Cockcroft, John D., 637
Coefficient of conductivity, 224
Coefficient of cubical expansion, 190–192
Coefficient of linear expansion, 188–189
Coefficient of resistance, 427
Coherent light, 580–581
Cohesion, 125
Coincidence counter, 698
Color, 290ff.
  temperature and, 366–367; wavelengths of, 308
Color photography, 294
Color television, 294
Color vision, 293–294
Columbus, Christopher, 390
Compass, magnetic, 381–382
  properties of, 388–389
Compensation pendulum, 185
Complementarity, principle of, 379
Compound, 487
  tagged, 645
Compound bar, 184
Compound ions, 509
Compound microscope, 287
Compression waves, 156–157
Compton, Arthur H., 378, 698
Compton effect, 378
Comte, Auguste, 300
Concave lens, 272
Condenser, 409–414
  reactance and, 467; variable, 412
Conductance, electrical, 425
Conduction, 221
Conductivity, coefficient of, 224
Conductors, electrical, 398
Conservation of angular momentum, 81–83
Conservation of baryon number, 709
Conservation of electric charge, 401–402, 600
Conservation of electron family number, 718, 729–732
Conservation of energy, 100, 222, 348, 715
Conservation of mass, 347, 488
Conservation of mass-energy, 348
Conservation of matter, 234, 488
Conservation of momentum, 67–72
  center of mass and, 76
Conservation of muon family number, 731–732
Conservation of parity, 734–735
Conservation of strangeness, 734
Contact potential difference, 419
Continuous spectrum, 299
Convection, 221
Converging lens, 271, 283–284
Converging series, 62, 483
Convex lens, 271
Convex mirror, 261–262
Copernicus, Nicholas, 38
Copper, 214–215, 224, 386, 419–421, 489–490
  electric wiring and, 430–431; resistivity of, 426
Copper-64, 668
Cornea, 273
Coryell, Charles D., 656
Cosmic particles, 698
Cosmic rays, 696–701
  nature of, 698
Cosmon, 712
Coulomb, Charles A. de, 383, 403
Coulomb, 404, 505–506
  electrolysis and, 423
Counterclockwise, 73
Couple, 85
Cowan, Clyde L., 721
Crest, 150
Critical angle, 267
Critical field strength, 447
Critical size, 679
Critical temperature, 211
Crookes, William, 510, 515, 601
Crookes tube, 510
Cross-section, nuclear, 653
Cryotron, 452
Crystalline lens, 273–274
Crystals, 408–409, 542
  radio and, 575
Cubical expansion, coefficient of, 190–192
Curie, Jacques, 409
Curie, Marie S., 590, 602, 689–690
Curie, Pierre, 388, 409, 602
Curie, 649–651
Curie point, 388
Curium, 657
Curium-242, 674
Current, electric, 417ff.
  alternating, 460ff.; direct, 462–464; electrons and, 426
Current intensity, 425
  effective, 464
Curved space, 361
Cyclotron, 638–640

Daguerre, Louis J. M., 280
Dalton, John, 146, 423, 490–491, 503
D'Arsonval, Jacques A., 452
D'Arsonval galvanometer, 452
Daughter elements, 601
Davisson, Clinton J., 582
Davy, Humphry, 422, 438, 486, 498, 504
Dawn, 309

Debierne, André L., 602
Deceleration, 28
Decibel, 161
Declination, magnetic, 389–390
Definite proportions, law of, 489
DeForest, Lee, 523
Degrees (of angles), 73
Degrees (of temperature), 187
Delayed neutrons, 681
Delta particles, 736
Democritus, 135–136, 483–484
Dempster, Arthur, J., 611
Density, 117
  optical, 266; pressure and, 144; temperature and, 143–144
Descartes, René, 264–265, 291
Deuterium, 623
Deuteron, 647–649
  acceleration of, 651–652
Deutsch, Martin, 707
Dewar, James, 226
Dewar flask, 226
Diamagnetism, 395–396
  superconductivity and, 447
Diamond, 266, 315, 397
Dielectic constant, 405
Dielectrics, 405
  condensers and, 410–411
Diffraction, light, 304–305
Diffraction gratings, 305
Diffuse reflection, 254
Diffusion, gas, 204ff.
Diffusion cascades, 677–678
Dimensional analysis, 18
Diode, 523
Diopters, 272
Dip, magnetic, 389
Dirac, Paul A. M., 585n, 702, 707, 728
Direct current, 462–464
Disorder, 239–240
Displacement, 27
Distance, 27
Diverging lens, 272, 285
Dollond, John, 297
Domains, magnetic, 384–388
Doppler, Christian J., 174, 316, 318
Doppler effect, 174, 316, 318
Doppler-Fizeau effect, 318, 364
Dorn, Friedrich E., 602
Double refraction, 320
Dry cell, 437
DuFay, Charles F., 399
Dulong, Pierre J., 215
Dulong and Petit, law of, 215
Dynamics, 88
Dyne, 32
Dysprosium, 388
Earth, 151, 445
  age of, 630–632; conservation of momentum and, 71–72; gravitational pull of, 37, 41–46; gravitational variations on, 55; magnetic properties of, 388–391, 445; mass of, 52; radius of, 46; rotation of, 55; shape of, 55–56; weight of, 60
Echo, 177–178
Echolocation, 180
Eclipse, 251, 278, 362–363
Eddington, Arthur S., 687
Edison, Thomas A., 281, 450, 463, 469, 522
Edison effect, 522
Effective current intensity, 464
Effective potential difference, 464
Eightfold way, 736
Einstein, Albert, 24n, 26, 26n, 47n, 100, 341ff., 365, 374–375, 478, 534, 587, 657, 669, 676, 706, 707, 728
Einstein equation, 351–352
Einsteinium, 657
Einthoven, Willem, 452
Elasticity, 50
Electric battery, 422, 436–440
Electric bell, 448
Electric charge, 397ff.
  conservation of, 401–402, 600
Electric circuits, 431–436
Electric current, 417ff.
  electrons and, 426
Electric field, 405
Electric force, 403–405
Electric fuse, 433
Electric generators, 455–459
Electric insulation, 435
Electric lines of force, 405
Electric motor, 472–475
Electric poles, 421
Electric potential difference, 407
Electric power, 428–431
Electric relay, 449
Electric shock, 412
Electric transformer, 470–472
Electric wiring, 430–431
Electrical cell, 422
Electrical conductance, 425
Electrical conductors, 399
Electrical current intensity, 423, 452–455
Electrical friction machine, 398
Electrical insulators, 399
Electrical resistance, 424ff.
  temperature coefficient of, 427
Electricity, 381, 507–509
  charging of, 397; conduction of, 507–509; discharging of, 398; element

formation by, 422; element isolation by, 504–505; galvanic, 422; lightning and, 414–415; magnetism and, 441ff.; motion as source of, 455–459; speed of, 426n; static, 397ff.; transmission of, 469–472; types of, 399–400; vacuum(s) and, 245, 509–511
Electrochemical cell, 422
Electrochemistry, 422
Electrode(s), 421, 504
Electrodynamics, 422
Electrolysis, 422, 505
  laws of, 505–507
Electrolytes, 505
Electromagnet, 446–452
  applications of, 448–452
Electromagnetic field, 477
Electromagnetic induction, 457
Electromagnetic radiation, 511ff.
  energy of, 636–637
Electromagnetic waves, 477
Electromagnetism, 441ff.
Electromotive force, 408
Electron(s), 372, 402, 509
  acceleration of, 640–641; annihilation of, 705; atom(s) and, 532ff.; atomic sharing of, 549–550; atomic shifts of, 547–548; beta rays and, 590; charge of, 520–521; chemical reactions and, 547–549; discovery of, 517; electric charge of, 404; electric current and, 426; elements and, 548–550; energy levels of, 570ff.; light quanta and, 374–375; mass of, 519; moving, 551–552; muon and, 726–727; nuclear, 593–595; nuclear atom and, 540; photoelectric effect and, 534–536; positive, 702; quantum theory and, 556; shells of, 546ff.; shifts of, 406; size of, 517–519; spin of, 559; subshells of, 560; wave properties of, 582ff.
Electron family number, conservation of, 718, 729–732
Electron microscope, 583–584
Electron shells, 560
Electron synchrotrons, 642
Electronics, 521ff.
Electron-neutrinos, 731–732
Electron-positron pair, 707
Electron-volt, 635
Electroplating, 505
Electroscope, gold-leaf, 589
Electrostatic generator, 638
Electrostatic induction, 398
Electrostatic unit, 404
Electrostatics, 397ff.
Element(s), 485ff.
  alphabetical list of, 488–489; atomic number of, 544–545; atomic weight of, 492–494; combinations of, 487; daughter, 601; definition of, 486; distribution of, 667; electrical production of, 504–505; electron distribution in, 548–550, 560ff.; equivalent weight of, 506–507; families of, 496–498, 549–550; Greek notions of, 486; isotopes of, 606ff.; periodic table of, 495–498; radioactive, 602–603, 629; radioactive changes in, 604–606; radioactive series of, 601ff.; rare earth, 498; stable isotopes of, 617–622; symbols of, 487–489; synthetic, 654–659; transition, 563ff.; transuranium, 658
Elementary particles, 727n
Emission spectrum, 298–299
Energy, 94
  atomic, 669n; available, 227–229; conservation of, 99–100, 222, 348, 715; forms of, 98–99; heat and, 98–99; internal, 213n; kinetic, 94–97; mass equivalence to, 348–351; mechanical, 97; nuclear, 669ff.; potential, 96–97; surface, 126
Energy levels, electronic, 570ff.; nuclear, 591
Entropy, 231ff.
  disorder and, 238–240
Epicurus, 484
Equilibrium, radioactive, 616
Equator, magnetic, 391
Equatorial bulge, 56
Equilibrium, 86–87
Equipotential surface, 407
Equivalent weight, 506
Erg, 90
Escape velocity, 63
Ether, 6, 245, 326ff., 365, 478
  properties of, 327
Ether wind, 332
Even-even nuclei, 662
Even-odd nuclei, 662
Exchange forces, 723
Exchange particles, 728
Excited state, 556
Exclusion principle, 560
Experiments, 8–9
Exponential notation, 52
Extraordinary ray, 320
Eye, 273–276
  defects of, 274–276; resolving power of, 285–286
Eye-glasses, 275, 285

Fahrenheit, Gabriel D., 186
Fahrenheit scale, 187–188
Falling bodies, 7–8, 10ff., 19–20, 141
Fallout, 692
Farad, 410
Faraday, Michael, 392–393, 396, 410, 421–423, 456–459, 465, 475–476, 504, 509
Faraday, 423, 506
Faraday's laws of electrolysis, 505–507
Farsightedness, 274–275
Fast reactors, 684
Fechner, Gustav T., 160n
Fermi, Enrico, 653, 656–657, 672, 679, 682, 716, 728
Fermi-Dirac statistics, 728
Fermions, 728
Fermium, 657
Ferrimagnetism, 388
Ferrites, 386, 388
Ferromagnetism, 386–388
Fessenden, Reginald A., 527
Field, Cyrus W., 450
Field, electric, 405
  electromagnetic, 477; magnetic, 393, 394
Field emission microscope, 501–502
Field glasses, 287
Field-ion microscope, 502
First law of motion, 24, 33, 329
First law of thermodynamics, 222, 229n, 233
Fission, nuclear, 672–674
  spontaneous, 674
Fission bombs, 685
Fission products, 673–674
FitzGerald, George F., 337
FitzGerald contraction, 336–340, 356
  size of, 338–339
Fixed proportions, law of, 489
Fleming, John A., 458, 522
Fleming's rule, 458
Flerov, G. N., 674
Fluid(s), 115–116
Fluid mechanics, 120
Fluorescence, 589
Flux, magnetic, 394
Fly-wheel, 81
F-number, 280
Focal length, 272
Focus, 257
  cirtual, 262
Foot-candle, 249–250
Foot-pound, 99
Force, 24
  acceleration and, 29n, 30; centrifugal, 55; centripetal, 55; definition of, 26; distance and, 89–90; electric, 403–405; electromotive, 408; gravitational, 25, 404; magnetic, 383–384, 391ff.; magnetic lines of, 392–394; mass and, 30; movement of, 75; multiplication of, 89; parallelogram of, 40–41; time and, 65–66; torque-producing, 75; units of, 32–33
Foucault, Jean B. L., 314
Fourier, Jean B. J., 168, 226
Frame of reference, 342
Francium, 536n, 610, 655
Franklin, Benjamin, 276, 399–403, 414–416
Fraunhofer, Joseph von, 298, 305
Fraunhofer lines, 298
Free fall, 10, 59–60
Frequency, 153–154
  of musical notes, 171–173; of sound waves, 163–164, 169ff.
Frequency modulation, 528
Fresnel, Augustin J., 179, 321
Friction, 25, 35, 130–131
  heat and, 98
Fuel cell, 439n
Fulcrum, 85
Functions, trigonometric, 111n
Furnace, solar, 260
Fuse, electric, 433
Fusion, latent heat of, 217
Fusion, nuclear, 686–689
  energy produced by, 694; positrons and, 704; sun and, 687–688; temperatures for, 694
Fusion bomb, 688
  fallout and, 692
Fusion power, 693–695

Gadolinium, 386
  Curie point of, 388
Galaxy(ies), 446, 700
  red shift of, 319–320
Galaxy (Milky Way), 316
Galilei, Galileo, 288, 310–311
Galileo, 9ff., 38–39, 88, 108, 113, 191, 486
Gallium, 448
Galvani, Luigi, 417–418
Galvanic electricity, 422
Galvanism, 422
Galvanometer, 452–454
Gamma rays, 363–364, 590–591, 605–606
  absorption of, 719; electron-positron annihilation and, 705–706; tracks of, 597
Gas(es), 116, 136ff., 484–485
  atomism and, 143ff.; coefficient of cubical expansion in, 192; diffusion of, 204ff.; gravity and, 198; inert,

496–497, 544, 546–547, 631–632; kinetic theory of, 197ff.; liquefaction of, 210; perfect, 196; permanent, 210; pressure of, 138ff.; sound velocity in, 206–207; specific gravity of, 136; temperature and, 191–193
Gassendi, Pierre, 484
Gauss, Karl F., 395
Gauss, 395
Gay-Lussac, Joseph L., 191
Gay-Lussac's law, 192, 195
kinetic theory of gases and, 200
Geiger, Hans, 595
Geiger-Müller counter, 596
Geissler, Heinrich, 509
Geissler tubes, 509
Gell-Mann, Murray, 733, 735–736
General theory of relativity, 354, 357, 363
testing of, 361–364
Generators, electric, 455–459
Geometric optics, 247ff.
Geophysics, 3
Germanium, 426, 572–573
Germer, Lester H., 582
Getting, Ivan A., 699
Gev, 636n
Giauque, William F., 628
Gilbert, William, 381, 388–389, 397
Gladstone, William E., 457n
Glass, 132n, 297
light velocity in, 315; temperature change and, 183
Goeppert-Mayer, Maria, 664
Gold, 426
Goldhaber, Maurice, 712
Gold-leaf electroscope, 589
Goldstein, Eugen, 510, 591
Graham, Thomas, 206
Graham's law, 206
Gram, 31
Gram-atomic weight, 499
Gram-molecular weight, 499
Gravitation, 478
light and, 362–363; relativity and, 359–364; vacuum and, 245
Gravitational constant, 48
value of, 52
Gravitational force, 404
Gravitational mass, 357
Gravitational potential difference, 406
Graviton, 701, 711, 728
Gravity, 20, 25, 44–45
center of, 77; distance and, 46, 61; pendulum and, 109–111
Gray, Stephen, 298–417
Grid, 523
Grimaldi, Francesco M., 304
Ground state, 556
Group displacement law, 606
Guericke, Otto von, 140, 398, 414
Gyroscope, 81

Hafnium, 545
Hahn, Otto, 603, 608, 672, 675
Half-life, 614–617
biologic, 692
Hallwachs, Wilhelm, 534
Halogens, 497
Harmonic analysis, 168
Harmonic motion, 102
H-bomb, 688
Heat, energy and, 98–99
entropy and, 238–239; flow of, 220–226, 235–238; friction and, 98; insulator of, 225; kinetic theory of, 241; latent, 215–218; mechanical equivalent of, 99; motion and, 234–240; specific, 214–215; temperature and, 212ff.; units of, 99
Heat rays, 317
Heaviside, Oliver, 525
Heavy hydrogen, 623
Heavy water, 623
Heisenberg, Werner, 570, 585–586, 598
Helium, 211–212, 427–428, 593
isotopes of, 607; sun and, 687; universe and, 667
Helium-4, 693
packing fraction and, 667–668
Helmholtz, Hermann L. F. von, 100, 230, 293, 686
Henry, Joseph, 446, 449, 465, 474–475
Henry, 466
Hero, 9
Herschel, William, 47, 317, 512
Hertz, Heinrich R., 479, 512, 515–517, 524–525, 534
Hess, Victor F., 697
Hevesy, Georg von, 646
Hiero, 123–124
Highlight, 254
Hittorf, Johann W., 510
Hofstadter, Robert, 710, 727
Hooke, Robert, 49, 113
Hooke's law, 50, 102
Horsepower, 93, 430
Huggins, William, 319
Huygens, Christiaan, 113, 301–302, 320
Huygens' principle, 302
Hydraulic press, 121–122
Hydrodynamics, 120, 134
Hydrogen, 136–137, 266, 491
atomic weight of, 492; buoyancy of, 141; critical temperature of, 211; electron

orbits of, 557ff.; isotopes of, 607, 623–624; melting point of, 212; molecular velocity of, 206; molecular weight of, 498; spectral lines of, 553–555, 558; sun and, 687; universe and, 667; weight of atom of, 501
Hydrogen-1, 666
fusion and, 693
Hydrogen-2, 623–624, 645–646, 648–649, 677
fusion of, 686, 693–694
Hydrogen-3, 649, 693
Hydrogen bomb, 688
Hydrogen ion, 507
Hypernucleus, 733
Hyperon, 733
Hyperopia, 275

Ice, 144
melting of, 216–217
Iceland spar, 320, 322
Iconoscope, 530
Ideal gas, 196
Ideal gas equation, 196
Image, 254ff., 277–281
mirror, 255; real, 259; size of, 263, 283ff.; virtual, 254
Impedance, 467
Impulse, 66–67
Incidence, angle of, 253
Inclined plane, 10ff., 88, 91–92
Index of refraction, 265, 315
Indium, 652
Inductance, 465
Induction, electromagnetic, 457
electrostatic, 398; magnetic, 385
Inductive reactance, 466
Inert gases, 496–497, 544, 546–547, 631–632
Inertia, 24
mass and, 30; moment of, 81; rotational, 79
Inertial mass, 357
Infrared radiation, 317, 512–513
Infrasonic waves, 180
Insulation, electric, 435
Insulator, heat, 225
Insulators, 399
permittivity of, 405
Intensity, current, 423
measurement of, 452–455
Intensity, magnetic, 384
Interaction, 723n
strong, 728; weak, 728
Interface, 128, 264
Interference, 167
light, 306–307
Interferometer, 332–333
Intermediate boson, 728
Internal energy, 213n
Internal resistance, 437–438
International cradle, 249
International Prototype Meter, 333
Invar, 183
Inverse-square law, 46
Iodine, 495
Iodo-quinine sulfate, 325
Ion(s), 421, 505
compound, 509; electrons and, 520–521, 577; positive, 507; radiation and, 590; subatomic particles and, 596–597
Ion-drive, 592n
Ionic dissociation, theory of, 508
Ionosphere, 525
Iridium, 667
Iron, 144, 214–215, 380–381, 385–587, 667
Curie point of, 388; electromagnetism and, 446; magnetic domains in, 386–388; permeability of, 396
Iron oxide, 386
Isogonic lines, 391
Isotopes, 606ff.
atomic weight and, 625–629; biochemical uses of, 645–649; mass differences among, 623; production of, 682; separation of, 677; stable, 617–622, 668; tracer uses of, 645
Isotropic substances, 190n

Jansky, Karl, 532
Janssen, Zacharias, 287
Johnson, Thomas H., 698
Joliot-Curie, Frédéric, 643–644, 704
Joliot-Curie, Irène, 643–644, 690, 704
Joule, James P., 99–100, 230, 234
Joule, 90, 99
Junction diode, 574
Jupiter, 207–208, 445
satellites of, 311–312

Kamerlingh-Onnes, Heike, 427
Kaons, 732, 734
K-capture, 622–624
Kelvin, Lord, 193, 450, 469
Kennelly, Arthur E., 525
Kennelly-Heaviside layer, 525
Kepler, Johannes, 378
Kerst, Donald W., 641
Kev, 635
Kilocalorie, 214
Kilogram, 31
Kilometer, 21
Kilowatt, 430

Kilowatt-hour, 430
Kinetic energy, 94–97, 348
  temperature and, 203
Kinetic theory of gases, 197ff.
Kinetic theory of heat, 241
Kirchhoff, Gustav R., 298–299, 366
Kleist, Ewald G. von, 413
K-mesons, 732
Kölliker, Rudolf A. von, 515
Krypton, 333

Lambda particles, 732–733
Land, Edwin H., 325
Langmuir, Irving, 547
Lanthanides, 498, 544, 658
  electron structure of, 566–567
Lanthanum-138, 622
Laser, 579–581
Latent heat, 215–218
Laue, Max von, 542–543
Lavoisier, Antoine L., 234, 487
Lawrence, Ernest O., 638, 655
Lawrencium, 658
Lead, 214–215, 439, 447, 627
  earth's age and, 630–631; isotopes of, 608–609, 612–613, 627, 630–631
Lead-204, 630–631
Lead-206, 669
Lead-210, 646
Lead-212, 646
Lead peroxide, 439
Lead sulfate, 439
Lebedev, Peter N., 377
Leclanché, Georges, 437
Leclanche cell, 437
Lee, Tsung-Dao, 734
Leeuwenhoek, Anton van, 286–287
Leibniz, Gottfried W., 94
Lenard, Philipp, 373, 516, 535, 537–538
Lenard rays, 516
Length, velocity variation of, 336–340
Lens, 271ff.
  achromatic, 297; camera and, 279; concave, 272; converging, 271, 283–285; convex, 271; crystalline, 273–274; diverging, 272, 285; magnification by, 284; types of, 271–272
Lens formula, 273
Lenz, Heinrich F. E., 465
Lenz's law, 465
Leptons, 702
Leukemia, 689
Lever, 85–90
Leverrier, Urbain J. J., 242, 562
Lewis, Gilbert N., 547
Leyden jar, 413–415
Light, 247ff.
  aberration of, 312–313; absorption of, 250–251; chemical reactions and, 317; coherent, 580–581; color and, 290ff.; color changes in, 316; constant velocity of, 342ff.; diffraction of, 304–305; electromagnetic nature of, 477; energy content of, 372–375; ether and, 326ff.; gravity and, 362; intensity of, 248–250; interference of, 306–307; invisible, 317; length standards and, 333; monochromatic, 295; nature of, 246, 301–305, 375–379; nature of waves of, 320–321; particle theory of, 375–379; photoelectric effect and, 372–375; photography and, 280–281; photons of, 375–379; plane-polarized, 322; polarization of, 322; polarized, 320–325; radiation of, 248; rectilinear propagation of, 247; reflection of, 252ff.; refraction of, 263–268, 302–303; scattering of, 308–310; spectral discussion of, 297; spectrum of, 291–292, 297–300; transmission of, 250–252; unpolarized, 321; vacuum and, 244–245; value of velocity of, 315–316; velocity of, 310–316, 331–336; visible, 318; vision and, 273–276; wave theory of, 301–305, 306ff.; wavelength of, 303, 307–308; white, 292
Light bulbs, 431, 522
Lightning, 166, 414–415
Lightning rod, 415
Light-year, 316
Lime, 486
Limiting sum, 62
Linac, 638
Linear accelerator, 638–639, 641
Linear expansion, coefficient of, 188–189
Linear velocity, 73
Lines of force, electric, 405
  magnetic, 392–394
Liquid(s), 116
  density of, 136–137
Lister, Joseph J., 297
Lithium, 667
Loadstone, 380, 386, 389
Logarithms, 160n
Longitudinal wave(s), 156–157, 320, 327
Lorentz, Hendrik A., 339
Lorentz-FitzGerald contraction, 339, 344
Loschmidt, J., 500
Loudness, 158–161
Lucretius, 484
Lumen, 249–250
Luminiferous ether, 326

Lunar eclipse, 251
Lyman, Theodore, 555
Lyman series, 555, 557–558

Mach, Ernst, 165
Mach number, 165
Machine, 88
Magic numbers, 664
Magnesium, 422, 668
Magnet(s), 380ff.
attracting properties of, 382; permanent, 386; poles of, 381–384; repelling properties of, 382ff.; temporary, 386
Magnetic axis, 390
Magnetic declination, 389–390
Magnetic dip, 389
Magnetic domains, 384–388
Magnetic equator, 391
Magnetic field, 393–394
Magnetic flux, 394
Magnetic flux density, 394
Magnetic force, 383–384, 391ff.
Magnetic induction, 385
Magnetic intensity, 384
Magnetic lines of force, 392–393
Magnetic permeability, 395–396
Magnetic poles (Earth), 390
Magnetic quantum number, 559
Magnetism, 380ff.
disruption of, 387; electricity and, 441ff.; superconductivity and, 447–448; vacuum and, 245
Magnification, 281ff.
Maiman, Theodore H., 579
Malus, Étienne L., 321
Manganese, 386
Manganese dioxide, 437
Marconi, Guglielmo, 525
Mariotte, Edme, 191
Mariotte's law, 191
Mars, 49
Maser, 578–579
Mass, 30
center of, 76; conservation of, 347, 488; energy equivalence to, 348–351; gravitational, 357; increase with velocity of, 339–340, 347–348; inertial, 357; measurement of, 57; units of, 31; weight and, 53ff.; weightlessness and, 58–60
Mass defect, 665
Mass spectrograph, 619
Mass-energy, conservation of, 348
Mathematics, 17
Matrix mechanics, 570
Matter, conservation of, 234, 488
Matter-waves, 582ff.
Maxwell, J. Clerk, 200, 235, 237, 241, 378, 394, 476–479, 500, 511–512
Maxwell, 394
Maxwell's Demon, 237–238
Maxwell's equation, 203, 477
Mayer, Julius R. von, 100
McMillan, Edwin M., 640, 655
Mechanical energy, 97
Mechanical equivalent of heat, 99
Mechanics, 88
fluid, 120
Mechanism, 242
Meitner, Lise, 603, 608, 672, 675
Mendeleev, Dmitri I., 495, 535, 657
Mendelevium, 657
Meniscus, 128, 271
Mercury (element), 128
barometer and, 138; thermometer and, 186
Mercury (planet), 47n, 361–362
Meson, 725
Mesonic atom, 726
Mesotron, 725
Metals, 398
Meter, 20–21
Metric system, 20–21
Mev, 636
Mho, 425
Michelson, Albert A., 315, 331–333, 336
Michelson-Morley experiment, 331–336
Microcuries, 651
Microfarad, 410
Micrometer, 307n
Micromicrofarad, 410
Micron, 307n
Microscope, 286–287
achromatic, 297; electron, 583; field-emission, 501–502; field-ion, 502
Microwatts, 160
Microwaves, 513, 530–532, 577–579
Millicuries, 651
Millikan, Robert A., 375, 520–521, 697
Millimeter, 47
Millimicrons, 307
Millipoise, 133
Mirages, 268
Mirror, 255ff.
convex, 261–262; parabolic, 257–261; plane, 255; spherical, 256
Mirror image, 255
mks system, 32
Moderators, 654
Modern physics, 372
Mole, 499
Molecular biology, 3
Molecular weight, 206, 498
Molecules, 147

Molybdenum, 655
Moment of force, 75
Moment of inertia, 81
Momentum, 66ff.
angular, 81; conservation of, 67–72; vectors and, 67
Moon, 420
apparent size of, 282–283; atmosphere of, 207; distance of, 46; gravitational force on, 57–58; magnetic properties of, 444; mass of, 49; motion of, 41ff.; shadow of, 250–251; shadows on, 310
Morley, Edward W., 336
Morse, Samuel F. B., 449
Moseley, Henry G.-J., 543–545
Mössbauer, Rudolf L., 363
Mössbauer effect, 363–364
Motion, 3ff.
absolute, 329–331; circular, 104–105; combined, 37–41; component, 40–41; electricity and, 457–459; energy and, 94–97; falling, 7–8; first law of, 24, 33, 329; Greek view of, 5–9; harmonic, 102; heat and, 234–240; heavenly, 6–7; laws of, 23ff.; moon's, 41ff.; periodic, 104; perpetual, 229; quantity of, 65–66; random, 198, 235–238; relative, 328ff.; rotational, 72; second law of, 30–31, 347; third law of, 34–36, 70; translational, 72; vibratory, 101ff.
Motion pictures, 281
Motor, electric, 472–475
Mueller, Erwin W., 501–502
Müller, S., 596
Multiplication factor, 681
Mu-meson, 726
Muon, 726–727
neutrinos and, 729–732
Muon family number, conservation of, 731–732
Muon-neutrinos, 731–732
Musical notes, 167ff.
frequency of, 171–173
Musschenbroek, Pieter van, 413
Mutations, 690
Myopia, 275, 279

Nagaoka, Hantaro, 552
Natural philosophy, 1
Near-point, 274
Nearsightedness, 274–275
Ne'eman, Yuval, 735
Negative electricity, 400–401
Negative pole, electric, 421
Neon, 618, 626
Neptunium, 656
Neptunium-237, 658
Neptunium-238, 657
Neptunium-239, 684
Neptunium series, 659
Neumann, John von, 585
Neutral atom, 545
Neutretto, 729
Neutrino, 716ff.
absorption of, 719; electron family and, 718; muons and, 729–732; spin of, 717
Neutron, 598
bombardment with, 651–654; breakdown of, 599–600, 711, 716–718; chain reactions and, 671–674; delayed, 681; magnetic field of, 710; mass of, 665; packing fraction of, 667; prompt, 681; slowing of, 653–654; structure of, 727; thermal, 654
Newton, Isaac, 23ff., 37ff., 242, 245, 291, 295, 296, 301, 303, 305, 320–321, 357, 359, 379
Newton, 33
Nicholson, William, 422
Nichrome, 426, 431
Nickel, 386, 495, 627
Nickel-64, 668
Nicol, William, 322
Nicol prism, 322–323
Niepce, Joseph N., 280
Niobium, 428, 447–448, 451–452
Nishijima, Kazuhiko, 733
Nitrogen, 431, 492
nuclear structure of, 594–595, 599
Nitrogen-13, 646
Nitrogen-14, 663
Nitrogen-15, 646, 667
Nobelium, 658
Nodes, 150
Nonelectrolytes, 505
Normal, 252
North magnetic pole (Earth), 390
North pole (magnet), 381
N-rays, 696n
Nuclear atom, 536–540
Nuclear batteries, 683–684
Nuclear bomb, 351
Nuclear chain reaction, 671
Nuclear cross-section, 653
Nuclear energy, 669ff.
Nuclear fission, 672–674
Nuclear force, 723ff.
Nuclear fusion, 686–689
energy produced by, 694; temperatures for, 694
Nuclear isomers, 608
Nuclear power, 682
Nuclear reactions, 663ff.

Nuclear reactor, 351, 680–683
Nuclear ships, 682–683
Nuclear spin, 594–595
Nucleon, 599
Nucleon shells, 664
Nucleus, atomic, 539–540
charge of, 540ff., 544–545; energy levels of, 591; masses of, 665–667; neutrons in, 598–600; proton/neutron structure of, 660–664; spin of, 594–595; structure of, 588ff., 664, 673, 722ff.
Nuclide, 619

Oblate spheroid, 56
Ockham, William of, 5
Ockham's razor, 5, 235
Octave, 171
Odd-odd nuclei, 662–663
Oersted, Hans C., 384, 441–442, 452
Oersted, 384
Ohm, Georg S., 424
Ohm, 425
Ohm's law, 424–425
Oliphant, Marcus L. E., 648
Omega-minus particle, 736
Opera glasses, 287
Oppenheimer, J. Robert, 651, 688
Optical activity, 324
Optical density, 266
Optical maser, 579
Optics, 247ff.
physical, 301
Orbital quantum number, 559
Orbital velocity, 64
Ordinary ray, 320
Ortho-positronium, 707
Osmium, 137
Overtones, 175
Oxygen, 205, 211, 486, 489–490, 491
atomic weight of, 492; isotopes of, 628; molecular weight of, 498; weight of molecule of, 500
Oxygen-15, 647
Oxygen-16, 667
atomic weight standard, 628
Oxygen-18, 646
Ozone, 498–499

Packing fraction, 666
Parabola, 39
Parabolic mirror, 257–261
Parachute, 142
Parallelogram of force, 40–41
Paramagnetism, 395–396
Para-positronium, 707
Parity, conservation of, 734–735
Particles, acceleration of, 635ff.
detection of, 595–598; elementary, 727n; exchange, 728; magnetic fields of, 710; resonance, 733; spin of, 728; stable, 599; subatomic, 520; unstable, 599; virtual, 724; wave properties of, 581ff.
Pascal, Blaise, 119, 139
Pascal's principle, 119
Paschen, Friedrich, 555
Paschen series, 555, 557–558
Pasteur, Louis, 324
Pauli, Wolfgang, 100, 560, 715, 722, 728
Pauling, Linus, 586
PC conservation, 735
Peltier, Jean C. A., 576
Peltier effect, 576
Pendulum, 108–114
compensation, 185
Peregrinus, Peter, 382, 392
Perey, M., 655
Perfect gas, 196
Period, 104n
Periodic motion, 104
Periodic table, 495–498, 536
atomic number and, 544–545; electron arrangements and, 549–550; gaps in, 545
Permanent magnet, 386
Permeability, magnetic, 395–396
Permittivity, 405
Perpetual motion, 229
Perspiration, 218–219
Petit, Alexis T., 215
Petrjak, K. A., 674
Philosophy, 1–2
Phosphorus-30, 644, 704
Phosphorus-31, 644
Phosphorus-32, 703
Photoelectric cell, 536
Photoelectric effect, 372–375, 534–536
Photography, 280–281
color, 294
Photon(s), 375–379, 728
absorption of, 719; antiparticles and, 711
Physical atomic weight, 628
Physical chemistry, 3
Physical optics, 301
Physical philosophy, 2
Physics, 2–3
Pi, 74, 74n
Piezoelectricity, 408–409
Pile, atomic, 678–682
Pi-meson, 727
Pinhole camera, 279
Pion, 727–730
breakdown of, 729–732; spin of, 728
Pitch, 162ff.
motion and, 173–174

Planck, Max K. E. L., 369, 375
Planck's constant, 371, 556, 587
Plane, inclined, 10ff., 88, 91–92
Plane mirror, 255
Plane-polarized light, 322
Planté, Gaston, 439
Plasma, 695
Plasma physics, 695
Plate, 522
Platinum, 137
Plücker, Julius, 509–510
Plutonium, 657
Plutonium-239, 684
Pneumatics, 120
Poise, 133
Poiseuille, Jean L. M., 133
Polariscope, 323–324
Polarized light, 320–325
  plane shift of, 324–325
Polaroid, 325
Poles, electric, 421
Polonium, 602
Polonium-209, 629
Polonium-212, 670
  half-life of, 617
Polonium-281, 610
Porta, Giambattista della, 279
Positive electricity, 400–401
Positive ions, 507
Positive pole, electric, 421
Positive rays, 592
Positron, 702
  annihilation of, 705; hydrogen fusion and, 704; radioactivity and, 703–704
Positronium, 707
Potassium, 422, 495, 661
Potassium-40, 622, 668
  earth's age and, 631
Potassium-41, 667
Potential difference, 407
  contact, 419; effective, 464; gravitational, 406; measurement of, 453
Potential energy, 96–97
Powell, Cecil F., 727
Power, 93
Power, electric, 428–431
Power stations, nuclear, 683
Presbyopia, 274
Pressure, 116ff., 138ff.
  gases and, 145–146; radiation, 377–378; units of, 116–117
Primary radiation, 699–700
Principal axis, 256
Principal quantum number, 557
Prism, 269–270
Projector, light, 280
Promethium, 656, 674
Prompt neutrons, 681
Proportionality, 14, 14n, 18–19
Proportionality constant, 19
Proton, 402
Protactinium, 545, 603
Protactinium-231, 611
Protactinium-233, 685
Protactinium-234, 608
Proton(s), 592–593
  acceleration of, 637–642; cosmic rays and, 700; mass of, 665; nuclei and, 661–663; structure of, 727
Proton synchrotrons, 642
Proust, Joseph L., 488–489
Proust's law, 489
Prout, William, 503
Prout's hypothesis, 504, 626
Pull, 26
Pulley, 88
Push, 26
Pyrex, 183
Pythagoras, 102, 154, 170, 171

Quantum, 370
  size of, 374
Quantum mechanics, 585
Quantum numbers, 557ff.
Quantum theory, 370ff., 555ff.
  photoelectric effect and, 374–375
Quartz, 183, 251, 424, 426
Quasars, 713

Rad, 691
Radar, 530–532
Radial velocities, 319
Radiation, 222, 248, 511ff.
  background, 691; black-body, 368–371; Cerenkov, 699; infrared, 317; ionizing, 590; pressure of, 377–378; primary, 699–700; radioactive, 590, 689; secondary, 699; temperature and, 366–368; ultraviolet, 318
Radiation sickness, 689–692
Radio, 524–528, 575
Radio astronomy, 532
Radio telescopes, 532
Radio tubes, 527
Radio waves, 479, 512, 526–528
Radioactive elements, 629
Radioactive equilibrium, 616
Radioactive series, 601ff.
Radioactivity, 589–590
  artificial, 644; atomic changes in, 604–606; units of, 649–651
Radiocarbon dating, 648
Radioisotopes, 644
Radium, 73–75, 602

Radium-226, 615
breakdown of, 649
Radium-228, 608
Radon, 602
Rainbow, 290–291
Random motion, 198, 235–238
Rankine, William J. M., 96
Rare earth elements, 498
Ray, light, 248
Rayleigh, Lord, 309, 368
Reactance, capacitive, 467–468
Reactance, inductive, 466
Reactions, nuclear, 633ff.
Reactor, nuclear, 680–683
Real image, 259
Reciprocal seconds, 154
Rectifier, 523
Red shift, 319, 363
Reflecting telescope, 294–297
Reflection, 177, 252ff.
angle of, 253; diffuse, 254; specular, 255; total, 267
Refracting telescope, 295
Refraction, 263–268, 302–303
angle of, 264; double, 320; index of, 265, 315; law of, 265
Refrigeration, 218–219
Regnault, Henri V., 208
Reines, Frederick, 721
Reinforcement, 168
Relative time, 352–356
Relativity, 341ff.
general theory of, 24n, 47n, 354, 357–364; gravitation and, 359–364; special theory of, 342; time and, 352, 356
Relay, electric, 449
Rem, 691
Rep, 690
Repulsion, magnetic, 382ff.
Resistance, electrical, 424ff.
in parallel, 434; in series, 431–432; internal, 437; measurement of, 454–455; temperature and, 426–428; temperature coefficient of, 427; variable, 455
Resistivity, 425–426
Resolving power, 286
Resonance, 176–177
Resonance particles, 733
Resonance theory 586
Rest, 6, 24n
Rest, absolute, 329–331
Rest-length, 337
Rest-mass, 339
Retina, 273, 293–294
Reverberation, 178
Revolutions per minute, 73
Rhenium, 545
Richards, Theodore W., 627
Ritter, Johann W., 317, 513
Roemer, Olaus, 311–312
Roentgen, 690
Roentgen equivalent man, 691
Roentgen equivalent physical, 690
Röntgen, Wilhelm K., 479, 514–515, 541
Röntgen rays, 514
Roosevelt, Franklin D., 676
Root mean square, 203n
Rossi, Bruno, 698
Rotational motion, 72–73
torques and, 75
Rubidium, 299
Rubidium-87, 622
earth's age and, 631
Rumford, Count, 234
Ruske, Ernst A. F., 583
Rutherford, Ernest, 538–540, 590, 592–593, 595, 601, 613, 614, 632–634
Rutherford, 651
Rydberg, Johannes R., 553
Rydberg constant, 554

Satellites, 60, 64
Saturn, 445
Scalar quantity, 27
Scattering, light, 308–310
Schmidt, Bernard, 277
Schmidt camera, 277
Schockley, William B., 575
Schoenheimer, Rudolf, 645–646
Schrödinger, Erwin, 583–585
Schrödinger wave equation, 585
Schweigger, Johann S. C., 452
Science, 1
Scintillation counter, 633
Screw, 88
Seaborg, Glenn T., 656–658, 684
Second law of motion, 30–31, 347
Second law of thermodynamics, 226ff., 233
Secondary radiation, 699
Seebeck, Thomas J., 420, 575
Seebeck effect, 575
Segrè, Emilio, 655, 709
Self-induction, 465
Semiconductors, 426, 572–573
Seneca, Lucius A., 291
Series, converging, 62, 483
Shadow, 250
Shear, 155
Shell, electron, 560ff.
Shell numbers, 664
Shock, electric, 412
Shock wave, 166

Short circuit, 435
Sigma particles, 732–733
Silicon, 426, 572, 643
Silver, 214–215, 423, 506
resistivity of, 426
Silver nitrate, 513
Simple harmonic motion, 102
Simple machine, 88
Simple microscope, 286–287
Sine, 111, 265
Sine curve, 150, 462
Sirius, 319
companion of, 363
Size, critical, 679
Sky, 310
Snell, Willebrord, 264
Soddy, Frederick, 601, 606, 613
Sodium, 299–300, 308, 422, 668
Solar battery, 576–577
Solar eclipse, 251, 278, 362–363
Solar furnace, 260
Solar spectrum, 300, 318
Solenoid, 444–445
Solids, 115, 125
density of, 137
Solid-state devices, 573–577
Sommerfeld, Arnold, 558
Sonar, 180
Sonic boom, 166
Sound, 154ff., 244
intensity of, 159; interference and, 167; loudness of, 158–161; pitch of, 162ff.; power of, 160; quality of, 175–177; reflection of, 177–180; reinforcement of, 168; temperature and, 164; velocity of, 164
Sound barrier, 165
Sound waves, 156–157
frequency of, 163–164, 169ff.
South magnetic pole (Earth), 390
South pole (magnet), 381
Space, absolute, 331
curved, 361
Special theory of relativity, 342
Specific activity, 650
Specific gravity, 137
Specific heat, 214–215
Specific rotation, 324
Spectacles, 275–276, 285
Spectral lines, 298–300
Spectroscopy, 298–299
Spectrum, 291–292
absorption, 299–300; continuous, 299; dispersion of, 297; emission, 298–299; lines in, 297–300; solar, 300, 318
Specular reflection, 255
Speed, 29
angular, 73
Spherical aberration, 257, 276–277
Spherical mirror, 256
Spin quantum number, 559
Spontaneous fission, 674
Spring balance, 53
Stainless steel, 386
Standing wave, 583
Stars, 288–289
color of, 316–317, 368; temperature of, 368
Static, 528
Static electricity, 397ff.
Statics, 88
Steam, 144
Steel, 166, 381
stainless, 386
Stefan, Josef, 366
Stefan-Boltzmann law, 366
Stefan's law, 366
Steinmetz, Charles P., 469–475
Stevinus, Simon, 11
Stokes, George G., 520
Stoney, George J., 509
Storage battery, 430–440
Stradivarius, Antonius, 177
Strain, 50
Strangeness, conservation of, 733–734
Streamlining, 131
Stress, 50
String galvanometer, 452
Strong interactions, 728
Strontium, 422
Strontium-87, 631
Strontium-90, 684
fallout and, 692
Strutt, John W., 309
Sturgeon, William, 446
Subatomic particles, 520, 728
acceleration, 635ff.; magnetic fields of, 710
Sulfur, 398, 703
resistivity of, 426
Sulfur dioxide, 210
Sulfuric acid, 439
Sum, limiting, 62
Sun, 445–446, 686
neutrinos and, 720–721; nuclear fusion in, 687–688
Sunset, 310
Superconductivity, 428, 451–452
magnetism and, 447–448
Superheterodyne receiver, 527
Supernova, 721
Supersonic velocity, 166
Surface energy, 126
Surface tension, 127–130

Synchrocyclotron, 640
Szilard, Leo, 675–676

Tagged compound, 645
Talbot, William H. F., 280
Tantalum, 451–452
Technetium, 654
Telegraph, 449–450
Telephone, 450–451
Telescope, 287–289
  reflecting, 294–297; refracting, 295
Television, 528–530
  color, 294
Teller, Edward, 676, 688
Tellurium, 495
Temperature, 181ff.
  absolute, 193–195; color and, 366–367; critical, 211; electrical resistance and, 426–428; gases and, 191–193; heat and, 212ff.; heat flow and, 223–224; magnetism and, 387–388; molecular kinetic energy and, 203; molecular velocity and, 205; radiation and, 366–368; sound and, 164
Temperature scales, 187–188, 193
Temporary magnet, 386
Terminal velocity, 131
  in air, 141–142
Tesla, Nikola, 471, 475
Thales, 380
Thermal neutrons, 654
Thermocouple, 420
Thermodynamics, 222ff.
  first law of, 222, 229n, 233; second law of, 226ff., 233
Thermoelectricity, 420–421, 575
Thermometer, 185–188
Thermonuclear reaction, 687
Thermos bottle, 226
Thermostat, 184–185
Third law of motion, 34–36, 70
Thompson, Benjamin, 234
Thomson, George P., 582
Thomson, Joseph J., 517, 534, 537, 592, 618
Thomson, William, 193, 450
Thorium, 590
Thorium-231, 617
Thorium-232, 606–608, 630, 685
  half-life of, 617; series of, 610–611
Thorium-234, 606, 608
Thorium series, 610–611
Threshold frequency, 373
Threshold value, 535
Thunder, 166
Time, measurement of, 107–108, 113–114
Time dilatation, 352–356
Tin, 428, 447–448, 667
Titanium, 428
Torque, 75
  angular acceleration and, 80; lever and, 85
Torricelli, Evangelista, 138, 140
Torricelli, 138
Torsion balance, 50
Total reflection, 267
Townes, Charles H., 578–579
Tracers, isotopic, 645
Transformer, electric, 470–472
Transistors, 575
Transition elements, 563ff.
Translational motion, 72
Transparency, 251–252
Transuranium elements, 658
Transverse wave(s), 150, 155–156, 320, 327
Triangular prism, 269n
Trigonometric functions, 111n
Tritium, 649
Triton, 649
Trough, 150
Tungsten, 212, 431
Turbulence, 131
Twilight, 309
Twin paradox, 355
Tyndall, John, 309
Tyndall effect, 309

Ultrasonic waves, 180
Ultraviolet radiation, 318, 513–514, 689
Uncertainty, principle of, 587, 723–724
Unified field theory, 478
Unit poles, 384
Units, 17
  common, 20, 31; systems of, 32
Universal gravitation, 44–45
Universe, 711–713
  Newtonian, 242; running down of, 233
Unpolarized light, 321
Uranium, 589–590
  earth's age and, 630–631; energy produced by, 669–690; fission of, 672–674; neutron bombardment of, 656
Uranium-233, 684–685
Uranium-235, 611–612
  fission of, 677–682; half-life of, 616–617; series of, 611–612
Uranium-238, 630, 677–678
  breakdown of, 606, 663–664, 669–674; fission half-life of, 674; half-life of, 615–616; packing fraction of, 667; series of, 608–609
Uranium hexafluoride, 137, 678
Uranium series, 608–609

Uranus, 242–243
Urbain, Georges, 544
Urey, Harold C., 624

Vacuum, 7, 245, 395, 509–511
heat flow and, 225–226; Torricellian, 140
Vacuum tube, 523
Van de Graaf, Robert J., 638
Van der Waals, Johannes D., 210
Van der Waals equation, 208–210
Van der Waals forces, 210
Vanadium, 448
Van't Hoff, Jacobus H., 325
Vapor, 211n
Vaporization, latent heat of, 218
Variable condenser, 412
Variable resistance, 455
Vector, 27–28
Veksler, Vladimir I., 640
Velocity, 14
angular, 73; escape, 63; linear, 73; molecular, 205; orbital, 64; supersonic, 166; terminal, 131, 141–142; units of, 15; vectors and, 28
Venus, 420
magnetic properties of, 445
Vibratory motion, 101ff.
period of, 103ff.
Vinci, Leonardo da, 279
Violet catastrophe, 369
Virtual focus, 262
Virtual image, 254
Virtual particles, 724
"Vis viva," 94
Viscosity, 130–133
Visible light, 318
Vision, 273–276
color, 293–294
Visual angle, 281–282
Vitruvius Pollio, Marcus, 155
Volt, 408
Volta, Alessandro, 408, 418–419, 422, 533
Voltage, 408
Voltage drop, 432n
Voltage multiplier, 638
Voltmeter, 453
Volume, temperature and, 182ff.

Wallis, John, 70
Walton, Ernest T. S., 637
Water, 116, 251, 491
boiling of, 217–218; coefficient of conductivity of, 224; critical temperature of, 211; density of, 136, 221; dielectric constant of, 405; index of refraction of, 266; light velocity in, 315; molecular weight of, 499; sound velocity in, 166; waves in, 148–149
Watson-Watt, Robert A., 531–532
Watt, James, 93, 218, 429
Watt, 93, 429
Watt-second, 430
Wave(s), 148ff.
compression, 156–157; longitudinal, 156–158; sound, 154ff.; transverse, 150, 155–156
Wave analysis, 169
Wave mechanics, 585
Wavelength, 152–153
light, 303; reflection and, 179–180
Weak interactions, 728
Weather, 139
Weber, Ernst H., 160n
Weber, Wilhelm E., 394, 395n
Weber, 394–395
Weber-Fechner law, 160n
Wedge, 88
Weight, 43, 53
pressure and, 116; units of, 54; variations of, 55–60
Westinghouse, George, 463, 471
Wheatstone, Charles, 453, 455
Wheatstone bridge, 455
Wheel and axle, 88
Whispering galleries, 178
White dwarf, 363
White light, 292
Wien, Wilhelm, 367–368, 592
Wien's law, 368
Wigner, Eugene P., 676, 735
Wilson, Charles T. R., 596, 697
Wiring, electric, 430–431
Wollaston, William H., 298
Wood, 224, 426
Work, 90–92
"stored," 93–94
Wu, Chien-Shiung, 735

Xi particles, 732–733
X-ray(s), 378–379, 541ff.
burns from, 689; characteristic, 542–545; discovery of, 514–516; muons and, 726; production of, 541–542; radioactivity and, 588–591; tracks of, 597; wavelength of, 543
X-ray tube, 542

Yang, Chen Ning, 734
Young, Thomas, 94, 293, 306–307, 321
Yukawa, Hideki, 723–727

Zeno, 483
Zinc, 419–421, 437–440, 668
Zinc chloride, 437
Zinc sulfide, 633
Zworykin, Vladimir K., 530